SHORT PROTOCOLS IN NEUROSCIENCE

CELLULAR AND MOLECULAR METHODS

A Compendium of Methods from
Current Protocols in Neuroscience

EDITORIAL BOARD

Charles R. Gerfen

Michael A. Rogawski

David R. Sibley

Phil Skolnick

Susan Wray

WILEY

Published by John Wiley & Sons, Inc.

Cover illustration: © Mediscan/Corbis

Published by John Wiley & Sons, Inc., Hoboken, New Jersey.
Published simultaneously in Canada.

While the authors, editors, and publisher believe that the specification and usage of reagents, equipment, and devices, as set forth in this book, are in accord with current recommendations and practice at the time of publication, they accept no legal responsibility for any errors or omissions, and make no warranty, express or implied, with respect to material contained herein. In view of ongoing research, equipment modifications, changes in governmental regulations, and the constant flow of information relating to the use of experimental reagents, equipment, and devices, the reader is urged to review and evaluate the information provided in the package insert or instructions for each chemical, piece of equipment, reagent, or device for, among other things, any changes in the instructions or indication of usage and for added warnings and precautions. This is particularly important in regard to new or infrequently employed chemicals or experimental reagents.

NOTE: All protocols using live animals must be reviewed and approved by an Institutional Animal Care and Use Committee (IACUC) and must follow officially approved procedures for the care and use of laboratory animals.

This book was edited by Drs. Gerfen, Rogawski, Sibley, and Wray in their private capacities. No official support or endorsement by the National Institutes of Health is intended or should be inferred.

Library of Congress Cataloging-in-Publication Data

Short protocols in neuroscience : cellular and molecular methods / editorial board, Charles R. Gerfen ... [et al.].

p. ; cm.

Includes bibliographical references and index.

ISBN-13: 978-0-471-78399-2 (cloth : alk. paper)
ISBN-10: 0-471-78399-4 (cloth : alk. paper)

1. Molecular neurobiology—Laboratory manuals. 2. Cytology—Laboratory manuals. 3. Neurons—Laboratory manuals. I. Gerfen, Charles R.

[DNLM: 1. Nervous System Physiology—Laboratory Manuals. 2. Gene Expression Regulation—physiology—Laboratory Manuals. 3. Neurobiology—methods—Laboratory Manuals. 4. Neurons—physiology—Laboratory Manuals. 5. Neurotransmitter Agents—physiology—Laboratory Manuals. WL 25 S559 2006]

QP356.2.S56 2006

573.8'4—dc22

2006016055

CIP

Printed in the United States of America

10 9 8 7 6 5 4 3 2 1

CONTENTS

3 Cellular and Developmental Neuroscience

4 Gene Cloning, Expression, and Mutagenesis

5 Molecular Neuroscience

Appendices

Preface

Neuroscience is one of the most dynamic disciplines in biology. The inherent interest in the nervous system is not sufficient to explain the explosion of work in this field. The last decade has seen considerable conceptual progress, but technical advances leading to the introduction of new methods in laboratories across the world have also played a major part in driving the field forward. The techniques for investigating the nervous system that have been developed require familiarity with electrophysiology, mouse genetics, optics, brain anatomy, and many other methods. More than most areas of biology, neuroscience is a multidisciplinary activity.

The rapid transfer of technical information and experience is the partner of profound conceptual change. Although it is hard to be comprehensive in this diverse field, our goal in creating this manual is to provide a single source for core techniques in the cellular and molecular aspects of neuroscience.

The methods in *Short Protocols in Neuroscience: Cellular and Molecular Methods* are shortened versions of methods published in *Current Protocols in Neuroscience* (CPNS). This compendium includes step-by-step descriptions for a broad range of cellular and molecular neuroscience methods. The methods are complete and easy to follow and the book is easier to use at the bench than the parent volumes. *Short Protocols* is ideal for graduate students and postdoctoral fellows who are familiar with the background and theory found in CPNS, and sufficient detail is provided to allow experienced investigators to use it as a stand-alone bench guide.

Although mastery of the techniques presented in this manual, and its companion, *Short Protocols in Neuroscience: Systems and Behavioral Methods*, will allow the reader to design and complete experiments in neuroscience, this manual is not intended to be a substitute for an advanced training program in neuroscience, nor for any number of comprehensive textbooks in the field. In addition to becoming thoroughly knowledgeable of the terms and concepts of neuroscience, readers should gain first-hand experience in basic techniques and safety procedures by working in a supervised neuroscience laboratory.

HOW TO USE THIS MANUAL

Format and Organization

Subjects in this manual are organized by chapters and protocols are contained in units. Protocol units, which constitute the bulk of the book, include one or more protocols with listings of materials and steps. Recipes for unique reagents and solutions are included in *APPENDIX 1* and references are included in the *References* section at the end of the book. The book also includes several overview units which contain theoretical discussions that lay the foundation for subsequent protocol units.

Certain units that describe commonly used techniques and recipes (e.g., neural tissue sectioning and mounting, solutions for preparing coated tissue culture dishes) are cross-referenced in other units that describe their application. Thus, whenever it is necessary to prepare mounted rat brain sections for microscopy, the appropriate unit in Chapter 1—describing various procedures for brain preparation, sectioning, and mounting—is cross-referenced (i.e., *UNIT 1.1*). For some widely used molecular biological techniques, *Current Protocols in Molecular Biology* (*CPMB*) is cross-referenced throughout the manual.

Protocols

Most units in this manual contain groups of protocols, each presented with a series of steps. One or more *Basic* Protocols are presented first in each unit and generally cover the recommended or most universally applicable approaches. *Alternate* Protocols are provided where different equipment or reagents can be employed to achieve similar ends, where the starting material requires a variation in approach, or where requirements for the end product differ from those in the Basic Protocol. *Support* Protocols describe additional steps that are required to perform the Basic or Alternate Protocols; these steps are separated from the core protocol because they might be applicable to other uses in the manual, or because they are performed in a time frame separate from the Basic Protocol steps.

Reagents and Solutions

Reagents required for a protocol are itemized in the materials list before the procedure begins. Many are common stock solutions, others are commonly used buffers or media, while others are solutions unique to a particular protocol. All recipes are presented in *APPENDIX 1*. It is important to note that the names of some of these special solutions might be similar from unit to unit (e.g., SDS sample buffer) while the recipes differ. To minimize confusion, the appropriate unit for a given recipe, except for those for commonly used solutions, is given parenthetically in the appendix.

NOTE: Deionized, distilled water should be used in the preparation of all reagents and solutions and for all protocol steps. When sterile solutions are required, that will be indicated in the protocol.

Equipment

When special equipment is required for an experiment, that equipment is listed in the Materials list for the protocol. The Materials list does not include every item used, rather it includes items that might not be readily available in the laboratory, items that have particular specifications, and items that require special preparation or temperature. Standard pieces of equipment used in the modern neuroscience laboratory are listed in the accompanying box. These items are used frequently in the protocols in this manual and are not necessarily listed in the Materials list.

Standard Equipment

Applicators, cotton-tipped and wooden
Autoclave
Balances, analytical and preparative
Beakers
Bench protectors, plastic-backed (including "blue pads")
Biohazard disposal containers and bags
Bottles, glass and plastic
Bunsen burners
Cell harvester, for determining radioactivity uptake in 96-well microtiter plates
Centrifuges, low-speed (6,000 rpm) and high-speed (20,000 rpm) refrigerated centrifuges and an ultracentrifuge (20,000 to 80,000 rpm) are required for many procedures. At least one microcentrifuge that holds standard 0.5- and 1.5-ml microcentrifuge tubes is essential. It is also useful to have a tabletop swinging-bucket centrifuge with adapters for spinning 96-well microtiter plates.
Clamps
Cold room or cold box, 4°C
Computer (IBM-compatible or Macintosh) and printer
Containers, assortment of plastic and glass dishes for gel and membrane washes
Coplin jars or staining dishes, glass, for 75 × 25–mm slides
Cryovials, sterile—e.g., Nunc
Cuvettes, plastic disposable, glass, and quartz
Darkroom and developing tank, or X-Omat automatic X-ray film developer (Kodak)
Desiccators (including vacuum desiccators) and desiccant
Dry ice
Filtration apparatus, for collecting acid precipitates on nitrocellulose filters or membranes
Flasks, glass (e.g., Erlenmeyer, beveled shaker)
Forceps
Fraction collector
Freezers, −20° and −80°C
Geiger counter
Gel dryer
Gel electrophoresis equipment, at least one full-size horizontal apparatus and one horizontal minigel apparatus, one vertical full-size and minigel apparatus for polyacrylamide protein gels, and specialized equipment for two-dimensional protein gels
Gloves, plastic and latex, disposable and asbestos
Graduated cylinders
Heating blocks, variable temperature up to 100°C; these thermostat-controlled metal heating blocks that hold test tubes and/or microcentrifuge tubes are very convenient for carrying out enzymatic reactions
Heat-sealable plastic bags and sealing apparatus
Hemacytometer
Hoods, chemical (fume), microbiological safety, and laminar flow (tissue culture)
Hot plates, with or without magnetic stirrer
Ice buckets
Ice maker
Incubators, 37°C for bacteria and humidified 37°C, 5% CO_2 for tissue culture
Kimwipes, or equivalent lint-free tissues
Lab coats
Light box, for viewing gels and autoradiograms
Liquid nitrogen and Dewar flask
Lyophilizer
Magnetic stirrers, (with heater is useful)
Markers, including indelible markers and china-marking pencils
Microcentrifuge, Eppendorf-type, maximum speed 12,000 to 14,000 rpm

continued

Commercial Suppliers

Throughout the manual, we have recommended commercial suppliers of chemicals, biological materials, and equipment. In some cases, the noted brand has been found to be of superior quality or it is the only suitable product available in the marketplace. In other cases, the experience of the author of that protocol is limited to that brand. In the latter situation, recommendations are offered as an aid to the novice neuroscientist in obtaining the tools of the trade. Experienced investigators are therefore encouraged to experiment with substituting their own favorite brands.

Addresses, phone numbers, and facsimile numbers of all suppliers mentioned in this manual are provided in *APPENDIX 3*.

References

Short Protocols gives only a limited number of the most fundamental references as background. These are listed at the end of the unit. Full bibliographic information for these and references cited in the unit are given in the *References* at the end of this book, Readers who would like a more extensive entry into the literature for background and

Standard Equipment, *continued*

Microcentrifuge tubes, 1.5-ml and 0.5-ml
Microscope, standard optical model (optionally with epifluorescence or phase-contrast illumination) and inverted microscope for tissue culture
Microscope slides and coverslips
Mortar and pestle
Ovens, drying, hybridization, vacuum, and microwave
Paper cutter, large size, for 46 × 57–cm Whatman paper sheets
Paper towels
Parafilm
pH meter
pH paper
Pipet bulbs, or battery-operated pipetting devices—e.g., Pipet-Aid (Drummond Scientific)
Pipets, Pasteur and graduated, glass and plastic, serological (1-ml to 25-ml)
Pipettors, adjustable delivery, volume ranges 0.5 to 10 µl, 10 to 200 µl, and 200 to 1000 µl. It is best to have one set of these three sizes for each full-time researcher and sets dedicated for radioactive and PCR experiments.
Plastic wrap, UV-transparent (e.g., Saran Wrap)
Pliers, needle nose
Polaroid camera
Power supplies, 300-V power supplies are sufficient for polyacrylamide gels; 2000- to 3000-V is needed for some applications
Racks, for test tubes and microcentrifuge tubes
Radiation shield, Lucite or Plexiglas
Radioactive waste containers, for liquid and solid waste
Razor blades
Refrigerator, 4°C
Ring stands and rings
Rotator, end-over-end
Rubber bands
Rubber policemen
Rubber stoppers
Safety glasses
Scalpels and blades
Scintillation counter
Scissors
Sectioning equipment, cryostat microtome, sliding microtome (with stage and knife), Vibratome
Shakers, orbital and platform, room temperature or 37°C. An enclosed shaker (e.g., New Brunswick Controlled Environment Incubator Shaker) that can spin 4-liter flasks is essential for growing 1-liter *E. coli* cultures. A rotary shaking water bath (New Brunswick R76) is useful for growing smaller cultures in flasks.
Spectrophotometer, UV and visible
Speedvac evaporator (Savant)
Stir-bars, assorted sizes
Surgical equipment, scale for weighing animals, electric razor, syringes, hypodermic needles, dissection instruments (scissors, scalpels, forceps, hemostats, retractor, bone drill, sutures), operating microscope (optional), heating pad, sterile gauze
Tape, masking and electrician's
Thermometers
Timer
UV cross-linker (e.g., Stratalinker from Stratagene)
UV light sources, long- and short-wave, stationary or hand-held
UV transilluminator
Vacuum aspirator
Vacuum line
Vortex mixers
Wash bottles, plastic and glass
Water baths, variable temperature up to 80°C
Water purification equipment, e.g., Milli-Q system (Millipore) or equivalent
X-ray film cassettes and intensifying screens

application of methods are referred to the appropriate unit in *Current Protocols in Neuroscience*.

Safety Considerations

Anyone carrying out these protocols may encounter the following hazardous or potentially hazardous materials: (1) radioactive substances, (2) toxic chemicals and carcinogenic or teratogenic reagents, (3) pathogenic and infectious biological agents, and (4) recombinant DNA constructs. Check the guidelines of your particular institution with regard to use and disposal of these hazardous materials. Only limited cautionary statements are included in this manual. Users are responsible for understanding the dangers of working with hazardous materials and for strictly following the safety guidelines established by manufacturers (e.g., MSDS) as well as local and national regulatory agencies. Radioactive substances must be used only under the supervision of licensed users, following the guidelines of the National Regularoty Commission (NRC).

Animal Handling

Many protocols call for use of live animals (usually rats or mice) for experiments. Prior to conducting any laboratory procedures with live subjects, the experimental approach must be submitted in writing to the appropriate Institutional Animal Care and Use Committee (IACUC). Written approval from this committee is absolutely required prior to undertaking any live-animal studies. Some specific animal care and handling guidelines are provided in the protocols where live subjects are used, but check with your IACUC guidelines to obtain more extensive guidelines.

ACKNOWLEDGMENTS

This manual is the product of dedicated efforts by many of our scientific colleagues who are acknowledged in each unit and by the hard work by the Current Protocols editorial staff at John Wiley and Sons. We are extremely grateful for the critical contributions Gwen Crooks (Series Editors) who kept the editors and the contributors on track and played a key role in bringing the entire project to completion. Other skilled members of the Current Protocols staff who contributed to the project include Kathy Morgan, Sheila Kaminsky, and Joseph White. The extensive copyediting required to produce an accurate protocols manual was ably handled by Susan Lieberman, Linda Plappinger, and Allen Ranz, and electronic illustrations were prepared by Gae Xavier Studios.

RECOMMENDED BACKGROUND READING

Bedcer, J.B., Breedlove, M.S., Crews, D., and McCarthy, M.M. (eds.) 2002. Behavioral Endocrinology, 2nd ed. MIT Press, Cambridge, Mass.

Cooper, J.R. 2002. The Biochemical Basis of Neuropharmacology, 8th ed. Oxford University Press, New York.

Charles R. Gerfen, Michael A. Rogawski, David R. Sibley, Phil Skolnick, and Susan Wray

Contributors

Evelyn Abordo-Adesida
University of Manchester
Manchester, United Kingdom

Roger L. Albin
University of Michigan and Ann Arbor Veterans' Administration Medical Center
Ann Arbor, Michigan

Phillipe I. H. Bastiens
European Molecular Biology Laboratory
Heidelberg, Germany

David Baulcombe
The Sainsbury Laboratory
The John Innes Center
Norwich, United Kingdom

Alain Bernard
Ares-Serono Pharmaceutical Research Institute
Geneva, Switzerland

Juan S. Bonifacino
National Institute of Child Health and Human Development
Bethesda, Maryland

Andrew Bradbury
Los Alamos National Laboratory
Los Alamos, New Mexico and International School for Advanced Studies (SISSA)
Trieste, Italy

Roger Brent
The Molecular Sciences Institute
Berkeley, California

Terry Brown
University of Manchester Institute of Science and Technology
Manchester, United Kingdom

Oliver Brüstle
University of Bonn
Bonn, Germany

James R. Bunzow
Oregon Health Sciences University
Portland, Oregon

Michael C. Byrne
Genetics Institute/Wyeth Research
Cambridge, Massachusetts

Heather A. Cameron
National Institute of Neurological Disorders and Stroke, NIH
Bethesda, Maryland

Maria G. Castro
University of Manchester
Manchester, United Kingdom

Jonathan D. Chesnut
Invitrogen Corporation
Carlsbad, California

Piotr Chomczynski
University of Cincinnati College of Medicine
Cincinnati, Ohio

John E. Coligan
National Institute of Allergy and Infectious Diseases, NIH
Bethesda, Maryland

Donald Coling
University of California
Berkeley, California

Peter Crino
University of Pennsylvania Medical Center
Philadelphia, Pennsylvania

Miles G. Cunningham
National Institute of Neurological Disorders and Stroke, NIH
Bethesda, Maryland

Esteban C. Dell'Angelica
National Institute of Child Health and Development
Bethesda, Maryland

Joseph DeRisi
University of California
San Francisco, California

Robert L. Dorit
Yale University
New Haven, Connecticut

James Eberwine
University of Pennsylvania Medical Center
Philadelphia, Pennsylvania

Markus U. Ehrengruber
University of Zurich
Zurich, Switzerland

Elaine A. Elion
Harvard Medical School
Boston, Massachusetts

JoAnne Engebrecht
State University of New York
Stony Brook, New York

R. Douglas Fields
National Institute of Child Health and Human Development, NIH
Bethesda, Maryland

Russell L. Finley, Jr.
Wayne State University School of Medicine
Detroit, Michigan

Maximillian T. Follettie
Genetics Institute/Wyeth Research
Cambridge, Massachusetts

Cornel Fraefel
Institute of Virology
University of Zurich
Zurich, Switzerland

Kirk A. Frey
University of Michigan
Ann Arbor, Michigan

Steven Fuller
Nabi
Rockville, Maryland

Sean Gallagher
UVP, Inc.
Upland, California

Christian A. Gerdes
University of Manchester
Manchester, United Kingdom

Charles R. Gerfen
National Institute of Mental Health, NIH
Bethesda, Maryland

Claudia Gerwin
National Institute of Neurological Disorders and Stroke, NIH
Bethesda, Maryland

Daniel H. Geschwind
UCLA School of Medicine
Los Angeles, California

Michael Gilman
Cold Spring Harbor Laboratory
Cold Spring Harbor, New York

Michelle Gilmor
Emory University
Atlanta, Georgia

Marianna Goldrick
Ambion, Inc.
Austin, Texas

Erica A. Golemis
Fox Chase Cancer Center
Philadelphia, Pennsylvania

Shiaoching Gong
Rockefeller University
New York, New York

David K. Grandy
Oregon Health Sciences University
Portland, Oregon

Jeno Gyuris
Mitotix, Inc.
Cambridge, Massachusetts

Rebecca A. Haberman
University of North Carolina
Chapel Hill, North Carolina

Thomas Hazel
National Institute of Neurological Disorders and Stroke, NIH
Bethesda, Maryland

J. S. Heilig
University of Colorado
Boulder, Colorado

Craig Heilman
Emory University
Atlanta, Georgia

James P. Hoeffler
Invitrogen Corporation
Carlsbad, California

Hennie Hoogenboom
Department Pathologie Academisch
Ziekenhius Maastricht and TargetQuest
Maastricht, The Netherlands

Richard L. Huganir
The Johns Hopkins University School of Medicine
Baltimore, Maryland

John G. R. Hurrell
FluorRx
Carmel, Indiana

Bechara Kachar
National Institute on Deafness and Other Communication Disorders, NIH
Bethesda, Maryland

Stanislav L. Karsten
UCLA School of Medicine
Los Angeles, California

Donald Kessler
Genomics Corp.
Foster City, California

Paul A. Kingston
University of Manchester
Manchester, United Kingdom

Robert E. Kingston
Massachusetts General Hospital and Harvard Medical School
Boston, Massachusetts

Mikhail G. Kolonin
Wayne State University School of Medicine
Detroit, Michigan

Phillip R. Kramer
NIH, NINDS, CDNS
Bethesda, Maryland

Gabriele Kroner-Lux
University of North Carolina
Chapel Hill, North Carolina

Miriam Leenders
National Institute of Neurological Disorders and Stroke, NIH
Bethesda, Maryland

Allan Levey
Emory University
Atlanta, Georgia

Filip Lim
Universidad Autonoma de Madrid
Madrid, Spain

Pedro R. Lowenstein
University of Manchester
Manchester, United Kingdom

Kenneth Lundstrom
F. Hoffmann-LaRoche Research Laboratories
Basel, Switzerland

Tricia C. Maleniak
University of Manchester
Manchester, United Kingdom

Roberto Marzari
Università di Trieste
Trieste, Italy

Thomas J. McCown
University of North Carolina
Chapel Hill, North Carolina

Eva Mezey
National Institutes of Health
Bethesda, Maryland

David Moore
Baylor College of Medicine
Houston, Texas

Richard Mortensen
University of Michigan Medical School
Ann Arbor, Michigan

Thomas Müller
National Institute of Neurological Disorders and Stroke, NIH
Bethesda, Maryland

Norman Nash
Emory University
Atlanta, Georgia

Kim A. Neve
Oregon Health Sciences University and Veterans Affairs Medical Center
Portland, Oregon

Rachael L. Neve
Harvard Medical School and McLean Hospital
Belmont, Massachusetts

Michael J. O'Donovon
National Institute of Neurological Disorders and Stroke, NIH
Bethesda, Maryland

Shigeo Okabe
National Institute of Bioscience and Human Technology
Ibaraki, Japan

Mark Payton
Glaxo-Wellcome
Stevenage, United Kingdom

Mary C. Phelan
Thompson Children's Hospital
Chattanooga, Tennessee

Kathryn R. Radford
Australian Membrane and Biotechnology Research
Sydney, Australia

Louise Rem
Department Pathologie Academisch
Ziekenhius Maastricht and TargetQuest
Maastricht, The Netherlands

Randall Ribaudo
National Institute of Allergy and Infectious Disease
Bethesda, Maryland

Katherine W. Roche
National Institute on Deafness and Other Communication Disorders, NIH
Bethesda, Maryland

John K. Rose
Yale University School of Medicine
New Haven, Connecticut

Nicoletta Sacchi
Johns Hopkins University
Baltimore, Maryland

R. N. S. Sachdev
The University of Texas at San Antonio
San Antonio, Texas

Richard Jude Samulski
University of North Carolina
Chapel Hill, North Carolina

Daniele Sblattero
International School for Advanced Studies (SISSA)
Trieste, Italy

John A. Schetz
University of North Texas Health Science Center
Fort Worth, Texas

Michael I. Schimerlik
Oregon State University
Corvallis, Oregon

Christine E. Seidman
Harvard Medical School
Boston, Massachusetts

Ilya Serebriiskii
Fox Chase Cancer Center
Philadelphia, Pennsylvania

Eswar P. N. Shankar
University of North Texas Health Science Center
Fort Worth, Texas

Jun Shao
Vanderbilt University
Nashville, Tennessee

Zu-Hang Sheng
National Institute of Neurological Disorders and Stroke, NIH
Bethesda, Maryland

Carolyn L. Smith
National Institute of Neurological Disorders and Stroke, NIH
Bethesda, Maryland

Thomas D. Southgate
University of Manchester
Manchester, United Kingdom

Timothy A. Springer
Center for Blood Research
Harvard Medical School
Boston, Massachusetts

Daniel Stone
University of Manchester
Manchester, United Kingdom

William M. Strauss
Harvard Medical School and Beth Israel Deaconess Medical Center
Boston, Massachusetts

Kevin Struhl
Harvard Medical School
Boston, Massachusetts

Lorenz Studer
National Institute of Neurological Disorders and Stroke, NIH
Bethesda, Maryland

Vivian Tabar
National Institute of Neurological Disorders and Stroke, NIH
Bethesda, Maryland

James P. Tam
Vanderbilt University
Nashville, Tennessee

Clare E. Thomas
University of Manchester
Manchester, United Kingdom

Didier Trono
University of Geneva
Geneva, Switzerland

Carlos Vicario-Abejón
Centro de Investigaciones Biológicas
Censejo Superior de Investigaciones Cientificas
Madrid, Spain

Laura A. Volpicelli-Daley
Emory University
Atlanta, Georgia

David E. Weinstein
GliMed Inc.
New York, New York

Maryann Z. Whitley
Genetics Institute/Wyeth Research
Cambridge, Massachusetts

Charles J. Wilson
The University of Texas at San Antonio
San Antonio, Texas

Scott E. Winston
Nabi
Rockville, Maryland

Fred S. Wouters
Imperial Cancer Research Fund
London, United Kingdom

X. William Yang
Neuropsychiatric Institute
David Geffen School of Medicine
at UCLA
Los Angeles, California

W. Scott Young, III
National Institutes of Health
Bethesda, Maryland

Romain Zufferey
University of Geneva
Geneva, Switzerland

CHAPTER 1

Neuroanatomical Methods

UNIT 1.1 describes some of the basic procedures for preparation of the brain for histologic processing that are common to most of the methods described in subsequent units. These include methods for preparing the brain in either a fixed or unfixed state as well as for sectioning the brain for subsequent histologic processing. The choice among these basic procedures depends on the type of histologic processing that is to be performed.

UNIT 1.2 presents methods for localization of messenger RNA (mRNA) by in situ hybridization histochemistry (ISHH). ISHH involves the specific binding of a labeled probe to a complementary portion or all of the RNA in brain cells. Methods are described for the localization of mRNA using oligonucleotide cDNA probes and using ribonucleotide probes which are detected using either radioactive or nonradioactive methods. Techniques described provide for both regional localization (localization to brain regions) and cellular-level resolution. In addition, combined methods provide for localization of multiple mRNAs in the same tissue section and for the use of ISHH localization in association with other techniques such as tract tracing or immunohistochemistry.

UNIT 1.3 describes localization of biochemicals by immunohistochemistry. Immunohistochemistry involves the localization of specific proteins, peptides, or glycoproteins by the binding of antibodies to specific antigenic sites on these biochemicals, and the visualization of the bound antibodies using histochemical processes. Methods are described for the histochemical procedures used on brain tissue sections. The production and characterization of antibodies are described in *UNITS 5.5–5.6*.

UNIT 1.4 describes localization of neurotransmitter receptor binding sites in brain sections. These methods involve the use of radioactively tagged ligands that are used to bind specifically to neurotransmitter receptor binding sites in brain tissue sections. In many cases, these ligands have been characterized by pharmacologic binding assays, and demonstrated to have receptor-specific binding. Obtaining similar receptor-specific binding in brain tissue sections requires that conditions be determined to block nonspecific binding while retaining receptor-selective binding.

UNIT 1.5 provides details for labeling individual or small numbers of neurons with the markers biocytin or Neurobiotin, which label all processes of the neuron, including both dendritic and axonal projections. Two methods are described, one in which neurons are injected intracellularly, and the other in which the marker is ejected extracellularly, to allow for uptake by nearby neuron cell bodies. These methods are most often used as part of neurophysiologic studies to enable identification of the neurons from which recording is taking place. As such labeling provides nearly complete filling of both dendrites and axonal projections, important information is obtained about the connections of the neuron from which recording is taking place.

Contributor: Charles Gerfen

UNIT 1.1

Basic Neuroanatomical Methods

BASIC PROTOCOL 1

PREPARATION OF UNFIXED FRESH-FROZEN BRAIN TISSUE

Materials

Isopentane
Dry ice
Rat or mouse for study
Anesthetic

Dissection instruments:
- Scissors
- Spatula
- Forceps

Plexiglas or metal sieve-like basket and metal container large enough to hold it

1. Place ~300 to 500 ml isopentane in a metal container large enough to hold a corresponding sieve-like basket. Place the metal container with the isopentane in dry ice for 15 to 30 min, until the temperature of the isopentane reaches −70°C.
2. Kill animal(s) with an overdose of anesthesia. Remove brain from skull.
3. Place brain on the mesh bottom of the sieve-like basket in a manner that preserves the normal shape of the brain. Immerse brain in the cooled isopentane for 20 to 30 sec (long enough to result in complete freezing of brain but not so long that it cracks).
4. Rapidly remove frozen brain from isopentane, detach from the mesh, and place briefly on absorbent paper to remove excess isopentane. Wrap dried, frozen brain in foil and store at –20° to –70°C until sectioning is performed.

BASIC PROTOCOL 2

PERFUSION FIXATION

Materials (*see* APPENDIX 1 *for items with* ✓)

Saline (0.9% w/v NaCl), 4°C
✓ Fixative solution for perfusion, room temperature
Rat or mouse for study
Anesthetic
✓ Sucrose-infiltration solution, 4°C

Peristaltic perfusion pump (e.g., Masterflex with variable-speed standard drives from Cole-Parmer)
Masterflex Tygon tubing (0.25-in.)
Blunt 13-G and 15-G hypodermic needles
Surgical instruments (Roboz Surgical or Fine Science Tools) including:
- Scalpel
- Scissors
- Clamps

Hegenbarth clip-applying forceps
Hemostats
Bone rongeur

1. Place cold saline (0.9% NaCl) and room temperature fixative solution in separate flasks and set up the peristaltic pump, Tygon tubing, and perfusion instruments according to the manufacturer's instructions in such a manner that the saline is first drawn through the pump into tubing that is to be connected to the animal. Use a valve system or other device to allow the fixative solution to be drawn through the tubing at a later point (set up such that air is not introduced during the switch between the two fluids). Fill the system with saline. Attach tubing primed with saline to a blunt 15-G hypodermic needle that will be used for perfusion.

2. Prepare rat or mouse for infusion by administering a lethal dose of anesthesia. Monitor until the point when the animal fails to respond to pinching of the foot. Make an incision through the abdomen just below the rib cage to expose the diaphragm. Make an incision in the diaphragm to expose the beating heart. Open the thoracic cavity with two horizontal cuts through the rib cage on either side of the heart. Clamp the sternum with a hemostat and fold the cut rib flap headward to expose the heart.

3. Make a small incision at the bottom apex of the left ventricle. Quickly insert a blunt 13-G hypodermic needle upward through the ventricle past the aortic valve so that it may be visualized ~5-mm inside the ascending aorta. Clamp needle in place with a Hegenbarth clip-applying forceps or hemostat across the ventricle.

4. Begin perfusion of cold saline very slowly (i.e., 20 to 40 ml/min). Immediately after the peristaltic pump begins pumping the saline, cut the right atrium to allow an escape route for the blood and perfusion fluid. Perfuse saline at a moderate to rapid rate (~40 ml/min) and continue until the effluent runs clear, which may require 200 to 500 ml of solution.

5. After effluent runs clear, stop pump and introduce fixative into the peristaltic pump line running into the animal. Perfuse fixative at a moderate to slow rate (~20 ml/min) such that ~500 ml fixative is perfused over 10 to 20 min.

6. Following perfusion of ~500 ml of fixative, remove brain from skull using a bone rongeur. If desired, prior to sucrose infiltration, post-fix brain for a variable period of time at room temperature in the same fixative as that used for perfusion.

7. Transfer brain to a vessel containing 4°C sucrose-infiltration solution. Incubate 24 to 48 hr at 4°C, until brain sinks into the sucrose solution, indicating that sucrose has infiltrated the brain. Section the brain (see Basic Protocols 3, 4, and 5).

BASIC PROTOCOL 3

CRYOSTAT SECTIONING OF FROZEN BRAIN TISSUE

Materials

Dry ice (powdered or pellets)
Embedding matrix (M-1, Shandon/Lipshaw or OCT compound, Miles Labs)
Brain tissue: fresh-frozen (see Basic Protocol 1) or perfusion-fixed (see Basic Protocol 2) and frozen on dry ice just prior to sectioning

Cryostat microtome
Specimen holder (cryostat chuck; metal platform for supporting specimen during sectioning)

Gelatin-subbed microscope slides (see Support Protocol)
Clean, soft paint brush (optional)
40°C warming plate (optional)
Small zip-lock bags

1. Place specimen holder/cryostat chuck on dry ice and place embedding matrix or water on the surface of the specimen holder/chuck. As the embedding matrix or water begins to freeze, place the frozen brain, base-down, into it so that the brain adheres to the specimen holder/chuck. Pour embedding matrix over the frozen brain to provide a thin coat that aids in maintaining the integrity of the brain sections during cutting. Place the brain, mounted on the specimen holder/cryostat chuck, in the cryostat microtome. Section the brain.

2a. *Method 1:* Carefully transfer the fragile sections from the specimen holder/chuck to the exact desired position on a warm (40°C) gelatin-subbed slide (section will adhere to the site that it touches).

2b. *Method 2:* Prechill slide and keep it cold in the cryostat. Position the cut brain section on the cold gelatin-subbed slide with a clean, soft brush. After the brain section is in position, warm slide by running a warm (40°C) steel bar on the back side of the slide, or place the slide on a 40°C warming plate, to allow the section to adhere.

3. Dry slides for 1 min on a 40°C warming plate. Place slides in storage bags (e.g., small zip-lock bags) and store at −20° to −80°C pending further processing.

BASIC PROTOCOL 4

SLIDING-MICROTOME SECTIONING OF FIXED BRAIN TISSUE

Materials *(see* APPENDIX 1 *for items with ✓)*

Sucrose-infiltrated fixed brains (see Basic Protocol 2)
Dry ice
✓ KPBS

Sliding microtome with knife (Leica or American Optical) and sliding microtome stage
Small brush
Container for collecting brain tissue sections (e.g., 24-well Costar tissue culture plate)
Gelatin-subbed microscope slides (see Support Protocol)

1. Freeze sucrose-infiltrated fixed brain in dry ice.

2. Cool the microtome stage according to the manufacturer's instructions. Attach the brain to the stage by placing water on the prechilled stage and placing the brain on the water to allow it to freeze in place.

3. Cut 10- to 50-μm sections of the brain by sliding the microtome knife across the surface of the frozen brain. Carefully transfer the single cut sections from the knife surface with a brush to a dish containing a suitable buffer solution (e.g., KPBS).

4. Place a gelatin-subbed slide in the dish underneath the floating sections and position the section of interest onto the slide with a brush. Slowly raise the slide out of the dish and allow the section to dry onto the slide. Once section is dried, reimmerse slide in dish and affix other sections by repeating the process. If desired, stain dried, slide-mounted sections with thionin (see Basic Protocol 7).

BASIC PROTOCOL 5

VIBRATOME SECTIONING

A vibratome (TPI) is an instrument that provides a vibrating knife to cut tissue. The advantage of this device is that the vibrating knife makes it possible to section brain tissue without the need for freezing and the consequent ultrastructural damage which is only partially averted via sucrose infiltration. While the Vibratome method is essential for electron microscopy, it also affords superior morphologic integrity of immunohistochemical labeling at the light-microscopic level.

BASIC PROTOCOL 6

POST-SECTIONING PROCEDURES I: DEFATTING

For some procedures, such as in situ hybridization histochemistry, brain sections may be fixed and defatted prior to histochemical processing.

Materials *(see* APPENDIX 1 *for items with* ✓*)*

Fresh-frozen (unfixed) slide-mounted brain sections (see Basic Protocol 3)
✓ 4% (w/v) formaldehyde in saline
Acetic anhydride
✓ Triethanolamine/saline solution
70%, 95%, and 100% ethanol
Chloroform

Metal 30-slide rack (optional for small numbers of slides)
500-ml (or appropriate-sized) staining dishes

1. Thaw slides with sections at room temperature in the storage bags. Fix mounted sections by immersing slides for 10 min in a staining dish with 4% formaldehyde/saline.
2. Add 1.25 ml acetic anhydride to 500 ml triethanolamine/saline solution in a beaker and stir rapidly for 20 sec. Immediately transfer solution to a staining dish containing the slides and immerse slides for 10 min.
3. Dehydrate sections by immersing slides successively for 1 min each in staining dish containing 70% and 95% ethanol, then twice for 1 min in 100% ethanol. Defat sections by immersing slides twice for 5 min in chloroform. Remove chloroform by immersing slides twice for 1 min in 100% ethanol, then once for 1 min in 95% ethanol. Allow slides to air dry and store in boxes at −20°C.

BASIC PROTOCOL 7

POST-SECTIONING PROCEDURES II: THIONIN STAINING

Thionin staining may be effectively used to determine the extent of neurotoxic lesions—e.g., 6-hydroxydopamine lesions of the dopamine neurons in the midbrain, substantia nigra, pars compacta, or excitotoxic lesions of the cortex and striatum. Most of its staining appears to be of RNA, which is particularly concentrated in the cell body and nucleus.

Materials *(see* APPENDIX 1 *for items with* ✓*)*

Brain sections mounted on slides, preferably fixed
✓ Thionin solution

50%, 70%, 95%, and 100% ethanol
Xylene
95% ethanol/1% acetic acid (optional)

Coverslips
Permount histological mounting fluid (e.g., Fisher)

1. Place slides with brain sections in staining racks. Dip successively in the following solutions for the indicated periods of time:

1 min	thionin solution
2 min	distilled, deionized H_2O
1 to 2 min	50% ethanol
1 to 2 min	70% ethanol
1 to 2 min	95% ethanol
1 to 2 min	100% ethanol
1 to 2 min	100% ethanol
2 min	xylene
2 min	xylene.

Store indefinitely in xylene until slides are to be coverslipped.

2. Allow to dry, then apply coverslip over section using Permount. Examine by bright-field microscopy for blue-purple staining with white matter relatively unstained.

3. If white matter has not destained sufficiently, pry off coverslip (if already applied) and dip slide in the following solutions for the indicated periods of time:

2 min	xylene
2 min	xylene
2 min	100% ethanol
2 min	100% ethanol
1 to 2 min	95% ethanol/1% acetic acid
2 min	100% ethanol
2 min	100% ethanol
2 min	xylene
2 min	xylene

Apply new coverslip as in step 2. Examine the slide.

BASIC PROTOCOL 8

POST-SECTIONING PROCEDURES III: PHOTOGRAPHIC-EMULSION COATING OF SLIDE-MOUNTED SECTIONS FOR AUTORADIOGRAPHY

Materials

Emulsion: Kodak NTB-3 or Amersham LM-1 (thaw 30 min prior to use)
0.1% (w/v) Dreft detergent in H_2O
Slide-mounted radioactively labeled tissue sections
Dektol developer (Kodak)
Stop bath: H_2O *or* 1.5% (v/v) acetic acid in H_2O
Rapid Fix (Kodak): prepare according to manufacturer's instructions without hardener

Darkroom with amber/red sodium photographic safelight and humidifier
Slide mailer (2-slide or 5-slide, Shandon/Lipshaw or Thomas Scientific)

40° to 42°C water bath
Blank microscope slides
Light-tight slide boxes
Staining racks and dishes
Glass staining dishes (Thomas Scientific)
Metal slide racks (Thomas Scientific)

1. In darkroom under safelight conditions, melt emulsion completely in a 40° to 42°C water bath in a slide mailer or other suitable vessel (shaped to minimize the volume of emulsion needed) in which slides may be immersed. Melt some emulsion separately in a beaker, which will be used to replenish the emulsion in the slide mailer as slides are dipped. For Kodak NTB-3 emulsion, dilute gel form of emulsion 1:1 with 0.1% Dreft detergent. For Amersham LM-1 emulsion, proceed to step 2 without dilution.

2. Remove bubbles from emulsion (in both slide mailer and beaker) by dipping blank slide(s) in it up to 100 times. Let emulsion sit 15 min to settle.

3. Dip each slide into emulsion, wipe back of slide with a paper towel to remove emulsion, then stand slides upright on absorbent paper towels to allow excess emulsion to flow off slide. Allow emulsion to dry for 1 to 3 hr in upright position. Store slides in light-tight box at −20°C until they are to be developed.

4. In darkroom under safelight conditions, dilute 1 part Dektol developer with 2 parts water in a staining dish. Cool developer solution on ice and maintain at 17°C during developing. If necessary, thaw slides for 30 min prior to developing. Place stop bath and Kodak Rapid Fix solutions in staining dishes.

5. Process slides in staining racks by passing them through the following solutions:

2 min	Dektol developer, 17°C
1 min	stop bath
2 to 3 min	Kodak Rapid Fix.

6. Rinse slides in cold running tap water for 15 to 30 min. Air dry. Counterstain if desired (Basic Protocol 7).

SUPPORT PROTOCOL

PREPARATION OF GELATIN-SUBBED MICROSCOPE SLIDES

Materials *(see* APPENDIX 1 *for items with* ✓*)*

✓ Gelatin-subbing solution
Glass slides
Slide racks
40°C glassware-drying oven

1. Place slides in slide racks and dip for 1 min in gelatin-subbing solution.

2. Remove racks containing slides from subbing solution and shake to facilitate removal of excess subbing solution. For twice-subbed slides (preferred), repeat steps 1 and 2.

3. Dry slides in a 40°C oven overnight.

Reference: Bolam, 1992

Contributor: Charles R. Gerfen

UNIT 1.2

Immunohistochemical Localization of Proteins in the Nervous System

BASIC PROTOCOL 1

IMMUNOHISTOCHEMISTRY FOR LIGHT MICROSCOPY

This method is based on the formation of an avidin-biotin complex (see Fig. 1.2.1).

Materials *(see APPENDIX 1 for items with ✓)*

✓ Tris-buffered saline (TBS), pH 7.2, 4°C
Brain tissue, fixed and sectioned (*UNIT 1.1*)
✓ NAS/avidin blocking solution
Monoclonal or polyclonal primary antibody against antigen of interest
✓ Primary antibody diluent
Vectastain ABC peroxidase kit (Vector Labs) containing:
  Reagent A (avidin)
  Reagent B (biotinylated horseradish peroxidase)
Biotin-conjugated secondary antibody (Vector Labs)
✓ Secondary antibody diluent
✓ DAB substrate solution (prepare immediately before use)
0.1 M sodium nitrate
10% (v/v) Triton X-100
0.1% (w/v) cresyl violet in 0.08% (v/v) acetic acid
70%, 95%, and 100% ethanol
Histo-Clear (National Diagnostics) or xylenes
Mounting medium: e.g., Permount (Fisher) or DPX (Electron Microscopy Sciences)

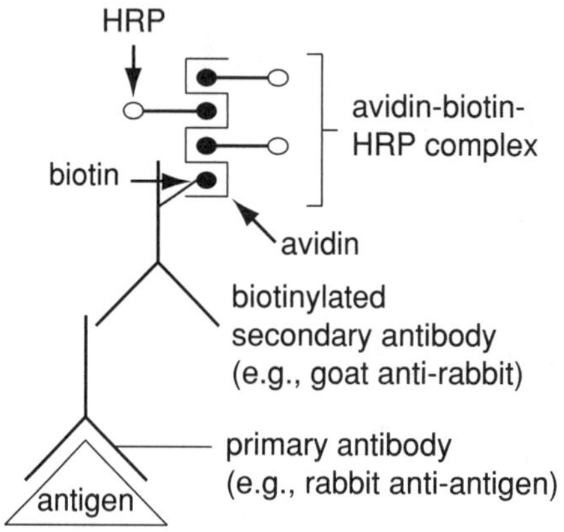

Figure 1.2.1 Immunoperoxidase immunohistochemistry. Abbreviation: HRP, horseradish peroxidase. The method utilizes secondary antibodies conjugated to biotin. The biotin is recognized by an avidin-biotinylated HRP complex (ABC complex). Enzymatic reaction of HRP with a substrate (e.g., DAB) results in deposition of a colored reaction product. Because avidin has multiple binding sites, many biotinylated enzymes are complexed, resulting in amplification of the signal.

6-, 12-, or 24-well tissue culture plates
Glass transfer pipets with the ends heat-sealed and bent to form a hook
Platform shaker
0.22-μm syringe filters (Millipore) and appropriate-sized syringes
Mounting dish or large petri dish
Superfrost Plus microscope slides (Fisher)
Fine-tipped paint brush
60°C water bath
Slide rack
24 × 60–mm coverslips

1. Place 4°C TBS into wells of a tissue culture plate. Transfer the fixed, sectioned brain tissue into the wells containing TBS, using the tip of a brush or a glass transfer pipet, the end of which has been sealed and bent in a flame to form a hook. Rinse six times, each time by incubating 5 min in TBS at 4°C with gentle shaking on a platform shaker, then transferring the section to another well containing TBS and repeating the incubation. Briefly rinse transfer tool in one to two containers of distilled water before placing it into a solution containing a different reagent.

2. Incubate tissue 30 min in NAS/avidin blocking solution with shaking at 4°C. Rinse sections three times in TBS, 5 min each, 4°C, then transfer to NAS/biotin blocking solution and incubate an additional 30 min as before.

3. Rinse the sections three times in TBS, 10 min each, 4°C, with slow shaking

4. Dilute primary antibody (against the antigen of interest) to the desired concentration in primary antibody diluent and place in wells of a tissue culture plate. Transfer tissue sections to the primary antibody–containing wells and incubate ~16 to 48 hr with shaking at 4°C. Rinse the tissue sections three times in TBS, 10 min each, 4°C, with slow shaking.

5. Dilute secondary antibody 1:200 in secondary antibody diluent and place in wells of a tissue culture plate. Transfer tissue sections to the secondary antibody–containing wells and incubate 1 hr at 4°C with shaking. Rinse the tissue sections three times in TBS, 10 min each, 4°C, with slow shaking.

6. Mix, 30 min prior to use, 1 drop each of reagent A (avidin) and B (biotinylated horseradish peroxidase) per 10 ml TBS and place on ice to allow formation of the avidin-biotinylated horseradish peroxidase complex (tertiary complex). Place this solution in the wells of a tissue culture plate. Transfer (biotinylated) secondary antibody–labeled tissue sections to these wells and incubate 1 hr at room temperature with shaking. Rinse the tissue three times in TBS, 10 min each, 4°C, with slow shaking.

7. Place the freshly prepared DAB substrate solution into an appropriate-size syringe and filter through a 0.22-μm syringe filter into the wells of a new tissue culture plate. Transfer the tissue sections into the wells of the plate. Incubate at room temperature with shaking (typically for 3 to 15 min) while carefully monitoring the reaction to ensure that background staining remains at a minimum (i.e., that negative controls remain white) whereas specific staining is high (i.e., that there is dark brown staining in regions of experimental tissue where the antigen is known to be present).

8. Once the desired level of staining has been reached, transfer the tissue to the wells of a tissue culture plate containing TBS to terminate the reaction. Rinse the tissue four times in TBS, 5 min each, 4°C, with slow shaking.

9. Place 0.1 M sodium nitrate in a mounting dish or a large petri dish. Add several drops of 10% Triton X-100 to facilitate the movement of tissue onto the slides. Transfer the tissue sections into the dish and, if desired, arrange in sequential order. Place a slide into the dish

1

and use a fine-tipped paint brush to move the sections onto the slide. Mark the frosted end of the slide with the relevant information for the tissue it contains. Dry slides overnight in a dust-free place. Proceed to counterstaining (steps 10 to 13) if desired, or proceed to dehydration, mounting, and application of coverslip (steps 14 to 15).

10. If a counterstain is desired (e.g., to aid in detection of immunohistologically unlabeled cells and/or neuroanatomical landmarks), heat 0.1% mg/ml cresyl violet to 60°C. Place the tissue-containing slides in a slide rack. While the cresyl violet is heating, dehydrate the tissue by immersing slides for 30 sec in water, then successively for 3 min each in:

    70% ethanol
    95% ethanol
    95% ethanol
    100% ethanol
    100% ethanol.

11. Immerse slides three times in Histo-Clear or xylene, each time for 5 min, then rehydrate the tissue by immersing slides successively for 3 min in each of the following to rehydrate:

    100% ethanol
    100% ethanol
    95% ethanol
    95% ethanol
    70% ethanol
    water.

12. Immerse slides for 15 to 30 sec in the 60°C cresyl violet solution. Rinse slides in distilled water and determine if a darker stain is desired. If the staining is acceptable, immerse slides in 70% ethanol (several minutes) until the desired level of differentiation of stain is achieved.

13. Immerse slides once for 2 min in 90% ethanol, then twice, each time for 2 min, in 100% ethanol. Immerse slides three times in Histo-Clear or xylene, each time for 3 min. Mount tissue and apply coverslip as described in step 16.

14. Place the tissue-containing slides (from step 9) in a slide rack. Place the racks in tap water for 30 sec to dissolve the salts remaining from the buffers previously used. Dehydrate tissue by immersing slides for 3 min each in:

    70% ethanol
    95% ethanol
    95% ethanol
    100% ethanol
    100% ethanol.

15. Clear the tissue by immersing slides three times for 5 min each time in Histo-Clear or xylene. After the third immersion, leave the slides in the Histo-Clear or xylene to avoid having the tissue dry out.

16. Lay a coverslip on a flat surface and place a line of mounting medium along the edge of one of the long axes. Using forceps, remove a slide from the Histo-Clear or xylene, allowing excess clearing agent to drip off. Align an edge of the slide with the axis of the coverslip containing the mounting medium. Bring the slide down so that the mounting medium "runs up" between the slide and coverslip without leaving air bubbles in between. Let slides dry overnight at room temperature and examine.

BASIC PROTOCOL 2

DOUBLE-LABELING IMMUNOFLUORESCENCE

Materials *(see* APPENDIX 1 *for items with ✓)*

✓ Tris-buffered saline (TBS), pH 7.2, 4°C
Brain tissue, fixed and sectioned (*UNIT 1.1*)
3% (v/v) H_2O_2 (dilute 30% H_2O_2 in TBS)
✓ NAS/avidin blocking solution (if amplification is not required, avidin can be omitted)
Monoclonal and/or polyclonal primary antibody against antigen of interest
✓ Primary antibody diluent (if amplification is not required, biotin can be omitted)
Donkey anti-species fluorophore-conjugated secondary antibodies for multiple labeling (Jackson Immunoresearch): these antibodies have been preabsorbed to multiple species, resulting in minimal cross-reactivity and minimal background staining; fluorescein, rhodamine-red X, and Cy5 (Cy5 can be used for triple labeling experiments) provide fluorophores with minimal overlap of emission/excitation spectra
✓ Secondary antibody diluent
Donkey anti-species biotin-conjugated secondary antibodies for multiple labeling (Jackson Immunoresearch)
Vectastain ABC peroxidase kit (Vector Labs) containing:
Reagent A (avidin)
Reagent B (biotinylated horseradish peroxidase)
Tyramide signal amplification (TSA) fluorescence system kit (PerkinElmer) including:
Tyramide-fluorophore solution
Amplification diluent
✓ Cupric sulfate solution
10% (v/v) Triton X-100
Vectashield mounting medium for fluorescence (Vector Labs)
Clear nail polish

6-, 12-, or 24-well tissue culture plates
Glass transfer pipets with the ends heat-sealed and bent to form a hook
Platform shaker
Mounting dish or large petri dish
Superfrost Plus microscope slides (Fisher)
Fine-tipped paint brush
60°C water bath

1. Place 4°C TBS into wells of a tissue culture plate. Transfer fixed, sectioned brain tissue into the wells containing TBS using a brush or a glass transfer pipet. Rinse six times, each time by incubating 5 min in TBS at 4°C with slow shaking on a platform shaker.

2. Incubate tissue 10 min in 3% H_2O_2 in TBS. Rinse three times with TBS or until most of the bubbles are gone. Incubate tissue 30 min in NAS/avidin blocking solution with shaking at 4°C. Rinse the sections twice in TBS, 5 min each, 4°C, with slow shaking.

3. Dilute primary antibodies to the desired concentration in primary antibody diluent and place in wells of a tissue culture plate.

   The optimal antibody concentration should be determined empirically. Two primary antibodies of divergent species can be incubated together. For example, an antibody raised in rabbit and an antibody raised in mouse can be incubated together. However, antibodies raised in mouse can cross-react with antibodies raised in rat.

4. Transfer tissue sections to wells containing the primary antibody and incubate ~16 to 48 hr with shaking at 4°C. Rinse the tissue sections three times in TBS, 5 min, 4°C, with slow shaking.

5. Dilute fluorophore-conjugated secondary antibody 1:200 in secondary antibody diluent and place in wells of a tissue culture plate.

   Secondary antibody can be diluted 1:1 in 80% glycerol and stored at −20°C. Wrap the vial containing the antibody in foil to protect from light. If the antibody is diluted in glycerol, use at 1:100.

   If both primary antibodies work well with a secondary antibody directly conjugated to a fluorophore, secondary antibodies can be incubated together (for example, FITC-conjugated donkey anti-mouse and rhodamine red-X-conjugated donkey anti-rabbit). In this case, perform step 6 and proceed directly to step 10.

6. Transfer tissue sections to the secondary antibody–containing wells and incubate 1 hr at 4°C with shaking. Rinse the tissue sections three times in TBS, 5 min each, 4°C, with slow shaking.

7. Dilute biotin-conjugated secondary antibody 1:200 in secondary antibody diluent and place in wells of a tissue culture plate. Incubate 1 hr at 4°C with shaking. Rinse the tissue sections three times in TBS, 5 min, 4°C, with slow shaking.

8. At a time point 30 min prior to use, mix 1 drop each of reagent A (avidin) and B (biotinylated horseradish peroxidase), from the Vectastain ABC peroxidase kit, per 10 ml TBS and place on ice to allow formation of the avidin-biotinylated horseradish peroxidase complex. Place this solution in the wells of a tissue culture plate. Transfer biotinylated secondary antibody–labeled tissue sections to these wells and incubate 1 hr at 4°C with shaking. Rinse the tissue three times in TBS, 5 min, 4°C, with slow shaking.

9. Dilute tyramide-fluorophore 1:100 in amplification diluent (supplied with TSA kit) and place this solution in the wells of a new tissue culture plate. Transfer the tissue sections into the wells of the plate containing the tyramide-fluorophore solution. Make sure that the sections are not folded and that all areas have access to the solution. Incubate at room temperature with shaking for 10 min. Transfer the tissue to the wells of a tissue culture plate containing TBS. Rinse the tissue three times in TBS, 5 min. 4°C, with slow shaking.

10. Place cupric sulfate solution into wells of tissue culture plate. Transfer the tissue sections into the wells and incubate at 4°C with shaking for 30 min. Transfer the tissue to the wells of a tissue culture plate containing TBS. Rinse the tissue three times in TBS, 5 min, 4°C, with slow shaking.

11. Place TBS in a mounting dish or a large petri dish. Add several drops of 10% Triton X-100 to facilitate the movement of tissue onto the slides. Transfer the tissue sections into the dish and, if desired, arrange in sequential order. Place a slide into the dish and use a fine-tipped paint brush to move the sections onto the slide. Mark the frosted end of the slide with the relevant information for the tissue it contains.

12. Carefully blot excess TBS from around the section with a Kimwipe. Place a drop of Vectashield mounting medium onto each section. Slowly overlay the coverslip onto the Vectashield/section without leaving air bubbles in between. Seal the coverslip by aspirating excess Vectashield around the coverslip (a seal will be formed between the coverslip and slide when there is no movement of the coverslip). Paint the edges of the coverslip with clear nail polish and allow to dry. Examine the sections by epifluorescence or confocal microscopy.

BASIC PROTOCOL 3

PRE-EMBEDDING IMMUNOHISTOCHEMISTRY FOR ELECTRON MICROSCOPY

Materials *(see* APPENDIX 1 *for items with ✓)*

✓ 0.1 M sodium phosphate buffer, pH 7.4
2% glutaraldehyde in sodium phosphate buffer, pH 7.4
✓ 1% (w/v) osmium tetroxide
0.05 M sodium acetate, pH 7.3 (store up to several weeks at 4°C)
2% (w/v) uranyl acetate in 0.05 M sodium acetate, pH 7.3 (prepare fresh)
Propylene oxide (Electron Microscopy Sciences)
Epoxy resin (e.g., Epon, Ted Pella, or Durcupan, Electron Microscopy Sciences)
✓ Release-agent-coated microscope slides

Platform shaker
Glass transfer pipets
Shell vials or scintillation vials with stoppers
Wooden applicators
✓ Resin stubs
60°C oven

CAUTION: Osmium tetroxide is a highly reactive reagent. Wear appropriate protective gear such as gloves and use only in a chemical fume hood. Used osmium tetroxide should be stored in a specifically designated waste bottle and disposed of according to institutional procedures.

1. Block nonspecific sites, incubate tissue in primary and secondary antibodies, form tertiary complex, and visualize antibodies with DAB (see Basic Protocol 1, steps 1 to 8), with the following modifications.
   a. Omit the Triton X-100 from the blocking solutions and the antibody diluent.
   b. Incubate tissue with secondary antibody ~24 hr instead of 1 hr (see Basic Protocol 1, step 5).
   c. Incubate tissue with tertiary complex solution 4 hr instead of 1 hr (see Basic Protocol 1, step 6).
2. After visualization of the antibodies with DAB, rinse tissue three times in TBS, then three times in 0.1 M sodium phosphate buffer, pH 7.4, 5 min each, 4°C, with slow shaking.
3. *Optional:* Incubate the tissue in 2% glutaraldehyde on a platform shaker for at least 1 hr (up to overnight) at 4°C. Rinse the tissue three times in 0.1 M sodium phosphate buffer, pH 7.4, 5 min, 4°C, with slow shaking.
4. Transfer the tissue sections to shell vials or glass scintillation vials containing 0.1 M sodium phosphate buffer, pH 7.4. Remove the buffer from the vials using a glass transfer pipet and add enough 1% osmium tetroxide to completely immerse the tissue. Place stoppers on vials and agitate on a shaker 30 min at room temperature. Remove the osmium tetroxide from the vials using a glass transfer pipet and dispose of it properly.
5. Rinse tissue in vials twice with 0.1 M sodium phosphate buffer, pH 7.4, then twice with 0.05 M sodium acetate buffer, each time by covering the section with buffer, incubating 5 min, then removing the buffer with a glass transfer pipet.
6. Cover the tissue in the vials with 2% uranyl acetate and incubate on a platform shaker overnight at 4°C. Remove the uranyl acetate from the vials using a glass transfer pipet.

Rinse tissue in vials three times with 0.5 M acetate buffer, each time by covering the section with buffer, incubating 5 min, then removing the buffer with a glass transfer pipet.

7. Dehydrate the tissue in increasing concentrations of ethanol as follows: rinse twice in 50% ethanol, each time by covering the section with the alcohol, incubating 5 min on a platform shaker, then removing the alcohol. Repeat with 70% and 95% ethanol (twice with each concentration), then repeat three times with 100% ethanol.
8. Cover the section with 100% propylene oxide, incubate 5 min on a platform shaker, then remove the propylene oxide and repeat the incubation with fresh 100% propylene oxide.
9. Prepare a mixture of 1 part epoxy resin and 1 part 100% propylene oxide. Remove the 100% propylene oxide from each vial containing a tissue section and cover the section with this mixture of propylene oxide and resin. Agitate on a shaker 30 min at room temperature.
10. Replace the propylene oxide/resin mixture with 100% epoxy resin and shake overnight at room temperature.
11. Use the beveled end of a wooden applicator to delicately remove the sections from the resin (the tissue will be brittle from the osmium tetroxide) and place them on release-agent-coated slides. Add fresh 100% epoxy resin on top of the sections, then place another coated slide on top so that the tissue and resin are sandwiched in between. Place in a 60°C oven for 2 to 3 days to harden the resin.
12. Pry the two slides apart carefully using a single-edged razor blade. Examine the slide containing the section with a microscope to identify areas of intense labeling, then use a scalpel and dissecting microscope to cut a 2-mm^2 area of the labeled tissue. Place the tissue sample onto a release-agent-coated slide. Place a drop of liquid resin onto one flat end of a hardened resin stub, then place that same end of the stub on top of the tissue sample. Place a drop of resin on the other end of the stub and affix a small piece of paper (e.g., "waste" paper from a hole puncher) containing information about the tissue sample. Place the slide with the "stubbed" tissue in a 60°C oven for 2 days, after which the tissue is ready to be cut on an ultramicrotome.

References: Cuello, 1993; Dawes, 1971; Levey et al., 1991, 1993; Sternberger et al., 1970

Contributors: Laura A. Volpicelli-Daley and Allan Levey

UNIT 1.3

Hybridization Histochemistry of Neural Transcripts

Controls for specificity in hybridization histochemistry include: (1) observing the same distribution of signal with probes directed against different portions of the same transcript; (2) observing blockage of signal by prior hybridization with unlabeled probe; (3) correlation of signal with immunocytochemical results; (4) observing different distribution of signal with probes against unrelated transcripts; and (5) observing that northern analysis using the probe under the same degrees of stringency shows bands of expected sizes.

BASIC PROTOCOL 1

HYBRIDIZATION HISTOCHEMISTRY WITH OLIGODEOXYNUCLEOTIDE PROBES

Materials *(see APPENDIX 1 for items with ✓)*

✓ 4× SSPE containing 50% (v/v) formamide
Slides containing 12-μm tissue sections prepared by cryostat, mounted on Superfrost Plus slides (Fisher), and defatted (*UNIT 1.1*)
✓ Hybridization solution containing ~1×10^6 dpm/50 μl ^{35}S-labeled oligodeoxynucleotide probe or 1 to 5 μl/50 μl digoxigenin-labeled oligonucleotide probe (see Support Protocol 1)
✓ 1× SSPE/1 mM DTT, room temperature and 55°C
70% ethanol

Sterile Bio-Assay dishes (Nunc; 245 × 245 × 30–mm)
Whatman 3MM chromatography paper
18 × 30–mm glass coverslips
Staining dishes and tubs
Slide rack
55°C water bath

NOTE: Use DEPC-treated water (*APPENDIX 1*) for all reagents in pretreatment and hybridization steps.

1. Cover the inside of the Bio-Assay dish top (which has a larger surface area than the bottom) with a piece of Whatman 3MM chromatography paper. Wet paper with 4× SSPE containing 50% formamide and lay the slides containing the tissue sections on the paper with the sections facing up (20 slides/dish). Place 45 μl hybridization solution containing ~1×10^6 dpm/50 μl ^{35}S-labeled (or 1 to 5 μl/50 μl digoxigenin-labeled) oligodeoxynucleotide over the tissue section on each slide and cover with an 18 × 30–mm dust-free coverslip. Cover the slides with the bottom of the Bio-Assay dish and incubate 20 to 24 hr at 37°C.

2. Remove slides from incubator and place with frosted ends up into a staining dish containing 1× SSPE/1 mM DTT. Gently slide the coverslip until it partially overhangs the slide, then pry off the coverslip with a forceps. Place the slide without a coverslip in a slide rack and immerse in a tub containing 1× SSPE/1 mM DTT. Repeat with each slide until all the slides have had their coverslips removed and have been placed in the same rack. Do not allow the sections to dry until step 4.

3. Wash the slides by immersing successively for 15 min each time in four changes of 55°C 1× SSPE/1 mM DTT, then for 5 min each time in two changes of room temperature 1× SSPE/1 mM DTT.

4a. *For ^{35}S-labeled probes:* Rapidly dip the slides into water and then into 70% ethanol, then blow them dry while the slides are oriented with their frosted ends down. Proceed to detection (see Basic Protocol 5).

4b. *For digoxigenin-labeled probes:* Using slides directly from the SSPE/DTT wash in step 3, proceed to detection (see Basic Protocol 3 or Alternate Protocol).

1

BASIC PROTOCOL 2

HYBRIDIZATION HISTOCHEMISTRY WITH RNA PROBES

Materials *(see APPENDIX 1 for items with ✓)*

✓ 4× SSPE containing 50% (v/v) formamide
Slides containing 12-μm tissue sections prepared by cryostat, mounted on Superfrost Plus slides (Fisher), and defatted (*UNIT 1.1*)
✓ Hybridization solution containing ~1 × 10^6 dpm/50 μl ^{35}S-labeled riboprobe or 3 to 10 μl/100 μl digoxigenin-labeled riboprobe (see Support Protocol 2)
✓ 4× SSPE/1 mM DTT
✓ RNase A solution, 37°C
✓ 0.1× SSPE/1 mM DTT, room temperature and 65°C
✓ 1×SSPE
70% ethanol

Sterile Bio-Assay dishes (Nunc; 245 × 245 × 30–mm)
Whatman 3MM chromatography paper
Glass coverslips
Staining dishes and tubs
Slide rack
55°C incubator
65°C water bath

NOTE: Use DEPC-treated water (*APPENDIX 1*) for all reagents in pretreatment and hybridization steps.

1. Cover the inside of the Bio-Assay dish top (which has a larger surface area than the bottom) with a piece of Whatman 3MM chromatography paper. Wet paper with 4× SSPE containing 50% formamide and lay the slides containing the tissue sections on the paper with the sections facing up (20 slides/dish). Place 45 μl hybridization solution containing ~1 × 10^6 dpm/50 μl ^{35}S-labeled (or 1 to 5 μl/50 μl digoxigenin-labeled) riboprobe onto the tissue section on each slide and cover with an 18 × 30–mm coverslip. Cover the slides with the bottom of the Bio-Assay dish and incubate 20 to 24 hr at 55°C.

2. Remove slides from incubator and place with frosted ends up into a staining dish containing 1× SSPE/1 mM DTT. Gently slide the coverslip until it partially overhangs the slide, then pry off the coverslip with a forceps. Place the slide without a coverslip in a slide rack and immerse in a tub containing 4× SSPE/1 mM DTT. Repeat with each slide until all the slides have had their coverslips removed and have been placed in the same rack. Do not allow the sections to dry until step 5.

3. Wash the slides by immersing successively for 5 min each time in four changes of room temperature 4× SSPE/1 mM DTT. Incubate slides in RNase A solution 30 min at 37°C. Rinse twice, each time by immersing 5 min in 0.1× SSPE/1 mM DTT at room temperature.

4. Wash the slides by immersing successively for 30 min each time in two changes of 65°C 0.1× SSPE/1 mM DTT (kept at 65°C in a water bath), then for 5 min each time in two changes of room temperature 1× SSPE.

5a. *For ^{35}S-labeled probes:* Rapidly dip the slides into water and then into 70% ethanol, then blow them dry while the slides are oriented with their frosted ends down. Proceed to detection (see Basic Protocol 5).

5b. *For digoxigenin-labeled probes:* Using slides directly from the 1× SSPE wash in step 8, proceed to detection (see Basic Protocol 3 or Alternate Protocol).

BASIC PROTOCOL 3

DETECTION OF DIGOXIGENIN-LABELED PROBES USING AP-CONJUGATED ANTI-DIGOXIGENIN ANTIBODIES

Materials *(see* APPENDIX 1 *for items with ✓)*

Digoxigenin-labeled sections on slides (see Basic Protocol 1, step 3 or Basic Protocol 2, step 4)
✓ TBS, pH 7.5 (room temperature)
✓ Detection buffer
Alkaline phosphatase–conjugated sheep polyclonal anti-digoxigenin antibody (Roche)
✓ Development buffer
✓ NBT/BCIP substrate working solution
✓ 1× SSPE
Cytoseal 60 (Stephens Scientific) or similar organic-basic mounting medium

Staining dishes or tubs
Slide warmer

1. Transfer slides from the final SSPE wash after digoxigenin labeling (see Basic Protocol 1, step 3, or Basic Protocol 2, step 4) to a staining dish or tub containing TBS, pH 7.5. Incubate 5 min at room temperature, then replace the solution with fresh TBS, pH 7.5, and incubate an additional 5 min. Transfer slides to a staining dish or tub containing detection buffer and incubate 30 min.

2. Prepare a 1:2000 dilution of alkaline phosphatase–conjugated anti-digoxigenin antibody in detection buffer. Incubate slides with this solution 5 to 16 hr at room temperature with gentle rocking.

3. Transfer slides to a staining dish or tub containing TBS, pH 7.5, and incubate 10 min, then replace the solution with fresh TBS, pH 7.5, and incubate an additional 10 min.

4. Transfer slides to a staining dish or tub containing substrate buffer and incubate 5 min, then incubate slides (1 to 2 hr with most probes) at room temperature in the dark with NBT/BCIP substrate working solution. Wash slides four times, each time by immersing 30 min in fresh 1× SSPE.

5. Dip slides briefly in water and blow dry. Thoroughly dry slides on a slide warmer. Apply coverslips to slides using Cytoseal 60 or similar organic-based mounting medium, or proceed to autoradiographic detection (see Basic Protocol 5).

    Color-development artifacts with alkaline phosphatase may be due to endogenous peripheral-type enzyme, which may be blocked with levamisole. Intestinal alkaline phosphatase is more refractory and requires a 10-min treatment with 0.1 M HCl at room temperature. DTT present during enzymatic development can impart a strong purplish color. The use of nonhybridized sections should reveal whether adventitious color formation is occurring. Loss of AP staining occurs with exposure to ethanol.

ALTERNATE PROTOCOL

TYRAMIDE SIGNAL AMPLIFICATION (TSA) DETECTION OF DIGOXIGENIN-LABELED PROBES

Additional Materials *(also see Basic Protocol 3; see* APPENDIX 1 *for items with ✓)*

Peroxidase-conjugated sheep polyclonal anti-digoxigenin antibody (Roche)

Renaissance TSA-Indirect kit (PerkinElmer) containing:
- 2× diluent
- Blocking reagent
- Biotinylated tyramide

Streptavidin-Texas red *or* streptavidin-fluorescein conjugate (PerkinElmer)
Streptavidin-HRP conjugate (PerkinElmer)
✓ TBS, pH 8.0 (room temperature)
Diaminobenzidine (DAB) tablets (Sigma)
Urea/hydrogen peroxide tablets (Sigma)
Streptavidin–alkaline phosphatase conjugate (Jackson Immunoresearch)
Cytoseal 60 (Stephens Scientific) or similar organic-basic mounting medium

1. Wash slides with TBS, pH 7.5, then with detection buffer, as for AP detection (see Basic Protocol 3, step 1).

2. Prepare a 1:300 to 1:600 dilution of peroxidase-conjugated anti-digoxigenin antibody in detection buffer. Incubate slides with this solution 2 hr at room temperature with gentle rocking, and then overnight at 4°C. Wash slides three times, each time by immersing for 5 min in fresh TBS, pH 7.5.

3. Incubate slides 5 min with 1 diluent (from Renaissance TSA-Indirect kit). Prepare a 1:50 to 1:100 dilution of biotinylated tyramide in 1× diluent. Incubate slides 10 min with this solution. Wash slides three times, each time by immersing 5 min in TBS, pH 7.5. Prepare a 0.5% solution of blocking reagent in TBS, pH 7.5. Incubate slides 30 min with this solution.

4a. *To label with Texas red or fluorescein (least sensitive):* Prepare a 1:2000 dilution of streptavidin-Texas red or streptavidin-fluorescein in TBS, pH 7.5. Transfer slides to this solution and incubate 60 min at room temperature. Wash slides four times, each time by immersing 5 min in fresh TBS, pH 7.5.

4b. *For detection with HRP and substrate (intermediate sensitivity):* Prepare a 1:3000 dilution of streptavidin-HRP conjugate in TBS, pH 7.5 containing 0.5% blocking reagent. Transfer slides to this solution, incubate 60 min at room temperature, then wash slides four times, each time by immersing 5 min in fresh TBS, pH 8.0. Dissolve one DAB tablet and one urea/hydrogen peroxide tablet in 15 ml of TBS, pH 8.0. Transfer slides to this solution and incubate until a signal develops (∼5 to 10 min). Wash slides twice, each time by immersing 5 min in fresh TBS, pH 8.0.

4c. *For detection with alkaline phosphatase and substrate (most sensitive):* Prepare a 1:25,000 dilution of streptavidin–alkaline phosphatase conjugate in detection buffer. Transfer slides to this solution and incubate 60 min at room temperature. Develop with NBT/BCIP substrate and wash with SSPE (see Basic Protocol 3, steps 3 to 4).

5. Dip slides briefly in water and blow dry. Thoroughly dry slides on a slide warmer. Apply coverslips to slides using Cytoseal 60 or similar organic-based mounting medium, or proceed to autoradiographic detection (see Basic Protocol 5).

Color-development artifacts with alkaline phosphatase may be due to endogenous peripheral-type enzyme, which may be blocked with levamisole. Intestinal alkaline phosphatase is more refractory and requires a 10-min treatment with 0.1 M HCl at room temperature. DTT present during enzymatic development can impart a strong purplish color. The use of nonhybridized sections should reveal whether adventitious color formation is occurring. Loss of AP staining occurs with exposure to ethanol.

BASIC PROTOCOL 4

DETECTION OF CHROMOSOMAL DNA (Y CHROMOSOME) USING RIBOPROBES

Materials *(see* APPENDIX 1 *for items with* ✓*)*

Superfrost (positively charged) slides containing defatted 12-µm fresh-frozen or paraffin-embedded sections prepared by cryostat (*UNIT 1.1*) or whole fixed cells
✓ Paraformaldehyde-picric acid solution
✓ Phosphate-buffered saline (PBS)
✓ 0.1×, 0.5×, 1×, and 2× SSC
Citrisolv (Fisher) or xylene
70%, 80%, 95%, and 100% ethanol
Citra Plus, pH 6 (BioGenex)
Reagents for combined immunocytochemistry/immunohistochemistry (optional) including:
  Blocking solution: PBS containing 1% (w/v) BSA and 0.6% (v/v) Triton X-100
  Biotinylated secondary antibody (from Vectastain ABC kit, Vector Laboratories, or Jackson Immunoresearch)
  ✓ 0.1 M Tris·Cl, pH 8
  Streptavidin-HRP conjugate (PerkinElmer)
  Biotinylated tyramide (PerkinElmer)
  Fluorochrome-labeled streptavidin of different color from that used to develop the hybridization histochemical signal
✓ Hybridization solution containing 3 to 10 µl/100 µl digoxigenin-labeled riboprobe (Support Protocol 2) to a mouse repeat sequence (Bishop and Hatat, 1987; Mezey et al., 2000) or human Y-chromosomal marker DYZ1 (Mezey et al., 2003)
70% formamide/2× SSC
DAKO Peroxidase Blocking Reagent or hydrogen peroxide

Plastic Coplin jar
Staining dishes and tubs
Slide rack
Sterile Bio-Assay dishes (Nunc, 245 × 245 × 30–mm)
55°, 65°, and 80°C incubators
65° and 80°C water bath

1a. *For fresh-frozen sections and whole fixed cells:* Immerse slides for 5 min in 1% paraformaldehyde-picric acid solution. Wash slides three times, each time for 5 min, in 1× PBS, then once for 5 min in 1× SSC. For fresh-frozen tissues, go to step 3 for combined immunocytochemistry/immunohistochemistry or to step 4 for hybridization histochemistry alone. For whole fixed cells, go to step 2.

1b. *For paraffin-embedded sections:* Using a slide rack, immerse slides containing tissue sections in staining dishes containing:

  Citrisolv (or xylene) three times, 7 min each, to deparaffinize
  100% ethanol three times, 3 min each
  95% ethanol, 5 min each
  80% ethanol, 5 min each
  70% ethanol, 5 min each.

  Rinse briefly in water. Go to step 2.

2. Place slides containing deparaffinized sections or whole fixed cells in a plastic Coplin jar filled with Citra Plus and transfer to a microwave oven. Bring to a boil at 600 W and then adjust power to 450 W for 15 min.

3. Immerse slides in 70% ethanol at 4°C for 5 min, then rinse quickly in water. Carry out the following procedures if combined immunocytochemistry/immunohistochemistry is to be performed:

   a. Incubate slides in blocking solution for 10 min at room temperature.

   b. Incubate slides in appropriately diluted primary antibody 1 hr at 37°C or overnight at 4°C.

   c. Wash three times in PBS, each time for 5 min at room temperature.

   d. Dilute biotinylated secondary antibody 1:1000 in blocking solution. Incubate slides in diluted secondary antibody 1 hr at room temperature.

   e. Wash three times in PBS, each time for 5 min at room temperature.

   f. Dilute streptavidin-HRP conjugate 1:250 in blocking solution. Incubate slides in diluted streptavidin-HRP conjugate 30 min at room temperature.

   g. Wash three times in 0.1 M Tris·Cl, pH 8, each time for 5 min at room temperature.

   h. Prepare a 1:50 to 1:100 dilution of biotinylated tyramide in 1× diluent (provided in Renaissance TSA-Indirect kit). Incubate slides in the biotinylated tyramide solution 5 min at room temperature.

   i. Wash three times in 0.1 M Tris·Cl, pH 8, each time for 5 min at room temperature.

   Continue with step 4. After step 7, add fluorochrome-labeled streptavidin (of a different color from that used to develop the hybridization histochemical signal) at a dilution of 1:500 to 1:2500 (see Alternate Protocol, step 4a) to visualize the antigen of interest.

4. Transfer to 0.2 N HCl at room temperature for 20 min, then rinse quickly in water. Place slides in 2× SSC at 80°C (in a water bath) for 25 min and then transfer them to 2× SSC at room temperature to cool.

5. Heat the hybridization buffer containing the probe at 80°C for 10 min. Simultaneously, place the slides in 70% formamide/2× SSC at 80°C for 10 min. Incubate the sections in 50 to 80 μl of hybridization buffer containing the probe at 80°C for 12 min, then transfer the slides to a Bio-Assay dish at 55°C for 30 min.

6. Remove the coverslips in 2× SSC. Wash slides by immersing successively in 1×, 0.5×, and 0.1× SSC at room temperature, each time for 5 min. Transfer the slides to 0.1× SSC at 65×C (in a water bath) for 30 min.

7. To block endogenous peroxidase prior to tyramide staining, use DAKO Peroxidase Blocking Reagent (usually 3% hydrogen peroxide will also suffice) for 5 min at room temperature. Perform tyramide signal amplification (TSA) detection of digoxigenin-labeled probes (see Alternate Protocol).

   With human brain tissue, the autofluorescence of lipofucsin granules is a serious problem in fluorescence hybridization histochemistry. A 2-min staining in 70% ethanol/Sudan Black at the end of the hybridization procedure will eliminate this.

BASIC PROTOCOL 5

DETECTION OF RADIOLABELED PROBES

Materials

Ilford K5.D or Kodak NTB-3 nuclear emulsion
7.5 M ammonium acetate
^{35}S-labeled samples, slide-mounted (see Basic Protocols 1 and 2)
Kodak D-19 photographic developer, 17°C
Kodak Rapid Fix (without hardener), 17°C
Counterstain (optional): 0.4% toluidine blue, 2 μg/ml ethidium bromide, hematoxylin/eosin, or other stain of choice
Cytoseal 60 (Stephens Scientific) or similar organic-based mounting medium

Darkroom with safelight
Coplin jars
Spatula
40°C water bath
Black slide boxes
Desiccant capsules (e.g., Humi-caps from United Desiccants-Gates)
Black photography tape
Slide racks
Slide warmer

1. In a darkroom under safelight conditions, scoop out 40 ml of emulsion with a spatula and transfer into a Coplin jar containing 1.6 ml of 7.5 M ammonium acetate (for final ammonium acetate concentration of 300 mM). Place the Coplin jar in a 40°C water bath for 20 to 30 min to allow air bubbles to rise. Mix gently and test for the complete elimination of bubbles by examining a clean slide after dipping it into the emulsion.

2. Dip ^{35}S-labeled, slide-mounted sections into the emulsion and allow to dry for several hours standing up. Typically, place five slides in red plastic slide grips and dip them five at a time into the emulsion, then hang them from a custom-made Plexiglas holder.

3. Place the emulsion-coated slides in black slide boxes with desiccant capsules. Tape the edges of the box with black photography tape and store the boxes at 4°C in the dark for the appropriate time.

4. Put the emulsion-coated slides in slide racks and develop as follows.
   a. Immerse 2 min in 17°C Kodak D-19 developer with agitation every 30 sec.
   b. Immerse 15 sec in running tap water with slight agitation.
   c. Immerse 2 min in Kodak Rapid Fix (without hardener) with agitation every 30 sec.

5. Rinse slides 8 min in running tap water. Counterstain, if desired, for 30 sec in 0.4% toluidine blue, 2 μg/ml ethidium bromide, hematoxylin/eosin, or other stain of choice, then rinse again briefly to remove excess stain. Dip slides very briefly into deionized water, then in 70% ethanol, and place on slide warmer to thoroughly dry. Apply coverslips to slides using Cytoseal 60 or similar organic-based mounting medium.

   Some stains may obscure colorimetric detection of the digoxigenin probe or destroy silver grains (e.g., periodic acid/Schiff reagent).

   Potential artifacts that arise from autoradiography include the spurious creation and destruction of grains (positive and negative chemography, respectively). Assess positive chemography using sections that are not hybridized or that are hybridized to a sense probe. Grains are especially susceptible to loss during staining or after coverslipping if moisture remains in the tissue sections.

SUPPORT PROTOCOL 1

PREPARATION OF OLIGONUCLEOTIDE PROBES FOR HYBRIDIZATION HISTOCHEMISTRY

Materials *(see* APPENDIX 1 *for items with ✓)*

- ✓ 5× tailing buffer
- Oligonucleotide to be used as probe (∼48 bases long)
- [α-^{35}S]dATP (>1000 Ci/mmol; PerkinElmer)
- Terminal deoxynucleotidyl transferase (TdT; Roche or Invitrogen)
- 250 μM digoxigenin-dUTP/1 mM dNTP (or dATP) mix (ingredients available from Roche; store mix indefinitely at –20°C)
- ✓ TE buffer, pH 7.6
- 4 M NaCl
- 25 μg/μl yeast tRNA
- 70% and 100% ethanol
- 5 M dithiothreitol (DTT)

For radiolabeled probes

1a. Prepare the following reaction mix on ice:

10 μl 5× tailing buffer
Oligonucleotide to 0.1 μM
[^{35}S]dATP to 1 μM
70 to 100 U TdT
H_2O to 50 μl.

The [^{35}S]dATP should not be added in the form of a solution containing EDTA or similar chelator that will interfere with the tailing buffer (a color change indicates that this has happened).

2a. Incubate reaction ∼2 to 5 min at 37°C to add 10 to 15 bases.

For digoxigenin-labeled probes

1b. Prepare the following reaction mix on ice:

10 μl 5× tailing buffer
Oligonucleotide to 0.1 μM
1 μl 250 μM digoxigenin-dUTP/1 mM dNTP (or dATP) mix
70 to 100 U TdT
H_2O to 50 μl.

2b. Incubate 2 hr at 37°C.

3. Add the following to the reaction mix:

375 μl TE buffer, pH 7.6
25 μl of 4 M NaCl
50 μg yeast tRNA
1 ml ethanol.

Mix thoroughly, incubate 10 min on wet ice, then microcentrifuge 10 min at maximum speed.

4a. *For radiolabeled probes:* Remove the supernatant and rinse the pellet with 1 ml of 70% ethanol. Microcentrifuge briefly at maximum speed, remove the ethanol, and add 50 μl TE buffer, pH 7.6, and 1 μl of 5 M DTT. Dissolve pellet with gentle vortexing and count

1 μl of the radiolabeled probe in a scintillation counter and record the number of dpm (expect ~500,000 dpm). Store probe up to 3 months at 4°C or at –80°C for longer periods.

4b. *For digoxigenin-labeled probes:* Remove the supernatant and rinse the pellet with 1 ml of 70% ethanol. Microcentrifuge briefly at maximum speed, remove the ethanol, and add 50 μl TE buffer, pH 7.6. Dissolve pellet with gentle vortexing. Store indefinitely at 4°C.

Use 3 to 10 μl of digoxigenin-labeled oligodeoxynucleotide probe to 100 μl hybridization solution (see Basic Protocol 1).

SUPPORT PROTOCOL 2

PREPARATION OF RNA PROBES (RIBOPROBES) FOR HYBRIDIZATION HISTOCHEMISTRY

Longer riboprobes offer greater sensitivity. Some researchers also employ alkaline hydrolysis of riboprobes to facilitate tissue penetration, but the authors have not found this helpful, and such a step may result in inconsistency in probe sizes.

Materials *(see APPENDIX 1 for items with ✓)*

Linearized plasmid containing cDNA to be transcribed
[α-^{35}S]UTP (>1000 Ci/mmol; PerkinElmer)
✓ 5× transcription buffer
100 mM dithiothreitol (DTT)
10 mM ATP, CTP, and GTP
RNasin (Promega)
10 to 20 U/μl RNA polymerase (SP6, T3, or T7)
1 mM UTP/4 mM digoxigenin-11-UTP mix (ingredients available from Roche; store mix indefinitely at –20°C)
✓ 1 U/μl RNase-free DNase I
✓ TE buffer
4 M NaCl
25 μg/μl yeast tRNA
70% and 100% ethanol
5 M DTT
✓ 10% (w/v) sodium dodecyl sulfate (SDS)
TE buffer/5% (w/v) SDS

NOTE: Use DEPC-treated water (*APPENDIX 1*) for all reagents.

For radiolabeled probes

1a. Combine ~250 ng of linearized plasmid (or 10 to 100 ng of PCR product) with [^{35}S]UTP at a final concentration of 20 to 30 μM in a microcentrifuge tube. Dry in a Speedvac evaporator at room temperature.

2a. Add, at room temperature, to the tube with the dried [^{35}S]UTP and DNA:

1 μl 5× transcription buffer
0.5 μl 100 mM DTT
0.5 μl 10 mM ATP
0.5 μl 10 mM CTP
0.5 μl 10 mM GTP
0.5 μl RNasin
0.5 μl of 10 to 20 U/μl SP6, T3, or T7 RNA polymerase
H_2O to 5 μl.

1

Mix thoroughly, especially where the $[^{35}S]UTP$ and DNA have been pelleted.

For digoxigenin-labeled probes

1b. Combine at room temperature in a microcentrifuge tube:

~250 ng linearized plasmid
0.5 µl 1 mM UTP/4 mM digoxigenin-11-UTP mix
1 µl 5× transcription buffer
0.5 µl 100 mM DTT
0.5 µl 10 mM ATP
0.5 µl 10 mM CTP
0.5 µl 10 mM GTP
0.5 µl RNasin
0.5 µl SP6, T3, or T7 RNA polymerase
H_2O to 5 µl.

2b. Mix thoroughly and proceed to step 3.

3. Incubate reaction mixture 30 min at 37°C and add another 0.5 µl of the appropriate (SP6, T3, or T7) RNA polymerase. Incubate 30 min at 37°C, then add 0.5 µl RNasin and 0.5 µl of 1 U/µl RNase-free DNase and continue incubating 10 min at 37°C.

4. Add 420 µl TE buffer, 25 µl of 4 M NaCl, 2 µl of 25 µg/µl tRNA, and 1 ml of 100% ethanol. Mix thoroughly, place on wet ice for 10 min, then microcentrifuge 10 min at maximum speed. Remove the supernatant and rinse the pellet with 1 ml of 70% ethanol. Microcentrifuge briefly at maximum speed.

5a. *For radiolabeled probes:* Remove the ethanol and add 485 µl TE buffer, 10 µl of 10% SDS, and 5 µl of 5 M DTT to dissolve the pellet. Count 1 µl of radiolabeled probe in a scintillation counter and record the number of dpm (expect ~1×10^6 dpm per µl). Freeze at –80°C until used.

5b. *For digoxigenin-labeled probes:* Remove the ethanol and add 50 µl TE buffer/0.5% SDS to dissolve the pellet. Store in aliquots indefinitely at −80°C.

Use 3 to 10 µl of digoxigenin-labeled riboprobe to 100 µl of hybridization solution (see Basic Protocol 2).

References: Albertson et al., 1995; Bishop and Hatat, 1987; Bobrow et al., 1989; Bradley et al., 1992; Hunyady et al., 1996; Mezey et al., 2003; Wilkinson, 1992; Young et al., 1986; Young, 1992

Contributors: W. Scott Young III and Éva Mezey

UNIT 1.4

Receptor Binding Techniques

In the past, a variety of clinically useful agents were discovered empirically through behavioral screening, but without clear indication as to their biochemical actions in the brain. Subsequently, it was recognized that many drugs interacted with neurotransmitter receptors or other minor protein constituents by means of stereospecific, saturable binding. The in vivo distribution of radioligands was then applied to detailed mapping of the binding sites in brain. It was later determined that in vitro assays employing radioligands could be performed for the same purpose. The potential advantages of the in vitro design became immediately apparent—e.g., the ability to directly control and measure aspects of the binding reaction, the use of reduced amounts of radioligand, and the possibility of employing radioligand binding to evaluate the properties of unlabeled drugs in competition assays. In vitro autoradiographic binding-site assays are now important tools in contemporary neuroscience and pharmacology, with strengths and limitations that are often complementary to those of homogenate binding assays. Performance of rapid kinetic association and dissociation assays and detailed multisite saturation or competition assays can be more difficult by autoradiography than with the use of tissue homogenates. Conversely, the very high sensitivity and spatial resolution of autoradiography affords the ability to study structures too small to accurately dissect for homogenate assays. The ability to combine binding-site autoradiography with other anatomical imaging procedures, including routine histological staining, histochemical staining, immunohistochemistry, nucleic acid hybridization assays, and some in vivo radioligand distribution assays, permits powerful and flexible multidisciplinary protocols not possible with homogenate assays.

Pharmaceutical development may be assisted by both in vitro and in vivo ligand autoradiography. Although in vitro methods are preferred for the detailed characterization of binding-site pharmacology, in vivo distribution studies continue to be of use in determining the detailed biodistributions of new drugs, affording important data on pharmacokinetics and regional distribution in preclinical animal studies.

RADIOLIGAND SELECTION

Isotopic Label

Identification of candidate radioligands for autoradiography of binding sites first necessitates consideration of the possible radioisotopes for their labeling. The most widely employed isotope, tritium (^3H), has the advantages of low cost, ease of safe handling and disposal, and a wide variety of approaches for introducing label into ligands. Tritium decays by the conversion of a neutron to a proton with emission of a β^- particle and an antineutrino, thus increasing the number of protons in the nucleus and resulting in an atom of helium. Tritium is unique among the isotopes employed for autoradiography because its average β^- energy permits travel through only a few micrometers of tissue before its absorption (Table 1.4.1). Typical frozen histological tissue sections are much thicker than the average ^3H β^- particle path length; sections of 15 to 20 μm or greater are infinitely thick with respect to ^3H emissions. Autoradiograms of tritium-ligand distribution in most tissue sections are thus insensitive to minor inconsistencies in sectioning precision. The advantage afforded by this insensitivity to tissue-section thickness is counterbalanced by the effects of tissue self-absorption and by the need for specialized films and emulsions to detect of the low-energy ^3H β^- emissions.

As depicted in Table 1.4.1, radionuclides employed for autoradiography differ in mode of decay as well as in physical half-life, each of which affect the feasibility of their use in binding-site autoradiography. As a general guide, if binding sites are present at an average tissue concentration of 10 to 100 nM, a single tritium atom per ligand molecule will permit autoradiographic imaging with exposure times of 5 to 7 days and specialized tritium-sensitive X-ray films. Increasing the number of labeled atoms per molecule or increasing the exposure time will each independently lower the detection threshold. If the known target binding-site density is substantially lower, use of a shorter-half-life nuclide at higher specific activity per labeled atom may be necessary.

The most commonly employed nuclide with very high specific activity is iodine-125 (^{125}I), obtainable at almost 100-fold higher activity per atom on the basis of its shorter half-life. However, additional features, including abundance of charged particulate emissions per decay, effective film-exposure index per particle, and the number of nuclide atoms per ligand molecule

Table 1.4.1 Properties of Some Nuclides Useful in Autoradiography

Nuclide	Emission	Max. energy (keV)	Avg. energy (keV)	Max. tissue range (mm)	Avg. tissue range (mm)	Half-life	Max. sp. act. (Ci/mol)	Particles per transition
^{14}C	β^-	156	49	0.301	0.042	5370 years	6.7×10^1	1.00
^{18}F	β^+	635	250	2.440	0.637	1.83 hr	1.7×10^9	1.00
^{3}H	β^-	18.6	5.7	0.008	0.004	12.3 years	2.9×10^4	1.00
^{123}I	ICE[a]	127	127	0.214	0.214	13.2 hr	2.4×10^8	0.14
^{125}I	AuE[b]	30	24[c]	0.019	0.012	60 days	2.2×10^6	0.20
^{131}I	β^-	606	192	2.300	0.421	8.04 days	1.6×10^7	0.89
^{32}P	β^-	1710	695	8.210	2.750	14.3 days	9.2×10^6	1.00
^{33}P	β^-	250	77	0.637	0.092	25 days	5.1×10^6	1.00
^{35}S	β^-	167	49	0.336	0.042	87 days	1.5×10^6	1.00
^{99m}Tc	ICE[a]	120	120	0.195	0.195	6.01 hr	5.2×10^8	0.09

[a]Internal conversion electron decay.

[b]Auger electron decay.

[c]For ^{125}I (which exhibits Auger electron decay), the average energy is calculated from the abundance-weighted mean of electron energies resulting from ten distinct transition possibilities with energies ranging from 22 to 30 keV.

additionally influence the relative sensitivities of ligands labeled with ^{125}I versus those labeled with ^{3}H. Labeling of ligands with ^{125}I may be accomplished by replacement of stable ^{127}I or by iodination of a parent ligand at a position distant from its site of interaction with the binding site of interest. Iodine-bearing ligands are often more challenging to employ in vitro than their noniodinated counterparts because of the effects of increased molecular size and greater lipophilic character conferred by iodine. These properties tend to slow the kinetics of ligand diffusion, binding, and dissociation—and also promote increased nonspecific tissue binding when assays employ polar, saline-based buffers. Emitted electrons, termed "Auger electrons," are the principal means of autoradiographic detection of ^{125}I. The energy of the Auger electron is characteristic of the difference in energy states of the initial electron transition as well as the energy of the electron absorbing the transition energy. In the decay of ^{125}I, Auger electrons of ten individual energies within the range of 20 to 30 keV are emitted with a cumulative probability of 20%. Thus, only one in five decaying ^{125}I atoms emits an Auger electron likely to interact with autoradiographic film. The energies of these Auger emissions, however, are sufficiently higher than those of ^{3}H β^- particles so that their tissue-penetration distances are similar to typical frozen-section thicknesses. Also, the effective film exposure per particle is much greater, and traditional anti-scratch-coated X ray film can be used for autoradiography. Overall, autoradiographic detection sensitivity per ^{125}I-decay particle may be 5- to 10-fold higher than for ^{3}H, and when this is combined with the differences in specific activity and particle yield per decay, autoradiographic detection may be enhanced by 50- to 100-fold over ^{3}H at the same tissue concentrations.

Other radionuclides have been employed successfully for autoradiography involving the emissions of β^- particles, including ^{14}C, ^{32}P, ^{33}P, ^{35}S, and ^{131}I. Of these nuclides, all but ^{14}C have specific activities adequate for routine autoradiographic imaging of low-abundance binding sites such as neurotransmitter receptors and transporters. The emitted energies of these tracers are all high enough to penetrate typical frozen-section tissue thicknesses well; furthermore these isotopes are not subject to tissue self-absorption concerns and do not require specialized films or emulsions for detection. However, the range of ligands that can be labeled with these nuclides is more limited than with ^{3}H, based on their infrequent occurrence in active drugs.

The β^+ emitter ^{18}F has proven useful in positron emission tomographic (PET) imaging of neuroreceptors and other binding sites when incorporated in ligands to replace stable ^{19}F, when substituted for OH groups, or when incorporated in an alkyl substituent (e.g., when employing fluoromethyl or fluoroethyl substitutions for methyl or ethyl groups). ^{18}F decay to stable ^{18}O involves emission of a β^+ particle with maximum energy of 635 keV, but as in the instance of β^- decay, division of energy between the β^+ particle and neutrino results in a spectrum of emitted β^+ energies averaging ~250 keV. The emitted particles are effective in exposing autoradiographic film, although they are of such high energy that anatomic resolution is considerably less than that typically achieved with either ^3H or ^{125}I. An additional consideration is the ultimate fate of the β^+ particle, because after loss of its initial kinetic energy it undergoes annihilation with an electron in the absorbing medium whereby the β^+ particle and electron are converted to 511-keV photons. These electromagnetic radiations, indistinguishable in properties from γ rays and X rays, require the use of shielding in the laboratory to reduce exposure of personnel during in vitro assays. Although the short half-life of ^{18}F can present challenges in tissue processing, this isotope has been successfully employed for ex vivo detection of ligand distributions.

The nuclides technetium-99m (^{99m}Tc) and iodine-123 (^{123}I) can decay by emission of internal conversion electrons (orbital electrons ejected after absorbing energy from a nuclear transition) with probabilities of 9% and 14%, respectively, per transition. In ~89% of ^{99m}Tc to ^{99}Tc transitions, energy is released as a γ ray of 140 keV. In one of several alternative transition routes, however, energy is ultimately released as an internal conversion electron of 120 keV. In the case of ^{123}I, initial transition by electron capture results in conversion of a proton to a neutron, and ~14% of transitions lead subsequently to the emission of an internal conversion electron of 127 keV that is suitable for autoradiographic imaging. Because ^{99m}Tc is a transition metal, it can be incorporated by chelation or coordination chemistry into organic ligands; however, as in the instance of iodination, the resulting ligands generally have much larger molecular sizes than related parent drugs.

Radionuclides with complicated transition schemes—e.g., ^{99m}Tc, ^{123}I, ^{125}I, and ^{131}I—most often give rise to multiple forms of emitted energy, including γ rays employed in medical diagnostic imaging. While these radionuclides are useful in tissue autoradiographic assays, this aspect necessitates consideration of additional shielding during in vitro assays, which is not necessary with the use of pure, low-energy β^- emitters such as ^3H, ^{14}C, ^{33}P, and ^{35}S.

Pharmacological and Chemical Ligand Profiles

Ligands selected for in vitro autoradiographic assays share many characteristics with ligands employed for homogenate binding assays. The most useful ligands have high affinity for a well-defined population or populations of binding sites, exhibit relatively low nonsaturable or nonspecific binding, and have kinetic properties facilitating convenient equilibrium-binding assay development. Generally speaking, antagonist ligands that do not distinguish different receptor states on the basis of effector coupling provide the best tools for quantification of binding-site numbers. As with homogenate binding assays, ligand polarity may influence the ability to interact with the full tissue complement of binding sites. The use of highly polar ligands, such as quaternary amines, may limit detection to receptors exposed on membrane surfaces, and in intact cells, to cell-surface receptors. Lipophilic ligands often detect additional binding sites presumed to reside in more lipophilic domains. In the example of binding to intact cells, these may include intracellular or internalized receptors. In homogenate assays and other broken-cell preparations, lipophilic ligands may still interact with a larger number of binding sites than polar ligands, suggesting that intramembranous receptor location or adjacent lipophilic membrane domains may be important. Ligand polarity/lipophilicity is most readily estimated on the basis of partitioning between aqueous and organic phases. A widely employed system uses octanol and saline phases. This system has been sufficiently studied to enable the estimation of the partitioning properties of new ligands by extrapolation from measurements made on related compounds (i.e., from the partitioning behavior of the parent molecular structure and its substituents). Ligands with octanol:saline partition coefficients less than unity are sufficiently polar that they do not cross cell membranes in intact cells and are likely to detect binding sites only in hydrophilic domains. Ligands with partition coefficients >10 are sufficiently lipophilic that they enter membranes readily, and are likely to detect binding sites in both lipophilic and hydrophilic domains. Ligands with partition coefficients >1000 pose additional problems in vitro assays, these result from nonspecific binding. These ligands, encountered frequently among iodinated compounds, have very poor solubility in saline-based buffers and tend to distribute avidly into tissue sections in binding assays. The addition of carriers in the assay buffer (e.g., albumin) or reduction in buffer polarity by isoosmotic substitution of sucrose for sodium chloride may be necessary to reduce nonspecific binding in some instances.

CHARACTERIZATION OF LIGAND-BINDING ASSAYS

General Aspects of Ligand-Binding Autoradiography

A tissue source must be selected and preprocessed, then tissue samples are incubated in the presence of labeled ligand under carefully controlled conditions, the labeled tissue is washed to remove nonspecifically bound ligand, and specifically bound ligand is quantified. In the following sections, general aspects and guidelines for performance of autoradiographic assays will be considered.

Tissue Preparation

In addition to specimens obtained at biopsy or necropsy of experimental animals, human surgical resection and post-mortem samples are often sufficiently preserved for autoradiography. In experimental animal studies, simple gross organ or tissue dissection followed by freezing is preferred. Perfusion fixation (*UNIT 1.1*), while sometimes affording improvements in histological, histochemical, or in situ nucleic acid hybridization studies, should generally be avoided in tissues that are to be used in ligand-binding experiments. Even very mild aldehyde or other tissue-fixation schemes may modify kinetic and pharmacological properties of binding sites. If fixed tissues must be employed, it is important to conduct all initial characterizations of the binding site(s) in similarly processed tissues. Effects of post-mortem tissue handling and delay prior to dissection and freezing can theoretically influence the recovery of binding sites and other biochemical tissue markers. These considerations are particularly important in assays of human post-mortem tissues, where intervals of many hours may pass before specimens are collected. Potential post-mortem artifacts may be reduced by careful matching of sample histories between experimental groups. This approach, however, does not entirely guarantee that important information and distinctions have not been lost prior to tissue collection and storage. An additional approach to this problem is to conduct controlled experiments emulating human post-mortem conditions and intervals in experimental animals, thereby permitting direct assessment of possible differences between immediate and delayed tissue dissection and freezing.

After dissection, small specimens can be readily frozen by covering them in crushed dry ice. This method is applicable to tissue blocks of up to 3 to 5 g with good results. Blocks of considerably larger size require more efficient heat transfer during freezing to avoid microscopic ice-crystal formation or macroscopic fissuring. Large tissue blocks should be frozen by intermittent, repetitive immersion in isopentane chilled with dry ice (*UNIT 1.1*). Blocks should be dipped for several seconds at a time, followed by withdrawal for several seconds to permit equalization of temperature between the block surfaces and center. Otherwise, the surfaces will freeze first, and when the center subsequently freezes and expands the tissue may fracture. Frozen tissue blocks can be maintained for years if stored at −70°C and protected from desiccation. Coating the surfaces of tissue blocks with a frozen-section embedding medium and packaging in air-tight plastic freezer bags permits routine storage with good tissue preservation and maintenance of sectioning properties. Storage at higher temperatures or repeated temperature cycling from −70°C to cryostat-sectioning temperature may hasten loss of some tissue biochemical reactivities. If there is concern that aging or repeated temperature cycling may alter or degrade binding-site properties, this can be explored directly in control tissues.

Tissue is prepared typically for in vitro autoradiographic binding assays by sectioning in a cryostat microtome (*UNIT 1.1*). Usual frozen-section thicknesses of 10 to 30 μm are acceptable for autoradiographic purposes. It is important to bear in mind when selecting tissue-sectioning conditions that generation of ^3H autoradiograms will not be appreciably affected by section thicknesses within this range, nor will section-to-section imprecision affect the autoradiograms. However, direct scintillation spectrometry assays of sections and autoradiographic binding assays employing radionuclides with higher particulate-energy emissions will each be sensitive both to the selected tissue thickness and to the precision of sectioning. In the instance of ^3H autoradiography, it is typical to collect duplicate adjacent tissue sections to guard against the intrusion of artifacts in processing. With other radionuclides, triplicate sections should be considered, to reduce additional section-thickness error effects, unless the microtome precision is known to be excellent. Microtome precision can be assessed experimentally by comparison of autoradiographic densities of adjacent sections after incubation reactions with radionuclides of moderate to high penetration, or in ex vivo autoradiograms after the accumulation of tracers labeled with penetrating radionuclides. Motorized cryostat microtomes offer the benefit of improved section-thickness precision as compared with manual sectioning, but are considerably more costly than comparable manually driven models.

Tissue sections for use in autoradiographic binding assays are usually mounted frozen on microscope slides, thawed, and dried. Slides should be prewashed and subbed to promote section adhesion during binding assays. Subbing solutions employing gelatin (*UNIT 1.1*) are generally inferior to those containing poly-L-lysine for use in lengthy in vitro assays. High-heat conditions should be avoided during section drying to prevent binding-site denaturation. After complete drying, slide-mounted sections may generally be stored at −70°C for up to several weeks without loss of binding activity. If a particular binding site is known to be fragile or is of unknown stability, effects of post-sectioning storage should be investigated directly in control experiments.

Detection of Specific Binding

Specific binding is by definition saturable; thus, a minimum criterion is that the radioligand demonstrate a component of binding that does not increase linearly with free ligand concentration. This is most easily demonstrated by the effect of adding unlabeled ligand to the assay. More precise pharmacological characterization of specific binding is achievable when the radioligand and the unlabeled competitor are chemically dissimilar, minimizing the likelihood of multiple sites of overlapping tissue interaction. It is helpful to begin the search for specific binding of a new radioligand with knowledge of one or more unlabeled competitors of the ligand at the desired binding site as well as the inhibitory equilibrium binding constant (K_i) of each. It is best to employ an unlabeled competitor at concentrations not above $1000 \times K_i$, resulting in occupancy of 99.9% of binding sites. Use of unnecessarily higher inhibitor concentrations increases the possibility of detecting unwanted, secondary sites rather than the binding site of interest. In the sections to follow, it is assumed that a definition of nonspecific radioligand binding has been formulated and that its basis is the use of a competitive inhibitor at an appropriate concentration relative to its K_i, the radioligand concentration, and the radioligand's equilibrium binding constant (K_d).

Determination of Incubation and Washing Parameters

After establishing an initial definition of specific radioligand binding, it is important to determine critical assay parameters to maximize its separation from nonspecific binding. Preliminary studies to determine optimal incubation and post-incubation washing conditions are often performed without the use of autoradiography for radioligand measurement. In this instance, it is common to conduct binding studies identically to autoradiographic assays with the exception that after post-incubation washing, sections are wiped from the microscope slide with glass-fiber filter paper and assayed by liquid scintillation counting or by γ counting, depending on the radionuclide involved. In this way, rapid results can be obtained and multiple potential assay modifications and refinements tested in logical sequence. However, a series of tissue sections bearing reproducible numbers of binding sites is necessary for these assays.

Aspects of the basic binding incubation—including choice of buffer system, pH, time, temperature, and duration of incubation—should be explored to maximize specific binding. Often, important initial estimates of these quantities can be derived from prior in vitro homogenate binding assays. It is generally advisable in autoradiographic assays, however, to employ incubation buffers that are isotonic, so as to avoid osmotic stresses on tissue sections. Hypotonic buffers are particularly problematic if tissue-section adhesion is marginal, and their use may result in tissue fragmentation and losses from the slide.

An advantage of autoradiographic binding assays over homogenate designs is the ability to sample the binding incubation buffer directly and analyze it during the course of the assay. In the early phases of assay development, direct measures of free ligand concentration and chromatography of the radioligand after tissue exposure can assist with detection of unanticipated technical artifacts. If ligand depletion resulting from metabolism is noted, a lower incubation temperature should be considered. If ligand depletion resulting from nonspecific binding is noted, alteration of the buffer properties to promote ligand solubility may help. If depletion is due to specific binding, the volume of incubation buffer per tissue section must be increased.

An initial parameter for assay optimization is the post-incubation washing time that best distinguishes specific from nonspecific binding. Comparable tissue sections are usually incubated in solutions containing labeled radioligand with and without unlabeled competitor to block specific binding. Sections are then assayed for radioligand activity after varying post-incubation washing times, including an initial measurement without washing (i.e., a "zero wash time," or $t = 0$ sample). These initial sections are processed by very brief (1 to 2 sec) dipping in buffer lacking both the ligand and competitor to remove binding assay buffer from the slide surface. This step is important for assessing the possibility of loss of labeled sites during prolonged washing. Varying wash times are assessed, comparing total, specific, and nonspecific binding to the initial $t = 0$ estimates. Optimal post-incubation washing procedures should reduce nonspecific binding to a minimum, but not promote loss of any specific binding detected in $t = 0$ samples. Prolonged washing procedures that enhance specific over nonspecific binding, but at the expense of lost specific binding, run the risk of preferentially revealing a subset of binding sites. Unless the nature of specific binding that remains versus that potentially lost in these processes is understood, experimental results may be biased and results of studies employing distinct radioligands targeting the same site(s) may be discordant.

Estimates of specific-to-nonspecific labeling of tissues derived from whole-section assays may actually provide minimum values in comparison with autoradiographic assays when

^3H-labeled ligands are employed. This phenomenon is due to the likelihood that washing is most effective at the section surface, and most protracted at the section-slide interface where the diffusion distance to reach the buffer solution is greatest. Thus, the limited tissue-depth penetration of ^3H β^- particles may result in autoradiographic images of the most favorable specific-to-nonspecific binding strata.

The appropriate tissue incubation time to achieve equilibrium in the radioligand assay should be determined prior to conducting saturation assays or other pharmacological characterizations. Since the forward ligand-receptor binding rate is a second-order process, it is sensitive to ligand and receptor concentration effects. The time required to achieve equilibrium is, therefore, a function of radioligand concentration, and will be longest at the lowest ligand concentration studied. Thus, it is useful to have initial estimates of radioligand K_d so that equilibration time for ligand concentrations below the half-maximal binding level can be safely used in saturation analyses. In practice, equilibration-time determinations are straightforward in design, involving incubations of tissue with radioligand for progressively increasing durations, followed by routine post-incubation washing and assay of bound tracer. A caveat, however, is the possible loss of intact binding sites in tissue sections during prolonged incubations. This may arise, in theory, because of solubilization of some sites into the binding buffer. Alternatively, losses of high-affinity ligand recognition sites may reflect proteolytic degradation or dissociation of multiple proteins or cofactors that are required together for ligand binding. If ligand-binding sites are lost over the interval of the assay, apparent equilibration may take place, but the data will actually reflect competition between increasing ligand binding and decreasing total binding capacity (B_{max}). This aspect of ligand equilibrium can be controlled by conducting parallel assays at the lowest ligand concentration of interest and at a higher, saturating concentration where binding equilibrates more rapidly. The low-concentration data are used to assess equilibration rate, while lack of decrease in binding over time at the high concentration will exclude the possibility of binding-site losses during incubation. Desirable assay characteristics include incubation times sufficiently short that no detectable losses of binding sites occur, but which readily permit equilibration of sub-K_d ligand concentrations. Ligands that do not readily equilibrate at low concentrations may, nevertheless, be useful for assays of regional B_{max} distribution alone. However, it is recommended that use of a near-saturating ligand concentration for this purpose be used to guard against effects of possible unrecognized affinity differences that cannot be explored directly because of protracted equilibration times at low ligand concentration.

Saturation Studies

In addition to regional mapping of ligand-binding sites, assays to determine the pharmacological properties of saturable binding sites can be conducted by quantitative autoradiography. In particular, saturation binding assays are indispensable for evaluation of new radioligands in autoradiography.

Autoradiographic ligand-binding saturation analyses should be designed and conducted in parallel with in vitro homogenate assays. A range of ligand concentrations is employed in incubations of slides bearing a series of adjacent tissue sections. At each concentration, or at regular, less frequent intervals, assessment of nonspecific binding is made in additional sections. A 100-fold range of ligand concentrations centered on the K_d constitutes an ideal selection for estimation of both affinity and binding capacity. As detailed previously, it is important to verify technical aspects of the assay conditions so that artifacts do not result in erroneous saturation curves. It is important to verify by direct assay of the binding buffer that ligand depletion has not occurred during the incubation. It is also important that sufficient time be allowed to permit equilibration of ligand and binding sites. Each of these considerations is more important the lower the ligand concentrations used, and failure to properly design assays will result in an underestimation of binding. These effects result in curvilinear Rosenthall (Scatchard) plots, mimicking effects of cooperativity at interacting binding sites.

Competition Assays

Competition assays often require six or more distinct incubation conditions per binding site to assess binding affinity and capacity. In instances of multiple receptor subtypes that are distinguished by the competing ligand, as many as 15 to 20 parallel conditions may be necessary. Homogenate binding studies are adapted to these designs, permitting a large number of individual samples with highly similar tissue and binding-site compositions. In autoradiographic assays, however, anatomic gradients and variability from section to section make identification of truly identical tissue samples difficult. As a result, there is a methodological trade-off between the detail of the experimental design (number of replicate sections per condition and number of conditions) and the ability to select sufficient adjacent tissue sections bearing similar receptor types and numbers. Success of detailed multisite competition assays is enhanced if large, relatively invariant brain regions are selected (e.g., cerebral or cerebellar cortical regions, striatum, or major thalamic nuclei). In order to guard against unanticipated anatomic receptor gradients, repetition of the total binding condition (absence of competitor) should be performed at regular anatomic intervals throughout a series of sections. If the total binding does not change over the anatomic extent of the sections, differences may be safely ascribed to effects of the competitor. If subtle gradients in total binding are detected, it may be possible to interpolate between the control sections and use an adjusted B_{max} for each anatomic level.

QUANTIFICATION OF LIGAND BINDING

Basic Autoradiographic Principles

All autoradiographic techniques are based on the apposition of a photographic film or emulsion to materials containing a radioactive source. The passage of charged particulate emissions such as electrons, β^-, and β^+ particles through the silver halide crystals of photographic emulsions results in the conversion of silver ions to metallic silver. During development, these metallic silver ions catalyze the transformation of whole silver halide crystals into metallic silver, with consequent opacification of the emulsion. While electromagnetic radiation in the form of γ rays, X rays, or annihilation photons is also capable of exposing X-ray film, direct interactions between these radiations and photographic emulsions are much less efficient than those involving charged particulate emissions. Thus, little if any contribution to tissue autoradiography with the radionuclides discussed previously is attributable to γ- or X-ray emissions.

Factors influencing the response of autoradiographic emulsions to a radioactive source include the type of ionizing radiation emitted by the source, the energy of the ionizing particles, the thickness of the specimen containing the radioactive source, the distribution of radioactivity within the specimen, the distance from the specimen to the emulsion, the thickness of the emulsion, and the density of silver halide crystals within the emulsion. In quantitative receptor autoradiography, many of these issues are not a major concern as long as experiments are performed in a standardized manner.

Contact Autoradiography

The reversible binding of most radioligands implies that exposure to aqueous solutions after completion of the binding assay may result in unwanted losses in both the amount and anatomic localization of bound ligand. This means that the traditional method for detection and high-resolution localization of labeled macromolecules (proteins labeled by amino acid incorporation or nucleic acids labeled by incorporation of labeled nucleotides)—i.e., dipping slide-mounted specimens directly in liquid photographic emulsions, may result in migration or loss of label in binding-site autoradiography. As a result, methods have been developed to allow dry contact between labeled tissue sections and autoradiographic emulsions. In the first technique, which is an extension of the traditional emulsion-dipping method, acid-washed coverslips may be coated with an appropriate nuclear emulsion. Coverslips are dried, affixed to microscope slides bearing the labeled tissue sample fixed at one end with contact cement, and then fastened with a removable clip at the opposite end, apposing the emulsion to the sections. After exposure, the clips are removed and the coverslips are gently raised from the tissue opposite the side of permanent attachment, to permit development and fixation of the emulsion.

The more popular and more readily quantifiable autoradiographic method is to appose tissue sections to an appropriate acetate-backed film. Using double-sided tape or contact cement, sections on microscope slides or coverslips are mounted on paper or cardboard and placed in a rigid, light-tight cassette. Film is placed immediately on top of the sections and the cassette sealed. The cassettes are stored in a cool, dry environment for an appropriate period, and then the film is removed and developed. In this design, technical aspects of autoradiogram generation are simplified; however, the structural registration of the image with corresponding tissue sections is less precise than with the emulsion-dipping or coverslip-apposition techniques.

In film-contact autoradiography applications, the radionuclide used has a marked influence on the choice of film and subsequent data analysis. By far the most common radionuclide is the β^- emitter ^{3}H, on the basis of its previously-discussed favorable properties. The relatively low energy of its β^- emissions means that these emissions travel shorter distances both through tissue sections and within film emulsions, necessitating specialized autoradiographic techniques. Typical photographic and autoradiographic films are protected by a gelatinous "anti-scratch" layer overlaying the silver halide matrix, which is designed to reduce mechanical damage as well as chemographic and contact-induced silver reductions. In the instance of ^{3}H, these protective layers are too thick for β^- particles to effectively penetrate them, and the resulting autoradiographic sensitivity is extremely low. Thus, specialized autoradiographic films lacking the protective antiscratch layer are necessary for contact autoradiography. These films are more costly than conventional X-ray or other photographic films useful in autoradiography of higher-energy emitters, and must be handled with great care to prevent mechanical damage and exposure of the emulsion. Additional consideration must be given to the possibility of direct chemical interactions between these emulsions and the section and slide, as is also possible with liquid-emulsion and dipped-coverslip autoradiography. While specialized tritium-sensitive films will certainly work with ^{14}C, ^{35}S, and ^{32}P, use of such film in detection of the higher-energy emissions of these isotopes is unnecessary. Conventional X-ray films or finer-grained mammography films work well for these radioisotopes; the finer-grained emulsions afford optimal anatomic resolution with ^{14}C and ^{35}S, but at the expense of longer exposure times.

Film Densitometry and Radioligand Quantification

Data analysis by film densitometry has become straightforward through the availability of commercial computer-assisted densitometric systems. The relative ease of quantitative densitometry conceals some potential pitfalls in the analysis of autoradiographic data. It is important to recall that the relationship between incident-radiation exposure and optical density on developed film is not linear over an indefinite range of exposures. At higher levels of exposure, optical density reaches a ceiling (film saturation) beyond which it no longer reflects additional incident-radiation exposure. Even before saturation is reached, the relationship between

1

exposure and film optical density deviates significantly from linearity, with reduced resolution of differences in incident-radiation intensity. It is important to prepare and analyze autoradiograms under conditions in which there is a strong relationship between incident-radiation exposure and developed-film optical density. This precaution necessitates careful attention to the duration of film exposure and uniformity of development conditions.

A related problem is that of radioactivity standards. To perform truly quantitative receptor autoradiography, tissue sections must be coexposed together with standards that provide an external reference for the relationship between radiation exposure and film optical density. Plastic standards with known amounts of radioisotopes are now commercially available and widely used. Standards containing short-lived radioisotopes must be replenished and recalibrated frequently. This is true even for ^{3}H standards if they are to be used over a period of years. Standards for ^{14}C have the advantage of not requiring recalibration because of the very slow rate of decay, and can be cross-calibrated against other radioisotopes to provide a universal set of standards. This concept, however, has been challenged by evidence that ^{14}C and ^{3}H plastic standards may not always produce parallel optical-density changes. In the authors' experience, this may occur when film reciprocity is violated in either of two ways. If the intensity of one, but not both, of the exposures has saturated, the exposures will no longer vary proportionately. In the instance of ^{14}C and ^{3}H, most film emulsions permit ^{14}C access to deeper levels, resulting in higher maximal optical density and later saturation than with ^{3}H. Additionally, if exposure times are very protracted (more than several months), fading of the latent images resulting from ^{14}C and ^{3}H may occur at distinct rates that are due to differing degrees of lattice distortion produced by deposition of high versus low β^{-} particle energies. Good correlations between relative ^{14}C and ^{3}H exposures are obtained if films are exposed for less than 3 months time and if care is taken to avoid saturation of the ^{3}H exposure. In addition, ^{14}C standards have been used to control for exposures arising from ^{18}F, ^{35}S, ^{99m}Tc, ^{125}I, and ^{123}I. On the basis of these experiences, it may be predicted that extension of this method of standardization to other nuclides listed in Table 1.4.1 should also succeed.

Another important consideration in the interpretation of [^{3}H]ligand autoradiography is that differential absorption of emissions within tissue, known as tissue quenching, may alter imaging results. In brain tissue, lipid-rich white matter absorbs ^{3}H emissions to a greater extent than gray matter. This absorption differential gives rise to apparent ligand-concentration differences between brain regions that may in part reflect differing gray versus white-matter compositions. This effect may confer as much as a 2-fold difference in the autoradiographic image exposures of comparably labeled gray and white-matter structures. Differential effects of ^{3}H quenching may be advantageous for anatomic localization of binding signals through contrast-enhancing effects. However, in between-group experimental comparisons, alterations in tissue composition rather than binding-site numbers may confound autoradiographic measures.

Three approaches have been employed to address the ^{3}H quenching problem. One design is to determine the relative changes in a reference binding site (a "control" site known to be experimentally unaffected) in comparison to the binding site of interest. If both sites apparently change by the same fraction between experimental groups, altered tissue composition is the most likely explanation. Some groups have developed procedures to eliminate lipids from dried tissue sections, thus, eliminating the major source of regional quenching differences. Unfortunately, most delipidation procedures result in denaturation or loss of ligand-binding sites.

The final approach to evaluating and correcting for differential quenching effects is to make direct measures of attenuation of a diffusely introduced label in parallel sections to those in the binding assays. For example, terminal in vitro labeling of proteins followed by their autoradiography has been employed successfully for analysis and correction of tissue quenching effects. Quenching is not a significant source of bias or error in typical 20-μm-thick tissue sections when labeled with the higher-energy-emitting radionuclides.

ANALYSES OF AUTORADIOGRAPHIC DATA

Anatomic Distribution

Interpretation of autoradiograms requires the ability to identify reliably the correct regions of interest. Regional anatomy can be identified in several complementary ways. The most direct of these is use of the autoradiogram itself. With many ligands, there is sufficient detail and contrast in the images to establish regional boundaries, even for small regions. Fiber tracts typically express low levels of synaptic binding sites, and in the instance of tritium-labeled ligands, the higher lipid levels of fiber tracts will further reduce emulsion exposure. Consequently, structures such as the corpus callosum, the deep cerebellar white matter, the internal capsule, and other myelinated fiber tracts are identified readily and may serve as anatomic landmarks for structural identification. Many binding sites also have regional gray-matter distribution patterns that permit ready identification of selected regions.

Conventional histology can also be used to guide interpretation of autoradiograms. Sections exposed to film can be processed subsequently for Nissl staining and then used for regional anatomic identification. Some image-processing systems (e.g., the MCID System family from Imaging Research) support parallel processing and analysis of multiple images, allowing superposition of Nissl-stained sections and autoradiograms and direct transposition of structural boundaries from the histologic to the binding images. Sections immediately adjacent to those chosen for autoradiography can alternatively be used for histology. In addition to conventional histological tissue staining, sections can be processed for histochemistry or other techniques that may reveal boundaries or tissue distinctions not appreciated with conventional histology.

Saturation and Competition Isotherms

Analyses of binding curves to estimate affinities and densities of sites are identical to the post-processing analyses conducted on data from in vitro homogenate binding studies. As mentioned previously, care should be taken to ensure that radioligand K_d estimates used to make estimations of unlabeled ligand K_i values are appropriate to intact tissue sections. Ideally, all important binding parameters would be determined in each individual brain and region studied. This is often impractical, however, and K_d values determined in representative samples of an experimental group must often be extrapolated to regional mapping and unlabeled ligand competition experiments performed in other brains. Computer-assisted analyses are of particular value when multiple binding-site estimations are undertaken to assure that unbiased parameter estimates are made and to permit proper statistical comparisons of alternative binding models. With typical autoradiographic measurement variances and difficulties in assuring stable anatomic receptor expression throughout a large series of adjacent tissue sections, it is unusual to identify and separate more than two sites in most binding experiments.

Combination with Other Anatomic Methods

Receptor autoradiography can be combined with almost any neuroanatomical method for which tissue preparation is compatible with receptor studies, using alternate sections derived from the same blocks of tissue for individual assays. Unfixed, frozen tissue sections can be used in a wide variety of neuroanatomic methods. In addition, some comparison measures can be made on the same tissue sections used for receptor autoradiography. Routine histology with Nissl staining is readily accomplished on sections used for receptor binding after film exposure, by post-fixing the sections over paraformaldehyde vapors. Sections are placed in a tightly sealed container in a rack or on a platform suspended above paraformaldehyde

1

powder. The sections are exposed to paraformaldehyde vapors by warming of the chamber for a minimum of 48 hr, and then stained with 0.5% cresyl violet in the conventional manner (see *UNIT 1.2*). This method produces sections with excellent regional histology and surprisingly good preservation of perikaryal cellular morphology. The quality of these sections is more than adequate to establish regional boundaries and to confirm the presence of experimental lesions or cannula placements.

Another technique that can be readily combined with receptor autoradiography and that can employ either identical or alternate sections is [^{14}C]2-deoxyglucose (2DG) autoradiography of cerebral glucose metabolism. The same tissue sections utilized for 2DG studies can be processed for receptor autoradiography after the metabolic autoradiograms are initially obtained, or, more often, adjacent sections from [^{14}C]2DG-labeled brains can be processed for ligand binding and autoradiograms exposed in parallel. The latter design is favored, since rapid desiccation of tissue sections on a slide warmer enhances the anatomic resolution of the 2DG method, but may diminish the recovery of binding sites. In any event, it is important to incorporate additional prewashes into the receptor autoradiography protocol to completely remove the [^{14}C]2DG from the tissue prior to binding assays with lower-energy radionuclides.

A number of additional techniques can be applied to alternate sections harvested during cryostat sectioning. Many histochemical techniques can be applied to fresh-frozen tissue sections. While somewhat lacking in cellular resolution, histochemistry on fresh-frozen sections can give valuable information about regional anatomy. Some investigators, for example, have used cytochrome oxidase histochemistry on fresh-frozen sections to estimate the extent of experimental lesions. It is also possible to quantitate the intensity of many histochemical reaction products with densitometric methods analogous to those used for receptor autoradiograms. The absence of fixation eliminates a confounding variable for measurement of reaction-product density and can allow at least semiquantitative estimation of enzyme activity.

Another useful neuroanatomical method that can be employed in fresh-frozen sections is in situ mRNA hybridization (*UNIT 1.3*). Use of this technique together with receptor autoradiography on alternate sections allows complementary evaluations of regional mRNA and protein expressions; the ligand-binding assay provides the index of gene expression.

Tissue preparations for receptor autoradiography and immunohistochemistry are not generally compatible. Fixation necessary for the optimal visualization of many antigens often makes receptor autoradiography impossible.

Three-Dimensional Reconstructions

An emerging technique, developed initially for analysis of image data from positron emission tomography (PET) of blood flow in the human brain, involves the analysis of binding-site distribution data throughout the entire brain volume at high resolution. This approach has been explored for its extension to experimental-animal autoradiographic imaging, with encouraging preliminary results after reconstructions of the entire brain from autoradiograms at regular intervals. Screening of the entire neuroaxis without the need for predefined regions of interest or preconceptions regarding regional boundaries may permit detection of unanticipated focal changes or of significant gradients within larger regions, which are not evident in qualitative examinations of autoradiograms. Conversely, these parallel analyses of large numbers of independent regions (voxels) require conservative statistical adjustments against false-positive results that may limit detection sensitivity. Experimental designs that employ limited numbers of a priori regional hypotheses, but which additionally include omnibus whole-brain survey techniques, may afford the greatest overall utility.

CONCLUSION

Ligand-binding autoradiography is a powerful technique with many strengths and advantages. Unlike most neuroanatomic techniques, it is readily quantifiable and provides numerical data that can be subjected to parametric statistical analyses. In addition to providing maps of binding-site distribution, it can also yield data about binding-site pharmacology and regulation that is often comparable to that obtained from homogenate binding techniques with less anatomical resolution. While receptor autoradiography does not typically provide cellular-level resolution, it can provide excellent resolution on a brain-regional basis; with careful technique and selection of ligands it can distinguish and quantify small nuclear regions and subregions. Binding-site autoradiography is also very efficient with respect to numbers of experimental subjects needed in assays. Autoradiography often permits analyses of small brain structures that would require pooling of tissue from several subjects to obtain sufficient tissue for homogenate assays. Furthermore, it is technically easier to screen the entire neuroaxis in autoradiographic assays than to dissect and analyze a comparable number of discrete regions by homogenate binding methods. Tissue harvested for receptor autoradiography may also be used for routine histology, in situ mRNA hybridization, and histochemical techniques—allowing use of tissues for multiple complementary analyses.

Contributors: Kirk A. Frey and Roger L. Albin

UNIT 1.5

Intracellular and Juxtacellular Staining with Biocytin

Intracellular and juxtacellular methods for staining neurons allow microscopic visualization of neurons during electrophysiological experiments. These methods label all processes of a neuron, including the cell body, axon, dendrites, and dendritic spines. When sufficient time is allowed for transport of biocytin into the cell, the axon may be followed for several millimeters, permitting complete visualization of axonal arborizations in some cell types.

CAUTION: Osmium tetroxide is a hazardous chemical; wear protective gear, use in hood, avoid spilling, and store in an airtight container. Also, use glutaraldehyde, picric acid, and paraformaldehyde in the hood.

BASIC PROTOCOL 1

INTRACELLULAR BIOCYTIN STAINING WITH SHARP ELECTRODES

Biocytin or Neurobiotin

A solution of 1% to 4% (w/v) biocytin (372.5 mol. wt., Sigma) or Neurobiotin (322.8 mol. wt., Vector Laboratories) in 0.5 to 3 M potassium acetate, potassium chloride, or potassium methylsulfate is prepared. Low salt concentrations (0.5 to 1.0 M) are recommended for more effective transfer of biocytin into cells, and to reduce alteration of the potassium equilibrium potential of the cell under study. Higher salt concentrations yield lower electrode resistances and may be required when using very fine-tipped sharp electrodes. Heat solutions (to ~50°C) to facilitate dissolution of biocytin or Neurobiotin. Biocytin is used by more investigators than is Neurobiotin, but Neurobiotin is more soluble and may be more effectively ejected by iontophoresis. Both forms work well for intracellular staining. A typical solution in a microelectrode is 1 M KCl and 2% biocytin or 2% Neurobiotin.

Micropipets and filling
Glass micropipets are selected to obtain optimal intracellular recordings for the physiological experiment. Filling micropipets with solutions containing biocytin or Neurobiotin is slightly more difficult than filling with simple salt solutions, primarily because the biocytin-containing salt solution is slightly more viscous, so air bubbles are more likely to become trapped in the electrode. Gentle tapping on the outer wall of microelectrodes dislodges most trapped bubbles. If the micropipets are pulled from filament-containing glass, this will probably not be noticeable. Electrodes can be filled from the shaft, replacing the usual pipet solution with one containing biocytin (or Neurobiotin).

Biocytin ejection
Biocytin is ejected by application of either positive or negative currents using a constant-current device such as that provided with an active bridge intracellular recording amplifier (e.g., from Cygnus technologies; also known as a Neurodata amplifier, Ir183 or 283). Recommended currents range from 0.25 to 5 nA, with a duty cycle of 50% (e.g., 300 msec on and 300 msec off), for 15 to 90 min. The ejection current should be adjusted so that it does not damage the cell. The most reliable predictor of good staining is stable and secure intracellular recording throughout the biocytin ejection period. It is important to terminate ejection current if there is a substantial deterioration of the intracellular recording. Continuing to eject biocytin into an obviously damaged cell (this is sometimes called kill-and-fill) is likely to result in staining of multiple neurons. After filling a cell, the survival period can be anywhere between 1 and 24 hr. Survival times >24 hr are counterproductive due to intracellular degradation of biocytin.

ALTERNATE PROTOCOL

INTRACELLULAR BIOCYTIN STAINING WITH PATCH ELECTRODES

Biocytin or Neurobiotin
Add 0.1% to 0.5% (w/v) biocytin (Sigma) or Neurobiotin (Vector Laboratories) to the usual patch solution (see Basic Protocol 1).

Micropipets and filling
To prevent inadvertent staining of the surrounding tissue, it is useful to fill the tip with biocytin-free patch solution by application of negative pressure to the electrode while immersing the tip in biocytin-free solution. Fill the electrode shaft with biocytin-containing solution, and remove bubbles.

Biocytin ejection
No ejection is required, as the biocytin exchanges into the cell by diffusion.

BASIC PROTOCOL 2

JUXTACELLULAR BIOCYTIN STAINING

Juxtacellular staining is initiated from an extracellular position, and so employs the micropipets and filling solutions used for extracellular recording. The most difficult step in the juxtacellular method is recognizing when the electrode is in the correct position relative to the recorded neuron. Extracellular recording is not always done using glass micropipets or active bridge amplifiers. For juxtacellular staining, glass micropipets are required and the use of an active bridge amplifier is strongly recommended.

Biocytin or Neurobiotin
Prepare a solution of 1% to 4% (w/v) biocytin or Neurobiotin in 1 M NaCl.

Micropipets and filling
The sizes and shapes of micropipets used in extracellular recording experiments are not as critical as those for intracellular recording. Pipets should be constructed in a way consistent with obtaining good extracellular recordings of the neurons under study. There should be a good signal-to-noise ratio and the electrode characteristics should make it easy to distinguish between action potentials recorded from two adjacent neurons. It is common to use pipets with diameters of 1 to 5 μm. Filling such electrodes is not difficult, although construction of pipets from filament-containing glass capillaries is recommended. There are no special precautions to prevent leakage of biocytin from the tip. Because biocytin can be ejected by either negative or positive current, application of a holding current while searching for cells is not recommended.

Biocytin ejection
Juxtacellular staining requires movement of the electrode into a position that facilitates staining. The ideal position of the micropipet for staining the cell is not the common configuration for extracellular recording. For juxtacellular labeling, the micropipet must be positioned as closely to the cell as possible. To achieve the correct position, the electrode is advanced until the extracellular action potential amplitude is at its maximum. At this point, the active bridge amplifier is used to pass positive (anodal) current pulses, with a 200- to 500-msec duration, 50% duty cycle (e.g., 300 msec on 300 msec off), starting at an amplitude of 1 nA. The amplitude of this current is gradually increased (should not exceed ~10 nA) while the neuron is monitored for changes in electrical activity. The electrical current pulses induce the juxtacellular configuration of the electrode, indicated by a sudden increase in background noise. The microscopic details of this phenomenon are not known, but it is probably comparable to a loose on-cell patch configuration. In some neurons, this is accompanied by a sudden increase in sensitivity of the cell to the current pulses. That is, the cell may begin to fire in response to the current pulse. Action potentials at this point will be of large amplitude (1 to 10 mV). Regardless of whether the cell begins to fire, the current amplitude should be quickly reduced when the change in background noise is observed, otherwise the cell is likely to be irreversibly damaged. Ejection currents in the range of 1 to 5 nA can then be continued for 15 to 90 min to stain the neuron while monitoring the background noise and cell firing to ensure that the micropipet has remained in the juxtacellular configuration. Occasionally, when the juxtacellular configuration is lost during the injection, it can be recovered by momentarily increasing the current amplitude as before.

Not every attempt to achieve the juxtacellular configuration is successful. Simply ejecting biocytin into the extracellular space almost never stains a single neuron. Either several cells and processes are stained or no processes are stained.

SUPPORT PROTOCOL

TISSUE PROCESSING

Tissue handling varies somewhat between in vivo and slice preparations. For slices, it is recommended that the brain be perfused (*UNIT 1.1*) with slice medium before slices are prepared, to reduce background staining. This is most important with the DAB method for visualizing biocytin because of the peroxidase activity of hemoglobin. In vivo, perfusion fixation is preferred for the same reason. Slices should be carefully removed from the recording chamber and fixative poured over them carefully as they lay flat on the bottom of a glass vial. The biocytin method is compatible with a variety of fixatives, but it is important to use a cross-linking fixative to immobilize the biocytin and prevent it from washing out of the cell during tissue processing.

Biocytin diffuses readily within cells. The soma, dendrites, and local axonal branches of most cells will be stained at all practical survival times (from 30 min up to 20 hr after staining). If staining is done to visualize these parts of the neuron, there is no reason to extend the survival time beyond the time required for ejection of biocytin. If long axonal branches must be stained, increasing survival time to 12 to 16 hr can be desirable. There is little value in going much beyond 12 to 16 hr, as staining intensity is compromised by clearance of biocytin from the cytoplasm.

Materials *(see* APPENDIX 1 *for items with ✓)*

Tissues
0.1% (v/v) and 1% Triton X-100
✓ PBS, pH 7.2
ABC kit or avidin-conjugated fluorescent marker (e.g., avidin-Texas Red)
0.05% (v/v) 3′,3′-diaminobenzidine (DAB)
3% (v/v) hydrogen peroxide

Vibratome or freezing microtome
Shaker

1. To facilitate penetration of labeled avidin into the tissue containing the biocytin-labeled cell, section tissue on a Vibratome to a 50-μm thickness so that it makes it likely that the cell is near one surface. Keep sections in serial order. Permeabilize the tissue sections by incubating in 0.1% Triton X-100 in PBS for 1 hr on a shaker at room temperature. Increase Triton X-100 concentration to 1% and the incubation time to 8 to 12 hr for thicker (up to 250-μm) sections.

2. Incubate in ABC solution (1:100 in PBS containing 0.1% to 0.2% Triton-X 100 mixed according to kit manufacturer's instructions) or avidin-Texas Red or other avidin-conjugated fluorophore (1:100 in PBS containing 0.1% Triton-X 100) overnight at room temperature for sections that are 150- to 200-μm thick or 2 to 4 hr at room temperature for thinner sections.

3. Wash sections several times in PBS, pH 7.2. For sections treated with ABC solution, incubate in 0.05% 3′,3′-diaminobenzidine and 0.003% hydrogen peroxidase in PBS ~20 min, monitoring staining by microscopic examination of wet sections. If additional time in the DAB solution is required, use repeated 20-min incubations, using freshly mixed solution each time. Wash repeatedly in PBS.

   For sections treated with avidin-Texas Red or other fluorescent avidin conjugate, it is possible to view immediately. If desired, subsequently treat these sections in ABC and react with DAB to make more permanent preparations (see steps 2 to 3).

4. Dehydrate and mount (*UNIT 1.1*), or embed for electron microscopy.

Reference: Horikawa and Armstrong, 1988

Contributors: Charles J. Wilson and R.N.S. Sachdev

CHAPTER 2

Imaging

This chapter serves as an introduction to the basic facts and key concepts for areas of microscopy that are in wide use. The fluorescence microscope is a standard instrument that is required in many experimental designs. *UNIT 2.1* provides an introduction to the physical chemistry of fluorescence and a simple overview of immunofluorescence. *UNIT 2.2* describes approaches for visualization, identification, and characterization of synaptic vesicle membrane proteins. Using biochemical as well as electron and light microscopic techniques, this unit provides protocols for addressing molecular mechanisms relevant to synaptic vesicle exocytosis and modulation of neurotransmission. The increasing use of confocal microscopes is placing new demands on the ability of researchers to acquire an appropriate technical background; *UNIT 2.3* explains the advantages and physical basis of confocal microscopy. This unit also discusses constraints on image acquisition from living specimens. The trend toward acquiring data from living cells continues with the various optical recording methods presented in *UNIT 2.4*. These techniques provide data on patterns of brain activity at high temporal resolution. Together these units provide conceptual and technical background that will assist the reader in choosing an appropriate imaging system—from microscope to dye—required to resolve the structures of interest.

Contributor: Susan Wray

UNIT 2.1

Theory and Application of Fluorescence Microscopy

Fluorescence microscopy allows selective examination of a particular component. A specimen labeled with fluorescent dye(s) is illuminated with filtered light of the absorbing wavelength; it is viewed by using a barrier filter that is opaque to the absorbing wavelength but transmits the longer wavelength of the emitted light. The structures marked with the fluorescent molecules will light up against the black background.

Many biological molecules fluoresce when illuminated with ultraviolet light. This effect is called autofluorescence. The fluorescent amino acid tryptophan, for instance, occurs in a large number of proteins. This molecule is of limited value in cell biology because of the lack of selectivity. Fortunately, autofluorescence of cells and biological tissues and fluids can be minimized by using probes that can be excited at >500 nm.

The most common use of multiband filters is to investigate the relative spatial localization of two antigens. This is typically done with red and green dyes, which form a yellow image from spectral summation when the antibodies colocalize. Colocalization of antigens on organelles as small as stress fibers and vesicles can be detected. Multidye filter sets with more than two-color capability are useful for illuminating intracellular organization of organelles and organization of cells within a tissue. To this end, nucleic acid dyes that stain the nucleus blue, such as DAPI and Hoechst 33342, are among the most commonly used dyes to complement red and green.

Usually high-pressure mercury lamps are used. A 2-hr "burning-in" period should be allowed after lamp replacement before adjusting the optics. Although an arc lamp may remain stable for 6 months to a year, it should be checked for even illumination with each use.

IMMUNOLABELING: GENERAL STEPS FOR LABELING FIXED CELLS AND TISSUES

Several excellent reference books are devoted to immunocytochemistry. The most often cited, and still quite applicable, is Sternberger (1979). Other sources include Pawley (1995) and Rost (1992).

Commercial Antibody Selection

Linscott's Directory (see *APPENDIX 3*) is a useful database of commercial sources of monoclonal and polyclonal antibodies with 40,000 different products and reagents. Printed and floppy disk versions are available. This resource is an invaluable aid in the face of the growing number of large and small companies that supply antibodies.

Selection and Preparation of Starting Material

This section is focused on fluorescence microscopy of cultured animal cells. After fixation, subsequent steps for labeling and visualization are essentially the same as might be used for any fixed specimen.

For tissue culture applications, cells are generally grown on glass coverslips. In a sterile hood, coverslips are sterilized with 70% (v/v) ethanol and flame. Cells may be grown directly on the glass or on glass coated with extracellular matrix proteins. The most popular coating agents are collagen, fibronectin, and laminin. References and protocols for the coating of coverslips may be readily obtained from suppliers (e.g., Life Technologies, Sigma); also see Chapter 3 for descriptions of various coverslip coating procedures.

Rinsing Cells (For Tissue Culture)

The rinsing step removes debris and serum components that may mask certain epitopes as a result of fixation. For many antigens, elimination of this step is possible without consequence. Most protocols use PBS at 37°C for rinsing. Dulbecco's PBS (Dulbecco and Vogt, 1954) with added calcium (Life Technologies) is a sensible choice for two reasons. First, calcium stabilizes cell-cell contact and membrane structure. Second, the ionic composition of electrolytes in Dulbecco's PBS mimics that of extracellular fluid and various culture media more closely than does PBS. Alternatives are fresh medium for debris removal and serum-free medium when serum proteins may interfere with antigenicity. To preserve cell morphology, it is important to follow temperature constraints, as microtubules may rapidly depolymerize at temperatures lower than cell culture temperature.

Fixation

The choice of fixation method will depend on the structure or protein being labeled. A common practice with tissue culture is to use 2% to 4% (w/v) formaldehyde in PBS. The fixative is warmed to 37°C and applied to coverslips immediately after the rinse. Usual incubation conditions are 10 min at room temperature. A stock formaldehyde solution may be prepared from the crystalline polymer paraformaldehyde as an 8% (w/v) aqueous solution and stored at 4°C for up to 1 week (see *UNIT 1.1*). (*CAUTION:* Preparation of such stock solutions is discouraged unless large volumes are needed—e.g., for cardiac perfusion. Formaldehyde vapors are toxic. Paraformaldehyde is insoluble in water at room temperature; preparation requires heating to 80°C in a fume hood, tedious pH adjustment with solid sodium hydroxide pellets, and filtering.) For fixation of tissue culture cells and local perfusion of tissues, readily available 16% (w/v) formaldehyde solutions are recommended. As an alternative fixation procedure, rinsed coverslips may be quickly immersed in −20° to −30°C acetone or methanol with 1% to 2% (w/v) formaldehyde. This fixative is especially useful when the question of artifactual reorganization of soluble antigen pools is being considered. Acetone and methanol fixatives are prepared by adding the appropriate volume of a 16% (w/v) aqueous formaldehyde stock to the organic solvent. Most protocols specify the use of this fixative at −20°C. In practice, it is easy to achieve −30°C by setting 10-ml glass coverslip containers on the surface of dry ice.

Wash

After fixation, further manipulations are carried out at room temperature. Three washes in PBS are normally performed to remove fixative and to rehydrate samples fixed in organic solvents.

Permeabilization

Permeabilization may be done as a separate step following fixation or by combining detergent with an aqueous fixation step. A detergent permeabilization step is often omitted in order to assess whether an epitope is located within a cell or on the extracellular surface. Usually 0.3% to 1% (v/v) Triton X-100 is used. For tissues, the authors suggest using 1% (v/v) Triton X-100 in PBS for a 30-min incubation followed by three 10-min washes in PBS. For monolayers of tissue culture cells fixed with formaldehyde in PBS, 0.3% (v/v) Triton X-100 is included with the fixative. For cells fixed with organic solvents, no further permeabilization is required because the solvent extracts lipids from the bilayer of the plasma membrane.

Blocking

Nonspecific binding can be blocked by using either BSA or serum. The authors conservatively use both 2% (w/v) BSA and 5% (v/v) goat serum for 1 hr at room temperature or overnight at 4°C.

Primary Antibody

Primary antibodies are diluted in blocking solution and used to probe tissues and cells. The appropriate dilution depends on antibodies and antigens. If a supplier recommends a particular dilution, say, 1:200, then it is best to initiate experiments by bracketing that dilution (i.e., 1:100, 1:200, and 1:1000). To remove aggregates, solutions are centrifuged 10 min at 14,000 × *g* just before use. For tissue culture cells, one can typically place coverslips in a moist chamber, face up, on Parafilm. The Parafilm helps to keep primary and secondary antibody solutions on the coverslip and aids in transferring coverslips to six-well dishes for washing. A 22 × 22–mm coverslip can be covered with 150 µl of antibody solution. Incubation times vary from 1 to 3 hr. Alternatively, coverslips may be placed, face down, on Parafilm on 30- to 50-µl droplets. The surface tensions that develop in this procedure are very large. Thus, although this method results in satisfactory images for most antibodies, its use should be limited to cases of extreme shortage of antibody.

2

Wash

The intermediate wash step is more critical than previous ones. For high-quality images, use five 10-min washes in PBS. For other goals, such as routine screening of monoclonal antibodies, this step can be reduced to three 5-min washes.

Secondary Antibody

Secondary antibodies coupled to fluorophores must be screened for titer with each new batch or lot. Dilutions are made in blocking serum, followed by centrifugation. Centrifugation is especially important for fluorescein isothiocyanate (FITC)–coupled antibodies, to reduce particulate background. Conditions for centrifugation are the same as for primary antibody. The supernatant should be carefully transferred to a fresh tube without disturbing the pellet. Titers can usually be adjusted for a 1-hr room temperature incubation.

Final Wash

The final wash step is as critical as the previous wash. The same considerations apply.

Mounting

A number of antifade mounting solutions can be made at the bench or purchased. One that is particularly easy to prepare and use is made by mixing 10 mg *p*-phenylenediamine (Sigma) with 1 ml PBS and 9 ml glycerol. Store the solution in aliquots at −20°C in the dark. Discontinue using when a ruby red color develops. Four dots of colorless nail polish, one at each corner of the coverslip, or four dots of High Vacuum Grease (Dow Chemical), one under each corner, will suffice to hold the coverslip in place. The silicone-based grease may be loaded into a syringe equipped with a wide bore needle for easy application.

Many other solutions are available to delay fading of fluorescent dyes. The commercial mounting medium ProLong (Molecular Probes) is recommended. In addition to an antifade reagent, the product contains a hardening agent that circumvents the need for nail polish or grease. It works well with cells or tissue. It requires a final wash with water and more careful handling. However, ProLong gives results that are noticeably superior to those achieved with phenylenediamine/glycerol solutions.

Double Labeling

Two basic methods may be used for double labeling. The two primary antibodies are typically generated in mouse and rabbit. They may, however, be from any two distinct host species, so long as neither of the secondary antibodies cross-reacts with the primary antibody from the other host. In the first method, the processing of the first primary antibody and its respective secondary antibody is completed, and then the second primary antibody and its respective secondary antibody are processed separately. The time for the complete operation is effectively doubled. This method offers the advantage that interactions between the two primary antibodies are avoided. By the same logic, it is absolutely necessary to prepare controls to ensure that the pattern of labeling is not dependent on the order of binding of the two primary antibodies.

An alternative method is to mix the two primary antibodies together and the two secondary antibodies together. This method produces satisfactory results for many pairs of antibodies and offers significant time savings. Additionally, it provides an opportunity to test for specificity in preadsorption controls (see below). Because of the possibilities of steric hindrance or other interactions in the binding of antibodies to closely situated epitopes, any double-labeling experiments must always be accompanied by controls using each primary antibody separately. This is especially important when one of the antibodies is monoclonal, a typical situation.

Controls

Mixing antigen, in the form of protein or synthetic peptide, with primary antibody provides the best control for specificity. This so-called preadsorption control may also be accomplished in the solid phase by coupling antigen to chromatography resins or by blotting large amounts of antigen on nitrocellulose. Solid-phase adsorption controls are especially useful with antigens like myosins that normally require high salt concentrations for solubilization. Some protocols call for overnight incubation. However, complete blocking may be obtained after a 1-hr incubation of antipeptide antibodies with 1 mg/ml immunizing peptide, whereas the same concentration of irrelevant peptide has no effect. A good way to implement this control is to titrate the antibody with antigen and combine the fluorescence microscopy results with immunoblotting. If titration is not performed, it is still important to show that the antigen is not so concentrated that it is blocking sample labeling by nonspecific mechanisms. This may be accomplished by performing double labeling with two primary antibodies simultaneously. The requirement is that the antigen specifically block labeling by the appropriate antibody only. Other control conditions introduced at this step are the use of preimmune serum, serum from the same host, or blocking solution alone.

References: Dulbecco and Vogt, 1954; Pawley, 1995; Rost, 1992; Sternberger, 1979

Contributors: Donald Coling and Bechara Kachar

UNIT 2.2

Multidisciplinary Approaches for Characterizing Synaptic Vesicle Proteins

NOTE: It is imperative to retain synaptic structural and molecular integrity while isolating and fractionating rat brain synaptosomes. The following precautions must be taken: (1) Dissect and homogenize cortex rapidly. (2) Perform all isolation and fractionation steps between 0° and 4°C, and keep all solutions, centrifuge tubes, and centrifuge rotors chilled prior to and during use. (3) Do not use excessive force on tissue; homogenize no faster than 900 rpm, and use a pipet (not a vortexer) to resuspend pellets.

NOTE: For Basic Protocols 1 and 2, use only freshly isolated synaptosomes (i.e., within 4 hr after preparation). Keep on ice until use.

BASIC PROTOCOL 1

SUBCELLULAR FRACTIONATION OF RAT BRAIN SYNAPTOSOMES

See flow chart (Fig. 2.2.1) for overview of procedure.

NOTE: All fractionation steps must be performed between 0° and 4°C, and all solutions and centrifuge tubes and rotors should be precooled below 4°C and kept on ice.

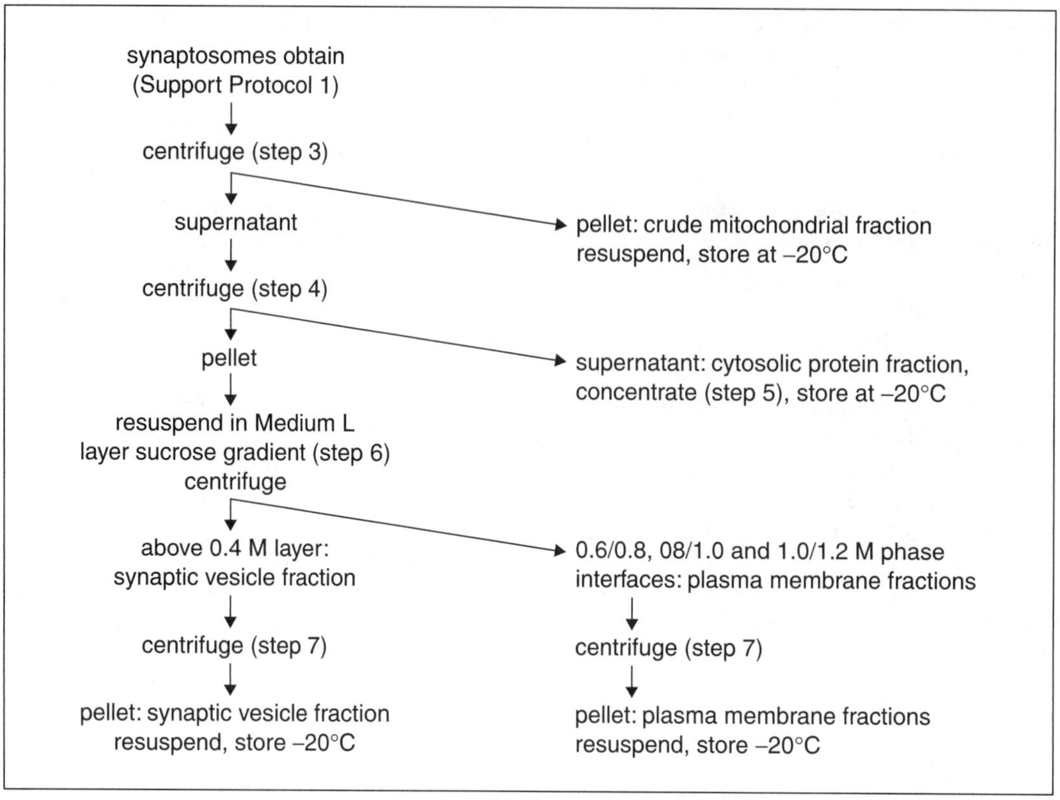

Figure 2.2.1 Flow chart for fractionation of synaptosomes.

Materials *(see* APPENDIX 1 *for items with ✓)*

Rat brain synaptosomes (see Support Protocol 1)
✓ Medium L
1 M KOH
1.0 M sucrose in Medium L (store up to 5 days at 4°C)
✓ 20 mM Tris·Cl, pH 7.4
✓ Protease inhibitors
✓ Sucrose gradients
Synaptic vesicle protein-specific antibodies for immunoblotting

55-ml Potter-Elvehjem homogenizer (Teflon-glass)
Ultracentrifuge
Ultracentrifuge rotors: swinging-bucket and fixed-angle rotors
Ultracentrifuge tubes: thick-walled polycarbonate, 31–ml capacity, 25 × 89–mm; thin-walled ultraclear, 38.5-ml capacity, 25 × 89–mm; and thick-walled polycarbonate, 10-ml capacity, 16 × 76–mm
10,000 MWCO dialysis tubing
Centriprep 10 centrifugal concentrators (Amicon)

1. Resuspend synaptosome pellet in ice-cold Medium L, 4 ml per rat cortex.

IMPORTANT NOTE: Combine synaptosomal preparations from six to eight rats to obtain a concentrated synaptic vesicle fraction for biochemical studies. Save a 0.5-ml aliquot of this synaptosomal suspension for validation of the fractionation procedure.

2. Manually homogenize four times slowly up and down in a 55-ml Potter-Elvehjem homogenizer. Incubate suspension on ice for 15 min, then repeat homogenization one more time. Adjust pH to 8.0 with 1 M KOH, then incubate on ice for another 45 min.

3. Layer 5 ml synaptosomal suspension onto 5 ml of 1.0 M sucrose in Medium L in a thick-walled, 31-ml capacity polycarbonate ultracentrifuge tube (using a total of six to eight tubes). Centrifuge 30 min at 96,300 × *g* (swinging-bucket rotor), 0° to 4°C. Collect the supernatant and set aside on ice for use in step 4 (Fig. 2.2.1). To use the pellet (a crude mitochondrial fraction) for further biochemical analysis, resuspend in 20 mM Tris·Cl (pH 7.4) containing protease inhibitors and store at −20°C.

4. Mix supernatant until homogeneous by gently inverting tubes covered with Parafilm, and centrifuge 14 hr at 25,000 ×*g* (swinging-bucket rotor), 4°C. Collect the supernatant (the cytosolic protein fraction) and set aside for use in step 5. Save the pellet (on ice) for further fractionation (step 6).

5. Dialyze the supernatant against 10 vol Medium L overnight at 4°C, using the 10,000 MWCO dialysis tubing. The next day, centrifuge the dialysate 1 hr at 140,000 × *g* in a fixed-angle rotor, 4°C. Collect supernatant and concentrate with a Centriprep 10 centrifugal concentrator by microcentrifuging 30 min at maximum speed, 4°C. Collect concentrated product, load onto a fresh Centriprep 10, and repeat microcentrifugation. Collect final concentrated cytosolic protein fraction from the Centriprep 10 and add protease inhibitors to the following final concentrations: 2 μg/ml aprotinin, 1 μg/ml leupeptin, and 1 mM PMSF. Store at −20°C up to several years in 50- to 100-μl aliquots (to avoid repeated freezing/thawing).

6. After setting up the overnight dialysis in step 5, resuspend the pellet from step 4 in Medium L (1.0 to 2.0 ml for each cortex). Layer 5 ml of the suspension onto each sucrose gradient. Mark phase interfaces on the tubes. Centrifuge sucrose gradients 90 min at 68,000 × *g* in a swinging-bucket rotor, 0° to 4°C. Collect each fraction carefully with a Pasteur pipet.

   The fraction of synaptic vesicles is located at the layer above the 0.4 M sucrose phase; the fractions enriched with the presynaptic plasma membrane proteins are located at the phase interfaces between 0.6/0.8, 0.8/1.0 (usually the purest plasma membrane fraction), and 1.0/1.2 M.

7. Transfer each fraction to a thick-walled 10-ml ultracentrifuge tube. Add Medium L to fill 3/4 of each tube. Centrifuge 45 min at 106,500 × *g* (fixed-angle rotor), 0° to 4°C. Discard each supernatant. Resuspend each pellet in 0.5 to 1 ml of 20 mM Tris·Cl, pH 7.4, containing protease inhibitors. Store at −20°C up to several years in 50- to 100-μl aliquots for further biochemical analysis.

8. Determine protein concentration (*CPMB UNIT 10.1A*) of each fraction: synaptosomes (step 1), cytosol (step 5), plasma membranes (step 7), and synaptic vesicles (step 7).

9. Load samples of equal amounts of each fraction on SDS-polyacrylamide gels (5 to 10 μg total protein for each fraction should be enough to detect most synaptic vesicle proteins) and perform electrophoresis (*CPMB UNIT 10.2*).

10. Detect protein of interest with specific antibody by immunoblotting (*CPMB UNIT 10.8*). To determine relative purity of each fraction, perform sequential immunoblotting with antibodies against well-characterized specific markers, such as synaptophysin (Chemicon) or VAMP (StressGen) for synaptic vesicles, Na^+/K^+-ATPase (Transduction Laboratories) for plasma membranes, and LDH for cytosol.

BASIC PROTOCOL 2

VISUALIZATION OF SYNAPTIC VESICLE PROTEINS BY PRE-EMBEDDING IMMUNOGOLD ELECTRON MICROSCOPY

Although this protocol, adapted from Tanner et al. (1996), is for synaptosomal preparations, it can be applied to cultured neuronal cells or freshly dissected brain tissue as well.

NOTE: All steps are performed at room temperature.

Materials *(see APPENDIX 1 for items with ✓)*

Rat brain synaptosomes (see Support Protocol 1)
✓ Wash buffer 0.1 M sodium phosphate buffer
4% (w/v) paraformaldehyde in 0.1 M sodium phosphate buffer
✓ PBS
PBS with 5% (w/v) nonfat dry milk
5% (v/v) normal goat serum/0.1% (w/v) saponin in PBS
Primary antibody against synaptic vesicle protein of interest
Nanogold-conjugated secondary goat antibody (Nanoprobes)
2% (w/v) glutaraldehyde in PBS
HQ silver enhancement kit (Nanoprobes)

12-well tissue culture plates
Electron microscope

1. Resuspend synaptosome pellet purified from one rat cortex in 1 ml wash buffer, divide into a minimum of four equal volumes, and microcentrifuge 5 min at maximum speed, room temperature.

   IMPORTANT NOTE: When using cultured neuronal cells, omit this step. Instead, remove culture medium from cells, wash cells with 0.1 M sodium phosphate buffer, then proceed to step 2.

2. Discard supernatant and fix pellet by adding 0.5 ml of 4% paraformaldehyde in 0.1 M sodium phosphate buffer and incubating 30 min at room temperature.

3. Wash fixed pellet four times, each time for 5 min with 1 ml of 0.1 M sodium phosphate buffer. Place each pellet on filter paper soaked with 0.1 M sodium phosphate buffer and divide into four to six pieces, using a razor blade (total, 16 to 24 pieces). Transfer the pieces into separate wells of 12-well tissue culture plates.

4. Block nonspecific binding sites and permeabilize synaptosomes by adding 1 ml of 5% (v/v) normal goat serum/0.1% (w/v) saponin in PBS to each of the pieces of fixed pellet and incubating for 1 hr.

5. Dilute the primary antibody against the synaptic vesicle protein of interest appropriately in 5% normal goat serum/0.1% saponin/PBS. Add 0.5 ml of the diluted primary antibody to each of the pieces of pellet and incubate for 1 hr at room temperature, or longer at 4°C.

6. Wash each piece of pellet four times, each time for 5 min with 1 ml of PBS containing 5% nonfat dry milk.

7. Dilute Nanogold-labeled secondary antibody 1:400 in PBS containing 5% nonfat dry milk. Add 400 µl of the diluted secondary antibody to each of the pieces of fixed pellet and incubate for 1 hr at room temperature.

8. Wash four times as described in step 6. Post-fix by adding 1 ml of 2% glutaraldehyde in PBS to each of the pieces of pellet and incubating for 30 min at room temperature.

9. Wash four times, each time for 5 min with 1 ml PBS, then wash an additional four times, each time for 5 min with distilled water.

10. Incubate the sample with 1 ml of silver enhancement (from HQ silver enhancement kit) for 4 min, then immediately perform four to six washes over a 10-min period, each with 1 ml water, followed by three washes, each for 5 min with 1 ml of 0.1 M sodium phosphate buffer.

11. Process further for electron microscopy as described in *UNIT 1.2*, starting with osmium tetroxide treatment, but using 1 ml of 0.2% (instead of 1%) osmium tetroxide in 0.1 M sodium phosphate buffer. Examine samples with electron microscope.

2

SUPPORT PROTOCOL 1

ISOLATION OF SYNAPTOSOMES FROM RAT BRAIN CORTEX

The authors of this unit have found that the following protocol, first described by Dunkley et al. (1988), yields the purest synaptosomal preparations.

NOTE: All isolation steps must be performed at 0° to 4°C, and all solutions and centrifuge tubes and rotors should be precooled below 4°C and kept on ice.

Materials *(see APPENDIX 1 for items with ✓)*

3- to 4-week old male rat (Wistar or Sprague-Dawley)
Anesthetic
✓ 1× sucrose buffer, ice-cold
1× sucrose buffer
✓ Percoll gradients, ice-cold
✓ Wash buffer, ice-cold

Rat guillotine
25-ml Potter-Elvehjem homogenizer (Teflon-glass)
High-speed centrifuge with fixed-angle rotor for eight 16 × 100–mm tubes)
Polycarbonate high-speed centrifuge tubes: 11-ml capacity, 16 × 100–mm *and* 50-ml capacity, 29 × 102–mm

1. Anesthetize the rat (*CPNS APPENDIX 4B*) and decapitate on guillotine. Remove brain and place in ice-cold 1× sucrose buffer.

2. Remove cerebellum, brain stem, midbrain, and white matter (Fig. 2.2.2). Put the remaining cortex in 9 ml ice-cold 1× sucrose buffer, and transfer to 25-ml Potter-Elvehjem homogenizer. Homogenize the cortex tissue at a maximum of 900 rpm by moving slowly 10 times up and down.

3. Centrifuge homogenate 10 min at 1000 × *g*, 4°C. Collect supernatant and carefully load 2 ml of supernatant onto each Percoll gradient. Centrifuge gradients for exactly 5 min at 32,500 × *g*, 4°C, using a timer to start counting the time when 32,500 × *g* is reached.

4. Using a Pasteur pipet, remove and discard the gradient material above the 15%/23% interface (the bottom interface). Then, using a clean Pasteur pipet, collect the synaptosomes (visible as a cloud of material concentrated within the 15%/23% interface) and transfer to a 50-ml polycarbonate centrifuge tube. Add 30 ml chilled wash buffer to the centrifuge tube. Centrifuge 15 min at 15,000 × *g*, 4°C.

5. Remove supernatant carefully (synaptosomes form a loose pellet), resuspend the pellet in an appropriate buffer, and proceed with analysis (see Basic Protocols 1 and 2) or keep on ice until used. Centrifuge again in 20 ml chilled wash buffer if a more solid pellet is preferred.

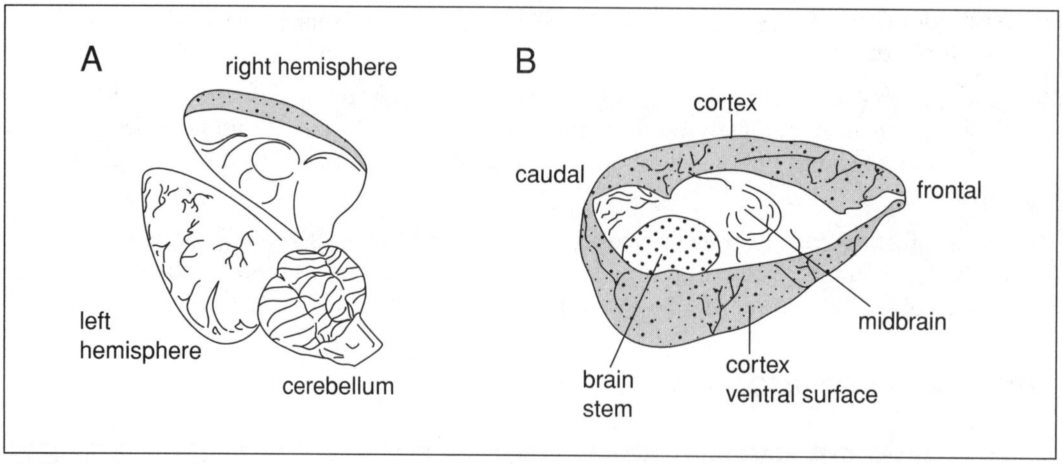

Figure 2.2.2 Illustration of rat brain anatomical features mentioned in Support Protocols 1 and 2.

BASIC PROTOCOL 3

IMMUNOFLUORESCENCE STAINING OF SYNAPTIC VESICLE PROTEINS ON CULTURED HIPPOCAMPAL NEURONS

NOTE: All steps are performed at room temperature.

Materials *(see* APPENDIX 1 *for items with ✓)*

Hippocampal neurons cultured on coverslips (see Support Protocol 2)
✓ PBS, pH 7.4
Fixative solution: 4% (w/v) paraformaldehyde/4% (w/v) sucrose in PBS
✓ Blocking/permeabilization buffer
Primary antibody against protein of interest
✓ Antibody dilution buffer
Secondary antibody labeled with fluorophore (e.g., TRITC, FITC, Cy2, Cy3, or Cy5; store protected from light with aluminum foil)
Mounting medium

12-well tissue culture plates
Microscope slides
Forceps
Fluorescence microscope with 40× and/or 63× oil-immersion lenses

1. Transfer each coverslip containing cultured hippocampal neurons to a separate well of a 12-well tissue culture plate containing 1 ml PBS. Aspirate the PBS.

2. Fix cells by adding 1 ml fixative solution to each well and incubating 10 to 15 min at room temperature *or* by adding 1 ml ice-cold methanol to each well and keeping at −20°C for 10 min.

3. Remove fixative and wash coverslips three times, for 5 min each time, with 1 ml PBS. Remove the final wash of PBS and add 1 ml blocking/permeabilization buffer to each of the fixed coverslips. Incubate for 1 hr at room temperature to block nonspecific sites and permeabilize the cells.

4. Dilute primary antibody appropriately (as determined empirically) in antibody dilution buffer. Add 400 μl of the diluted primary antibody to each fixed coverslip and incubate 2

to 3 hr at room temperature *or* overnight at 4°C. Following incubation, wash the coverslips as in step 3.

5. For each coverslip, dilute a fluorophore-conjugated secondary antibody appropriately (typically, 1:100 to 1:500 for commercial preparations) in 400 μl of antibody dilution buffer. Microcentrifuge 5 min at maximum speed to remove fluorophore aggregates.

6. Add 400 μl of the diluted secondary antibody to each coverslip and incubate at room temperature for 1 to 2 hr. Wash the coverslips as in step 3. After last wash, add 1 ml PBS to each coverslip. Remove the PBS, add 1 ml of water, and leave the coverslip in water for 10 min before mounting on slide.

7. Label microscope slides and place 1 drop of mounting medium on each slide. Using forceps, pick up the coverslip from the well, gently blotting off excess water by touching the edge to a Kimwipe. Invert the coverslip cell-side-down onto the drop of mounting medium. Gently blot mounted coverslip with paper towel. Allow to air dry. Store 6 months to a year at 4°C in the dark.

8. View the coverslip, using a fluorescence microscope with a 40× and/or 63× oil immersion objective.

2

SUPPORT PROTOCOL 2

CULTURE OF HIPPOCAMPAL NEURONS FROM EMBRYONIC RAT BRAIN

NOTE: All incubations are carried out in a humidified, 37°C, 5% CO_2 incubator unless otherwise specified. All reagents and equipment coming into contact with live cells must be sterile.

NOTE: Coverslips used in culturing must be stripped using concentrated HNO_3, rinsed well, and autoclaved prior to use. Chamber slides, coverslips, and dishes must be coated with polyornithine and fibronectin before the dissection day (see *UNIT 3.4*).

Materials *(see APPENDIX 1 for items with ✓)*

19-day pregnant rat
Anesthetic
70% (v/v) ethanol
HNO_3
Polyornithine
Fibronectin
✓ Dissection buffer, ice-cold
✓ Chopping buffer, ice-cold
Trypsin (Invitrogen)
✓ Glial feed
✓ Neuronal feed, with and without GlutaMAX supplement

Surgical instruments:
- Straight surgical scissors
- Straight fine scissors
- Straight spring scissors
- Curved forceps
- Straight, wide-tipped forceps
- Dumont no. 5-45 forceps (45° tip)

35-mm tissue culture dishes and 10-cm dissection dishes
Dissecting microscope and fiber-optic lights
15-ml tubes
9-in. cotton-plugged borosilicate glass pipets with flame-polished tips
Polyornithine/fibronectin-coated (*UNIT 3.4*) 12- to 15-mm-diameter circular coverslips and/or two-well Lab-Tek chamber slides (Nalge Nunc International)
12-well tissue culture plate

2

1. Anesthetize pregnant rat (typically, use E18 to E19 embryos) with CO_2 or other approved method (*CPNS APPENDIX 4B*). Sacrifice rat by cervical dislocation.

2. Wipe off abdomen with 70% ethanol. Grab abdominal skin with straight, wide-tipped forceps. Cut skin and muscle along the midline with the surgical scissors. Make two cuts to the sides and pull out both horns of the uterus. Cut uterus and place it in a 10-cm dish.

3. Open the uterus with the straight, fine scissors. Remove the embryos and decapitate with sharp scissors. Place the heads in a dish of cold dissection buffer.

4. Working under a dissecting microscope with fiber-optic lights, place one head on clean bench paper. Using curved forceps, hold the head firmly by inserting the tips of the forceps into the orbits. Using spring scissors, carefully cut through the skin and skull from the point of decapitation up towards the orbits. Next, holding the head with one pair of forceps, peel back the skin and skull with another pair, to expose the brain (work from beneath the olfactory bulbs, caudally). Using the curved forceps, carefully remove the brain *intact* and place it in a 35-mm dish containing a few drops of dissection buffer. Then rapidly dissect out the hippocampi, as described in steps 5 through 7.

5. Using spring scissors, remove the cerebellum, pons, and cervical spinal cord (Fig. 2.2.2). Next, separate the hemispheres of the brain by moving the Dumont forceps through the midline.

6. One hemisphere at a time, remove the septum, thalamus, and hypothalamus from the cortex, exposing the hippocampus curved along the medial edge of the cortex. Remove the olfactory bulb and striatum, to further expose the hippocampus for easy removal. Finally, carefully remove the meninges and choroid plexus.

7. Using the sharp Dumont forceps, remove the hippocampus and place it in a 35-mm dish containing dissection buffer.

8. Repeat steps 6 and 7 with the other hemisphere. Repeat this entire procedure (steps 4 through 7) with each head. Place all hippocampi in the same 35-mm dish filled with dissection buffer.

9. Under a laminar flow hood, remove the dissection buffer in which the hippocampi have been isolated. Add 3 to 4 ml of fresh chopping buffer. Mince tissue into fine pieces. Pipet the minced tissue into a 15-ml tube. Rinse the dish with ~1 ml of chopping buffer to collect any remaining tissue. Next, add an appropriate amount of trypsin to the tube and place tube in 37°C incubator for 6 to 7 min.

   Because the activity of trypsin varies from lot to lot, both the amount of trypsin and the length of time needed to trypsinize will vary. Generally, when switching to a new lot of trypsin, test it at a 1:10 dilution with a 6-min trypsinization period. Then adjust the amount of trypsin used and the time of incubabation on the basis of the results of plating and the appearance of the cells.

10. After trypsinization, remove supernatant carefully and discard. Add ~10 ml of ice-cold dissection buffer, being sure to gently disturb the settled tissue, thus permitting the trypsin to be washed out of the pelleted tissue. Once the tissue has settled to the bottom of the tube, remove supernatant. Repeat the wash two more times, gently disturbing the pelleted tissue each time to optimize the wash.

11. After final wash, add 5 to 6 ml of chopping buffer to tissue pellet and triturate six to ten times with a flame-polished, 9-in., cotton-plugged, borosilicate glass pipet. Pipet slowly to avoid formation of bubbles and to facilitate maximum trituration. Use approximately half of these cells to set up glial beds (step 12) and the other half (step 13) to plate neurons onto already existing glial beds (7 to 10 days old).

12. Just before plating cells, remove water from polyornithine/fibronectin-coated coverslips or chamber slides. Count cells (*APPENDIX 21*). Resuspend half of the cells from step 11 in glial feed (which selects for the growth of astrocytic-type glial cells), at 1–2 × 10^5 cells/ml, apply the suspension to coverslips (in 12-well plates) or chambers, and place in incubator. Allow the glial cells to grow for 7 to 10 days, performing 1/3 medium changes every 4 to 5 days. Change the medium very slowly, to avoid disturbing the cells, using medium that has been warmed in a 37°C incubator.

13. Use approximately half of the hippocampal cells from step 11 to generate a neuronal culture. Resuspend the cells in neuronal feed containing GlutaMAX supplement, at a density of 1–2 × 10^5 cells/ml. Remove ~80% of the glial feed volume from a glial bed that has been growing 7 to 10 days. Plate the cells suspended in neuronal feed on the glial bed, using a sufficient volume of suspended cells to replace the glial feed that was removed. Return plates to incubator. After 3 days, remove 1/3 of the feed and replace with neuronal feed without GlutaMAX supplement. Every 4 days thereafter, perform a 1/3 to 1/2 feed change, always omitting the GlutaMAX.

References: Dunkley et al., 1988; Tanner et al., 1996

Contributors: Miriam Leenders, Claudia Gerwin, and Zu-Hang Sheng

UNIT 2.3

Basic Confocal Microscopy

Confocal microscopy produces sharp images of structures within relatively thick specimens (up to several hundred microns). It is particularly useful for examining fluorescent specimens. Thick fluorescent specimens viewed with a conventional widefield fluorescent microscope appear blurry and lack contrast because fluorophores throughout the entire depth of the specimen are illuminated and fluorescence signals are collected not only from the plane of focus but also from areas above and below. Confocal microscopes selectively collect light from thin (~1 μm) optical sections representing single focal planes within the specimen. Structures within the focal plane appear more sharply defined than they would with a conventional microscope because there is essentially no flare of light from out-of-focus areas. A three-dimensional view of the specimen can be reconstructed from a series of optical sections at different depths.

The information presented herein is intended to provide background and practical tips needed to get started with confocal microscopy. An excellent source of theoretical and technical information is the *Handbook of Biological Confocal Microscopy* (Pawley, 1995).

THE BASIS OF OPTICAL SECTIONING

Confocal microscopes accomplish optical sectioning by scanning the specimen with a focused beam of light and collecting the fluorescence signal from each spot via a spatial filter (generally a pinhole aperture) that blocks signals from out-of-focus areas of the specimen. A pinhole aperture in the image plane allows fluorescence from the illuminated spot in the specimen to pass to the detector but blocks light from out-of-focus areas. The diameter of the pinhole determines how much of the fluorescence emitted by the illuminated spot in the specimen is detected, and the thickness of the optical section. The separation of the in-focus signal from the out-of-focus background achieved by a properly adjusted pinhole is the principle advantage of confocal microscopy for examination of thick specimens.

TYPES OF CONFOCAL MICROSCOPES

Several types of confocal microscopes are available, each having unique features and advantages. The types most commonly used for examining fluorescence specimens are laser-scanning confocal microscopes. These microscopes collect images by scanning a laser beam across the specimen. Lasers provide intense illumination within a narrow range of wavelengths. The emission wavelengths of several types of lasers, together with the excitation spectra of familiar fluorophores, are illustrated in Figure 2.3.1.

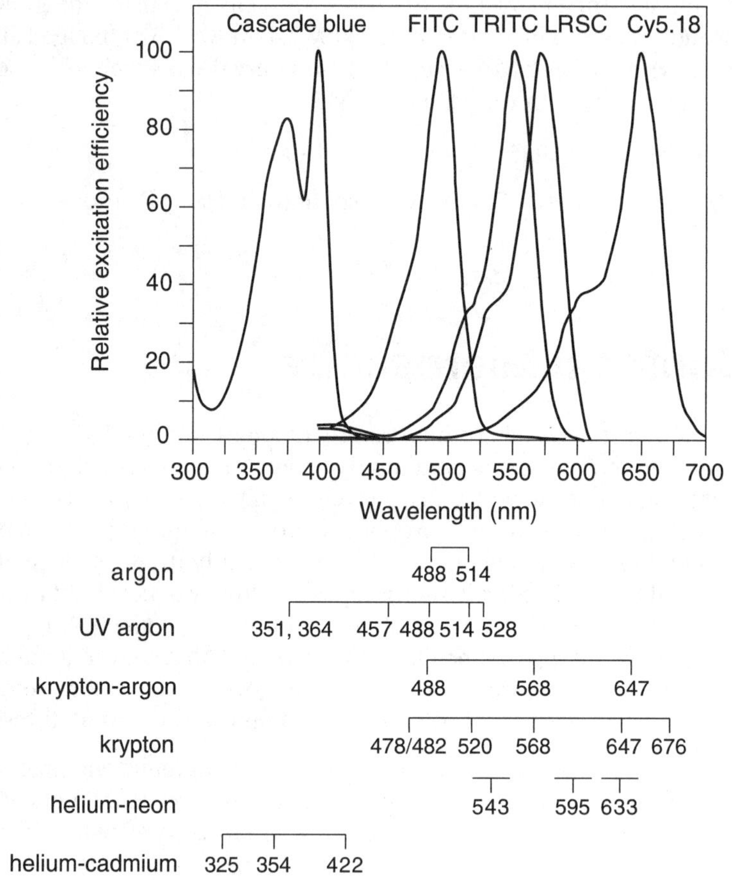

Figure 2.3.1 Comparison of the emission wavelengths of various lasers and the excitation spectra of representative fluorophores. The lasers most commonly used for laser-scanning confocal microscopy are air-cooled argon, krypton-argon, and helium-neon lasers. UV argon lasers generally require water cooling and are more expensive. They may be configured to provide only UV wavelengths (351 nm and 364 nm) or both UV and longer wavelengths. Data for the excitation spectra of Cascade blue, fluorescein (FITC), tetramethylrhodamine (TRITC), lissamine rhodamine (LRSC), and cyanine 5.18 (Cy5.18) are from Wessendorf and Brelje (1993). Modified from Brelje et al. (1993).

PRACTICAL GUIDELINES

Sample Preparation: Immunofluorescence in Fixed Specimens

Fixation

The best fixative is one that accurately preserves the three-dimensional geometry of the specimen. The standard fixative for fluorescence microscopy (2% to 4% formaldehyde in PBS) is not ideal because it can cause blebbing of the plasma membrane, vesiculation of intracellular membrane compartments, and other alterations in cellular morphology. Moreover, some commercial preparations of formaldehyde contain methanol, which shrinks cells. Techniques for optimizing formaldehyde fixation are described by Bacallao et al. (1995; also see UNIT 2.1). The buffer should be chosen to match the osmolality and pH of the specimen. Fixatives containing 0.125% to 0.25% glutaraldehyde in addition to formaldehyde preserve cellular morphology better than formaldehyde alone. Some investigators avoid using glutaraldehyde for fluorescence microscopy because it induces autofluorescence. However, autofluorescence can be reduced by treating the sample after fixation with $NaBH_4$ (1 mg/ml in PBS, pH 8.0; two treatments of 5 min each for dissociated cells, longer for thicker samples). A more serious drawback of glutaraldehyde for immunofluorescence studies is that it destroys the antibody recognition sites of some antigens. An alternative fixation technique that preserves tissue better than chemical fixation is rapid freezing followed by freeze substitution (Bridgman and Reese, 1984).

Choices of fluorophores

Good choices for multiwavelength imaging with a krypton-argon laser are: FICT/Bodipy/Oregon green for excitation at 488 nm; lissamine rhodamine/Cy3/Texas red for excitation at 568 nm; and Cy5 for excitation at 647 nm. UV fluorophores also are good for multicolor imaging (some of the best dyes for DNA are UV fluorophores) if a UV laser is available.

Control samples

Confocal microscopy uses electronic image enhancement techniques that can make even a dim autofluorescence signal or nonspecific background staining look bright. In order to be able to distinguish a real signal from background it is essential to prepare appropriate control samples. For immunofluorescence experiments with one primary antibody, the appropriate control samples are unstained specimens and specimens treated with the secondary antibody but no primary antibody. Experiments with two primary and secondary antibodies require additional controls to test whether the secondary antibodies cross-react with the "wrong" primary antibody. Other control experiments may be required to verify the specificity of labeling (see UNIT 2.1).

Mounting the specimen

The mounting medium should preserve the three-dimensional structure of the specimen. PBS or a mounting medium consisting of 50% glycerol/50% PBS preserves the shapes of cells quite well, but Mowiol and gelvatol cause a 10% decrease in height (Bacallao et al., 1995). Adding an antioxidant to the mounting medium helps to alleviate photobleaching. One of the best antioxidants is 100 mg/ml 1,4-diazabicyclo[2,2,2]octane (DABCO; Sigma; Bacallao et al., 1995). *n*-propyl gallate (Giloh and Sedat, 1982) and *p*-phenylenediamine (PPD; Johnson et al., 1982) also are effective antibleaching agents, but the former may cause dimming of the fluorescence, while the latter may damage the specimen (Bacallao et al., 1995).

The choice of mounting medium should take into account the type of microscope objective that will be used to observe the specimen. In order for an objective to perform optimally, the mounting medium should have the same refractive index as the objective immersion medium. Mismatches in the refractive indexes produce spherical aberration leading to loss of light at the detector, as well as decreased *z* axis resolution and incorrect depth discrimination. Image deterioration caused by spherical aberration increases with depth into the specimen. Significant

Table 2.3.1 Refractive Indexes of Common Immersion and Mounting Media

Medium	Refractive index
Immersion media	
Air	1.00
Water	1.338
Glycerol	1.47
Immersion oil	1.518
Mounting media	
50% glycerol/PBS/DABCO	1.416[a]
5% *n*-propyl gallate/0.0025% *p*-phenylene diamine (PPD) in glycerol	1.474[a]
0.25% PPD/0.0025% DABCO/5% *n*-propyl gallate in glycerol	1.473[a]
VectaShield (Vector Labs)	1.458[a]
Slow Fade (Molecular Probes)	1.415

[a]Data from Bacallao et al. (1995).

losses of signal intensity and axial resolution are apparent at distances of just 5 to 10 μm when an oil immersion objective is used to examine a specimen in an aqueous medium (Keller, 1995). Table 2.3.1 gives the refractive indexes of standard objective immersion media and mounting media.

Most microscope objectives are designed for viewing specimens through a glass coverslip of a specific thickness (typically 0.17 μm, a no. 1 1/2 coverslip). Correct coverslip thickness is especially critical for high-NA (>0.5) dry objectives and water immersion objectives (Keller, 1995). Use of a coverslip that differs from the intended thickness by only 5% causes significant spherical aberration. High-NA dry and water immersion objectives typically have an adjustable collar to correct for small variations in coverslip thickness.

The specimen should be mounted as close to the coverslip as possible, especially for observation with immersion objectives, which have short working distances (~100 to 250 μm, depending on the type of objective). This also helps to avoid image deterioration due to spherical aberration. Fragile specimens should be protected by supporting the coverslip, with, for example, a thin layer of nail polish, strips of coverslips, or a gasket made from a sheet of silicon rubber (Reiss). Sealing the edges of the coverslip—with nail polish or silicon vacuum grease (Dow Corning)—helps to prevent specimen desiccation and movement.

Living Specimens

Confocal microscopy of living preparations is challenging for several reasons. The specimen must be mounted in a chamber that keeps it healthy and immobile while at the same time providing access for the objective. For high-resolution transmitted-light imaging (e.g., by laser-scanning differential interference contrast microscopy), the chamber must be thin enough to accommodate a high-NA (oil immersion) condenser. Fluorescence signals in living specimens generally are weak, and the light levels needed to detect them can be damaging to the specimen. Photobleaching inevitably is a problem for experiments that require collecting many images. Temperature fluctuations in specimens kept at nonambient temperatures make it difficult to maintain adequate focus.

A simple chamber for culture preparations grown on glass coverslips can be made by forming a well on a glass slide with a gasket cut from a sheet of silicon rubber or a plastic ruler. To prevent the well from leaking, it should be sealed with silicon vacuum grease, a mixture of melted paraffin and petroleum jelly, or Sylgard (Dow Corning). The well is filled with medium, and the coverslip with attached cells is placed, cell side down, on top of the well. The preparation can be kept warm during observation on the microscope with a heated air blower (e.g., a hair dryer with variable power source) or with infrared lamps. An important factor to consider in choosing a chamber is whether it maintains the desired temperature while in contact with an immersion objective that acts as a heat sink. One solution to this problem is to heat the objective as well as the chamber. A heated chamber and objective warmer designed for microscopy with a high-NA objective and condenser are available from Bioptechs.

Addition of an oxygen quencher to the medium can help to alleviate photobleaching of the fluorophores. Photobleaching not only leads to dimming of the signal but also to generation of oxygen radicals that can damage cells. Several oxygen quenchers have been reported to be effective, including oxyrase (0.3 U/ml; Oxyrase; Waterman-Storer et al., 1993); ascorbic acid (0.1 to 3.0 mg/ml; Sigma; Terasaki and Dailey, 1995); a mixture of Trolox (10 μM; Aldrich) and *N*-acetylcysteine (50 μM; Sigma; M. Burack and G. Banker, pers. comm.); and crocetin (Tsien and Waggoner, 1995).

Optimizing Imaging Parameters

Choice of objectives

High-NA objectives generally are preferable for fluorescence microscopy because they collect more light than low-NA objectives (brightness is proportional to NA^4). Most high-quality high-NA objectives have >80% transmission at visible wavelengths, but some have low transmission at UV wavelengths (Keller, 1995).

Water immersion objectives are the best choice for visualizing specimens in aqueous solutions (e.g., living specimens). High-NA water immersion objectives specifically designed for confocal microscopy of biological specimens can have working distances of ~250 μm.

Oil immersion objectives can have higher NAs than water immersion objectives. Most have fairly short working distances (~100 μm) although some recently introduced oil objectives have working distances of ~200 μm. A long-working-distance oil objective will be useful only if the specimen is mounted in a medium that matches the refractive index of immersion oil ($\eta = 1.518$). If an aqueous mounting medium is used, images from depths at more than ~20 μm into the specimen will be noticeably degraded by spherical aberration. Also, distance measurements in the *z* axis will need to be corrected.

Zoom factor

The zoom setting on a confocal microscope determines the size of the scan region and the apparent magnification of the image. A zoom factor of 2 will scan an area half as long and wide as a zoom factor of 1. Images are made up of the same number of samples (points along the horizontal axis, lines along the vertical axis) and are displayed on the image monitor by a fixed number of pixels, regardless of the zoom factor. Therefore, the pixels in a zoom-2 image will represent areas within the specimen half as large in each dimension as the areas represented by the pixels at zoom 1. If the pixel size for an objective at zoom 1 represents 0.25 μm × 0.25 μm, then the pixel size at zoom 2 will be 0.125 × 0.125 μm. For each objective, there is an optimal zoom setting which yields pixel dimensions small enough to take advantage of the full resolution of the objective but large enough to avoid oversampling. In order for the minimum resolvable entity to be visible on the display monitor, the pixel dimensions need to be smaller than (less than one-half) the optical resolution. However, if the pixel size is made too small by using a higher-than-optimal zoom factor, the specimen is subjected to more irradiation than necessary with an increased risk of photobleaching. The rate of photobleaching increases proportionally to the square of the zoom factor (Centonze and Pawley, 1995). A guideline for selecting an appropriate zoom factor derived from information theory (the Nyquist Sampling

Theorem) states that the pixel dimensions should be equal to the optical resolution divided by 2.3 (see Webb and Dorey, 1995). However, pixel dimensions smaller than this may produce more informative images.

Z axis sectioning interval

In order to study the three-dimensional structure of a specimen, images are collected at a series of focal levels at intervals determined by the settings sent to the focus motor. The most straightforward way to ensure that the reconstructed images have correct proportions in the x, y, and z axes is to collect optical sections at z axis intervals equal to the x, y pixel dimension. However, the interfocal plane interval needed to adequately sample the specimen in the z axis is not as small as the x, y pixel dimension because the axial resolution is poorer than the lateral resolution. The optimal interfocal plane interval (according to the Nyquist Sampling Theorem) is equal to the axial resolution divided by 2.3. Collecting images at shorter intervals results in oversampling with an increased risk of photobleaching.

Illumination intensity

Fluorescence emission increases linearly with illumination intensity up to a level at which emission saturates. Optimal signal-to-background and signal-to-noise ratios are obtained with illumination levels well below saturation (Tsien and Waggoner, 1995). The illumination intensity on a laser-scanning microscope can be adjusted by inserting neutral-density filters into the light path and/or by operating the laser at submaximal power. In general, the best images are obtained with illumination levels as high as possible without producing unacceptable rates of photobleaching.

PMT black level and gain

The contrast and information content of confocal images are influenced by the black level and gain of the photomultiplier tube (PMT) amplifiers. To obtain maximal information, the black level and gain should be adjusted to take advantage of the full dynamic range of the PMTs. The appropriate black level setting can be found by scanning while the light path to the PMT is blocked. The image that appears on the display monitor should be just barely brighter than the background, which is black (gray level = 0). To set the gain, scan the specimen and adjust the gain so that the brightest pixel in the image is slightly below white (gray level = 255). Selecting black level and gain settings that ensure that all signals fall within the dynamic range of the PMT is important for quantitative imaging experiments. The software provided with many confocal microscopes includes a pseudocolor image display mode that facilitates selection of appropriate black level and gain settings by highlighting pixels with intensity values near absolute black and absolute white.

Averaging

Confocal images of dimly fluorescent specimens captured at typical scan rates (1 to 2 sec/frame for a slow-scan confocal microscope) appear noisy because of the small numbers of photons collected from each spot. In some instances, it may be possible to improve the signal-to-noise ratio by scanning the specimen at slower rates. Another way to obtain a better image is by summing and averaging the signals obtained in multiple scans (frame averaging). Some confocal microscopes provide a second averaging method (line averaging), in which individual lines are repeatedly scanned and averaged. Line averaging generally produces sharper images than frame averaging (which averages full frames) because there is less risk of blurring due to movements or changes in the specimen.

Image display

Commercial confocal microscope packages provide software for some types of image enhancement and display. The display options for three-dimensional datasets typically include "z projections," which are two-dimensional displays formed by superimposition of stacks of optical sections, and stereoscopic views, which are made by combining two image stacks, one aligned in the z axis and the other with a displacement between successive images. Many systems also have the capability to compute cross-sections and projections of the specimen from varying angles. Computed projections for a sequence of view angles can be played as a movie in which the specimen appears to rotate around an axis. Such movies give the viewer a striking impression of the three-dimensional geometry of the specimen. Additional display options are available in various integrated software/hardware packages specifically designed for visualization and analysis of three-dimensional images.

Resources Available via Internet

NIH Image, a powerful image analysis program for Macintosh computers developed by W. Rasband (Research Services Branch, National Institute of Mental Health, NIH), has many useful tools for analysis of confocal images. It can be downloaded from *http://rsb.info.nih.gov/nih-image/* or obtained via FTP from *zippy. nimh.nih.gov*. A version of NIH Image modified for operation under Windows also is available. Confocal Assistant, a program designed for operating on images from Biorad confocal microscopes, is available via FTP from *genetics.bio-rad.com/public/confocal/cas/cas255.zip.*

References: Inoue, 1986; Matsumoto, 1993; Pawley, 1995; Russ, 1995; Shotton, 1993; Wilson, 1990

Contributor: Carolyn L. Smith

UNIT 2.4

Imaging Nervous System Activity

The optical imaging of nervous system activity outlined in this unit involves monitoring changes in fluorescence signals on an epifluorescence microscope, typically an inverted microscope. High numerical aperture (NA) lenses are essential, requiring that tissue preparations be viewed through glass coverslips with oil immersion lenses. Xenon-arc and mercury-arc lamps can generally be used for illumination (but mercury-arc lamps are not appropriate for use with fura-2 indicator dyes). Electronic shutters in the illumination pathway are necessary to minimize exposure of the specimen to illumination.

BASIC PROTOCOL 1

BATH APPLICATION OF MEMBRANE-PERMEANT CALCIUM-SENSITIVE FLUORESCENT INDICATORS (ACETOXYMETHYL-AM-ESTER DYES) TO CULTURED NEURONS

Materials *(see APPENDIX 1 for items with ✓)*

Cultured neurons (see Chapter 3)
Bathing solution: Tyrode's solution at room temperature for frog or chick neurons (see *APPENDIX 1*); Ringers solution prewarmed to 37°C for rat or mouse neurons (see *APPENDIX 1*)
✓ Calcium-sensitive membrane-permeant indicator dye solution (~1 mM; see Table 2.4.1)
BSA (electrophoresis grade)
Saline
Balanced salt solution lacking serum, such as artificial cerebrospinal fluid (see *APPENDIX 1*)

Perfusion chamber
Sonicator

2

Table 2.4.1 Characteristics of Calcium-Sensitive Indicators Used to Monitor Neuronal Activity

Indicator	Excitation λ (nm) $[Ca^{2+}]$ Low	Excitation λ (nm) $[Ca^{2+}]$ High	Emission λ (nm) $[Ca^{2+}]$ Low	Emission λ (nm) $[Ca^{2+}]$ High	K_d	Comments
Calcium green 1	488	488	525	525	190	Single-wavelength indicator that is more fluorescent than Fluo-3 at resting calcium levels. Can be coupled to dextran for retrograde tracing studies.
Fluo-3	500	500	525	525	390	Single-wavelength indicator for nonquantitative calcium measurements. Large signal of nearly 100-fold increase in fluorescence upon binding calcium (5- to 10-fold increase from the normal cytosolic calcium concentration of neurons). AM form is not fluorescent.
Fura-2	360	335	510	505	145	May be used as a ratiometric quantitative indicator or as a single-wavelength indicator for nonquantitative measurements. AM form is fluorescent.
Indo-1	345	330	475	400	230	Emission ratiometric indicator for quantitative measurements. Can be excited by argon ion laser for confocal microscopy. AM form is fluorescent.
Mag-Fura-2	370	330	510	510	25,000	Useful for measuring high levels of intracellular calcium concentration, which would be beyond the sensitivity of indicators with lower K_d values (e.g., Fura-2). AM form is fluorescent.

1. Wash neurons 10 min in bathing solution to remove serum-containing culture medium.

2. Just before use, dilute 7.5 µl of the 1 mM calcium-sensitive indicator dye solution in 1.5 ml saline containing 1% BSA (5 µM final concentration of indicator). Sonicate to help disperse indicator in the saline solution. Protect the diluted dye solution from light.

3. Incubate neurons in 1.5 ml calcium indicator solution for 20 min to 1 hr, away from light.

   The exact indicator concentration and incubation time and temperature will differ for different types of neurons. Adjust these variables empirically, to optimize the dynamic response to stimulation, and to limit sequestration of the indicator into intracellular compartments.

4. Wash neurons three times with fresh balanced salt solution at room temperature to remove excess calcium indicator. Allow at least 30 min for deesterification of the calcium indicator before acquiring data. Protect samples from light.

Table 2.4.2 Typical Filter Configurations for the Most Commonly Used Calcium-Sensitive Indicators[a]

	Wavelength (nm)					
Indicator	Ex #1	Ex #2	Dichroic 1	Dichroic 2	Em #1	Em #2
Fluo-3/calcium green 1	485 ± 11	None	505	None	535 ± 17.5	None
Fura-2[b]	340 ± 7.5	380 ± 7.4	30	None	510 ± 20	None
Indo-1[b]	355 ± 7.5	None	390	455	405 ± 21.5	495 ± 10

[a] Adapted from Haugland, 1996.

[b] Fura-2 and Indo-1 are UV-excitable ratiometric calcium indicators, which are the most commonly used probes for ratio imaging microscopy. Ex, excitation; em, emission.

5. Monitor the fluorescence signal on an epifluorescence microscope equipped with excitation and emission filters that are appropriate for the specific fluorescent probes (see *UNIT 2.1* and Table 2.4.2).

ALTERNATE PROTOCOL 1

BATH APPLICATION OF MEMBRANE-PERMEANT CALCIUM-SENSITIVE DYES TO EN-BLOC NERVOUS SYSTEM PREPARATIONS OR TISSUE SLICES

This protocol, adapted from Yuste et al. (1992), O'Donovan et al. (1994), and Wong et al. (1995), provides a dye-loading method suitable for neural slices or blocks of tissue.

***Additional Materials** (also see Basic Protocol 1)*

Neural tissue: brain slice or spinal cord preparation
Calcium-sensitive membrane-permeant indicator dye (fura-2 AM, fluo-3, or calcium green 1, available in 1-mg batches; Molecular Probes)
DMSO (anhydrous) containing 2.5 mg/ml pluronic F-127 detergent
Tyrode's solution

Superfusion system:
- Pump
- 95% O_2/5% CO_2
- Regulators
- Perfusion bath
- Heater (if necessary)

Fine-tipped microdissection scissors or vibrating blade mounted on a micromanipulator

1. Prepare dye by adding 100 µl DMSO/pluronic F-127 mixture to 1 mg calcium-sensitive indicator dye (fura-2 AM, calcium green, or fluo-3). Pipet this solution into 100 ml Tyrode's solution, which will be used to perfuse the tissue (10 µM final dye concentration). Sonicate the perfusate while adding the dye to facilitate its dispersion, but avoid heating. To improve uptake of dye, add 1% purifed BSA.

2. Cut the brain or spinal cord tissue with fine-tipped scissors or a vibrating blade mounted on a micromanipulator (see O'Donovan et al., 1994) to expose neurons in en-bloc preparations.

3. Oxygenate perfusate by gently bubbling with O_2. Avoid bubbling too vigorously; the perfusate may bubble out of the container due to the presence of the pluronic F-127. Load the cells with dye by perfusing the tissue with the perfusate solution containing dye for 1 to 2 hr at ~10 to 20 ml/min, recirculating the perfusate.

   En-bloc or tissue slices are generally superfused continuously during electrical and optical recording. For the isolated spinal cord, superfusion is maintained during dye loading, although tissue slices may be immersed in the dye solution as long as adequate oxygenation (with 95% O_2/5% CO_2) is maintained.

2

4. Following dye loading, perfuse the tissue with bathing solution—initially without recirculation—to wash away excess extracellular dye. Then perfuse with fresh bathing solution for another hour before making recordings.

5. Monitor the fluorescence signal on an epifluorescence microscope equipped with excitation and emission filters that are appropriate for the specific fluorescent probes (see *UNIT 2.1* and Table 2.4.2).

BASIC PROTOCOL 2

RETROGRADE LOADING OF CALCIUM-SENSITIVE DYES INTO EN-BLOC OR SLICE PREPARATIONS BY MICROINJECTION

Retrograde loading methods have been developed to overcome the limitations found to be associated with the bath-applied-dye techniques (Yuste and Katz, 1991). This protocol is adapted from O'Donovan et al. (1993; 1994).

Materials *(see APPENDIX 1 for items with ✓)*

En-bloc or slice preparation of neural tissue
✓ Calcium-sensitive membrane-impermeant indicator dye solution
Bathing solution:Tyrode's solution at room temperture for frog or chick tissue (see *APPENDIX 1*); Ringers solution prewarmed to 37°C for rat or mouse neurons (see *APPENDIX 1*)

Glass microelectrodes pulled on a standard puller from 1- to 1.5-mm filament glass to have ~50-MΩ impedance when filled with KCl (i.e., suitable for intracellular recording)
Micromanipulator

1. Back-fill glass microelectrodes with a small volume (~1 μl) of the dye solution and then mount in a micromanipulator for injection into the tissue. Under a dissecting microscope, pressure-inject the dye into the tissue or identifiable nerve tracts using a pressure pulse of ~5 to 20 psi for 100 to 500 msec.

2. Set up the perfusion flow with bathing solution to wash excess dye away from the preparation, or do not recirculate the perfusate during the injection procedure. Do not inject a significant amount of the dye into the perfusate around the tissue. Repeatedly wash tissue with a Pasteur pipet filled with the perfusate to avoid extracellular accumulation of the dye.

3. Leave the tissue with continuous perfusion (see Alternate Protocol 1, step 3) for 6 to 18 hr at room temperature to allow sufficient time for loading neuronal cell bodies and dendrites.

4. Monitor the fluorescence signal on an epifluorescence microscope equipped with excitation and emission filters that are appropriate for the specific fluorescent probes (see *UNIT 2.1* and Table 2.4.2).

ALTERNATE PROTOCOL 2

RETROGRADE LOADING OF CALCIUM-SENSITIVE DYES INTO EN-BLOC OR SLICE PREPARATIONS USING A SUCTION ELECTRODE

This loading method (adapted from O'Donovan and Ritter, 1995) is superior to microinjection because a very high concentration of the dye is in constant contact with the axons to be filled. However, it can only be used if a nerve tract can be drawn into an electrode.

Additional Materials (also see Basic Protocol 2)

Suction electrode: 1- to 3-in. (2.5- to 7.5-cm) length of polyethylene (PE) 50-90 tubing pulled over a flame to provide a taper with a 100- to 200-μm tip diameter
PE 50-160 tubing
Tygon tubing, 1- to 1.5-mm i.d.

1. Attach a length of thin PE tubing to the back of the suction electrode and apply suction. Pull the nerve into the electrode. Gently detach the PE tubing, insert the Tygon tubing into the electrode and evacuate the Tyrode's solution (to prevent the dye from being diluted), leaving just enough for the tissue to remain moist.
2. Draw up a small amount of the dye solution into another fine tube that can be introduced into the back of the suction electrode. Introduce the dye into the back of the electrode, being careful not to dislodge the nerve.
3. Leave the tissue with continuous perfusion (see Alternate Protocol 1, step 3) for 6 to 18 hr at room temperature to allow sufficient time for loading neuronal cell bodies and dendrites.
4. Monitor the fluorescence signal on an epifluorescence microscope equipped with excitation and emission filters that are appropriate for the specific fluorescent probes (see *UNIT 2.1* and Table 2.4.2).

BASIC PROTOCOL 3

RETROGRADE LOADING OF VOLTAGE-SENSITIVE DYES INTO EN-BLOC OR SLICE PREPARATIONS BY MICROINJECTION

Materials

En-bloc or slice preparation of neural tissue
10 to 20 mg/ml voltage-sensitive dye (di-8-ANEPPQ or di-12-ANEPEQ; Molecular Probes) in ethanol, chloroform, or DMSO

Glass microelectrodes pulled on a standard puller from 1- to 1.5-mm filament glass to have ~50-MΩ impedance when filled with KCl (i.e., suitable for intracellular recording)
Micromanipulator

1. Back-fill a glass microelectrode with the dye.
2. Pressure-inject the dye into the nerve tracts (see Basic Protocol 2, step 1).
3. Leave the tissue at room temperature with continuous perfusion (see Alternate Protocol 1, step 3) for 12 to 24 hr to allow sufficient time for the dye to reach cell bodies.

4. Monitor the fluorescence signal on an epifluorescence microscope equipped with excitation and emission filters that are appropriate for the specific fluorescent probes (see *UNIT 2.1* and Table 2.4.2).

References: Cohen, 1993; O'Donovan and Ritter, 1995; O'Donovan et al., 1993; 1994; Tsau et al., 1996; Wenner et al., 1996; Wong et al., 1995; Yuste et al., 1992

Contributors: R. Douglas Fields and Michael J. O'Donovan

CHAPTER 3

Cellular and Developmental Neuroscience

Since the early part of this century, neuroscientists have devoted considerable effort to developing tissue culture methods that would allow them to dissociate the complex interactions between cells that occur in the nervous system.

Four units present methods that permit the culture of "pure" populations of the five major cell types of the nervous system. *UNIT 3.1* presents protocols for clonal analysis of the multipotential stem cells that are present in both the developing and adult nervous system. These methods permit analysis of proliferating stem cells and the first stages of their differentiation. There are many distinct neurons in each region of the nervous system and specialized knowledge is required to culture these many cell types. *UNITS 3.4 & 3.5* illustrate common features of these culture methods as applied to the hippocampus and the substantia nigra, and *UNIT 3.6* provides a standard procedure for culturing astrocytes.

3

Most of the protocols take advantage of developmental periods at which large numbers of specific precursor cells are present in vivo. The efficient differentiation of embryonic stem (ES) cells into neurons has a number of important applications. ES cells can be made to differentiate into glia as described in *UNIT 3.2*. This in vitro differentiation model using ES cells facilitates the study of the signaling system involved in the initial step of neural induction and can also be used for rapid analysis of the neural phenotype of genetically modified ES cells.

UNIT 3.3 is concerned with in vivo studies of cell proliferation and differentiation. There is considerable interest in the proliferation of neuronal precursor cells in the adult CNS. *UNIT 3.3* presents details of methods for labeling dividing cells in the strain. *UNIT 3.7* provides protocols for cell transplantation into the embryonic, neonatal, and adult nervous systems.

Contributor: Susan Wray

UNIT 3.1

Culture of Neuroepithelial Stem Cells

NOTE: All tissue culture incubations should be performed in a humidified 37°C, 5% CO_2 incubator unless otherwise specified. All reagents and equipment coming into contact with live cells must be sterile.

BASIC PROTOCOL

FETAL RAT BRAIN NEUROEPITHELIAL STEM CELL CULTURE

This protocol yields cultures that are relatively homogeneous and consist almost exclusively of multipotent precursors expressing the intermediate filament nestin, a marker for CNS stem cells.

Materials *(see* APPENDIX 1 *for items with ✓)*

- Pregnant rat
- ✓ HBSS/HEPES, room temperature and 37°C
- ✓ N2 medium
- 0.2% (w/v) trypan blue
- 10-cm tissue culture dishes coated with poly-L-ornithine and fibronectin (see Support Protocol)
- Basic fibroblast growth factor (bFGF)

- Pasteur pipets
- 15-ml centrifuge tubes
- 10-ml pipets
- Cell lifters (Costar)

3

1. Sacrifice a pregnant rat by CO_2 asphyxiation and as quickly as possible, harvest tissue from the desired region of the embryonic CNS into ice-cold sterile HBSS/HEPES. Keep on ice.

   The gestational stage of the embryos should be selected such that the CNS region of interest is proliferating, e.g., E-14 to E-15 for cortex, E-13 to E-14 for striatum, E-15 to E-16 for hippocampus. Rats yield more proliferating cells and the survival of these cells is significantly better than for those obtained from mice.

   Avoid exposure of cells to serum at any stage of culture.

2. Using a sterile Pasteur pipet maintained vertically, consolidate the tissue from all embryos in a sterile 15-ml centrifuge tube and centrifuge 5 min at a maximum of $1000 \times g$, 4°C.

3. Aspirate HBSS/HEPES from the tube, taking care not to dislodge the pellet. Add 1 ml sterile HBSS/HEPES to the tissue pellet and dissociate the cells by slowly pipetting up and down (no more than ten times) using a 1000-μl micropipettor. Allow the pipet tip to rest gently against the bottom of the centrifuge tube to assist in dispersing the tissue. Add ~10 ml sterile HBSS/HEPES to the dispersed cells and slowly pipet up and down ten to twenty times to disperse any remaining clumps of tissue. Centrifuge as in step 2.

4. Aspirate the supernatant, taking care not to disturb the cell pellet. Resuspend the cells in 5 to 10 ml N2 medium by gently pipetting until the pellet is completely resuspended. Assess the yield of cells and their overall viability by mixing a small aliquot 1:1 with 0.2% trypan blue and counting the live (generally 70% to 80%) and dead cells on a hemacytometer (*APPENDIX 2I*). Inoculate $1–1.5 \times 10^6$ cells per 10-cm dish coated with poly-L-ornithine and fibronectin in 5 to 10 ml N2 medium, and add 10 ng/ml bFGF as a mitogen. Culture 4 to 5 days (until cells are 50% to 70% confluent), adding fresh bFGF daily and replacing the medium every other day.

5. Harvest cells for passage into new precoated dishes to remove unwanted cells. Wash cells two to three times with 5 ml HBSS/HEPES, 37°C, then incubate cells 15 min at 37°C in 7 to 10 ml HBSS/HEPES. Using a 10-ml pipet, gently spray HBSS/HEPES over the surface of the dish to wash off lightly adherent cells. Remove any remaining cells by scraping the surface of the dish with a cell lifter. Pipet cells gently to disperse any clumps and centrifuge as in step 2.

6. Aspirate the supernatant and resuspend the cells in 5 to 10 ml N2 medium, then count cells and assess viability as in step 4. Inoculate precoated dishes at the desired density in N2 medium containing 10 ng/ml bFGF.

ALTERNATE PROTOCOL

ADULT RODENT BRAIN NEUROEPITHELIAL STEM CELL CULTURE

Materials *(see* APPENDIX 1 *for items with ✓)*

Adult rat or mouse
L15 medium
✓ aCSF, rodent, supplemented with 0.2 mg/ml kynurenic acid and aerated with 95% O_2/5% CO_2; *or* HBSS/HEPES supplemented with 10 mM glucose and 0.2 mg/ml kynurenic acid
Trypsin (tissue culture grade, ~10,000 BAEE U/mg, from bovine pancreas; Sigma)
Hyaluronidase (1500 U/mg Type V; Sigma)
✓ DMEM/F-12 medium supplemented with 0.2 mg/ml kynurenic acid, 0.05% (w/v) DNase I (Worthington), and 0.7 mg/ml trypsin inhibitor from chicken egg white (Boehringer Mannheim)
✓ HBSS/HEPES
10-cm tissue culture dishes coated with poly-L-ornithine and fibronectin (see Support Protocol)
✓ N2 medium
B-27 supplement (Life Technologies; optional)
Basic fibroblast growth factor (bFGF)

Tissue culture dishes
Dissection microscope
Dissection instruments, including forceps and a sharpened tungsten needle, *or* a 25-G needle attached to 1-ml syringe
Scalpels
50- and 15-ml Falcon tubes
32° to 35°C shaking water bath

1. Sacrifice an adult rat or mouse by CO_2 intoxication. Aseptically remove the brain, then use a razor blade to cut the forebrain coronally into a series of ~1-mm sections. Collect the sections in a tissue culture dish containing L15 medium.

2. Using a dissection microscope, distinguish the subventricular zone (SVZ) lining the lateral ventricles and overlying the striatum in these coronal sections. Cut out the SVZ from the ventral tip of the lateral ventricle to the corpus callosum, using forceps and a sharpened tungsten needle (or a 25-G needle attached to 1-ml syringe). See Paxinos and Watson (1982) to identify the appropriate regions for dissection.

3. Mince tissue fragments finely using two scalpels. Collect the minced tissue from five animals in a 50-ml Falcon tube in 10 ml rodent aCSF supplemented with kynurenic acid and aerated with 95% O_2/5% CO_2, or with HBSS/HEPES/glucose. Add trypsin and hyaluronidase, close the tube tightly, and incubate 30 min at 32° to 35°C in a shaking water bath.

4. Wash the digested tissue fragments by centrifuging 5 min at 1000 × *g*, 4°C, and resuspending in 10 ml DMEM/F-12 supplemented with kynurenic acid, DNase I, and trypsin inhibitor. Repeat wash and perform two more similar washes in HBSS/HEPES. Centrifuge 5 min at 1000 × *g*, 4°C, to sediment tissue pieces.

5. Resuspend the tissue fragments in 2 ml HBSS/HEPES. Triturate with ten slow strokes using a a 1000-μl micropipettor. Let undissociated tissue pieces settle out for a few minutes. Transfer 1 ml of the supernatant to a 15-ml conical Falcon tube. Add 1 ml HBSS/HEPES to the undissociated tissue and repeat trituration and sedimentation two more times. Pool the supernatants containing dissociated cells.

6. Pellet the dissociated cells by centrifugation. Discard the supernatant and gently resuspend the cells in N2 medium with or without B-27 supplements (at manufacturer's recommended concentration) for the first 24 hr. Count viable cells (see Basic Protocol, step 4).

7. Plate ~1–5 × 10^5 cells into a 10-cm tissue culture dish coated with poly-L-ornithine and fibronectin in 5 ml N2 medium supplemented with 10 ng/ml bFGF. Incubate the dishes, adding fresh bFGF daily and replacing the medium every other day. Continue cultivation until a relatively homogeneous precursor cell population is obtained (see Basic Protocol, steps 5 and 6).

SUPPORT PROTOCOL

PREPARATION OF POLYORNITHINE/FIBRONECTIN-COATED TISSUE CULTURE DISHES

Materials *(see* APPENDIX 1 *for items with ✓)*

✓ 15 μg/ml poly-L-ornithine
✓ PBS, sterile
1 mg/liter bovine fibronectin in sterile PBS
10-cm tissue culture dishes

1. Add 5 ml of 15 μg/ml poly-L-ornithine to 10-cm tissue culture dishes and incubate several hours to overnight at 37°C.

2. Aspirate the solution and add enough PBS to cover the surface of the dish (~5 to 7 ml per 10-cm dish). Allow the dishes to incubate at least 1 hr, then aspirate the solution and repeat the PBS wash.

3. Using a volume sufficient to completely cover the dish, add 1 mg/liter fibronectin in PBS and incubate 1 hr to overnight at 37°C. Aspirate the solution and wash once with PBS. Store any unused dishes in PBS and discard dishes within 3 to 4 days.

Reference: Johe et al., 1996

Contributors: Thomas Hazel and Thomas Müller

UNIT 3.2

Differentiation of Embryonic Stem Cells

Embryonic stem (ES) cells are pluripotent cells whose developmental state is equivalent to cells of the inner cell mass in the blastocyst-stage embryo.

NOTE: All tissue culture incubations should be performed in a humidified 37°C, 5% CO_2 incubator unless otherwise specified. All reagents and equipment coming into contact with live cells must be sterile.

BASIC PROTOCOL

DIFFERENTIATION OF EMBRYONIC STEM CELLS

Materials *(see* APPENDIX 1 *for items with ✓)*

- Feeder cells prepared from embryonic fibroblasts (see Support Protocol 1), frozen
- ✓ DMEM/10% (v/v) FBS (e.g., Life Technologies)
- ES cells (e.g., J1, D3, R1, or CJ7 cells available from ATCC, Genome Systems, or individual laboratory), frozen
- ✓ ES proliferation (ES PRO) medium
- ✓ Dissociation solutions A and B (Dis A and Dis B)
- ✓ 2× frozen stock (FS) solution
- 0.2% (w/v) gelatin, autoclaved
- ✓ ES differentiation medium (ES DIF)
- ✓ DMEM/F-12 (e.g., Life Technologies)
- ✓ ITSFn medium
- ✓ DMEM/F-12/5% (v/v) FBS (e.g., Life Technologies)
- ✓ N3FL medium
- Polyornithine- and laminin-coated 6-cm tissue culture dishes: prepare as for polyornithine- and fibronectin-coated dishes in *UNIT 3.4*, but in place of fibronectin use 1 μg/ml laminin (Life Technologies) in DMEM/F-12, filtered through 0.2-μm filter
- Basic fibroblast growth factor (bFGF)
- Glial cell monolayer cultures in 35-mm plates (see Support Protocol 2), between 3 and 14 days after passage
- ✓ Neurobasal medium/B-27/5% (v/v) FBS
- Cytosine-β-D-arabinofuranoside (Ara-C)

- 15-ml conical tubes
- 10- and 6-cm tissue culture dishes
- Pasteur pipets
- Freezing container (Nalgene)
- 6-cm bacteriological petri dishes
- Disposable pipets with wide openings

1. Thaw one 1-ml frozen vial (1×10^7 cells) of feeder cells and transfer cell suspension into a sterile 15-ml conical tube containing 10 ml DMEM/10% FBS. Centrifuge 5 min at 300 × *g*, 20°C, aspirate supernatant, resuspend cells in 2 ml DMEM/10% FBS, and plate in several 10-cm tissue culture dishes, using 10 μl of the cell suspension per plate.

2. On the next day, thaw a vial of ES cells in a 37°C water bath. Mix the cells with 5 ml ES PRO medium in a sterile 15-ml conical tube and centrifuge 5 min at 300 × *g*, 20°C to collect the cells. Aspirate the supernatant and resuspend the cells in 10 ml ES PRO medium. Remove medium from feeder cells by aspiration and add ES cell suspension to the 10-cm feeder plate. Incubate cells 2 to 4 days until confluent.

   Keep quality of undifferentiated ES cells constant by using lower-passage-number ES cells, minimizing unnecessary passage, and optimizing culture conditions.

3. When ES cells become confluent, remove the growth medium by aspiration, and wash the plate once with 5 ml Dis A. After aspirating Dis A, add 1.5 ml Dis B and incubate 5 min at 37°C (or until cells start to detach). Dissociate cells completely by gently passing the cell suspension through a Pasteur pipet ensuring homogeneous single-cell suspensions. Check for the presence of cell aggregates under a microscope. If a large number of cell aggregates remain, repeat this step.

4. Transfer cell suspension to 15-ml conical tube, add 3.5 ml ES PRO medium, and centrifuge 5 min at 300 × g, 20°C. Aspirate the supernatant, resuspend cells in 10 ml ES PRO medium, and count cells using a hemacytometer.

5. Add ~3 × 10^6 cells in 10 ml ES PRO medium to a new 10-cm feeder plate. Alternatively, to make a frozen stock, adjust the cell density to 5–10 × 10^6 cells/ml, add 1 vol of 2× FS solution, and divide into 1-ml aliquots in individual labeled cryovials. Place cryovials in a freezing container and incubate overnight at −70°C to achieve −1°C/min rate of cooling. The next day, transfer frozen vials into a liquid nitrogen freezer. Keep frozen stocks up to 2 years.

   Optimize the cell density and the degree of trypsinization. Under optimal conditions, a 10-cm plate can hold up to 6 × 10^7 cells.

6. Spread 1 ml of 0.2% gelatin on 6-cm tissue culture dishes and incubate 10 to 60 min. Aspirate gelatin from petri dishes, and immediately plate ES cells onto the dishes at 2 × 10^6 cells/ml in 3 ml ES PRO medium. Incubate until the culture becomes confluent, usually 2 to 3 days.

7. Add 2 ml Dis A solution and aspirate. Add 1 ml Dis B solution and incubate at room temperature until cells start to detach from the substrate (~5 min). Gently rock the plate until most of the cells detach from the substrate. Add 11 ml ES DIF medium to the cell aggregates to neutralize the trypsin. Dispense 3 ml of the cell aggregates to each of four 6-cm uncoated bacteriological dishes using a disposable pipet with a wide opening. Incubate the suspension cultures for 3 to 4 days to allow EB formation.

8. If the culture contains a lot of cell debris on the day after plating, eliminate debris by placing the suspension in a 15-ml conical tube for 5 min to allow the EBs to settle to the bottom of the tube, then aspirating the supernatant and resuspending EBs in 3 ml fresh ES DIF medium. Plate onto a 6-cm uncoated bacteriological petri dish.

   Feeder cells can be eliminated from ES cells by passaging ES cells once on gelatin-coated dishes without feeder cells.

9. Combine the suspension of EBs from four 6-cm bacteriological dishes into a 50-ml conical tube. Let the solution sit 5 min to settle the EBs at the bottom of the tube, aspirate the medium, and resuspend EBs in 12 ml ES DIF medium. Plate 3 ml of the EB suspension onto each of four 6-cm tissue culture dishes.

   The day after plating, EBs should spread onto the substrate to make a monolayer of cells at the periphery; ideally, 50% of the surface area should be covered with spreading EBs.

10. The day after plating, wash plates with DMEM/F-12 three times and add 3 ml ITSFn medium to each dish. Replace the ITSFn medium every 2 days and check cell morphology every day for the appearance (generally 5 to 7 days after plating) of small, cluster-forming cells, which are nestin-positive neuronal precursor cells.

11. When cluster-forming cells are observed, wash 6-cm plates once with 3 ml Dis A solution and incubate with 1 ml Dis B solution for 5 to 10 min at room temperature. When cells start to round up, gently flush the solution on the surface using a Pasteur pipet until most of the cells detach. Further dissociate cells by gentle trituration through a Pasteur pipet several times. Add 5 ml DMEM/F-12/5% FBS to neutralize the solution, transfer cells to a 15-ml conical tube, and centrifuge 5 min at 300 × g, 20°C.

12. Resuspend in 3 ml N3FL medium and let sit for 5 min to settle remaining cell clumps at the bottom of the tube. Count the cell density of the supernatant and replate the cells at a density of 3 × 10^5 cells/cm^2 onto polyornithine- and laminin-coated 6-cm dishes.

Incubate cells, adding bFGF every day at a final concentration of 5 ng/ml and replacing medium every 2 days with 3 ml fresh N3FL medium.

A critical parameter for the transfer of precursor cells from ITSFn medium to N3FL medium is plating density. The proliferation index (PI) is calculated as follows:

$$PI = \frac{\text{cells/microscope field at day 3}}{\text{cells/cm}^2 \text{ at plating}}$$

13. When the cells reach confluence, wash cells once with Dis A solution. Add 1 ml Dis A solution per 6-cm dish again and incubate 5 to 10 min. Gently spread the solution onto the surface to detach cells from the plate. Dissociate cells by passing through a Pasteur pipet several times until cell clumps disappear. Add 4 ml DMEM/F-12, transfer to 15-ml conical tube, and centrifuge 5 min at 300 × *g*, 20°C. Remove supernatant and resuspend cells in N3FL medium.

14. To expand cells from one 6-cm dish to a larger surface area, resuspend cells in 9 ml N3FL medium and plate onto three 6-cm dishes coated with polyornithine and laminin. Add bFGF every day at a final concentration of 5 ng/ml and replace medium every 2 days. Grow cells until they reach confluence (2 to 5 days).

    In this proliferation phase, 90% of cells express the neuronal precursor cell marker nestin and 10% of cells express the neuronal cell marker MAP2. Few cells are positive for glial cell markers, such as GFAP (see UNIT 3.6) or Gal-C.

15. Aspirate medium from a 6-cm dish containing ES cell–derived neuronal precursor cells. Add 1 ml N3FL medium. Dissociate cells by gently flushing the medium onto the surface. Aspirate serum-containing medium from preprepared 35-mm plates containing glial monolayer cells. Add $1-5 \times 10^5$ ES cell–derived neuronal precursor cells in 2 ml N3FL medium, plate, and incubate.

    Plating desity of precursor cells onto gial monolayers can range from $1–20 \times 10^5$ cells per 35-mm dish.

16. Replace the medium with 2 ml Neurobasal medium/B-27/5% FBS the day after plating. To prevent glial cell proliferation, add 10 µM Ara-C 3 to 5 days after plating. Change one-fifth of the medium every 4 days.

    Dense synapse formation can be observed after 3 weeks.

3

SUPPORT PROTOCOL 1

PREPARATION OF FEEDER CELLS FROM EMBRYONIC FIBROBLASTS

It is generally desirable to prepare a large amount of γ-irradiated cells at one time and determine the optimal plating density before starting the actual experiments with that batch.

Materials *(see APPENDIX 1 for items with ✓)*

Pregnant mice at 13 to 14 gestational days
✓ 1.25× PBS
✓ Trypsin/EDTA solution
✓ DMEM/10% FBS (e.g., Life Technologies)
✓ 2× FS solution

Dissection instruments: forceps and scissors
50-ml disposable tubes

10-cm tissue culture dishes
10-ml disposable pipets
Pasteur pipets
γ-irradiation apparatus

1. Euthanize pregnant mice at 13 to 14 gestational days, remove embryos, and place in a 10-cm dish containing 1.25× PBS. Using forceps and scissors, remove head, extremities, tail, and visceral organs (liver, heart, intestine) from embryos. Wash the remaining trunk two times with 1.25× PBS.

2. Place embryo carcasses in a 10-cm dish, add 1 ml trypsin/EDTA solution per embryo, and mince tissue with small scissors. Transfer contents of the dish to a 50-ml disposable tube. Incubate 15 min with agitation at 80 rpm, 37°C.

3. Add 5 ml DMEM/10% FBS per embryo to the tube and dissociate tissue by passing through a 10-ml disposable pipet. Allow the tube to stand for 5 min to settle debris at the bottom. Remove supernatant into a new 50-ml tube and adjust the volume to 10 ml per embryo. Plate 10 ml of cell suspension to each 10-cm dish. Incubate 24 hr. Replace medium with 10 ml fresh DMEM/10% FBS and continue incubation until cells reach confluency (~3 to 6 days after plating).

4. Wash the plate of confluent cells once with 10 ml of 1.25× PBS, add 2 ml trypsin/EDTA solution, and incubate 5 min at 37°C. Dissociate cells by gently passing through a sterile Pasteur pipet. Add 5 ml DMEM/10% FBS and centrifuge 5 min at 300 × *g*, 20°C. Aspirate supernatant, resuspend cells in 10 ml DMEM/10% FBS per plate, and plate cells in four to six 10-cm plates. Incubate 3 to 5 days until cells reach confluency.

5. Dissociate cells as in step 4, and resuspend in 10 ml DMEM/10% FBS. Subject cell suspension to γ irradiation at 3500 to 5500 R. Alternatively, stop proliferation of embryonic fibroblasts by treating cell suspension with mitomycin C (Robertson, 1987). Adjust cell density to 2×10^7 cells/ml, add 1 vol of 2× FS solution, and divide the cell suspension into separate cryovials in 1-ml aliquots. Freeze the vials according to standard procedure.

6. Before using cells, thaw one vial, prepare a set of serial dilutions of the cells (ranging from 1/2 to 1/16), and plate in several 6-cm dishes. Choose the lowest density of mycoplasma-free cells which still form a monolayer to use in the actual experiments.

SUPPORT PROTOCOL 2

PREPARATION OF GLIAL MONOLAYER CULTURE

Materials *(see* APPENDIX 1 *for items with ✓)*

Postnatal day 2 (P-2) mouse pups
✓ 1.25× PBS
✓ Trypsin/EDTA solution
✓ DMEM/5% (v/v) FBS (e.g., Life Technologies)

10-cm tissue culture dishes
Dissection scissors sterilized by autoclaving
35-mm tissue culture plates
200-μm nylon mesh, autoclaved (optional)

1. Euthanize P-2 mouse pups and isolate their cerebral hemispheres aseptically. Place them in 1.25× PBS in a 10-cm dish and remove and discard the meninges. Mince the tissue with small scissors. Transfer the tissue to a 50-ml disposable tube, add 2 ml trypsin/EDTA solution per brain, and shake 10 min at 80 rpm, 37°C.

2. Dissociate tissue by passing through a 10-ml disposable pipet several times. Leave the tube for 10 min to settle cell aggregates at the bottom of the tube. If large cell aggregates form and do not settle, pass the cell suspension through autoclaved 200-μm nylon mesh. Remove the supernatant and add an equal volume of DMEM/5% FBS. Collect cells by centrifuging 5 min at 300 × *g*, 20°C.

3. Resuspend the pellet in 1 ml DMEM/5% FBS per brain. Calculate the cell density and plate $1-3 \times 10^6$ cells in DMEM/5% FBS per 10-cm plate. Incubate cells, replacing the medium every 3 days.

4. When the culture becomes confluent, wash cells once with 1.25× PBS, add 3 ml trypsin/EDTA solution to a 10-cm plate, and incubate at 37°C until cells start to round up. Dissociate cells by gently passing through a Pasteur pipet, add 3 ml DMEM/5% FBS, and collect cells by centrifuging 5 min at 300 × *g*, 20°C. Plate cells in 2 ml DMEM/5% FBS in 35-mm dishes that will be used for plating ES cell–derived neuronal cells and incubate.

   ES cell–derived neuronal precursor cells can be plated onto the glial monolayers between 3 and 14 days after passage.

References: Cole and de Vellis, 1989; Okabe et al., 1996

Contributor: Shigeo Okabe

3

UNIT 3.3

Quantitative Analysis of In Vivo Cell Proliferation

BASIC PROTOCOL 1

IN VIVO [^{3}H]THYMIDINE AUTORADIOGRAPHY

CAUTION: Potentially contaminated waste, such as unused [^{3}H]thymidine, animal bedding, perfusate, and animal carcasses, may have to be disposed of as radioactive waste, depending on their levels of radioactivity and local policies on radioactivity. Consult local radiation safety authorities for guidance.

Materials

Tissue sections (see Support Protocol 1)
Gelatin-subbed slides (*UNIT 1.1*)
Photographic emulsion suitable for detecting tritium (e.g., NTB-2, Eastman Kodak)
Photographic developer (e.g., Dektol, Eastman Kodak)
Fixer (e.g., Polymax T, Eastman Kodak)
0.1% (w/v) cresyl violet acetate in distilled water (store indefinitely at room temperature; filter before use)
70% and 100% (v/v) ethanol
0.25% (v/v) acetic acid in 95% ethanol (store indefinitely at room temperature)
Histological clearing agent (e.g., Histoclear, National Diagnostics)
Permanent histological mounting medium (e.g., Permount, Fisher)

Slide boxes
Darkroom with red safelight (e.g., 15-W bulb with Eastman Kodak safelight filter no. 2)

40- to 80-ml new (unused) glass beaker
43° to 45°C water bath
Emulsion dipping vessel (e.g., Dip Miser, EM Science)
Plain (unsubbed) slides

1. Mount one or two tissue sections on each gelatin-subbed slide. Air dry overnight at room temperature.
2. Place slides in a uniform orientation in slide boxes for easy handling in the dark room. Perform the following steps 3 to 6 in a darkroom under a red safelight.
3. Combine a small amount of photographic emulsion with an equal amount of distilled water in a new (unused) glass beaker (~20 ml of emulsion for 40 ml of diluted emulsion, which will coat 100 slides). Place the beaker in a water bath at 43° to 45°C. Slowly mix with a glass rod until the emulsion has liquified and combined with the water, avoiding air bubbles in the emulsion mixture. Pour diluted emulsion into a dipping vessel to a depth that will cover all but the frosted area of a slide.
4. Wearing clean gloves, dip a clean plain glass slide into the emulsion mixture, and hold it up to the safelight to check for air bubbles. Repeat this with three to four additional test slides until no air bubbles are visible.
5. Place two slides back to back (with the sections facing out), hold by the frosted end, and dip the slides in to the emulsion once using a slow, steady motion. Separate the slides and place them vertically against the wall of a counter, in a sectioned box, or a slide drying tray to dry for 1 hr.
6. Place slides back in the slide boxes with a packet of dessicant, and wrap the boxes with three layers of aluminum foil to keep out light. Leave slide boxes at 4°C for 2 to 4 weeks, preferably in a dry atmosphere.
7. Set up three processing solutions in staining dishes:

   1:2 (v/v) Dektol/distilled water (developer), ~15°C
   Distilled water (rinse)
   1:2 (v/v) Polymax T/distilled water (fix solution).

8. In a darkroom under the safelight, remove slides from boxes and place in slide racks. Place slide racks in developer solution for 90 sec. Dip briefly in distilled water and place in fix solution for 10 min. Rinse slides in a large beaker or other container under cold running tap water for 20 to 30 min.
9. Remove slide racks from the darkroom to counterstain for Nissl. Incubate slides (still in racks) in 0.1% cresyl violet for 5 min. Dip in distilled water. Place in 70% ethanol for ~1 min. Incubate in 0.25% acetic acid/95% ethanol until background staining is cleared from acellular regions. Dehydrate two times in 100% ethanol, 5 min each time. Incubate 2 to 5 min in histological clearing agent. Apply 2 to 3 drops permanent histological mounting medium and cover with a coverslip, gently pressing out any air bubbles over sections. Assign numbers to slides before analysis.
10. Select a labeling criterion (i.e., the minimum number of silver grains a cell must have over its nucleus in order to be counted as labeled; a cutoff of five grains is generally a good compromise). Count the number of labeled cells using 1000× magnification of a brightfield microscope. Keep track of the number of individual silver grains over each labeled cell to estimate the number of times the cell has divided.

    Because tritium can only be detected within 4 μm of the emulsion, grains that represent [^{3}H]thymidine incorporated by dividing cells' DNA should be directly over, or <4 μm away from, the nucleus.

3

11. Express labeled cell counts as either percentages of total cell number or number of labeled cells per unit area (densities) to normalize the data.

BASIC PROTOCOL 2

5-BROMO-2′-DEOXYURIDINE (BrdU) IMMUNOHISTOCHEMISTRY

This protocol (adapted from Cai et al., 1997) should be used when an immunohistochemical method for identification of dividing cells is preferable to an autoradiographic method. It is always a good idea to process several sections together on the same slide or in the same well.

Materials *(see* APPENDIX 1 *for items with* ✓*)*

Tissue sections (see Support Protocol 1)
0.3% (v/v) hydrogen peroxide in PBS (prepare in advance and store at 4°C)
Silanized slides (see Support Protocol 2) or Superfrost Plus slides (Fisher)
✓ Trypsin buffer
2 N HCl in PBS (prepare in advance and store at room temperature)
✓ PBS
✓ Blocking solution
✓ Antibody buffer
Mouse monoclonal anti-BrdU antibody (Becton Dickinson Immunocytometry)
Normal serum from species in which secondary antibody was produced
Biotinylated anti-mouse secondary antibody (Sigma or Vector Labs)
0.5% (w/v) cresyl violet in distilled water
Immunohistochemistry enhancement (e.g., Elite Peroxidase ABC Kit, Vector Labs)
Histological clearing agent (e.g., Americlear, Baxter, or Hemo-De, Fisher)
Permanent histological mounting medium (e.g., Permount, Fisher)
Peroxidase substrate for BrdU, preferably strong chromogenic substrate: e.g., cobalt-enhanced diaminobenzidine (e.g., DAB-Ni-Co, Sigma) or VIP (Vector Labs)
PAP pen

1. Mount sections on silanized slides. Incubate tissue sections in 0.3% hydrogen peroxide in PBS for 30 min to block endogenous peroxidase activity. Let slides dry 1 to 2 hr at room temperature, do not overdry. Circle tissue with a PAP pen.

   All of the following incubations on slide-mounted sections should be done with slides lying flat in a humidified chamber. Approximately 300 to 500 μl of solution is sufficient to cover an entire slide (when the PAP pen is used).

2. Incubate sections in trypsin buffer for 10 min. Rinse in PBS either three times or for 5 min. Incubate in 2 N HCl in PBS for 30 min. Rinse in PBS.

3. Incubate sections in blocking solution for 20 min and remove excess blocking solution.

4. Incubate sections with anti-BrdU antibody (diluted 1:100 in antibody buffer) overnight (or less if working with thin sections) at 4°C. Rinse in PBS.

5. Incubate 1 hr with biotinylated secondary antibody (diluted 1:100 in PBS containing 1.5% normal serum). Rinse in PBS.

6. Amplify signal by incubating 1 hr with peroxidase-conjugated Elite ABC. Incubate in peroxidase substrate solution for 2 to 10 min. Dry for 1 to 2 hr.

7. Dip slides (still in racks) briefly in distilled water, then incubate in 0.5% cresyl violet for 5 min. Dip in distilled water and place in 70% ethanol for ~1 min. Incubate in

0.25% acetic acid/95% ethanol until background staining is cleared from acellular regions (~1 to 2 min). Dehydrate two times in 100% ethanol, 5 min each time.

8. Incubate in histological clearing agent until sections are transparent (~2 to 5 min). Apply 2 to 3 drops permanent histological mounting medium and cover with a coverslip, gently pressing out any air bubbles over sections.

SUPPORT PROTOCOL 1

IN VIVO CELL LABELING AND TISSUE PREPARATION

CAUTION: Because they are DNA precursors, thymidine and BrdU are teratogenic and mutagenic. Read the safety information provided by the manufacturers for guidelines on handling, storage, and disposal.

Materials *(see* APPENDIX 1 *for items with ✓)*

Rat or mouse
Label: [^{3}H]thymidine in sterile aqueous solution (e.g., methyl-[^{3}H]thymidine, 1 mCi/ml, 50 to 90 Ci/mmol, DuPont NEN), *or* 5-bromo-2′-deoxyuridine (BrdU) labeling reagent (*APPENDIX 1*)
Anesthetic: e.g., Metofane (A.J. Buck & Son) or pentobarbital
✓ 4% (w/v) paraformaldehyde

Syringes with small-gauge (e.g., 25-G) needles
Cryostat or oscillating tissue slicer (e.g., OTS, EM Science; Vibratome, TPI)

1. Inject animal intraperitoneally (i.p.) with the selected label using a small-gauge needle.
   a. For autoradiography, use [^{3}H]thymidine in water (5.0 μCi/g body weight).
   b. For immunohistochemistry, use 50 to 200 μg/g body weight BrdU (as labeling reagent).
2. Allow animal to survive for the desired amount of time following injection, usually at least 1 to 2 hr to allow available [^{3}H]thymidine or BrdU to be incorporated into cells.
3. Perfuse heavily anesthetized animal through the heart with 4% paraformaldehyde (see *UNIT 1.1*). Remove the tissue of interest and postfix it overnight in a small vial of 4% paraformaldehyde.
4. Cut sections of the tissue on a cryostat or oscillating tissue slicer and mount on appropriate slides. Mount thin sections right away (on silanized slides for BrdU processing and either gelatin-subbed or silanized slides for [^{3}H]thymidine autoradiography) so they are completely flat.

SUPPORT PROTOCOL 2

SILANIZED SLIDES

Materials

2% (w/v) 3-aminopropyltriethoxysilane (Aldrich) in acetone
Acetone

Glass slides
42°C oven

1. Wash slides 30 min in detergent and hot water. Rinse slides 45 min in running tap water. Rinse slides briefly in distilled water. Air dry.

2. Incubate dry slides in 2% 3-aminopropyltriethoxysilane in acetone for 15 sec at room temperature. Wash two times in acetone, 1 min each time. Dip two times in distilled water. Dry slides at 42°C. Store indefinitely at room temperature.

References: Baserga and Malamud, 1969; Feinendegen, 1967

Contributor: Heather A. Cameron

UNIT 3.4

Long-Term Culture of Hippocampal Neurons

In culture, hippocampal cells can develop to express neuronal antigens and acquire mature neuronal morphologies, including axons, complex dendritic trees, and synapses that can be visualized at the electron microscopy level and are electrophysiologically active. This system is suitable for studying different events in neuronal differentiation and, in particular, relatively late events, such as synaptogenesis (Vicario-Abejon et al., 1998). It is also a valuable model for investigating synaptic plasticity and exploring the mechanisms of neuronal degeneration. Although embryos at day 18 (E-18) are used in this protocol, cultures from E-19 or E-20 embryos can give similar results. Cells prepared from younger (e.g., E-16 or E-17) embryos are more difficult to grow into mature neurons in vitro.

NOTE: Gloves, and a mask if desired, should be worn during this procedure. Surgical instruments should be autoclaved for 30 min, and surgical instruments and hands should be cleaned frequently with 70% ethanol. Dissection can be performed on a clean laboratory bench covered with paper.

3

BASIC PROTOCOL

CULTURE OF HIPPOCAMPAL NEURONS FROM EMBRYONIC DAY 18 RAT OR MOUSE

Materials *(see APPENDIX 1 for items with ✓)*

Pregnant rat (e.g., Sprague-Dawley, Taconic) or mouse, ~18 days pregnant (E16 to E18)
CO_2, metophane, or ether to anesthetize animal
70% ethanol
✓ Dissection solution, ice cold (keep in ice bucket)
✓ Chopping solution
✓ Trypsinization solution
✓ Supplemented DMEM/F-12/N2/10% (v/v) FBS
✓ Trituration solution
0.2% to 0.4% (w/v) trypan blue
✓ PBS with antibiotics
✓ Supplemented DMEM/N2/10% (v/v) FBS (see recipe for N2 supplements)
1 mM cytosine β-D-arabinofuranoside (Ara-C; Sigma)

Surgical instruments:
- Scissors
- Microdissecting scissors
- Curved forceps (medium and small sizes)
- Dumont no. 5 titanium forceps
- Microdissecting scissors with angled blades (Castroviejo style)

Dissecting microscope and optic fiber lights
Neubauer hemacytometer
Inverted microscope
12-mm-diameter polyornithine/fibronectin-coated circular coverslips for 24-well microtiter plate wells (see Support Protocol)

1. Anesthetize pregnant rat or mouse with CO_2, metophane, or ether. Sacrifice animal by cervical dislocation. Wipe off abdomen with 70% ethanol.

2. Open abdomen by grabbing skin with curved forceps and cutting skin and muscle along the midline with scissors. Make two cuts to the sides and pull out both horns of the uterus. Place uterus in a 10-cm dish containing dissection solution. Open uterus using small scissors pointed upward to avoid damaging the embryos.

3. Remove embryos (an average pregnant rat has 10 to 12 embryos), decapitate them, and place the heads in a fresh dish of dissection solution. If they continue to bleed, move them to a new dish of fresh dissection solution.

4. Under a dissecting microscope, place one head in a 6-cm dish containing dissection solution. Remove skin and skull using two pairs of Dumont forceps. Holding the head firmly with one pair of forceps, carefully remove the skin and skull with the other. Begin on the dorsal side over lambda and continue forward to bregma (see Fig. 3.4.1A). Separate the ventral side of the brain from the rest of the head.

5. Pick the brain up with curved small forceps and place into a new 6-cm dish of fresh dissection solution. Using Dumont forceps, remove the caudal parts of the central nervous system (CNS), such as the cerebellum, pons, and cervical spinal cord (see Fig. 3.4.1B). With the ventral aspect of the brain facing up, separate both hemispheres by moving the forceps through the midline (see Fig. 3.4.1B).

6. Working on one hemisphere, carefully remove the septum, thalamus, and hypothalamus from the cortex (see Fig. 3.4.1C), revealing the hippocampus as a slightly thicker portion lining the curved medial edge of the cortex (see Fig. 3.4.1D). Make a cut to separate the cortex from the striatum and olfactory bulbs. Then cut longitudinally through the boundaries between the hippocampus and cortex. Gently remove the meninges and choroid plexus, both of which are highly vascularized (see Fig. 3.4.1E). Using a Pasteur pipet, move the hippocampus to a fresh 6-cm dish containing dissection solution.

7. Repeat step 6 on the other hemisphere and then steps 4 to7 on the remaining heads. Collect the hippocampi in a single 6-cm dish and begin working under a laminar flow hood.

8. Put the tissue in a 35-mm dish containing chopping solution. Thoroughly mince the tissue into small pieces using angled scissors. Pipet the minced tissue into a 15-ml tube. Rinse the dish well with more chopping solution and collect the remaining tissue. Centrifuge 5 min at $\sim$200 $\times$ g, 4°C, and carefully remove the supernatant with a pipet.

9. Add 2 to 4 ml trypsinization solution to the 15-ml tube containing the minced tissue. Pipet into a 50-ml tube. Swirl in a 37°C water bath for 5 to 7 min (or up to 10 min if hippocampi are from older embryos, i.e., E-19 or E-20 rat embryos).

10. Quench trypsin activity by adding 10 ml DMEM/F-12/N2/10% FBS as quickly as possible. Pipet into a 15-ml tube. If clumping occurs, pipet tissue up and down two to four times. Centrifuge 5 min at $\sim$200 $\times$ g, 4°C.

11. Remove the supernatant and add 1.5 ml trituration solution to the tissue pellet. Triturate the tissue by pipetting it five to eight times through a 1000-μl pipet tip. Pipet slowly to avoid formation of bubbles. Add 5 ml dissection solution and centrifuge 5 min at $\sim$200 $\times$ g, 4°C.

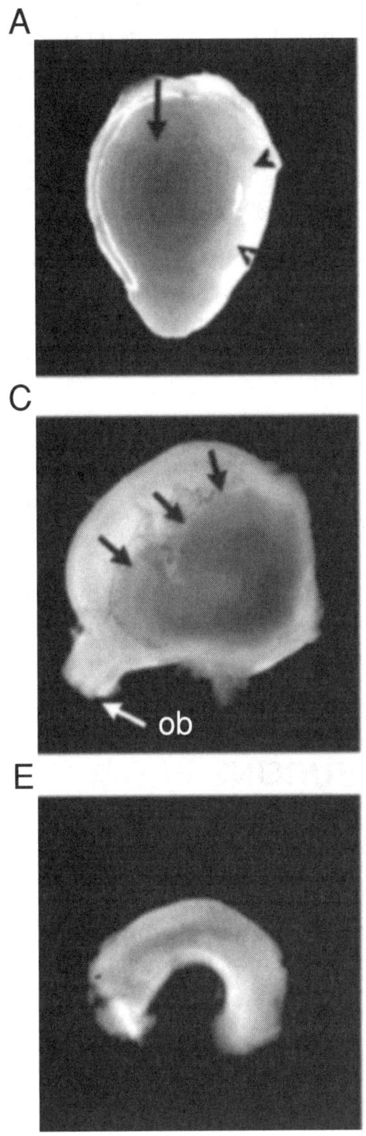

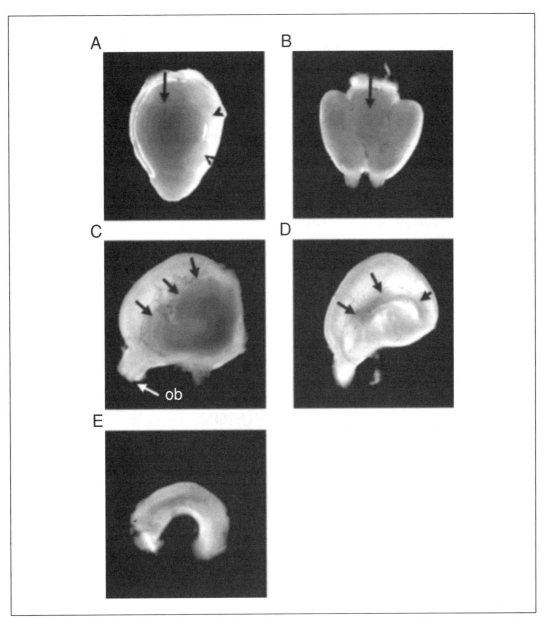

Figure 3.4.1 Dissection of the embryonic rat hippocampus. (**A**) With the head firmly grasped with a pair of Dumont forceps, the skin and skull are removed using another pair of forceps, beginning from lambda (arrow) and continuing forward to bregma. The tissue surrounding the head (arrowheads) is also removed. (**B**) The caudal parts of the CNS, such as the cerebellum, pons, and cervical portions of the spinal cord, are then removed. With the ventral aspect of the brain facing up, the two hemispheres are separated by moving the forceps through the midline (arrow). (**C**) In the next step, the septum and the diencephalic tissue (thalamus and hypothalamus) of one hemisphere are removed (arrows). The olfactory bulb (ob) is indicated. (**D**) The hippocampus is visible as a slightly thicker portion lining the curved medial edge of the cortex (arrows). (**E**) The hippocampus is dissected out by making a longitudinal cut through the border and the cortex. This can be done from either the ventral or dorsal side. Finally, the meninges and choroid plexus surrounding the hippocampus are carefully removed.

12. Remove the supernatant and add 500 to 1000 μl dissection solution, depending on the size of the pellet. Resuspend the pellet by tapping the tube and pipetting through a 1000-μl pipet tip. Pipet slowly to avoid formation of bubbles. Dilute 15 μl of cell suspension in 15 μl of 0.2% to 0.4% trypan blue (or 10 μl of cell suspension in 20 μl of trypan blue), and count both viable cells (those that exclude dye, ideally ~90%) and nonviable cells (those that take up the dye and become blue) using a hemacytometer (*APPENDIX 2I*).

13. Prepare cell suspension of desired density of viable cells by diluting the cell suspension in DMEM/F-12/N2/10% FBS. Plan to plate 2–2.5 × 10^5 cells/cm^2.

14. Just before plating the cells, aspirate fibronectin from the wells of 24-well microtiter plates containing polyornithine/fibronectin-coated circular coverslips. Wash once with PBS, if desired. Add 1 ml cell suspension to each well. Place the multiwell plate into a humidified 37°C, 5% CO_2 incubator. Let the cells sit undisturbed for several hours after plating to allow them to attach and distribute evenly on the coverslips.

15. Four to five days after plating, change 300 μl of the medium (per well) as follows: Place DMEM/N2/10% FBS in the incubator for 30 min to allow it to equilibrate to 37°C and 5% CO_2. Add 6 to 8 μM Ara-C to prevent proliferation of non-neuronal cells, primarily astrocytes. Then remove 300 μl medium from each well and replace it with 300 μl of the freshly prepared medium, using a micropipettor and 1000-μl pipet tip. Continue to incubate plates for up to 3 to 4 weeks, changing 250 to 300 μl of the medium every 7 to 10 days. Use fresh DMEM/N2/10% FBS medium.

3

ALTERNATE PROTOCOL

CULTURE OF HIPPOCAMPAL NEURONS FROM TRANSGENIC MOUSE EMBRYOS

Neurons (and glia) from transgenic mouse embryos can be cultured so that their in vitro development and responses to different factors can be studied. Embryos within a litter of transgenic mice have different genotypes. As a result, the hippocampi from different embryos cannot be pooled, and cell dissociation for the hippocampal hemispheres from the brain of each embryo must be performed individually.

Additional Materials *(also see Basic Protocol; see APPENDIX 1 for items with ✓)*

Pregnant mouse bearing transgenic embryos, 16 to 18 days pregnant
✓ DMEM/F-12/N2
FBS (Sigma)

1. Isolate embryos (see Basic Protocol, steps 1 to 3). Once the embryos are removed from the uterus, keep them in separate dishes and tubes throughout the procedure. Collect the tail and a leg from each embryo for DNA extraction and genotyping.

2. Dissect the hippocampus in dissection solution (see Basic Protocol, steps 4 to 7). Working in the hood, pipet the two hippocampi from a single brain into a 15-ml tube containing 1 ml DMEM/F-12/N2. Triturate the tissue by squirting medium against it three or four times using a micropipettor and 1000-μl tip to separate it into small pieces, then slowly pipetting it up and down five to eight times.

3. Add 800 μl DMEM/F-12/N2 to the triturated solution and plate 900 μl of the cell suspension into each of two wells of 24-multiwell plates containing polyornithine/fibronectin-coated coverslips (~2 × 10^5 cells/cm^2; 50% to 60% viability). Add 100 μl FBS to each well so that the final medium is DMEM/F-12/N2/10% FBS.

4. The day after plating aspirate all of the medium, rinse the cells with DMEM/F-12/N2 to remove debris, and add 1 ml DMEM/F-12/N2/10% FBS.
5. Perform subsequent changes of medium and treatment with Ara-C (see Basic Protocol, steps 13 and 15).

SUPPORT PROTOCOL

POLYORNITHINE/FIBRONECTIN-COATED COVERSLIPS

This Support Protocol uses a combination of polyornithine and fibronectin, although polylysine and laminin are also commonly used to grow hippocampal neurons (Banker and Goslin, 1998). Note that once the astrocytes develop, they become the real substrate for neuronal growth.

Materials *(see* APPENDIX 1 *for items with ✓)*

Concentrated HNO_3
✓ 15 μg/ml polyornithine
✓ PBS with antibiotics
✓ 1 μg/ml fibronectin

12-mm-diameter circular coverslips for 24-well microtiter plate wells (Bellco Glass)
Staining rack (cover-glass staining outfit, Thomas Scientific)
Oven, 225°C
24-well microtiter plates

1. Using small forceps clean coverslips in concentrated HNO_3, rinse with excess water, and sterilize in dry heat for 6 hr at 225°C.
2. In a laminar-flow hood, flame the coverslips, place them in 24-well plates, and cover them with 15 μg/ml polyornithine solution at least 24 hr before cells are plated. Wrap the multiwell plates with plastic or aluminum foil and leave them up to 7 to 10 days at room temperature.
3. On the day of use, remove the polyornithine solution using a vacuum aspirator. Wash three times with PBS. Add enough 1 μg/ml fibronectin to cover the coverslips and leave the plates in the incubator for 2 to 6 hr at 37°C. Immediately before plating the cells, remove fibronectin and, optionally, rinse once with PBS.

References: Altman and Bayer, 1990; Amaral and Witter, 1989; Banker and Goslin, 1998; Kaufman, 1995

Contributor: Carlos Vicario-Abejon

UNIT 3.5

Culture of Substantia Nigra Neurons

The culture system should be chosen based on the following criteria: (1) the level of organotypicity that has to be achieved, (2) whether direct observation of single cells during the culture period is needed, (3) how long the tissue must be maintained in vitro, and (4) what parameters will be assessed at the end of the experiment. In addition, the equipment available and the experience of the investigator must be taken into account.

Protocols described in *UNITS 3.1 & 3.4* can be adapted for use with tissue derived from the substantia nigra, but differences in the developmental stages should be considered. Neurogenesis of the dopaminergic cell population in the substantia nigra takes place between embryonic days 12 and 15 (E-12 and E-15); cells in the cortex or hippocampus have much longer developmental windows. The donor age should be chosen appropriately.

NOTE: All tissue culture incubations should be performed in a humidified 37°C, 5% CO_2 incubator unless otherwise specified. All reagents and equipment coming into contact with live cells must be sterile.

BASIC PROTOCOL

PREPARATION OF DISSOCIATED NIGRAL CELL CULTURES

Materials *(see APPENDIX 1 for items with ✓)*

Pregnant Sprague-Dawley rat
✓ Dissection medium, ice cold
✓ DMEM/F-12/N2 medium
0.4% (w/v) trypan blue, *or* (if inverted fluorescence microscope is available) two-color fluorescence cell viability assay (Molecular Probes)

Dissection tools:
- Small and medium-sized scissors
- 2 small spatulas
- 2 sets of Dumont no. 5 forceps
- 30.5-G needles (Thomas) *or* sharpened tungsten needles, attached to 1-ml Luer syringes, to be used as sterile microknives
- Dissecting microscope

10-cm polystyrene tissue culture dishes (Falcon)
5-ml glass pipet (Falcon)
15-ml (17 × 120–mm) conical-bottom polystyrene tubes (Falcon)
6- or 10-cm tissue culture dishes or 8-well glass chamber slides, coated with poly-D-lysine, polyornithine/fibronectin (*UNIT 3.1*), or entactin-collagen IV-laminin (UPI)

1. Euthanize pregnant rat with CO_2 (dry ice) and remove uterus (see *UNIT 3.4*). Separate yolk sack from embryos and place embryos in new petri dish with dissection medium.

2. Cut out mesencephalic region (see Fig. 3.5.1A). Microdissect the anlage of the substantia nigra (see Fig. 3.5.1B-D). Transfer all dissected pieces from 10-cm petri dish to 15-ml tube. Centrifuge 5 min at 200 × *g*, 4°C. Aspirate dissection medium and add 1 ml fresh dissection medium.

3. Resuspend pellet and triturate five to eight times in 1 ml dissection medium with micropipettor and 1000-μl pipet tip. Add 4 ml dissection medium and further disperse solution with a 5-ml glass pipet. Centrifuge 5 min at 200 × *g*, 4°C, and aspirate dissection medium.

4. Add 1 to 5 ml DMEM/F-12/N2, depending on the number of embryos dissected, and disperse cells with a 2- or 5-ml glass pipet. Remove 10 μl cell suspension and mix with 10 μl of 0.4% trypan blue. Calculate percentage (which should be >80%) and total number of living cells using a standard hemacytometer (*APPENDIX 2I*). Add additional DMEM/F-12/N2 to obtain a cell density of 0.5×10^6/ml.

5. Plate 25-μl droplets of cell suspension (1.25×10^4 cells each) onto tissue culture dishes or 8-well glass chamber slides precoated with poly-D-lysine. Place cultures in a humidified

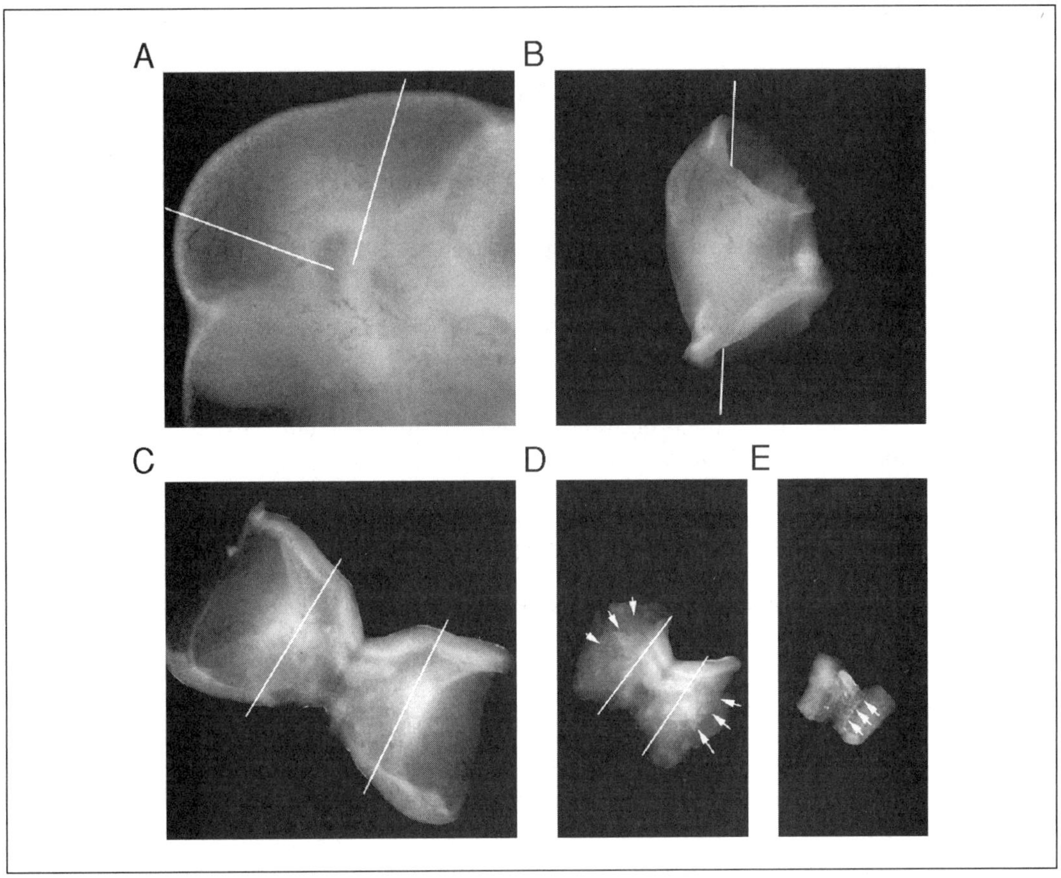

Figure 3.5.1 Dissection of fetal ventral mesencephalon. (**A**) Overview of mesencephalic flexure of E-12 rat embryo. Cuts are made along the indicated lines to isolate the mesencephalic region encompassing the anlage of the substantia nigra. (**B**) The dorsal mesencephalon (tectum anlage) is slit medially. (**C**) Same piece as in B now exhibits butterfly-like shape. Lateral parts are cut away along the indicated lines trying to tear meninges to the side. (**D**) Loosened meninges can be readily recognized (arrows) and are carefully separated from the CNS tissue using a Dumont no. 5 forceps to hold the meninges and a 30.5-G needle used as a microknife. (**E**) The final size of the dissected piece is ~1 × 5 × 0.3 mm. The arrows point to the border between the floor plate and the anlage of the substantia nigra.

37°C, 5% CO_2 incubator for 1 hr to allow cells in the droplets to attach to the matrix. Add DMEM/F-12/N2 medium (375 µl for 8-well chamber slides, 3 ml for 6-cm, and 5 ml for 10-cm culture dishes) and, if required by the experimental protocol, appropriate growth factors.

Cell survival can be greatly improved by supplementing DMEM/F-12/N2 with 1% FBS during the first 12 hr after plating.

6. Continue to incubate cells, changing medium every other day by carefully aspirating the old medium with a Pasteur pipet and adding the respective amount of fresh DMEM/F-12/N2 medium and growth factors.

7. After 4 to 9 days in vitro quantify the cells by in situ immunohistochemistry. Perform all wash steps carefully to minimize detachment. Use PBS/0.01–0.1% (w/v) BSA for fixation and staining procedures.

ALTERNATE PROTOCOL 1

PREPARATION OF FREE-FLOATING ROLLER TUBE (FFRT) CULTURES

The size and shape of these cultures allows them to be easily loaded into 18- to 21-G stereotactic needles after being taken up with a sterile Pasteur pipet and collected in a petri dish. These cultures can subsequently be used for transplantation without disrupting internal neuronal circuits. Cultures are usually used for transplantation after in vitro periods of 4 to 16 days.

Additional Materials *(also see Basic Protocol; see* APPENDIX 1 *for items with ✓)*

✓ Neurobasal medium/2% B27 (Neurobasal/B27)
Dry 37°C, 5% CO_2 incubator
15-ml (17 × 120–mm) conical-bottom polystyrene tubes (Falcon), stored in 5% CO_2 incubator for at least 24 hr prior to use
Roller drum (Bellco)

1. Dissect the mesencephalic region from rat embryos (see Basic Protocol, steps 1 to 3). Cut dissected region into pieces ~0.5 to 1 mm in size. Transfer pieces with sterile Pasteur pipet into a fresh petri dish filled with dissection medium and leave for 5 min to wash out enzymes.

   This will result in two pieces/embryo for E-12, four pieces/embryo for E-13 or E-14, and eight pieces/embryo for E-15 donors.

2. Preincubate 15-ml tubes in a dry 5% CO_2 incubator for 24 hr and fill tubes with 1 ml Neurobasal/B27 medium. Place one piece of tissue in each tube. Incubate tubes in roller drum, tilted 5° and turning at 1 to 2 rpm, placed in a dry 37°C, 5% CO_2 incubator.

3. Change medium at least two times a week by slowly pouring out old medium (so that the culture remains trapped in the small amount of medium left in the vortex of the tube) and adding 1 ml fresh Neurobasal/B27 medium.

ALTERNATE PROTOCOL 2

PREPARATION OF MATRIX-BASED ORGANOTYPIC CULTURES

This culture system can be used to prepare mesencephalic tissue for long-term maintenance (up to 60 days) in vitro. The cultures generally show a high degree of organotypicity. Direct observation in matrix-based organotypic cultures is dependent on the degree of flattening exhibited by the tissue over time. The amount and type of serum used affects the flattening of the tissue, possible overgrowth of glial cells (which might require the use of an antimitotic), and the long-term survival of dopaminergic neurons.

Other materials, such as polyornithine, polylysine, or collagen, have been used for organotypic slice and explant cultures (Wray, 1992). Another simple organotypic culture technique is growing explants on Millicell membranes (Millicell-CM, Millipore). This technique, based on the work of Stoppini (Stoppini et al., 1991), does not require a rolling movement and can be combined with the use of various matrices, including chicken plasma (Plenz and Kitai, 1996). If the observation of single cells in long-term cultures is required, however, the roller tube technique is preferable because it produces more pronounced flattening of the tissue.

NOTE: Enough cleaned coverslips for the procedure should be prepared before starting the dissection.

Additional Materials *(also see Basic Protocol; see* APPENDIX 1 *for items with ✓)*

5 ml lyophilized chicken plasma (Sigma or Cocalico) reconstituted in 5 ml H_2O
✓ 200 U/ml thrombin
✓ Supplemented DMEM/HBSS/10% (v/v) FBS

Standard tissue chopper (McIlwain, Brinkman)
✓ 12 × 24–mm glass coverslips, cleaned and sterilized

1. Dissect mesencephalic region from rat embryos (see Basic Protocol, steps 1 to 3).

2. For E-14 or younger embryos, cut the anlage of the substantia nigra into 0.5- to 1-mm pieces. For E-15 to E-17 or older embryos, first chop the mesencephali in 0.3-mm slices using a standard tissue chopper, then further cut slices into pieces not bigger than 0.3 × 1 × 1–mm.

3. Place five washed 12 × 24–mm glass coverslips onto a 10-cm petri dish. Add two drops (100 μl) of chicken plasma to one end of each coverslip and one drop (50 μl) of thrombin to the other end.

4. Transfer tissue, with a minimum amount of dissection medium, into chicken plasma using a sterile Pasteur pipet. Mix thrombin with chicken plasma by careful rotatory movements using a small spatula. Let coverslips stand 30 min at room temperature or up to 2 hr at 4°C.

5. Fill CO_2-preincubated 15-ml tubes (see Alternate Protocol 1, step 4) with 1 ml supplemented DMEM/HBSS/10% FBS medium. Remove coverslips from the petri dish with Dumont no. 5 tweezers and place into the prepared tubes. Incubate tubes in roller drum, tilted 5° and turning at 1 to 2 rpm, placed in a dry 37°C 5% CO_2 incubator.

6. Change medium at least two times a week by pouring out old medium and adding 1 ml fresh DMEM/10% FBS medium, pouring it along the wall and taking care to avoid directly hitting the culture on the coverslip.

References: Gähwiler, 1981; Spenger et al., 1994; Takeshima et al., 1996

Contributor: Lorenz Studer

UNIT 3.6

Isolation and Purification of Primary Rodent Astrocytes

BASIC PROTOCOL

ISOLATION AND CULTURE OF ASTROCYTES FROM RODENT CEREBELLUM

Materials *(see* APPENDIX 1 *for items with ✓)*

Mice aged P-0 to P-7
✓ PBS/glucose
✓ 60% and 30% (v/v) Percoll
✓ Trypsin/DNase solution
✓ DNase solution
✓ PBS/glucose containing 5 mg/liter $MgSO_4$

✓ DMEM/10% (v/v) FBS containing 2% (w/v) glucose
0.4% (w/v) trypan blue solution (e.g., Life Technologies)
Tissue culture plates coated with poly-L- (or poly-D-) lysine
✓ PBS
Horse serum (optional)
0.05% (w/v) trypsin/0.53 mM EDTA (Life Technologies)
Anti-Thy 1.1 or 1.2 antibody; depending on the species and strain of animals used as the source of astrocytes (if monoclonal, IgM antibodies are preferable)
✓ DMEM
Rabbit or guinea pig complement (Life Technologies)
Fetal bovine serum (FBS; optional)
Dimethyl sulfoxide (DMSO; optional)

Dissecting instruments:
- 1 pair microdissecting scissors (~3.5-in.)
- 1 pair dissecting scissors (~6.5-in)
- 2 pairs no. 5 Dumont forceps (~4.75-in.)
- 1 pair Semkin dissecting forceps (~6-in)
- 1 pair curved no. 5 Dumont forceps (optional)

100-mm petri dishes (Falcon)
5-ml and 15-ml conical tubes (Falcon)
9-in. Pasteur pipets (one flame polished) and bulb
Swinnex 13-mm filter units (Millipore) with 20-μm nylon mesh filters (Fisher)
100-mm tissue culture plates (Falcon)
1-ml syringes
18-, 20-, and 23-G needles
Humidified 37°C, 7% CO_2 incubator
Bright-field microscope equipped for epifluoresence

1. Sterilize instruments in an autoclave, or by dipping in ethanol and flaming. Autoclave the Swinnex filters after they have been assembled with nylon mesh in place.

2. In the lid of a petri dish, make several small puddles (approximately five times more than the number of pups to be harvested) of ~100 μl PBS/glucose, and a single puddle of ~400 μl.

3. Place each size needle (18-, 20-, and 23-G) on a separate 1-ml syringe.

4. For discontinuous Percoll gradients (one gradient for each two to three cerebella to be dissected), add 3 ml of 60% Percoll to the bottom of a 15-ml conical tube. Carefully and slowly overlay 2 ml of 30% Percoll such that there is a sharp separation between the two concentrations. Use gradients up to ~1 hr after pouring.

5. Place one 1-ml aliquot of trypsin/DNase and two to six 1-ml aliquots of DNase (depending on the number of animals to be taken; about one aliquot per five animals) in a 37°C water bath for use in steps 12 and 14.

6. Anesthetize the pups by cooling them on ice one at a time. Sacrifice one animal at a time and remove the cerebellum from each before moving onto another animal. Once the animal has stopped moving, decapitate with the large scissors, drop the head into a 100-mm petri dish, and grasp the snout with large forceps using the left hand (for right-handed individuals). Slide the closed points of the small scissors between the scalp and the cranium and separate the scalp from the underlying tissue by sliding the scissors back and forth. Cut away the skin thus loosened and discard.

7. Open the small scissors, insert one point through the foramen magnum (at the base of the skull), angling the blade to the extreme side of the head, and cut along the side of the skull

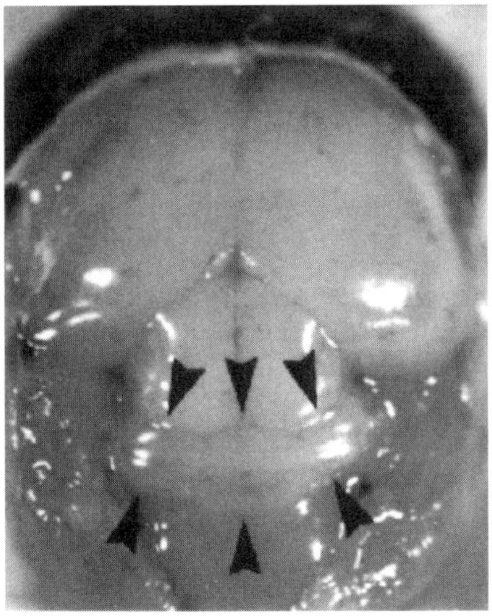

Figure 3.6.1 Neonatal rodent cerebellum. The arrowheads indicate the rostral and caudal borders of the developing cerebellum. Harvesting of this tissue is accomplished by passing Dumont forceps under the cerebellum at the sites of the lateral arrowheads and gently lifting the tissue up, separating the cerebellum from the mid- and hindbrain.

3

until the coronal suture is reached. Repeat the process on the other side, then cut across the cranium and lift it away from the underlying tissue with a fine forceps.

8. Place the blades of either a straight or curved forceps under the cerebellum, and gently lift it away from the remainder of the brain (see Fig. 3.6.1). Place each cerebellum harvested into a separate small puddle in the lid of the petri dish.
9. Move the petri dish lid with the cerebella to the stage of a dissecting microscope. Using the Dumont forceps, hold the tissue with one forceps, and peel away the meninges and pia mater from both sides of the cerebellum. If blood vessels are observed on the surface of the tissue, remove them along with the associated membranous material to reduce the number of contaminating fibroblasts.
10. Move each stripped cerebellum without the dissociated membranes to the larger puddle in the dish lid.
11. Transfer the cleaned cerebella to a 5-ml tube, rinse with PBS/glucose, and remove the fluid. Immediately add 1 ml trypsin/DNase solution, making sure that the fluid is in contact with all of the tissue in the tube. Cap the tube and transfer it for 3 min to a 37°C water bath.
12. Remove the trypsin/DNase solution and carefully wash the tissue three times by adding ~2 ml PBS/glucose, allowing the tissue to settle to the bottom of the tube, and then removing PBS/glucose with a Pasteur pipet and bulb, making sure that the pellet is undisturbed.
13. After removing the last wash solution, add 1 ml DNase solution. Beginning with the 18-G needle, draw up the tissue and DNase solution and expel it back into the tube. Repeat this procedure for a total of fifteen times, then repeat (fifteen times each) with the 20- and 23-G needles.
14. Draw up the cell suspension and discard the needle. Using sterile technique, attach the Swinnex filter unit onto the syringe and pass the cell suspension through the filter into a fresh 5-ml tube. Centrifuge the tube 1 min at 150 × *g*, 4°C, in a tissue culture or clinical centrifuge.

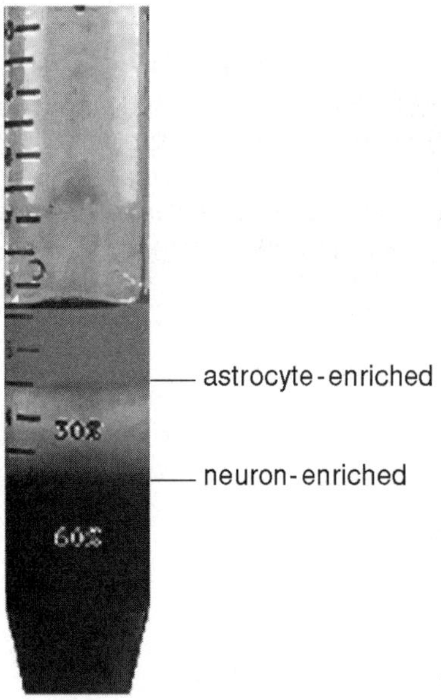

Figure 3.6.2 Discontinuous Percoll gradients. Enrichment of astrocytes from a heterogeneous cerebellar cell suspension is achieved by passing the cells over a Percoll gradient. The small, dense granule cell neurons pass through the 30% Percoll and come to rest on a cushion of 60% Percoll. The larger, membranous astrocytes stop at the medium/30% Percoll interface. The 60% Percoll will be blue, the 30% Percoll will be white, and the loading medium will be red.

15. For each gradient (two to three cerebella) used, resuspend the pellet in 1 ml of a 1:1 mixture of DNase solution and PBS/glucose/$MgSO_4$ (e.g., 5 ml for five gradients). Without disturbing the gradient interfaces, overlay the gradients with 1 ml of the cell suspension, and cap tube. Transfer the gradients to a centrifuge and spin 10 min at 150 × *g*, 4°C. Observe the astrocyte-enriched fraction at the buffer/30% Percoll interface, and the neuron-enriched fraction at the 30%/60% interface (see Fig. 3.6.2).

16. Transfer the gradient back to a laminar flow hood. Place a rubber bulb on a flame-polished Pasteur pipet and expel all of the air from the bulb (do not introduce bubbles into the gradient). Insert the pipet into the gradient just above the astrocyte-enriched layer and carefully aspirate the cells at the interface; the cells will adhere to one another, so lift them as a carpet of material. Transfer the cell suspension to a fresh 15-ml conical tube, top off the tube to 15 ml with PBS/glucose, and centrifuge 10 min at 150 × *g*, 4°C.

17. Resuspend the pellet in DMEM/10% FBS/2% glucose, determine the cell number and viability using trypan blue and a hemacytometer (*APPENDIX 2I*), and plate at 2.5–5 × 10^4 cells/ml on polylysine-coated plates, using 1 ml DMEM/10% FBS/2% glucose for every 10 mm^2 tissue culture area. Incubate all cultures in a 37°C, 7% CO_2 humidified incubator.

18. Two days after the establishment of astrocyte-enriched cultures, trypsinize the cells with trypsin/EDTA solution. Wash the cells three times with PBS/glucose, remove the last wash, and add 1 ml trypsin/EDTA for every 30 mm^2 tissue culture surface area. Check frequently under the microscope. As soon as the cells begin to loosen from the plate, tap the plate firmly against a hard surface to facilitate lifting of the cells. Once the majority of cells are in suspension, add 3 vol DMEM/10% FBS/2% glucose, transfer to 15-ml conical tube, and centrifuge 5 min at ~150 × *g*, 4°C.

19. Resuspend the pellet in 5 ml DMEM/10% FBS/2% glucose and centrifuge 5 min at ~150 × *g*, 4°C. Resuspend the pellet in 1 ml monoclonal antibody supernatant, or antibody diluted in DMEM/10% FBS/2% glucose as recommended by the manufacturer for cytolysis. Incubate the cells 1 to 2 hr on ice.

20. Spin cells in a refrigerated centrifuge 5 min at ~150 × *g*, 4°C. Resuspend in 3 ml DMEM and add 1 ml complement. Incubate 1 hr at 37°C. Add DMEM to full volume of the tube (15 ml) and centrifuge the cells 5 min at ~150 × *g*, 4°C. Resuspend the pellet in DMEM and centrifuge 5 min at ~150 × *g*, 4°C. Repeat again.

21. Plate the cells at 5 × 10^4 cells/ml in DMEM/10% FBS/2% glucose on polylysine-coated plates and incubate to produce cultures of essentially pure astrocytes.

22. Passage the cells when they reach confluence, splitting them 1:4. For use in the future, harvest the cells as in step 19, resuspend the pellet in 1 ml of 95% heat-inactivated FBS/5% DMSO, and freeze overnight at −80°C in a cryotube. The following day, move the cryotube to liquid nitrogen for long-term storage.

ALTERNATE PROTOCOL

ENRICHMENT OF ASTROCYTES BY DIFFERENTIAL ADHESION

Additional Materials *(also see Basic Protocol; see* APPENDIX 1 *for items with ✓)*

✓ PBS, 37°C

1. Complete Basic Protocol steps 1 to 14. Then resuspend the cells in 1 to 2 ml DMEM/10% FBS/2% glucose and determine the concentration and viability using trypan blue and a hemacytometer (*APPENDIX 2I*). Plate the cells at 5 × 10^4 cells/ml in DMEM/10% FBS/2% glucose on poly-L- (or poly-D-) lysine-coated plates and incubate in a humidified 37°C, 7% CO_2 incubator, as in Basic Protocol, step 17.

2. Two to three hours after plating, move the plates to a laminar flow tissue culture hood and dislodge the loosely adherent neurons by aggressively agitating the plate, taking care not to spill the medium over the sides of the plate or otherwise contaminate the culture. Firmly tap the plate against a benchtop, then gently wash the surface of the culture with PBS warmed to 37°C. Monitor the culture after each wash to be sure that astrocytes are not inadvertently removed. Repeat washes until most of the contaminating neurons are removed.

3. Following the final wash, add DMEM/10% FBS/2% glucose to the cultures and return them to the incubator.

SUPPORT PROTOCOL

IMMUNOHISTOCHEMICAL DETECTION OF GLIAL FIBRILLARY ACIDIC PROTEIN (GFAP)

Materials *(see* APPENDIX 1 *for items with ✓)*

Astrocyte culture (see Basic Protocol or Alternate Protocol)
Glass coverslips or 8-well slides (Lab-Tek, Nalge Nunc) coated with poly-L- (or poly-D-) lysine
✓ PBS
100% methanol, ice-cold
Blocking reagent: 10% (v/v) goat serum/0.1% (v/v) Triton X-100 in PBS
Anti-GFAP monoclonal antibody (clone GA-5; Boehringer Mannheim)
Anti-mouse antiserum conjugated with chromofluor or enzyme

1. Before passaging cells (Basic Protocol, step 19), coat either glass coverslips or a Lab-Tek 8-well slide with polylysine. Plate 3×10^3 to 1×10^4 cells per slide well or 1×10^4 cells per coverslip in DMEM/10% FBS/2% glucose and culture overnight under standard conditions.
2. Rinse cells two times with PBS. Cover the cells with ice-cold 100% methanol and let sit 8 min at −20°C. Rinse the cells three times with PBS or CMF-PBS and add 300 μl blocking reagent. Incubate 1 hr at room temperature.
3. Dilute anti-GFAP monoclonal 1/20 in blocking reagent. Add to slides in blocking reagent and let sit 1 hr at room temperature. Wash slides three times for 5 min each in PBS.
4. Dilute anti-mouse antiserum conjugated to a chromofluor or enzyme per the manufacturer's instructions in blocking reagent, add to slides, and incubate in the dark 1 hr at room temperature. Wash three times for 5 min each in PBS.
5. Coverslip and view under fluorescence optics (*UNIT 2.1*).

References: Hatten, 1985; Weinstein et al., 1990

Contributor: David E. Weinstein

3

UNIT 3.7

Experimental Transplantation in the Embryonic, Neonatal, and Adult Mammalian Brain

STRATEGIC PLANNING

Animal Care

Animal procedures described in these protocols serve only as general guidelines. It is essential to consult the local animal health and care unit (Institutional Animal Care and Use Committee) before initiating any transplant experiments. Transplant protocols are subject to legal regulations and require approval by the local animal care and health committee.

Fresh bedding should not be given to adults or newborn litters for at least 3 days prior to surgery or delivery of a litter.

Transplant Material

The choice of transplant material (tissue fragments versus single-cell suspensions) is especially important for studies addressing the developmental potential of neural precursors in rodents. Compared with isolated cells, cells transplanted within a multicellular context are less amenable to the external stimuli of their new environment and tend to maintain their original phenotype when transferred to a heterotopic location. These observations have led to the view that exposure to the new environment at the single-cell level is crucial for regional cellular integration. Transplantation of tissue fragments is, however, an excellent method for the study of axonal plasticity (Stanfield and O'Leary, 1985; O'Leary and Stanfield, 1989; Barbe and Levitt, 1992, 1995). In reconstructive transplantation, solid grafts allow transfer of differentiated neurons that, due to their complex morphology, would not survive preparation of a single-cell suspension.

Cell Tracing

Selection of a reliable cell labeling system is crucial for the correct interpretation of transplant experiments. Today, genetic labeling systems have become the state of the art. There is a steadily growing number of mice carrying reporter genes under the control of various cell type–specific promoters as well as *lacZ*-expressing "knockout" mice. It is well worth considering these strains as potential donor sources. Alternatively, donor cells can be transduced with adenoviruses and replication-deficient retroviruses carrying suitable reporter genes. Because the interpretation of the results will depend largely on the quality of the cell labeling procedure, this aspect requires particular attention during the planning of each transplant experiment.

BASIC PROTOCOL 1

TRANSPLANTATION INTO THE ADULT RODENT BRAIN

A two-person team approach to transplantation procedures is preferable.

Materials

- 70% ethanol
- Adult rodent
- Anesthetic
- Betadine solution
- Material to be injected: transplant, such as cell suspension (see Support Protocol 2) or solid tissue fragment; *or* neurotoxin solution or other substance
- Saline (0.9% w/v NaCl), sterile

- Animal scale
- Surgical instruments:
  - Scalpel (surgical blade) with no. 10 or no. 15 blade
  - Tissue forceps
  - Fine scissors
  - Sterile gauze
  - Small self-retaining retractor (e.g., ALM, Tieman)
  - Surgical spears or ophthalmic lancets (Merocel, Xomed)
  - 26-G needle and 1-ml syringe
  - Needle holder or skin staples
  - Sutures (nonabsorbable nylon monofilament 5-0 Ethilon)
- Surgical gloves and masks
- Stereotactic frame (e.g., David Kopf Instruments)
- Operating microscope (optional)
- Fiber optic light source (optional)
- Injection instruments:
  - Hamilton needles and syringes (1, 5, or 10 μl) for conventional implantation of cell suspensions
  - Injection device (see Support Protocol 1) or glass capillaries (50- to 100-μm orifice) attached to a Hamilton needle via a polyethylene cuff for microtransplants
  - Spinal needle (preferably blunt-tipped) for placement of solid grafts
- Dental drill or bone drill with small (1- to 1.5-mm) cutting burrs
- Heating pad (Aquamatic, American Medical Systems), 37°C, or heat lamps

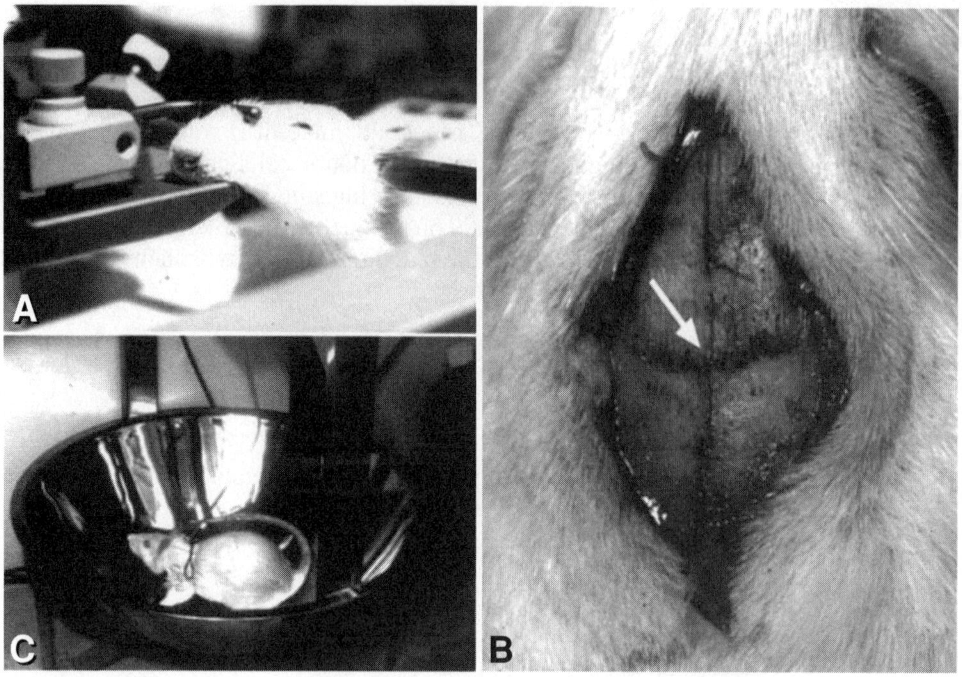

Figure 3.7.1 Stereotactic transplantation into the 6-hydroxydopamine-lesioned striatum of an adult rat, a rodent model of Parkinson's disease. (**A**) Positioning of the animal in the stereotactic frame. (**B**) Identification of the bregma at the intersection of the coronal and sagittal sutures (arrow). Drug-induced rotations are analyzed before and after transplantation, using an automated rotometer (**C**).

1. Sterilize surgical instruments and instruments used for injection by autoclaving or immersion in ethanol. Wear surgical gloves, mask, and lab coat throughout the surgical procedure, and follow appropriate sterile technique.
2. Set up surgical field over a large, flat surface in a quiet room (preferably a designated operating area). Position stereotactic frame, operating microscope (optional), and fiber optic light source (optional). Spray all parts to be handled during the procedure with 70% ethanol. Arrange sterilized instruments over sterile towels.
3. Weigh animal and determine required amount of anesthetic. Deeply anesthetize animal by inhalation or injection prior to handling; absence of a paw reflex upon pinching is usually a good sign of adequate anesthesia.
4. Shave the head of the animal, preferably with an electric razor and position animal in the stereotactic frame (see Fig. 3.7.1A). Open the mouth of the animal and insert incisors across the tooth bar. Holding the head firmly in one hand, gently advance a cupped or 45° earbar into the auditory canal, avoiding rupturing the tympanic membrane. Lock the earbar and, while continuing to support the head, insert and lock the opposite earbar.
5. Partially unlock both earbars and align head first in the horizontal and then in the coronal plane. Adjust orientation in the sagittal plane by setting the toothbar at the appropriate position and gently tighten the nose clamp over the snout, without exerting undue pressure or using it as a restraining device.
6. Clean the skin with a gauze soaked in 70% ethanol or a betadine solution avoiding contact with the eyes.

7. Perform a midline incision ~1 to 1.5 cm with a scalpel (no. 10 or no. 15 blade). Dissect the underlying fascia and pericranium using forceps and fine scissors, starting in the midline and moving laterally. Remove remaining pericranium off the skull using a piece of dry gauze held with forceps. Identify bregma at the intersection of the coronal and the sagittal sutures (see Fig. 3.7.1B).

   Stereotactic coordinates usually include toothbar position, reference point (bregma or interaural line), anteroposterior (AP), mediolateral (ML), and ventral (V) or depth (D) readings. For easy reference, post these coordinates in the surgical area prior to the procedure.

8. Load the Hamilton syringe or alternative injection device with the transplant material (cell suspension or solid tissue) or other material (e.g., neurotoxin solution) that is to be injected. Attach syringe (or alternative injection device) to the holder on the stereotactic frame.

9. Move the tip of the injection needle to the bregma and record the AP and ML bregma coordinates. Calculate the AP and ML target coordinates by adding or subtracting the reference coordinates from the bregma readings.

10. Move injection needle to the newly calculated coordinates and drill a burr hole the size of the injection needle at the target site. Avoid penetrating the dura. Clean the burr hole site with the tip of the surgical spear.

11. Lower injection needle to the level of the dura, stretching it minimally, and record the vertical coordinate. Calculate the vertical target coordinate using the level of the dura as zero.

12. Open the dura with the tip of a 26-G needle. Avoid injury to underlying brain tissue. Lower the needle through the dura in a continuous slow motion down to the target depth. Start injection of the transplant material at the rate specified in the protocol, typically 0.5 to 1 μl/min. Leave the needle in place for 3 to 5 min, then start slow withdrawal, e.g., at a rate of 1 mm/min.

    Slow injection and a subsequent waiting period are crucial to avoid backflow along the needle tract, penetration into the ventricular system, and tissue destruction.

13. Dry the area with the spear or a cotton swab. Irrigate with saline if necessary. Close skin with one or two nylon sutures or clips. Remove the animal from the frame and place it on a warming pad.

BASIC PROTOCOL 2

TRANSPLANTATION INTO THE NEONATAL RODENT BRAIN

This method allows transplantation into a variety of neonatal brain regions, including the dentate gyrus (Fig. 3.7.2).

Materials

70% ethanol
Neonatal rodent (P-0 to P-8)
50% ethanol and dry ice for cooling the stereotactic frame
Material to be transplanted: cell suspension (see Support Protocol 2) or neurotoxin solution or other substance
Saline (0.9% w/v NaCl), sterile

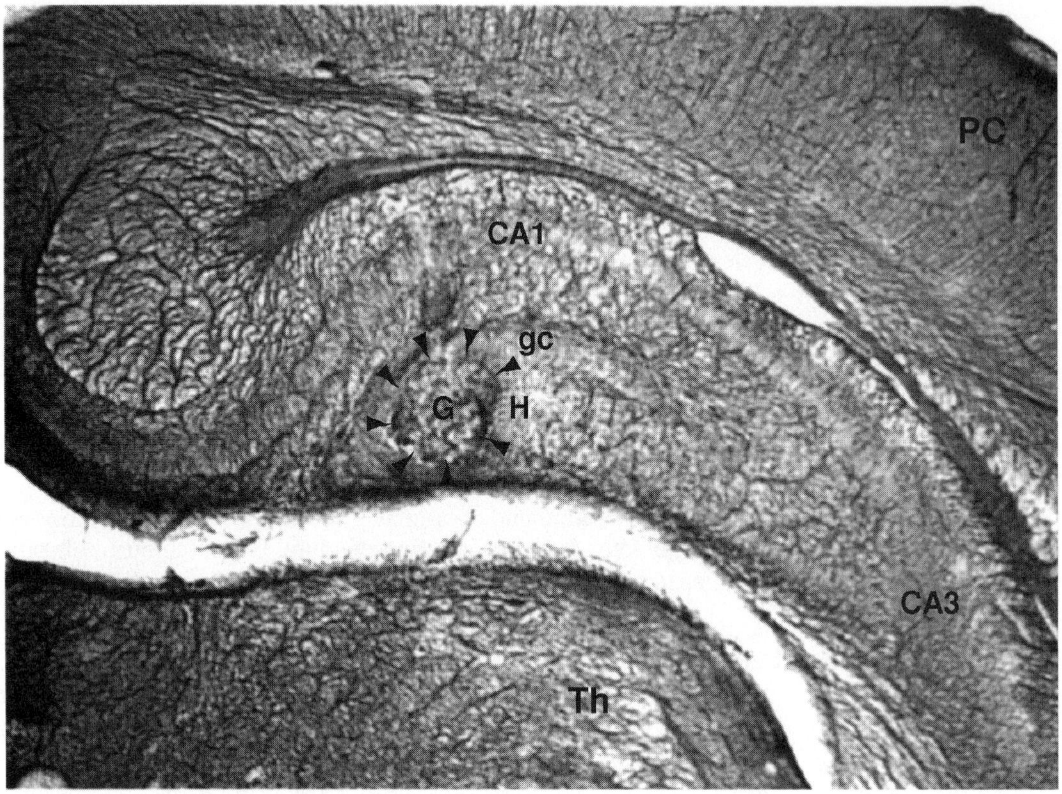

Figure 3.7.2 Micrograft of a 250-nl cell suspension into the hippocampus of a neonatal rat. The graft (G) has been stereotaxically placed into the hilus (H) of the dentate gyrus. The animal was sacrificed at the day of transplantation. Abbreviations: gc, granule cell layer; PC, parietal cortex; Th, thalamus; CA1 and CA3, sectors of the pyramidal cell layer.

Animal scale
Surgical instruments:
- Scalpel with small blade
- Fine tissue forceps
- Tweezers or watchmaker forceps
- Fine scissors
- Sterile gauze
- Surgical spears or ophthalmic lancets (Merocel, Xomed)
- 27-G hypodermic needle
- Needle holder
- 7-0 monofilament nylon suture

Surgical gloves and mask
Miniaturized stereotactic frame allowing continuous hypothermia (e.g., Stoelting)
Operating microscope
Fiber optic light source
Dental drill or bone drill with 0.5-mm carbide dental burrs
Injection device (see Support Protocol 1) or glass capillary (50- to 100-μm orifice) attached to the needle of a Hamilton syringe using a cuff of polyethylene tubing
Heating pad (Aquamatic, American Medical Systems), 37°C, or heat lamps

Figure 3.7.3 Hypothermic miniaturized stereotactic instrument designed for transplants into neonatal rodents (Cunningham and McKay, 1993). The device is cooled by a mixture of dry ice and ethanol and can be mounted onto a conventional stereotactic frame.

1. Sterilize equipment and set up surgical field (see Basic Protocol 1, steps 1 to 3).
2. Weigh animal. Induce hypothermic anesthesia by covering the animal with ~6 cm of crushed ice for ~7 min (1 min/g body weight is usually appropriate).
3. Place the animal on the cooled surface (cooled by dry ice/ethanol) of the stereotactic device (Fig. 3.7.3).

   Hypothermic anesthesia can be maintained safely for ~30 min by adding 10 g of dry ice every 10 min to a 50% ethanol bath in the instrument's reservoir. This will maintain the temperature of the instrument at ~5°C (under normal atmospheric conditions). Higher alcohol content of the solution in the reservoir will result in colder temperatures. For longer surgical procedures, the animal should be removed from the apparatus after 20 to 30 min and warmed to the point of being slightly responsive to a pinch to the tail or paw. This allows the animal respite from extended hypothermia and also allows the researcher to check that the animal is surviving the procedure. The animal can then be returned to the ice for ~4 min and repositioned in the apparatus for the next phase of the surgical procedure. For 90-min-long procedures using P-0 to P-3 rat pups, mortality using this method is ~5%.

4. Clean skull with 70% ethanol. Make a midline skin incision from just behind the eyes over the length of the cranium. With fine forceps, loosen the connective tissue and retract the skin, exposing the cartilaginous external acoustic meatus. Reflect skin and the membranous meatus and gently insert the earbars into the cartilaginous meatus. Do not apply excessive force because the external acoustic meatus is still immature and ossification is not yet complete, and the skull itself is very thin and elastic. Stabilize the animal's snout by inserting the mouthpiece and tightening the nose bridge.

5. Focus operating microscope and fiberoptic light source on operating field. Standardize head position in all three axes. To do so, adjust earbars and the mouthpiece so that:
   a. an anteroposterior movement of the pipet precisely traces the midline of the skull;
   b. bregma and lambda have the same vertical coordinates; and
   c. points 3 mm bilateral to lambda (for 7-g rats; 2 mm for 3-g mice) have the same vertical coordinate.
6. Clear skull surface of loose connective tissue using fine forceps, followed by gentle abrasion with a slightly dampened cotton swab. Measure the coordinates of the appropriate landmark (e.g., bregma or lambda), calculate the anteroposterior and lateral coordinates for the target site, and mark target site with permanent ink.
7. Using a drill equipped with a 0.5-mm carbide dental burr, drill a hole with a 0.5- to 1.0-mm diameter without damaging the underlying dura. Standardize the vertical coordinate by slowly advancing the pipet through the drill hole until it makes contact with the dura. Note the vertical coordinate and calculate depth coordinate of target site.
8. Withdraw pipet and carefully incise dura using the bevel of a high-gauge (e.g., 27-G) hypodermic needle. Slowly lower pipet to the site of interest. Leave pipet in this position for ~1 min before beginning the injection. Inject cell suspensions at a rate of 0.25 µl/min. Leave the pipet in place for another 2 min before beginning a slow withdrawal (1 mm/min).
9. After removal of the pipet, clean the exposed area with sterile saline and close skin incision using 7-0 (10-0 for P-3 mice) monofilament, small-diameter nylon suture (one interrupted knot every 3 mm; 1.5 mm in mice).
10. Remove animal from frame, clean, and warm on a heating pad before returning to its mother.

BASIC PROTOCOL 3

TRANSPLANTATION INTO THE EMBRYONIC RODENT BRAIN

NOTE: The entire procedure is performed on a 37°C heating pad.

Materials *(see APPENDIX 1 for items with ✓)*

70% ethanol
Pregnant rodent at day 14 to 18 of gestation
Inhalation anesthetic: e.g., methoxyflurane (Metofane, Pittman-Moore)
Injection anesthetic: 80 mg/kg ketamine (Ketaset; Fort Dodge Labs)/10 mg/kg xylazine (Rompun 20; Miles)
✓ PBS, sterile
Cells to be transplanted (see Support Protocol 2)
Betadine solution

Surgical gloves and masks
Heating pad, 37°C
Injection device (see Support Protocol 1)
Operating microscope
Fiber optic light source
Animal scale
Inhalation device for anesthesia
Sterile surgical gauze and cotton swabs
Sterile surgical instruments (scalpel, scissors, hemostats, needle holder, medium and fine forceps)

Suture material (e.g., 5-0 monofilament)
10-ml syringe

1. Sterilize surgical instruments by autoclaving or immersion in ethanol.
2. Cover heating pad with clean benchtop paper and position injection device, fiber optic system, and operating microscope appropriately. Spray arms of fiber optic light source and all parts of the microscope that may be touched during the operation with 70% ethanol.
3. Weigh animals.
4. Prepare inhalation chamber and place in vented area. Prepare 80 mg/kg ketamine/10 mg/kg xylazine mixture. Anesthetize pregnant rats lightly by brief exposure to Metofane, then remove quickly from inhalation chamber and inject intraperitoneally with ketamine/xylazine (0.1 to 0.2 ml/100 g body weight). Cover operating field with sterile gauze or sterile cloth, leaving only the operating area free.
5. Shave abdomen and wash with 70% ethanol. Place animal on its back on the heating pad and fix legs, slightly stretched out, with tape.
6. Perform a 3-cm long midline skin incision above lower abdomen; retract skin using sutures held by hemostats. Cover operating field with sterile gauze or sterile cloth, leaving only the operating area free. Open abdominal cavity by a 2.5-cm-long midline incision along the tendon between the abdominal muscles. Retract muscles using sutures held by hemostats.
7. Using sterile cotton swabs or medium-sized blunt forceps, very carefully extract one uterine horn and place on wet gauze. Be sure to prevent any kinking of blood vessels. Keep exposed uterine horn and intestines wet using a 10-ml syringe filled with PBS.
8. Using transillumination with one or both fiber optic arms, locate the head of the first fetus (see Fig. 3.7.4.). Do not leave fiber optics in direct contact with uterine horn for longer than a few seconds.
9. Front-load injection pipet with cells, leaving a 1-mm air cushion between cell suspension and buffer. Holding the fetus in place with the first two or three fingers, quickly penetrate uterine wall opposite placenta, skull, and ventricle with the capillary. Inject cells quickly (over the course of 2 sec) and remove capillary immediately (see Fig. 3.7.4). Do not apply too much pressure while holding the fetus.

   To access the telencephalic vesicle, use a frontal approach. Care must be taken not to damage any uterine or fetal blood vessels. Injecting part of the air cushion is an easy method to confirm intraventricular injection. Alternatively, color dyes might be added to the cell suspension.

10. Following injection, inspect the fetus closely. There should be no bleeding and a regular heartbeat. Confirm presence of air bubbles or color dye within the ventricles. Avoid applying any pressure to the injected fetus to avoid loss of amniotic fluid, increasing the risk of resorption.
11. Proceed to the next fetus, consecutively injecting all fetuses in one uterine horn.
12. Gently return injected uterine horn to abdominal cavity and repeat procedure on the other side. Close muscle with 5-0 suture, placing stitches every 2 mm.

    To prevent loss of amniotic fluid, the abdominal cavity may be filled with sterile saline.

13. Working from the inside out, apply betadine solution to closed muscle and skin. Close skin using single stitches and 5-0 suture. Apply betadine solution and place animal back in warmed cage. Place water and food in bottom of cage to allow operated animals easy access.

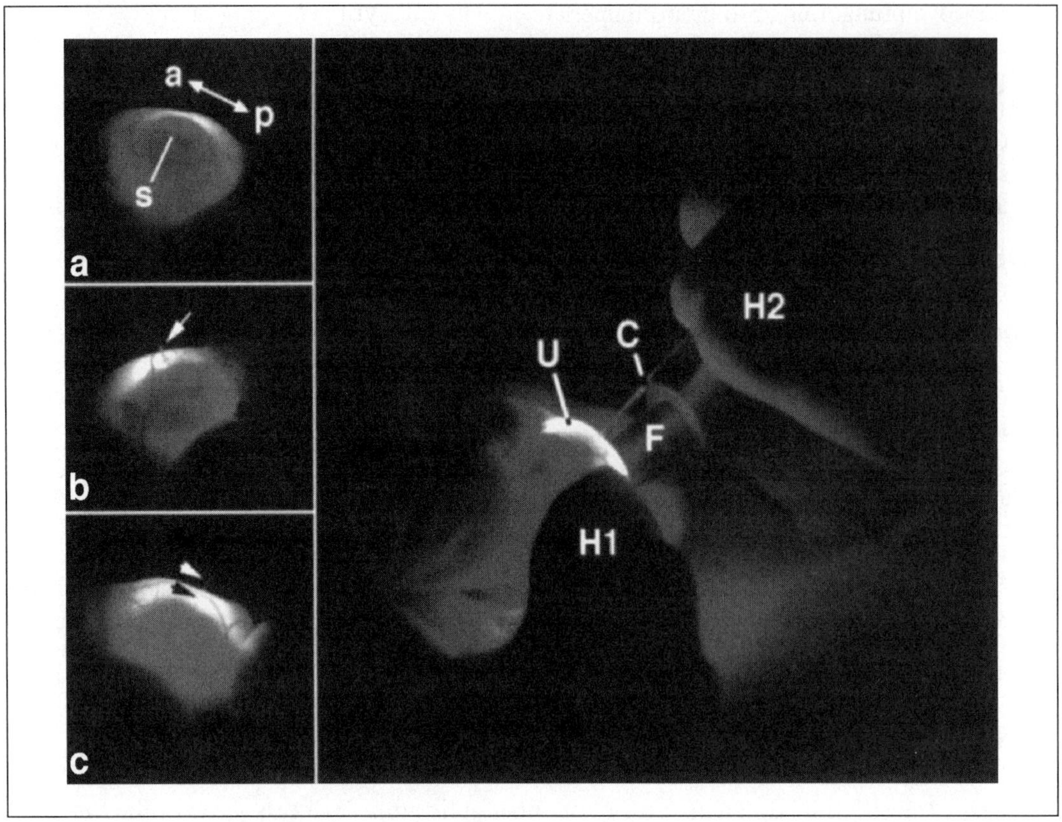

Figure 3.7.4 Transuterine injection into the telencephalic vesicle of an E-17 rat fetus. Insets on the left show orientation of the cranial sutures before injection (**A**) and the path of the injection capillary (arrows) from a dorsal (**B**) and a lateral (**C**) view. Abbreviations: a, anterior; C, injection capillary; F, fiber optic light source; H, experimenter's hands; p, posterior; s, sagittal suture; U, uterus. Photograph by Thomas Muller.

SUPPORT PROTOCOL 1

PREPARATION OF RODENT BRAIN INJECTION SYSTEM

Materials *(see* APPENDIX 1 *for items with ✓)*

Super glue
70% ethanol
✓ PBS, sterile

10-μl Hamilton syringe
30-cm segment of rigid polyethylene tubing, i.d. 0.38 mm, o.d. 1.09 mm (e.g., Intramedic, Becton Dickinson)
Glass injection pipet with a 50- to 100-μm orifice, pulled (e.g., Microcaps, Drummond Scientific Company)
2-ml syringe
27-G hypodermic needle

1. Insert the needle of the Hamilton syringe 2 to 3 mm into one end of the rigid polyethylene tubing, then attach the injection pipet to the other end of the tubing (the glass pipet should fit snugly *over* the plastic tubing, and the tubing itself should fit snugly *over* the needle of the syringe). Secure connections with Super glue.

2. Remove plunger from injection syringe, and back-fill syringe, tubing, and injection capillary with 70% ethanol. Using a 2-ml syringe and a hypodermic needle, insert the needle into the back of the injection syringe and squeeze ethanol through the syringe and tubing. Reattach plunger, submerge injection capillary in 70% ethanol, and let sit until use.

3. Before transplantation and after each injection, flush system several times with sterile PBS, as described in step 2.

SUPPORT PROTOCOL 2

PREPARATION OF SINGLE-CELL SUSPENSIONS FOR TRANSPLANTATION

This protocol may be modified depending on the donor tissue and the specific requirements of the experiment. In particular, early embryonic brain tissue often does not require trypsinization. It should be noted that removal of surface antigens by trypsinization may influence the cell-cell interaction of the transplanted cells.

Materials *(see APPENDIX 1 for items with ✓)*

Rat or mouse
✓ CMF-HBSScontaining 2 g/liter glucose, ice cold
1% (w/v) trypsin (e.g., Worthington) in CMF-HBSS (store frozen in small aliquots at −20°C)
1% (w/v) DNase (e.g., Worthington) in CMF-HBSS (store in small aliquots at −20°C)
Four flame-polished Pasteur pipets with decreasing orifice size (0.8, 0.6, 0.4, and 0.2 mm)

1. Dissect desired brain region and collect the fragments in ice-cold CMF-HBSS. See *UNITS 3.1 & 3.4* for details of dissection procedures.

2. Wash dissected fragments three times in ice-cold CMF-HBSS. In most cases, collect fragments by sedimentation without centrifugation.

3. Aspirate buffer and add a solution of 0.1% trypsin and 0.1% DNase in CMF-HBSS (prepared from 1% stock solutions of each). Incubate 10 min at room temperature with occasional gentle agitation. Wash dissected fragments three times in ice-cold CMF-HBSS.

4. Add 3 ml of 0.05% DNase in CMF-HBSS and triturate fragments with the four flame-polished Pasteur pipets. Move from large to smaller orifice size, pipetting up and down five to ten times with each pipet.

5. Add ice-cold CMF-HBSS to 10 ml and centrifuge cells 5 min at 200 × *g*, 4°C. Resuspend pellet in another 10 ml ice-cold CMF-HBSS and remove 10 μl of the suspension for trypan blue staining and cell counting. Centrifuge cells 5 min at 200 × *g*, 4°C, and resuspend in ice-cold CMF-HBSS at the desired concentration. Keep cell suspension on ice until use. If the cells are kept on ice for extended periods of time (>2 hr), count cells intermittently to ensure sufficient viability. After completion of the experiments, grow residual cells in culture to assay long-term survival.

SUPPORT PROTOCOL 3

6-HYDROXYDOPAMINE LESIONING

NOTE: 6-OHDA is hazardous. Wear protective mask and gloves when handling the substance.

Additional Materials *(also see Basic Protocol 1; see* APPENDIX 1 *for items with ✓)*

✓ 6-OHDA solution, sterile

1. Prepare operating area and animal for injection procedure, make incision, and determine injection position (see Basic Protocol 1, steps 1 to 12).
2. Fill a 10-μl Hamilton syringe with 5.5 μl of the sterile 6-OHDA solution. Inject 2.5 μl at the following coordinates: AP −4.4 mm, ML −1.2 mm, and V (from the dura) −7.8 mm, toothbar set at −2.4. Inject at a rate of 1 μl/min and wait 5 min before withdrawing the needle.
3. In the same way, inject 3.0 μl at the following coordinates: AP −4.0 mm, ML −0.8 mm, and V −8.0 mm, toothbar set at +3.4.
4. Follow routine guidelines for postoperative care.

3

SUPPORT PROTOCOL 4

IBOTENIC ACID LESIONING

NOTE: Ibotenic acid is hazardous. Wear protective mask and gloves when handling the substance.

Additional Materials *(also see Basic Protocol 1; see* APPENDIX 1 *for items with ✓)*

✓ Ibotenic acid solution

1. Prepare operating area and animal for injection procedure, make incision, and determine injection position (see Basic Protocol 1, steps 1 to 12).
2. Fill a 10-μl Hamilton syringe with 1.5 μl ibotenic acid solution. Carefully clean the outside of the needle with a gauze soaked in sterile saline to prevent exposure of cortical tissue to ibotenic acid. Inject 0.5 μl at the following coordinates: AP +0.2 mm, ML −3.0 mm, and V (from the dura) −5.5 mm, toothbar set at −2.3. Inject at a rate of 0.1 μl/min and wait 2 min before withdrawing the needle.
3. To perform the next injection at the same AP and ML coordinates but at a shallower depth, gently withdraw the needle up to the newly calculated depth at V −4.0 mm. Inject 0.5 μl at a rate of 0.1 μl/min. Keep the toothbar set at −2.3 mm. Wait an additional 2 min.
4. Similarly, inject 0.5 μl at the following coordinates: AP +1.5 mm, ML −2.5 mm, and V −4.7 mm, toothbar set at −2.3.
5. Follow routine guidelines for postoperative care.

References: Cunningham and McKay, 1993; Dunnett and Björklund, 1992; Paxinos and Watson, 1982

Contributors: Oliver Brustle, Miles G. Cunningham, Vivian Tabar, and Lorenz Studer

CHAPTER 4

Gene Cloning, Expression, and Mutagenesis

Chapter 4 presents a variety of methods and techniques for the cloning, expression, and structural analysis of neural genes and proteins, primarily through mutagenesis. It is not meant to provide comprehensive coverage of all of these topics, but rather to present selected techniques that will be of use to neuroscientists who have at least some introductory background in recombinant DNA technology or are working with collaborators who do. More comprehensive coverage of these topics can be found in the sister manual to this present volume, *Current Protocols in Molecular Biology* (CPMB; Ausubel et al., 2006), which is extensively referred to in the units that follow. For further background, the reader may also wish to consult any of the wide variety of textbooks that are specifically devoted to recombinant DNA techniques.

In order to characterize a gene, and its corresponding protein product, at the molecular level, one must first isolate or "clone" that gene using any of a number of approaches. A selection of cloning techniques likely to be useful to neuroscientists is presented here. One cloning approach, in which the polymerase chain reaction (PCR) is performed using primers derived from conserved sequences of related genes, is presented in *UNIT 4.1*. In this case, the idea is to amplify a small segment of the target gene and then to use this fragment to obtain a full-length clone either through the screening of an appropriate DNA library or through the rapid amplification of cDNA ends (RACE).

In cases where no preliminary sequence information is available, it may be necessary to take another approach to isolating a desired cDNA or gene. This approach, known as expression cloning, has the limitation that it requires that a biological assay exist (or be developed) for detecting the expression of the target protein. *UNIT 4.2* presents the novel yeast two-hybrid system, which allows clones to be isolated based solely on protein-protein interactions. *UNIT 4.3* provides protocols for constructing cDNA libraries from single cells. This novel approach allows for the isolation and cloning of genes that are selectively expressed within a specific cell type. Once the library is constructed, it can be screened using a variety of approaches described elsewhere in this chapter.

Once a DNA clone has been successfully isolated and characterized via sequence analysis and other techniques, it is frequently desirable to express the clone in order to study its protein product. Generally, however, one must subclone the cDNA or gene into an expression vector suitable for the cell or tissue system in which the expression is to take place. *UNIT 4.4* provides an overview of expression and vector considerations. If one is using mammalian cells for expression, then this can typically be done either transiently, meaning that expression of the protein occurs for only a brief period of time, or stably, meaning that the expression construct is integrated into the genome of the cell and the protein is made continuously. Considerations for stable transfection and expression and selection of stable transformants are covered in *UNIT 4.5*. A particularly useful system for overexpression of proteins is presented in *UNIT 4.6*: the *Drosophila* S-2 cell system, which allows stable expression of proteins at high levels. The S-2 cells grow in suspension culture to high densities and therefore represent useful starting material for protein purification efforts. Large-scale production of proteins using baculovirus-derived vectors is also covered in *UNIT 4.7*. This is an extremely popular and useful method to prepare large quantities of proteins for antibody production, biochemical analyses, or purification efforts.

UNITS 4.8–4.15 introduce expression technologies that utilize viral-based vectors for the expression of genes in cultured cells, as well as for gene delivery to the brain. Given the terminally differentiated state of most neurons in the brain, it is essential that the gene-transfer vectors used persist stably in post-mitotic cells and can be targeted both spatially and temporally in the nervous system in vivo. *UNITS 4.13–4.15* present protocols for the use of herpes simplex virus 1 (HSV-1)-based vectors for gene delivery into cells and intact organisms. *UNIT 4.11* presents protocols for the use of adeno-associated viral vectors for neuronal gene expression. *UNIT 4.10* presents protocols for using lentiviral vectors in gene transfer while *UNIT 4.12* covers alphavirus-mediated gene transfer into neurons. Finally adenoviral-mediated gene transfer into neurons and the central nervous system of rodents is covered in *UNITS 4.8 & 4.9*, respectively.

UNITS 4.16 & 4.17 present an emergent and rapidly growing technology used to monitor gene expression. Nucleic acid arrays, or DNA microarrays, provide a high throughput method to analyze the expression, or changes in expression, of multiple genes simultaneously. *UNIT 4.16* provides an overview of nucleic acid arrays covering their uses and methods of analyses. *UNIT 4.17* provides protocols needed for the first step of array analysis, the preparation of mRNA for expression profiling. Specific experimental protocols for analyzing gene expression using cDNA microarrays are presented in *UNIT 4.18*.

An overview of gene targeting using homologous recombination is discussed in *UNIT 4.19*. Initially, this technique was used for gene inactivation or deletion (knock-out); it has now also been used for gene replacement (knock-in), in which the replacement gene may contain a mutation of the protein under study.

Contributor: David Sibley

UNIT 4.1

PCR Cloning of Neural Gene Products

The protocols in this unit assume that either total RNA or poly(A)$^+$ RNA has been purified from tissue culture cells or some tissue of interest (*APPENDIX 2E*) and demonstrated to be intact (i.e., undegraded). Although G protein–coupled receptors (GPCRs) are used as the focus of this unit, the strategies and techniques described are applicable to the cloning of a wide variety of neuronal gene products.

NOTE: All Reagents and Solutions used for RNA procedures should be treated with DEPC to inhibit RNase activity.

CAUTION: DEPC is a suspected carcinogen and should be handled carefully.

BASIC PROTOCOL 1

AMPLIFICATION USING DEGENERATIVE PRIMERS

NOTE: Experiments involving PCR require extremely careful technique to prevent contamination. See *CPMB UNIT 15.1* and *APPENDIX 2A* of this manual for guidelines.

NOTE: When designing primers for PCR, it may be worthwhile to divide the oligonucleotides into two domains: one for cloning purposes and the other for annealing to the target sequence. When designing the annealing portion of a degenerate primer, clusters of conserved amino acids are typically targeted; for this purpose it is recommended that primers not be extremely degenerate.

Materials *(see* APPENDIX 1 *for items with* ✓*)*

0.5 μg/μl oligo$(dT)_{12-18}$ *or* 50 ng/μl random hexamers
1 μg/μl total *or* poly$(A)^+$ RNA (*APPENDIX 2E*)
✓ Diethylpyrocarbonate (DEPC)-treated H_2O
5× first-strand cDNA buffer: 250 mM Tris·Cl /375 mM KCl/15 mM $MgCl_2$
0.1 M dithiothreitol (DTT)
✓ 10 mM 4dNTP mix: 10 mM each dATP, dCTP, dGTP, dTTP in H_2O or TE buffer, pH 7.5
200 U/μl SuperScript II reverse transcriptase (RT; Life Technologies)
2 U/μl *E. coli* RNase H
10× degenerate PCR amplification buffer (usually supplied by the manufacturer; also see recipe)
10 to 20 μM degenerate oligonucleotide primers in H_2O
2.5 mM 4dNTP mix: 2.5 mM each dATP, dGTP, dCTP, dTTP in TE buffer, pH 7.5
5 U/μl *Pfu* DNA polymerase (Taq Extender, Stratagene)
5 U/μl *Taq* DNA polymerase

37° or 42°, 55°, and 70°C water baths or constant-temperature heating blocks
Programmable thermal cycler

1. To a sterile 1.5-ml microcentrifuge tube add:

   2 μl 50 ng/μl random hexamers or 2 μl 0.5 μg/μl oligo$(dT)_{12-18}$
   1 to 5 μl 1 μg/μl total RNA (up to 5 μg for rare message) *or* 0.5 to 1.0 μg poly$(A)^+$ RNA
   DEPC-treated H_2O to 12 μl.

   Heat the mixture 10 min at 70°C in a water bath or constant-temperature heating block. Immediately place tube in wet ice and chill 5 min. Microcentrifuge briefly at maximum speed, room temperature, to collect any condensate.

2. To each tube add (19 μl total):

   4 μl 5× first-strand cDNA buffer
   2 μl 0.1 M DTT
   1 μl 10 mM 4dNTP mix.

   Place tube in 42°C water bath for 2 min to equilibrate contents.

3. Add 1 μl of 200 U/μl SuperScript II RT into reaction (total volume 20 μl). Draw the solution into the pipet and gently expel it repeatedly to assist mixing. Be careful to avoid foaming: do not vortex. Collect all of the reaction volume in the bottom of the tube by brief microcentrifugation at maximum speed, room temperature. Incubate tube 30 to 60 min in a 37° or 42°C (42° advisable for cDNAs with >60% GC content) water bath or constant-temperature heating block.

4. Incubate 5 min at 55°C. Add 2 U *E. coli* RNase H. Incubate 10 min at 55°C. Directly amplify the resulting first-strand cDNA by PCR, according to the following steps, *or* store it at −20°C for future use.

5. Prepare the following PCR reaction mixture (50 μl total; scale up or down proportionally, if necessary):

   29 μl sterile H_2O
   5 μl 10× degenerate PCR amplification buffer
   5 μl each 10 to 20 μM degenerate oligonucleotide primer (1 to 2 μM final, vary the concentration of each according to the total degeneracy of the mix)
   5 μl 2.5 mM 4dNTP mix
   1 μl first-strand cDNA.

For a rare mRNA, increase the amount of first-strand cDNA by using up to 5 μl of the first-strand reaction mixture. If nonspecific PCR products become a problem, reduce the amount of first-strand cDNA by using less of the cDNA mixture. If the amount of first-strand DNA is changed, adjust the volume of water to give a total reaction volume of 50 μl.

6. Heat the mixture 5 to 10 min at 95°C in a programmable thermal cycler. Lower the sample temperature to 80°C. Maintain the temperature at 80°C and add 1 μl of 5 U/μl *Pfu* DNA polymerase and 1 μl of 5 U/μl *Taq* DNA polymerase to each tube in the thermal cycler, removing the tube from the cycler briefly, if necessary, to add the enzymes.

   For very small products (50 to 250 bp), Pfu polymerase may be omitted.

7. Program the thermal cycler with the following settings:

35 cycles:	60 to 90 sec	94°C	(denaturation)
	90 to 120 sec	45° to 55°C	(annealing)
	90 to 150 sec	72°C	(extension)
1 cycle:	10 min	72°C	(final extension).

8. Analyze 10% to 20% of the PCR product by electrophoresis on agarose (see *CPMB UNIT 2.5A*) or nondenaturing polyacrylamide (see *CPMB 2.7*) gels depending on the expected size of the desired amplification product(s). Store the remaining PCR product at 4°C or indefinitely at −20°C.

4

BASIC PROTOCOL 2

ANCHORED AMPLIFICATION OF REGIONS DOWNSTREAM (3′) OF KNOWN SEQUENCE

Typically, 100 to 300 ng of total poly(A)$^+$ RNA prepared from vertebrate tissue is sufficient for conducting this procedure. Three PCR primers are required: an oligo(dT) primer and two sequence-specific primers, one for the original amplification and one for reamplification. The internal sequence-specific primer can be immediately adjacent (3′) to or can partially overlap primer 1.

Materials *(see APPENDIX 1 for items with ✓)*

≥100 ng/μl poly(A)$^+$ RNA (*APPENDIX 2E*) *or* 1 μg/μl total RNA
✓ 5× MMLV RT buffer
5 μg/μl BSA
10 mM 4dNTP mix: 10 mM each dNTP in TE buffer, pH 7.5
500 ng/μl actinomycin D
200 U/μl MMLV reverse transcriptase
100 ng/μl oligo$(dT)_{20}$ primer
✓ TE buffer, pH 7.5
✓ 10× downstream PCR amplification buffer
100 pmol/μl each of sequence-specific primers 1 and 2
2.5 mM 4dNTP mix: 2.5 mM each dNTP in TE buffer, pH 7.5
2.5 U/μl *Taq* DNA polymerase
Mineral oil

37° and 70°C water baths or constant-temperature heating blocks
Programmable thermal cycler

1. Place ~100 ng poly(A)$^+$ RNA or 1 μg of total RNA (≥1 μl) in a 1.5-ml microcentrifuge tube and incubate 2 min at 70°C. Microcentrifuge briefly and place immediately on ice.

2. In a separate 1.5-ml microcentrifuge tube, prepare on ice (10 μl total):

   2 μl 5× MMLV RT buffer
   1 μl 5 μg/μl BSA
   1 μl poly(A)$^+$ RNA (from step 1)
   1 μl 10 mM 4dNTP mix
   1 μl 500 ng/μl actinomycin D
   1 μl 200 U/μl MMLV reverse transcriptase
   1 μl 100 ng/μl oligo(dT)$_{20}$ primer
   2 μl sterile H_2O.

   Mix gently, microcentrifuge briefly, and incubate 1 hr at 37°C. Add 40 μl TE buffer.

3. Prepare the following on ice (100 μl final):

   1 μl cDNA template (from step 2)
   10 μl 10× downstream PCR amplification buffer
   1 μl 100 ng/μl oligo(dT)$_{20}$ primer
   1 μl 100 pmol/μl sequence-specific primer 1
   6 μl 2.5 mM 4dNTP mix
   1 μl 2.5 U/μl *Taq* DNA polymerase
   80 μl sterile H_2O.

   Overlay reaction mixture with mineral oil. Carry out 30 to 40 cycles of amplification.

35 cycles:	45 to 60 sec	94°C	(denaturation)
	30 to 60 sec	55°C	(annealing)
	60 to 120 sec	72°C	(extension)
1 cycle:	10 min	72°C	(final extension).

   $MgCl_2$ concentration, and annealing temperature should be optimized for the specic template. Primer concentrations can be reduced to 30 pmol of each primer.

4. Analyze an aliquot by agarose gel electrophoresis (see CPMB UNIT 2.5A).

5. Remove a 1-μl aliquot of product from step 3 to serve as the template for reamplification. Carry out a second round of PCR as in step 3, using 40 to 100 pmol each of the oligo(dT)$_{20}$ primer and the internal sequence-specific primer 2.

6. Analyze an aliquot of second amplification by agarose gel electrophoresis. If desired, characterize product by cloning into an appropriate vector or by direct sequencing.

BASIC PROTOCOL 3

ANCHORED AMPLIFICATION OF REGIONS UPSTREAM (5′) OF KNOWN SEQUENCE

Materials *(see* APPENDIX 1 *for items with ✓)*

100 pmol/μl sequence-specific primers 3 and 4
100 ng/μl oligo(dT)$_{20}$ primer
1 M NaCl
✓ 200 mM Tris·Cl, pH 7.5

✓ 25 mM EDTA
100 ng/μl poly(A)$^+$ RNA or 1 μg/μl total RNA
100% and 70% ethanol, ice cold
✓ 5× MMLV RT buffer
5 μg/μl BSA
10 mM 4dNTP mix: 2.5 mM each dNTP in TE buffer, pH 7.5
500 ng/μl actinomycin D
200 U/μl MMLV reverse transcriptase
25:24:1 (v/v/v) phenol/chloroform/isoamyl alcohol
24:1 (v/v) chloroform/isoamyl alcohol
✓ 3 M sodium acetate
✓ TE buffer
✓ 5× TdT buffer
15 mM $CoCl_2$
1 mM dATP
Terminal deoxynucleotidyl transferase

40°C and 37°C water baths
Programmable thermal cycler

1. Prepare a 5-μl annealing mix containing:

   1 μl 1 pmol/μl sequence-specific primer 3
   1 μl 1 M NaCl
   1 μl 200 mM Tris·Cl, pH 7.5
   1 μl 25 mM EDTA
   1 μl 100 ng/μl poly(A)$^+$ RNA or 1 μg of total RNA.

   Incubate 3 min at 65°C. Microcentrifuge briefly and place immediately on ice. Incubate 3 to 4 hr at 40°C.

   As a general rule, a 5- to 10-fold molar excess of oligomer primer relative to target template yields the best results.

2. Add 15 μl ice-cold 100% ethanol, place 10 min in dry ice/ethanol bath, and microcentrifuge at high speed 10 min at 4°C. Add 50 μl ice-cold 70% ethanol to pellet and gently invert tube several times. Microcentrifuge 2 min, dry pellet briefly under vacuum, and resuspend the annealed primer/template in 10 μl water.

3. Prepare mix (25 μl total) on ice containing the following ingredients (in order) and incubate 1 hr at 37°C to synthesize cDNA strand:

   5 μl 5× MMLV RT buffer
   2.5 μl 5 μg/μl BSA
   2.5 μl 10 mM 4dNTP mix
   2.5 μl (500 ng/μl) actinomycin D
   10 μl annealed primer/template (from step 2)
   1.5 μl sterile H_2O
   1 μl 200 U/μl MMLV reverse transcriptase.

4. Using 1.5-ml microcentrifuge tubes, extract twice with 1 vol of 25:24:1 (v/v/v) phenol/chloroform/isoamyl alcohol and then once with 1 vol of 24:1 (v/v) chloroform/isoamyl alcohol. Transfer supernatant to new microcentrifuge tube.

5. Add 2.5 μl of 3 M sodium acetate (0.3 M final) and 75 μl ice-cold 100% ethanol, place 5 min in dry ice/ethanol bath, and microcentrifuge 20 min at high speed, 4°C. Resuspend

pellet in 25 μl TE buffer. Add 2.5 μl of 3 M sodium acetate (0.3 M final) and repeat ethanol precipitation.

If necessary, add 1 to 5 μg carrier tRNA to facilitate precipitation.

6. Add 100 μl ice-cold 70% ethanol to pellet and rinse by gently inverting tube. Microcentrifuge 5 min, dry pellet briefly under vacuum, and resuspend pellet in 5 μl water. Boil 2 min, microcentrifuge briefly, and place immediately on ice.

7. Prepare mix (10 μl final) on ice containing the following ingredients (in order) and incubate 30 min at 37°C, to add poly(A) tail.

   2 μl 5× TdT buffer
   1 μl 15 mM $CoCl_2$ (1.5 mM final)
   1 μl 1 mM dATP (100 μM final)
   5 μl cDNA mix (from step 6)
   1 μl 25 U/μl terminal deoxynucleotidyl transferase.

   Inactivate enzyme by heating 2 min at 65°C. Add 1 μl of 3 M sodium acetate and 30 μl ice-cold 100% ethanol, and precipitate as in step 5. Wash and dry pellet as in step 6, and resuspend pellet in 10 μl water.

8. Carry out 40 amplification cycles as described (see Basic Protocol 2, step 3) using 40 to 100 pmol each of sequence-specific primer 3 and oligo$(dT)_{20}$ primer.

9. Remove a 1-μl aliquot of the product to serve as template for a new round of amplification. Carry out 35 cycles of PCR (Basic Protocol 2, step 3), using 40 to 100 pmol each of primer 4 and the oligo(dT) primer.

   Primer 4 can be immediately adjacent to or even partially overlapping with primer 3. Because anchored PCR procedures are carried out independently in the 3′ and 5′ directions, sequence-specific primers 3 and 4 can be complements of primers 1 and 2 used in the previous protocol.

10. Analyze an aliquot by agarose gel electrophoresis (see CPMB UNIT 2.5A). If desired, characterize the PCR product by cloning into an appropriate vector and/or by direct sequencing.

Contributors: David K. Grandy, James R. Bunzow, and Robert L. Dorit

UNIT 4.2

Interaction Trap/Two-Hybrid System to Identify Interacting Proteins

To understand the function of a particular protein, it is often useful to identify other proteins with which it associates. This can be done by a selection or screen in which novel proteins that specifically interact with a target protein of interest are isolated from a library. One particularly useful approach to detect novel interacting proteins—the two-hybrid system or interaction trap (Figs. 4.2.1 and 4.2.2)—uses yeast as a "test tube" and transcriptional activation of a reporter system to identify associating proteins. This approach can also be used specifically to test complex formation between two proteins for which there is a prior reason to expect an interaction.

NOTE: All solutions and equipment coming into contact with cells must be sterile, and proper aseptic technique should be used accordingly.

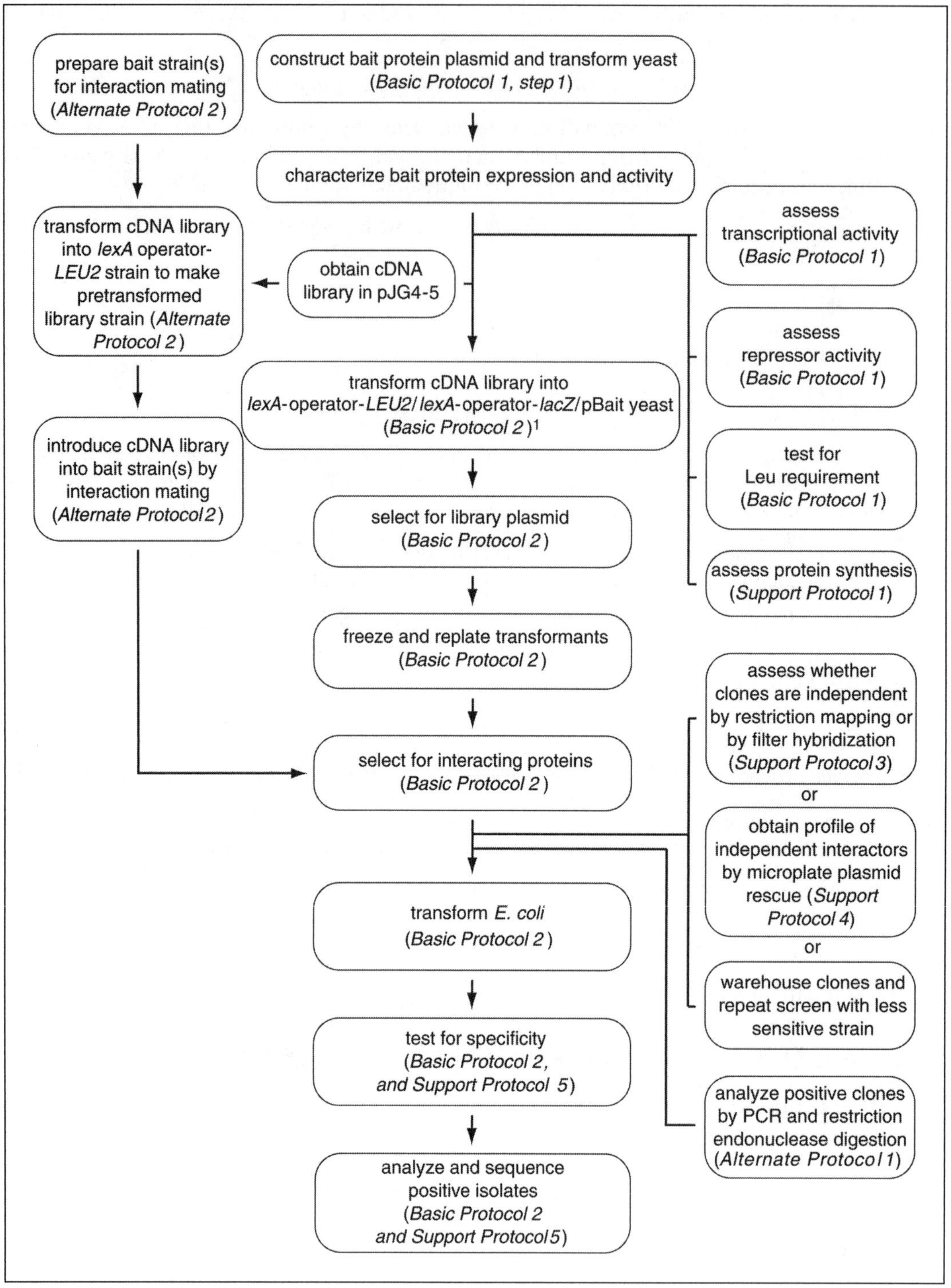

Figure 4.2.1 Flow chart for performing an interaction trap.

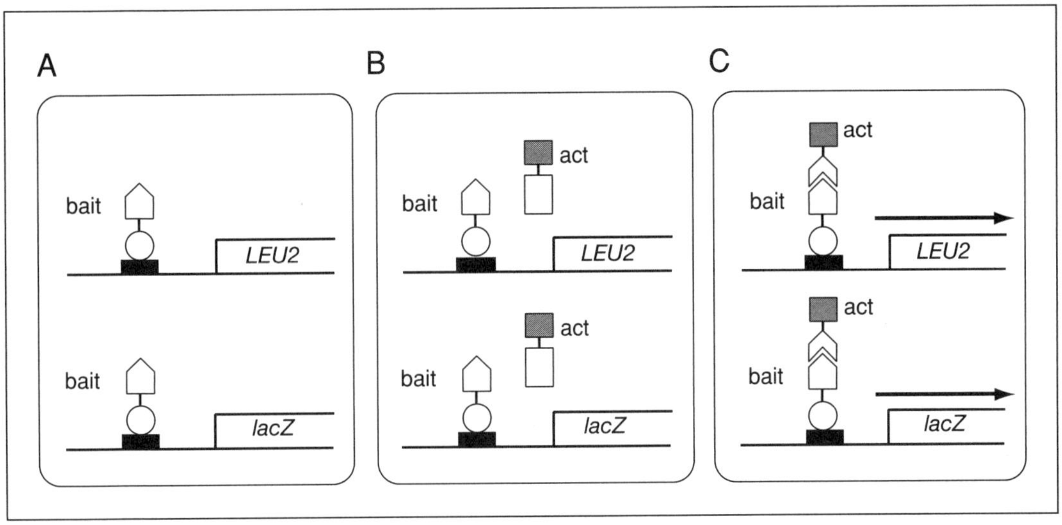

Figure 4.2.2 The interaction trap. (**A**) An EGY48 yeast cell containing two *LexA* operator-responsive reporters, one a chromosomally integrated copy of the *LEU2* gene (required for growth on −Leu medium), the second a plasmid bearing a *GAL1* promoter-*lacZ* fusion gene (causing yeast to turn blue on medium containing Xgal). The cell also contains a constitutively expressed chimeric protein, consisting of the DNA-binding domain of LexA fused to the probe of bait protein, shown as being unable to activate either of the two reporters. (**B**) and (**C**), EGY48/pSH18-34/pbait-containing yeast have been additionally transformed with an activation domain (act)-fused cDNA library in pJG4-5, and the library has been induced. In (**B**), the encoded protein does not act specifically with the bait protein and the two reporters are not activated. In (**C**), a positive interaction is shown in which the library-encoded protein interacts with bait protein, resulting in activation of the two reporters (arrow), thus causing growth on medium lacking Leu and blue color on medium containing Xgal. Symbols: black rectangle. *LexA* operator sequence; open circle, LexA protein; open pentagon, bait protein; open rectangle, library protein; shaded box, activator protein (acid blob in Fig. 4.2.6).

4

BASIC PROTOCOL 1

CHARACTERIZING A BAIT PROTEIN

The first step in an interactor hunt is to construct a plasmid that expresses LexA fused to the protein of interest. This construct is transformed into reporter yeast strains containing *LEU2* and *lacZ* reporter genes, and a series of control experiments is performed to establish whether the construct is suitable as is or must be modified, and whether alternative yeast reporter conditions should be used. These controls establish that the bait protein is made as a stable protein in yeast, that it is capable of entering the nucleus and binding *LexA* operator sites, and that it does not appreciably activate transcription of the *LexA* operator-based reporter genes. The LexA-fused bait protein must not activate transcription of either reporter—the EGY48 strain (or related strain EGY191) that expresses the LexA fusion protein should not grow on medium lacking Leu, and the colonies should be white on medium containing Xgal. The characterized bait protein plasmid is used for Basic Protocol 2 to screen a library for interacting proteins.

Materials *(see* APPENDIX 1 *for items with* ✓*)*

DNA encoding the protein of interest

Plasmids (Table 4.2.1):

pEG202 (Fig. 4.2.3), pSH18-34 (Fig. 4.2.4), pSH17-4, pRFHM1, and pJK101 for basic characterization; other plasmids for specific circumstances as described (Clontech, Invitrogen, OriGene, or R. Brent)

Table 4.2.1 Interaction Trap Components[a,b]

Plasmid name/source	Selection		Comment/description
	In yeast	In *E. coli*	
LexA fusion plasmids			
pEG202[c,d,e]	*HIS3*	Ap^r	Contains an *ADH* promoter that expresses LexA followed by polylinker
pJK202	*HIS3*	Ap^r	Like pEG202, but incorporates nuclear localization sequences between LexA and polylinker; used to enhance translocation of bait to nucleus
pNLexA[e]	*HIS3*	Ap^r	Contains an *ADH* promoter that expresses polylinker followed by LexA; for use with baits where amino-terminal residues must remain unblocked
pGilda[d]	*HIS3*	Ap^r	Contains a *GAL1* promoter that expresses same LexA and polylinker cassette as pEG202; for use with baits whose continuous presence is toxic to yeast
pEE202I	*HIS3*	Ap^r	An integrating form of pEG202 that can be targeted into *HIS3* following digestion with *Kpn*I; for use where physiological screen requires lower levels of bait to be expressed
pRFHM1[e,f] (control)	*HIS3*	Ap^r	Contains an *ADH* promoter that expresses LexA fused to the homeodomain of bicoid to produce nonactivating fusion; used as positive control for repression assay, negative control for activation and interaction assays
pSH17-4[e,f] (control)	*HIS3*	Ap^r	*ADH* promoter expresses LexA fused to GAL4 activation domain; used as a positive control for transcriptional activation
pMW101[f]	*HIS3*	Cm^r	Same as pEG202, but with altered antibiotic resistance markers; basic plasmid used for cloning bait
pMW103[f]	*HIS3*	Km^r	Same as pEG202, but with altered antibiotic resistance markers; basic plasmid used for cloning bait
pHybLex/Zeo[f,g]	Zeo^r	Zeo^r	Bait cloning vector compatible with interaction trap and all other two-hybrid systems; minimal ADH promotor expresses LexA followed by extended polylinker
Activation domain fusion plasmids			
pJG4-5[c,d,e,f]	TRP1	Ap^r	Contains a *GAL1* promoter that expresses nuclear localization domain, transcriptional activation domain, HA epitope tag, cloning sites; used to express cDNA libraries
pJG4-5I	*TRP1*	Ap^r	An integrating form of pJG4-5 that can be targeted into *TRP1* by digestion with *Bsu*36I (New England Biolabs); to be used with pEE202I to study interactions that occur physiologically at low protein concentrations
pYESTrp[g]	*TRP1*	Ap^r	Contains a *GAL1* promoter that expresses nuclear localization domain, transcriptional activation domain, V5 epitope tag, multiple cloning sites; contains f1 ori and T7 promoter/flanking site; used to express cDNA libraries (Invitrogen)
pMW102[f]	*TRP1*	Km^r	Same as pJG4-5, but with altered antibiotic resistance markers; no libraries yet available
pMW104[f]	*TRP1*	Cm^r	Same as pJG4-5, but with altered antibiotic resistance markers; no libraries yet available

continued

Table 4.2.1 Interaction Trap Components[a,b], continued

Plasmid name/source	Selection: In yeast	Selection: In *E. coli*	Comment/description
LacZ reporter plasmids			
pSH18-34[d,e,f]	*URA3*	Ap^r	Contains 8 *LexA* operators that direct transcription of the *lacZ* gene; one of the most sensitive indicator plasmids for transcriptional activation
pJK103[e]	*URA3*	Ap^r	Contains two *LexA* operators that direct transcription of the *lacZ* gene; an intermediate reporter plasmid for transcriptional activation
pRB1840[e]	*URA3*	Ap^r	Contains 1 *LexA* operator that directs transcription of the *lacZ* gene; one of the most stringent reporters for transcriptional activation
pMW112[f]	*URA3*	Km^r	Same as pSH18-34, but with altered antibiotic resistance marker
pMW109[f]	*URA3*	Km^r	Same as pJK103, but with altered antibiotic resistance marker
pMW111[f]	*URA3*	Km^r	Same as pRB1840, but with altered antibiotic resistance marker
pMW107[f]	*URA3*	Cm^r	Same as pSH18-34, but with altered antibiotic resistance marker
pMW108[f]	*URA3*	Cm^r	Same as pJK103, but with altered antibiotic resistance marker
pMW110[f]	*URA3*	Cm^r	Same as pRB1840, but with altered antibiotic resistance marker
pJK101[e,f] (control)	*URA3*	Ap^r	Contains a *GAL1* upstream activating sequence followed by two *lexA* operators followed by *lacZ* gene; used in repression assay to assess bait binding to operator sequences

[a]All plasmids contain a 2 μm origin for maintenance in yeast, as well as a bacterial origin of replication, except where noted (pEE202I, pJG4.5I).

[b]Interaction Trap reagents represent the work of many contributors: the original basic reagents were developed in the Brent laboratory (Gyuris et al., 1993). Plasmids with altered antibiotic resistance markers (all pMW plasmids) were constructed at Glaxo in Research Triangle Park, N.C. (Watson et al., 1996). Plasmids and strains for specialized applications have been developed by the following individuals: E. Golemis, Fox Chase Cancer Center, Philadelphia, Pa. (pEG202); J. Kamens, BASF, Worcester, Mass. (pJK202); cumulative efforts of I. York, Dana-Farber Cancer Center, Boston, Mass. and M. Sainz and S. Nottwehr, U. Oregon (pNLexA); D.A. Shaywitz, MIT Center for Cancer Research, Cambridge, Mass. (pGilda); R. Buckholz, Glaxo, Research Triangle Park, N.C. (pEE2021, pJG4-51); J. Gyuris, Mitotix, Cambridge, Mass. (pJG4-5); S. Hanes, Wadsworth Institute, Albany, N.Y. (pSH17-4); R.L. Finley, Wayne State University School of Medicine, Detroit, Mich. (pRFHM1); S. Hanes, Wadsworth Institute, Albany, N.Y. (pSH18-34); J. Kamens, BASF, Worcester, Mass. (pJK101, pJK103); R. Brent, The Molecular Sciences Institute, Berkeley, Calif. (pRB1840). Specialized plasmids not yet commercially available can be obtained by contacting the Brent laboratory at (510) 647-0690 or *brent@molsci.org*, or the Golemis laboratory, (215) 728-2860 or *EA_Golemis@fccc.edu*.

[c]Sequence data are available for pEG202 (pLexA) accession number pending.

[d]Plasmids commercially available from Clontech and OriGene; for Clontech pEG202 is listed as pLexA, pJG4-5 as pB42AD, and pSH18-34 as p8op-LacZ.

[e]Plasmids and strains available from OriGene.

[f]In pMW plasmids the ampicillin resistance gene (Ap^r) is replaced with the chloramphenicol resistance gene (Cm^r) and the kanamycin resistance gene (Km^r) from pBC SK(+) and pBK-CMV (Stratagene), respectively. The choice between Km^r and Cm^r or Ap^r plasmids is a matter of personal taste; use of basic Ap^r plasmids is described in the basic protocols. Use of the more recently developed reagents would facilitate the purification of library plasmid in later steps by eliminating the need for passage through KC8 bacteria, with substantial saving of time and effort. Ap^r has been maintained as marker of choice for the library plasmid because of the existence of multiple libraries already possessing this marker. These plasmids are the basic set of plasmids recommended for use.

[g]Plasmids commercially available from Invitrogen as components of a Hybrid Hunter kit; this kit also includes all necessary positive and negative controls (not listed in this table).

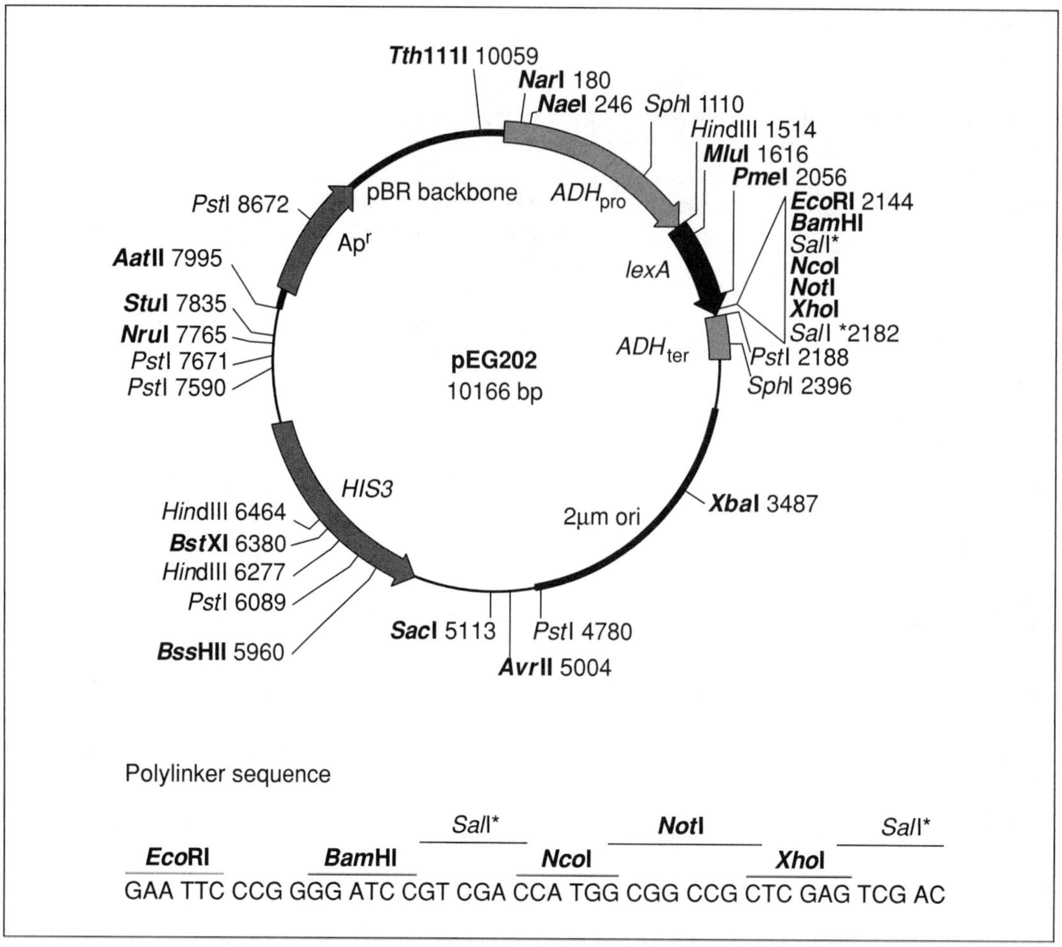

Figure 4.2.3 LexA-fusion plasmids: pEG202. The strong constitutive *ADH* promoter is used to express bait proteins as fusions to the DNA-binding protein LexA. Restriction sites shown in this map are based on recently compiled pEG202 sequence data and include selected sites suitable for diagnostic restriction endonuclease digests. A number of restruction sites are available for insertion of coding sequences to produce protein fusions with LexA; the polylinker sequence and reading frame relative to LexA are shown below the map with unique sites marked in bold type. The sequence 5′-CGR CAG CAG AGC TTC ACC ATT G-3′ can be used to design a primer to confirm correct reading frame for LexA fusions. Plasmids contain the *HIS3* selectable marker and the 2 μm origin of replication to allow propagation in yeast, and the Ap^r antibiotic resistance gene and the pBR origin of replication to allow propagation in *E. coli.* In the recently developed LexA-expression plasmids pMW101 and pMW103, the ampicillin resistance gene (Ap^r) has been replaced with the chloramphenicol resistance gene (Cm^r) and the kanamycin resistance gene (Km^r), respectively (see Table 4.2.1 for details).

Yeast strain EGY48 (*ura3 trp1 his3 3LexA*-operator-*LEU2*), or EGY191 (*ura3* trp1 his3 1LexA-operator-*LEU2*; Table 4.2.2)

✓ Complete minimal (CM) medium dropout plates, supplemented with 2% (w/v) of the indicated sugars (glucose or galactose), in 100-mm plates:
   Gul/CM, −Ura, −His
   Gal/CM, −Ura, −His
   Gal/CM, −Ura, −His, −Leu

✓ Z buffer with 1 mg/ml 5-bromo-4-chloro-3-indolyl-β-D-galactosidase (Xgal)

Gal/CM dropout liquid medium, supplemented with 2% Gal

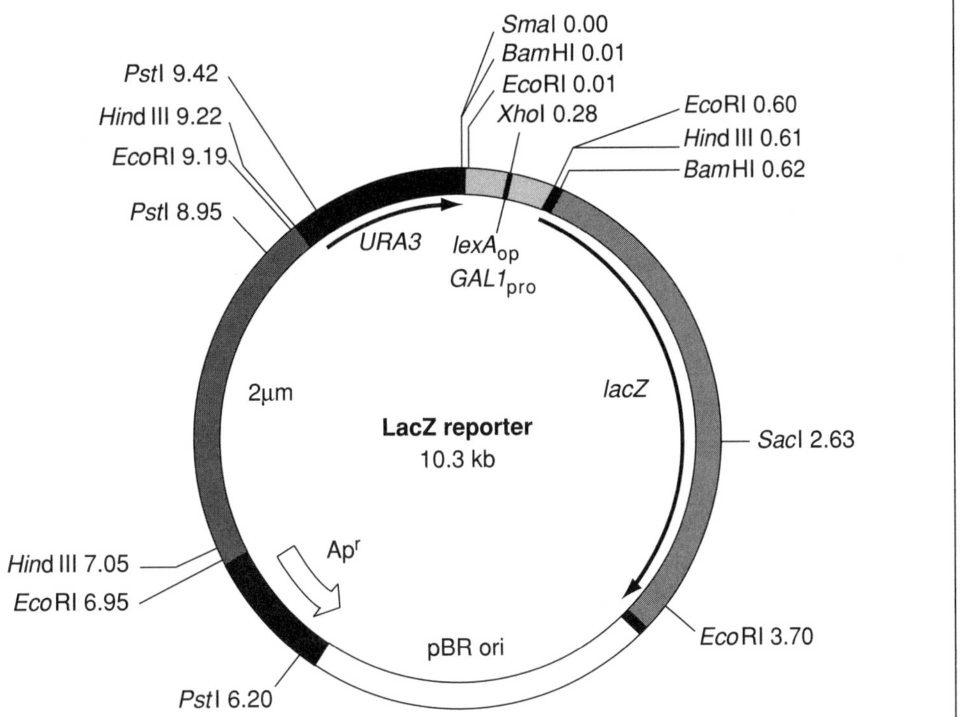

Figure 4.2.4 LacZ reporter plasmid. pRB1840, pJK103 and pSH18-34 are all derivatives of LR1Δ1 (West et al., 1984) containing eight, two, or one operator for LexA (*LexA*$_{op}$) binding inserted into the unique *Xho*I site located in the minimal *GAL1* promoter (*GAL1*$_{pro}$; 0.28 on map). the plasmid contains the *URA3* selectable marker, the 2 μm origin to allow propatation in yeast, the ampicillin resistance (Ap^r) gene, and the pBR322 origin (ori) to allow propagation in *E. coli.* Numbers indicate relative map positions. In the recently developed derivatives, the ampicillin resistance gene (Ap^r) has been replaced with the chioramphenicol or kanamycin resistance genes (see Table 4.2.1 for details).

Antibody to LexA or fusion domain: monoclonal antibody to LexA (Clontech, Invitrogen) or polyclonal antibody to LexA (available by request from R. Brent or E. Golemis)
H_2O, sterile

30°C incubator
Nylon membrane
Whatman 3MM filter paper

1. Using standard subcloning techniques (*CPMB UNIT 3.16*), insert the DNA encoding the protein of interest into the polylinker of pEG202 (see Fig. 4.2.3) or other LexA fusion plasmid to make an in-frame protein fusion.
2. Perform three separate lithium acetate transformations (*CPMB UNIT 13.7*) of EGY48 using the following combinations of plasmids:

   pBait + pSH18-34 (test)
   pSH17-4 + pSH18-34 (positive control for activation)
   pRFHM1 + pSH18-34 (negative control for activation).

3. Plate each transformation mixture on Glu/CM −Ura, −His dropout plates. Incubate 2 days at 30°C to select for yeast that contain both plasmids.

Table 4.2.2 Interaction Trap Yeast Selection Strains[a]

Strain	Relevant genotype	Number of operators	Comments/description
EGY48[b,c,d]	*MATα trp1, his3, ura3,* lexAops-LEU2	6	*lexA* operators direct transcription from the *LEU2* gene; EGY48 is a basic strain used to select for interacting clones from a cDNA library
EGY191	*MATα trp1, his3, ura3,* lexAops-LEU2	2	EGY191 provides a more stringent selection than EGY48, producing lower background with baits with intrinsic ability to activate transcription
L40[c]	MATα *trpl, leu2, ade2,* GAL4, *lexAops-HIS34, lexAops-lacZ8*		Expression driven from *GAL1* promoter is constitutive in L40 (inducible in EGY strains); selection is for *HIS* prototrophy. Integrated *lacZ* reporter is considerably less sensitive than pSH18-34 maintained in EGY strains

[a]Interaction Trap reagents represent the work of many contributors: the original basic reagents were developed in the Brent laboratory (Gyuris et al., 1993). Strains for specialized applications have been developed by the following individuals: E. Golemis, Fox Chase Cancer Center, Philadelphia, Pa. (EGY48, EGY191); A.B. Vojtek and S.M. Hollenberg, Fred Hutchinson Cancer Research Center, Seattle, Wash. (L40). Specialized strains not yet commercially available can be obtained by contacting the Brent laboratory at The Molecular Sciences Institute, Berkeley, (510) 647-0690 or *brent@molsci.org*, or the Golemis laboratory, (215) 728-2860 or *EA_Golemis@fccc.edu*.

[b]Strains commercially available from Clontech.

[c]Strains commercially available from Invitrogen as components of a Hybrid Hunter kit; the kit also includes all necessary positive and negative controls (not listed in this table).

[d]Strains commercially available from OriGene.

4

4. Streak a Glu/CM −Ura, −His master dropout plate with at least five or six independent colonies obtained from each of the three transformations in step 3 (test, positive control, and negative control) and incubate overnight at 30°C.

Perform filter assay for β-galactosidase activity

5a. Lift colonies by gently placing a nylon membrane on the yeast plate and allowing it to become wet through. Remove the membrane and air dry 5 min. Chill the membrane, colony side up, 10 min at −70°C.

6a. Cut a piece of Whatman 3MM filter paper slightly larger than the colony membrane and soak it in Z buffer containing 1 mg/ml Xgal. Place colony membrane, colony side up, on Whatman 3MM paper, or float it in the lid of a petri dish containing ∼2 ml Z buffer with 1 mg/ml Xgal.

7a. Incubate at 30°C and monitor for color changes.

Perform Xgal plate assay for lacZ activation

5b. Prepare Z buffer Xgal plates as described in *CPMB UNIT 13.1*.

6b. Streak yeast from master plate to Xgal plate and incubate at 30°C.

7b. Examine plates for color development at intervals over the next 2 to 3 days.

For LexA fusions that do not activate transcription, confirm by performing a repression assay (Brent and Ptashne, 1984) that the LexA fusion protein is being synthesized in yeast (some proteins are not) and that it is capable of binding *LexA* operator sequences (Fig. 4.2.5). The following steps can be performed concurrently with the activation assay.

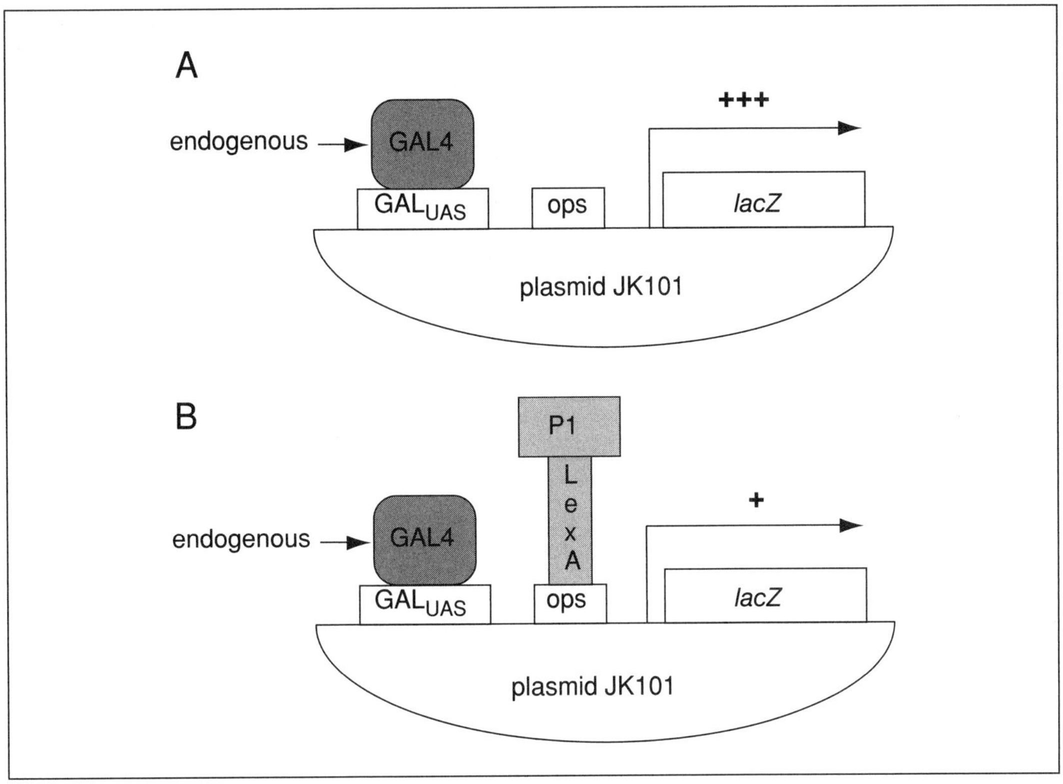

Figure 4.2.5 Repression assay for DNA binding. (**A**) The plasmid JK101 contains the upstream activating sequence (UAS) from the *GAL1* gene followed by *LexA* operators upstream of the *lacZ* coding sequence. Thus, yeast containing pJK101 will have significant β-galactosidase activity when grown on medium in which galactose is the sole carbon source because of binding of endogenous yeast GAL4 to the GAL_{UAS}. (**B**) LexA-fused proteins (P1-LexA) that are made, enter the nucleus, and bind the *LexA* operator sequences (ops) will block activation from the GAL_{UAS}, repressing the β-galactosidase activity (+) 3- to 5-fold. On glucose/Xggal medium, yeast containing pJK101 should be white because GAL_{UAS} transcription is repressed.

8. Transform EGY48 yeast with the following combinations of plasmids (three transformations):

   pBait + pJK101 (test)
   pRFHM1 + pJK101 (positive control for repression)
   pJK101 alone (negative control for repression).

9. Plate each transformation mix on Glu/CM −Ura, −His dropout plates or Glu/CM −Ura dropout plates, as appropriate to select yeast cells that contain the indicated plasmids. Incubate 2 to 3 days at 30°C until colonies appear.

10. Streak colonies to a Glu/CM −Ura, −His or Glu/CM −Ura dropout master plate and incubate overnight at 30°C.

11. Assay β-galactosidase activity of the three transformed strains (test, positive control, and negative control) by liquid assay (using Gal/CM dropout liquid medium), filter assay (steps 5a to 7a, first restreaking to Gal/CM plates to grow overnight), or plate assay (steps 5b to 7b, using Gal/CM −Ura XGal plates). Run the assay 1 to 2 hr for membranes and up to 24 to 36 hr for Xgal plates.

12. If a bait protein neither activates nor represses transcription, perform immunoblot analysis by probing an immunoblot of a crude lysate with antibodies against LexA or the fusion domain to test for protein synthesis (see Support Protocol 1).

These steps, which test for Leu requirement, can be performed concurrently with the *lacZ* activation and repression assays.

13. Disperse a colony of EGY48 containing pBait and pSH18-34 reporter plasmids into 500 μl sterile water. Dilute 100 μl of suspension into 1 ml sterile water. Make a series of 1/10 dilutions in sterile water to cover a 1000-fold concentration range.

14. Plate 100 μl from each tube (undiluted, 1/10, 1/100, and 1/1000) on Gal/CM −Ura, −His dropout plates and on Gal/CM −Ura, −His, −Leu dropout plates. Incubate overnight at 30°C.

Actual selection in the interactor hunt is based on the ability of the bait protein and acid-fusion pair, but not the bait protein alone, to activate transcription of the LexA operator-LEU2 gene and allow growth on medium lacking Leu. Thus, the test for the Leu requirement is the most important test of whether the bait protein is likely to have an unworkably high background. The LEU2 reporter in EGY48 is more sensitive than the pSH18-34 reporter for some baits, so it is possible that a bait protein that gives little or no signal in a β-galactosidase assay would nevertheless permit some level of growth on −Leu medium. If this occurs, there are several options for proceeding, the most immediate of which is to substitute EGY191 (see Table 4.2.2), a less sensitive screening strain, and repeat the assay.

BASIC PROTOCOL 2

PERFORMING AN INTERACTOR HUNT

An interactor hunt involves two successive large platings of yeast containing LexA-fused probes and reporters and libraries in pJG4-5 (Fig. 4.2.6 and Table 4.2.3) with a cDNA expression cassette under control of the *GAL* promoter. A list of libraries currently available for use with this system is provided in Table 4.2.3. Variations on the system are listed in Table 4.2.4.

Materials *(see APPENDIX 1 for items with ✓)*

Yeast containing appropriate combinations of plasmids (see Tables 4.2.1 and 4.2.2):
- EGY48 containing *LexA*-operator-*lacZ* reporter and pBait (see Basic Protocol 1)
- EGY48 containing *LexA*-operator-*lacZ* reporter and pRFHM-1
- EGY48 containing *LexA*-operator-*lacZ* reporter and any nonspecific bait

Complete minimal (CM) dropout liquid medium (*CPMB UNIT 13.1*) supplemented with sugars (glucose, galactose, and/or raffinose) as indicated [2% (w/v) Glu, or 2% (w/v) Gal + 1% (w/v) Raff]:
- Glu/CM −Ura, −His
- Glu/CM −Trp
- Gal/Raff/CM −Ura, −His, −Trp

H_2O, sterile

✓ TE buffer, pH 7.5, sterile (optional)

TE buffer (pH 7.5)/0.1 M lithium acetate

Library DNA in pJG4-5 (Table 4.2.3 and Fig. 4.2.6)

High-quality sheared salmon sperm DNA (see Support Protocol 2)

40% (w/v) polyethylene glycol 4000 (PEG 4000; filter sterilized)/0.1 M lithium acetate/TE buffer (pH 7.5)

Dimethyl sulfoxide (DMSO)

✓ Complete minimal (CM) medium dropout plates supplemented with sugars and Xgal (20 μg/ml) as indicated [2% (w/v) Glu, and 2% (w/v) Gal + 1% (w/v) Raff]:

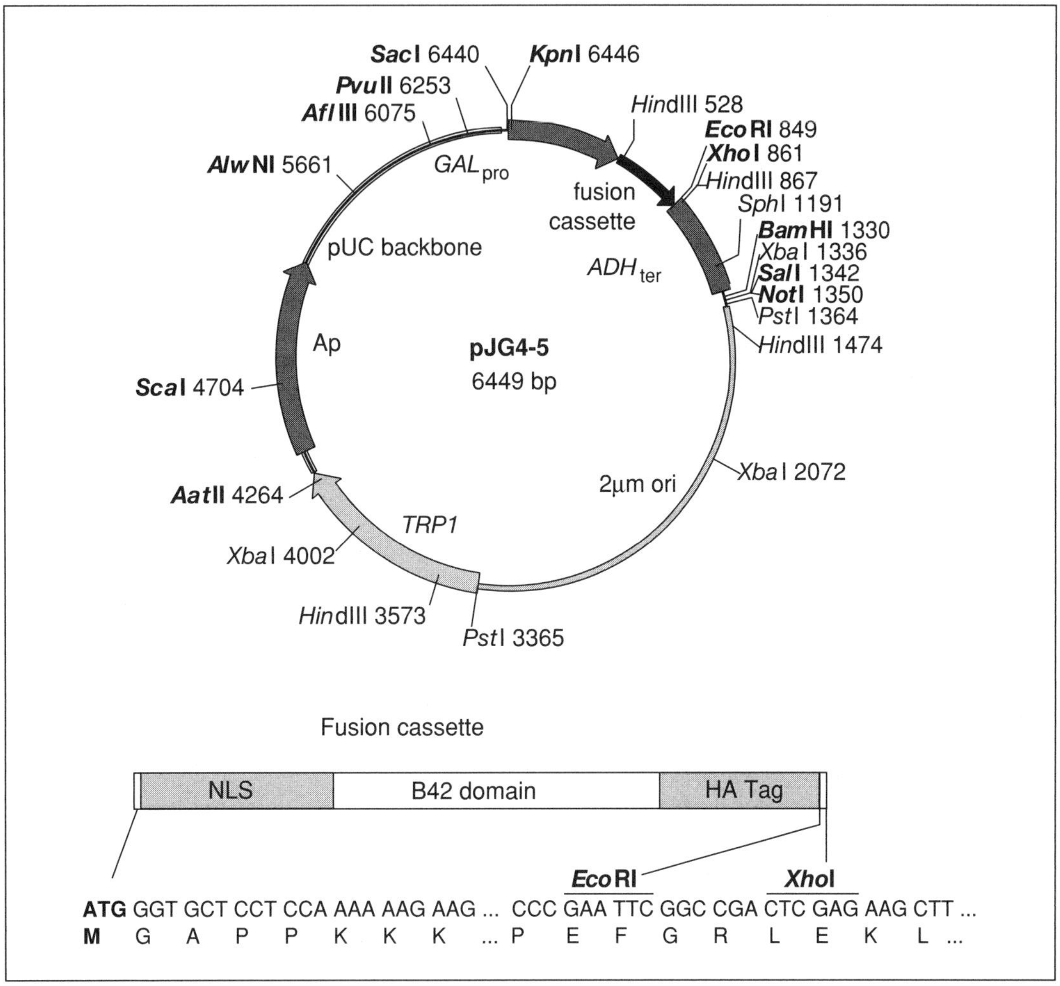

Figure 4.2.6 Library plasminds: pJG4-5. Library plasmids express cDNAs or other coding sequences inserted into unique *Eco*RI and *Xho*I sites as a translational fusion to a cassette consisting of the SV40 nuclear localization sequence (NLS; PPKKKRKVA), the acid blob B42 domain (Ruden et al., 1991), and the hemagglutinin (HA) epitope tag (YPYDVPDYA). Expression of cassette sequences is under the control of the *GAL1* galactose-inducible promoter. This map is based on the sequence data available for pJG4-5, and includes selected sites suitable for diagnostic restriction digests (shown in bold). Thye sequence 5′-CTG AGT GGA GAT GCC TCC-3′ can be used as a primer to identify inserts or to confirm correct reading frame. The pJG4-5 plasmid contains the *TRP1* selectable marker and the 2 μm origin to allow propagation in yeast, and the antibiotic resistance gene and the pUC origin to allow propagation in *E. coli.* In the recently developed pJG4-5 derivative plasmids pMW104 and pMW102, the ampicillin resistance gene (Ap^r) has been replaced with the chloramphenicol resistance gene (Cm^r) and the kanamycin resistance gene (Km^r), respectively (see Table 4.2.2 for details). Currently existing libraries are all made int he pJG4-5 plasmid (Gyuris et al., 1993) shown on this figure. Unique sites are marked in bold type.

Glu/CM −Ura, −His, −Trp, 24 × 24-cm (Nunc) and 100-mm
Gal/Raff/CM −Ura, −His, −Trp, 100-mm
Gal/Raff/CM −Ura, −His, −Trp, −Leu, 100-mm
Glu/Xgal/CM −Ura, −His, −Trp, 100-mm
Gal/Raff/Xgal/CM −Ura, −His, −Trp, 100-mm
Glu/CM −Ura, −His, −Trp, −Leu, 100-mm
Glu/CM −Ura, −His, 100-mm
Gal/CM −Ura, −His, −Trp, −Leu, 100-mm

Table 4.2.3 Libraries Compatible with the Interaction Trap System[a]

Source of RNA/DNA	Vector	Independent clones	Insert size (average)[b]	Contact information
Cell lines				
HeLa cells (human cervical carcinoma)	JG	9.6×10^6	0.3-3.5 kb (1.5 kb)	R. Brent, Clontech, Invitrogen, OriGene
HeLa cells (human cervical carcinoma)	Y	3.7×10^6	0.3-1.2 kb	Invitrogen
WI-38 cells (human lung fibroblasts), serum-starved, cDNA	JG	5.7×10^6	0.3-3.5 kb (1.5 kb)	R. Brent, Clontech, OriGene
Jurkat cells (human T cell leukemia), exponentially growing, cDNA	JG	4.0×10^6	0.7-2.8 kb (1.5 kb)	R. Brent
Jurkat cells (human T cell leukemia)	Y	3.2×10^6	0.3-1.2 kb	Invitrogen
Jurkat cells (human T cell leukemia)	Y	3.0×10^6	0.5-4.0 kb (1.8 kb)	Clontech
Jurkat cells (human T cell leukemia)	JG	5.7×10^6	(>1.3)	OriGene
Jurkat cells (human T cell leukemia)	JG	2×10^6	0.7-3.5 kb (1.2 kb)	S. Witte
Be Wo cells (human fetal placental choriocarcinoma)	Y	5.4×10^6	0.3-0.8 kb	Invitrogen
Human lymphocyte	JG	4.0×10^6	0.4-4.0 kb (2.0 kb)	Clontech
$CD4^+$ T cell, murine, cDNA	JG	$>10^6$	0.3-2.5 kb (>0.5 kb)	R. Brent
Chinese hamster ovary (CHO) cells, exponentially growing, cDNA	JG	1.5×10^6	0.3-3.5 kb	R. Brent
A20 cells (mouse B cell lymphoma)	Y	3.11×10^6	0.3-1.2 kb	Invitrogen
Human B cell lymphoma	JG	—	—	H. Niu
Human 293 adenovirus–infected (early and late stages)	JG	—	—	K. Gustin
SKOV3 human Y ovarian cancer	Y	5.0×10^6	(>1.4 kb)	OriGene
MDBK cell, bovine kidney	JG	5.8×10^6	(>1.2 kb)	OriGene
MDCK cells	JG	—	—	D. Chen
HepG2 cell line cDNA	JG	2×10^6	—	M. Melegari
MCF7 breast cancer cells, untreated	JG	1.0×10^7	(>1.5 kb)	OriGene
MCF7 breast cancer cells, estrogen-treated	JG	1.0×10^7	(>1.1 kb)	OriGene
MCF7 cells, serum-grown	JG	1.0×10^7	0.4-3.5 kb	OriGene
LNCAP prostate cell line, untreated	JG	2.9×10^6	(>0.8 kb)	OriGene
LNCAP prostate cell line, androgen-treated	JG	4.6×10^6	(>0.9 kb)	OriGene
Mouse pachytene spermatocytes	JG	—	—	C. Hoog
Tissues				
Human breast	Y	9×10^6	0.4-1.2 kb	Invitrogen
Human breast tumor	Y	8.84×10^6	0.4-1.2 kb	Invitrogen
Human liver	JG	$>10^6$	0.6-4.0 kb (>1 kb)	R. Brent
Human liver	Y	2.2×10^6	0.5-4 kb (1.3 kb)	Clontech
Human liver	JG	3.2×10^6	0.3-1.2 kb	Invitrogen
Human liver	JG	1.1×10^7	(>1 kb)	OriGene
Human lung	Y	5.9×10^6	0.4-1.2 kb	Invitrogen

continued

Table 4.2.3 Libraries Compatible with the Interaction Trap System[a], continued

Source of RNA/DNA	Vector	Independent clones	Insert size (average)[b]	Contact information
Human lung tumor	Y	1.9×10^6	0.4-1.2	Invitrogen
Human brain	JG	3.5×10^6	0.5-4.5 kb (1.4 kb)	Clontech
Human brain	Y	8.9×10^6	0.3-1.2 kb	Invitrogen
Human testis	Y	6.4×10^6	0.3-1.2 kb	Invitrogen
Human testis	JG	3.5×10^6	0.4-4.5 kb (1.6 kb)	Clontech
Human ovary	Y	4.6×10^6	0.3-1.2 kb	Invitrogen
Human ovary	JG	4.6×10^6	(>1.3 kb)	OriGene
Human ovary	JG	3.5×10^6	0.5-4.0 kb (1.8 kb)	Clontech
Human heart	JG	3.0×10^6	0.3-3.5 kb (1.3 kb)	Clontech
Human placenta	Y	4.8×10^6	0.3-1.2 kb	Invitrogen
Human placenta	JG	3.5×10^6	0.3-4.0 kb (1.2 kb)	Clontech
Human mammary gland	JG	3.5×10^6	0.5-5 kb (1.6 kb)	Clontech
Human peripheral blood leucocyte	JG	1.0×10^7	(>1.3 kb)	OriGene
Human kidney	JG	3.5×10^6	0.4-4.5 kb (1.6 kb)	Clontech
Human fetal kidney	JG	3.0×10^6	(>1 kb)	OriGene
Human spleen	Y	1.14×10^7	0.4-1.2 kb	Invitrogen
Human prostate	Y	5.5×10^6	0.4-1.2 kb	Invitrogen
Human normal prostate	JG	1.4×10^6	0.4-4.5 kb (1.7 kb)	Clontech
Human prostate	JG	1.4×10^6	(>1 kb)	OriGene
Human prostate cancer	JG	1.1×10^6	(>0.9 kb)	OriGene
Human fetal prostate	JG	—	—	OriGene
Human fetal liver	JG	3.5×10^6	0.3-4.5 kb (1.3 kb)	Clontech
Human fetal liver	Y	2.37×10^6	0.3-1.2 kb	Invitrogen
Human fetal liver	JG	8.6×10^6	(>1 kb)	OriGene
Human fetal brain	JG	3.5×10^6	0.5-1.2 kb (1.5 kb)	R. Brent, Clontech, Invitrogen, OriGene
Mouse brain	JG	6.1×10^6	(>1 kb)	OriGene
Mouse brain	JG	4.5×10^6	0.4-4.5 kb (1.2 kb)	Clontech
Mouse breast, lactating	JG	1.0×10^7	0.4-3.1 kb	OriGene
Mouse breast, involuting	JG	1.0×10^7	0.4-7.0 kb	OriGene
Mouse breast, virgin	JG	1.0×10^7	0.4-5.5 kb	OriGene
Mouse breast, 12 days pregnant	JG	6.3×10^6	0.4-5.3 kb	OriGene
Mouse skeletal muscle	JG	7.2×10^6	0.4-3.5 kb	OriGene
Rat adipocyte, 9-week-old Zucker rat	JG	1.0×10^7	0.4-5.0 kb	OriGene
Rat brain	JG	4.5×10^6	0.3-3.4 kb	OriGene
Rat brain (day 18)	JG	—	—	H. Niu
Rat testis	JG	8.0×10^6	(>1.2 kb)	OriGene
Rat thymus	JG	8.2×10^6	(>1.3 kb)	OriGene
Mouse liver	JG	9.5×10^6	(>1.4 kb)	OriGene
Mouse spleen	JG	1.0×10^7	(>1 kb)	OriGene

continued

Table 4.2.3 Libraries Compatible with the Interaction Trap System[a], continued

Source of RNA/DNA	Vector	Independent clones	Insert size (average)[b]	Contact information
Mouse ovary	JG	4.0×10^6	(>1.2 kb)	OriGene
Mouse prostate	JG	—	—	OriGene
Mouse embryo, whole (19-day)	JG	1.0×10^5	0.2-2.5 kb	OriGene
Mouse embryo	JG	3.6×10^6	0.5-5 kb (1.7 kb)	Clontech
Drosophila melanogaster, adult, cDNA	JG	1.8×10^6	(>1.0 kb)	OriGene
D. melanogaster, embryo, cDNA	JG	3.0×10^6	0.5-3.0 kb (1.4 kb)	Clontech
D. melanogaster, 0-12 hr embryos, cDNA	JG	4.2×10^6	0.5-2.5 kb (1.0 kb)	R. Brent
D. melanogaster, ovary, cDNA	JG	3.2×10^6	0.3-1.5 kb (800 bp)	R. Brent
D. melanogaster, disc, cDNA	JG	4.0×10^6	0.3-2.1 kb (900 bp)	R. Brent
D. melanogaster, head	JG	—	—	M. Rosbash
Miscellaneous				
Synthetic aptamers	PJM-1	$>1 \times 10^9$	60 bp	R. Brent
Saccharomyces cerevisiae, S288C, genomic	JG	$>3 \times 10^6$	0.8-4.0 kb	R. Brent
S. cerevisiae, S288C, genomic	JG	4.0×10^6	0.5-4.0 kb	OriGene
Sea urchin ovary	JG	3.5×10^6	(1.7 kb)	Clontech
Caenorhabditis elegans	JG	3.8×10^6	(>1.2 kb)	OriGene
Agrobacterium tumefaciens	JG	—	—	—
Arabidopsis thaliana, 7-day-old seedlings	JG	—	—	H.M. Goodman
Tomato (*Lycopersicon esculentum*)	JG	8×10^6	—	G.B. Martin
Xenopus laevis embryo	JG	2.2×10^6	0.3-4 kb (1.0 kb)	Clontech

[a]Most libraries are constructed in either the pJG4-5 vector or the pYESTrp vector (JG or Y in the Vector column); the peptide aptamer library is made in the pJM-1 vector. Libraries available from the public domain were constructed by the following individuals: (1) J. Gyuris; (3) C. Sardet and J. Gyuris; (4) W. Kolanus, J. Gyuris, and B. Seed; (39) D. Krainc; (50-52) R. Finley; (55) P. Watt; (54) P. Colas, B. Cohen, T. Jessen, I. Grishina, J. McCoy, and R. Brent (Colas et al., 1996). All libraries mentioned above were constructed in conjunction with and are available from the laboratory of Roger Brent, (510) 647-0690 or *brent@molsci.org*. The following individual investigators must be contacted directly: (18) J. Pugh, Fox Chase Cancer Center, Philadelphia, Pa.; (8,9) Vinyaka Prasad, Albert Einstein Medical Center New York, N.Y.; (57, 58) Gregory B. Martin, *gmartin@dept.agry.purdue.edu*; (11) Huifeng Niu, *hn34@columbia.edu*; (16) Christer Hoog, *christer.hoog@cmb.ki.se*; (12) Kurt Gustin, *kgus@umich.edu*; (6) Stephan Witte, *Stephan.Witte@nimbus.rz.uni-konstanz.de*.

[b]Insert size ranges for pJG4-5 based libraries originally constructed in the Brent laboratory, which are now commercially available from Clontech, were reestimated by the company.

Table 4.2.4 Two-Hybrid System Variants[a]

System	DNA-binding domain	Activation domain	Selection	Reference
Two-hybrid	*GAL4*	*GAL4*	Activation of *lacZ*, *HIS3*	Chien et al., 1991
Interaction trap	*LexA*	*B42*	Activation of *LEU2*, *lacZ*	Gyuris et al., 1993
"Improved two-hybrid"	*GAL4*	*GAL4*	Activation of *HIS3*, *lacZ*	Durfee et al., 1993
Modified two-hybrid	*LexA*	VP16	Activation of *HIS3*, *lacZ*	Vojtek et al., 1993
KISS	*GAL4*	VP16	Activation of *CAT*, hyg^r	Fearon et al., 1992
Contingent replication	*GAL4*	VP16	Activation of T-Ag, replication of plasmids	Vasavada et al., 1991

[a]Abbreviations: *CAT*, chloramphenicol transferase gene; hyg^r, hygromycin resistance gene; T-Ag, viral large T antigen.

4

✓ Glycerol solution
E. coli KC8 (*pyrF leuB600 trpC hisB463*; constructed by K. Struhl and available from R. Brent)
✓ LB/ampicillin plates
E. coli DH5α or other strain suitable for preparation of DNA for sequencing
Bacterial defined minimal A medium plates: 1× A medium plates containing 0.5 μg/ml vitamin B1 and supplemented with 40 μg/ml each Ura, His, and Leu

30°C incubator, with and without shaking
Low-speed centrifuge and rotor
50-ml conical tubes, sterile
1.5-ml microcentrifuge tubes, sterile
42°C heating block
Glass microscope slides, sterile

1. Grow an ~20-ml culture of EGY48 or EGY191 containing a *LexA*-operator-*lacZ* reporter plasmid and pBait in Glu/CM −Ura, −His liquid dropout medium overnight at 30°C.

2. In the morning, dilute culture into 300 ml Glu/CM −Ura, −His liquid dropout medium to 2 × 10^6 cells/ml (OD_{600} = ~0.10). Incubate at 30°C until the culture contains ~1 × 10^7 cells/ml (OD_{600} = ~0.50).

3. Centrifuge 5 min at 1000 to 1500 × *g* in a low-speed centrifuge at room temperature to harvest cells. Resuspend in 30 ml sterile water and transfer to 50-ml conical tube.

4. Centrifuge 5 min at 1000 to 1500 × *g*. Decant supernatant and resuspend cells in 1.5 ml TE buffer/0.1 M lithium acetate.

5. Add 1 μg library DNA in pJG4-5 and 50 μg high-quality sheared salmon sperm carrier DNA to each of 30 sterile 1.5-ml microcentrifuge tubes. Add 50 μl of the resuspended yeast solution from step 4 to each tube.

   The total volume of library and salmon sperm DNA added should be <20 μl and preferably <10 μll. A typical library transformation will result in 2 to 3 × 10^6 primary transformants. This transformation requires a total of 20 to 30 μg library DNA and 1 to 2 mg carrier DNA.

6. Add 300 μl of sterile 40% PEG 4000/0.1 M lithium acetate/TE buffer, pH 7.5, and invert to mix thoroughly. Incubate 30 min at 30°C.

7. Add DMSO to 10% (~40 μl per tube) and invert to mix. Heat shock 10 min in 42°C heating block.

8a. *For 28 tubes:* Plate the complete contents of one tube per 24 × 24-cm Glu/CM −Ura, −His, −Trp dropout plate and incubate at 30°C.

8b. *For two remaining tubes:* Plate 360 μl of each tube on 24 × 24-cm Glu/CM −Ura, −His, −Trp dropout plate. Use the remaining 40 μl from each tube to make a series of dilutions in sterile water. Plate dilutions on 100-mm Glu/CM −Ura, −His, −Trp dropout plates. Incubate all plates 2 to 3 days at 30°C until colonies appear.

   The dilution series gives an idea of the transformation efficiency and allows an accurate estimation of the number of transformants obtained.

9. Cool all of the 24 × 24-cm plates containing transformants for several hours at 4°C to harden agar.

10. Wearing gloves and using a sterile glass microscope slide, gently scrape yeast cells off the plate. Pool cells from the 30 plates into one or two sterile 50-ml conical tubes.

11. Wash cells by adding a volume of sterile TE buffer or water at least equal to the volume of the transferred cells. Centrifuge ~5 min at 1000 to 1500 × *g*, room temperature, and discard supernatant. Repeat wash.

12. Resuspend pellet in 1 vol glycerol solution, mix well, and store up to 1 year in 1-ml aliquots at −70°C.

13. Remove an aliquot of frozen transformed yeast and dilute 1/10 with Gal/Raff/CM −Ura, −His, −Trp dropout medium. Incubate with shaking 4 hr at 30°C to induce the *GAL* promoter on the library.

14. Make serial dilutions of the yeast cells using the Gal/Raff/CM −Ura, −His, −Trp dropout medium. Plate on 100-mm Gal/Raff/CM −Ura, −His, −Trp dropout plates and incubate 2 to 3 days at 30°C until colonies are visible.

15. Count colonies and determine the number of colony-forming units (cfu) per aliquot of transformed yeast.

16. Thaw the appropriate quantity of transformed yeast based on the plating efficiency, dilute, and incubate as in step 13. Dilute cultures in Gal/Raff/CM −Ura, −His, −Trp, −Leu medium as necessary to obtain a concentration of 10^7 cells/ml (OD_{600} = ~0.5), and plate 100 µl on each of as many 100-mm Gal/Raff/CM −Ura, −His, −Trp, −Leu dropout plates as are necessary for full representation of transformants. Incubate 2 to 3 days at 30°C until colonies appear.

17. Carefully pick appropriate colonies to a new Gal/Raff/CM −Ura, −His, −Trp, −Leu master dropout plate. Incubate 2 to 7 days at 30°C until colonies appear.

Pick a master plate with colonies obtained on day 2, a second master plate (or set of plates) with colonies obtained on day 3, and a third with colonies obtained on day 4. If many apparent positives are obtained, it may be worth making master plates of the much larger number of colonies likely to be obtained at day 4 (and after).

18. Restreak from the Gal/Raff/CM −Ura, −His, −Trp, −Leu master dropout plate to a 100-mm Glu/CM −Ura, −His, −Trp master dropout plate. Incubate overnight at 30°C until colonies form.

19. Restreak or replica plate from this plate to the following plates:

Glu/Xgal/CM −Ura, −His, −Trp
Gal/Raff/Xgal/CM −Ura, −His, −Trp
Glu/CM −Ura, −His, −Trp, −Leu
Gal/Raff/CM −Ura, −His, −Trp, −Leu.

At this juncture, colonies and the library plasmids they contain are tentatively considered positive if they are blue on Gal/Raff/Xgal plates but not blue or only faintly blue on Glu/Xgal plates, and if they grow on Gal/Raff/CM −Leu plates but not on Glu/CM −Leu plates.

The following steps test for Gal dependence of the Leu$^+$ *insert and lacZ phenotypes to confirm that they are attributable to expression of the library-encoded proteins. The GAL1 promoter is turned off and −Leu selection eliminated before reinducing.*

20a. Transfer yeast plasmids directly into *E. coli* by following the protocol for direct electroporation (*CPMB UNIT 1.8*). Proceed to step 22.

20b. Isolate plasmid DNA from yeast by the rapid miniprep protocol (*CPMB UNIT 13.11*) with the following alteration: after obtaining aqueous phase, precipitate by adding sodium acetate to 0.3 M final and 2 vol ethanol, incubate 20 min on ice, microcentrifuge 15 min at maximum speed, wash pellet with 70% ethanol, dry, and resuspend in 5 µl TE buffer.

21. Use 1 µl DNA to electroporate (*CPMB UNIT 1.8*) into competent KC8 bacteria, and plate on LB/ampicillin plates. Incubate overnight at 37°C.

22. Restreak or replica plate colonies arising on LB/ampicillin plates to bacterial defined minimal A medium plates containing vitamin B1 and supplemented with Ura, His, and Leu but lacking Trp. Incubate overnight at 37°C.

23. Purify library-containing plasmids using a bacterial miniprep procedure (*APPENDIX A.2F*).

    Do not sequence DNAs obtained at this stage, as it is still possible that none of the isolated clones will express bona fide interactors. Complete the specificity tests before committing the effort to sequencing.

    Because multiple 2 µm plasmids with the same marker can be simultaneously tolerated in yeast, a single yeast may contain two or more different library plasmids, only one of which encodes an interacting protein. Work up at least two individual bacterial transformants for each yeast positive. Digest (CPMB UNIT 3.1) with EcoRI + XhoI to release cDNA inserts, and run an agarose minigel (CPMB UNIT 2.5A) to confirm that both plasmids contain the same insert.

24. In separate transformations, use purified plasmids from step 23 to transform yeast that already contain the following plasmids and are growing on Glu/CM −Ura, −His plates:

    EGY48 containing pSH18-34 and pBait
    EGY48 containing pSH18-34 and pRFHM-1
    EGY48 containing pSH18-34 and a nonspecific bait (optional).

25. Plate each transformation mix on Glu/CM −Ura, −His, −Trp dropout plates and incubate 2 to 3 days at 30°C until colonies appear.

26. Create a Glu/CM −Ura, −His, −Trp master dropout plate for each library plasmid being tested. Streak adjacently five or six independent colonies derived from each of the transformation plates. Incubate overnight at 30°C.

27. Restreak or replica plate from this master dropout plate to the same series of test plates used for the actual screen:

    Glu/Xgal/CM −Ura, −His, −Trp
    Gal/Raff/Xgal/CM −Ura, −His, −Trp
    Glu/CM −Ura, −His, −Trp, −Leu
    Gal/CM −Ura, −His, −Trp, −Leu.

    True positive cDNAs should make cells blue on Gal/Raff/Xgal but not on Glu/Xgal plates, and should make them grow on Gal/Raff/CM −Leu but not Glu/CM −Leu dropout plates only if the cells contain LexA-bait. cDNAs that meet such criteria are ready to be sequenced (see legend to Fig. 19.1.3 for primer sequence) or otherwise characterized. Those cDNAs that also encode proteins that interact with either RFHM-1 or another nonspecific bait should be discarded.

    It may be helpful to cross-check the isolated cDNAs with a database of cDNAs thought to be false positives. This database is available on the World Wide Web as a work in progress at http://www.fccc.edu:80/research/labs/golemis/InteractionTrapInWork.html.

28. If appropriate, conduct additional specificity tests. Analyze and sequence positive isolates.

    The primer sequence for use with pJG4-5 is provided in the legend to Figure 4.2.4. DNA prepared from KC8 is generally unsuitable for dideoxy or automated sequencing even after use of Qiagen columns and/or cesium chloride gradients. Library plasmids to be sequenced should be retransformed from the KC8 miniprep stock (step 23) to a more amenable strain, such as DH5α, before sequencing is attempted.

ALTERNATE PROTOCOL 1

RAPID SCREEN FOR INTERACTION TRAP POSITIVES

Additional Materials *(also see Basic Protocol 2; see* APPENDIX 1 *for items with ✓)*

Yeast plated on Glu/CM −Ura, −His, −Trp master plate (see Basic Protocol 2, step 19)
✓ Lysis solution
10 μM forward primer (FP1): 5′-CGT AGT GGA GAT GCC TCC-3′
10 μM reverse primer (FP2): 5′-CTG GCA AGG TAG ACA AGC CG-3′

Toothpicks or bacterial inoculating loops, sterile
96-well microtiter plate
Sealing tape, e.g., wide transparent tape
✓ 150- to 212-μm glass beads, acid-washed
Vortexer with flat plate

1. Perform an interactor hunt (see Basic Protocol 2, steps 1 to 19).

2. Use a sterile toothpick or bacterial inoculating loop to transfer yeast from the Glu/CM, −Ura, −His, −Trp master plate into 25 μl lysis solution in a 96-well microtiter plate. Seal the wells of the microtiter plate with sealing tape and incubate 1.5 to 3.5 hr at 37°C with shaking.

3. Remove tape from the plate, add ∼25 μl acid-washed glass beads to each well, and reseal with the same tape. Firmly attach the microtiter plate to a flat-top vortexer, and vortex 5 min at medium-high power.

4. Remove the tape and add ∼100 μl sterile water to each well. Swirl gently to mix, then remove sample for step 5. Press the tape back firmly to seal the microtiter plate and place in the freezer at −20°C for storage.

5. Amplify 0.8 to 2.0 μl of sample by standard PCR (*CPMB UNIT 4.1*) in a ∼30-μl volume using 3 μl each of the forward primer FP1 and the reverse primer FP2. Perform PCR using the following cycles:

Initial step:	2 min	94°C
31 cycles:	45 sec	94°C
	45 sec	56°C
	45 sec	72°C.

These conditions have been used successfully to amplify fragments up to 1.8 kb in length; some modifications, such as extension of elongation time, are also effective.

6. 6.Load 20 μl of the PCR reaction product on a 0.7% low melting temperature agarose gel (*CPMB UNIT 2.5A*) to resolve PCR products. Based on insert sizes, group the obtained interactors in families, i.e., potential multiple independent isolates of identical cDNAs. Reserve gel until results of step 7 are obtained.

7. While the gel is running, use the remaining 10 μl of PCR reaction product for a restriction endonuclease digestion with *Hae*III in a digestion volume of ∼20 μl (*CPMB UNIT 3.1*). Based on analysis of the sizes of undigested PCR products in the gel (step 6), rearrange the tubes with *Hae*III digest samples so that those thought to represent a family are side by side. Resolve the digests on a 2% to 2.5% agarose gel (*CPMB UNIT 2.5A*).

Most restriction fragments will be in the 200-bp to 1.0-kb size range, so using a long gel run is advisable.

8. Isolate DNA fragments from the low melting temperature agarose gel (step 6).

9. Remove the microtiter plate of lysates from the freezer, thaw it, and remove 2 to 4 μl of lysed yeast for each desired positive. Electroporate DNA into either DH5α or KC8 *E. coli* as appropriate, depending on the choice of bait and reporter plasmids (see Table 4.2.1). Refreeze the plate as a DNA reserve in case bacteria fail to transform on the first pass.

10. Prepare a miniprep of plasmid DNA from the transformed bacteria (*APPENDIX 2F*) and perform yeast transformation and specificity assessment (see Basic Protocol 2, steps 24 to 28).

ALTERNATE PROTOCOL 2

PERFORMING A HUNT BY INTERACTION MATING

An alternative way of conducting an interactor hunt is to mate a strain that expresses the bait protein with a strain that has been pretransformed with the library DNA, and screen the resulting diploid cells for interactors (Bendixen et al., 1994; Finley and Brent, 1994). Interaction mating allows several interactor hunts with different baits to be conducted using a single high-efficiency yeast transformation with library DNA. This can be a considerable savings, since the library transformation is one of the most challenging tasks in an interactor hunt. Strain combinations other than those described below can also be used in an interaction-mating hunt. The key to choosing the strains is to ensure that the bait and prey strains are of opposite mating types and that both have auxotrophies to allow selection for the appropriate plasmids and reporter genes. Also, once the bait plasmid and *lacZ* reporter plasmid have been introduced into the bait strain and the library plasmids have been introduced into the library strain, the resulting bait strain and library strain must each have auxotrophies that can be complemented by the other, so that diploids can be selected.

4

Additional Materials *(also see Basic Protocols 1 and 2; see APPENDIX 1 for items with ✓)*

Yeast strains: either RFY206 (Finley and Brent, 1994), YPH499 (Sikorski and Hieter, 1989; ATCC #6625), or an equivalent *MATa* strain with auxotrophic markers *ura3*, *trp1*, *his3*, and *leu2*
✓ YPD liquid medium
Glu/CM -Trp plates: CM dropout plates −Trp (*APPENDIX 1*) supplemented with 2% glucose
pJG4-5 library vector (Fig. 4.2.6), empty
✓ 100-mm YPD plates

1. Perform construction of the bait plasmid (pBait; see Basic Protocol 1, step 1).

2. Cotransform the *MATa* yeast strain (e.g., either RFY206 or YPH499) with pBait and pSH18-34 using the lithium acetate method (*CPMB UNIT 13.7*). Select transformants on Glu/CM −Ura, −His plates by incubating plates at 30°C for 3 to 4 days until colonies form. Combine 3 colonies for all future tests and for the mating hunt.

   The bait strain will be a MATa yeast strain (mating type opposite of EGY48) containing a lacZ reporter plasmid like pSH18-34 and the bait-expressing plasmid, pBait.

3. *Optional:* Assay *lacZ* gene activation in the bait strain (see Basic Protocol 1, steps 4 to 7).

   If the bait activates the lacZ reporter, a less sensitive lacZ reporter plasmid (Table 4.2.1), or an integrated version of the lacZ reporter should be tried.

4. Perform a large-scale transformation of EGY48 with library DNA using the lithium acetate method (see Basic Protocol 2, steps 1 to 8, except start with EGY48 bearing no other plasmids). To prepare for transformation, grow EGY48 in YPD liquid medium. Select library transformants on Glu/CM −Trp plates by incubating 3 days at 30°C.

5. Collect primary transformants by scraping plates, washing yeast, and resuspending in 1 pellet vol glycerol solution (see Basic Protocol 2, steps 9 to 12). Freeze 0.2 to 1.0 ml aliquots at −70° to −80°C.

6. Transform EGY48 grown in YPD liquid medium with the empty library vector, pJG4-5, using the lithium acetate method (*CPMB UNIT 13.7*). Select transformants on Glu/CM −Trp plates by incubating 3 days at 30°C.

7. Pick and combine three transformant colonies and use them to inoculate 30 ml of Glu/CM −Trp medium. Incubate 15 to 24 hr at 30°C (to OD_{600} >3).

8. Centrifuge 5 min at 1000 to 1500 × *g*, room temperature, and remove supernatant. Resuspend in 10 ml sterile water to wash cells.

9. Centrifuge 5 min at 1000 to 1500 × *g*, room temperature, and remove supernatant. Resuspend in 1 pellet vol glycerol solution and freeze 100-µl aliquots at −70° to −80°C.

10. After freezing (at least 1 hr), thaw an aliquot of each pretransformed strain (from step 5 and step 9) at room temperature. Make several serial dilutions in sterile water, including aliquots diluted 10^5-fold, 10^6-fold, and 10^7-fold. Plate 100 µl of each dilution on 100-mm Glu/CM −Trp plates and incubate 2 to 3 days at 30°C.

11. Count the colonies and determine the number of colony-forming units (cfu) per aliquot of transformed yeast.

    The plating efficiency for a typical library transformation and for the control strain will be ~1 × 10^8 cfu per 100 il.

12. Grow a 30-ml culture of the bait strain in Glu/CM −Ura, −His liquid dropout medium to mid to late log phase (OD_{600} = 1.0 to 2.0, or 2 to 4 × 10^7 cells/ml).

13. Centrifuge the culture 5 min at 1000 to 1500 × *g*, room temperature, to harvest cells. Resuspend the cell pellet in sterile water to make a final volume of 1 ml.

14. Set up two matings. In one sterile microcentrifuge tube mix 200 µl of the bait strain with 200 µl of a thawed aliquot of the pretransformed control strain from step 9. In a second microcentrifuge tube mix 200 µl of the bait strain with ~1 × 10^8 cfu (~0.1 to 1 ml) of the pretransformed library strain from step 5.

15. Centrifuge each cell mixture for 5 min at 1000 to 1500 × *g*, pour off medium, and resuspend cells in 200 µl YPD medium. Plate each suspension on a 100-mm YPD plate. Incubate 12 to 15 hr at 30°C.

16. Add ~1 ml of Gal/Raff/CM −Ura, −His, −Trp to the lawns of mated yeast on each plate. Mix the cells into the medium using a sterile applicator stick.

17. Transfer each slurry of mated cells to a 500-ml flask containing 100 ml of Gal/Raff/CM −Ura, −His, −Trp dropout medium. Incubate with shaking 6 hr at room temperature to induce the *GAL1* promoter, which drives expression of the cDNA library.

18. Centrifuge the cell suspensions 5 min at 1000 to 1500 × *g*, room temperature, to harvest the cells. Wash by resuspending in 30 ml of sterile water and centrifuging again. Resuspend each pellet in 5 ml sterile water. Measure OD_{600} and, if necessary, dilute to a final concentration of ~1 × 10^8 cells/ml.

19. For each mating make a series of dilutions in sterile water, at least 200 µl each, to cover a 10^6-fold concentration range. Plate 100 µl from each tube (undiluted, 10^{-1}, 10^{-2}, 10^{-3}, 10^{-4}, 10^{-5}, and 10^{-6} dilution) on 100-mm Gal/Raff/CM Ura, −His, −Trp, −Leu plates. Plate 100 µl from the 10^{-4}, 10^{-5}, and 10^{-6} tubes on 100-mm Gal/Raff/CM −Ura, −His, −Trp plates. Incubate plates at 30°C. Count the colonies on each plate after 2 to 5 days.

20. For the mating with the pretransformed library, prepare an additional 3 ml of a 10^{-1} dilution. Plate 100 µl of the 10^{-1} dilution on each of 20 100-mm Gal/Raff/CM −Ura, −His, −Trp, −Leu plates. Also plate 100 µl of the undiluted cells on each of 20 100-mm Gal/Raff/CM −Ura, −His, −Trp, −Leu plates. Incubate at 30°C. Pick Leu^+ colonies after 2 to 5 days and characterize them beginning with step 17 of Basic Protocol 2.

To screen all of the pretransformed library, it will be necessary to pick a sufficient number of Leu^+ colonies in addition to background colonies produced by the transactivation potential of the bait itself. Thus, the minimum number of Leu^+ colonies that should be picked in step 20 of this protocol is given by:

(transactivation potential, Leu^+/cfu) × (# library transformants screened).

For example, if 10^7 library transformants were obtained in step 2 (and at least 10^8 cfu of these transformants were mated with the bait strain in step 14, since only ~10% will form diploids), and the transactivation potential of the bait is 10^{-4} Leu^+/cfu, then at least 1000 Leu^+ colonies must be picked and characterized. In other words, if the rarest interactor is present in the pretransformed library at a frequency of 10^{-7}, to find it one needs to screen through at least 10^7 diploids from a mating of the library strain. However, at least 1000 of these 10^7 diploids would be expected to be Leu^+ due to the bait background if the transactivation potential of the bait is 10^{-4}. The true positives will be distinguished from the bait background in the next step by the galactose dependence of their Leu^+ and $lacZ^+$ phenotypes.

SUPPORT PROTOCOL 1

4

PREPARATION OF PROTEIN EXTRACTS FOR IMMUNOBLOT ANALYSIS

Materials *(see* APPENDIX 1 *for items with ✓)*

Master plates with pBait-containing positive and control yeast on Glu/CM −Ura, −His dropout medium (see Basic Protocol 1, step 4)
Glu/CM −Ura, −His dropout liquid medium: CM dropout plates −Ura, −His (*APPENDIX 1*) supplemented with 2% glucose
✓ 2× Laemmli sample buffer
Antibody to fusion domain or LexA: monoclonal antibody to LexA (Clontech, Invitrogen) or polyclonal antibody to LexA (available by request from R. Brent or E. Golemis)

30°C incubator
100°C water bath

1. From the master plates, start a 5-ml culture in Glu/CM −Ura, −His liquid medium for each bait being tested and for a positive control for protein expression (i.e., RFHMI or SH17-4). Incubate overnight at 30°C.

For each construct assayed, it is a good idea to grow colonies from at least two primary transformants, as levels of bait expression are sometimes heterogenous.

2. From each overnight culture, start a fresh 5-ml culture in Glu/CM −Ura, −His at OD_{600} = ~0.15. Incubate again at 30°C.

3. When the culture has reached OD_{600} = 0.45 to 0.7 (~4 to 6 hr), remove 1.5 ml to a microcentrifuge tube. Microcentrifuge cells 3 min at 13,000 × *g*, room temperature. If a pellet is not visible after 3 min, microcentrifuge another 3 min. When the pellet is visible, remove the supernatant.

4. Working rapidly, add 50 μl of 2× Laemmli sample buffer to the visible pellet in the tube, vortex, and place the tube on dry ice or freeze at −80°C.

5. Transfer frozen sample directly to a boiling water bath or a PCR machine set to cycle at 100°C. Boil 5 min, then microcentrifuge 5 sec at maximum speed to pellet large cellular debris.

6. Perform SDS-PAGE (*CPMB UNIT 10.2A*) using 20 to 50 μl sample per lane. To detect the protein, immunoblot and analyze (*CPMB UNIT 10.8*) using antibody to the fusion domain or LexA.

SUPPORT PROTOCOL 2

PREPARATION OF SHEARED SALMON SPERM CARRIER DNA

Materials *(see APPENDIX 1 for items with ✓)*

High-quality salmon sperm DNA (e.g., sodium salt from salmon testes, Sigma or Boehringer Mannheim), desiccated
✓ TE buffer, pH 7.5, sterile
TE-saturated buffered phenol
1:1 (v/v) buffered phenol/chloroform
Chloroform
3 M sodium acetate, pH 5.2
100% and 70% ethanol, ice cold

Magnetic stirring apparatus and stir-bar, 4°C
Sonicator with probe
50-ml conical centrifuge tube
High-speed centrifuge and appropriate tube
100°C and ice-water baths

1. Dissolve desiccated high-quality salmon sperm DNA in TE buffer, pH 7.5, at a concentration of 5 to 10 mg/ml, by pipetting up and down in a 10-ml glass pipet. Place in a beaker with a stir-bar and stir overnight at 4°C to obtain a homogenous viscous solution.

   It is important to use high-quality salmon sperm DNA. Sigma Type III sodium salt from salmon testes has worked well, as has a comparable grade from Boehringer Mannheim. Generally it is convenient to prepare 20- to 40-ml batches at a time.

2. Shear the DNA (to 2 to 15 kb, ∼7 kb avg.) by sonicating briefly using a large probe inserted into the beaker.

   The goal of this step is to generate sheared salmon sperm DNA (sssDNA) with an average size of 7 kb, but ranging from 2 to 15 kb.

3. Once DNA of the appropriate size range has been obtained, extract the sssDNA solution with an equal volume of TE-saturated buffered phenol in a 50-ml conical tube, shaking vigorously to mix.

4. Centrifuge 5 to 10 min at 3000 × *g*, room temperature, or until clear separation of phases is obtained. Transfer the upper phase, containing the DNA, to a clean tube.

5. Repeat extraction using 1:1 (v/v) buffered phenol/chloroform, then chloroform alone. Transfer the DNA into a tube suitable for high-speed centrifugation.

6. Precipitate the DNA by adding vol of 3 M sodium acetate and 2.5 vol of ice-cold 100% ethanol. Mix by inversion. Centrifuge 15 min at ∼12,000 × *g*, room temperature.

7. Wash the pellet with 70% ethanol. Briefly dry either by air drying, or by covering one end of the tube with Parafilm with a few holes poked in and placing the tube under vacuum. Resuspend the DNA in sterile TE buffer at 5 to 10 mg/ml.

   Do not overdry the pellet or it will be very difficult to resuspend.

8. Denature the DNA by boiling 20 min in a 100°C water bath. Then immediately transfer the tube to an ice-water bath.

9. Place aliquots of the DNA in microcentrifuge tubes and store frozen at −20°C. Thaw as needed.

   DNA should be boiled again briefly (5 min) immediately before addition to transformations.

   Before using a new batch of sssDNA in a large-scale library transformation, it is a good idea to perform a small-scale transformation using suitable plasmids to determine the transformation efficiency. Optimally, use of sssDNA prepared in the manner described will yield transformation frequencies of >10^5 colonies/ig input plasmid DNA.

Internet Resources

http://www.clontech.com
http://www.invitrogen.com
http://www.origene.com
Commercial sources for basic plasmids, strains, and libraries
http://proteome.wayne.edu/finlabindex.HTML
Source of two-hybrid information, protocols, and links.

References: Finley and Brent,1994; Gyuris et al., 1993

Contributors: Erica A. Golemis, Ilya Serebriiskii, Russell L. Finley, Jr., Mikhail G. Kolonin, Jeno Gyuris, and Roger Brent

4

UNIT 4.3

cDNA Library Construction from Single Cells

This unit describes procedures for producing cDNA libraries from individual cells and comparing the products between cells, as outlined in Fig. 4.3.1.

BASIC PROTOCOL 1

PRODUCTION AND AMPLIFICATION OF cDNA FROM SINGLE CELLS

NOTE: To prevent large changes in the proportions of specific cDNA transcripts between the starting material (mRNA) and the final cDNA library, the reverse transcription reaction is truncated by including a limiting amount of DTT within the reaction and shortening the reaction time from the usual >1 hr to 15 min. Small cDNA products of a similar length result using this procedure. Thus, any variation in the amplification step due to cDNA size is minimized (i.e., small cDNAs are amplified at a greater rate than longer cDNA transcripts) and the proportion of specific messages is maintained relative to the starting cDNA. This rationale can be verified by Southern analysis of the cDNA products.

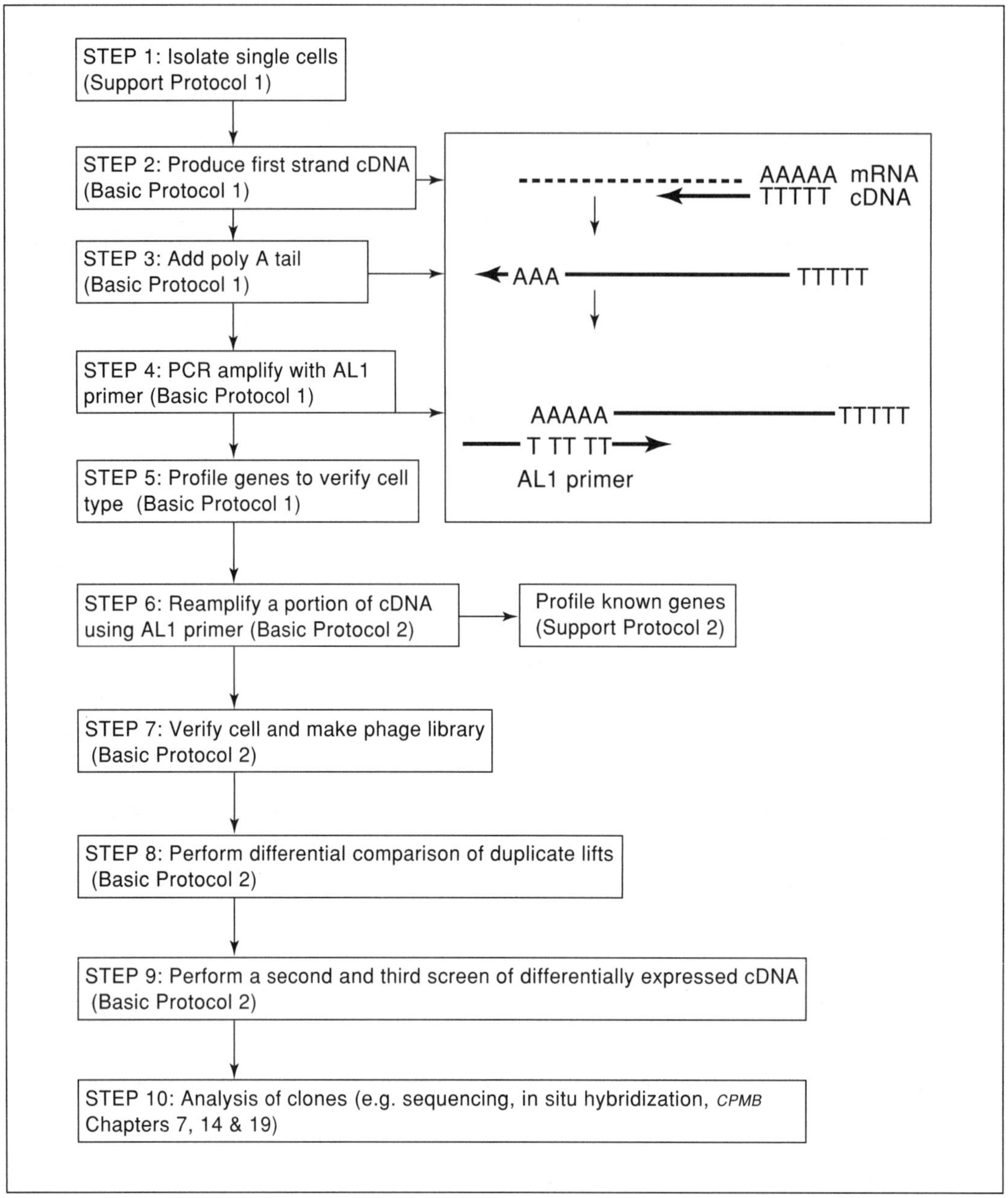

Figure 4.3.1 Flowchart for producing cDNA libraries from individual cells using the protocols provided in this unit.

NOTE: The first round of amplification takes ~5.5 hr; starting early is therefore recommended.

Materials *(see* APPENDIX 1 *for items with ✓)*

Isolated cells in cDNA lysis buffer (see Support Protocol 1)
200 U/μl MMLV reverse transcriptase with 5× buffer (Life Technologies)
1 to 5 U/μl AMV reverse transcriptase (Life Technologies)
10× PCR buffer II (PE Biosystems)
25 mM $MgCl_2$ (Perkin Elmer)
20 mg/ml BSA, molecular grade

100 mM solutions of each dATP, dGTP, dCTP, dTTP
5% (v/v) Triton X-100, RNase- and DNase-free (e.g., Sigma)
PAGE-purified AL1 primer: ATT GGA TCC AGG CCG CTC TGG ACA AAA TAT GAA TCC $(T)_{24}$ (Life Technologies)
AmpliTaq DNA polymerase (PE Biosystems)
5× terminal transferase buffer (Life Technologies)
25 U/μl terminal transferase (Boehringer Mannheim)
Mineral oil, molecular biology grade
✓ TE buffer, pH 7.5
Tris-buffered phenol
24:1 (v/v) chloroform/isoamyl alcohol
100% ethanol
✓ 10 mM Tris·Cl, pH 7.0
1.5% (w/v) agarose gel

37°C and 65°C water baths
0.2-ml thin-walled reaction tubes and adapters if necessary
DNA thermal cycler (e.g., Perkin Elmer or equivalent with extended cycle time feature)
1.5-ml microcentrifuge tubes

1. Lyse the isolated cells in the cDNA lysis buffer in a 65°C water bath for 1 min.

   A negative control reaction, in which a cell has not been added to the lysis buffer, and a positive control, which contains 20 pg of purified total RNA (obtained from any tissue source or commercial supplier), should both be included.

2. Remove tubes from the 65°C water bath and cool for 1 to 2 min at room temperature (22°C), while the oligo$(dT)_{19-24}$ anneals to the RNA poly-A tail. Place tubes on ice for 1 min, then microcentrifuge for 2 min at maximum speed, 4°C. Add 0.5 μl of a 1:1 (vol:vol) mix of AMV and MMLV reverse transcriptases, mix, and incubate for exactly 15 min at 37°C. Incubate 10 min at 65°C to inactivate the reverse transcriptases. Place inactivated reverse-transcription reaction tubes on ice.

3. Prepare the PCR reaction buffer on ice:

   10 μl 10× PCR buffer II
   10 μl 25 mM $MgCl_2$
   0.5 μl 20 mg/ml BSA
   1 μl each 100 mM dNTP
   1 μl 5% Triton X-100
   5 μl 1 μg/μl AL1 primer (0.05 μg/ml final)
   67.5 μl H_2O
   2 μl AmpliTaq DNA polymerase.

4. Prepare the 2× tailing buffer:

   800 μl of 5× terminal transferase buffer
   30 μl of 100 mM dATP
   1.17 μl non-DEPC-treated water.

5. On ice, combine 4.5 μl of 2× tailing buffer and 0.5 μl of 25 U/μl terminal transferase for each reaction. Add 5 μl of this tailing mixture to each thin-walled tube, mix, and incubate by placing tube in a 37°C water bath for 15 min. Inactivate the terminal transferase by placing tubes in a 65°C water bath for 10 min. Microcentrifuge 2 min at maximum speed, 4°C.

6. Add 90 μl of ice-cold PCR reaction buffer (from step 3) to each tube and layer 1 or 2 drops of mineral oil over the reaction.

7. Perform 25 cycles of PCR, programming the thermal cycler as follows:

1 min	94°C	(denaturation)
2 min	42°C	(annealing)
6 min	72°C	(extension).

Add 10 sec to the extension time at each cycle.

8. Immediately after the completion of the first 25 cycles, add 1 μl of AmpliTaq DNA polymerase directly into each tube and perform 25 additional cycles with the same program as in step 7 without the addition of 10 sec to the extension step after each cycle.

9. Transfer the samples from the thin-walled PCR tubes to 1.5-ml microcentrifuge tubes and add 100 μl of TE buffer. Extract protein from the completed PCR reaction by adding 100 μl Tris-buffered phenol per tube and vortexing vigorously for 1 min. Microcentrifuge samples for 5 min at maximum speed, room temperature, and carefully transfer the upper aqueous layer to new 1.5-ml microcentrifuge tubes. Repeat the Tris–buffered-phenol extraction.

10. To remove residual phenol, add 50 μl of 24:1 chloroform/isoamyl alcohol to each tube and shake or vortex vigorously for 1 min. Microcentrifuge samples for 5 min at maximum speed, room temperature, remove upper aqueous layer and transfer to new 1.5-ml microcentrifuge tubes, avoiding the interface. Repeat chloroform/isoamyl alcohol extraction.

11. Add 2.5 vol of 100% ethanol, mix well (critical), chill samples at −20°C for >1 hr before centrifugation, and microcentrifuge for 30 min at maximum speed, room temperature. Decant solution and let air dry or dry under vacuum. Resuspend the samples in 100 μl of sterile 10 mM Tris·Cl, pH 7.0. Store 25- to 50-μl aliquots at −80°C for up to 1 year.

12. Fractionate 5 μl of the cDNA sample from step 11 on a 1.5% agarose gel (*CPMB UNIT 2.5A*).

13. Transfer the cDNA onto nylon membrane and perform Southern analysis (*CPMB UNITS 2.9A & 2.10*).

    Probes for hybridization or primers for PCR should be made against the 3′-end of the genes due to the fact that this technique produces a truncated cDNA starting at the 3′-end of the mRNA transcript. Alternatively, the PCR described in Support Protocol 2 can be used, although all controls must be performed.

14. Identify several cells that yield a positive signal to the specific gene of interest and the housekeeping gene(s), and verify that the no-cell control produces no hybridization signal. Proceed to Basic Protocol 2 or Support Protocol 2 for further analysis.

SUPPORT PROTOCOL 1

ISOLATION OF INDIVIDUAL CELLS FROM ACUTE TISSUE AND TISSUE EXPLANTS GROWN IN VITRO

NOTE: After removal of the tissue from the organism, dissections can be completed on ice, but the specifics of individual protocols will have to be determined empirically.

NOTE: Tissue is fragmented before enzymatic lysis, in order to decrease the time for obtaining individual cells by enzyme treatment, but the fragmentation itself should not induce cell lysis.

Materials *(see APPENDIX 1 for items with ✓)*

Tissue sample
✓ CMF-PBS or CMF-HBSS
✓ 0.75 mM EDTA

CMF-PBS or CMF-HBSS containing 0.025% trypsin/0.75 mM EDTA, 37°C
5× MMLV buffer (e.g., Life Technologies)
✓ DEPC-treated, sterile, deionized, RNase-free H_2O
NP-40 (e.g., USB)
RNase inhibitors: Prime RNase inhibitor (Eppendorf) and RNA guard (Amersham Pharmacia Biotech)
✓ Stock primer mix

Dissecting microscope
Sterile 60- or 100-mm polystyrene or glass petri dish
Sterile scalpel or razor blade
5-ml plastic pipets
37°C incubator
0.5-ml thin-walled tubes, sterile, RNase-free
15-ml polypropylene tubes
Inverted microscope and micromanipulator (e.g., Sutter Instruments)
0.5 (i.d.) × 130-mm glass capillary tubing
Micropipet puller (e.g., Sutter Instruments)
Syringe with attached tubing fitted for the glass capillary tube *or* microinjector (e.g., Nanoject II, Drummond Scientific)
Sterile 1.5-ml polypropylene tubes

1. Prepare several glass microcapillary pipets from conventional 130-mm capillary tubing using a micropipet puller (settings determined empirically), or pull them by hand over a small flame with an overall taper length of ~1 cm. Touch the end of the tip with a sterile pointed metal object (e.g., fine-tipped forceps), to break off the tip of the micropipet and create a beveled opening slightly larger than a cell diameter (10- to 20-μm). Discard those with smaller openings.

2. Dissect tissue in CMF-PBS or CMF-HBSS. Fragment the tissue in a 60-mm petri dish under a dissecting microscope into pieces as small as possible, using a sterile scalpel or razor blade.

3. Remove all buffer with a pipet. Remove potential residual RNA by two or three repeated cycles of rinsing, in CMF-PBS, followed by centrifugation. Then add 2 ml CMF-PBS or CMF-HBSS containing 0.025% trypsin/0.75 mM EDTA prewarmed at 37°C. Gently pipet up and down two or three times with a 5-ml plastic pipet to mix tissue and trypsin.

4. Transfer the tissue to a fresh 60-mm petri dish and place in a 37°C incubator for 1 to 15 min (until several hundred cells are in suspension). Pipet very gently, as in step 3, every 5 min, if required. If <100 single cells are observed after 15 minutes, proceed.

5. While the cells are being dissociated, prepare the cDNA-lysis buffer (4 μl/cell) on ice:

   20 μl 5× MMLV buffer
   75.5 μl RNase-free H_2O
   0.5 μl NP-40
   1 μl Prime RNase inhibitor
   1 μl RNA guard
   2 μl freshly made 1:24 dilution of the stock primer mix in water.

   Place 4 μl of this buffer in each 0.5-ml, thin-walled PCR reaction tube. Keep on wet ice.

6. Examine the tissue for separated individual cells. If individual cells are not observed, remove solution and add fresh trypsin solution. Incubate 10 min at 37°C, then check again for dissociated cells.

4

7. Transfer cells, tissue, and trypsin solution into 10 ml of prewarmed CMF-HBSS or CMF-PBS in 15-ml polypropylene tubes, and gently invert 2 times. Centrifuge 10 min at 800 × *g*, room temperature. Remove the supernatant with a pipet.

8. Add 5 ml CMF-PBS and triturate very gently by pipetting up and down four to five times, slowly and carefully. Allow clumps of tissue to settle and transfer a small volume (~1 ml) of the top layer (suspended cells) to a 60-mm plate/petri dish. Place on an inverted microscope and allow cells to settle.

9. Place a microcapillary pipet (from step 1) in the holder of a micromanipulator. Attach a 3-ml syringe or microinjector to the micromanipulator apparatus holding the microcapillary pipet. Position the apparatus so that both the cell and microcapillary can be visualized together under the inverted microscope.

10. Before approaching the cell, very gently blow out residual buffer from within the microcapillary (a small amount of buffer per cell is critical for successful RT-PCR). Advance the open end of the microcapillary pipet so the cell is partially within the opening. Gently apply vacuum with a syringe or other suction-producing device until the cell just enters the opening and the total volume is ~0.2 to 0.5 µl.

11. Once the cell is in the pipet, bring the tip of the pipet out of the PBS solution to a height of ~4 to 7 cm. Place the tip into the 4 µl of ice-cold lysis buffer in the PCR reaction tube, and carefully but quickly inject the microcapillary contents into the lysis buffer.

12. Microcentrifuge the tube immediately for 30 sec at maximum speed to make sure the cell is completely within the cDNA lysis buffer. Place the tube on dry ice up to several hours.

4

BASIC PROTOCOL 2

PHAGE LIBRARY CONSTRUCTION AND DIFFERENTIAL SCREENING

Differential screens between two cells or groups of cells can utilize phage libraries or bacterial plasmid libraries. Bacterial cDNA libraries are used due to the ease in plating the cells, but the technical advantage to using phage libraries (*UNIT 5.9*) is that, when taking multiple lifts, the amount of cDNA attached to the membrane is more homogeneous from lift to lift.

Materials *(see APPENDIX 1 for items with ✓)*

Cell-amplified cDNA (see Basic Protocol 1, step 12)
PCR buffer (see Basic Protocol 1, step 3)
Tris-buffered phenol
24:1 (v/v) chloroform/isoamyl alcohol
✓ 10 mM, 150 mM, and 1 M Tris·Cl, pH 7.0, pH 8.0 and pH 7.5, respectively
*Eco*RI restriction enzyme and appropriate buffer (*CPNS APPENDIX 1M*)
Low-melting agarose (melts at 65°C at 1.5%; e.g., Seaplaque GTG or NuSieve GTG agarose)
0.5 µg/ml ethidium bromide
PCR prep DNA purification resin (e.g., Promega) or equivalent commercially available DNA gel purification method
100% ethanol
Phage vector DNA (e.g., Lambda ZapII arms, *Eco*RI cut and dephosphorylated; Stratagene)
200 U/µl high-concentration ligase and 10× ligase buffer
High-efficiency phage packaging extract (e.g., Gigapack II gold extract; Stratagene) containing host bacteria (e.g., XL-1-Blue MRF′; Stratagene)

10× *Taq* PCR buffer (150 mM Tris·Cl, pH 8.0, 500 mM KCl)
25 mM $MgCl_2$ in dH_2O
100 mM solutions of each of dATP, dTTP, dCTP, dGTP
5 μg/ml T7 primer (5′-TAATACGACTCACTATAGGG-3′) in TE buffer (Life Technologies)
5 μg/ml T3 primer (5′-TAACCCTCACTAAAGGGA-3′) in TE buffer (Life Technologies)
AmpliTaq DNA polymerase (PE Biosystems)
0.5 M NaOH
0.5× sodium phosphate buffer containing 1% (w/v) BSA and 4% (w/v) SDS
✓ APH solution
1-5 × 10^7 cpm/ml radiolabeled probe for specific gene or control genes such as tubulin, actin, GAPDH (e.g., labeled as described in *CPMB UNIT 3.5*)
✓ 0.5% SDS in 0.5× SSC
10× PCR buffer II (PE Biosystems)
20 mg/ml BSA
[α-^{32}P]dCTP
5% Triton X-100
PAGE-purified AL1 primer: ATT GGA TCC AGG CCG CTC TGG ACA AAA TAT GAA TCC $(T)_{24}$ (Life Technologies)

Thermal cycler
0.2-ml thin-walled PCR tubes
37°C and 50°C water baths
Razor blades
1.5-ml microcentrifuge tubes
137-mm nylon membranes (e.g., Hybond N+, NEN Genescreen)
20-G needle
Plastic wrap
Paper towels or Whatman paper
65°C incubator
X-ray film

1. Add 10 μl of cell-amplified cDNA from Basic Protocol 1 step 12 to 90 μl of the PCR buffer described in Basic Protocol 1, step 3.

2. Perform one round of PCR (to fill in any single-strand areas) with the following parameters:

5 min	94°C	(denaturation)
5 min	42°C	(annealing)
30 min	72°C	(extension).

3. Perform Tris-buffered phenol and 24:1 chloroform/isoamyl alcohol extractions as described in Basic Protocol 1, steps 10 to 12. Resuspend the samples in 100 μl sterile 10 mM Tris·Cl, pH 7.0.

   cDNAs from multiple cells can be combined at this step (in sterile water) to create a more representative sample of heterogeneous gene expression in a particular cell type. Thus, samples can be dried or precipitated to reduce the total volume in which the cDNA is dissolved.

4. Digest the sample with 40 to 90 U *Eco*RI for 2 to 3 hr at 37°C in a total volume of 50 to 100 μl (*CPNS APPENDIX 1M*).

5. Electrophorese the digested samples through a 2% low-melting agarose gel (*CPMB UNIT 2.5A*) and visualize bands by staining gel with 0.5 μg/ml ethidium bromide. Using a new razor blade, isolate the region of the gel between 300 and 1000 bp. Mince the gel slice into cubes, <0.5 mm on a side. Place the gel material in a 1.5-ml microcentrifuge tube.

6. Purify the DNA from the gel material, with any commercial DNA gel purification procedure *or* according to the following directions:

   Add ~300 μl TE to every 300 μl of gel material. Vortex briefly and place at 4°C for 24 hr.

   Clean the DNA by two phenol and two chloroform/isoamyl alcohol extractions as described in Basic Protocol 1, steps 10 and 11.

7. Precipitate DNA by adding 2.5 vol of 100% ethanol, mix well, and microcentrifuge for 30 min at maximum speed, producing a very small pellet. Decant solution carefully. Resuspend the pellet in 5 μl of water and pipet to completely dissolve the pellet. If the pellet had been quite large (indicating residual gel material), remove the residual gel material by centrifuging the tube at high speed for 15 min. Remove the supernatant (containing the DNA, ~100–150 ng/5 μl) and transfer to a new tube.

8. Mix the following on ice to ligate the purified *Eco*RI-cut cDNA into phage arms:

   1 μl *Eco*RI-digested and dephosphorylated phage arms
   3 μl (~100 ng) *Eco*RI-cut cell cDNA
   0.5 μl 10× ligase buffer
   0.2 μl H_2O
   0.3 μl 200 U/μl high-concentration ligase.

   Incubate the ligation reaction overnight at 12°C or over the weekend at 4°C.

9. Package 1 μl of the ligation reaction according to the phage packaging extract's manufacturer's instructions.

10. Titer the phage (use Stratagene's recommended protocol and calculations *or* see UNIT 5.9).

    If using the Stratagene protocol, do not use the X-gal colorometeric assay to determine the presence of inserts.

    From 100 ng of cDNA, 10^6 to 10^7 independent recombinants should contain clones of very rare mRNA transcripts (see UNITS 5.8 and 5.9).

11. Determine the ligation efficiency of PCR inserts into the phage arms by amplifying the inserts from 10 to 20 plaques. Using a 20- to 200-μl pipet tip, stab the phage plaque (~0.5 μl) and place the tip in a 1.5-ml microcentrifuge tube containing the following:

    25 μl dH_2O
    4 μl 10× *Taq* PCR buffer
    8 μl 25 mM $MgCl_2$
    0.4 μl each 100 mM solution of dNTP
    0.2 μl 100 μM T7 primer
    0.2 μl 100 mM T3 primer
    0.5 ml AmpliTaq DNA polymerase
    Total volume 40 μl.

12. Perform PCR using the following parameters:

1 cycle:	5 min	93°C	(denaturation)
	5 min	48°C	(annealing)
	3 min	72°C	(extension)
30 cycles:	1 min	93°C	(denaturation)
	1 min	48°C	(annealing)
	2 min	72°C	(extension)
Final step:	5 min	72°C	(final extension).

13. Fractionate the PCR reactions on a 1% agarose gel (*CPMB UNIT 2.5A*). If 80% to 90% of the phages have inserts, then package the remainder of the ligation reaction (step 9). Amplify part (<75%) of the obtained library.

14. Plate 1000 to 5000 plaques at a density of 500 pfu/82-mm plate (e.g., use Stratagene's protocols, with the suggested modification of using 8 ml of top agar for each 150-mm plate).

15. Place a 137-mm nylon membrane on the plate for 1.5 min and mark orientation by poking a 20-G needle through the membrane into the agar several times near the edge of the plate. Place the membrane in a 0.5-ml pool of 0.5 M NaOH on plastic wrap for 2 min. Make sure that the NaOH solution soaks the entire membrane but that it does not pool to a visible amount since this will smear the DNA of the plaque. Transfer the membrane to a 0.5-ml pool of 1 M Tris·Cl, pH 7.5, for 2 min and then a second pool of 0.5 ml of 1 M Tris·Cl, pH 7.5, for 2 min. If performing duplicate lifts, place a second membrane on the plate for 6 min, mark the orientation and perform the lysis and neutralization steps.

16. Dry the membranes briefly on paper towels or Whatman paper. Crosslink the DNA to the moist membranes with a UV crosslinker (120 mJ/cm2). Air dry the membrane until dry to the touch. Soak for 5 to 10 min in 100 mM Tris·Cl, pH 7.5, and remove residual agar or phage material by rubbing gently with a gloved finger. Prehybridize the membrane for 2 hr at 65°C in APH solution (~10 ml/150 cm^2).

17. Add 1-5 $\times$ 10^7 cpm/ml labeled probe directly to prehybridization buffer and hybridize overnight at 65°C.

18. Wash the membranes by adding 0.5% SDS in 0.5× SSC (20 ml/50 cm^2). Wash three times, 15 to 20 min each wash, at 65°C. Wrap membranes in plastic wrap and expose to film.

    Tubulin should be present at 0.2% to 4.0% of the total mRNA content within the cell (calculation based on the number of plaques that hybridized the tubulin probe/total number of plaques plated).

19. Plate the library (Stratagene's protocol or *UNIT 5.9*) at a density of 8,000 to 10,000 pfu/150-mm plate and incubate the plates at 37°C until the plaques are 2 mm in diameter.

20. Perform filter lifts on duplicate filters as described in steps 15 and 16.

21. Assemble the following PCR mixture on ice:

    1 µl original cell cDNA
    5 µl 10× PCR buffer II
    5 µl 25 mM $MgCl_2$
    0.25 µl 20 mg/ml BSA
    0.5 µl each 0.2 µM dGTP, dATP, dTTP
    5 µl 20 mCi/ml [α-^{32}P]dCTP (2 mCi/ml final)
    0.5 µl 5% Triton X-100
    2.5 µg AL1 primer
    1 µl AmpliTaq DNA polymerase.

    Bring to a total volume of 50 µl with deionized water. Reamplify for 10 cycles as described in Basic Protocol 1, step 7 without the 10-sec addition to each extension step.

22. Add 2.5 vol of 100% ethanol to each PCR reaction, mix well, microcentrifuge for 30 min at maximum speed, 4°C, air dry, resuspend in 300 µl water, and determine the specific activity.

23. Prehybridize and hybridize the membranes as in steps 16 and 17. Hybridize at least 16 hr with a probe concentration of 1–5 $\times$ 10^7 cpm/ml. Use cDNA from cells that have different

4

gene expression than the cell from which the phage library is made (cell B) for the first lift and probes from the cell from which the library was made (cell A) for the second lift.

24. Wash the membranes by adding 0.5% SDS in 0.5× SSC (20 ml/50 cm^2). Perform wash three times, 15 to 20 min each, at 65°C. Wrap membranes in plastic wrap and expose to film.

25. Carefully compare the film from the lifts and pick plaques that exhibit a positive signal from cell A and not cell B. If differential expression is not clear, pick the plaque anyway. Amplify the insert of all the plaques that were picked using the T3 and T7 primers from step 11. Electrophorese these amplified inserts on a 1.5% agarose gel (*CPMB UNIT 2.5A*) and perform a Southern transfer (*CPMB UNIT 2.9A*) of the PCR products.

26. Hybridize each membrane with a probe from cell B and strip the membrane and hybridize with probe from cell A.

27. Sequence cDNA (*CPMB CHAPTER 7*) that was confirmed to be differentially expressed by Southern hybridization of the PCR products, and perform database searches (*CPMB CHAPTER 19*) as well as in situ hybridizations (*UNIT 1.3*) to analyze cellular localization of the mRNA.

4

SUPPORT PROTOCOL 2

GENE EXPRESSION PROFILING USING PCR AMPLIFICATION OF TRUNCATED cDNA LIBRARIES

Both PCR and Southern hybridization produce the same results.

Materials *(see APPENDIX 1 for items with ✓)*

Amplified cDNA from a single cell or pooled cellular cDNAs (see Basic Protocol 1)
10× PCR buffer (e.g., *Taq* Gold PCR buffer)
25 mM $MgCl_2$
✓ TE buffer
100 mM each dNTP
100 μM PCR primer X dissolved in water or TE buffer
100 μM PCR primer Y dissolved in water or TE buffer
5 U/μl DNA polymerase (e.g., AmpliTaq Gold; PE Biosystems)
Gel loading dye
0.5 μg/ml ethidium bromide

Thermal cycler
0.5-ml thin-walled PCR tubes

1. For each amplification reaction, mix the following on ice:

    0.5 to 2 μl amplified cDNA from a single cell or pooled cellular cDNAs
    4 μl 10× PCR buffer
    3 to 8 μl 25 mM $MgCl_2$
    0.4 μl each dNTP
    0.2 μl 100 μM primer X
    0.2 μl 100 μM primer Y
    1.0 μl 5 U/μl AmpliTaq Gold DNA polymerase
    Deionized water to a total volume of 40 μl.

    As a starting point, try 2 μl cDNA and 8 μl $MgCl_2$, a $MgCl_2$ concentration that has worked with multiple primer sets.

The primersd must be complimentary to the last 500 bp at the 3′ end of the transcript for the gene of interest.

2. Mix by briefly vortexing reaction, microcentrifuge for 1 min at maximum speed, room temperature. Add a drop of mineral oil if necessary and place in thermal cycler.
3. Program the thermal cycler as follows:

1 cycle:	10 min	94°C	(initial denaturation)
40 cycles:	45 sec	94°C	(denaturation)
	1 min	55°C	(annealing)
	2 min	72°C	(extension)
	10 min	72°C	(final extension).

The optimum annealing temperature for each primer set should be determined empirically.

4. Add loading dye and electrophorese 15 to 20 μl of PCR reaction on a 1% agarose gel with molecular weight markers, stain with ethidium bromide, and visualize under UV light (*CPMB UNIT 2.5A*).

Reference: Dulac and Axel, 1995

Contributor: Phillip R. Kramer

4

UNIT 4.4

Overview of Neural Gene Expression

Genetically engineered mammalian cell lines provide a system for studying the biochemical, pharmacological, and physiological functions of neural gene products in homogeneous populations of cells expressing an exogenous cDNA. The cell lines chosen for neural cDNA expression normally do not express the gene for the transfected cDNA, so the investigator can analyze point mutations and deletions in a milieu uncluttered by expression of the wild-type gene. Use of transient, high-expression vectors such as vaccinia virus or baculovirus permits production of large amounts of gene product for purification and biochemical characterization as well as for development of antibodies.

The expression of neural genes in genetically engineered mammalian cells requires consideration of the following: (1) What type of expression vector should be used? (2) What type of cell should be used? (3) How will the recombinant vector be introduced into the cells? (4) Should the analyses be done on transiently transfected/infected populations of cells, or on clonal stable transfectants? (5) Which promoter should be used?

CHOICE OF EXPRESSION VECTOR

The expression vector is chosen on the basis of the cell type to be used, the type of expression desired (transient or stable), and the type of experiment to be performed (e.g., purification of the recombinant protein or functional assays). All other factors being equal, a vector with convenient restriction endonuclease sites for cloning the cDNA is preferable to one without such sites. Without restriction sites cDNA cohesive ends would need to be altered by adding adaptors or by filling in and ligating on synthetic restriction site linkers.

For maximally efficient transient expression assays of cloned neural genes, an SV40-based vector can be transfected into cells such as COS cells, which have an integrated copy of the early region of SV40 and which express SV40 T antigen. These cells support the replication of vectors that carry the SV40 origin of replication. The resultant multiple copies of recombinant vector in the cell allow for high levels of transient expression of the recombinant gene, but multiple vector copies are ultimately deleterious to the survival of the transfected cells. Alternatively, Epstein Barr virus (EBV)–based vectors are maintained episomally in primate and canine lines and give high levels of expression of recombinant proteins. Expression systems that direct the transient synthesis of large amounts of recombinant protein for purification or for other biochemical studies are also available. Two of the most widely used expression systems are vaccinia virus (Fuerst et al., 1986; *CPMB UNITS 16.15–16.19*) and baculovirus (Luckow and Summers, 1988; *CPMB UNITS 16.9–16.11*).

Stable expression in other types of cells can be achieved with a wide variety of vectors. Most useful are those vectors that include a ColE1 origin of replication for growth in *Escherichia coli* and the ampicillin resistance gene, encoding β-lactamase, for selection in *E. coli*. With respect to expression in mammalian cells, the vector must contain a promoter (see Choice of Promoter), transcription-termination and RNA-processing signals, and a selectable marker. RNA-processing signals, which are necessary for stability of the transgene mRNA, are most commonly taken from SV40, although polyadenylation and transcription-termination signal sequences from the bovine growth hormone gene are gaining widespread use. Both are effective, but because there may be subtle cell line-specific differences in their efficiencies, it is useful to test both in a given cell line. The availability of multiple selectable markers makes it possible to coexpress two or more cDNAs in a cell line. One of the most popular selectable markers is resistance to G418, conferred by the *neo* gene; Zeocin (Invitrogen) and blastocidin (ICN Biomedicals) resistance genes also offer flexibility in that they confer resistance over a reasonably wide range of drug concentrations. Vectors carrying genes for resistance to hygromycin, puromycin, and bleomycin are also useful, but the window of resistance conferred by these genes is narrow and must be determined empirically for each cell line. As an alternative to using two different selectable markers, given the tendency for multiple plasmids to enter a single cell together, more than one cDNA can be transfected with a single selectable marker.

Ideally, neural genes should be expressed in primary neuronal cultures. However, neurons do not yield easily to genetic intervention. The terminally differentiated state of most neurons precludes the use of vectors, such as conventional retroviruses, that are dependent on cell replication for stable maintenance in the cell. Vectors for gene delivery into neurons must be able to infect and persist stably in postmitotic cells. A number of such gene delivery systems have been developed over the last decade (reviewed in Neve and Geller, 1996). These include herpes simplex virus (HSV), adenovirus, adeno-associated virus, and retrovirus vectors, as well as DNA-liposome complexes. Table 4.4.1 compares the advantages and disadvantages of each vector type.

CHOICE OF CELL TYPE

The cell type into which neural genes will be introduced depends very much on the nature of the assay for the genetically engineered cell. If the goal is simply to produce large amounts of protein for biochemical characterization or for production of antibodies, cells that are appropriate for infection by baculovirus or vaccinia virus recombinants can be used. If functional assays will be performed, nonneuronal or neuronal cell lines or primary neuronal cultures can be used. The advantage of using nonneuronal cell lines to study the function of a neural gene is that the function of the exogenous gene can be analyzed on a null background, in the absence of the endogenous gene product. However, it is prudent to check nonneuronal lines for expression, because neuronal genes are occasionally expressed ectopically in nonneuronal cell lines. Valuable information can be gained from pharmacologic and electrophysiologic analysis of neural receptors, morphologic analysis of neural cytoskeletal proteins, and biochemical analysis of signal transduction proteins in nonneuronal cell lines. Nevertheless, these studies

Table 4.4.1 Comparison of Vectors for Delivery of Genes into Primary Neurons

Vector	Advantages	Disadvantages
Herpesvirus	Broad host and cell-type range Episomal (no possibility of insertional activation of host genes) Easy genetic manipulation of amplicon vectors Can accommodate up to 15 kb of foreign DNA	Occasional cytotoxicity Lack of persistence of expression Long-term safety unknown
Adenovirus	Broad mammalian host and cell-type range Episomal (no possibility of insertional activation of host genes) Growth to high titers ($\sim 10^{10}$/ml) High level of expression of foreign genes	Elicits host immune response Stability of expression unknown Can accommodate only ~7.5 kb of foreign DNA Genetic manipulation is unwieldy Long-term safety unknown
Adeno-associated virus	Broad mammalian host and cell type range Nonpathogenic Helper virus–free stocks possible	Can accommodate only 4.7 kb of foreign DNA Lack of persistence of expression Long-term safety unknown
Retrovirus	Useful for cell marking/lineage analysis and ablation of brain tumors	Low efficiency of infection of nondividing cells Variable expression of foreign genes Maximum titres $\sim 10^6$/ml Can accommodate only 6-8 kb of foreign DNA
DNA-liposome complexes	Biologically safe Easy to manipulate genetically: unmodified plasmid vectors without eukaryotic replicans can be used Virtually no size limitations Probably extrachromosomal	Stability of expression unknown Cytotoxicity of cationic lipids Low efficiency of transfection

are inherently limited by the fact that the functions of the neuronal genes are being investigated in a nonneuronal milieu.

As an alternative, multiple well-characterized neuronal cell lines have been used for genetic intervention. A primary defining feature of neurons, their terminally differentiated state, is not recapitulated in neuronal lines, but often can be generated conditionally with differentiating agents such as nerve growth factor or retinoic acid. Ideally, information obtained about neural genes using cell lines should be confirmed in primary neuronal cultures.

INTRODUCING RECOMBINANT VECTORS INTO CELLS

The most commonly used techniques for introducing neural genes into mammalian cells are calcium phosphate transfection, DEAE-dextran transfection, electroporation, liposome-mediated transfection, and infection. The parameters for DNA transfer into cells by these techniques are very dependent on cell type and must be optimized for each cell line.

Calcium phosphate and DEAE-dextran transfections create a chemical environment in which the DNA attaches to the cell surface, after which it is taken up into the cell by an as-yet-undefined mechanism. In electroporation, an electric field is used to open up pores in the cell, through which the DNA presumably diffuses. In liposome-mediated transfection, liposomes containing cationic and neutral lipids mediate the entry of DNA into the cell. Negatively charged phosphate groups of DNA bind to the positively charged surface of the liposome, and the residual positive charge is thought to facilitate binding of the complex to charged sialic acid residues on the cell surfaces.

All four of these methods can be used for transient transfection of DNA into cell lines, and all but the DEAE-dextran transfection method are also applicable to stable transfections. Because there are so many effective liposome reagents commercially available, liposome-mediated transfection may be most reproducible for cell lines. Calcium phosphate transfection has been used to introduce DNA successfully into primary neuronal cultures (Xia et al., 1996), but infection of neurons with viral vectors (reviewed in Neve and Geller, 1996) is potentially the most efficient and the most consistent means of gene transfer into neurons.

TRANSIENT OR STABLE GENE EXPRESSION

In deciding whether to express exogenous neural proteins transiently or stably, the following considerations may be helpful. For many purposes, such as the expression of neural gene products for purification, for development of antibodies, or for certain types of radioligand binding studies, the primary consideration is to express recombinant products at as high a concentration as possible. Most systems for transient expression of transgenes attain higher levels of expression than can be attained by stable expression of transgenes. Transient expression may yield results more rapidly than stable expression, because it eliminates the 4 to 6 weeks required to generate, isolate, and characterize transfected cell lines. On the other hand, when the ratio of exogenous gene product to other cellular components is important, such as for studies of signal transduction or regulation of receptor responsiveness, stable expression of recombinant neural genes provides two advantages. First, when a clonal cell line is used, the ratio of transgene product to other cellular components can be maintained at a consistent level throughout a set of experiments. In contrast, transient expression typically involves the use of material in which an unknown percentage of the cells express recombinant gene products at variable and unknown levels. Furthermore, transfection efficiency can vary from one preparation of DNA to another. The second advantage is that the lower level of expression typically observed for stably expressed exogenous neural genes may mimic endogenous expression more closely than does the often very high expression of transient systems.

CHOICE OF PROMOTER

Some neuronal cells, such as PC12 cells, do not support high levels of expression of most of the commonly used mammalian vector promoters. The simian cytomegalovirus (CMV) and Rous sarcoma virus (RSV) promoters were found to be almost an order of magnitude more efficient than any of five other widely used viral promoters tested (Donis et al., 1993). These two promoters also direct robust expression in many other cell lines, although the clear advantage of CMV and RSV promoters that is observed in PC12 cells is not as prominent in nonneuronal cell lines.

Inducible promoters are optimal because they allow the investigator to express a transgene either acutely or chronically. The most popular inducible promoters have included the cadmium- and zinc-activated metallothionein promoters and glucocorticoid-inducible promoters, such as the mouse mammary tumor virus (MMTV) promoter. However, cadmium and glucocorticoids have significant effects on cellular metabolism and can affect the very pathways in the cell that are being examined. Therefore, there has been a trend towards using heterologous promoters from bacteria and insects that are regulatable by agents that have minimal effects on mammalian cells. These include transactivation systems that are regulated by lactose (Baim et al., 1991), tetracycline (Gossen and Bujard, 1992; *CPMB UNIT 16.21*; also reviewed in Shockett and Schatz, 1996), the insect steroid hormone ecdysone (No et al., 1996), the synthetic steroid 4-hydroxytamoxifen (OHT), which binds to a transcriptionally inactive estrogen receptor (Littlewood et al., 1995), and rapamycin (Rivera et al., 1996). These systems require vectors that express two transcriptional units, but that disadvantage is more than offset by the tight control over gene expression that they make possible.

Contributors: Rachael L. Neve and Kim A. Neve

UNIT 4.5

Selection of Transfected Mammalian Cells

Analysis of gene function frequently requires the formation of mammalian cell lines that contain the studied gene in a stably integrated form. Since only one cell in 10^4 will stably integrate DNA (the efficiency can vary depending on the cell type), a dominant selectable marker is used to permit isolation of stable transfectants.

4

STRATEGIC PLANNING

Constructs

A typical construct is depicted in Fig. 4.5.1.

Promoters. The promoter and accompanying enhancers drive expression of the coding regions of both the gene of interest and the selectable marker. Promoters are usually short DNA sequences (<1 kb) that bind endogenous transcription factors. They may or may not have a TATA box. Cytomegalovirus (CMV) and simian virus 40 (SV40) promoters have been favorite general mammalian promoters for years, and many derivatives have been made and are commercially available.

Translational start site. Most selectable markers are based on a bacterial resistance gene, and in their native state may give low levels of expression. Whereas bacterial systems translate

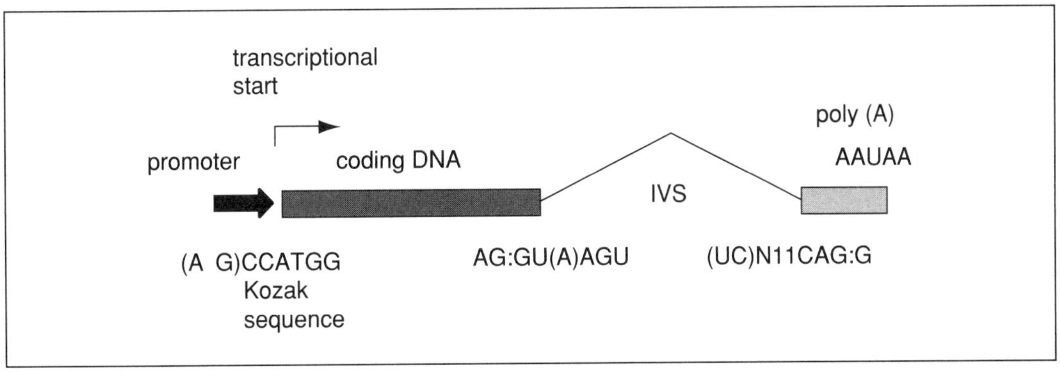

Figure 4.5.1 Generalized construct for expression of genes in mammalian cells. Each integrated construct must contain a promoter region with a transcriptional start site and a coding sequence. Features that increase expression include translational start sites according to Kozak's rules (Kozak, 1989), intervening sequences (IVS) with splice donor and acceptor sites, and polyadenylation signals.

polycistronic messages well, translation of most mammalian mRNA is greatly decreased. In addition, the mammalian start site is most efficient if it follows Kozak's rules (Kozak, 1989). Many commonly used selectable markers have been modified to give good mammalian translation.

Polyadenylation site. A sequence in the 3′ end of the expression vector that signals the addition of a poly(A) tail often stabilizes the mRNA and leads to better expression. *Intervening sequence and splice sites.* Addition of an intervening sequence (intron) that is spliced out in the final mRNA can also increase expression.

Introduction of DNA into Cells

For the creation of stable lines, the results can vary depending on the transfection method used. Calcium phosphate (*CPMB UNIT 9.1*), DEAE-dextran (*CPMB UNIT 9.2*), and liposome (*CPMB UNIT 9.4*) methods all tend to introduce multiple copies of transfected genes, usually at a single site. These methods aggregate the DNA and are suitable for co-transfection of separate plasmids, typically in the ratios of five to ten parts DNA of interest to one part DNA of selectable marker. Electroporation (*CPMB UNIT 9.3*) usually introduces DNA in a single copy. If only a single copy is desired, then electroporation should be used with a single plasmid that contains both the gene of interest and the selectable marker.

4

Removal of Selectable Markers

The selectable marker can be removed using recombinases. As illustrated in Fig. 4.5.2, this is accomplished by transiently expressing Cre recombinase, which removes the sequence between two identical recognition sequences (*lox* sites). Removal of the selectable marker is desirable if more than one sequential genetic manipulation is planned. Because tandem repeats are eliminated by recombinase expression, single-copy integration is desirable, and electroporation is thus the preferred method for transfection.

Amplification of Expressed Gene

Some selectable markers (most notably DHFR) can be used to amplify the expression of the integrated DNA. Amplification is accomplished by gradually increasing the selection pressure with higher concentrations of selection medium. This results in tandem duplication of the integrated DNA.

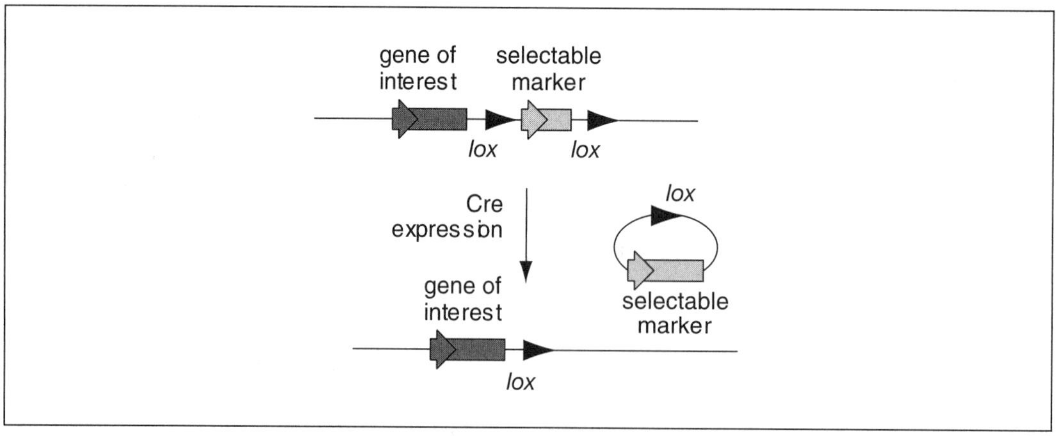

Figure 4.5.2 Using the Cre-*lox* system to remove a selectable marker after integration at a unique site. Transient expression of Cre recombinase causes the excision of sequences between the *lox* sites.

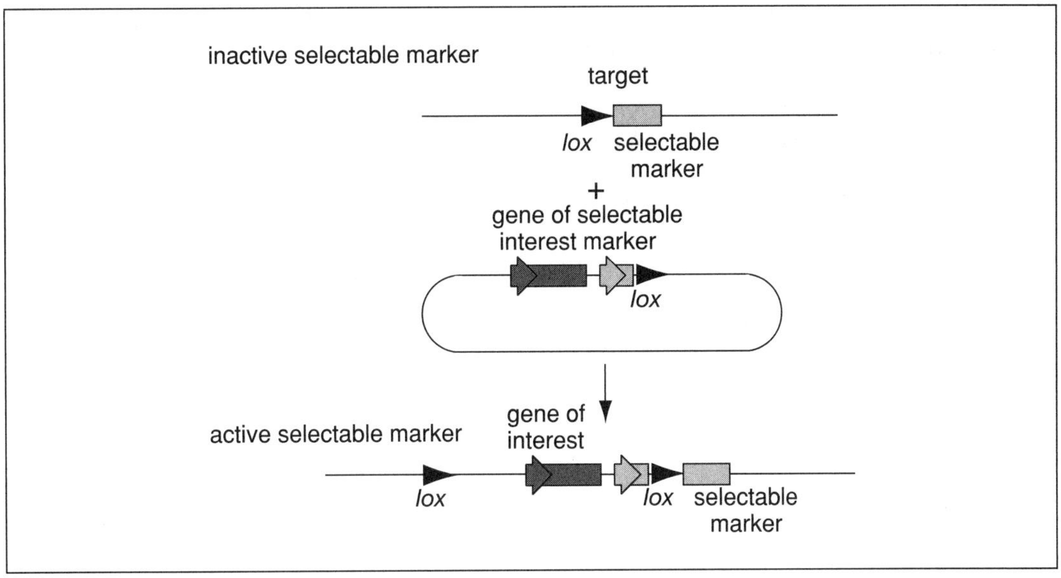

Figure 4.5.3 Using the Cre-*lox* system to integrate an expression construct at a specific reproducible site. The target site is a previously integrated defective form of a selectable marker (lacking a promoter and ATG start codon). Transfection with a plasmid containing the gene of interest leads to integration at the *lox* site and adds a promoter and translational start site, restoring expression of the selectable marker.

4

Introduction of DNA at Specific Sites

One of the problems with production of stable lines is that the transfected DNA integrates at a random site and many of the integrations will not give adequate expression. To address this problem, homologous recombination can be used to introduce DNA at a specific site. A cell line can then be created where the gene of interest is expressed by an endogenous promoter in its normal genomic location. The Cre recombinase system can also be used to introduce DNA at a selected site (Fig. 4.5.3).

BASIC PROTOCOL 1

STABLE TRANSFER OF GENES INTO MAMMALIAN CELLS

Materials *(see* APPENDIX 1 *for items with ✓)*

Cells to be transfected
✓ Complete medium
Selective medium (see Basic Protocol 2)
Cloning cylinders

1. Ensure that cells can grow as isolated colonies. For adherent cells, plate ~100 cells on a 10-cm tissue culture dish, feed every 4 days for 10 to 12 days, and count the number of viable colonies.

   A cell line that cannot grow as isolated cells cannot be stably transfected. To determine how many colonies are present on a tissue culture dish, replace medium with ~2 ml of 2% methylene blue (50% ethanol stock), wait 2 min, pour off methylene blue, and wash off residual dye in cold water. Cell lines that survive poorly as single cells may be selected by using a feeder layer (Robertson, 1987).

2. Determine selection conditions for the parental cell line. For example, if hygromycin-B selection is to be used, determine the minimum level of hygromycin-B that must be added to the medium to prevent cell growth. Split a confluent dish of cells 1:15 into medium containing various levels of the drug. Incubate 10 days, feeding with the appropriate selective medium every 4 days, and examine the dishes for viable cells.

   It is essential to split cells at least 1:15 so that the cells don't reach confluence before the selection can take effect.

3. Determine the most efficient means of transfecting the parental cell type (*see* APPENDIX 2A).

4. Co-transfect desired gene and marker gene into parental cell line. Use a 5:1 molar ratio of plasmid containing the gene of interest to plasmid containing the selective marker. Include a control, for which carrier DNA (e.g., pUC13) is used instead of the plasmid containing the gene of interest. Do several separate transfections containing the gene of interest in case one transfection fails.

   If selection requires transfection of a large amount of the selection plasmid, a single plasmid should be constructed that contains both the selectable marker and the gene of interest.

   Efficient transfection by electroporation requires linearization of the construct.

5. Allow cells to double twice under nonselective conditions, then split 1:15 into selective medium. Plate at least five dishes with selective medium from each transfected dish to maximize the number of colonies that can be picked and expanded into cell lines.

6. Feed cells with selective medium every 4 days (or as needed). After 10 to 12 days, inspect plates for colonies by holding them up to a light at an angle. If desired, stain one plate as in step 1 (staining kills cells). Pick large, healthy colonies (with ~500 to 1000 cells).

SUPPORT PROTOCOL 1

PICKING STABLE COLONIES USING CLONING CYLINDERS

Additional Materials *(also see Basic Protocol 1)*

Cloning cylinders
0.05% trypsin/0.06 mM EDTA in PBS, 37°C

1. Ten to twelve days after placing the cells in selective medium, check the dishes for colonies. Circle the opaque patches to make colonies easier to detect under a microscope.

2. Select the colonies to be picked. Circle the chosen colonies with a laboratory marker to determine where to place the cloning cylinders. Select large, healthy colonies with ~500 cells.

   It is possible at this point to pool large numbers of transformants that have integrated the vector into different sites. Because different integration sites have quite different potentials for amplification, use sequential increases in MTX resistance to rapidly select cells that have amplified the gene to high copy number.

3. Coat one end of a cloning cylinder with a thin film of sterile vacuum grease by touching the cylinder to grease that has been autoclaved in a glass petri dish. Gently place the cylinder around the colony to be picked.

4. Using a sterile Pasteur pipet, rinse the colony with 37°C trypsin/EDTA by filling and emptying the cloning cylinder.

5. Add 3 drops of 37°C trypsin/EDTA to the cloning cylinder. Wait 1 min. Fill the cloning cylinder with medium and repeatedly run the contents of the cylinder in and out through

a Pasteur pipet in order to remove the trypsinized cells from the dish and disperse them. Plate the cells in a 40-mm dish.

6. As the cells grow out, split them frequently (every 4 to 5 days or so) so that they do not form large colonies because central cells in large colonies do not fare well.

BASIC PROTOCOL 2

SELECTABLE MARKERS FOR MAMMALIAN CELLS

The desired concentration for many of the drugs used for selection will vary depending on susceptibility of a particular cell line to the drug. For these drugs, approximate ranges are given, and precise ranges for a given cell line can be determined by characterizing cell viability across a range of concentrations of the drug (see Basic Protocol 1, step 2).

Positive Selection

Adenosine deaminase (ADA)

Selection conditions. Medium supplemented with 10 μg/ml thymidine, 15 μg/ml hypoxanthine, 4 μM 9-β-D-xylofuranosyl adenine (Xyl-A), and 0.01 to 0.3 μM 2′-deoxycoformycin (dCF). Fetal bovine serum (FBS) contains low levels of ADA, which will detoxify the medium, so serum should be added immediately prior to use.

Basis for selection. Xyl-A can be converted to Xyl-ATP and incorporated into nucleic acids, resulting in cell death. Xyl-A is detoxified to its inosine derivative by ADA. dCF is a transition state analog inhibitor of ADA, and is needed to inactivate ADA endogenous to the parental cell type. As the level of endogenous ADA varies with cell type, the appropriate concentration of dCF for selection will vary as well.

Comments. High levels of expression of the transfected ADA gene are necessary to achieve selection in cells with high endogenous ADA levels. ADA-deficient CHO cells are available. ADA can be used in amplification systems, as increasing the level of dCF selects for cells that have amplified the ADA gene.

Reference: Kaufman et al., 1986

Aminoglycoside phosphotransferase (neo, G418, APH)

Selection conditions. 100 to 800 μg/ml G418 in complete medium. G418 should be prepared in a highly buffered solution (e.g., 100 mM HEPES, pH 7.3) so that addition of the drug does not alter the pH of the medium.

Basis for selection. G418 blocks protein synthesis in mammalian cells by interfering with ribosomal function. It is an aminoglycoside, similar in structure to neomycin, gentamicin, and kanamycin. Expression of the bacterial APH gene (derived from Tn*5*) in mammalian cells therefore results in detoxification of G418.

Comments. Varying concentrations of G418 should be tested, as cells differ in their susceptibility to the drug. Different lots of G418 can have different potencies, causing many investigators to buy a large amount of one lot to standardize selection conditions. Cells will divide once or twice in the presence of lethal doses of G418, so the effects of the drug take several days to become apparent.

Reference: Southern and Berg, 1982

Bleomycin (bleo, phleo, zeocin)

Selection conditions. Complete medium supplemented with 0.1 to 50 μg/ml bleomycin.

Basis for selection. This gene encodes a 13,000-Da protein that stoichiometrically binds the drug (bleomycin, phleomycin, or zeocin) and inactivates it.

Comments. Three homologous genes that confer bleomycin resistance have been isolated from the gram-positive bacterial plasmid pUB110, from the central region of the gram-negative transposon Tn*5*, and from resistant strains of *Actinomycetes.*

References: Mulsant et al., 1988; Sugiyama et al., 1994

Cytosine deaminase (CDA, CD)
Selection conditions. Medium containing 1 mM *N*-(phosphonacetyl)-L-aspartate, 1 mg/ml inosine, and 1 mM cytosine.

Basis for selection. Cytosine deaminase converts cytosine to uracil. *N*-(Phosphonacetyl)-L-aspartate blocks the de novo synthesis pathway of pyrimidines and forces the cells to rely on a cytosine deaminase scavenger pathway.

Comment. Originally defined as a negative selectable marker.

Reference: Wei and Huber, 1996

Dihydrofolate reductase (DHFR)
Selection conditions. α^- medium supplemented with 0.01 to 300 μM methotrexate (MTX) and dialyzed FBS.

4

Basis for selection. DHFR is necessary for purine biosynthesis. In the absence of exogenous purines this enzyme is required for growth. Dialysis of serum to remove endogenous nucleosides and use of media devoid of nucleosides is therefore necessary for selection. MTX is a potent competitive inhibitor of DHFR, so increasing MTX concentration selects for cells that express increased levels of DHFR.

Comments. Extremely high levels of expression of the transfected normal DHFR gene are needed for selection in cell lines with high endogenous DHFR levels. A mutant DHFR gene is available that encodes an enzyme resistant to MTX. This gene can be used for dominant selection in most cell types. DHFR can be used to amplify transfected genes. This is most efficiently accomplished using a DHFR-deficient CHO cell line and a normal DHFR gene for selection.

Reference: Simonsen and Levinson, 1983

Histidinol dehydrogenase (hisD)
Selection conditions. Complete medium supplemented with 2.5 mM histidinol or medium lacking histidine and containing 0.125 mM histidinol.

Basis for selection. This gene encodes an enzyme that catalyzes the oxidation of L-histidinol to L-histidine. Thus, medium deficient in histidine but containing histidinol will not support cell growth. In addition, histidinol is toxic in the presence of histidine because of inhibition of histidyl-tRNA synthase, so *hisD*-containing cells can be selected by their ability to detoxify histidinol.

Comment. Histidinol in complete medium usually works well.

Reference: Hartman and Mulligan, 1988

Hygromycin-B-phosphotransferase (HPH)
Selection conditions. Complete medium supplemented with 10 to 400 μg/ml hygromycin-B.

Basis for selection. Hygromycin-B is an aminocyclitol that inhibits protein synthesis by disrupting translocation and promoting mistranslation. The HPH gene (isolated from *E. coli* plasmid pJR225) detoxifies hygromycin-B by phosphorylation.

Comments. While the level of hygromycin-B needed for selection can vary from 10 to 400 μg/ml, many cell lines require 200 μg/ml.

References: Gritz and Davies, 1983; Palmer et al., 1987

Puromycin-N-acetyl transferase (PAC, puro)
Selection conditions. Complete medium supplemented with 0.5 to 10 μg/ml puromycin.

Basis for selection. Puromycin inhibits protein synthesis. The gene encodes an enzyme for the inactivation of puromycin by acetylation.

Comment. This selectable marker is comparable to *neo* in selection efficiency. Many cell types are selected well at 2.0 μg/ml.

Reference: de la Luna et al., 1988

Thymidine kinase (TK)
Selection conditions. Forward (TK^- to TK^+): Complete medium supplemented with 100 μM hypoxanthine, 0.4 μM aminopterin, 16 μM thymidine, and 3 μM glycine (HAT medium).

Basis for selection. Under normal growth conditions, cells do not need thymidine kinase, as the usual means for synthesizing dTTP is through dCDP. Selection of TK^+ cells in HAT medium is primarily due to the presence of aminopterin, which blocks the formation of dTDP from dCDP. Cells therefore need to synthesize dTTP from thymidine, a pathway that requires TK.

Comments. Thymidine kinase is widely used in mammalian cell culture because both forward and reverse selection conditions exist. Unlike markers such as ADA and DHFR, however, it is not possible to select for variable levels of TK, so the gene cannot be used for amplification. Like ADA and DHFR, most mammalian cell lines express TK, removing the possibility of using the marker in those lines unless BrdU is used to select a TK^- mutant.

Reference: Littlefield, 1964

Xanthine-guanine phosphoribosyltransferase (XGPRT, gpt)
Selection conditions. Medium containing dialyzed FBS, 250 μg/ml xanthine, 15 μg/ml hypoxanthine, 10 μg/ml thymidine, 2 μg/ml aminopterin, 25 μg/ml mycophenolic acid, and 150 μg/ml L-glutamine.

Basis for selection. Aminopterin and mycophenolic acid both block the de novo pathway for synthesis of GMP. Expression of XGPRT allows cells to produce GMP from xanthine, allowing growth on medium that contains xanthine but not guanine. It is therefore necessary for selection to use dialyzed FBS and a medium that does not contain guanine.

Comments. XGPRT is a bacterial enzyme that does not have a mammalian homolog, allowing XGPRT to function as a dominant selectable marker in mammalian cells. The amount of mycophenolic acid necessary for selection varies with cell type and can be determined by titration in the absence and presence of guanine.

Reference: Mulligan and Berg, 1981

Negative Selection

Cytosine deaminase (CDA, CD)
Selection conditions. Complete medium supplemented with 50 to 250 μg/ml 5-fluorocytosine.

Basis for selection. Cytosine deaminase converts the 5-fluorocytosine to 5-fluorouracil, resulting in inhibition of proliferation.

Comment. If CD^+ cells are a minority (<1%), there is no detectable bystander killing; however, if the majority of cells contain CD, then CD^- cells survive only at higher dilutions (≤10,000 plated cells per 100-mm plate).

Reference: Mullen et al., 1992

Diptheria toxin (DT)
Selection conditions. The expression of the diptheria toxin gene is itself toxic.

Basis for selection. Diptheria toxin inhibits protein synthesis by the NAD-dependent ADP-ribosylation of elongation factor 2.

Comment. Since this selection is not conditional, this marker is only useful for elimination of cells expressing DT. Stable cell lines expressing DT cannot be isolated. It has been used as a substitution for HSV-TK for homologous recombination or for the elimination of tissues in transgenic animals.

Reference: Yagi et al., 1990

Herpes simplex virus thymidine kinase (HSV-TK), + to −
Selection conditions. Complete medium supplemented with 2 μM 9-[(1,3-dihydroxy-2-propoxy)methyl]guanine (DHPG or ganciclovir) *or* 0.2 μM 1-(2′-deoxy-2′-fluoro-1-β-D-arabinofuranosyl-5-iodo)uracil (FIAU).

Basis for selection. The selection drugs (ganciclovir or FIAU) are nucleoside analogs and are phosphorylated by the HSV-TK gene, which leads to incorporation of the drug into the growing DNA chain during S phase, and subsequent cell death. Although the related compound 9-[(2-hydroxyethoxy)methyl]guanine (acyclovir) causes chain termination, ganciclovir does not. These drugs are ∼1000-fold poorer substrates for mammalian TK and are thus not harmful to non-HSV-TK-containing cells at these concentrations.

4

Comment. This system has been widely used for increasing the selection of homologous recombinants by including the HSV-TK gene outside the regions of homology. It can be used for selection of recombination after Cre recombinase expression if TK is placed between *lox* sites.

References: Cheng et al., 1983; Staschke et al., 1994

BASIC PROTOCOL 3

RAPID SELECTION OF TRANSFECTED MAMMALIAN CELLS

Materials *(see* APPENDIX 1 *for items with ✓)*

- Gene of interest
- Cell line for transfection and appropriate complete medium (*CPMB APPENDIX 3F*)
- Capture-Tec system (Invitrogen) consisting of pHook-1, pHook-2, or pHook-3 plasmid (Fig. 4.5.4) and Capture-Tec magnetic beads
- 3 mM EDTA in PBS
- ✓ PBS
- Cell scraper
- 60-mm tissue culture plates
- Magnetic stand (e.g., Invitrogen; other models may be used) or strong magnet
- End-over-end rotating mixer

1a. *If an expression construct containing the gene of interest already exists:* Cotransfect cells with construct containing gene of interest and pHook-1. Mock transfect several plates without added DNA to determine the amount of background selection (if any). Cotransfect with pHook-1 and control plasmid pcDNA3.1/His/lacZ to assess cotransfection efficiency.

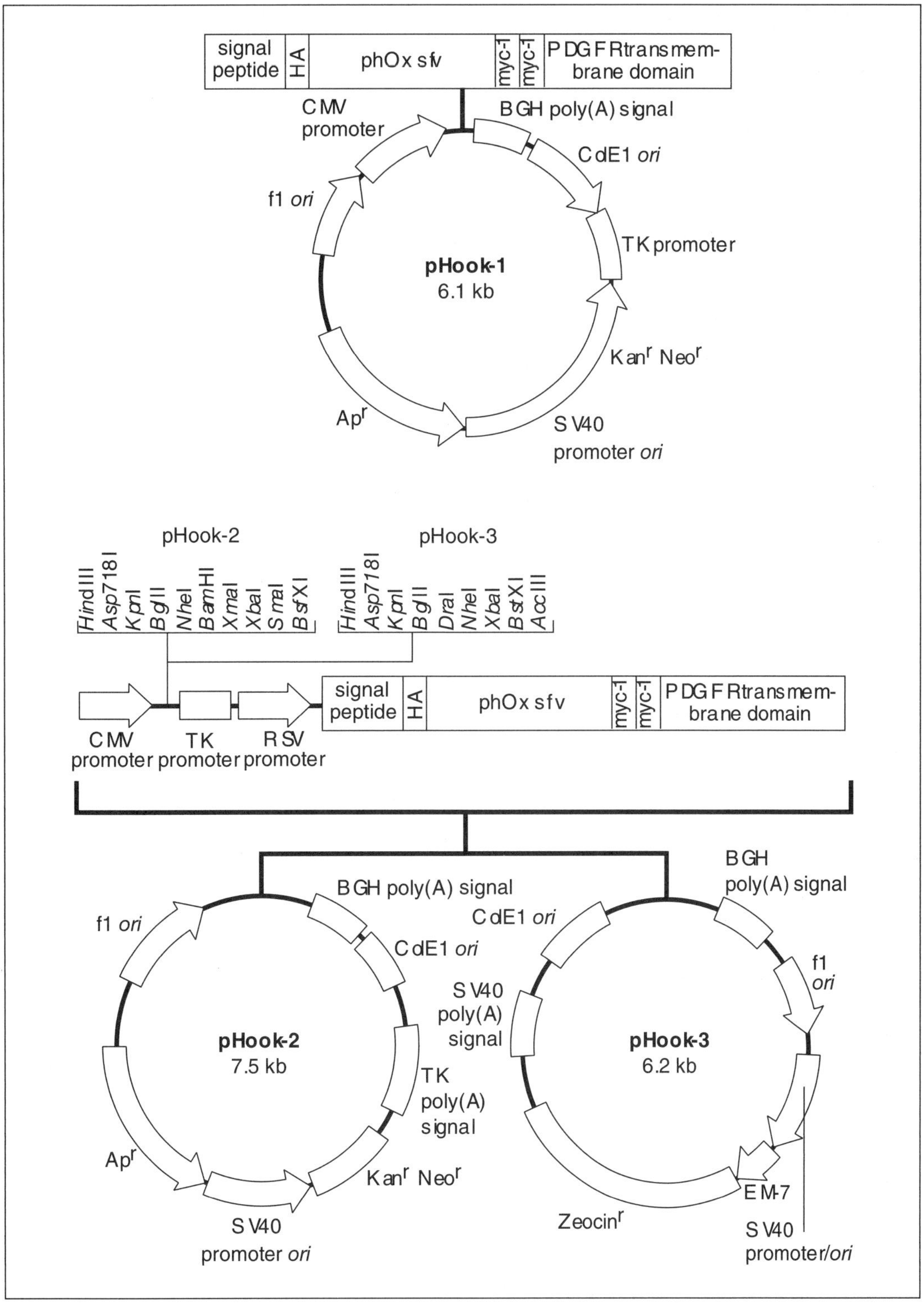

Figure 4.5.4 Plasmid maps of pHook-1, pHook-2, and pHook-3. pHook-1 expresses a fusion protein consisting of the signal peptide from Ig κ V-J2-C region fused with the anti-phOx sFv and the transmembrane domain from the PDGF receptor. pHook-2 and pHook-3 express the same fusion protein but also contain a separate expression cassette consisting of the CMV promoter, a multiple cloning site, and a polyadenylation signal. Definitions: EM-7, synthetic bacterial promoter; HA, hemagglutinin epitope tag; myc-1, myc epitope tag (see text for additional definitions).

1b. *If it is desirable to link expression of the gene of interest with the sFv hook:* Subclone gene of interest into the multicloning site of either pHook-2 or pHook-3, then transfect cells with the plasmid. Mock transfect several plates without added DNA to determine the amount of background selection (if any). Transfect cells with pHook-2/lacZ or pHook-3/lacZ control plasmids to assess transfection efficiency.

2. After transfection (or mock transfection), return cells to incubator for 2 to 48 hr.

3. After incubation, replace medium with 3 mM EDTA in PBS. Harvest cells, using a cell scraper if necessary, and transfer to a centrifuge tube. Centrifuge cells 5 to 10 min at 800 to 1000 × *g*, room temperature. Decant supernatant. Resuspend cells to a single-cell suspension in 1 ml complete medium per 60-mm plate (∼1 × 106 cells). Transfer each sample to a single 1.5-ml microcentrifuge tube.

4. Resuspend enough Capture-Tec magnetic bead slurry for three magnetic separations by vortexing, then place 10 μl (1.5×10^6 beads) into each of a series of microcentrifuge tubes.

Figure 4.5.5 summarizes the magnetic selection procedure.

5. Wash beads in each tube by resuspending in 1 ml complete medium, then separating them out into a pellet with a magnetic stand or strong magnet and removing the medium.

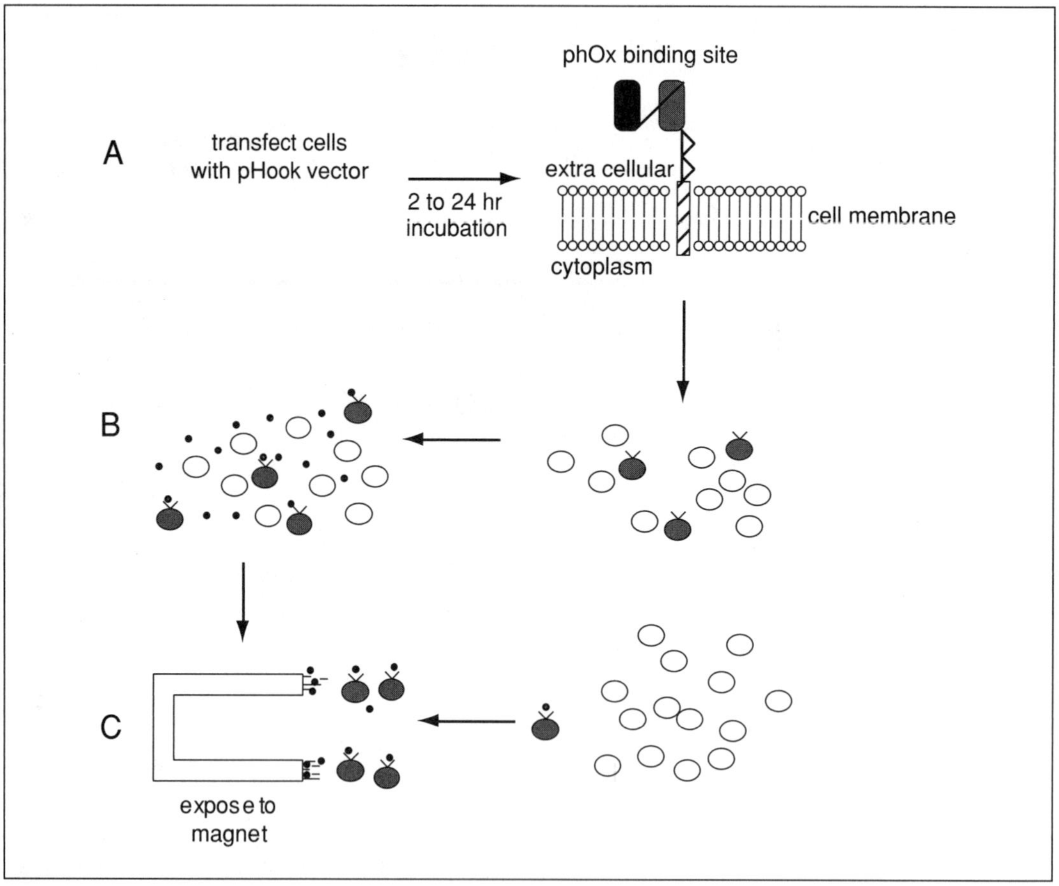

Figure 4.5.5 Magnetic cell-selection protocol. (**A**) Cells are transfected with one of the pHook vectors and returned to culture for 2 to 24 hr. (**B**) The cell population, including transfected cells displaying the sFv hook (dark circles) and untransfected cells (open circles), is then harvested and incubated with antigen-coated magnetic beads (small dots) for 30 min. (**C**) Selection of transfected cells is achieved by exposing the cell/bead mixture to a magnet and discarding cells that are not retained by the magnet.

6. Add each of the cell suspensions from step 3 to one of the microcentrifuge tubes containing washed Capture-Tec beads. Incubate 30 min at 37°C on an end-over-end rotator at 5 to 10 rpm.

7. Place the tubes containing the bead/cell mixture in a magnetic stand and mix for 30 sec to 1 min with gentle rocking. Remove supernatant containing unbound cells fron the transfected and control cell suspensions.

8. Remove the tubes from the magnetic stand and resuspend the beads and cells thoroughly in 1 ml complete medium. Vortex gently to resuspend cells.

9. Repeat magnetic separation on the cells (steps 8 to 10) twice more.

10. Resuspend selected cells in a small volume (50 to 100 ul) of complete medium or PBS and count cells.

SUPPORT PROTOCOL 2

OPTIMIZATION OF COTRANSFECTION CONDITIONS

The positive control plasmid pcDNA3.1/His/*lacZ*, provided with pHook-1, results in the expression of β-galactosidase when successfully cotransfected into cells along with pHook-1. Similarly, the plasmids pHook-1/*lacZ* and pHook-2/*lacZ* result in expression of β-galactosidase when successfully transfected. Hence, these plasmids can be used to check for cotransfection in selected cells and to assess transfection efficiencies. Optimal cotransfection conditions are defined as the pHook-1 to pcDNA3.1/His/*lacZ* ratio that gives the greatest enrichment of blue-stained cells in the selected population. For pHook-2/*lacZ* and pHook-3/*lacZ* controls, optimum transfection conditions are defined as the DNA concentrations that yield the highest number of blue cells selected.

Materials *(see* APPENDIX 1 *for items with ✓)*

Magnetically selected and unselected (supernatant) control cells (see Basic Protocol 3) transfected with: pHook-1 and control plasmid pcDNA3.1/His/*lacZ* (if using pHook-1) pHook-2/*lacZ* (if using pHook-2) pHook-3/*lacZ* (if using pHook-3)
Magnetically selected and unselected (supernatant) mock transfected cells
✓ PBS
✓ Xgal staining solution

Tissue culture dishes
Inverted microscope

1. Centrifuge the suspensions of cells 5 min at 800 to 1000 × *g*, room temperature. Decant supernatants. Wash by resuspending each pellet in 1 ml PBS, centrifuging again, then decanting the supernatants.

2. Resuspend each pellet of selected and unselected cells in 100 ul Xgal staining solution. Transfer to tissue culture dishes or incubate the cells in tubes and transfer to a hemacytometer for counting. Incubate cells 2 to 24 hr at 37°C.

3. Examine dishes under an inverted microscope for development of blue color. Count total cells and blue-stained cells. Normalize the cell counts to total cell number.

References: Cheng et al., 1983; Perucho et al., 1980; Robins et al., 1981; Southern and Berg, 1982; Staschke et al., 1994; Wigler et al., 1977

Contributors: Richard Mortensen, Jonathan D. Chesnut, James P. Hoeffler, and Robert E. Kingston

4

Protein Expression in the *Drosophila* Schneider 2 Cell System

The assorted limitations of bacterial, yeast, and mammalian protein systems have led to the exploitation of insect cell systems for the overexpression of functional proteins from a variety of sources. In contrast to the transient, baculoviral-aided overexpression of recombinant proteins in insect *Sf* 9 cells (Bernard et al., 1999; Murphy and Piwnica-Worms, 1999), Schneider's embryonic *Drosophila* cell line 2 (S2 cells; Schneider, 1972) can be used for stable expression of a variety of recombinant proteins using vectors specifically engineered for protein expression in *Drosophila* cells (Tables 4.6.1 and 4.6.2; also see *http://flybase.bio.indiana.edu;*). Thus, the S2-based system has all the advantages of the *Sf* 9-based system with the major additional advantage that S2 cells can be used to attain stable expression under the control of strong, inducible promoters.

NOTE: All solutions and equipment coming into contact with living cells must be sterile, and aseptic technique should be used accordingly.

4

BASIC PROTOCOL 1

PREPARATION OF STABLE POLYCLONAL S2 CELL LINES

Commercially available, quality control-tested transfection kits are recommended in order to obtain reproducible results.

Materials

- Schneider 2 (S2) cells at or near log phase (see step 2 or 8 of Support Protocol 1)
- Complete M3 medium (see Support Protocol 2)
- Plasmid DNA purified such that $OD_{260/280} \geq 1.8$
- *N,N*-Bis(2-hydroxyethyl)-2-aminoethanesulfonic acid (BES) buffer-based calcium phosphate transformation reagents (e.g., Stratagene)
- 70% (v/v) ethanol

Table 4.6.1 Plasmids Carrying Drug Resistance Genes for Use in Selection of S2 Cells

Plasmid name	Drug resistance gene	Drug resistance conferred	Vector backbone	Promoter	Poly(A)$^+$ signal	Reference
pHGCO	*E. coli* dihyrofolate reductase	Methotrexate	pBR322	*Drosophila* COPIA 5′ LTR	Early SV40	van der Straten et al. (1987)
pCodHygro	*E. coli* hygromycin B phosphotransferase	Hygromycin B	pBR322	*Drosophila* COPIA 5′ LTR	Early SV40	van der Straten et al. (1987)
pUChsneo	*E. coli* neomycin phosphotransferase	Geneticin (G418)	pUC8	*Drosophila* hsp70	None specified	Steller and Pirrotta (1985)
pCo*neo	*E. coli* neomycin phosphotransferase	Geneticin (G418)	pBR322	*Drosophila* COPIA 5′ LTR	Early SV40	van der Straten et al. (1987)

Table 4.6.2 Protein Expression Vectors for Use in S2 Cells

Plasmid name	Vector backbone	Promoter	Poly(A)$^+$ signal	Unique sites in MCS[a]	Other features	Reference
pRmHa-3	pUC18	*Drosophila* Mtn	*Drosophila* Adh	*Eco*R1, *Sac*1, *Kpn*1, *Sma*1, *Bam*H1, *Sal*1	None	Bunch et al. (1988)
pJACKS-3tag	pUC18	*Drosophila* Mtn	*Drosophila* Adh	*Bam*H1, *Apa*1, *Avr*2, *Stu*1, *Sal*1	*N-terminal features:* influenza virus hemagglutinin signal sequence; *Drosophila* Kozak's consensus sequence; FLAG tag; 6× polyhistidine tag; prothrombin protease cleavage site; S-protein tag; enterokinase protease cleavage site	Schetz et al. (2003)
pMT/V5-His A, B, and C	pBR322	*Drosophila* Mtn	Late SV40	*Kpn*1, *Spe*1, *Eco*R1, *Eco*R5, *Not*1, *Xho*1	*C-terminal features:* V5 antigen; 6× polyhistidine tag	SKB/ Invitrogen
pDS47/V5-His A, B, and C	pBR322	DS47	Late SV40	*Kpn*1, *Sac*1, *Bam*H1, *Eco*R1, *Eco*R5, *Not*1, *Xho*1	*C-terminal features:* V5 antigen; 6× polyhistidine tag	SKB/ Invitrogen
pMT/BiP/V5-His A, B, and C	pBR322	*Drosophila* Mtn	Late SV40	*Bgl*2, *Nco*1, *Sma*1, *Kpn*1, *Spe*1, *Eco*R1, *Eco*R5, *Not*1, *Xho*1	*N-terminal features*: *Drosophila* homolog of mammalian BiP signal sequence (HSC3, hsp72) *C-terminal features:* V5 antigen; 6× polyhistidine tag	SKB/ Invitrogen

[a]MCS, multiple cloning site.

Incomplete S2 medium (e.g., Life Technologies)
100× solution of selection drug (optional): e.g., 100 mg/ml geneticin (G418) or 20 to 30 mg/ml hygromycin B

150-cm^2 polystyrene tissue culture dishes
17 × 100-mm 14-ml tubes (Falcon #2059)
50-ml conical polypropylene centrifuge tubes with caps
Cell lifter
75-cm^2 screw-cap (nonaerating) polystyrene tissue culture flasks
27°C incubator

1. Seed 20 ml S2 cells at a density of 1 to 2 × 10^6 cells/ml in complete M3 medium at room temperature into 150-cm^2 polystyrene tissue culture dishes (see Support Protocol 1).

2. In a 17 × 100-mm Falcon tube, mix sufficient purified plasmid DNA and freshly prepared BES-based calcium phosphate solution to deliver 1 ml of solution containing 20 μg plasmid DNA to each dish. When cotransfecting with a drug resistance gene, use a 20:1 ratio of target gene plasmid to drug resistance gene plasmid, to increase the number of copies of target DNA in the drug-selected polyclonal population.

3. Evenly add 1 ml calcium phosphate/DNA precipitate solution dropwise to each dish, swirl lightly to mix, and replace the dish lid. Sterilize 2 × 12-in. strips of Parafilm by wiping with 70% ethanol and allowing them to dry in a sterile hood. Seal the lids of the dishes by carefully but tightly stretching the Parafilm around the perimeter of the dish. Alternatively, place dishes in an airtight container (e.g., Tupperware). Store the dishes in the dark at room temperature for at least 24, but not more than 48 hr.

4. Transfer cells and medium to a 50-ml conical centrifuge tube. Add 25 to 30 ml incomplete S2 medium to the dish, gently dislodge any adherent cells with a cell lifter, and add this rinse to the 50-ml tube. Centrifuge cells 10 min at 225 × *g* at room temperature. Decant and discard supernatant, and resuspend cells in 10 ml complete M3 medium supplemented with the appropriate concentration of selection drug from a 100× stock solution.

5. Add cell suspension to a 75-cm^2 screw-cap polystyrene tissue culture flask, cap the flask tightly, and incubate at 27°C.

6. Maintain cells under drug selection for 3 to 4 weeks before testing for expression of the target gene (see Basic Protocol 2). Once a week, add 5 ml complete M3 medium with the appropriate concentration of fresh selection drug. If cell densities are $\geq 5 \times 10^6$ cells/ml at the beginning of the third week, centrifuge cells and resuspend at 2×10^6 cells/ml in fresh drug selection medium.

7. Maintain selected polyclonal populations under constant drug selection for an additional 4 weeks, splitting cells 1:5 (v/v) about every 3 to 5 days. See Support Protocol 1 for storage and reculturing procedures.

4

BASIC PROTOCOL 2

PROTEIN EXPRESSION BY INDUCTION OF EXOGENOUS TARGET GENES IN S2 CELL LINES

The *Drosophila* metallothionein (Mtn) promoter and the *Drosophila* heat shock protein 70 (hsp70) promoter are strong, inducible promoters. Both can be used in combination for co-transfected cells. The hsp70 promoter has low levels of basal activity in S2 cells and can be activated by heat shock as well as other environmental stressors (e.g., >1 μM cadmium). In contrast, the basal activity of the Mtn promoter is virtually undetectable in S2 cells.

NOTE: Because activation of heat shock promoters temporarily inhibits RNA splicing and the synthesis of non-heat shock proteins, induction by heat shock in cotransfected cells should be performed 1 to 2 days prior to induction of the target gene under the control of a non-heat shock promoter.

NOTE: If the *Drosophila* vectors being used are under the control of a constitutively active promoter (e.g., actin 5C distal promoter), this protocol can be omitted.

Materials *(also see Basic Protocol 1 and Support Protocol 2)*

S2 cultures transfected with gene of interest (see Basic Protocol 1)
Complete M3 medium (see Support Protocol 2; for Mtn induction)
50 mM $CuSO_4$ *or* 1 mM $CdCl_2$ (for Mtn induction): ultrapure (e.g., Aldrich) and sterilized with a 0.22-μm filter
Aluminum foil (for hsp70 induction)
37°C incubator

1. To reduce the risk of contamination, check that the lid of the S2 culture flask is tightly secured and completely seal the flask with fresh aluminum foil. Wrap the foil twice around

the width of the flask, then crimp and roll the ends to seal. To heat shock the cells, place the foil-covered flask in a 37°C bacterial incubator for 30 to 40 min.

2. Remove the flask from the incubator, discard the foil, and return the S2 cells to their previous culture conditions. Repeat heat shock once in 5 to 7 days.

3. To induce the MTN promoter, split a near log-phase S2 culture 1:5 (v/v) in fresh complete M3 medium. Add 1/100th vol of 50 mM $CuSO_4$ (final 500 μM) or 1 mM $CdCl_2$ (final 10 μM). Allow 8 to 72 hr for induction of protein expression before measuring protein expression levels by radioligand binding, immunofluorescence (*UNIT 2.2*), or immunoblotting (*CPMB UNIT 10.8*).

   S2 cells can be cultured continually in the presence of these concentrations of copper for at least 20 days; prolonged exposure to cadmium tends to be cytotoxic.

SUPPORT PROTOCOL 1

CULTURE AND STORAGE OF *DROSOPHILA* S2 CELLS

Although S2 cells grow rapidly at high densities in conditioned medium, it may take several weeks to reculture frozen stocks to a sufficient volume and density for rapid expansion.

Materials *(also see Support Protocol 2)*

Schneider 2 (S2) cells (ATCC #CRL-1963)
Complete M3 medium (see Support Protocol 2)
Cell culture-grade DMSO
Incomplete S2 medium (e.g., Life Technologies)

25- and 150-cm^2 screw-cap (nonaerating) polystyrene tissue culture flasks
27°C incubator (or other dark, dry environment)
1.8-ml polypropylene cryotubes
Liquid nitrogen storage tank
15-ml conical polypropylene centrifuge tubes with caps
10-ml pipet

1. Dilute near log-phase active S2 cultures ~1:5 (v/v) in complete M3 medium to yield cell densities ranging from ~1 to 3 × 10^6 cells/ml and transfer the cells with their conditioned medium.

2. Add 25 ml diluted S2 cells to a 150-cm^2 screw-cap polystyrene tissue culture flask, tightly cap the flask to prevent aeration, and maintain at 27°C in a dark, dry environment. Continue to split actively growing cells 1:5 (v/v) about every 3 to 5 days.

3. For short term storage, split log-phase S2 cell cultures 1:5 (v/v) in complete M3 medium. Culture cells in a tightly capped flask in the dark at 18° to 25°C for up to 1 month.

4. For long-term storage, split log-phase S2 cell cultures 1:2 (v/v) in complete M3 medium and allow the cells to grow for an additional 24 hr at 27°C. Use a hemacytometer (*APPENDIX 2I*) to check that the cell density is ≥1 × 10^7 cells/ml.

5. Suspend cells by swirling and transfer several milliliters of the newly log-phase culture to a 15-ml polypropylene tube. Add ~10% (v/v) DMSO and then transfer 1.5-ml aliquots to 1.8-ml cryotubes. Place cells in liquid nitrogen for long-term storage (up to 2 years).

6. To reculture frozen stock, quick-thaw 1.5 ml frozen S2 stock by incubating 3 to 5 min in a clean 37°C water bath. Transfer cells to a 15-ml conical centrifuge tube and add 10 ml incomplete S2 medium. Centrifuge for 10 min at 225 × *g* at room temperature and decant the supernatant, to remove the DMSO.

7. Resuspend cells in 3 to 4 ml complete M3 medium by mild trituration with a 10-ml pipet, and then transfer to a 25-cm^2 tissue culture flask. Break the surface tension by lightly swirling so that the medium covers the entire bottom of the flask. Also make certain that the culture area is level.

8. Without splitting cells, continue to expand the cultures to ~30 ml by doubling the medium volume every 5 to 7 days with complete M3 medium. Use these cells for large-scale expansion. Dilute and passage the cells as in steps 1 and 2 to prepare stable cell lines.

SUPPORT PROTOCOL 2

PREPARATION OF A MODIFIED SHIELDS AND SANG COMPLETE M3 MEDIUM

In 12% FBS-supplemented M3 medium (as described here), S2 cells grow primarily in suspension and can be expected to double about every 20 hr at 27°C, with a plateau in cell density at ~10 to 20 million cells/ml. Comparable results are achieved using a simplified, finely ground powder form of Shields and Sang M3 medium, which contains no Bacto Peptone and twice the amount of yeast extract as the original formulations (Shields and Sang, 1977).

NOTE: Serum that has not been sufficiently heat inactivated or culturing under conditions favorable for mammalian cell culture (i.e., 37°C aerated with 5% CO_2) is cytotoxic to S2 cells.

NOTE: As some media components are temperature sensitive, it is important to store both dry and liquid media at 2° to 8°C.

4

Materials

Tissue culture-tested FBS
Powdered modified Shields and Sang M3 insect medium (e.g., Sigma)
Tissue culture-grade $KHCO_3$
Penicillin/streptomycin at 5000 U/ml and 5 mg/ml, respectively

Clear resealable plastic bag (e.g., Ziploc)
65°C circulating water bath
50-ml polypropylene centrifuge tubes
250-ml, 0.45-μm tissue culture-grade sterile filtration devices with glass fiber prefilters
1-liter, 0.22-μm tissue culture-grade sterile filtration device with glass fiber prefilter

1. Seal a 500-ml bottle of room temperature, tissue culture-tested FBS in a clear resealable plastic bag. Fill a circulating water bath such that the level of the water is a few centimeters above the serum level in the bottle, and preheat to 65°C. Place the bottle of serum in the water bath and heat for 30 min.

2. Allow the serum to cool by placing it in a refrigerator for ~30 min. Aliquot the cooled, heat-inactivated serum into 50-ml polypropylene centrifuge tubes and centrifuge at ≥1000 × *g* for 20 min at 4°C.

3. Place a glass fiber prefilter on top of a 250-ml, 0.45-μm sterile filtration device, covering the entire surface of the filtration device. Apply a strong vacuum. Gradually pour 250 ml of the FBS supernatants onto the center of the prefilter. Repeat with the remaining 250 ml of FBS supernatants, using a second filtration device. If heat-inactivated FBS is not to be used immediately, divide into 40-ml aliquots and freeze at −80°C for long-term storage (e.g., up to 2 years).

4. Carefully weigh out 39.3 g/liter powdered, modified Shields and Sang M3 insect medium without inhaling the powder. Dissolve in 900 ml distilled water by gently mixing in a 2-liter Erlenmeyer flask until the medium has completely clarified (do not use heat or

hot water to dissolve; avoid prolonged exposure to light). Mix in 0.5 g $KHCO_3$. Transfer medium to a 1-liter graduated cylinder and bring the total volume to 1000 ml.

5. Sterilize 875 ml medium, 125 ml heat-inactivated FBS, and 20 ml penicillin/streptomycin by filtration through a 1-liter, 0.22-µm sterile filtration device equipped with a glass fiber prefilter. Do not adjust pH, which should be between 6.5 and 6.9. Store complete M3 medium up to 1 month in the dark at 2° to 8°C.

References: Angelichio et al., 1991; Ashburner, 1989a,b

Contributors: John A. Schetz and Eswar P.N. Shankar

UNIT 4.7

Protein Expression in the Baculovirus System

This unit describes methods associated with the large-scale production of recombinant proteins in the baculovirus expression system. Performing the protocols in this unit requires that the desired protein has already been cloned into an appropriate vector for expression in cultured insect cells under the control of a strong baculoviral promoter; it is also assumed that both infection of cells by the recombinant virus and production of recombinant proteins have been analyzed in preliminary experiments (Murphy and Piwnica-Worms, 1994a,b; Summers and Smith, 1987).

NOTE: Although the expression of recombinant protein with the baculovirus system is not hazardous to humans, as the virus cannot infect cells of noninsect origin, handling of virus stocks and infected cultures should be restricted to well-defined equipment, materials, and rooms that are kept separate from those where the uninfected host cell lines are maintained. This helps prevent any uncontrolled latent viral infection of the host cell line stocks.

NOTE: All protocols should be carried out using strict aseptic technique in the absence of antibiotics.

BASIC PROTOCOL 1

LARGE-SCALE PRODUCTION OF VIRAL STOCK

Materials

Spodoptera frugiperda (*Sf*9) cells (ATCC, Pharminogen, Invitrogen)
Insect cell medium (Life Technologies): TC100 or IPL41 with 5% to 10% (v/v) fetal calf serum (FCS) *or* serum-free Sf-900 II
Pure recombinant virus (at 10^7 pfu/ml)
1 N NaOH

1- to 10-liter-capacity shaker flasks or spinner culture flasks
27°C shaker incubators, humidity controlled
Low-speed centrifuge
1-liter glass conical-bottom centrifuge bottles (Corning or Bellco), sterile
175-cm^2 tissue culture flasks

1. Prepare a 500-ml culture of *Sf*9 cells in a shaker flask or spinner culture flask containing prewarmed insect cell medium (with or without serum) in a 27°C shaker incubator under controlled humidity. Use a low liquid/total volume ratio (<0.4) and set agitation to 50 to 100 rpm to ensure good aeration of the culture while minimizing mechanical shear effects.

   Pluronic F-68 (BASF) should be included at 2 g/liter in media that lack a shear protectant.

The Sf9 cell line should be maintained and regularly passaged two to three times per week in either monolayer cultures or small shaker cultures in the same medium as the one chosen for the protocol.

2. Withdraw a sample of culture and monitor cell growth (by counting with a hemacytometer; *APPENDIX 2.1*), cell viability (by morphology and trypan blue exclusion), and nutrient levels to ensure an adequate growth environment. Use cultures with a minimum of 2–3 × 10^5 cells/ml, at a viability >98%, and glucose and glutamine levels that do not decrease below 0.2 g/liter and 0.2 mM, respectively.

3. When the culture reaches mid-exponential phase (~1–5 × 10^6 cells/ml), infect with pure recombinant virus to achieve a multiplicity of infection (MOI) of 0.01 to 0.1. Transfer the spinner or shaker flask to a biological hood reserved for handling virus-infected cultures. Carefully open the flask and use a sterile pipet to add the appropriate volume of virus to the culture. Mix by gentle swirling, close the flask, and return it to the incubator.

   The MOI is the ratio of infective virus particles per cell. The volume of virus solution (V_v in ml) is calculated as $V_v = MOI \times N \times V_c T$, where N is the cell density in the culture (cells/ml), V_c is the culture volume (ml), and T is the titer of the virus solution (pfu/ml).

   A low MOI (<0.1) allows some degree of postinfection cell growth, which leads to higher numbers of infected cells and the production of high-titer viral stocks.

4. Incubate 4 to 5 days in a 27°C shaker incubator under controlled humidity to allow virus production. Monitor cell growth daily.

   At 4 to 5 days postinfection a large proportion of the cells (>50%) display hallmarks of infection: large cell bodies, condensed nuclei, rough membranes, and, for some cell lines, sausage shapes. At this stage, viability should be <20%.

5. Harvest the culture by centrifugation in sterile conical-bottom bottles for 30 min at 200 × *g*, 4°C, to remove cell debris. Retain the supernatant as the viral stock. Before discarding cell debris, inactivate all viruses in the pellet by soaking 30 min in 1 N NaOH.

6. Withdraw a small sample (e.g., 5 ml) for viral titration and store remaining stock up to several months at 4°C, protected from light, in 175-cm^2 tissue culture flasks. Titer stocks before use.

   Stocks of <10^7 pfu/ml are useless and should be discarded. 1–5 × 10^7 pfu/ml are considered to be low titer. The expected average titer is 10^8 pfu/ml.

BASIC PROTOCOL 2

DETERMINATION OF EXPRESSION KINETICS

The following protocol can be used to compare host cell lines, media, virus constructs, and usual process parameters (e.g., cell density, MOI, pH, and agitation). The results are used to guide the timing of harvesting in large-scale production systems (see Basic Protocol 3 and Alternate Protocol).

Materials

Cell line: *Spodoptera frugiperda* (*Sf*9, *Sf*21; ATCC, Pharminogen, Invitrogen) *or* *Trichoplusia ni* (*Tn*5 line; High Five cell line, Invitrogen)
Insect cell medium: TC100 (GIBCO/BRL) or IPL41 (GIBCO/BRL), serum free or with 5% to 10% (v/v) fetal calf serum (FCS); *or* Ex-Cell 401 or Ex-Cell 405 (JRH Biosciences) *or* Sf-900 II (GIBCO/BRL), serum free
Recombinant viral stock (≥5 × 10^7 pfu/ml; see Basic Protocol 1)
1 N NaOH

1.5- to 3-liter shaker flasks or spinner culture flasks
27°C incubator, humidity controlled

NOTE: Generally, *Sf*9 cells are optimal for virus production, and *Tn*5 cells are optimal for secreted proteins; serum-free medium generally supports higher expression levels.

1. For each cell culture and infection variable to be analyzed, prepare a separate culture of cells in a 1.5- to 3-liter shaker flask or spinner culture flask containing 0.5 to 1 liter insect cell medium (with or without serum). Place flask in a 27°C incubator and allow cells to grow. Monitor cell growth, cell viability, and nutrient levels (see Basic Protocol 1, step 2).
2. Infect culture at high MOI (5 to 10; see Basic Protocol 1, step 3) by sterile addition of an aliquot of recombinant viral stock. For serum-containing media, use a cell density $\leq 1 \times 10^6$ cells/ml. For serum-free media, use cultures up to 5×10^6 cells/ml. Return culture to 27°C incubator and incubate under controlled humidity.
3. Withdraw 10-ml culture samples, measure nutrient levels, and prepare aliquots for protein analysis. Monitor nutrient levels and cell viability (using trypan blue) twice per day until 4 to 5 days postinfection.

   An increase in cell size is usually a better indicator of infection than trypan blue exclusion.
4. Determine the optimal harvest time from the kinetics of target protein expression, based on SDS-PAGE, immunoblotting, and biological assay, if available. For some proteins, it may be beneficial to harvest before the actual peak of expression because of progressive proteolytic degradation. Neutralize test cultures for 30 min in 1 N NaOH and discard.

4

BASIC PROTOCOL 3

PRODUCTION IN BIOREACTORS

Materials

Insect cell medium: TC100 or IPL41 (Life Technologies), serum free or with 5% to 10% (v/v) fetal calf serum (FCS); *or* Ex-Cell 401 (JRH Biosciences) or Sf-900 II (Life Technologies), serum free
Inoculum (15% to 20% bioreactor culture volume): spinner or shaker flask culture of uninfected insect cells at high viability and in full exponential growth phase
Recombinant viral stock (see Basic Protocol 1)

Bioreactor fitted with temperature, pH, agitation, air flow, and pO_2 controls
Millipak filter, sterile (Millipore)
Viral stock and inoculum transfer vessel fitted with transfer line
Peristaltic pump (optional)

1. Prepare bioreactor by inspecting for the absence of residual biological material on the reactor vessel walls, agitator shaft baffles, or other parts; calibrating sensors for temperature, pH, pO_2 and pressure; and sterilizing accessory lines and transfer vessels by autoclaving. Fill reactor vessel with water such that all probe tips are immersed, and sterilize. Empty reactor.
2. Connect medium addition line and add insect cell medium (75% to 80% of final culture volume) through a sterilizing filter. Disconnect medium addition line.
3. Set all parameters to the desired set points. If possible, leave the bioreactor in an uninoculated state for 24 to 48 hr and check sterility.

   Among strategies available to control dissolved oxygen levels in the culture (e.g., flow rate of gas, agitation, control through headspace), pure O_2 pulsing through the sparger is the easiest method *to monitor. A permanent airflow through the reactor headspace (at 1 liter air/liter culture/min) also damps possible oscillations around the set point. Generally, a set point of 30% to 50% air saturation should be optimal.*

4. Inoculate bioreactor with inoculum at a starting density of 2–3 × 10^5 cells/ml. Pool cultures if necessary and transfer into a sterile inoculation transfer vessel under a biological hood. Connect transfer vessel using a septum port that has not been previously used for emptying or filling the reactor.

5. Transfer the inoculum aseptically into the reactor either via a peristaltic pump or by pressurizing the inoculum vessel with sterile air. Avoid foaming at all times during transfer.

6. Monitor cell growth until cell density reaches half-maximal levels, as estimated from inoculum growth curves.

7. Infect culture at high MOI (5 to 10; see Basic Protocol 1, step 3) by aseptically adding recombinant viral stock. Allow virus infection, virus replication, and protein expression to proceed for 4 to 5 days, withdrawing samples twice daily to determine cell viability and nutrient levels and to perform protein analysis.

8. Terminate the process by harvesting at the optimal time point (see Basic Protocol 2). Harvest by centrifugation or tangential flow microfiltration (see Support Protocol). Keep the cell pellet or supernatant depending on the target protein localization. Collect representative samples before and after any subsequent procedures to monitor changes in target protein stability or conformation.

ALTERNATE PROTOCOL

PRODUCTION IN PERFUSION CULTURES

Perfusion cultures permit high cell densities and thus increase protein production, particularly if proteins are designed for secretion. Equipment setup is as in Basic Protocol 3 except that two tanks and pumps are used for adding and removing extra medium. Because of the large holding tanks required, this protocol is most easily implemented at reactor scales of 1 to 5 liters.

Additional Materials *(also see Basic Protocol 3)*

Cell retention device: spin filter, internal membrane filter, external membrane filter, or external centrifuge
Fermentor level controller (contact probe)
Medium feed and filtrate polypropylene holding tanks (10 to 15 times reactor volume; Nalgene)
Load cell or electronic balance for feed tank (Mettler or equivalent)
Peristaltic pumps (Watston-Marlow or equivalent)

1. Prepare the bioreactor and start the culture (see Basic Protocol 3, steps 1 to 6). Equip the bioreactor with a cell retention device and a fermentor level controller that activates addition of fresh medium to the reactor by a peristaltic pump as culture volume decreases.

2. Prepare a large volume of medium in a feed tank. Set tank on load cell or electronic balance and connect to the reactor.

3. Connect a filtrate holding tank to the filtrate port of the cell retention device. Set up peristaltic pumps for medium feed and filtrate removal.

4. Monitor cell growth. Once cell density has reached half-maximal density for a batch operation, start perfusion with insect cell medium at a rate of 0.5 reactor volume/day by

setting the filtrate withdrawal pump. Monitor feed rates by recording changes in the weight of the feed tank per unit time (taking into account the density of the feed medium).

5. Gradually increase perfusion rate in proportion to the increase in cell density. Allow 3 to 5 days of perfusion to reach the target cell density (1–3 × 10^7 cells/ml at >95% viability).

   Ideally, a preliminary perfusion process should be run without infection to determine the maximal cell density achievable with the given hardware, cell line, and medium.

6. Infect the culture by adding recombinant viral stock to the reactor (see Basic Protocol 3, step 7). Withdraw a 5-ml sample for a negative control for analysis of protein expression. Turn off the filtrate withdrawal pump to halt perfusion for a short time after infection (1 to 2 hr) to allow virus attachment and/or penetration.

   Because the cell density is unusually high, it may be difficult to achieve a high MOI. However, infection at very low MOI (0.01) is still as productive as a batch operation with a high MOI.

   To avoid nutrient limitation during the arrest of perfusion, the feed rate can be increased 2- to 3-fold during the hour preceding the infection.

7. Adjust filtrate withdrawal pump to a maximum perfusion rate (3 to 4 reactor volumes/day) and let infection proceed for at least 4 days. Monitor cell viability and protein expression.

8. To avoid dilution of the product in the filtrate (and if the protein is secreted), remove the accumulated filtrate every 24 hr and concentrate by ultrafiltration (see Support Protocol).

   If expression of the secreted protein is under control of a late promoter, the accumulated filtrate for the first 24 hr postinfection usually contains little product and can be discarded.

4

SUPPORT PROTOCOL

HARVESTING

Depending on the target protein localization, either the cell pellet or the supernatant fraction is kept for further purification. Protease inhibitors should be added in all cases and temperature should be kept at ~4°C to limit possible degradation. In practice, however, it becomes expensive to add protease inhibitors before the material is concentrated to a small volume.

Materials *(see APPENDIX 1 for items with ✓)*

✓ Protease inhibitor mix
1-liter conical-bottom centrifuge bottles (Bellco), sterile
Low-speed centrifuge, continuous feed centrifuge (Heareus Contifuge or equivalent), *or* tangential flow microfiltration system (Microgon KrosFloII or equivalent; ≥0.04 m^2/liter culture)

To harvest by batch centrifugation (≤5 liters)

1a. Transfer the reactor contents into 1-liter conical-bottom glass centrifugation bottles by pressurizing the headspace into the bottles.

2a. Centrifuge 30 min at 200 × *g*. Concentrate the cell-free supernatant by tangential flow ultrafiltration.

To harvest by continuous feed centrifugation (5 to 20 liters)

1b. Connect the inlet line of the centrifuge to the reactor harvest valve, then connect the filtrate outlet line from the centrifuge to a harvesting tank.

2b. Start centrifuge at 200 × *g*, 250 ml/min. Check that supernatant is clear and free of cells. Concentrate the cell-free supernatant by tangential flow ultrafiltration.

To harvest by tangential flow microfiltration (>20 liters)

1c. Connect the inlet of the filtration system to the reactor harvest valve. Connect the retentate line to a return port at the bottom of the reactor. Ensure that the return port is immersed in the culture to avoid foaming.

2c. Set the circulation rate and transmembrane pressure so that the average shear rate is $<2500\ sec^{-1}$, and the transmembrane pressure does not exceed 500 mbar.

3c. Start the circulation pump with the filtrate line closed. Operate without filtration for 3 to 5 min, then open the filtrate line. Carry out filtration quickly (<1 hr). When the reactor is almost empty, transfer the retentate to centrifuge bottles. Centrifuge and concentrate as described in step 2a.

References: King and Possee, 1992; Murphy and Piwnica-Worms, 1994a,b; O'Reilly et al., 1994; Summers and Smith, 1987; Tokashiki and Takamatsu, 1993

Contributors: Alain Bernard, Mark Payton, and Kathryn R. Radford

4

UNIT 4.8

Gene Transfer into Neural Cells In Vitro Using Adenoviral Vectors

Important features of adenovirus vectors include: the ability to infect post-mitotic cells, including neurons; the capacity to be grown to high titers (up to 10^{13} pfu/ml); high levels of transgene expression; long-term stable expression in the brain (described in rats and mice); and the ability to infect target cells with more than one vector. Moreover, adenoviruses are rarely associated causatively with brain pathology. Adenovirus vectors can be exploited for target expression in brain neocortical and glial cells in culture.

NOTE: It is assumed from the outset that readers will be familiar with the molecular biology techniques required for production and analysis of shuttle vectors containing the desired transgenes under the control of selected promoters, and with the techniques and skills necessary for culture of mammalian cell lines in vitro.

NOTE: All cell culture incubations should be carried out in a humidified 37°C, 5% CO_2 incubator unless otherwise stated. All solutions and equipment coming into contact with living cells must be sterile, and aseptic technique should be used accordingly.

BASIC PROTOCOL 1

PREPARATION OF RECOMBINANT ADENOVIRUS VECTORS

Ad vectors for gene transfer and therapy have been constructed using derivatives of human adenovirus serotypes 2 or 5 (Ad2 or Ad5). First-generation recombinant adenovirus vectors (RAds) for gene transfer typically have deletions in the E1 region to render the vector replication-defective, thereby preventing virus spread and target cell lysis. These vectors can be propagated in the E1-complementing human embryonic kidney cell-derived 293 line (Graham et al., 1977). The E3 region, which is dispensable for growth in culture, is also deleted in many Ad vectors to allow the cloning of larger inserts. Most of the Ad vectors in use today are derived from the Ad5 strain *dl*309 or its derivative plasmid, pJM17, which carries a deletion/substitution in E3 but can still express the E3 19-kD and 11.6-kD proteins. See Figure 4.8.1 for examples of transfer and shuttle vectors used to construct recombinant adenovirus.

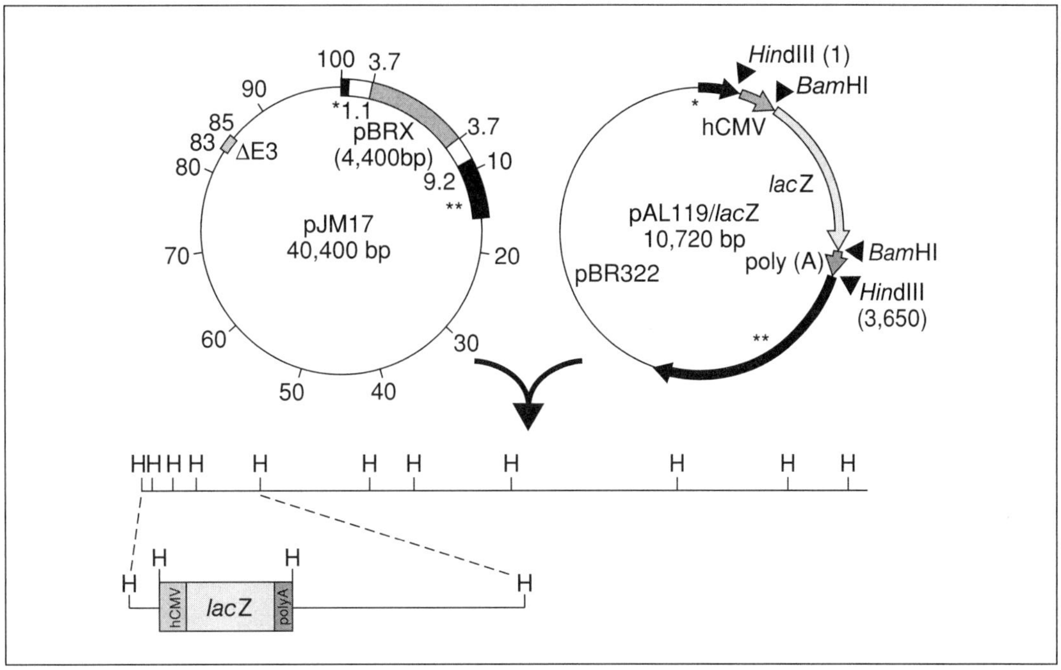

Figure 4.8.1 Schematic diagram of shuttle vectors used for construction of recombinant adenoviruses. The transfer vector pAL119/*lacZ* was constructed from plasmid pXCX2 (Spessot et al., 1989) with the addition of a linker (containing the *Hin*dIII cloning site) at the unique *Xba*I cleavage site. *lacZ* was cloned into pAL119 under hCMV promoter control and upstream of a polyadenylation signal. The transfer vector provides the expression cassette with flanking sequences from the Ad5 E1a gene region (* and **). The plasmid pJM17 contains the entire Ad5dl309 genome and the prokaryotic vector sequence pBRX inserted into the E1a gene at 3.7 map units. Insertion of the prokaryotic vector makes pJM17 too large to be packaged into Ad nucleocapsids. Cotransfection of both plasmids into 293 cells allows homologous recombination, resulting in the replacement of the entire E1a region and prokaryotic sequences with the *Hin*dIII expression cassette. The genome of the recombinant vector is reduced sufficiently in size to be packaged into nucleocapsids, allowing vector replication and resulting in plaque formation in the 293 cell monolayer. A linear diagram of the RAd genome is shown, illustrating the position of the expression cassettes and the *Hin*dIII restriction sites throughout the genome used for characterization of the recombinant adenovirus.

Materials *(see* APPENDIX *1 for items with ✓)*

Adenoviral transfection plasmid, e.g., pJM17 or pBHG10 (Microbix Biosystems), *and* shuttle plasmid pMV60 or pAL119 (constructed from plasmid pXCX2; Spessot et al., 1989, Lowenstein et al., 1996)
Plasmid purification system (e.g., Maxi-prep, Qiagen)
293 cells (ATCC #1573; ECACC # 85120602)
✓ 293 medium, prewarmed to 37°C
✓ Low-Tris/EDTA buffer
2 M $CaCl_2$
✓ 2× HEPES-buffered saline (HeBS)
1% Virkon solution (Anachem)
DPBS (Life Technologies), prewarmed to 37°C
Arklone P (Basic Chemical Company)
✓ 1.33 and 1.45 g/ml CsCl solution
Mineral oil
✓ TE buffer, pH 7.8
✓ Buffers A and B

25-cm^2 and 175-cm^2 plastic tissue culture flasks (Greiner Labortechnik)
15-ml and 50-ml sterile polypropylene conical tubes (Greiner Labortechnik)
1.5-ml screw-cap microcentrifuge tubes
24-well and 96-well tissue culture plates (Greiner Labortechnik)
14-ml ultracentrifuge tubes
Ultracentrifuge with swinging bucket rotor
Wide bore needles (18-G spinal needles, Sherwood Medical) and 5-ml syringes

CAUTION: The following procedures should be carried out in a Class 2 (B.L.2) tissue culture suite in a Class 2 laminar flow cabinet. All media used should be discarded into a solution of 1% Virkon and all plasticware which has been in contact with the recombinant adenovirus should be washed in a solution of 1% Virkon prior to autoclaving and incineration.

1. Prepare adenoviral transfection plasmids such as pJM17 or pBHG10 and the shuttle plasmids, such as pMV60 or pAL119 using a system such as Maxi-prep (Qiagen) and assay them for recombination as described in Support Protocol 1.

2. The day before cotransfection, thinly seed 293 cells in a fresh 25-cm^2 tissue culture flask in 7 ml of 293 medium. Incubate overnight to ~40% confluency on the day of transfection (see *APPENDIX 21*). The next day, immediately before transfection, aspirate the medium and add 5 ml of fresh, prewarmed 293 medium.

3. Pipet a volume containing 5 µg of the shuttle plasmid containing the transcription unit and a volume containing 5 µg pJM17 (or pBHG10) into a sterile 15-ml polypropylene conical tube. Add low-Tris/EDTA buffer to a final volume of 210 µl and mix gently by flicking the tube. Add 30 µl of 2 M CaCl2 and centrifuge gently 1 min at 200 × g, room temperature to ensure that all the solution is in the bottom of the tube.

4. Using a sterile glass Pasteur pipet and an automatic pipettor, add the DNA/$CaCl_2$ mixture to a second sterile tube containing 240 µl of 2× HeBS while very gently bubbling air into the mixture. Allow precipitate to form for 30 min at room temperature.

5. Add the HeBS/DNA/$CaCl_2$ mixture directly to 5 ml fresh pre-warmed 293 medium in a 25-cm^2 tissue culture flask containing the 293 cells, and tilt dish gently to ensure even distribution of the mixture across the cell layer. Check that a fine precipitate has formed on the cells by viewing at high magnification. Leave DNA precipitate on the cells while incubating for 16 hr.

6. Gently aspirate the medium and add 5 ml of DPBS to wash the cells. Aspirate the DPBS, gently wash the cells once with fresh prewarmed 293 medium, then add 6 ml fresh prewarmed 293 medium and return to the incubator. Use a nonviral transfer shuttle plasmid containing the transgene LacZ (e.g., pMV12; Shering et al., 1997) as a control to determine representative transfection efficiency (ideally >30%). Incubate the cells for 2 days then check the number of transfected cells using the method described for Xgal staining of infected cultures (Support Protocol 7).

7. Replenish medium every 3 days, or more often if it becomes acidic (medium changes color from red to orange/yellow). Discard all media and wash all plastics used in a solution of 1% Virkon, then autoclave. Beginning at day 6, examine the cultures for plaques. Continue to incubate to allow plaques to spread throughout the monolayer. Discard cultures that are negative after 25 days.

8. Harvest the cells into a sterile 15-ml polypropylene conical tube (the cells detach from the surface of the flask on gentle washing or tapping) and centrifuge 15 min at 300 × *g*, 4°C. Aspirate the supernatant and resuspend the cells in 100 µl of DPBS.

9. Lyse the cells by rapidly freeze-thawing 3 times, pellet the cellular debris by centrifugation for 15 min at 300 × *g*, 4°C, and transfer the supernatant to a sterile screw-cap

microcentrifuge tube. Store at −70°C. To ensure that the subsequent virus stock is derived from a single infectious viral particle, plaque purify the recombinant virus three times by endpoint dilution (steps 10 to 16).

10. Seed each well of a 96-well tissue culture plate with 5×10^3 293 cells in 100 µl pre-warmed 293 medium.

11. On the next day, place 10 µl of the recombinant adenovirus stock (from step 9) in 990 µl of fresh, prewarmed 293 medium. Using this 10^{-2} dilution, prepare a series of further 10-fold dilutions in prewarmed 293 medium, using sterile microcentrifuge tubes or a 24-well tissue culture plate. Continue a dilution factor of 10^{-11} is reached.

12. Aspirate the medium from the 96-well plate (from step 10) and rinse each well of cells gently with 100 µl prewarmed DPBS. Aspirate the DPBS rinse from each well and add 100 µl of each of the viral dilutions to triplicate wells, starting with the highest dilution (10^{-11}) and working down to the lowest (10^{-2}) dilution, without changing the tip on the pipet. Return the plate to the incubator, gently rocking the plate every 15 min for 90 min, to ensure even distribution of the viral inoculum.

13. At 24 hr postinfection, add an additional 100 µl of fresh, prewarmed 293 medium to each well. Monitor the wells daily for evidence of plaque formation, changing the medium every 3 days (sooner if the medium becomes acidic).

14. After 8 days, locate the endpoint wells (those wells with the highest dilution of viral stock that contains a plaque). Harvest the cells from the triplicate endpoint wells by scraping with the end of a 200-µl pipet tip. Place the cells from each well in separate, sterile screw-cap microcentrifuge tubes, gently microcentrifuge 15 min at 300 × *g*, room temperature, to pellet cells. Remove the supernatants, and resuspend the cells in 100 µl DPBS.

15. Lyse the cells by rapidly freeze-thawing 3 times. Pellet the cell debris by microcentrifuging 15 min at 300 × *g*, room temperature, and transfer the supernatant to a fresh, sterile screw-cap microcentrifuge tube. If necessary, store the supernatant at −70°C for up to several years.

16. Repeat steps 10 to 15 two more times, selecting the cell lysate from one well at random from the three endpoint wells from step 14. Store the supernatants from the other two wells at −70°C as backup.

Production and purification of the viral stock is undertaken in three stages: (1) production of a master stock of the endpoint dilution–purified recombinant adenovirus (It is recommended that at this stage the viral preparation be thoroughly characterized for the presence of the transcription unit within the RAd genome, as outlined in Support Protocol 1.); (2) production of the recombinant adenovirus at titers of 10^8 to 10^{10} pfu/ml, with a purity suitable for in vitro gene transfer into non-neural cells; (3) CsCl gradient centrifugation, to purify the recombinant virus to a level suitable for in vivo gene delivery, and to concentrate the virus to titers ranging from 10^{11} to 10^{12} pfu/ml, which are suitable for infection of neural cells in vitro.

17. Seed a 25-cm² tissue culture flask with 293 cells in 7 ml of prewarmed 293 medium, and culture until they reach 70% to 80% confluency.

18. In a sterile 15-ml polypropylene conical tube, gently mix 5 ml of fresh prewarmed 293 medium and 25 µl of the recombinant virus stock produced after the third round of plaque-purification (from step 16).

19. Aspirate the medium from the culture flask, wash the cells gently with prewarmed DPBS, then add the viral inoculum (step 18), and return cells to the incubator. Gently rock the flask every 15 min for 90 min to ensure even distribution of the inoculum. After 24 hr, increase volume in flask to 7 ml with fresh, prewarmed 293 medium. After 2 to 3 days,

check for a plaque that extends throughout the whole cell monolayer (cytopathological effect, CPE).

20. When the CPE occurs, detach cells from the flask by tapping the side of the flask gently with the hand and then harvest the cell suspension into a sterile 15-ml polypropylene tube. Pellet the cells 15 min at 300 × *g*, 4°C, discard the supernatant, and resuspend the cell pellet in 200 μl of DPBS.

21. Lyse the cells by rapidly freeze thawing 3 times, pellet the cells as in step 20, and transfer the supernatant to a fresh, sterile screw-cap microcentrifuge tube. If necessary, store the supernatant at −70°C for up to several years.

22. Repeat steps 17 to 21, using 75 μl of the new stock of recombinant virus isolated in step 21, to infect the 293 cells. Increase the final volume, in which the cells are resuspended in step 20, to 500 μl.

23. Titrate this stock by endpoint dilution (steps 10 to 16) before proceeding to a medium-scale stock production. If the titer is not $\geq 10^8$ pfu/ml, repeat steps 17 to 22, and re-titer.

24. Grow twenty-one 175-cm^2 flasks of 293 cells to 70% to 80% confluence in 293 medium. At confluency, count the number of cells in the 21st flask, using a hemacytometer (*APPENDIX 2I*), to determine the amount of inoculum needed in the next step.

25. Aspirate the medium and inoculate each flask with 25 ml fresh prewarmed 293 medium containing viral stock at a multiplicity of infection (MOI) of 3. Determine the volume of virus stock as follows:

$$\text{vol. virus stock to be added} = \frac{\text{MOI} \times \text{no. cells to be infected}}{\text{viral titer}}$$

26. Return the flask to the incubator and gently rock the flask every 15 min for 90 min to ensure even distribution of the viral inoculum. After 24 hr, add an additional 25 ml of fresh, prewarmed 293 medium to each flask.

27. When CPE occurs, detach the cells by tapping the side of each flask gently with your hand. Harvest the cell suspension from each flask into a sterile 50-ml polypropylene conical tube and pellet the cells 15 min at 300 × *g*, 4°C. Discard supernatant and resuspend the cell pellet in 0.5 ml prewarmed DPBS per tube.

28. Pool the cell suspension from each tube into one sterile 50-ml tube and add an equal volume of Arklone P to lyse and solubilize the cells. Mix thoroughly by repeated inversion of the tube for 2 to 3 min, then pellet cells 20 min at 500 × *g*, 4°C.

29. Harvest the top aqueous layer and divide into aliquots in sterile screw-cap microcentrifuge tubes. Take care to avoid disturbing the layer of cellular debris that forms at the aqueous/Arklone P interface. Snap freeze with liquid N_2 or dry ice and store at −70°C (up to several years). Alternatively, proceed directly to purification byCsCl gradient centrifugation (steps 30 to 36).

30. Pipet 2.5 ml of 1.33 g/ml CsCl solution into a 14-ml ultracentrifuge tube. Then, using a 5-ml syringe equipped with an 18-G wide bore needle, place the needle at the bottom of the tube and very slowly inject 1.5 ml of 1.45 g/ml CsCl solution, so the less dense layer "floats" on top of the more dense layer.

31. Carefully layer up to 7 ml of the Arklone P–purified virus (from step 29) onto the gradient. Layer mineral oil on top of the viral layer to 2 mm from the top of the tube. Weigh the

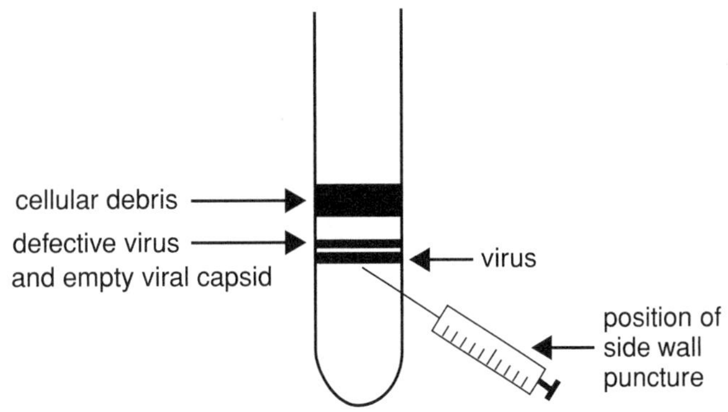

Figure 4.8.2 Banding on CsCl gradient. Schematic representation of relative size and position of each of the three bands present after ultracentrifugation. The size of the cellular debris band varies between preparations.

tube and prepare a balance tube of identical weight (weighed to ±0.02 g). Place the tubes in the centrifuge buckets and seal the buckets prior to leaving the Class 2 laminar flow cabinet. Centrifuge 2 hr at 90,000 × *g*, 4°C in an ultracentrifuge with a swinging-bucket rotor, using maximum acceleration and minimum deceleration.

32. Very carefully remove the tubes from the buckets, in the laminar flow cabinet, and remove the banded virus (Fig, 4.8.2) by side-wall puncture, ~1 cm below the level of the band, with a 5-ml syringe and wide-bore (18-G) needle. Dilute the virus fraction with 0.5 vol of sterile TE buffer, pH 7.8.

33. Prepare a second CsCl gradient as in step 30, using 1 ml of 1.45 g/ml and 1.5 ml of 1.33 g/ml. Carefully layer the diluted virus band onto the gradient and layer mineral oil on top of the virus solution as previously described. Weigh the tube and set up a balance tube, then seal into the buckets again. Centrifuge in the swinging bucket rotor, 18 hr at 100,000 × *g*, 4°C.

34. Remove the viral band as in step 33.

35. Dialyze the banded virus twice against buffer A for 1 hr and once against buffer B for 2 hr, at room temperature, using 12,000 to 14,000 MWCO dialysis tubing.

36. Divide the virus into 10- and 20-μl aliquots and store at −70°C. Titer the recombinant virus by endpoint dilution (Steps 37 to 41).

37. Seed each well of a 96-well tissue culture plate with 5×10^3 293 cells in 100 μl 293 medium/well.

38. On the following day, prepare a series of 10- and 2-fold dilutions of the recombinant adenovirus (from step 36) in sterile microcentrifuge tubes or using a 24-well tissue culture plate as shown in Table 4.8.1.

39. Aspirate the medium from the 96-well plate and rinse the cells gently with prewarmed DPBS. Add 100 μl of each of the viral dilutions to triplicate wells, starting with the highest dilution (7.63×10^{-12}) and working down to the 10^{-2} dilution, without changing the tip on the pipet. Return the plate to the incubator, gently rocking the plate every 15 min for 90 min to ensure even distribution of the viral inoculum. At 24 hr post infection, add an additional 100 μl of fresh, pre-warmed 293 medium to each well.

Table 4.8.1 Dilution Scheme for Titering Recombinant Adenoviruses by Endpoint Dilution

Mixture	Dilution
10 μl of recombinant adenovirus stock + 990 μl of fresh prewarmed 293 medium	10^{-2}
100 μl of 10^{-2} dilution + 900 μl of fresh prewarmed 293 medium	10^{-3}
100 μl of 10^{-3} dilution + 900 μl of fresh prewarmed 293 medium	10^{-4}
100 μl of 10^{-4} dilution + 900 μl of fresh prewarmed 293 medium	10^{-5}
100 μl of 10^{-5} dilution + 900 μl of fresh prewarmed 293 medium	10^{-6}
500 μl of 10^{-6} dilution + 500 μl of fresh prewarmed 293 medium	5×10^{-7}
500 μl of 5×10^{-7} dilution + 500 μl of fresh prewarmed 293 medium	2.5×10^{-7}
etc. to a final end dilution of	7.63×10^{-12}

Table 4.8.2 Titer of Recombinant Adenovirus Preparations Corresponding to Serial Dilution Endpoint Wells

Well	1	2	3	4	5	6
Dilution	10^{-2}	10^{-3}	10^{-4}	10^{-5}	10^{-6}	5.0×10^{-7}
pfu/ml	1.0×10^{3}	1.0×10^{4}	1.0×10^{5}	1.0×10^{6}	1.0×10^{7}	2.0×10^{7}
Well	7	8	9	10	11	12
Dilution	2.5×10^{-7}	1.25×10^{-7}	6.25×10^{-8}	3.12×10^{-8}	1.56×10^{-8}	7.81×10^{-9}
pfu/ml	4.0×10^{7}	8.0×10^{7}	1.6×10^{8}	3.2×10^{8}	6.4×10^{8}	1.28×10^{9}
Well	13	14	15	16	17	18
Dilution	3.91×10^{-9}	1.95×10^{-9}	9.77×10^{-10}	4.88×10^{-10}	2.44×10^{-10}	1.22×10^{-10}
pfu/ml	2.56×10^{9}	5.12×10^{9}	1.02×10^{10}	2.05×10^{10}	4.10×10^{10}	8.19×10^{10}
Well	19	20	21	22	23	24
Dilution	6.10×10^{-11}	3.05×10^{-11}	1.53×10^{-11}	7.63×10^{-12}	Uninfected controls	
pfu/ml	1.64×10^{11}	3.28×10^{11}	6.55×10^{11}	1.32×10^{12}		

40. Monitor the wells daily for evidence of plaque formation, changing the medium every 3 days, or earlier if the medium becomes acidic.

41. After 8 days, read the plate for the highest dilution in which all 3 wells contain plaques. Use Table 4.8.2 to determine the total number of plaque forming units/ml (PFU/ml).

SUPPORT PROTOCOL 1

CHARACTERIZATION OF RECOMBINANT ADENOVIRUS VECTORS

Each recombinant vector should be thoroughly characterized to assess the presence of the transcription unit—i.e., promoter-transgene-poly(A)—after cotransfection, after plaque purification, and after each subsequent CsCl purification procedure.

Materials *(see* APPENDIX 1 *for items with ✓)*

293 cells (ATCC #1573, ECACC # 85120602)
Recombinant adenovirus
DPBS (Life Technologies), prewarmed to 37°C
✓ Virion lysis buffer

5 M NaCl
25:24:1 (v/v/v) phenol/chloroform/isoamyl alcohol
100% and 70% ethanol
RNase Type I-A (Sigma)
Dig DNA labeling and detection kit (Roche)
Restriction enzymes appropriate for digesting plasmids of interest
Shuttle plasmid containing transgene
1% agarose gel containing ethidium bromide (*CPMB UNIT 2.5A*)
1-kb DNA step ladder (Promega)
0.25 M HCl
0.4 M NaOH
✓ 0.5× SSC (prepare from 20× stock)
✓ Pre-hybridization solution
Hybridization solution (pre-hybridization solution containing 5 µl denatured Dig DNA probe for every ml solution), prepare fresh just before use
0.2× SSC (prepare from 20× stock) containing 0.1% (w/v) SDS
0.02× SSC (prepare from 20× stock) containing 0.1% (w/v) SDS
0.01× SSC (prepare from 20× stock) containing 0.1% (w/v) SDS
✓ Color buffers 1, 2, 3, and 4

25-cm^2 tissue culture flasks
Screw-cap microcentrifuge tubes
Nytran nucleic acid and protein transfer media, pore size 0.45 µm (Schleicher & Schuell)
Plastic bags for hybridization
Hybridization oven
95°C water bath or heating block

NOTE: This protocol is a modified Hirt procedure (Gluzman and Van Doren, 1983). For steps 1 to 4, make sure that all solutions are disposed of in 1% Virkon and that all plasticware is autoclaved before disposal.

1. Culture a 25-cm^2 flask of 293 cells to 70% to 80% confluency and then inoculate with the recombinant adenovirus at an MOI of 3 (see Eq. 4.8.1) in 5 ml of fresh 293 medium and return the flask to the incubator. When the cells become rounded (after ~40 hr), collect them by gently tapping on the side of the flask and subsequently triturating the cell/medium suspension.

2. Pellet the cells 10 min at 300 × *g*, 4°C, rinse the cells with DPBS, and pellet again 10 min at 300 × *g*, 4°C. Resuspend the pellet in 400 µl virion lysis buffer. Transfer to a screw-cap microcentrifuge tube and incubate for 2 hr at 37°C. Add 100 µl of 5 M NaCl and store the tube overnight in an ice/water mixture in a Styrofoam box and refrigerate.

3. Centrifuge 1 hr at 15,000 × *g*, 4°C, and harvest the supernatant.

4. Extract residual proteins in the supernatant with an equal volume of 25:24:1 phenol: chloroform:isoamyl alcohol, then centrifuge 15 min at 15,000 × *g*, 4°C (*see APPENDIX 2C*).

5. Remove the aqueous layer. Precipitate the DNA by adding 2 vol of 100% ethanol; incubate for 1 hr at −70°C or overnight at −20°C. Centrifuge 10 min at 15,000 × *g*, room temperature. Carefully remove the supernatant, using a pipet, and discard. Wash the pellet in 1 ml 70% ethanol. Centrifuge 15 min at 15,000 × *g*, room temperature, and discard the supernatant. Allow the DNA pellet to air dry.

6. Resuspend the DNA pellet in 20 µl sterilized distilled water containing 0.1 mg/ml RNase Type I-A, and incubate for 30 min at 37°C. Store the resuspended DNA at −20°C for up to several years.

7. Label ssDNA probe with digoxigenin-dUTP using the Dig DNA kit and instructions supplied by Roche. Make a probe no larger than 2000 bp that encompasses at least part of the promoter/transgene sequence that has been inserted into the recombinant adenovirus.
8. Using suitable restriction enzymes, digest the purified recombinant adenovirus DNA (from step 6), the original adenoviral vector, (i.e., pJM17 or pBHG10) as a negative control, and the shuttle plasmids containing the transgene as a positive control.
9. Electrophorese the three samples along with a DNA ladder on a 1% agarose gel containing ethidium bromide until all the bands have separated and are clearly visible (*see CPMB 2.5A*). Note the relative positions of the DNA ladder fragments and the fragments of the recombinant adenovirus and its controls by photographing the gel or by measuring the gel directly on an UV illuminator. Rinse the gel with distilled water.
10. Depurinate the DNA by soaking the gel for 30 min in 10 vol of 0.25 M HCl, with gentle shaking. Rinse the gel once with distilled water for 2 min. Denature the DNA by soaking the gel twice in 5 vol of 0.4 M NaOH for 10 min each time with gentle shaking.
11. Meanwhile, prepare the Nytran nucleic acid and protein transfer media (sterile filter pore size 0.45 µm), following the manufacturer's instructions.
12. Prepare the blotting stack as shown in Figure 4.8.3, being careful to ensure that no air bubbles are trapped between layers, as this can significantly affect the success of the DNA transfer. Allow transfer to proceed overnight.
13. Carefully remove the nylon membrane and wet it in 0.5× SSC, then place it in a plastic bag containing enough pre-hybridization solution to cover the membrane completely. Seal the bag so it is airtight, with no bubbles, then incubate for 1 hr at 68°C.
14. Heat denature the probe for 10 min at 95°C and then immediately place it on ice.
15. Open the bag containing the membrane and exchange the pre-hybridization solution for fresh hybridization solution. Seal the bag again and hybridize overnight at 68°C.

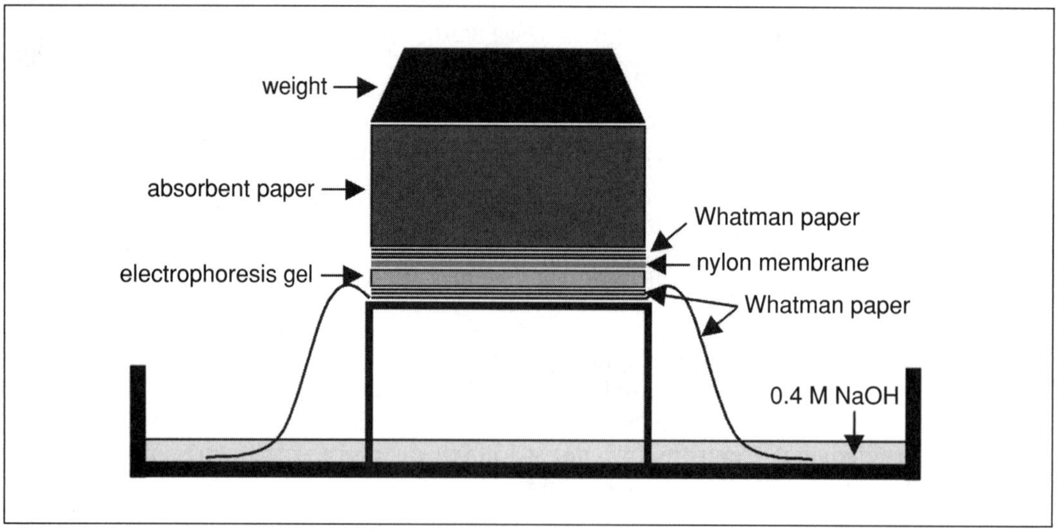

Figure 4.8.3 Transfer of DNA to nylon membrane.

16. Remove the membrane from the bag and wash 3 times (5 min/wash) in 0.2 SSC containing 0.1% SDS, at room temperature. Wash twice (5 min/wash) in 0.02× SSC containing 0.1% SDS, at room temperature. Wash for 30 min in 0.01× SSC containing 0.1% SDS, at 65°C. Wash membrane for 2 min in color buffer 1.

17. Block membrane by incubating for 30 min in color buffer 2 in a sealed plastic bag, with no bubbles. Remove the membrane from the bag and wash for 2 min in color buffer 1.

18. Dilute AP-conjugated anti-digoxigenin antibody (from the Dig DNA kit) 1:5000 in color buffer 1 and incubate it with the membrane in a sealed plastic bag, with no bubbles, for 30 min at room temperature.

19. Remove membrane from bag and wash twice, for 15 min each, with color buffer 1 to remove unbound antibody conjugate. Equilibrate membrane 2 min in color buffer 3.

20. Incubate membrane in freshly prepared color substrate solution (from the Dig DNA kit) in a sealed bag in the dark, again with no bubbles. Do not shake or mix solution while it is developing. Allow color to develop (occasionally up to 24 hr).

21. When the desired level of color is reached, stop the reaction by removing the membrane from the bag and washing it in color buffer 4 for 5 min. Photograph, scan, or photocopy the membrane while wet, as the color will fade slightly when dried.

SUPPORT PROTOCOL 2

PCR ANALYSIS OF RECOMBINANT ADENOVIRAL DNA

The design of primer pairs should encompass unique sections of the expression cassette and the adenoviral genome. Shown below are unique pairs of primers for the IVa2 transcription unit of the adenoviral type 5 genome present in both recombinant and wild-type adenovirus, located at 4071 to 4090 bp and 4738 to 4757 bp, which produce a PCR product of 0.69 kb, and the E1B transcription unit of the adenoviral type 5 genome present only in wild-type (replication competent) adenovirus, which produce a PCR product of 0.56 kb (Dewey et al., 1999).

Materials *(also see* UNIT 4.1*)*

Primers for IVa2:
  Forward: 5′-AAGCAAGTGTCTTGCTGTCT-3′
  Reverse: 5′-GGATGGAACCATTATACCGC-3′
Primers for E1B:
  Forward: 5′-CAAGAATCGCCTGCTACTGTTGTC-3′
  Reverse: 5′-CCTATCCTCCGTATCTATCTCCACC-3′
2% agarose gel
Thermalcycler

NOTE: With all PCR experiments, it is essential that samples and reagents be kept as clean as possible to minimize contamination and DNA degradation. Laboratory coats and clean latex gloves should be worn routinely and procedures to prepare the PCR samples should be undertaken in a laminar flow cabinet.

1. Dilute the recombinant viral DNA (positive for transcription unit and adenoviral genome), shuttle plasmid (positive control for expression cassette) and adenoviral genomic vector (positive control for adenoviral genome and negative for the expression cassette), i.e., pJM17 or pBHG10 (see Fig. 4.8.4), to a concentration of 100 ng/ml with sterile distilled water in separate tubes.

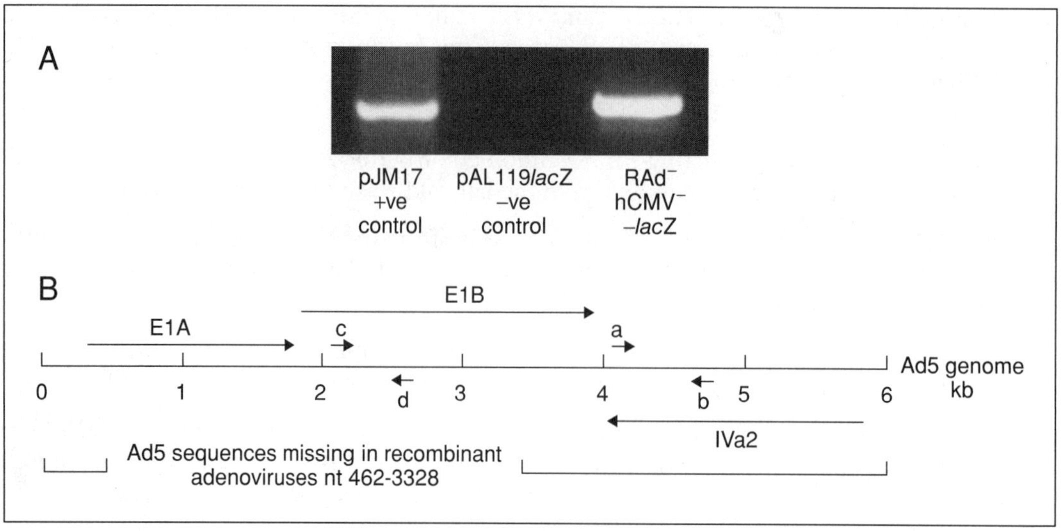

Figure 4.8.4 PCR detection of the IVa2 transcription unit from the Ad5 genomic sequence. (**A**) The extracted DNA from RAd-hCMV//*lacZ* was tested for the presence of the IVa2 transcription unit by PCR. (**B**) Schematic of first 6000 bp of the adenoviral type 5 genome showing the sites at which the primer pairs amplify their respective transcription units and the sequences removed by recombination. Primer pair (a,b) refer to the IVa2 transcription unit, while primer pair (c,d) refers to the E1B transcription unit.

2. In a sterile PCR tube add the following components (100 μl final volume):

   2 μl 10 mM dNTPs
   5 μl 20 ng/ml first primer of pair (final concentration 1 ng/ml)
   5 μl 20 ng/ml second primer of pair (final concentration 1 ng/ml)
   10 μl 10× *Taq* DNA polymerase buffer
   1 U *Taq* DNA polymerase
   1 μl 100 ng/μl template
   4 μl 25 mM $MgCl_2$ (1 mM final)
   72 μl sterile H_2O.

   Overlay reaction mixture with 60 μl mineral oil and then place tube in thermal cycler.

3. Carry out PCR amplification (the following example of a program is for the adenoviral type 5 genomic DNA primers under Materials, above; it will have to be optimized for other vectors):

1 cycle:	5 min	94°C	(initial denaturation)
	30 sec	56°C (IVa2) or 57°C (E1B)	(annealing)
	1 min	72°C	(extension)
35 cycles:	30 sec	94°C	(denaturation)
	30 sec	56°C (IVa2) or 57°C (E1B)	(annealing)
	1 min	72°C	(extension)
1 cycle:	10 min	72°C	(final extension)
	indefinite	4°C	(hold).

4. Run 10 μl of the PCR product on a 2% agarose electrophoresis gel (*see* CPMB 2.5A) and store the mainder at −20°C.

SUPPORT PROTOCOL 3

QUALITY CONTROL OF RECOMBINANT ADENOVIRUS VECTORS: ASSAYS FOR (1) LIPOPOLYSACCHARIDE CONTAMINATION, (2) REPLICATION-COMPETENT VIRUS, AND (3) DETERMINATION OF PURITY

One of the main problems with the standard strategies for Ad vector generation and large-scale vector production is the inadvertent generation of E1-positive replication-competent adenoviruses (RCA). In culture, contaminating RCA can be generated by recombination between the vector and the E1 sequences, extending from nucleotide 1 to 4344 (Louis et al., 1997), carried by the complementing 293 cell line. Another major source of concern is the contamination of the viral preparation with lipopolysaccharide (LPS). The source of LPS could be any material/reagent used in the preparation of the viral stocks. In sufficient levels, LPS is able to elicit an immune response in its own right when administered in vivo, thereby altering the host response to the recombinant adenoviral vector.

Materials *(see* APPENDIX 1 *for items with ✓)*

Recombinant adenovirus preparation to be tested
Limulus Amebocyte lysate pyrogen kit, (BioWhittaker)
HeLa cells (ATCC #CCL-2; ECACC #93021013)
✓ Maintenance medium, prewarmed to 37°C
DPBS (Life Technologies), prewarmed to 37°C
✓ Virion lysis buffer

10-cm sterile tissue culture dishes (Greiner Labortechnik)
24-well tissue culture plate
56°C water bath
Quartz cuvette

To assess LPS contamination

1a. Assay levels of LPS using the Limulus Amebocyte lysate test kit, following the manufacturer's instructions.

To assess of levels of RCA contamination within recombinant adenovirus preparation (Dion et al., 1996)

1b. Grow seven 10-cm tissue culture dishes of HeLa cells containing 6×10^6 cells/dish.

2b. Seed six culture dishes (use the remaining culture dish as a mock infection) with the recombinant adenovirus to be tested at an MOI of 30 in 4 ml of prewarmed maintenance medium, representing a test dose of 1×10^9 pfu. Incubate for 6 hr, then add an additional 4 ml prewarmed maintenance medium and return to the incubator.

3b. At 48 hr after infection, seed each well of a 24-well plate with 5×10^4 HeLa cells and incubate for 72 hr.

4b. After 72 hr, harvest the supernatant from the six 10-cm dishes and pellet the detached cells by centrifuging 15 min at 300 × *g*, 4°C. Keep the supernatant from each dish separate.

5b. Wash the cells in the 24-well plate gently with pre-warmed DPBS and then add 300 µl of each of the supernatants harvested from step 4b to three separate wells containing HeLa cells. Incubate 6 hr, then replace the medium with 500 µl of prewarmed maintenance medium. Incubate ~6 days, until plaques appear if the culture is contaminated with RCA.

To determine purity of recombinant adenovirus preparation (Burlingham et al., 1974, Liebermann.and Mentel, 1994)

1c. Thaw a 20-μl aliquot of the recombinant virus and dilute 1:20 in virion lysis buffer in a screw-cap microcentrifuge tube. Incubate for 10 min at 56°, mixing the tube by inversion every 2 min.

2c. Place the solution of disrupted virions in a quartz cuvette and measure the absorbance at 260 and 280 nm versus a blank of 400 μl virion lysis buffer in a UV spectrophotometer. Calculate the ratio of A_{260}/A_{280} and the virus particle concentration as follows:

$$\text{virus particle concentration} = \frac{A_{260} \times 20}{9.09 \times 10^{-13}}$$

where 20 is the dilution factor and 9.09×10^{-13} is the extinction coefficient for wild-type adenovirus.

3c. Calculate the ratio of defective to infectious viral particles (i.e., virus particles/plaque forming unit as assayed by serial dilution endpoint) as in the following typical example:

$A_{260} = 0.276$

$A_{280} = 0.204.$

Therefore the A_{260}/A_{280} ratio = 1.35. The virus particle concentration for this example is then calculated as:

$$\text{virus particle concentration} = \frac{0.276 \times 20}{9.09 \times 10^{-13}} = 6.07 \times 10^{12} \text{ viral particles}$$

If the titer of the recombinant adenovirus by endpoint dilution = 1.64×10^{11}, then the ratio of virus particles to infectious units is:

$$\text{ratio of virus particles to infectious units} = \frac{6.07 \times 10^{12}}{1.64 \times 10^{11}} = 37$$

and is therefore expressed as 37:1.

An A_{260} of 1.0 corresponds to a viral particle concentration of 1.1×10^{12}/ml, and the A_{260}/A_{280} ratio for pure virus is 1.3 (Burlingham et al., 1974). The A_{260}/A_{280} ratio is decreased for defective particles, which contain less DNA, and is increased by impurities in the preparation, notably cellular RNA.

The particle/infectious unit ratio of wild-type Ad5 is 20. Packing of recombinant genomes is less efficient than wild-type.

BASIC PROTOCOL 2

INFECTION OF NEURONAL AND GLIAL CELLS IN PRIMARY CULTURE

Materials *(see* APPENDIX 1 *for items with ✓)*

1% (v/v) Virkon
Neocortical glial culture (see Support Protocol 6)
Neocortical neuronal culture (see Support Protocol 4)

4

Ventral-mesencephalic (VM) culture (see Support Protocol 5)
DPBS (Life Technologies), prewarmed to 37°C
✓ Plating medium *or*
✓ VM culture medium, prewarmed to 37°C

1. Remove the viral aliquot(s) from the freezer and quickly thaw in a 37°C water bath until the ice just disappears, and then place on ice.

2. Move reagents to a suitable laminar flow cabinet. Prepare a beaker of 1% Virkon for disposal of contaminated plastic and liquids.

3. Dilute the virus in a minimum volume of plating medium or VM culture medium to the appropriate virus dose (MOI). For a 35-mm well, do not exceed a volume of 0.7 to 1 ml. Place the inoculum on ice. Just prior to infection (in step 4), gently warm it to 37°C.

4. Carefully remove all medium from the wells containing neuronal or glial cells in culture, (prepared in Support Protocols 4, 5, or 6), and gently wash the cells with prewarmed D-PBS. Remove the wash, then gently add the warmed inoculum to the culture. Return the plate to the incubator and gently rock the plate every 15 min for the next 2 hr to circulate the virus. After 12 hr, adjust the volume of each well to 2 ml with prewarmed plating medium or VM culture medium.

5. Sterilize all items that have been in contact with the recombinant adenovirus by autoclaving.

6. Process cultures to assess transgene expression or other physiological properties 48 to 72 hr post-infection (Support Protocol 7).

4

SUPPORT PROTOCOL 4

PREPARATION OF LOW-DENSITY PRIMARY NEOCORTICAL NEURONAL CULTURES

Materials *(see* APPENDIX 1 *for items with ✓)*

Pregnant rat (E17-18)
70% Ethanol
CMF-HBSS (Life Technologies), ice-cold and sterile
1 M HEPES buffer solution (Life Technologies)
0.05% (w/v) trypsin/0.02% (w/v) EDTA (Life Technologies)
✓ Plating medium, prewarmed to 37°C

Sterile dissecting equipment
Petri dishes
Dissection microscope (e.g., Leica)
Sterile flame-polished glass Pasteur pipets
70-μm nylon cell strainer
15-ml and 50-ml polypropylene conical tubes
35-mm well–diameter tissue culture plates
✓ Poly-L-lysine coated 22-mm^2 glass coverslips

1. Sacrifice a pregnant rat (E17-18) using an accepted procedure, and disinfect the animal's fur with 70% ethanol

2. Using sterile dissection tools, make a large opening in the skin of the abdomen, leaving the abdominal wall intact. Use fresh sterile tools to open a large area of the abdominal wall and reveal the embryos in the uterus.

3. Prepare ice-cold sterile CMF-HBSS containing 10 mM HEPES (add from 1 M HEPES stock). Remove both horns of the uterus, move to a laminar flow hood, and place each horn into a petri dish containing the ice-cold CMF-HBSS/10 mM HEPES. Remove the embryos from the uterus, decapitate them, and place the heads in a fresh petri dish with ice-cold CMF-HBSS/10 mM HEPES.

4. With the aid of a dissecting microscope, use fine curved forceps to remove the embryo's nose, and then peel away the skin layer to reveal the top of the skull. Using the point of the forceps, gently open the skull, remove the brain and transfer it to a petri dish containing ice cold CMF-HBSS/10 mM HEPES; repeat this procedure for all the embryos.

5. Remove the meninges from the brains and dissect out the cortex, cutting it into pieces of ~1 mm^3. Transfer the tissue pieces to a sterile 15-ml polypropylene conical tube, add 150 μl of 0.05% trypsin/0.02% EDTA, and adjust volume to 15 ml with fresh CMF-HBSS/10 mM HEPES to aid dispersion and avoid formation of clumps among the cells. Incubate for 15 min at 37°C. Mix by inversion every 5 min.

6. Allow the tissue to settle and remove as much of the medium as possible. Add fresh HBSS/10 mM HEPES to a total volume of 15 ml and leave for 5 min at room temperature. Repeat 3 times. Resuspend the last wash in 15 ml of plating medium, invert 3 times and allow tissue to settle.

7. Remove all but 5 ml of plating medium. Triturate the tissue in the remaining medium, and using sterile glass flame-polished 5-ml Pasteur pipets, dissociate the tissue. Once the suspension is homogenous, adjust volume to 15 ml with prewarmed plating medium. Filter through a 70-μm nylon cell strainer into a 50-ml polypropylene conical tube.

8. Count cells using a hemacytometer (*APPENDIX 21*) and then adjust the density to 1×10^6 per ml with prewarmed plating medium. (Twelve embryos yield $\sim 3.6 \times 10^7$ cells.)

9. Add 1 ml of the neuronal suspension to each well of a 35-mm well–diameter tissue culture plate, each containing a poly-llysine-coated 22-mm^2 coverslip. Place in incubator overnight.

10. Remove cellular debris by replacing the medium with 1 ml of fresh prewarmed plating medium and continue culture up to 5 to 9 days. For longer culture, coculture with glial cells.

The cells will take a couple of days to establish a neuronal-like morphology and become suitable for further studies. Refrain from changing the medium too frequently, as this will promote the growth of contaminating nonneuronal cells such as glia and fibroblasts.

SUPPORT PROTOCOL 5

PREPARATION OF LOW-DENSITY PRIMARY VENTRAL-MESENCEPHALIC CULTURES

***Materials** (see APPENDIX 1 for items with ✓)*

Pregnant rat (E14)
70% ethanol
CMF-HBSS (Life Technologies), ice-cold (sterile)
1 M HEPES buffer solution (Life Technologies)

0.05% (w/v) trypsin/0.02% (w/v) EDTA (Life Technologies)
✓ VM culture medium

Sterile dissecting equipment
Petri dishes
Dissection microscope (e.g., Leica)
29-G, 0.5-in needles
Sterile, flame-polished glass Pasteur pipets
15-ml conical polypropylene tube
70-μm nylon cell strainer
22-mm well–diameter tissue culture plates
✓ Poly-L-ornithine and laminin coated 16-mm–diameter glass coverslips

1. Sacrifice a pregnant rat (E14) using an accepted procedure, and disinfect the animal's fur with 70% ethanol. Using sterile dissection tools, make a large opening in the skin of the abdomen, leaving the abdominal wall intact. Use fresh sterile tools to open a large area of the abdominal wall to reveal the embryos in the uterus.

2. Prepare ice-cold sterile CMF-HBSS containing 10 mM HEPES (add from 1 M HEPES stock). Remove both horns of the uterus, move to a laminar flow hood, and place each horn into a petri dish containing the ice-cold CMF-HBSS/10 mM HEPES. Remove the embryos from the uterus.

3. With the aid of a dissecting microscope and consulting an atlas of the developing rat brain (Paxinos et al., 1994), remove the fetal brain, isolate the brainstem, and remove the meninges.

4. Make a cut at the junction of the diencephalon and mesencephalon, using the prominent diencephalic protuberance as a point of reference. Using 29-G, 0.5-in needles as micro-knives, use one needle to steady the tissue, position the brainstem on its dorsal surface, and split the tectum medially through the ventricular opening. Cut away the most rostral portions of the ventral brainstem and the tectum to leave the medial portion of the remaining rostral brainstem (~1 mm^3).

5. Dissect out the ventral mesencephalon from each embryo and place it into a 15-ml polypropylene tube containing 5 ml of CMF-HBSS/10 mM HEPES on ice.

6. Incubate the tissue with 50 μl 0.05% trypsin/0.02% EDTA for 15 min at 37°C, mixing by inversion every 5 min. Then, rinse the tissue 3 times, 5 min per rinse, with sterile CMF-HBSS/10 mM HEPES. Remove the HBSS and add 5 ml of fresh cold VM culture medium.

7. Gently disperse the cells by triturating with sterile flame-polished glass Pasteur pipets, then filter through a sterile 70-μm nylon cell strainer. Count cells using a hemacytometer (*APPENDIX 2I*), then adjust the cell density to 2×10^5 cells/500 μl with fresh prewarmed to 37°C VM culture medium. (Twelve embryos yield ~9.6×10^6 cells.)

8. Add 500 μl of the cell suspension to each well of a 22-mm-well–diameter plate, each containing a previously prepared 16-mm–diameter glass coverslip coated with poly-L-ornithine and laminin and place the plate in an incubator overnight.

9. Replace the medium with 500 μl of fresh, prewarmed VM culture medium to remove any cellular debris. Replace the medium twice the following day, using fresh prewarmed VM culture medium to remove cellular debris. Assess the cultures by immunocytochemistry (Support Protocol 7) after 1 to 3 days of culture.

SUPPORT PROTOCOL 6

PREPARATION OF NEOCORTICAL GLIAL CULTURES

Materials *(see* APPENDIX 1 *for items with ✓)*

- 0- to 3-day-old rat pups
- 100% ethanol
- CMF-HBSS (Life Technologies), ice-cold (sterile)
- 1 M HEPES buffer solution (Life Technologies)
- ✓ Plating medium, prewarmed to 37°C

- Sterile dissecting equipment, including fine curved forceps
- Petri dishes
- Dissection microscope (e.g., Leica)
- 15- and 50-ml conical polypropylene tubes
- Sterile flame-polished glass Pasteur pipets
- 70-μm nylon cell strainer
- 35-mm well–diameter tissue culture plates
- Poly-L-ornithine/laminin-coated 22-mm^2 coverslips

1. Sacrifice 0- to 3-day-old rat pups according to an accepted procedure, then immerse them in 100% ethanol in a laminar flow cabinet and decapitate with a sterile scalpel blade.

2. Prepare ice-cold sterile CMF-HBSS containing 10 mM HEPES (add from 1 M HEPES stock). Place the heads in the CMF-HBSS/10 mM HEPES.

3. Hold the head with curved forceps inserted into the eye sockets and using a fine pair of scissors make a cut into the skull at the base of the skull and open it to reveal the brain. Remove the brain, carefully remove the meninges with the aid of a dissecting microscope and an atlas of the developing rat brain (Paxinos et al., 1994), and then dissect out the cortex and cut it into 1-mm^3 pieces using sterile curved forceps.

4. Place the tissue fragments in 5 ml of fresh ice-cold CMF-HBSS/10 mM HEPES in a 15-ml polypropylene conical tube. Dissociate the tissue by triturating using flame-polished glass Pasteur pipets until no clumps are visible. Adjust volume to 15 ml with CMF-HBSS/10 mM HEPES and pass through a 70-μm nylon cell strainer into a 50-ml polypropylene conical tube.

5. Centrifuge for 15 min at 300 × *g*, 4°C, with no brake. Remove supernatant, resuspend in 10 ml plating medium, and count cells using a hemacytometer (*APPENDIX 21*).

6. Adjust density to 1 × 10^6 cell/ml with plating medium (12 embryos yield ≥3.6 × 10^7 cells) and place 1 ml of the cell suspension in each well of a 35-mm well–diameter tissue culture plate, each containing a poly-L-ornithine/laminin-coated 22 × 22–mm coverslip. Add an additional 1 ml of plating medium and transfer to incubator. Continue incubation until cultures are confluent (1 to 2 weeks), feeding the cells every 3 to 4 days.

   It is possible to passage the cells, but this will also increase the number of fibroblasts that contaminate the culture. This protocol will produce a glial culture enriched in astrocytes. Low numbers of cortical oligodendrocytes do grow within the astrocytic cultures. Detailed protocols are available for obtaining pure oligodendrocyte cultures (Noble and Mayer-Proschel, 1998).

SUPPORT PROTOCOL 7

THREE ASSAYS TO DETECT TRANSGENE EXPRESSION WITHIN INFECTED NEURONAL AND GLIAL CELLS IN PRIMARY CULTURE: (1) Xgal STAINING, (2) FLUORESCENCE IMMUNOCYTOCHEMICAL STAINING, AND (3) FACS ANALYSIS

The Xgal histological staining of infected cultures is used only for virus expressing the marker transgene *E. coli* β-galactosidase. The fluorescence immunocytochemical and fluorescence-activated cell sorting (FACS) procedures can be used for all other transgenes, if suitable antibodies are available.

Materials *(see* APPENDIX 1 *for items with* ✓*)*

Infected cells in primary culture
DPBS (Life Technologies), prewarmed to 37°C
1% (v/v) Virkon
✓ 4% (w/v) paraformaldehyde solution
✓ PBS
0.1% (v/v) Triton X-100 in PBS
✓ 1 mg/ml 5-bromo-4-chloro-3-indoyl β-D-galactopyranoside (Xgal; Sigma) in DMSO
✓ Tris-buffered saline (TBS)
✓ Mowiol 4-88 mounting solution
10% and 1% (v/v) normal blocking serum (preferably from species in which secondary antibody was raised) in PBS
Primary antibody
Fluorescent secondary antibody labeled with fluorescein, Texas red, or R-phycoerythrin (Jackson Immunoresearch; Dako)
0.05% (w/v) trypsin/ 0.02% (w/v) EDTA (Life Technologies)
✓ Plating medium, ice-cold
✓ FACS staining buffer
✓ FACS permeabilizing buffer

6-well tissue culture plates
Fine forceps
Humidified chamber: e.g., sealable box containing a damp tissue
Glass slides
Epifluorescence microscope
Polypropylene centrifuge tubes suitable for FACS (different machines require different tubes so check with FACS manufacturer)
FACS machine (Becton Dickenson)
15-ml and 50-ml polypropylene conical tubes
Shaking platform
35-mm tissue culture plates
50°C water bath

Xgal staining of infected cultures

1a. Gently aspirate medium from the cells and wash with DPBS. Dispose of medium and washes in 1% Virkon.

2a. Fix the cells for 15 min with 2 ml fresh 4% paraformaldehyde solution. Wash 3 times, 5 min each time, with 5 ml PBS.

3a. Add 2 ml of 0.1% Triton X-100 in PBS and incubate 10 min to permeabilize. Wash 3 times, 5 min each time, with 5 ml PBS.

4a. Incubate cells for 3 hr, in the dark at 37°C, with 2 ml Xgal staining solution.

5a. Wash cells 3 times, 5 min each time, with 5 ml PBS, then once for 5 min with 5 ml TBS. Allow to air dry or apply coverslip after addition of 100 to 200 μl of the Mowiol 4-88 mounting solution.

Fluorescent immunocytochemistry of infected cultures

1b. At the appropriate time (usually 2 to 3 days) after infection, working in the laminar hood, use forceps to transfer the coverslips with the infected cell culture into wells of a 6-well dish, each containing 2 ml of DPBS.

2b. Gently wash the cells twice with 2 ml DPBS, disposing of each wash in 1% Virkon.

3b. Fix cells for 15 min with 2 ml of fresh 4% paraformaldehyde solution. Wash cells gently 3 times, each time with 5 ml PBS.

4b. Permeabilize the cells by incubating for 10 min with 2 ml 0.1% Triton X-100 in PBS. Wash cells 3 times, each time with 5 ml PBS.

5b. Incubate for 1 hr with 10% normal blocking serum in PBS. Wash briefly with 1% normal blocking serum in PBS.

6b. Place primary antibody diluted in 1% normal blocking serum, (100 μl per coverslip) onto a strip of clean Parafilm in a humid chamber, and use fine forceps to invert the coverslips over the antibody dilutions. Incubate 4 hr at room temperature or overnight at 4°C.

7b. Replace coverslips into a 6-well plate and remove unbound primary antibody thoroughly with five 5-min washes, each with 5 ml 1% normal blocking serum in PBS.

8b. Incubate coverslips in 100 μl fluorescein-conjugated secondary antibodies diluted in 1% blocking serum in PBS according to manufacturer's instructions, using Parafilm as above (step 6b) in a dark humid box covered in foil for 1 hr at 4°C, and perform all subsequent washes under dimmed ambient light.

9b. Replace coverslips into a 6-well plate and wash thoroughly with PBS 6 times, each time for 5 min. Wash briefly in TBS, then mount by inverting onto 100 μl of Mowiol 4-88 mounting solution on glass slides. Allow to harden at 4°C overnight before viewing under epifluorescence.

FACS analysis

1c. Aspirate medium from cells and wash cells gently with DPBS. Dispose of the medium and wash in 1% Virkon. To each 35-mm well add 200 μl of 0.05% (w/v) trypsin/0.02% (w/v) EDTA and incubate for 2 min.

A single well from a 6-well plate will generally provide enough cells to run 2 to 3 test samples; however, it is best to use a single well per test.

2c. Lightly tap the tissue culture plate to dislodge the cells, then add 2 ml of ice cold plating medium supplemented with 20% FBS to each well. Dispense the cell suspension into a polypropylene centrifuge tube suitable for use on a FAC sorter, and pellet the cells for 15 min at 200 × *g*, 4°C.

3c. Aspirate all but 100 µl of the medium from the tube, resuspend the pellet by brushing the bottom of the centrifuge tube along a rack, and add 5 ml of FACS staining buffer. Centrifuge 15 min at 200 × *g* at 4°C, aspirate and resuspend, add 1 ml 4% paraformaldehyde solution, quickly vortex, and fix cells 10 min on ice.

If the cells do clump, pass the cell suspension through a 70-µm nylon cell strainer to eliminate the clumps.

4c. Centrifuge 10 min at 500 × *g*, room temperature, aspirate, and resuspend (as in step 3c), and then wash cells with 5 ml FACS staining buffer. Centrifuge 10 min at 500 × *g*, room temperature, then aspirate supernatant and resuspend pellet in 5 ml FACS permeabilizing buffer. Incubate for 20 min at room temperature, then centrifuge the cells for 10 min at 500 × *g*, room temperature, and aspirate the supernatant.

5c. Resuspend the cells in 100 µl of FACS permeabilizing buffer, add the primary antibody (diluted in FACS permeabilizing buffer) at the appropriate dilution, and mix by gentle vortexing. Incubate 1 hr at room temperature. Add 5 ml FACS permeabilizing buffer, centrifuge 10 min at 500 × *g*, room temperature, and resuspend in 100 µl of FACS permeabilizing buffer.

6c. Add fluorescent secondary antibody at the appropriate dilution. Mix by gently vortexing, and incubate for 45 min in the dark at room temperature.

For fluorescein conjugates use a 1:50 dilution; for R-phycoerythrin use 1:25; Texas red conjugates are not suitable for FACS.

7c. Add 5 ml FACS permeabilizing buffer, spin down as before, and resuspend in 1 ml FACS staining buffer.

8c. Sort the cells with respect to their size, (forward scatter) and granularity (side scatter). Set the level of fluorescent compensation using negative and single-color positive controls before reading the test samples (Morrelli et al., 1999; Cowsill et al., 2000).

Neurons are smaller and less granular than glial cells.

References: Berkner and Sharp, 1983; 1984; Graham et al., 1977

Contributors: Thomas D. Southgate, Paul A. Kingston, and Maria G. Castro

UNIT 4.9

Gene Transfer into Rat Brain Using Adenoviral Vectors

Adenoviruses are easily purified to the high titers required for in vivo administration, and they are efficient in transducing terminally differentiated cells such as neurons and glial cells, resulting in high levels of transgene expression. At low doses, adenoviral vectors delivered to the brain parenchyma cause minimal inflammation or toxicity, and vector-mediated transgene expression is relatively stable and generally, spatially restricted to the region of virus administration.

The majority of current gene therapy protocols utilizing adenoviral vectors have involved "first-generation" vectors. These are recombinant vectors which are rendered nonreplicative by deletion of the E1 region from the viral genome. Detailed protocols for the generation and purification of first-generation adenoviral vectors are presented in *UNIT 4.8*. Before any virus preparation is used in vivo, it must be accurately titered and subjected to the stringent quality control tests described in *UNIT 4.8*.

BASIC PROTOCOL 1

IN VIVO ADENOVIRUS-MEDIATED GENE TRANSFER INTO THE CNS OF ADULT RATS

UNIT 3.7 describes administration of inoculum to a precise location within the adult rat CNS using stereotaxic guidance. In the protocol below, the technique is adapted to gaseous anesthesia and viral vector injection (Fig. 4.9.1).

NOTE: The surgical procedure is relatively quick, and optimally the animal should be under anesthetic for no longer than 40 min. In most cases the animals recover from the surgery within 10 to 15 min and show no behavioral signs of neurological dysfunction.

NOTE: If several animals are to be operated on in succession, it is often not practicable to sterilize the surgical tools after each operation. In such cases, it is imperative to thoroughly clean the Hamilton needle and other tools between animals, first by soaking in an enzymatic cleaning solution (specially formulated for surgical tools, e.g., Endozyme) and then by rinsing several times with 70% ethanol, and, finally, sterile saline. The drill head can also be cleaned in this manner between operations.

4

Materials *(see APPENDIX 1 for items with ✓)*

Adenoviral vector (*UNIT 4.8*)
✓ Sterile PBS, pH 7.4
Adult rat (250 g body weight 70% (v/v) ethanol/isopropyl alcohol

Stereomicroscope (e.g., Zeiss Stemi 1000 zoom) equipped with 16× eyepieces and 0.4× auxiliary objective lens, and mounted on hinged coupling arm on a heavy foot stand (or equivalent)
Rubber balloon
Stereotaxic frame with rat adapter and blunt ear bars (Stoelting), modified for gas anesthesia (Fig. 4.9.1)
Gas anesthetic trolley with the following components:
- Halothane gas anesthetic
- Halothane vaporizer (e.g., Fluotec)
- Medical oxygen cylinder
- Medical nitrous oxide cylinder
- Induction chamber
- Halothane scavenger (e.g., Fluovac)

Electric drill with 1.75-mm drill bit (Stoelting)
Heat pad
Surgical shavers (Stoelting)
Scalpel and blades
Skin retractors
Cotton swabs
10-μl Hamilton syringe with needle (model 701RN, Fisher)
Fiber optic illuminator with twin goose-neck pipes (Leica)
Sterile 23-G or 25-G hypodermic needles
1-ml syringes
Curved and straight forceps
Holding scissors

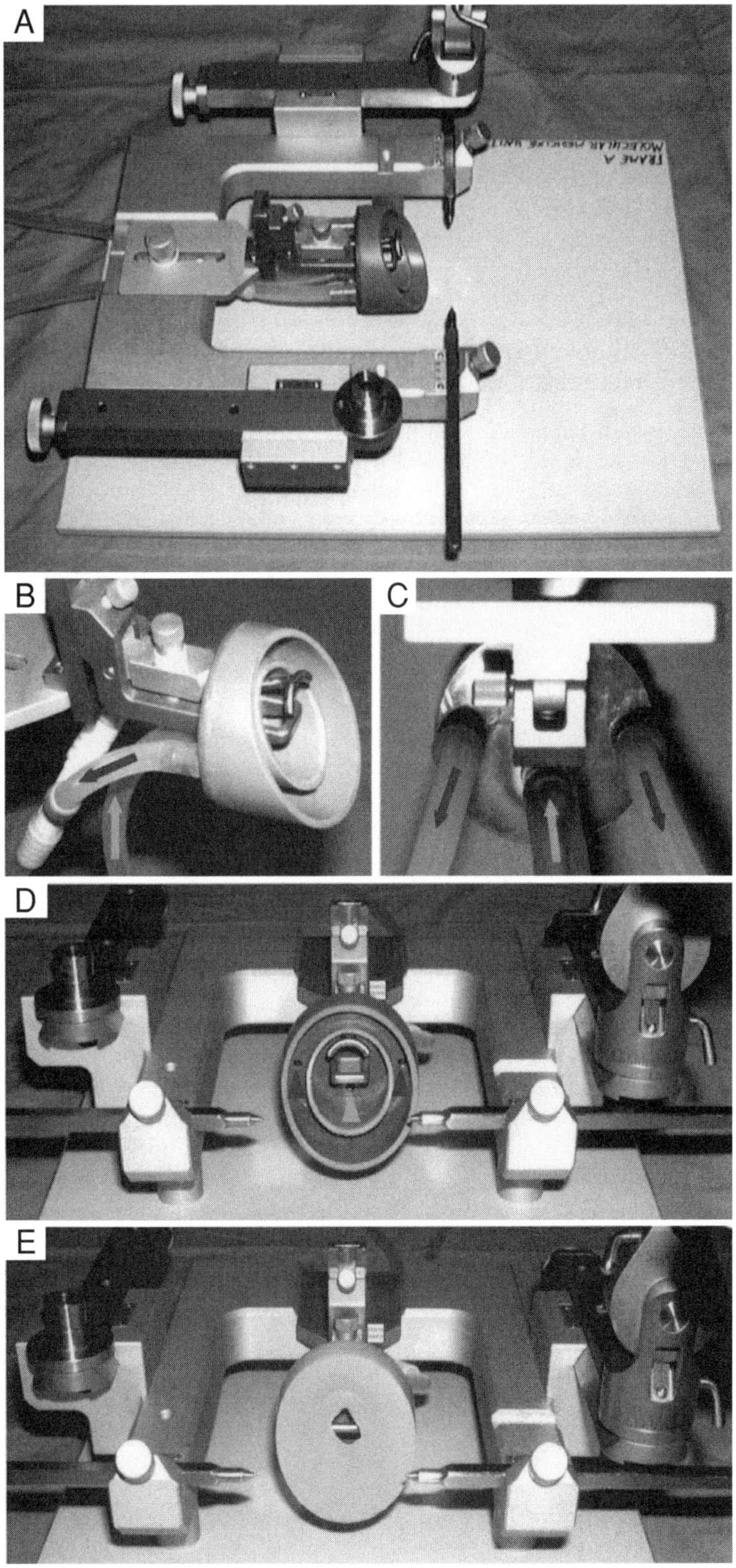

Figure 4.9.1 Stereotaxic frame modified for inhalational anesthesia. (**A**) View of the entire stereotaxic frame (model 51603 Stoelting) modified for inhalational anesthesia. (**B**) The modified mouthpiece. A stereotaxic anesthetic mask (model MkI, International Market Supplies) is attached to the incisor bar. The authors modified the MkI mask by cutting out a section of plastic at the back of the mask, allowing the nose bar to be moved up and down. (**C**) The back of the mask incorporates connections to one inlet tube bringing gas from the anesthetic machine to the animal (arrow) and 2 outlet tubes, which should be attached to a vacuum source or fluosorber for removing exhaled gas (arrows). (**D**) Frontal view of the modified mouthpiece showing the positions of the holes for the inlet tubing (arrow) and the outlet tubes (arrows). (**E**) A rubber balloon with a hole cut into the tip is stretched over the plastic frame to complete the mask.

Sharp scissors
Sterile gauze
Chromic catgut absorbable sutures

1. Position the stereotaxic frame relative to the light microscope and light box such that the microscope is focused on the ear bars of the frame. Stretch a rubber balloon over the mouthpiece of the stereotaxic unit and cut a small hole in the end to allow insertion of the animal's nose (Fig. 4.9.1). Position the anesthetic trolley such that the anesthetic tubing can be connected to the stereotaxic frame, and lay out the drill, heat pad, and sterilized surgical tools next to the frame.

2. Dilute the adenovirus preparation (with sterile saline solution or PBS, pH 7.4) to appropriate final injection concentration. Store on ice in a microcentrifuge tube, for up to several hours, until required. Do not refreeze the diluted virus.

   High doses of virus (i.e., above 10^8 infectious units) are generally associated with transient transgene expression and severe inflammation, while low doses (below 10^6 infectious units) result in minimal transduction (depending on promoter strength). Generally, administration of 10^7 infectious units of virus is optimal (when using vectors expressing transgenes from the major intermediate early human CMV promoter) in terms of providing relatively stable transgene expression with minimal inflammation.

3. Place the animal in the induction chamber and anesthetize with 4% halothane gas, vaporized with an oxygen/nitrous oxide mix (66% oxygen:33% nitrous oxide, i.e., O_2 set at a flow rate of 1500 ml/min and nitrous oxide at a flow rate of 750 ml/min).

4. When the animal is fully anesthetized, route the flow of halothane to the stereotaxic unit, remove the rat from the induction chamber, and quickly shave the fur on top of the head. Open the mouth of the rat and fit the mouthpiece of the stereotaxic frame such that the animal's nose is entirely enclosed within the rubber balloon mask. Reduce the halothane level to 1.5% of the carrier gas.

5. Slide the ear bars into each ear canal and tighten in place so that the head of the animal is firmly positioned and does not wobble. Check that the top of the head is lying horizontally. Place the heat pad underneath the animal to prevent hypothermia during surgical anesthesia. Monitor the animal's breathing rate throughout the operation.

6. When the animal is positioned firmly and correctly within the frame, place a drop of sterile saline solution into each of the eyes and swab the surgical site with a cotton swab dipped in 70% ethanol/isopropyl alcohol. Ensure that the animal is fully anesthetized by checking the lack of responses to footpad and tail pinching. Using a scalpel, make a midline incision into the skin, from above the eyes to the level of the ears. Use skin retractors to hold back the skin on either side of the incision, and determine that the bregma (the junction of the sagittal and transverse sutures) is visible on the exposed skull (see Fig. 3.7.1).

7. Remove the connective tissue covering the top of the skull by cleaning the cranium with a cotton swab.

8. Load the Hamilton syringe with the adenovirus solution and expel a small amount of the virus onto a cotton swab to verify the needle is not blocked. Clamp the needle into position on the frame.

   The volume of virus solution injected into the CNS will depend on the anatomical site. No more than 2 μl of virus solution should be injected into the brain parenchyma; however, up to 30 μl of virus can be administered to the ventricles.

9. Direct the light beams from the fiberoptic light pipes onto the exposed skull and focus the microscope onto bregma. (Turn light beams off or away from animal when not using microscope.) Position the syringe over the skull and tighten into place. While viewing the

brain through the microscope, position the needle using the 3 slide rules so that the needle bevel is directly over bregma.

10. Read the coordinates of bregma from the frame. Calculate the new coordinates of the site of injection by adding or subtracting the appropriate lateral and anterior/posterior values from bregma (see Paxinos and Watson, 1982). Move the needle to these new coordinates.

11. Lower the needle at the new coordinates, so that it is just touching the surface of the skull (be careful not to damage the needle point), and mark this position with a small dot using a very fine marker pen. Raise the needle to allow drilling.

12. Viewing the surface of the skull through the microscope, drill a small hole ~2 mm in diameter most of the way through the skull at the position marked by the small dot.

    Drilling into bone generates considerable heat. It is therefore advisable to drill in short bursts, intermittently bathing the hole with cold sterile saline solution, which can then be removed with a cotton swab before recommencing drilling. The hole may bleed during drilling if blood vessels coursing through the bone are ruptured. This should cause only very limited bleeding, which can be staunched with a cotton swab if necessary. It is sometimes necessary to refocus the microscope on the bottom of the deepening hole during drilling.

13. Stop drilling when the base of the hole becomes translucent. Perforate the remaining thin layer of skull with a sterile needle and, using a pair of sharp, curved forceps, carefully remove this remaining layer of bone to expose the dura matter. Using a cotton swab, soak up any clear CSF leaking into the hole.

14. Using a sterile, bent needle, carefully perforate the dura (whitish, opalescent, and elastic) and remove as much of the membrane as needed to expose the surface of the brain (darker and more yellow in color).

15. Lower the Hamilton needle into the hole until it just touches the surface of the brain and read the vertical coordinate of this position. Calculate the new vertical coordinate of the site of injection. Adjust coordinates slightly, if necessary, to avoid rupturing large blood vessels running across the surface of the brain.

16. Lower the needle into the brain to the site of injection and wait 2 to 3 min before slowly depressing the syringe plunger by 0.5 μl, over a further minute. Wait 1 min for the virus solution to infuse into the brain. Inject another 0.5 μl of virus and wait one moreminute. Repeat until the entire 2 μl of virus has been administered. Wait for 5 min after the final administration. Remove the needle from the brain very slowly and close the skin incision with sutures.

    Superfine glass needles can be fitted with wax over the Hamilton needle to reduce physical damage to the brain during injection.

17. Swab the sutured area with sterile saline, turn off the anesthetic and nitrous oxide while maintaining oxygen flow to the unit, and allow the animal to recover.

BASIC PROTOCOL 2

EVALUATION OF GENE TRANSFER, INFLAMMATORY RESPONSES, AND VECTOR-MEDIATED TOXICITY FOLLOWING IN VIVO ADMINISTRATION OF RECOMBINANT ADENOVIRUS VECTORS

There are many different methods for evaluating adenovirus-mediated gene transfer in an in vivo paradigm. Immunohistochemical detection of the transgene product within brain sections

4

Table 4.9.1 Antibodies Used to Detect Inflammation, Immune Responses and/or Adenovirus Vector-mediated Cytotoxicty in the Rat Brain

Antibody	Specificity within the brain	Dilution	Source
Inflammatory and immune responses:			
ED1	Infiltrating monocytes and activated macrophages/microglia	1:1000	Serotec
CD43	Infiltrating lymphocytes	1:500	Serotec
CD4	Helper T cells (perivascular microglial cells are also recognized)	1:200	Serotec
CD8	Cytotoxic T lymphocytes and natural killer cells.	1:500	Serotec
CD8b	Cytotoxic T lymphocytes	1:2000	Pharmingen
CD161	Natural killer cells	1:2000	Serotec
CD45RA	B cells	1:2000	Pharmingen
OX-6	MHC class II	1:200	Serotec
OX-18	MHC class I	1:200	Serotec
OX-62	Dendritic cells and $\gamma\delta$ T cells	1:20	Serotec
ICAM I	Perivascular microglia/activated endothelial cells	1:100	Serotec
Virus-mediated acute cytotoxicity:			
GFAP	Glial fibrillary acidic protein; activated astrocytes	1:200	Roche
NeuN	Neuronal nuclei	1:50	Autogen Bioclear
Myelin integrity:			
MBP	Myelin basic protein	1:2000	Dako

4

is a widely used procedure. UNIT 1.2 describes a method in which serial brain sections are processed using horseradish peroxidase (HRP)-based immunohistochemistry to detect the transgene product. In the protocol below, a modification of the method described in UNIT 1.2 is presented. Adenovirus vectors are inflammatory and immunogenic, and can be cytotoxic at high multiplicities of infection. It is therefore expedient to evaluate the inflammatory, immune, and cytotoxic responses to vector delivery, simultaneously with any evaluation of vector-mediated gene transfer. The HRP-based immunohistochemistry protocol described here can be used with appropriate antibodies to visualize the acute and/or chronic infiltration of immune and inflammatory cells, the activation of brain microglia and astrocytes, the loss of glial cell or neuronal markers (indicating cell loss through acute or chronic cytotoxicity), and/or the integrity of brain myelination after in vivo administration of adenovirus vectors to the brain. Table 4.9.1 shows some of the antibodies that have been used to investigate host responses to adenoviral vector administration in the rat striatum.

Materials *(see* APPENDIX 1 *for items with ✓)*

Brain sections (see Support Protocol 1)
✓ TBS
TBS/Triton: TBS containing 0.5% (v/v) Triton X-100
0.3% (v/v) H_2O_2 in PBSTBS/Triton containing 10% horse serum (TBS/Triton/10% HS)
TBS/Triton containing 1% horse serum (TBS/Triton/1% HS)
Primary antibody recognizing epitope of interest (e.g., transgene or immune cell marker)
Secondary biotinylated antibody
Vectastain ABC elite kit (Vector Laboratories)
✓ PBS

0.1 M sodium acetate, pH 6.0
✓ DAB staining solution

Soft-bristled paintbrush
Glass scintillation vials with plastic caps
Platform shaker
10-ml pipets and pipet aid
Glass Coplin jars
Gelatin-coated glass slides (*UNIT 1.1*)

1. Using a soft-bristled paintbrush, transfer the brain sections into glass scintillation vials containing approximately 5 ml of TBS/Triton. Wash the sections by shaking the vials at room temperature on a platform shaker for 5 min. Remove the TBS/Triton with a pipet aid and 10-ml pipet and discard. Add 3 ml of 0.3% H_2O_2 to each vial and incubate with shaking for 15 min to inactivate endogenous peroxidase. Wash the sections 3 times by incubating with 3 ml of TBS/Triton and shaking for 5 min for each wash.

2. Block nonspecific antibody binding sites and Fc receptors by incubating the sections with 1 ml of TBS/Triton/10% HS for 45 min with shaking. Ensure that all the sections are immersed in the blocking solution and are not stuck to the sides of the vial.

3. Wash the sections once for 5 min in TBS/Triton/1% HS, with shaking.

4. Incubate the sections with the primary antibody, diluted to the required extent in TBS/Triton/1% HS, overnight at room temperature, with shaking. Cap the tubes tightly.

   Before using an antibody for the first time, it is advisable to determine the optimal dilution factor by performing a small titration experiment incorporating 3 or 4 different dilutions. The data sheets provided with commercially available antibodies will often recommend a dilution factor and this can be used as a starting point. Primary antibodies can be reused if stored with sodium azide.

5. Wash the sections 5 times in TBS/Triton as in step 1. Incubate with the biotinylated secondary antibody, diluted appropriately in TBS/Triton/1% HS, for 4 hr with shaking.

6. Toward the end of the incubation period, prepare the avidin/biotinylated HRP complex (Solution AB) using the Vecta stain ABC Elite kit as follows:

   a. for each vial, prepare 1 ml of Solution AB by adding 10 µl of Solution A (avidin) and 10 µl of Solution B (biotinylated HRP) to 1 ml of PBS.

   b. Incubate with gentle mixing for at least 60 min before using.

7. Wash 5 times in TBS/Triton as in step 1. Incubate the sections for 3 hr in 1 ml of Solution AB, with shaking. Wash three times with 3 ml of PBS, 5 min each time. Wash twice with 3 ml 0.1 M sodium acetate (pH 6.0), 5 min each time.

8. Stain the sections by incubating in 1 ml of DAB staining solution with gentle shaking for 1 to 7 min.

   Carefully monitor the development of the stain; the sections will become faintly purple/black throughout. Specific staining of cells will sometimes be visible only under the microscope after mounting, dehydration, and coverslipping. To avoid overdeveloping, stain sections from only one or two vials at a time (the other sections can be left in sodium acetate until required), but ensure that all sections stained at different times are developed for the same length of time.

   CAUTION: *The staining solution is toxic and the DAB should be precipitated and inactivated with bleach before discarding.*

9. Wash the stained sections twice with 3 ml sodium acetate and twice with 3 ml of PBS. Store washed sections in the glass vials, at 4°C, for up to several days, or proceed directly to step 10.

10. Using a soft-bristled paintbrush, transfer the sections from the vial into a clear container (e.g., glass Coplin jar) filled with filtered PBS. Partially submerge a gelatin-coated slide in the PBS and float the sections onto the slide, using the paintbrush. Arrange the sections on the slide in anatomical order if possible (approximately 6 rat brain sections will fit on each slide).

11. Allow the sections to air dry in a dust-free environment for several hours to overnight, until they become fixed onto the gelatin-coated slides.

12. Dehydrate the sections and coverslip as described in *UNIT 1.2*, Basic Protocol 1.

ALTERNATE PROTOCOL 1

FLUORESCENCE-BASED IMMUNOHISTOCHEMICAL STAINING OF BRAIN SECTIONS

Basic Protocol 2 describes the immunohistochemical localization of a single marker antigen within tissue sections. The following protocol describes double-labeling of tissue sections using two different fluorescently-labeled antibodies. Confocal microscopy will greatly improve the accuracy and quality of data.

4

Additional Materials *(also see Basic Protocol 2)*

Two primary antibodies recognizing epitopes of interest, generated in different species)
Appropriate secondary antibodies labeled with two different fluorescent markers (e.g., fluorescein and Texas red)
Blocking serum (from the species in which the secondary antibodies were generated)
Fluorescence microscope

1. Using a soft-bristled paintbrush, transfer the brain sections (<30 μm thick) into glass scintillation vials containing ~5 ml of TBS/Triton. Wash the sections by shaking the vials at room temperature on a platform shaker for 5 min.

2. Block nonspecific antibody binding sites by incubating the sections for 45 min in TBS/Triton containing 10% blocking serum, from the species in which the first secondary antibody to be used was generated.

3. Wash the sections once for 5 min with TBS/Triton containing 1% blocking serum.

4. Incubate the sections overnight with the first primary antibody, diluted in the required amount of TBS/Triton/1% blocking serum.

5. Wash the sections 5 times in TBS/Triton. Incubate the sections for 4 hr with the appropriate fluorescent secondary antibody diluted in TBS/Triton/1% blocking serum, keeping sections away from strong light. Wash 5 times in TBS/Triton.

6. Repeat steps 2 to 5 for the second set of antibodies.

7. Mount the sections on gelatin-coated glass slides, dehydrate, and coverslip as described in Basic Protocol 1 of *UNIT 1.2*.

8. Visualize labeled cells using a fluorescence microscope.

BASIC PROTOCOL 3

SIMULTANEOUS EVALUATION OF VECTOR-INDUCED INFLAMMATION AND DEMYELINATION WITH LUXOL FAST BLUE AND CRESYL VIOLET

The following protocol describes a staining procedure wherein vibratome brain sections are histologically stained with both cresyl violet and Luxol fast blue for the simultaneous detection of cell nuclei (to visualize areas of inflammation) and myelin (to detect areas of demyelination). When examining a novel experimental paradigm, it is important to confirm actual demyelination using specific myelin antibodies (see Table 4.9.1).

Materials *(see APPENDIX 1 for items with ✓)*

Free-floating vibratome-cut brain sections (see Support Protocol 1)
50%, 70%, 80%, and 96% ethanol
95% methylated spirits
0.05% (w/v) aqueous lithium carbonate
✓ Luxol fast blue (LFB) solution
0.1% (w/v) cresyl violet in 1% (v/v) acetic acid

APES-coated glass slides (see Support Protocol 2)
Soft-bristled paintbrush
60°C oven

1. Mount free-floating vibratome-cut brain sections on APES-coated slides as described in step 17 of Basic Protocol 2 for gelatin-coated slides. Allow the sections to air dry and adhere to the slides.
2. Dehydrate the sections by sequentially immersing the slides for 10 min in each of the following dilutions of ethanol: 50%, 70%, 80%, and 96%.
3. Immerse the slides in filtered LFB solution in a glass jar with a lid and incubate the sections overnight (or for 16 hr) in a 60°C oven.

   Perform this staining section by section since it is highly variable.
4. Rinse the sections by immersion in 95% methylated spirits for 5 min, followed by distilled water for 5 min.
5. Start the differentiation process by agitating each slide for a few seconds in a jar of 0.05% lithium carbonate, using fresh lithium carbonate after agitation of several slides.
6. Continue to differentiate by agitation in 70% alcohol.
7. Rinse the slides in distilled water and examine the sections under the microscope. White matter (myelinated regions) should appear blue, while the rest of the brain should be pale. Repeat the differentiation steps if necessary until the gray and white matter are clearly distinguishable.
8. Counterstain by incubating the sections in 0.1% cresyl violet in 1% acetic acid for 15 min.
9. Wash the sections in distilled water, then dehydrate rapidly through graded alcohols and coverslip as described in Basic Protocol 1, *UNIT 1.2*. Nuclei and Nissl substance should appear purple.

BASIC PROTOCOL 4

EVALUATION OF VECTOR-MEDIATED TOXICITY BY TOLUIDINE BLUE STAINING

Toluidine blue staining of semi-thin plastic embedded sections (standard preparation for electron microscopy analysis; see *UNIT 1.2*) containing the injection site enables an evaluation of the integrity of the structure of neural cell bodies and processes, and allows a determination of the presence of extracellular edema within the tissue (Dewey et al., 1999).

Materials *(see APPENDIX 1 for items with ✓)*

Semi-thin plastic embedded 5-µm brain sections (see *UNIT 1.2*) mounted on gelatin-coated (see *UNIT 1.1*) or APES-coated (see Support Protocol 2) glass slides
✓ 1% (w/v) toluidine blue

1. Place a drop of 1% toluidine blue onto the brain section and leave for 1 to 2 min at room temperature. Wash off the stain by immersing the slide in distilled water.
2. Dehydrate and coverslip as described in Basic Protocol 2 of *UNIT 1.2*.
3. View using light microscopy.

BASIC PROTOCOL 5

DETECTION OF VECTOR GENOME IN BRAIN SECTIONS BY PCR

Levels of adenovirus-mediated transgene expression within the CNS generally decline over time. This protocol describes a nonquantitative PCR-based method for detecting adenoviral genomes within brain sections.

NOTE: Where possible, all steps described in this protocol should be performed in a tissue culture hood to avoid contamination. Particular care should be taken when setting up a PCR reaction to avoid cross-contamination between samples; use of aerosol-resistant filter pipet tips will reduce this risk. All tubes, tips and solutions (including phenol, chloroform, etc.) should be designated for this procedure only and should not be stored in close proximity to any virus stocks. Where possible, predesignated PCR-only pipets and racks should be used.

Materials *(see APPENDIX 1 for items with ✓)*

Free-floating vibratome-cut brain sections (see Support Protocol 1)
✓ Digestion buffer
25:24:1 (v/v/v) phenol/chloroform/isoamyl alcohol
✓ 3 M sodium acetate, pH 5.2
100% and 70% ethanol
Appropriate primer pairs (custom synthesized):
Adenovirus transcription unit IVa2 (primer pair amplifies a 687-bp fragment corresponding to residues 4071 to 4757, map units 11.31 to 13.21, of adenovirus type 5 genome):
Forward primer sequence: 5′-AAGCAAGTGTCTTGCTGTCT-3′
Reverse primer sequence: 5′-GGATGGAACCATTATACCGC-3′
Adenovirus transcription unit E1B primers (Easton et al., 1998; primer pair amplifies a 561-bp fragment, corresponding to residues 2100 to 2660, map units 5.83 to 7.38, of adenovirus type 5 genome):
Forward Primer: 5′-CAAGAATCGCCTGCTACTGTTGTC-3′
Reverse Primer: 5′-CCTATCCTCCGTATCTATCTCCACC-3′

Rat β-actin primers (modified from Lee and Cotanche, 1995; primer pair amplifies a 340-bp fragment within exon 4 of rat cytoplasmic β-actin gene, Genbank accession no. J00691):
Forward primer sequence: 5′-CCAGCCATGTACGTAGCCATCC-3′
Reverse primer sequence: 5′-GCAGCTCATAGCTCTTCTCCAGG-3′
HSV-1 thymidine kinase primers (Dewey et al., 1999; primer pair amplifies a 364-bp product of the HSV-1 TK gene):
Forward primer sequence: 5′-AAAACCACCACCACGCAACT-3′
Reverse primer sequence: 5′-GTCATGCTGCCCATAAGGTA-3′
10 mM dNTP mix (includes dATP, dTTP, dCTP and dGTP, each at 10 mM)
25 mM $MgCl_2$
Taq DNA polymerase and 10× buffer
2% (w/v) agarose gel containing ethidium bromide
✓ TAE buffer

95°C heating block
Sterile PCR tubes
Sterile pipet tips with filters
Thermal cycler

1. Transfer a single free-floating, unstained brain section from PBS into a sterile, screw-cap, 1.5-ml microcentrifuge tube, using a sterile pipet tip or scalpel blade.

   It is also possible to perform a modified version of the same procedure on tissue that has been immunohistochemically stained and mounted on glass slides, as described in UNIT 1.2.

2. Add 100 to 200 μl of freshly prepared digestion buffer (roughly twice the volume occupied by the brain section) and ensure the tissue is completely immersed. Screw the cap very tightly onto the tube. Incubate the sample for 24 hr in a shaking incubator at 37°C. Tape the tube horizontally to bottom of shaker, for greater agitation of tissue

3. Inactivate the proteinase K in the digestion buffer by heating the sample for 10 min at 95°C. Remove the proteinase K by two rounds of phenol/chloroform/isoamyl alcohol extraction. In each extraction, add an equal volume of 25:24:1 phenol/chloroform/isoamyl alcohol, briefly vortex to mix, and separate the aqueous phase from the organic phase by microcentrifugation for 2 min at 10,000 × *g*.

4. Precipitate the genomic DNA by adding 1/10 vol of 3 M sodium acetate, pH 5.2, and 2 vol of 100% ethanol. Microcentrifuge 10 min at maximum speed, room temperature, then discard the supernatant and wash the DNA pellet in 70% ethanol.

5. Air-dry the pellet in a tissue-culture hood and resuspend the DNA in 50 μl of deionized sterile filtered water. Use the DNA immediately for PCR, or add DNase-free RNase to a final concentration of 20 μg/ml and store at −20°C.

6. Prepare PCR reaction samples in a 50-μl reaction volume in appropriate sterile PCR tubes as follows:

   10 μl template DNA (step 5)
   200 nM each dNTP (add from 10 mM dNTP mix)
   2 ng/μl each primer
   2 mM $MgCl_2$(add from 25 mM stock)
   1 U *Taq* DNA polymerase.

   Remember to include negative controls: (1) a reaction sample prepared without template DNA and (2) a sample prepared with template DNA extracted from a noninfected brain. This latter sample should be positive for β-actin, but negative for adenoviral genome. Where possible, a positive control for adenoviral genome should also be included (this can be prepared with DNA extracted from cesium chloride-purified virus stock).

7. Perform PCR amplification (*see* CPMB 15.1), using a thermal cycler programmed with the appropriate cycle parameters for each primer pair:

   a. For adenovirus transcription unit IVa2 primers:

1 cycle:	5 min	95°C	(initial denaturation)
35 cycles:	30 sec	95°C	(denaturation)
	30 sec	56°C	(annealing)
	1 min	72°C	(extension)
1 cycle:	10 min	72°C	(final extension)

   b. For adenovirus transcription unit E1B primers:

1 cycle:	5 min	95°C	(initial denaturation)
35 cycles:	30 sec	95°C	(denaturation)
	30 sec	57°C	(annealing)
	1 min	72°C	(extension)
1 cycle:	10 min	72°C	(final extension)

   c. For rat β-actin primers:

1 cycle:	5 min	95°C	(initial denaturation)
30 cycles:	1 min	95°C	(denaturation)
	1 min	63°C	(annealing)
	1 min	72°C	(extension)
1 cycle:	10 min	72°C	(final extension)

   d. For HSV-1 thymidine kinase primers:

1 cycle:	5 min	95°C	(initial denaturation)
30 cycles:	30 sec	95°C	(denaturation)
	30 sec	63°C	(annealing)
	1 min	72°C	(extension)
1 cycle:	10 min	72°C	(extension)

8. Analyze the amplification products by agarose gel electrophoresis in TAE buffer using a 2% agarose gel containing ethidium bromide (*see* CPMB 2.5A).

BASIC PROTOCOL 6

ASSESSMENT OF BLOOD-BRAIN BARRIER PERMEABILITY IN THE RAT BRAIN

Mechanical or vector-induced disruptions to the integrity of the BBB are usually repaired within a few days. Any assessment of vector-induced BBB permeability should therefore be performed within the first few days after surgery.

Materials *(see* APPENDIX 1 *for items with* ✓*)*

Horseradish peroxidase (HRP; Type II; Sigma)
Normal saline: 0.9% (w/v) NaCl, sterile
Adult rat

✓ PBS, pH 7.4
✓ Hanker-Yates solution A
✓ Hanker-Yates solution B

Towel or other restraining device for rats
0.5-μl micro-fine disposable needle-fitted syringe (e.g., 0.3 mm × 12.7–mm, Becton, Dickinson)
6-well tissue culture dishes

1. Immediately before use, dissolve 17.5 mg HRP in 1 ml of sterile saline. Keep this solution on ice and out of the light as much as possible.

2. Restrain the rat by wrapping the animal in a towel so that only the tail protrudes and dilate the tail vein by placing the tail into a beaker of warm water (about 40°C) for ~5 min, or until the veins are visibly dilated.

3. Inject 400 μl of the freshly prepared HRP solution into dilated tail vein using a micro-fine needle-fitted syringe (e.g., insulin syringe). Return the animal to the cage and wait for 20 min, then perform perfusion fixation (see Support Protocol 1).

4. Section the brain using a vibratome as described in Support Protocol 1, and store in PBS, pH 7.4 (without sodium azide), until ready to perform the HRP detection assay (ideally, within a few days; no later than a few weeks).

5. Put the brain sections in the wells of a 6-well tissue culture dish containing 4 to 5 ml of Hanker-Yates solution A and incubate on a shaker for 15 min.

   CAUTION: *Hanker-Yates solutions are very toxic. Wear gloves.*

   If a positive control is desired, brain sections from a rat intrastriatally injected with 1 to 5 μg of lipopolysaccharide (LPS) can be included.

6. Remove and discard this solution and wash sections twice in PBS, pH 7.4, allowing 3 to 5 min for each wash.

7. Transfer the sections into another 6-well plate containing 4 to 5 ml of freshly prepared Hanker-Yates solution B. Incubate the plate in the dark (covered with foil) on a shaker for 15 min to develop the colored reaction product.

   If the blood-brain barrier has been damaged, a dark brown reaction product will be formed wherever the HRP has leaked into the brain.

8. Wash sections twice in PBS for a total of 5 min. Mount sections on gelatin-coated slides and allow to dry before coverslipping with DPX mountant as described in *UNIT 1.2*, Basic Protocol 1.

4

BASIC PROTOCOL 7

ACTIVATION OF A STRONG ANTI-VIRAL IMMUNE RESPONSE BY INTRADERMAL ADMINISTRATION OF ADENOVIRUS VECTOR

The relative persistence of vector-mediated expression in the brain, compared with that in peripheral organs, is attributed to the failure to elicit an effective anti-adenoviral T cell response following the intraparenchymal vector injection. After peripheral exposure to vector, activated anti-adenoviral T cells can cross the blood-brain barrier and facilitate the elimination of adenoviral vector-mediated transgene expression in the brain.

Materials

Adult rat
Normal saline: 0.9% (w/v) NaCl, sterile
Virus aliquots in 1.5-ml microcentrifuge tubes on ice (*UNIT 4.8;* each tube should contain 100 µl of virus diluted in saline such that each 100 µl contains 5×10^8 infectious units of vector)

Gas anesthetic trolley with the following components:
- Halothane gas anesthetic
- Halothane vaporizer (e.g., Fluotec)
- Medical oxygen cylinder
- Medical nitrous oxide cylinder
- Induction chamber
- Halothane scavenger (e.g., Fluovac)

Surgical shavers (Stoelting)
0.5-µl micro-fine disposable needle-fitted syringe (e.g., 0.3 mm × 12.7-mm, Becton, Dickinson)
Gauze surgical swabs

NOTE: soak used needles in bleach or 1% Virkon before discarding and autoclaving.

1. Place the animal in the induction chamber and anesthetize with 4% halothane gas, vaporized with oxygen at a flow rate of 1500 ml/min and nitrous oxide at a flow rate of 750 ml/min.

2. When the animal is fully anesthetized, remove it from the induction chamber and place it on a surgical table with its nose inserted into the outlet tubing of the anesthetic trolley. Reduce the halothane level to 1.5% of the carrier gas. Shave a patch of fur from the top of animal's back. Clean the area by swabbing with sterile saline.

3. Withdraw 100 µl of the virus solution (containing 5×10^8 infectious units of vector) into the 0.5-ml syringe.

4. Pinch the shaved patch of skin with one hand so that a ridge protrudes between the thumb and forefinger. With the bevel uppermost, insert the needle horizontally into the dermis, close to the top of the ridge of skin. Gently push the needle until at least 1 cm is inserted. With the needle still inserted, allow the skin to relax to its natural position. Slowly depress the syringe plunger to inject the virus into the dermis. Gently withdraw the needle from the skin before switching off the anesthetic gas and allowing the animal to recover.

SUPPORT PROTOCOL 1

PREPARATION OF BRAIN TISSUE FOR PROCESSING

UNIT 1.1 describes a method for perfusion fixation of animals using a peristaltic pump to perfuse the animal with saline and fixative. In the protocol below, the animal is instead perfused with oxygenated Tyrode's solution containing heparin (see recipe in *APPENDIX 1*) prior to being perfused with fixative to maintain the supply of oxygen to the brain and tissues and avoid clotting during blood clearance, thus preventing cellular death prior to perfusion of the fixative. Figure 4.9.2 depicts a perfusion apparatus that uses gravity to facilitate the flow of Tyrode's and fixative. Also, the descending aorta is clamped just below the liver prior to starting the perfusion of fixative, to improve the flow of fixative to the brain.

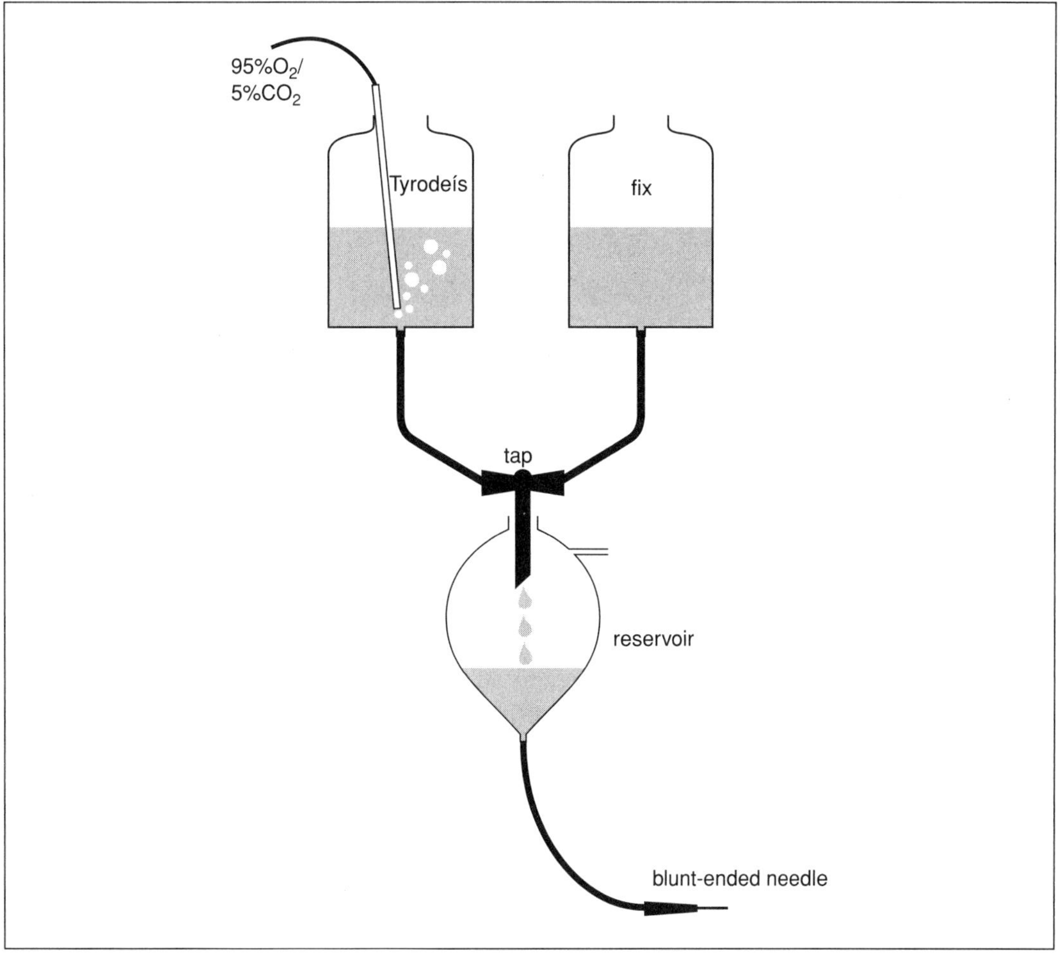

Figure 4.9.2 Apparatus for perfusion-fixation by gravity.

Materials *(see* APPENDIX 1 *for items with ✓)*

✓ PBS

Perfusion-fixed rat brain (*UNIT 1.1*; see introduction to this protocol, above, for modifications), post-fixed for 5 hr

Vibratome (e.g., model VT1000S, Leica)
Soft-bristled paintbrush
6-well or 12-well tissue culture plates

1. Add ~4 ml of PBS to each well of a fresh 6-well tissue culture dish, the storage container for the brain sections.

2. Mount the region of brain containing the injection site on a cutting platform using superglue. Slice the tissue at 20- to 70-μm intervals using a vibratome according to the manufacturer's instructions. Always immerse the brain in PBS (not water) during sectioning and use a fresh blade to section each brain. Use a soft-bristled paintbrush to transfer the brain sections from the cutting chamber to the 6-well plate.

3. Place serial sections in consecutive wells of the 6-well plate, so that each well contains a similar representation of sections throughout the injection site/area of interest. Place at least 10 sections in each well to obtain a good representation of a large area of the striatum. Store the sections at 4°C.

If the sections are to be stored for more than one week, it is essential to add sodium azide (to a final concentration of 0.02%) to each of the wells to prevent microbial growth. Sodium azide degrades spontaneously and needs to be replenished at regular intervals (e.g., every couple of months).

CAUTION: *Sodium azide is highly toxic and can cause death if ingested or injected.*

SUPPORT PROTOCOL 2

PREPARATION OF APES-COATED SLIDES

APES-coated slides are applicable for applications in which slide-mounted sections are stained (e.g., Luxol fast blue staining) because gelatin is water-soluble and the brain sections can become detached when the slides are incubated in solution for long periods of time.

Materials

95% ethanol
Acid alcohol: 1% (w/v) hydrochloric acid in 70% methylated spirits
3% (v/v) 3-aminopropyl-trioxysilane (APES; Sigma) in acetone
Acetone
Glass microscope slides
37°C or 50°C drying oven

1. Rack the glass microscope slides and degrease them by immersion in 95% alcohol for at least 2 min. Rinse three times in tap water, then once in deionized distilled water, and once in acid alcohol. Leave the slides to air dry.
2. Immerse the rack of slides in freshly prepared 3% APES in acetone for 2 min.
3. Rinse the slides by immersion in acetone for 2 min. Wash in distilled water and dry in the oven at 37°C or 50°C overnight. Store up to several months.

SUPPORT PROTOCOL 3

TRANSPORT OF RECOMBINANT ADENOVIRAL VECTORS ON DRY ICE

Clinical trials and collaborations between research groups using recombinant adenoviral vectors require the use of dry ice transport between laboratories in order to keep the virus stocks frozen. It is recommended that the virus be retitrated after shipment, as titration protocols (and therefore results) vary from laboratory to laboratory. This is especially important when comparing the effects of two or more viruses against each other. It has recently been observed that significant loss of viral titer occurs during dry ice transport. A combination of sealing the aliquot in Parafilm, placing this into a 50-ml centrifuge tube, then placing the tube into two polythene bags, maintains the viral titer after storage in dry ice for 48 hr. Alternative methods are described in the original manuscript by Nyberg-Hoffman and Aguilar-Cordova (1999).

Materials

Virus aliquoted into labeled 0.5-ml Treff Lab tubes or other suitable tubes
50-ml polypropylene centrifuge tubes (Greiner)

Minigrip resealable polythene bags (Fisher)
Parafilm, cut into squares 3 cm × 3 cm
Polystyrene box
Cardboard box (into which the polystyrene box snugly fits)
Brown tape

NOTE: Each step performed below is done individually at short intervals in the −80°C freezer to keep the aliquots frozen. Freeze/thawing of the aliquot may affect the viral titer.

1. Label the 50-ml polypropylene centrifuge tubes and Minigrip resealable polythene bags with the name and titer of the virus and the number of aliquots to be shipped. Place the tubes and bags in the −80°C freezer to cool.
2. Quickly wrap each virus aliquot completely with a 3 × 3–cm Parafilm square. Place the aliquots into the 50-ml centrifuge tubes and firmly screw the caps.
3. Place the centrifuge tube into two nested polythene bags. Remove excess air and seal completely. Keep the packaged viruses at −80°C until the dry ice container is ready.
4. Obtain enough dry ice pellets to completely fill the polystyrene box used for the shipment.

   Polystyrene boxes holding 3 kg of dry ice are sufficient for shipments lasting 48 hr. For longer shipments (i.e., U.S.A. to Europe, or between countries with strict customs regulations), use a larger box holding 5 to 7 kg of dry ice so that the dry ice will last at least 72 hr.

5. Fill the polystyrene box with 3 cm of dry ice. Take this box and the rest of the dry ice to the −80°C freezer. Place the virus aliquots on top of the 3-cm layer of dry ice. Immediately pour the rest of dry ice on top, filling the box completely. Shake the box to settle the pellets and top off the box if necessary, leaving no gaps.
6. Place the lid on the polystyrene box and seal with tape. Place the box in the cardboard box and seal for shipment.

References: Dewey et al., 1999; Easton et al., 1998; Gerdes et al, 2000; Thomas et al, 2001; Wood et al., 1996a,b

Contributors: Clare E. Thomas, Evelyn Abordo-Adesida, Tricia C. Maleniak, Daniel Stone, Christian A. Gerdes, and Pedro R. Lowenstein

UNIT 4.10

Production of High-Titer Lentiviral Vectors

Lentiviruses, such as the human immunodeficiency virus (HIV), are a subfamily of retroviruses that can infect both growth-arrested and dividing cells. Accordingly, lentiviral vectors efficiently transduce targets such as neurons and glial cells, both in tissue culture and in vivo (Naldini et al., 1996; Blomer et al., 1997; Miyoshi et al., 1997).

When producing vector stocks, it is mandatory to avoid the emergence of replication-competent recombinants. In one version of the lentiviral vector system described here, vector particles are generated from four separate plasmids (Fig. 4.10.1). This ensures that only replication-defective viruses are produced, because the four plasmids would have to undergo multiple and complex recombination events to regenerate a replication-competent entity. The pathogenic potential of HIV stems from the presence of nine genes that all encode for important virulence factors. Fortunately, six of these genes can be deleted from the HIV-derived vector system without altering its gene-transfer ability. The resulting multiply-attenuated design of currently used HIV vectors ensures that the parental virus cannot be reconstituted.

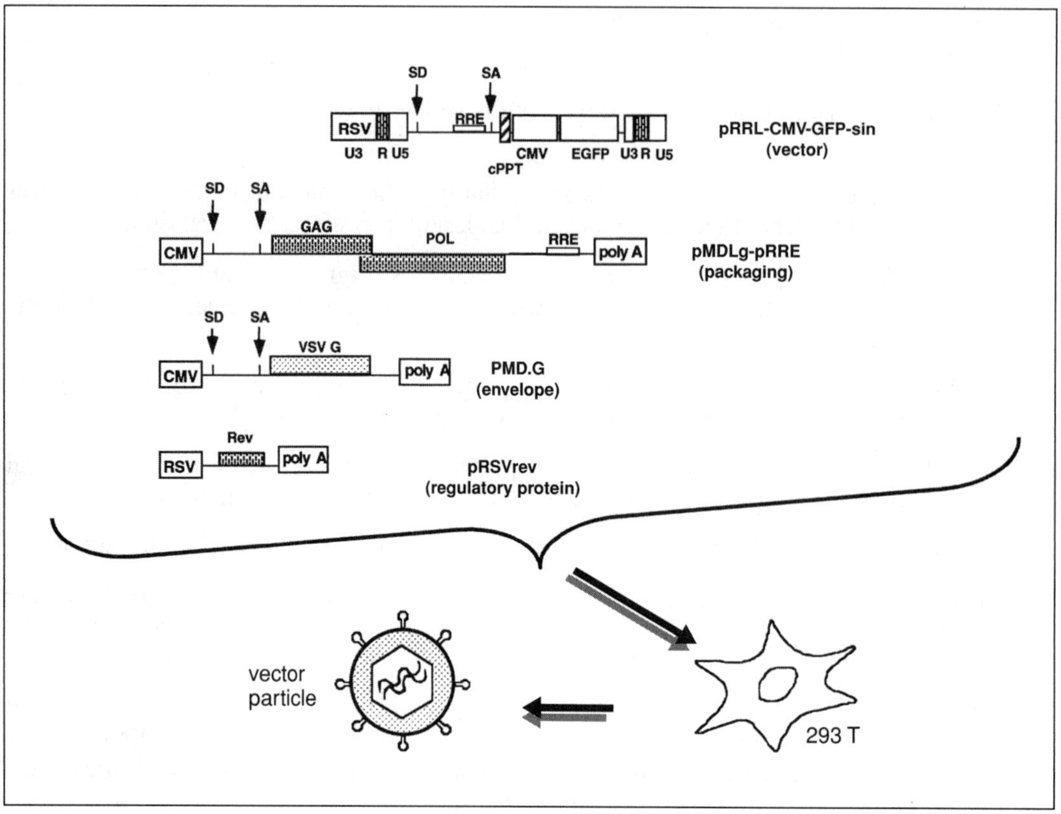

Figure 4.10.1 Production of HIV vectors: Schematic description of the protocol for HIV vector production. Vectors are produced by transfecting 293T cells with four plasmids, the main features of which are shown. The vector plasmid (top) is characterized by the presence of a chimeric RSV/HIV 5′ LTR and a partially deleted 3′ LTR. The Gag and Pol proteins of HIV-1 and the G protein of VSV are encoded by two independent plasmids (second and third from top). The fourth plasmid (bottom) encodes the Rev post-transcriptional regulator, essential for Gag/Pol expression and vector production. Abbreviations: CMV, immediate early promoter from the human cytomegalovirus; cPPT, central polypurine tract (Follenzi et al., 2000); PolyA, polyadenylation signal; RRE, Rev responsive element; RSV, U3 region (promoter) from the Rous Sarcoma Virus; SD and SA, splice donor and acceptor sites, respectively; sin, self-inactivating.

CAUTION: VSV G-pseudotyped lentiviral vectors have a broad tropism, both in vitro and in vivo; biosafety precautions need to take into account the nature of the transgene. A P2 laboratory is required. Procedures using lentiviral vectors must be reviewed and approved by the local biosafety committee of the institution where they are conducted.

NOTE: All solutions and equipment coming into contact with living cells must be sterile, and aseptic technique should be used accordingly.

BASIC PROTOCOL

PRODUCTION OF HIGH-TITER HIV-1-BASED VECTOR STOCKS BY TRANSIENT TRANSFECTION OF 293T CELLS

The VSV G-pseudotyped vector is the best choice for most gene-delivery experiments, both in vitro and in vivo. Detailed information on the sequence of the plasmids used in this protocol is available on the website *http://www.tronolab.unige.ch*. This website can also be used to request all the plasmids mentioned in this unit.

Materials *(see* APPENDIX 1 *for items with* ✓*)*

293T cells (information on laboratories providing 293T cells is available at *http://www.tronolab.unige.ch*)
✓ DMEM, supplemented with 10% FBS (DMEM-10)
pMD.G (encoding the VSV G envelope protein) in TE
pRRL-CMV-GFP-sin (vector) in TE
pMDLgag/polRRE (encoding the HIV-1 Gag and Pol proteins) in TE
pRSVrev (encoding the HIV-1 Rev protein) in TE
✓ TE buffer, pH 8.0
✓ 0.5 M $CaCl_2$
✓ 2× HeBS
70% ethanol in a spray bottle
✓ PBS, pH 7.4
14% (v/v) bleach

10-cm tissue culture dishes
37°C humidified incubators, 10% and 5% CO_2
0.5- and 1.5-ml microcentrifuge tubes, sterile, disposable
15- and 50-ml conical centrifuge tubes, sterile
125-ml filter bottles, 0.45-μm pore size
75-cm^2 tissue culture flasks
30-ml (25 × 89–mm) and 5-ml (13 × 51–mm) disposable conical polyallomer ultracentrifuge tubes (Beckman)
Ultracentrifuge, swinging bucket rotors

4

NOTE: All solutions and equipment coming into contact with living cells must be sterile, and proper aseptic technique should be used accordingly.

1. Maintain 293T cells in DMEM-10 medium, in 10-cm tissue culture dishes. Culture cells in a 37°C humidified incubator with a 10% CO_2 atmosphere, and split 1:4 to 1:6, three times per week.

2. Two days before the transfection, prepare 10 dishes of 2 × 10^6 293T cells. Incubate overnight in a 37°C humidified incubator with a 10% *or* 5% CO_2 atmosphere. On the following day, prepare 20 dishes of 293T cells by splitting the 10 dishes 1:2. Seed cells in 10 ml of DMEM-10. Incubate overnight in a 37°C humidified incubator with a 10% *or* 5% CO_2 atmosphere.

3. Adjust the DNA concentration of all plasmids (i.e., pMD.G, pRRL-CMV-GFP-sin, pMDLg/pRRE, and pRSVrev) to 1 mg/ml in TE buffer, pH 8.0 (see *APPENDIX 2G* for DNA quantitation method). For 20 plates, mix 100 μl of pMD.G, 400 μl of pRRL-CMV-GFP-sin, 240 μl of pMDLg/pRRE, and 60 μl of pRSVrev in a sterile 1.5-ml microcentrifuge tube.

4. Add 210 μl of sterile distilled water to each of 20 disposable 1.5-ml microcentrifuge tubes. Add 40 μl of the DNA mixture to each of these tubes, and vortex vigorously. Add 250 μl of 0.5 M $CaCl_2$ to each tube, and vortex.

   Since plasmids are stored in TE buffer, the water/DNA mixture contains 1.6 mM Tris. The increased buffer capacity is important for good transfection efficiency.

5. Add 500 μl of 2× HeBS to twenty 15-ml sterile conical tubes. To each of these tubes, then slowly transfer, dropwise, the 500 μl of DNA/$CaCl_2$ mixture from each tube prepared above, while vigorously vortexing to form a fine precipitate that can be taken up efficiently by cells. Leave at room temperature for 30 min. Then add the 1 ml of precipitate in these tubes, dropwise, to each of the 20 culture dishes prepared in step 2. Mix by gentle swirling until the medium has recovered a uniformly red color.

6. Place the dishes overnight in a 37°C humidified incubator with a 5% CO_2 atmosphere. Early the next morning, aspirate the medium and gently add 10 ml of fresh DMEM-10, prewarmed to 37°C. Incubate for 28 hr.

7. Transfer the culture medium from each plate to four 50-ml centrifuge tubes. Close the tubes, and spray them with 70% ethanol before taking them out of the hood. Centrifuge 2 min at 500 × *g*, 4°C, to pellet detached cells.

8. Connect a 125-ml filter bottle to a vacuum line. Filter 100 ml of culture medium through the 0.45-μm pore size membrane. Use a second filter bottle to clear the remaining 100 ml of culture medium. Pool the filtered medium in a 75-cm^2 tissue culture flask.

9. *Optional:* Save a 1-ml aliquot of the filtered vector stock in a 1.5-ml microcentrifuge tube for titration. Keep this aliquot at 4°C until used.

   VSV G-pseudotyped lentivectors have a halflife of 24 hr at 37°C. Aliquots of vector stocks should be stored at 4°C for up to 24 hr, or frozen at −80°C for longer storage.

Concentrate vector (first ultracentrifugation)

10. Transfer the filtered culture medium into six 30-ml disposable conical polyallomer ultracentrifuge tubes. Place the tubes in a swinging-bucket ultracentrifugerotor. Close the buckets before taking them out of the hood.

    P2 practice implies that vector suspensions are in a closed container when taken out of the hood.

11. Ultracentrifuge 90 min at 72,100 × *g*, 16°C.

12. Open the ultracentrifuge buckets in the hood. Use forceps to take the tubes carefully out of the buckets. Invert the tubes to transfer the supernatants to a 75-cm^2 tissue culture flask. Save a small aliquot of the supernatant for titration (<5% of the pre-centrifugation viral particles should be present in the supernatant). To the remaining supernatant, add 1/6 volume of 14% bleach, mix, wait 1 hr, and discard. Keep the tubes inverted and wipe the walls of the tubes with paper towels to eliminate as much as possible of the supernatant. Hold paper towels with forceps. Do not wipe the conical part of the tubes. Put the driedtubes on ice. Add 600 μl of PBS to each of the vector pellets.

13. Resuspend pellets by pipetting up and down 20 times with a 1000-μl pipet tip, avoiding foaming. Leave on ice 30 min.

14. Pipet up and down again 20 times. Pool the resuspended vector particles from the six tubes into one 5-ml polyallomer centrifuge tube. Add sufficient PBS to fill the tube.

15. Dilute a 5-μl aliquot of the vector suspension into 195 μl DMEM-10. Keep this aliquot at 4°C until used for titration (see Support Protocols 1 and 2).

    Comparing the vector particle numbers at steps 9 and 15 allows calculation of the recovery achieved after the first centrifugation. Similarly, particle recovery after the second centrifugation is calculated by comparing the particle number at steps 15 and 19. To calculate particle recovery, divide the particle number after the centrifugation by the particle number before the centrifugation. Particle number in a suspension is calculated by multiplying the titer (TU/ml) by the volume (ml) of the suspension.

Concentrate vector further (second ultracentrifugation)

16. Ultracentrifuge 90 min at 76,000 × *g*, 16°C.

17. Repeat step 12 through discarding supernatants and wiping wall of centrifuge tube. Thenadd 210 μl of PBS to the vector particles at the bottom of the tube.

18. Resuspend the vector particles with a 1000-μl pipet tip by pipetting up and down 20 times. Incubate on ice for 2 hr. Complete the resuspension by pipetting up and down 20 times again.

19. Dilute 5 μl of the concentrated vector stock in 495 μl of DMEM-10. Mix 100 μl of this first dilution with 400 μl of fresh DMEM-10. Keep these 1:100 and 1:500 diluted stocks at 4°C until used for titration (see Support Protocol 1 and 2). Compare aliquots saved at steps 9, 15, and 19 to calculate percent recovery).

20. Divide the concentrated stock (step 18) into 10 aliquots of 20 μl. Store the aliquots at −80°C.

SUPPORT PROTOCOL 1

TITRATION OF LENTIVIRUS GFP VECTOR STOCKS

Even though 293T cells are transduced as efficiently as HeLa cells, they have the disadvantage of poor adherence characteristics. While this is not a problem for the GFP detection protocol (Support Protocol 1), it is a point to keep in mind when using the LacZ staining procedure (Support Protocol 2). In either case, if using 293T cells, all reagents must be added by very gentle pipetting.

Titrating GFP-positive cells using a fluorescence-activated cell sorter (FACS) is more rapid, convenient, and sensitive than it is by counting green cells visualized with a fluorescent microscope.

Additional Materials *(also see Basic Protocol)*

HeLa cells (ATCC #CCL-2)
Vector (see Basic Protocol)
0.25% trypsin/0.53 mM EDTA (without dye; Life Technologies); dilute commercially available 10× stock in PBS

6-well tissue culture plates
Fluorescence-activated cell sorter (FACS; Becton Dickinson) and appropriate tubes

1. The day before titration, seed 0.5×10^5 HeLa cells in 2 ml of DMEM-10 into all wells of a 6-well culture plate, ensuring a uniform spread of cells on the bottom of the wells. Prepare one plate for each vector stock to be titrated. Incubate overnight at 37°C, 10% CO_2.

   0.5×10^5 cells are seeded to produce 1×10^5 cells per well the next day. The number of cells seeded must be accurate because it will be taken into account when calculating the titer (see Equation 4.10.1).

2. To five wells, add aliquots of the vector to be titrated: use 50 μl and 25 μl of the undiluted stock, and 100 μl, 50 μl and 25 μl of a 1:50 diluted stock (corresponding to 2.0, 1.0, and 0.5 μl of undiluted vector). Do not infect the cells in the last well; these are controls. Incubate 2 days.

3. Before the fluorescence-activated cell sorter (FACS) analysis, remove the culture medium, wash once with 2 ml PBS, and add 500 μl of 0.25% colorless trypsin/0.53 mM EDTA. Incubate 5 min at 37°C (to detach cells). Pipet up and down with a 1000-μl pipet tip to disrupt clumps. Transfer cells to a FACS tube containing 500 μl PBS.

   If desired, an aliquot of the cells can be kept in culture. If the FACS analysis is not done within 1 hr, cells can be fixed in 4% (w/v) paraformaldehyde solution for 30 min and kept for at least 1 week at 4°C.

4. Determine the percentage of GFP-positive cells by FACS analysis.

5. Calculate the titer in transducing units (TU)/ml, according to the formula:

$$\frac{(1 \times 10^5 \text{ seeded cells} \times \% \text{ GFP-positive cells}) \times 1000}{\mu\text{l of vector}}$$

For accurate titer calculations, the number of GFP-positive cells in 2 wells infected with 2 consecutive dilutions must be close to the expected 1:2 ratio. This linearity is observed when <15% of the target cells are transduced.

SUPPORT PROTOCOL 2

TITRATION OF LENTIVIRUS *LacZ* VECTOR STOCKS

Additional Materials *(also see Basic Protocol; see* APPENDIX 1 *for items with ✓)*

HeLa cells (ATCC #CCL-2)
Vector (see Basic Protocol)
✓ 4% (w/v) paraformaldehyde solution
✓ Xgal staining solution

96-well tissue culture plates
37°C humidified incubator with 10% CO_2 atmosphere
Multichannel pipettor with appropriate tips
Inverted microscope

4

1. On the day before titration, prepare a 96-well plate by seeding 5000 HeLa cells in 200 μl DMEM-10 medium in each of the 96 wells (sufficient for titrating four vector stocks, each titration in duplicate), ensuring a uniform spread of cells on the bottom of the wells. Incubate overnight in a 37°C, 10% CO_2 incubator.

2. Add 2 μl of vector diluted in 200 μl of DMEM-10 to the wells of the second column (400 μl total in wells), leaving the wells of the first column uninfected as a control.

   A 2-μl inoculum is chosen assuming a titer between 10^5 and 10^6 TU/ml. A higher inoculum would be required for titers lower than 10^3 TU/ml.

3. Prepare a serial dilution (1:2) by transferring 200 μl of DMEM-10 medium from the wells in one column to the wells of the following column with a multichannel pipettor. Start with the second column, and continue until all but the first column wells contain vector, mixing the medium by pipetting up and down several times at each passage. Incubate for 2 days at 37°C.

4. To detect *lacZ*-positive cells, remove the medium, wash once with 400 μl PBS, fix cells with 250 μl 4% (w/v) paraformaldehyde solution for exactly 5 min, wash twice with PBS, and add 250 μl of Xgal staining solution. Incubate at 37°C.

   Because the incubation time depends on the expression level of lacZ, check the plates regularly. Staining may be prolonged as long as the endogenous β-galactosidase activity remains undetectable in the control wells.

5. Replace the Xgal staining solution with 250 μl PBS.

   Plates can be stored at 4°C for at least 1 week.

6. Count transduction events in two wells containing between 10 and 100 blue foci, on an inverted microscope without phase contrast. Divide the number of transduction events by the dilution factor and multiply by 1000 to calculate the titer in TU/ml.

References: Cisterni et al., 2000; Déglon et al., 2000

Contributors: Romain Zufferey and Didier Trono

UNIT 4.11

Production of Recombinant Adeno-Associated Viral Vectors and Use in In Vitro and In Vivo Administration

Recombinant adeno-associated virus (rAAV) vectors are capable of delivering genes to both dividing and nondividing cells. This property, along with the physical stability of the virions, permits transduction of CNS cells both in tissue culture and in the whole animal. An overview of the rAAV production procedure is shown in Figure 4.11.1. See *www.med.unc.edu/genether/* for a map and sequence of the required plasmids.

4

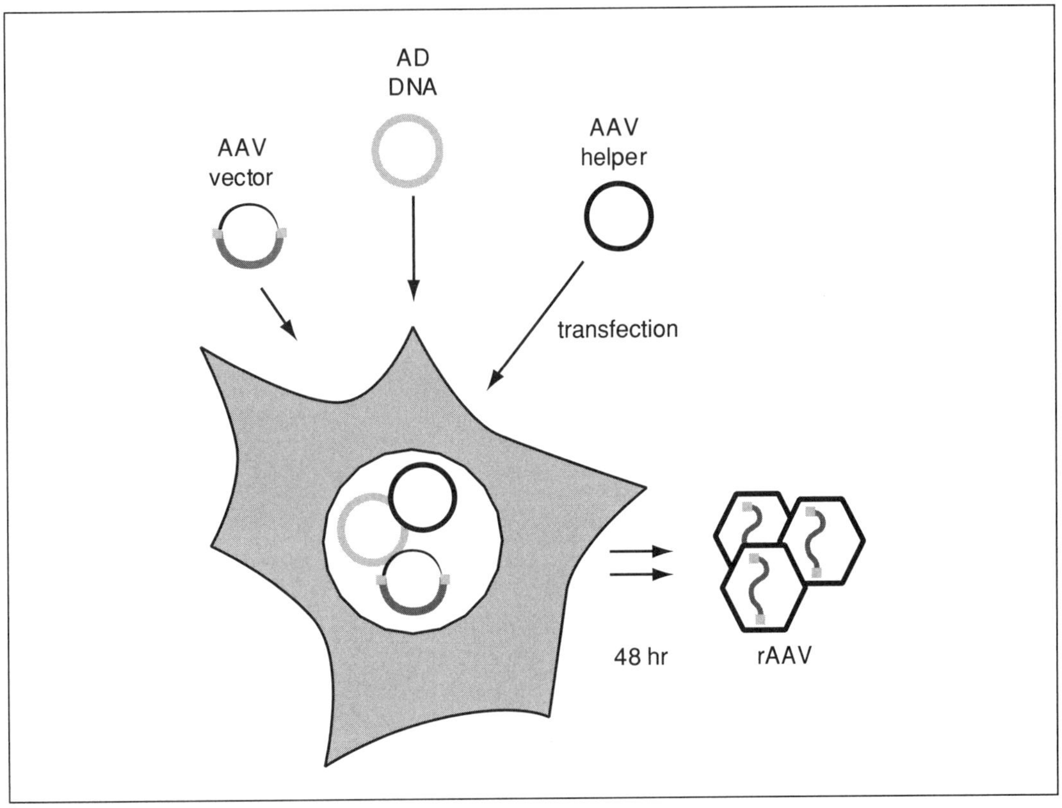

Figure 4.11.1 Generation of adenovirus-free recombinant adeno-associated virus. 293 cells (which supply the Ad *E1* gene) are transfected using three plasmids: the plasmid carrying the transgene (sub201-gene "*X*," AAV vector), the plasmid supplying the replication and capsid genes of AAV2 without terminal repeats (pXX2, AAV helper), and the plasmid supplying the adenovirus helper genes *E2*, *E4*, and *VA* RNA genes (pXX6, AD DNA), thereby generating Ad-free rAAV.

NOTE: All equipment and reagents coming in contact with tissue culture cells (uninfected and infected) and the virus following CsCl gradients should be sterile.

CAUTION: Virus work should be performed in a dedicated tissue culture hood and incubator separate from the hood and incubator used for the maintenance of laboratory cell lines. Proper disposal of virally contaminated materials should be performed.

BASIC PROTOCOL 1

PRODUCTION OF ADENOVIRUS-FREE rAAV BY TRANSIENT TRANSFECTION OF 293 CELLS

Materials *(see APPENDIX 1 for items with ✓)*

- psub201 plasmid: used to clone the transgene between AAV termini (Fig. 4.11.2; ATCC #68065)
- *Xba*I and *Hin*dIII restriction endonucleases, with appropriate buffers
- pXX2 plasmid: the AAV helper plasmid (UNC Vector Core Facility)
- pXX6 plasmid: the adenoviral helper plasmid (UNC Vector Core Facility)
- 293 tissue culture cell line (ATCC #CRL 1573)
- ✓ Complete DMEM/10% (v/v) FBS medium
- ✓ Complete IMDM/10% (v/v) FBS medium
- 0.05% (w/v) porcine trypsin/0.02% (w/v) EDTA
- ✓ 2.5 M $CaCl_2$
- ✓ 2× HEPES-buffered saline (HeBS)
- ✓ Complete DMEM/2% (v/v) FBS
- Dry ice/ethanol bath

- OPTI-MEM I (Life Technologies)
- ✓ Saturated ammonium sulfate, pH 7.0, 4°C
- ✓ 1.37 g/ml and 1.5 g/ml density CsCl
- 70% ethanol
- ✓ PBS
- 15-cm tissue culture plates
- 50-ml disposable polystyrene and polypropylene centrifuge tubes
- Cell scrapers
- 250-ml polypropylene centrifuge bottles
- Refrigerated high-speed centrifuge with appropriate fixed-angle rotors
- Sonicator with a 3-mm diameter probe
- Tabletop centrifuge
- 50-ml high-speed polypropylene centrifuge tubes
- Ultracentrifuge with swinging-bucket rotor and 12.5-ml clear tubes
- 21-G needles
- Pierce Slide-A-Lyzer dialysis cassettes (MWCO 10,000)

NOTE: All tissue culture incubations are performed in a humidified 37°C, 5% CO_2 incubator unless otherwise specified.

1. Digest the plasmid psub201 (Fig. 4.11.2) with *Xba*I and *Hin*dIII restriction endonucleases (*CPMB UNIT 3.1*) to remove the *rep* and *cap* fragments, and gel purify the 4000-bp plasmid backbone containing the AAV ITRs (*CPMB UNIT 2.6*). Insert the desired transgene cassette (<4400 bp) into the *Xba*I sites to construct the rAAV vector plasmid (*CPMB UNIT 3.16*). If the transgene cassette is less than 3000 bp, include filler sequence. Purify a large-scale

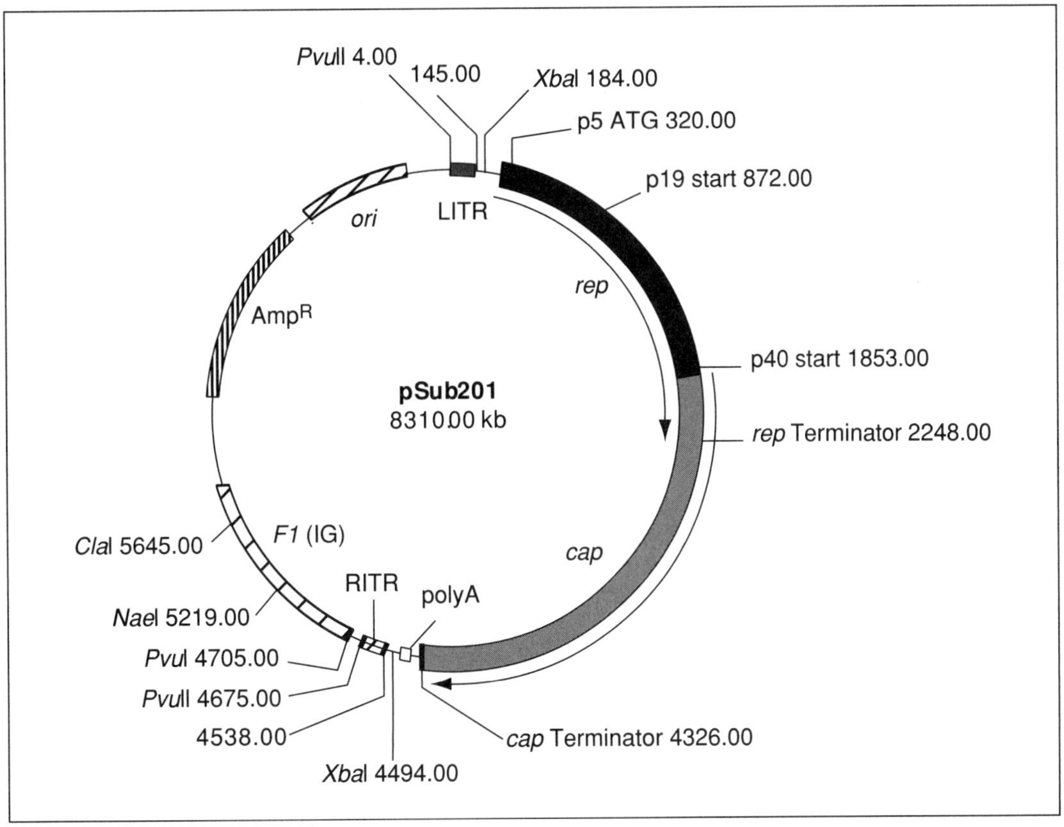

Figure 4.11.2 Plasmid sub201. The plasmid contains the terminal repeats and the replication (*rep*) and capsid (*cap*) genes of AAV Type 2. The *rep-cap* fragment can be replaced by the gene cassette of interest, as only the terminal repeats are needed for packaging.

plasmid preparation (at least 1 mg) of the rAAV vector and pXX2 and pXX6 plasmids by double CsCl gradient fractionation (*CPMB UNIT 1.7*).

The AAV ITRs are unstable in E. coli. When growing the rAAV vector plasmid, do not allow the culture to remain in stationary phase too long, as ITR deletions occur more rapidly after the culture has proceeded past log phase. The pXX6 plasmid is also unstable in bacteria. After receiving the plasmid, transform in SURE bacteria (Stratagene) and check the integrity of the plasmid by digesting and electrophoresing a miniprep. After the selection of an intact clone, individual laboratory stocks should be prepared. When growing the plasmid, grow the bacteria at 30°C for no longer than 12 hr. The plasmid preparations must be free of contaminants (ethidium bromide, CsCl, and RNA) before use.

2. Seed 2×10^7 293 cells in 15-cm dishes to achieve 70% to 80% confluency the next morning. Maintain 293 cell line in complete DMEM/10% FBS medium and split cells before they reach confluency by trypsinizing with 0.05% trypsin/EDTA (also see *APPENDIX 2I*). Split a near-confluent 15-cm dish into three to four 15-cm dishes. Prepare twenty 15-cm dishes.

The transfection procedure described in this protocol is for twenty 15-cm dishes.

3. At 12 to 16 hr following the split, replace the medium with 25 ml per 15-cm dish of complete IMDM/10% FBS. Incubate for 3 hr before the transfection.

4. For transfection of four 15-cm dishes, combine the following in a disposable 50-ml polystyrene tube:

   90 μg pXX6 helper plasmid (adenoviral helper genes)
   30 μg rAAV vector plasmid
   30 μg pXX2 helper plasmid (AAV helper genes)
   0.4 ml 2.5 M $CaCl_2$
   Deionized distilled H_2O to 4 ml.

   Add 4 ml of 2× HeBS to the mixture, and pipet up and down with a 10-ml plastic pipet three times to mix. Incubate 1 to 5 min at room temperature.

   The precipitate should be very fine and should be checked under the microscope at 40×. One should test making the precipitate on a small scale before doing the large-scale transfection; if the precipitate is too coarse, reduce the incubation time. The use of more DNA will slow the rate of precipitate formation. The precipitate becomes more coarse over time, so the transfection of 20 dishes is performed in smaller batches, that require less time.

5. Vortex the suspendedDNA/$CaCl_2$precipitate for 5 sec and then add 2 ml dropwise to the medium in each of four 15-cm plates of cells (from step 3). Swirl the plates to disperse evenly. Repeat from step 4, four dishes at a time until all 20 dishes have been transfected. Incubate 24 hr.

6. Aspirate the medium (two plates at a time so cells do not dry out) and add 25 ml of 37°C DMEM/2% FBS. Incubate until 48 hr post-transfection.

4

Ammonium sulfate precipitation removes most of the cellular debris from the virions before the CsCl gradient is performed. These recipes are for 20 to 40 15-cm plates, but can be scaled up two to four times.

7. Scrape the cells in the tissue culture plates, with a cell scraper, to collect cells and medium. Transfer suspension to 250-ml polypropylene centrifuge bottles.

8. To liberate virus particles from cells, freeze and thaw the cell suspension three times by transferring the tubes between a dry ice/ethanol bath and a 37°C water bath. Proceed to step 9 or store suspension at −20°C for up to 6 months.

9. Centrifuge the bottles for 10 min at 3000 × g, 4°C, to pellet the cells. Decant the supernatant into fresh 250-ml polypropylene centrifuge tubes. Retain both the supernatant and the cell pellet, separately.

10. To the supernatant, add 78.25 g $(NH_4)_2SO_4$ per 250 ml of supernatant, to precipitate the virus in the supernatant. Mix thoroughly to dissolve the ammonium sulfate and then precipitate on ice for 20 min. Centrifuge the tubes 10 min at 8300 × g, 4°C.

11. Slowly decant the supernatant into a container, separating it from the yellowish precipitate that has formed on the bottom and sides of the centrifuge vessel. Autoclave the supernatant, then discard. Maintain the centrifuge bottle containing the precipitate on ice until the cell pellet from step 9 has been processed.

    Do not use bleach to decontaminate ammonium sulfate supernatants because a noxious odor will be produced.

12. Resuspend the cell pellet from step 9 in 20 ml of OPTI-MEM I medium by pipetting up and down or vortexing, and transfer to a disposable 50-ml polypropylene centrifuge tube. Sonicate (40 bursts, 50% duty, power level 2) at room temperature in a tissue culture hood dedicated for virus work. Use ear protection when using the sonicator.

13. Centrifuge the tubes for 5 min at 3000 × *g* in a tabletop centrifuge at room temperature to pellet insoluble debris. Collect the clarified supernatant and transfer to a disposable 50-ml polypropylene tube.

14. Resuspend the cell pellet in 20 ml of OPTI-MEM I, and repeat steps 12 and 13.

15. Pool the supernatants from steps 13 and 14. Use these pooled supernatants to dissolve the ammonium sulfate precipitate in the centrifuge bottle (from step 11), by pipetting the solution down the side and the bottom of the bottle.

16. Measure volume of the liquid from step 15 and add 1 vol of 4°C saturated ammonium sulfate, pH 7.0, per 3 vol of the virus-containing suspension (25% saturation, to precipitate undesired proteins). Mix well, and put on ice for 10 min.

17. Centrifuge 10 min at 7700 × *g* (fixed-angle rotor), 4°C. Transfer each supernatant to a 50-ml high-speed polypropylene centrifuge tube, avoiding the yellow precipitate.

18. Using the volume of the lysate from step 15, add 2 vol of 4°C saturated $(NH_4)_2SO_4$ solution per 3 vol of the lysate (50% saturation, to precipitate the virions), and incubate on ice for 20 min. Centrifuge 20 min at 17,000 × *g* (fixed-angle rotor), 4°C. Remove supernatant and autoclave (do not add bleach; see above) before discarding.

19. Dissolve pellets in 20 ml total (for 20 to 40 plates) of 1.37 g/ml CsCl solution. Resuspend the pellets as completely as possible so that, after mixing, no undissolved material can be seen. If a large amount of undissolved material is present, add more 1.37 g/ml CsCl solution.

20. Add 0.5 ml of 1.5 g/ml CsCl solution to each of two 12.5-ml clear ultracentrifuge tubes. Overlay ~12 ml of virus sample, in 1.37 g/ml CsCl, in each tube by holding the pipet to the side of the tube and slowly applying the sample. Move the pipet up along the side of the tube as the sample is dispensed.

21. Centrifuge 36 to 48 hr at 288,000 × *g*, 15°C. Decelerate with brake to 500 rpm, then turn brake off.

22. Wipe the outside of the centrifuge tube with 70% ethanol, insert a 21-G needle ~1 cm from the bottom of the tube at a 90° angle, and allow the gradient to drip into sterile microcentrifuge tubes. Collect 15 to 20 fractions of 10 drops each. Collect multiple gradients of the same rAAV into one set of tubes.

23. Assay 1 to 5 μl of each fraction by dot blot, using an rAAV-specific probe to find the peak (see Support Protocol 1).

    The rAAV peak should band in the gradient with a density of 1.40 to 1.42 g/ml, and this can be checked with a refractometer (be sure to decontaminate the refractometer after use).

24. *Optional:* For increased purity and concentration of rAAV (but with some loss of material), add 1.37 g/ml CsCl solution to the pooled peak fractions to attain a final volume of 12 ml. Add 0.5 ml of 1.5 g/ml CsCl to one clear ultracentrifuge tube and overlay virus solution as in step 20, then reband, drip, and assay the gradient as in steps 21 to 23.

25. Dialyze the rAAV in MWCO 10,000 Slide-A-Lyzer dialysis cassettes against three 500-ml changes of sterile 1× PBS, for at least 3 hr each, at 4°C.

26. Divide the virus suspension into convenient aliquots (typically 100 to 200 μl) to avoid repeated freezing and thawing. Store at −20°C or −80°C (can be kept for ≥1 year).

ALTERNATE PROTOCOL

rAAV PURIFICATION USING HEPARIN SEPHAROSE COLUMN PURIFICATION

Additional Materials *(also see Basic Protocol 1; see* APPENDIX 1 *for items with ✓)*

✓ PBS-MK
✓ 15%, 25%, 40%, and 60% iodixanol
PBS-MK containing 1 M NaCl
0.5 M NaOH
20% (v/v) ethanol
Phenol red (Life Technologies)
Ethanol

Econo Pump peristaltic pump (Bio-Rad)
32.4-ml Optiseal tubes (Beckman)
50-μl disposable borosilicate glass capillary pipets (Fisher)
Ultracentrifuge with fixed-angle rotor
1-ml or 5-ml HiTrap heparin-Sepharose columns, (1-ml for virus prep from twenty 15-cm plates; 5-ml for larger preps; Amersham Pharmacia Biotech)
FPLC apparatus
Pierce Slide-A-Lyzer dialysis cassettes (MWCO 10,000)

1. Perform the transfection and carry out ammonium sulfate precipitation of the rAAV (see Basic Protocol 1, steps 1 to 18). Dissolve the ammonium sulfate pellet in PBS-MK to attain a total volume of 15 ml.
2. Set up the Econopump and insert the 50-μl capillary pipets into the tubing. Calibrate the pump according to the instructions. Place 7.5 ml of the ammonium sulfate pellet dissolved in PBS-MK (step 1) in each of two Optiseal tubes.
3. Insert one capillary pipet into each tube and, using the pump at a flow rate of 3 ml per minute, underlay the virus solution with 6 ml of the 15% iodixanol. Then underlay the 15% iodixanol with 5 ml of 25% iodixanol, and underlay the 25% with 5 ml of 40% iodixanol. Finally, underlay the gradient with 5 ml of 60% iodixanol. Do not introduce air bubbles.
4. Carefully remove the glass capillary without disturbing the gradient. Slowly add PBS to the viral solution, the uppermost layer in the tube, until the tube is filled to the top (Fig. 4.11.3).
5. Insert a plug in each tube and centrifuge 1 hr at 500,000 × *g*, 15° to 25°C.
6. Carefully remove the tubes and, in a viral hood, unplug the sealing cap. Insert an 18-G needle attached to a 5-ml syringe just above the 60% iodixanol fraction. Extract the 40% iodixanol solution (clear band, containing the virus), leaving ~0.5 ml, so that none of the 25% iodixanol band is also extracted. Proceed to step 7 or store virus overnight at 4°C.
7. Perform a pump wash with PBS-MK, then insert a heparin column in the FPLC.
8. Equilibrate the column in five column vol PBS-MK, then reduce the flow rate (0.2 ml/min for a 1-ml column; 1 ml/min for a 5-ml column) and inject the sample. Collect fractions throughout the column run (flowthrough, wash, and elution).
9. Wash the column with 5 column vol PBS-MK.
10. Elute the bound virus in a linear gradient over 5 column vol from 0% to 100% PBS-MK containing 1 M NaCl. Collect 0.5-ml fractions (for a 1-ml column) *or* 1-ml fractions (for a 5-ml column).

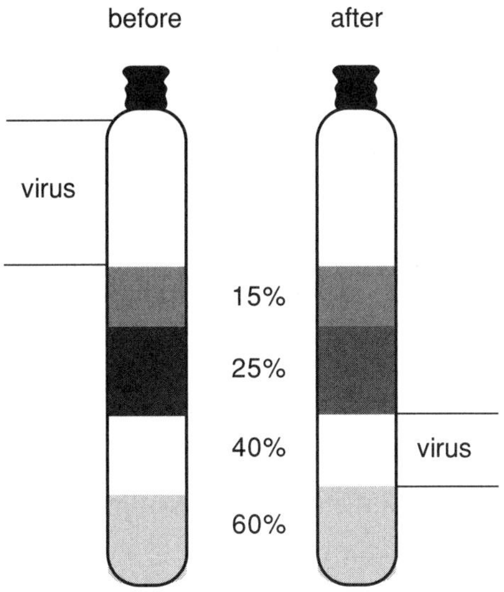

Figure 4.11.3 Iodixanol step gradient before and after centrifugation. The step gradient is generated by underlying the virus concentrate with 15%, 25%, 40% and 60% iodixanol solutions (see Alternate Protocol for instructions on forming the gradient). After centrifugation the virus resides exclusively in the 40% iodixanol (clear) layer. Care should be taken when removing this layer not to remove any of the 25% layer, as this contains cell debris.

4

11. Wash the column with two column vol 100% PBS-MK containing 1 M NaCl. Disconnect the heparin column and discard. Sterilize the FPLC by running 0.5 M NaOH through all the lines that came in contact with the virus. Fill all lines with 20% ethanol for storage.
12. Assay fractions (from throughout the purification process) for virus, using a dot blot with an AAV-specific probe to find the viral peak (see Support Protocol 1). Check flowthrough and wash as well as the fractions to ensure that all the virus was bound to the column and eluted in the gradient.
13. Pool the fractions that have the highest rAAV titer and dialyze them in a MWCO 10,000 Slide-A-Lyzer dialysis cassette against two 1-liter changes of PBS, 2 hr each, at 4°C.
14. Divide the virus suspension into 100- to 200-μl aliquots and store at −20°C or −80°C (viable for ≥1 year).

SUPPORT PROTOCOL 1

DETERMINATION OF rAAV TITERS BY THE DOT-BLOT ASSAY

A positive signal in this assay indicates that rAAV virions were produced, and quantitation yields a particle number in virions per ml.

Materials *(see* APPENDIX 1 *for items with ✓)*

Virus fractions or final virus preparation (e.g., from Basic Protocol 1 or Alternate Protocol)

✓ DNase digestion buffer

✓ 0.1 M EDTA
✓ Proteinase solution
0.5 M NaOH
rAAV plasmid used to make recombinant virus (see Basic Protocol 1, step 1)
✓ TE buffer, pH 7.5
0.5 M NaOH containing 1 M NaCl
✓ 0.4 M Tris·Cl, pH 7.5
0.5 M NaCl containing 0.5 M Tris·Cl, pH 7.5
Radiolabeled probe to transgene (made using Boehringer Mannheim random-primed DNA labeling kit according to manufacturer's instructions)

96-well plate
50°C water bath
Dot-blot apparatus
0.45-μm nylon membrane (Zeta Probe, Bio-Rad)

1. Prepare an experimental plan for the position of samples, controls, and DNA standards in duplicate in a 96-well format.

2. Place samples (usually 5 μl of CsCl- or column-purified fraction with 20 μl as the maximum if the virus is in CsCl) and controls in the appropriate wells of a 96-well plate. Add 50 μl DNase digestion buffer per well and incubate 1 hr at 37°C, to digest any DNA that may be present but not packaged into virions.

3. Stop the digestion by adding 10 μl of 0.1 M EDTA to each well. Mix very well. Release encapsulated virion DNA by adding 60 μl of proteinase solution. Incubate 30 min at 50°C.

4. Denature viral DNA by adding 120 μl of 0.5 M NaOH per well and incubate 10 min at room temperature.

5. Prepare linearized plasmid (from the plasmid used for transfection of this virus) at 0.1 μg/μl for DNA concentration standards. Make a 5-fold dilution series in 100 μl TE buffer by adding 12.5 μl DNA to 112.5 μl TE buffer in position 1 and successively mixing 25 μl from each well to the next (625 ng, 125 ng, 25 ng, 5 ng, 1 ng, 200 pg, 40 pg).

6. Denature DNA standards by adding 100 μl of 0.5 M NaOH/1 M NaCl to each well.

7. Equilibrate nylon membrane in 0.4 M Tris·Cl, pH 7.5. Prepare a dot-blot manifold apparatus with a prewetted nylon membrane. Add the denatured DNA samples without vacuum, and then apply the vacuum.

8. Disassemble the apparatus, remove membrane, and wash for 5 min with gentle shaking in 100 ml of 0.5 M NaCl/0.5 M Tris·Cl, pH 7.5.

9. Probe the filters with a radiolabeled probe specific for the rAAV sequences (*CPMB UNIT 2.10* and *APPENDIX 2A* in this manual). Use probe limited to the transgene cassette, not including plasmid backbone or ITR sequences.

10. Place the filters against film and expose (autoradiography; *CPMB APPENDIX 3A*). After developing the film, align the spots, excise the regions of the filter, and quantitate in a scintillation counter. Calculate how many molecules of the plasmid standards correspond to a given dilution of the rAAV stock, taking into account the fact that the plasmid standards are double-stranded, but the rAAV virions harbor only a single strand.

SUPPORT PROTOCOL 2

INFECTION OF CELLS IN VITRO WITH rAAV AND DETERMINATION OF TITER BY TRANSGENE EXPRESSION

Assaying for transgene expression is the most stringent method of determining rAAV titer. However, the lack of standardization between the different expression assays used for each gene makes the yields of rAAV with various transgenes difficult to compare.

Materials

Target cells
Tissue culture medium for target cells (see supplier's instructions for cells)
rAAV with appropriate transgene (see Basic Protocol 1 or Alternate Protocol)
Tissue culture plates (multiwell plates recommended)

NOTE: All tissue culture incubations are performed in a humidified 37°C, 5% CO_2 incubator unless otherwise specified. Any manipulations using virus should be carried out in a tissue culture hood dedicated for virus work.

NOTE: To assay transducing titer, cells can be coinfected with adenovirus (see Support Protocol 3), which acts as a helper virus and increases the transduction efficiency. The addition of adenovirus gives a better indication of the number of particles that are competent to transduce a cell by inducing an optimal environment for AAV infection. However, the adenovirus has a cytopathic effect because of its antigenicity and should not be used in vivo. Thus, a titer derived with use of adenovirus in vitro may not accurately reflect an in vivo competency.

1. Seed appropriate target cells into multiwell tissue culture plates with appropriate medium. Do not treat cells with DNA synthesis inhibitors such as Ara-C at any time before infection. (AAV, a single-stranded virus, requires second strand synthesis before gene expression is possible.)

2. Infect cell by adding rAAV directly to the medium of the cells or mix rAAV with fresh medium immediately before adding it to the cells. Infect with serial 5-fold dilutions of the rAAV stock. Alternatively, infect cells at the time of plating.

3. Carry out the specific assay for expression of the transgene (e.g., immunofluorescence, histological staining, drug resistance).

SUPPORT PROTOCOL 3

GROWING AN ADENOVIRUS HELPER STOCK

Materials *(see* APPENDIX 1 *for items with* ✓*)*

Adenovirus type 5 (ATCC #VR5)
293 tissue culture cell line (ATCC #CRL 1573) growing in complete DMEM/10% FBS medium in 15-cm plates and in 60-mm plates
✓ Complete DMEM/2% FBS and complete DMEM/10%
✓ TBS
✓ 1.2 g/ml, 1.3 g/ml, 1.4 g/ml, and 1.5 g/ml density CsCl
✓ 2× adenovirus storage buffer

4

Serum-free DMEM
✓ Plaque overlay solution, 39°C
3.3 g/liter neutral red stain (Sigma; store up to 6 months at room temperature)

500-ml polypropylene centrifuge bottle
Refrigerated high-speed centrifuge with fixed-angle rotor
50-ml disposable polypropylene centrifuge tubes
Dry ice/ethanol bath
Tabletop centrifuge
12.5-ml clear ultracentrifuge tubes
Ultracentrifuge with swinging-bucket rotor
5-ml syringe
21-G needle
Econo Pump peristaltic pump (Bio-Rad)
50-μl borosilicate glass capillary pipets (Fisher)

NOTE: All tissue culture incubations are performed in a humidified 37°C, 5% CO_2 incubator unless otherwise specified. Any manipulations using virus should be carried out in a tissue culture hood dedicated for virus work.

1. Add adenovirus dl309 to DMEM with 2% FBS at a multiplicity of infection (MOI) of 10. Replace the medium on ten 15-cm plates of 293 cells (70% to 80% confluent) with this Ad/DMEM solution. Incubate until full cytopathic effects (CPE) occur (36 to 56 hr), recognized by the presence of rounded cells that float and by the medium's turning an orange/yellow color.

2. Harvest the cells and the culture supernatant and transfer to a 500-ml polypropylene centrifuge bottle. Centrifuge 5 min at 4000 × *g*, 4°C.

3. Decant supernatant and autoclave before discarding. Resuspend the pellet in 10 ml TBS. Transfer suspension to a disposable 50-ml polypropylene tube. Freeze and thaw three times by transferring the tube between a dry ice/ethanol bath and a 37°C water bath. Centrifuge 5 min at 3000 × *g* in a tabletop centrifuge at 4°C and transfer the supernatant to a new tube.

4. Place 3.5 ml of 1.4 g/ml CsCl solution in each of two 12.5-ml ultracentrifuge tubes. Carefully layer 3.5 ml of 1.2 g/ml CsCl solution on top of the 1.4 g/ml CsCl. Finally, layer 5 to 6 ml of the supernatant from step 3 on top of each of the CsCl step gradients. Centrifuge 1 hr at 150,000 × *g*, 20°C.

5. Remove the lower band, containing the virus, with a 21-G needle and 5-ml syringe (see Basic Protocol 1, step 22), combine the bands from different tubes, then mix with 1.3 g/ml CsCl solution to a final volume of 12 ml. Place this in a 12.5-ml Ultra-Clear tube.

6. Using a glass capillary tube and setting the Econo Pump at a flow rate of 3 ml per minute, underlay the solution with 0.5 ml of 1.5 g/ml CsCl solution (see Alternate Protocol, steps 2 and 3). Centrifuge overnight at 150,000 × *g*, 20°C.

7. Remove the adenovirus band from the gradient with a 21-G needle and 5-ml syringe. Mix the adenovirus with an equal volume of 2× adenovirus storage buffer, divide into aliquots, and store up to 1 year at −20°C.

8. Make eight 10-fold serial dilutions of the adenovirus stock, each in 1 ml of serum-free DMEM. Infect 60-mm dishes of 293 cells (80% confluent) with 100 μl of each dilution and incubate 2 hr. Prepare the plaque overlay solution during this incubation.

9. Aspirate the medium from the cells and slowly overlay with 5 ml of 39°C plaque overlay solution. Allow the agar to harden at room temperature.

10. Incubate the plates at 37°C in 5% CO_2 for 5 days, then feed the plates by adding 2 ml of fresh 39°C plaque overlay solution.

11. When plaques become visible to the naked eye (after day 7) feed the plates again with 2 ml of plaque overlay solution containing a 1:100 dilution of 3.3 g/liter neutral red stain.

    Add the stain to 2× DMEM before combining with the agarose. The neutral red aids in the detection of the plaques, but kills the cells.

12. The next day, count the plaques at each dilution and multiply these numbers by the corresponding dilution factor to determine the titer of the undiluted adenovirus stock.

BASIC PROTOCOL 2

STEREOTACTIC MICROINJECTION OF rAAV INTO THE RAT BRAIN

Materials *(see* APPENDIX 1 *for items with* ✓*)*

Sprague-Dawley rats, 6 to 8 weeks old
Anesthetic: pentobarbital (50 mg/kg body weight)/atropine methyl nitrate (2 mg/kg body weight) in 0.9% NaCl *or* ketamine (150 mg/kg body weight)
Betadine scrub or 70% ethanol
1 mg/ml epinephrine
1% procaine·HCl
✓ PBS
rAAV preparation (see Basic Protocol 1or Alternate Protocol) diluted in PBS to desired concentration (centrifuge briefly to pellet any debris and keep at 4°C; see note below)

Rat brain atlas (e.g., Paxinos and Watson, 1986)
Dental drill or Dremel with #862 Dremel engraving cutter
Rat stereotaxic frame (Kopf) in sterile surgical area
Variable speed syringe pump
Surgical instruments: scalpel, forceps, needle holders, and small spatula
Sharp scissors or razor
30-G stainless steel tubing for injector (32-G or 33-G can also be used)
PE-10 intramedic tubing, Clay Adams
0 to 10-µl syringe with Teflon-tipped plunger (e.g., Precision Sampling glass Teflon tip plunger, gas-tight 0 to 10-µl syringe; Valco Instruments)
1-ml syringes with 26-G and 30-G needles
Animal clippers
Sterile gauze
Bonewax
4-0 silk suture
Polysporin ointment
Heating pad, heated to 37°C
Clean cage without bedding

NOTE: Before beginning injections with virus, it is important to verify the placement coordinates. This can be done by infusing dye (e.g., fast green) into one or two animals using the following procedure and then slicing the brain to verify injector placement.

NOTE: High titer of virus is necessary for this procedure. No fewer than 1×10^5 transducing units should be administered; 1 to 5×10^6 (usually equivalent to 10^{12} to 10^{13} particles/ml, although this varies with transgene) is typical. Fewer than 10^5 transducing units will most likely not be effective.

1. Using the rat atlas, determine the injection coordinates.
2. Set up the stereotaxic frame, dental drill, and syringe pump in a sterile area appropriate for surgery. Determine setting on syringe pump needed to inject 1 μl over 10 min.
3. Sterilize all surgical instruments (scalpel, forceps, needle holders, and spatula).
4. Using scissors, cut a 3- to 7-cm length of 30-G stainless steel tubing. Cut a 12- to 18-inch length of PE-10 tubing.
5. Insert 1 to 2 cm of the 30-G stainless steel tubing into one end of the PE-10 tubing, forming a water-tight seal. If 32- or 33-G tubing is used, place some Super Glue or contact cement on the stainless steel tubing, close to the end, then slide it onto the PE-10 tubing. Allow glue to dry for 30 min.
6. Test the integrity of the injector by filling a 1-ml syringe with water and fitting it with a 30-G needle. Insert the needle into the PE-10 tubing and slowly push the water through the injector. Check for leaks at the joint.
7. Fill the PE-10 tubing with sterile water. Place the open end of the PE-10 tubing on the 10-μl syringe (airtight seal). Withdraw the syringe plunger 0.3 to 0.5 μl, introducing an air bubble into the injector tubing.
8. Submerge the injector tip into the rAAV solution and slowly withdraw the plunger. Place the syringe into the holder of the pump.

   The air bubble should move back toward the syringe separating the water and the rAAV solution. During the infusion, movement of the air bubble should be monitored by marking the drug-air interface on the PE-10 tubing with a permanent marker before infusion. If the air bubble compresses instead of moving beyond the mark, then the injector tip is plugged.

9. Using two 1-ml syringes equipped with 26-G needles, anesthetize rat with an intraperitoneal (i.p.) injections of pentobarbital and atropine methyl nitrate (to prevent excess salivation). Wait until there is no toe pinch or eye blink response, indicating full anesthetization.
10. Shave hair from top of head, place the animal in the stereotaxic frame and clean the skin with betadine scrub or 70% ethanol. (See *UNIT 3.10* for more detail on stereotaxic frame and determination of injection coordinates.)
11. Make an incision no larger than 1.0 to 1.5 cm from the front of the skull to the back. Put several drops of 1 mg/ml epinephrine on the incision site, followed by a small amount of 1% procaine·HCl.
12. Using blunt dissection, remove facia on the top of the skull; wipe clean with sterile gauze and allow the skull to dry.
13. Plot the rostral/caudal and medial/lateral coordinates. Use the rostral suture intersection (bregma) as the starting point. Mark the point with a permanent marker and replot to verify accuracy (see Fig. 3.10.1). Drill a hole in the skull on the mark and lower injector the specified distance.
14. When injector is in place, turn on the infusion pump. Mark the air bubble at this point and monitor its progress. If the injector tip is plugged, try dislodging the plug by moving the injector slightly up and then back down again.
15. When specified volume (typically 1 to 2 μl) is infused, turn off pump and leave injector in place for 1 min before removing.
16. Remove injector, clean the skull, cover hole with a small amount of bone wax, and suture the incision with 4-0 silk suture. Swab the incision with polysporin ointment. Allow the

animal to recover in a clean cage, lined with paper towels on a heating pad, until fully mobile before returning it to normal housing. Remove sutures in 5 to 7 days. See Support Protocol 4 for methods of detecting expression of rAAV.

SUPPORT PROTOCOL 4

HISTOLOGICAL DETECTION OF β-GALACTOSIDASE OR GREEN FLUORESCENT PROTEIN EXPRESSION

GFP expression can be detected directly by fluorescence microscopy and *lacZ* is detected by histochemical staining. See *UNIT 1.1* for information regarding perfusion fixation and sectioning of brain.

Materials *(see APPENDIX 1 for items with ✓)*

Rats transduced with *GFP* or *lacZ* (Basic Protocol 2; AAV-*GFP* or AAV-*lacZ* virus available from UNC Vector Core Facility. pAAV-*GFP* or pAAV-*lacZ* plasmid constructs also available from UNC Vector Core Facility)
4% (w/v) paraformaldehyde in PBS, 4°C
✓ PBS, 4°C
2% agar (store at 4°C in solid form, heat to 70°C to liquefy before using)
✓ Fluorescent mounting medium
40 mg/ml Xgal in *N,N*-dimethylformamide (store up to several weeks in the dark at 4°C)
41 mg/ml potassium ferricyanide (store up to several weeks in the dark at 4°C)
53 mg/ml potassium ferrocyanide (store up to several weeks in the dark at 4°C)
2 M $MgCl_2$
Standard mounting medium, e.g., Permount or Accumount

Large forceps
Small forceps with hooks on the ends
50-ml conical centrifuge tubes
Single-edged razor blades
Super Glue
24-well tissue culture plates
Vibrating microtome with blades
Fine-bristled paint brush
Glass petri dish, shallow
Dark work surface
Glass slides and coverslips
Drying rack
Fluorescence microscope with GFP filter

1. Perform perfusion fixation of brains as described in *UNIT 1.1*, but place brain in a 50-ml conical tube containing 15 to 20 ml 4% paraformaldehyde in PBS and leave, at 4°C, overnight.

2. Use large forceps to remove brains from 50-ml conical tubes. Using small forceps, remove as much of the meninges as possible.

3. With a single-edge razor blade, block the brain by cutting in the coronal plane (parallel to the injection tract) ~5 mm both anterior and posterior to the infusion site. Make sure symmetry is maintained. Place on an absorbent paper towel to remove paraformaldehyde solution. Using a small amount of Super Glue, glue the brain to the mounting platform of the microtome. Press gently. Allow glue to dry.

4. Set up the microtome, using PBS to immerse the brain during sectioning. Fill several wells of the 24-well plate with PBS.

5. Coat the brain with a thin layer of melted (70°C) 2% agar and allow it to solidify. Place mounted brain in the vibrating microtome and set speed, vibration amplitude, and section thickness (40 to 60 μm).

6. Begin sectioning, using a fine-bristled paint brush to guide sections during slicing. As the section is cut, use the paint brush to lift the section gently from the blade and into a PBS-filled well of the 24-well plate. Tearing of the sections during slicing may indicate residual meninges; if all the meninges are removed and tearing still occurs, increase the frequency of blade vibration and decrease the speed. Replace the blade after each brain for optimal results. Mount sections immediately or store in PBS, at 4°, for up to several days.

For GFP

7a. Mount sections by placing them in a shallow glass petri dish set filled with 0.5× PBS solution and set on a dark surface. Using the paint brush, slide sections gently onto glass slides. Place slides upright to allow PBS to drain, and allow sections to dry until they become translucent (~5 to 10 min depending upon thickness).

8a. Rinse sections by gently dipping the slide in distilled water several times. Repeat twice. Dry another 10 to 30 min.

9a. To coverslip the sections, place a small amount of fluorescent mounting medium on the slide and place a glass coverslip carefully on top. Avoid producing bubbles. Remove excess mounting medium by draining briefly on a paper towel. Allow several minutes for medium to set (it will harden more over the next 24 hr).

Slides can be stored in the dark at 4°C for ~2 weeks with minimal loss of fluorescence. Longer storage times result in progressive loss of fluorescence. Bubbles in the mounted slides will disrupt viewing and impair fluorescence after storage.

10a. View GFP with a UV fluorescent microscope fitted with the correct filter.

For lacZ

7b. Prepare stain by adding the following to a 50-ml conical tube: 0.625 ml of 40 mg/ml Xgal, 1 ml of 41 mg/ml potassium ferricyanide, 1 ml of 53 mg/ml potassium ferrocyanide, and 25 μl of 2 M $MgCl_2$. Bring to 25 ml with PBS. Warm to 37°C. Make the stain fresh each time.

8b. Wash sections in PBS by moving sections with a paintbrush into a well filled with fresh PBS and shaking gently for 30 sec. Repeat twice.

9b. Add sections to a well filled with stain from step 7b. Incubate for 6 to 24 hr. Wash as in step 8b.

10b. Mount sections onto glass slides as in steps 7a and 8a.

11b. Place a small amount of standard mounting medium on sections and cover with glass coverslip, avoiding bubble formation as much as possible. View *lacZ* staining (blue) with light microscope.

References: McCown et al., 1996; Samulski et al., 1999; and Xiao et al., 1998

Contributors: Rebecca A. Haberman, Gabriele Kroner-Lux, Thomas J. McCown, and Richard Jude Samulski

UNIT 4.12

Alphavirus-Mediated Gene Transfer into Neurons

Kits for the production of packaged Semliki Forest virus (SFV) and Sinbis virus (SIN) replicons are commercially available (SFV, Life Technologies; SIN, Invitrogen). Comparable results are obtained using either SFV or SIN vectors in cultured rat hippocampal slices (Ehrengruber et al., 1999). The SFV and SIN systems (see Fig. 4.12.1) are similar, and most steps in the protocols for generating infectious SFV and SIN particles are identical. Additonal background information on the two systems is available at *www2.lifetech.com/catalog/techline/molecular_biology/Manuals_PPS/18448019.pdf* (SFV); *www.microbiology.wustl.edu/sindbis/sin_genes* (SIN); and *www.invitrogen.com/catalog.html* (SIN). Packaged replicons derived by this approach will infect target cells, but the absence of structural protein genes prevents the generation of new particles. They are therefore termed "suicide vectors."

CAUTION: Follow biosafety level 2 practices when using helper vectors in packaging alphaviral replicons. Recommended precautions are standard microbiological practices, laboratory coats, inactivation of all infectious waste, limited access to working areas, protective gloves, posted biohazard signs, and class I or II biological safety cabinets used for mechanical and manipulative procedures that cause splashes or aerosol. SFV and SIN vectors expressing a highly toxic protein should be treated as a special risk. NIH and CDC safety guidelines can be reviewed at *www.ehs.psu.edu/biosafety/bmbl-1.htm.*

NOTE: All solutions and equipment coming into contact with living cells must be sterile, and aseptic technique should be used accordingly. All cell culture incubations should be carried out in a humidified 37°C, 5% CO_2 incubator unless otherwise stated.

NOTE: See Troubleshooting Guide at end of this unit (Table 4.12.1) for summary of problems that may arise, and possible solutions.

BASIC PROTOCOL 1

PREPARATION OF PACKAGED SFV AND SIN REPLICONS

NOTE: The yield of high concentrations of packaged replicons depends on the integrity of both vector and helper RNA which are transfected into the BHK-21 cells. Normal standard precautions for the work with RNA must be used, i.e., protective gloves, pipet tips with filter inserts, and decontamination of the work bench from RNase (using chemical compounds, e.g., RNase AWAY, Molecular BioProducts).

Materials *(see* APPENDIX 1 *for items with ✓)*

- Vector plasmid pSFV2gen (Life Technologies, upon request; Fig. 4.12.2)
- Helper plasmid pSFV-Helper2 (Life Technologies)
- Restriction endonucleases *Spe*I, *Nru*I, and *Sap*I (including corresponding buffers)
- Vector plasmid pSINRep5 (Invitrogen; Fig. 4.12.3) or pSINrep504
- Linearized/nonlinearized helper plasmid DH-BB or DH(26S)5′SIN (Invitrogen)
- Restriction endonuclease *Xho*I (if the cDNA of interest contains an *Xho*I site, use either *Not*I or *Pac*I to linearize pSINRep5) including corresponding buffer
- 0.8% (w/v) agarose gel
- 25:24:1 (v/v/v) phenol/chloroform/isoamyl alcohol

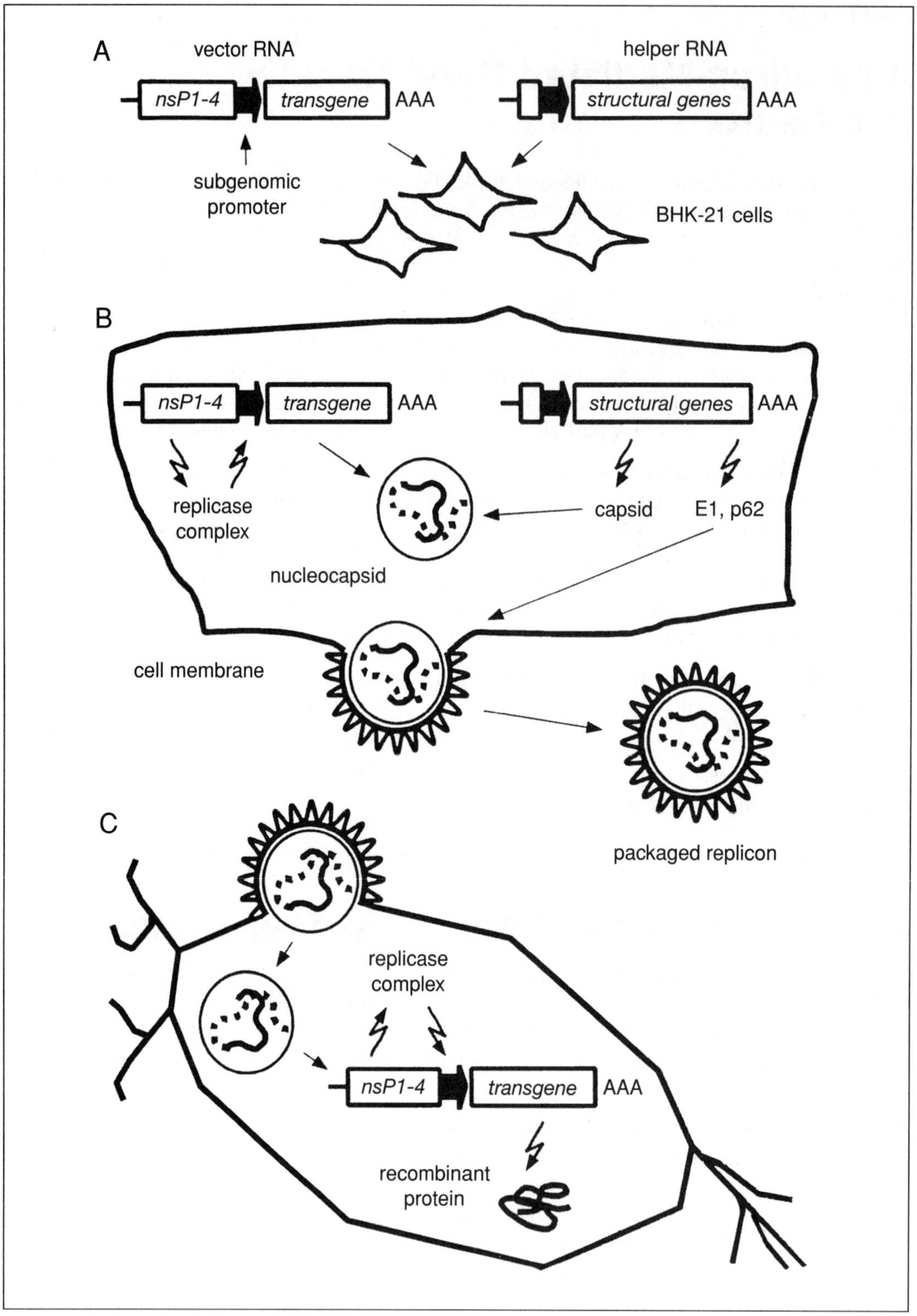

Figure 4.12.1 *(see legend at right)*

4

✓ 3 M sodium acetate
95% and 70% ethanol
✓ 10× transcription buffer (SFV system) *or* commercial 5× transcription buffer provided with the SP6 polymerase (SIN system)
10 mM $m^7G(5')ppp(5')G$ (sodium salt; Pharmacia or New England Biolabs)
50 mM dithiothreitol (DTT)
rNTP mix (10 mM rATP, 10 mM rCTP, 10 mM rUTP, 5 mM rGTP; Roche Molecular Biochemicals)
10 to 50 U/μl RNase inhibitor (Roche Molecular Biochemicals)
10 to 20 U/μl SP6 RNA polymerase (Roche Molecular Biochemicals, Life Technologies, Epicentre Technologies, or Promega)
✓ Gel loading buffer
Molecular weight marker (e.g., digested λ DNA)
BHK-21 cells (ATCC #CRL-6281, ~80% confluent)
✓ PBS, 37°CEDTA
Trypsin/EDTA (0.5 mg/ml trypsin and 0.2 mg/ml EDTA in PBS)
✓ Complete BHK-21 cell medium 1.5-ml microcentrifuge tubes

Two heating blocks or water baths (37°C and 80 to 90°C)
Sterile electroporation gap cuvettes (e.g., 0.2-cm gap, Bio-Rad)
Electroporator (e.g., Gene Pulser, Bio-Rad)
Tissue culture flasks or dishes (24-, 35-, 60-, or 100-mm)
Plastic syringes (10- or 20-ml) with attached 0.22-μm sterile filters

4

Subclone and linearize SFV system

1a. Subclone the gene of interest into the multiple cloning region of pSFV2gen (*CPMB UNIT 3.16*). Grow up resulting plasmid and pSFV-Helper2 and prepare purified DNA (*APPENDIX 2F*; it is preferable to utilize highly purified—e.g., Maxiprep, Qiagen—DNA).

2a. Linearize recombinant pSFV2gen with *Nru*I or *Sap*I, and pSFV-Helper2 with *Spe*I. If using pSFV1, linearize with *Spe*I or *Sap*I (*CPNS APPENDIX 1M*).

Subclone and linearize SIN system

1b. Subclone the gene of interest into the multiple cloning region of pSINRep5 or pSINrep504 (*CPMB UNIT 3.16*). Grow up resulting plasmid and prepare purified DNA (*APPENDIX 2F;* it is preferable to utilize highly purified—Maxiprep, Qiagen—DNA).

2b. Linearize recombinant pSINRep5 or pSINrep504 with *Xho*I (preferred), *Not*I, or *Pac*I (as appropriate for the inserted cDNA; *CPNS APPENDIX 1M*). Linearize DH-BB and DH(26S) 5′SIN with *Xho*I (or *Eco*RI), if not using their linearized forms, supplied by Invitrogen.

Figure 4.12.1 *(at left)* Generation of recombinant SFV and SIN particles. (**A**) Vector RNA encoding non-structural proteins 1-4 (*nsP1-4*) and the transgene (hatched) under the control of the subgenomic RNA promoter (broad arrow), and defective helper RNA (encoding the structural alphavirus proteins downstream of the subgenomic RNA promoter) are obtained by in vitro transcription from their respective cDNAs and cotransfected into BHK-21 cells. (**B**) Within the cytoplasm of BHK-21 cells, vector RNA replication occurs through the action of *nsP1-4*. In parallel, the defective helper RNA is also replicated and transcribed by *nsP1-4*, and the capsid protein, spike protein E1, and precursor protein p62 (later split into the E2 and E3 spike proteins) are translated from its subgenomic RNA. Capsid proteins package only vector RNA containing the transgene (dashed) while spike proteins are incorporated into the BHK-21 cell membrane. Nucleocapsids dock to the cell membrane where spike proteins have been incorporated, thus allowing the budding of SFV and SIN particles. (**C**) Upon infection of a neuron, nucleocapsids are released into the cytoplasm and the vector RNA is liberated. The replicase complex composed of *nsP1-4* amplifies the vector RNA, and the transgene is translated into the recombinant protein.

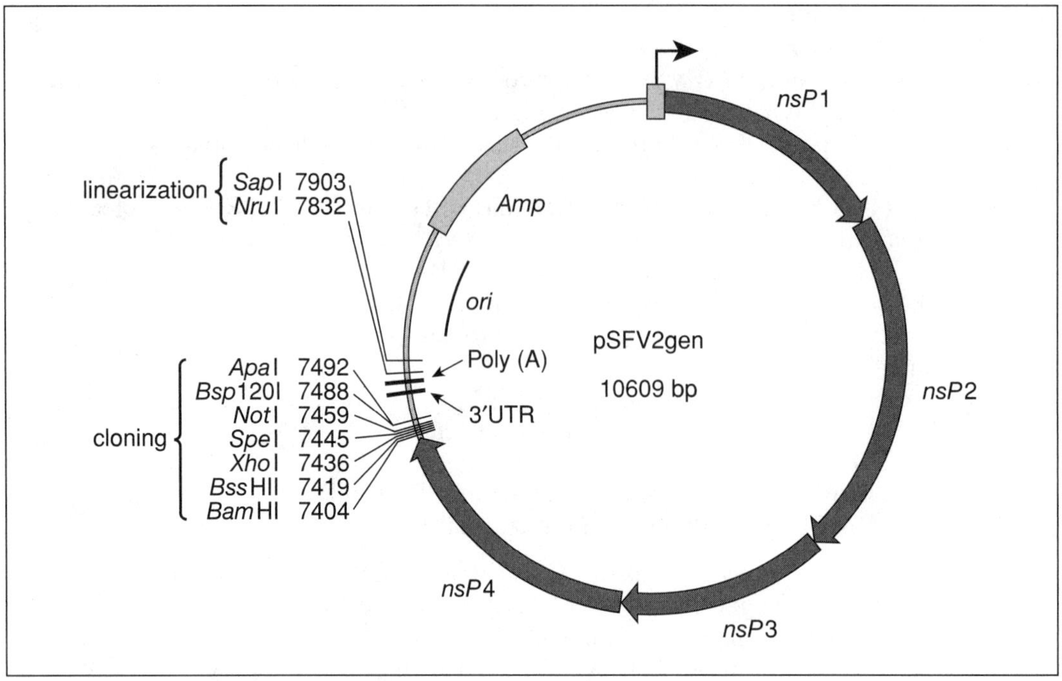

Figure 4.12.2 Map of the pSFV2gen vector plasmid. Numbering starts with the first nucleotide of the cDNA derived from SFV genomic RNA. Upstream from the SFV cDNA is the SP6 promoter (hatched box and arrow) for in vitro transcription by SP6 RNA polymerase. The regions encoding the nonstructural proteins are shown (*nsP1-4*). Unique restriction sites and their positions for the insertion of the cDNA of interest (cloning) and linearization for in vitro transcription are given. Regions corresponding to the 3′-untranslated region (3′UTR) of the cloning site, the poly(A) sequence, the origin of replication (*ori*), and the ampicillin resistance gene (*Amp*) are also shown.

3. Run 2- to 5-μl aliquots of plasmid linearization mixes on a 0.8% agarose gel together with undigested plasmid to confirm linearization (*CPNS APPENDIX 1N*).

4. Purify linearized DNA by extraction with phenol/chloroform as follows (also see *APPENDIX 2C*):

   a. Increase digestion volume to 100 μl with water and add 100 μl of 25:24:1 phenol/chloroform/isoamyl alcohol.

   b. Invert tubes a few times and microcentrifuge for 2 min at maximum speed, room temperature.

   c. Collect upper phase (∼100 μl), which contains the DNA.

5. Add 10 μl of 3 M sodium acetate and 250 μl 95% ethanol, mix well, and allow to precipitate for 15 to 20 min at −80°C or overnight at −20°C. Microcentrifuge 15 min at maximum speed, 4°C. Wash pellet with 70% ethanol, microcentrifuge 3 min at maximum speed, 4°C, discard supernatant, and air dry pellet for ∼5 min. Resuspend DNA in water at 0.5 μg/μl.

6. Set up 50-μl in vitro transcription reactions in 1.5-ml microcentrifuge tubes for vector and (separately) for helper RNA. Mix the following in the order given at room temperature (when planning more than one transcription reaction, prepare a master mix containing everything but the DNA and the SP6 RNA polymerase and then aliquot into individual tubes):

   5 μl (∼2 μg) linearized DNA
   5 μl 10× transcription buffer (for SFV system) *or* 10 μl commercial
   5× transcription buffer provided with the SP6 polymerase (for SIN system)

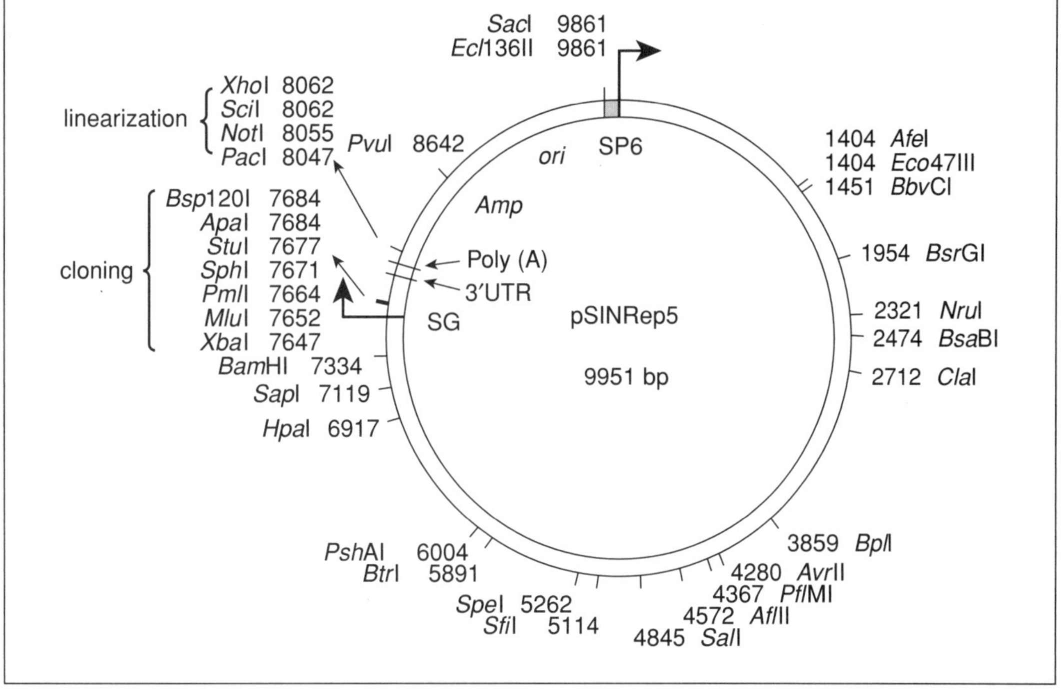

Figure 4.12.3 Map of the pSINRep5 vector plasmid. Numbering starts with the first nucleotide of the cDNA derived from SIN genomic RNA. Upstream from the SIN cDNA is the SP6 promoter (hatched box and arrow) for in vitro transcription by SP6 RNA polymerase. Unique restriction sites and their positions, including those for insertion of the cDNA of interest (cloning) and linearization for in vitro transcription, are indicated. The figure also indicates the subgenomic RNA promoter (SG; the arrow at position 7598 shows the start of the subgenomic RNA), the 3′ untranslated region (3′UTR) of the cloning site, the poly(A) sequence, the ampicillin resistance gene (*Amp*; nt 8227-9076), and the origin of replication (*ori*). Figure provided by Dr. Sondra Schlesinger; see Bredenbeek et al. (1993) for further details.

5 μl 10 mM $m^7G(5')ppp(5')G$
5 μl 50 mM DTT
5 μl rNTP mix
1.5 μl 50 U/μl RNase inhibitor
3.5 μl 10 to 20 U/μl SP6 RNA polymerase
20 μl (for SFV system) or 15 μl (for SIN system) sterile H_2O.

Mix and incubate for 60 min at 37°C (or 40°C when using the SIN system).

7. To examine the quality and quantity of the in vitro-transcribed RNA, transfer a 1- to 4-μl aliquot into 10 μl sterile water, add 3 μl gel loading buffer, and heat for 3 to 5 min at 80° to 90°C. Analyze sample on a 0.8% agarose gel by using, e.g., digested λ DNA as a molecular weight marker (see *CPNS APPENDIX 1N*). Store the RNA ∼8 kb) in aliquots at −80°C (stable for months).

To achieve maximal transfection efficiencies by electroporation, the BHK-21 cells must not be clumped. For the trypsinization, use cells grown in 150-mm petri dishes or 175-cm^2 flasks, ∼80% confluent. Avoid using cells at >2 days after passaging. For each electroporation, 10^7 cells are required.

8. Aspirate growth medium and wash cells with PBS, 37°C. Add 3 ml trypsin/EDTA per culture dish or flask, remove trypsin/EDTA, and incubate for 5 min at 37°C. Resuspend

cells in a small volume of complete BHK-21 cell medium (to stop the trypsin activity), remove cell clumps, and add medium up to the original volume.

9. Centrifuge cells for 5 min at 400 × *g*, room temperature. Resuspend cell pellet in the original volume of PBS. Centrifuge cells 5 min at 400 × *g*, room temperature. Resuspend cell pellet using 2 ml PBS per 175-cm^2 flask, resulting in 10^7 cells/ml. Keep cell suspension on ice, and use immediately for electroporation.

10. Transfer 0.4 ml cell suspension into a 0.2-cm gap electroporation cuvette, 0.8 ml cell suspension for a 0.4-cm gap cuvette.

11. Mix 20 to 40 μl recombinant vector RNA with 20 μl helper RNA (both from step 6) in a microcentrifuge tube, avoiding formation of air bubbles. Add RNA to cells in the cuvette. For SIN system, mix 5- to 10-/l each of vector and helper RNA.

12. Insert cuvette into electroporator holder and pulse the cell suspension twice with minimal interval between the two shocks. Use the following settings with a Bio-Rad Gene Pulser:

Capacitance extender	960 μF
Voltage	1500 V
Capacitor	25 μF
Resistance for the pulse controller	∞ Ω
Expected time constants	0.7 to 0.8 sec.

The new electroporator version from Bio-Rad (i.e., Gene Pulser II) requires some modifications to the settings.

13. Dilute electroporated cells 25-fold in complete BHK-21 cell medium. Transfer the cell suspension into tissue culture flasks and dishes. Use ~1 ml for a 24-mm dish, ~2 ml for a 35-mm dish, ~4 ml for a 60-mm dish, and ~10 ml for a 100-mm dish. Incubate electroporated cells 24 hr in a 37°C, 5% CO_2 incubator.

Cotransfected cells start to show cytopathic effects (i.e., round up, phase brightly, and stop growing) by 24 hr. Cells will express GFP fluorescence if this reporter gene is being used.

14. Harvest the released, packaged replicons by filtering the supernatant through a syringe with attached 0.22-μm filter, to remove cell debris and ensure that the viral stock is sterile. Dispense 0.1 to 1-ml aliquots and freeze at −80°C.

Stocks of packaged replicons can be activated (see Basic Protocol 2) immediately after harvesting or after storage at −20°C (short-term storage) or −80°C (storage for up to >1 year). It is generally convenient to activate the stocks prior to preparing aliquots for storage.

BASIC PROTOCOL 2

ACTIVATION OF PACKAGED SFV REPLICONS

Materials *(see APPENDIX 1 for items with ✓)*

20 mg/ml α-chymotrypsin (Sigma; freeze aliquots of 20 mg/ml for storage)
Packaged SFV replicons (see Basic Protocol 1)
10 mg/ml aprotinin (Sigma or Roche Molecular Biochemicals)
✓ PBS, 37°C
BHK-21 (ATCC #CRL-6281), CHO (ATCC #CCL-61), or HEK293 cells (ATCC #CRL-1573) in 6- or 12-well plates (70% to 80% confluent), plus the corresponding cell culture medium

1. Add 20 mg/ml α-chymotrypsin to the packaged replicon stock to achieve a final concentration of 500 μg/ml α-chymotrypsin. Incubate for 15 to 30 min at room temperature (30 min provides a slightly higher yield).

2. Stop the reaction by adding 10 mg/ml aprotinin to achieve a final concentration of 250 μg/ml aprotinin. Store aliquots of the activated particles, now ready for infection, at −20°C (short term) or −80°C (long term). When planning to infect cultures of dissociated neurons (see Basic Protocol 4), transfer activated particles into neuronal medium before preparing aliquots.

   Avoid more than a few cycles of freezing and thawing as this may inactivate recombinant SFV.

3. To test the expression of the transgene before infecting neurons (optional, but recommended), wash cells—that are 50% to 70% confluent—once with 37°C PBS.

4. Add 0.5 ml of the solution containing infectious replicons (from step 2, diluted in cell culture medium) to the cells. Test serial dilutions of up to 1000-fold for each stock. Incubate for ≥1 hr.

5. Completely replace the solution containing infectious replicons with fresh cell culture medium. Incubate the infected cells for the time required for transgene expression (see Support Protocols 1 and 2 for analyzing the titers of the viral stocks and for metabolic labeling, respectively; see *UNIT 4.15* for detecting β-galactosidase).

4

BASIC PROTOCOL 3

INFECTION OF HIPPOCAMPAL SLICES

NOTE: The virus injection procedure must be performed under sterile conditions in a biosafety level 2 cabinet. Unless otherwise stated, all steps in this protocol are carried out at room temperature.

Materials *(see APPENDIX 1 for items with ✓)*

✓ Cutting medium, pre-warmed to 37°C
Infectious SFV or SIN replicon stock (see Basic Protocols 1 and 2)
95% ethanol
Hippocampal slice cultures (roller-tube type, *CPNS UNIT 6.11*)
Roller tube culture medium (*CPNS UNIT 6.11*)

Glass capillaries (i.d. 1.5 mm, o.d. 1.17 mm; e.g., Clark capillaries, Harvard Apparatus)
Electrode puller
Autoclavable electrode holder
Micromanipulator with both coarse and fine controls (Narishige)
Metal plate containing a base for a 35-mm petri dish
3-way valve
1-ml syringe
Plastic tubing (i.d. 1 mm, o.d. 3 mm)
35-mm plastic petri dishes
Dissection microscope
Microloader pipet tips (autoclaved; Eppendorf)
Forceps

1. Pull glass capillaries on an electrode puller to obtain micropipets with a long tip, characteristic of sharp electrodes (*CPNS UNIT 6.1*).

2. Transfer micropipets into an autoclavable electrode holder, then autoclave the electrode holder with the micropipets.

3. Mount the electrode holder onto the micromanipulator, fix the micromanipulator onto a metal plate containing a base for a 35-mm petri dish, and place the set-up into the biosafety level 2 cabinet.
4. Connect the electrode holder to a 1-ml syringe with the plastic tubing, via a 3-way valve.
5. Insert a 35-mm plastic petri dish into the base on the metal plate, and add 2 to 3 ml cutting medium (prewarmed to 37°C) to the dish. (When infecting acute hippocampal slices, replace cutting medium with the normal saline solution of choice.)
6. Focus the dissection microscope onto the bottom of the 35-mm petri dish.
7. Dilute the stock of infectious replicons by 10- to 1000-fold in cutting medium.
8. Back-fill an autoclaved glass micropipet with ~30 µl of diluted viral solution, using an autoclaved microloader pipet tip. Insert micropipet into the electrode holder.
9. Sterilize forceps by prewetting in 95% ethanol, thenflaming. Break the micropipet tip with sterilized forceps to a final diameter of ~20 µm.
10. Verify that virus solution is exiting from the micropipet tip by applying positive pressure from the 1-ml syringe.
11. With sterilized forceps, transfer a hippocampal slice culture from the roller-tube into the petri dish containing cutting medium.
12. Lower the loaded micropipet into the pyramidal and/or granule cell layer by using the micromanipulator. Perform one injection of short duration (<2 sec) by applying pressure from the 1-ml syringe. Stop the injection by releasing the pressure via the 3-way valve.
13. Reposition the micropipet into an adjacent site and repeat the injection. Select 10 to 15 injection sites per slice. Apply a total of 1 to 2 µl virus solution per slice.
14. Wash the slice by bathing it, in a 35-mm petri dish, with 2 to 3 ml roller tube culture medium for 10 to 60 sec.
15. Return the slice to the roller-tube and culture for 1 to 5 days.

BASIC PROTOCOL 4

INFECTION OF DISPERSED NEURONS

For studies on certain recombinant proteins (e.g., metabotropic glutamate receptors), the medium from BHK-21 cells (in which the alphaviral particles are harvested) can be toxic to primary neurons. In these cases, it is recommended that the medium be exchanged by either virus pelleting, sucrose purification, or ultrafiltration. Virus pelleting is faster and simpler than sucrose purification, but can reduce infectious particle titers by 10- to 100-fold. Ultrafiltration is achieved with Centriprep concentrators. The Centriprep-50 device can be used for the concentration of a 15-ml sample to 0.7 ml within 20 min. Recovery of >90% of the virus is typical.

The infection rate is usually very high (70% to 95%). Test infections with a GFP or β-galactosidase reporter gene construct at different virus concentrationsare recommended. The neurons will tolerate SFV and SIN infection reasonably well (better than many cell lines), but it is recommended that fairly young neuronal cultures be used to obtain prolonged survival.

Materials *(see APPENDIX 1 for items with ✓)*

Stock of infectious SFV or SIN replicons (see Basic Protocols 1 and 2)
✓ BHK-21 cell medium
50% and 20% (w/v) sucrose solution

12- or 13-ml ultracentrifuge tubes, to fit particular rotor used
Ultracentrifuge with swingingbucket rotor
Centriprep-50 or Centriprep-100 centrifugal concentrators (Amicon)
50-ml disposable plastic tubes (Falcon)
Low-speed centrifuge (500 to 3000 × *g*), accommodating 50-ml plastic tubes
1.5-ml microcentrifuge tubes
1-ml disposable plastic pipet

Purify virus by pelleting

1a. Transfer viral stock into an ultracentrifuge tube. Add BHK-21 cell medium to fill the ultracentrifuge tube if the volume of the viral stock is <12 to 13 ml.

2a. Ultracentrifuge 2 hr at 217000 × *g*, 4°C.

3a. Resuspend the pellet containing the infectious replicons in neuronal culture medium at 1/10 of the original volume.

4a. Check the titer of the resuspended stock (see Support Protocol 1).

Purify virus by sucrose gradient centrifugation

1b. Prepare a step gradient in an ultracentrifuge tube by adding 1 ml of 50% sucrose solution (bottom) and 3 ml of 20% sucrose solution (top).

2b. Carefully layer 9 ml or 8 ml (depending upon the ultracentrifuge tube used) of viral stock onto the sucrose gradient.

3b. Ultracentrifuge 1.5 hr at 160000 ×, 4°C.

4b. Discard the medium fraction. Also discard the bottom 0.8 ml consisting of 50% sucrose. Transfer the resulting bottom 0.8 ml with the infectious replicon band (at the interface between the 20% and 50% sucrose) into a 1.5-ml microcentrifuge tube, using a 1-ml disposable plastic pipet. Check the titer of the virus stock (see Support Protocol 1).

Purify virus by ultrafiltration

1c. Load virus onto the sample container of the Centriprep concentrator (as described by the manufacturer).

2c. Centrifuge the assembled concentrator at an appropriate *g*-force (according to the manufacturer's recommendations) until the fluid levels inside and outside the filtrate collector equilibrate.

3c. Remove the device from the centrifuge, snap off the air-tight seal cap, decant the filtrate, replace the cap and centrifuge the concentrator a second time.

4c. Decant the filtrate. Loosen the twist-lock cap and remove the filtrate collector. Collect the concentrated virus sample with a 1-ml disposable plastic pipet. Repeat centrifugation if further concentration is desired.

5c. Infect the neurons by adding 2 to 10 viral particles per neuron, directly to the culture medium or after resuspending them in neuronal medium.

6c. Incubate the infected neurons for 1 to 2 hr, 37°C. Replace the medium containing infectious replicons with fresh or conditioned culture medium. Incubate the infected neurons at 37°C until transgene expression can be detected.

Expression of the gene of interest can be observed as early 6 hr post-infection. Infected neurons should remain alive for 2 to 4 days post-infection.

ALTERNATE PROTOCOL

LIPID-MEDIATED COTRANSFECTION OF RNA

Additional Materials *(also see Basic Protocol 1)*

Opti-MEM I reduced-serum medium (Life Technologies)
DMRIE-C reagent (Life Technologies)

1. Plate an appropriate number (i.e., 1.5–3 × 10^5) of BHK-21 cells in 35-mm dishes or 6-well plates in 2 ml complete BHK-21 cell medium.
2. The following day, when plated cells are ~80% confluent, wash cells once with 2 ml of Opti-MEM I reduced-serum medium at room temperature.
3. Prepare cationic lipid/RNA complexes.
   a. Add 1 ml Opti-MEM I reduced-serum medium at room temperature to each of six 1.5-ml microcentrifuge tubes.
   b. Add 0, 3, 6, 9, 12, and 15 μl DMRIE-C reagent to each labeled tube. Vortex briefly.
   c. Add 10 μl in vitro–transcribed vector RNA and 5 μl in vitro–transcribed helper RNA (~5 and 2.5 μg, respectively) to each tube (see Basic Protocol 1). Vortex briefly.
4. Immediately add lipid/RNA complexes to the washed cells from step 2. Incubate for 4 hr at 37°C, then replace the transfection medium with 37°C complete BHK-21 cell medium (>90% transfection efficiency).
5. Harvest the supernatant containing the released, packaged replicons. Sterile filter through a 0.22-μm syringe filter. Freeze aliquots at −80°C (up to >1 year) or at −20°C (short-term storage). Avoid more than a few cycles of freezing and thawing.

   Transferinfectious replicons into neuronal medium before preparing aliquots when planning to infect cultures of dissociated neurons (see Basic Protocol 4).

SUPPORT PROTOCOL 1

TITER DETERMINATION

Viral stocks expressing reporter genes (i.e., β-galactosidase, GFP) can be visualized directly. Indirect immunofluorescence detecting the recombinant protein is useful for other alphaviral constructs.

Additional Materials *(also see Basic Protocols 1 and 4)*

100% methanol
0.2% (w/v) gelatin in PBS (blocking solution)
Primary antibody directed against the recombinant protein
Secondary antibody coupled to detection system
2.5% (w/v) DABCO (reduces fading of FITC) in Mowiol 4-88 (Calbiochem)
Glass slides

Express reporter gene

1a. Define the number of cells (e.g., BHK-21 cells) grown in 6- or 12-well plates or on coverslips.

2a. Infect the cells of each well with a serially diluted stock of infectious replicons encoding GFP or β-galactosidase. Incubate overnight at 37°C.

3a. Analyze the GFP or β-galactosidase expression (<24 hr post-infection, if possible) as described in *UNIT 4.15*: Count the cells expressing β-galactosidase (blue) or GFP (green) for each dilution. Estimate the titer (infectious replicons/ml) based on the number of infected cells per well or the percentage of cells expressing the reporter gene, and the virus dilution.

Observe by immunofluorescence

1b. Define the number of cells grown to 70% confluency on coverslips.

2b. Infect cells with serial dilutions of the viral stock. Incubate overnight at 37°C.

3b. Rinse coverslips twice with PBS. Fix cells with 100% methanol for 5 min at −20°C.

4b. Remove methanol and wash cells three times with PBS.

5b. Soak coverslips in 0.2% gelatin in PBS (blocking solution) and incubate for 30 min at room temperature.

6b. Remove the blocking solution. Soak coverslips in primary antibody in blocking solution. Incubate for 30 min at room temperature.

7b. Remove primary antibody. Wash cells three times with PBS, room temperature. Soak coverslips in secondary antibody in blocking solution. Incubate for 30 min at room temperature.

8b. Wash cells three times with PBS and once with water at room temperature. Air dry.

9b. Mount coverslips on glass slides using 10 to 20 μl of 2.5% DABCO in Mowiol 4-88 as a mounting medium. Examine each coverslip, using conventional microscopy. Determine the titer (infectious replicons/milliliter) based on the number of cells per coverslip, the percentage of cells recognized by the primary antibody, and the virus dilution.

4

SUPPORT PROTOCOL 2

METABOLIC LABELING

The electroporated cells will produce large quantities of SFV or SIN structural proteins as early as 4 hr post-electroporation. For monitoring heterologous gene expression, it is advisable to label cells at ≥6 hr post-infection. High expression levels are generally observed at 12 to 16 hr post-infection.

CAUTION: Radioactive materials require special handling; all supernatants must be considered radioactive waste and disposed of accordingly.

Additional Materials *(also see Basic Protocol 1 and Alternate Protocol; see APPENDIX 1 for items with ✓)*

Infected or transfected cells (see Basic Protocols 1, 2, 4, and Alternate Protocol for cell lines and neurons)
✓ Starvation medium
>1000 Ci/mmol [^{35}S]-methionine (1 mCi/0.1 ml), added to starvation medium
✓ Chase medium
✓ Lysis buffer
✓ 2× SDS-PAGE sample buffer
Precast 10% (w/v) Tris-glycine polyacrylamide gels (Novex/Invitrogen)
✓ Fixing solution
Amplify solution (Amersham Pharmacia Biotech)
Hyperfilm-MP (Amersham Pharmacia Biotech)

SDS-PAGE apparatus
Power unit
Bio-Rad gel dryer

1. Remove the medium from infected cells grown in 6- or 12-well plates. Wash cells once with 37°C PBS. Add 1 to 2 ml starvation medium and incubate cells for 30 min at 37°C.

2. Remove the medium and add 250 or 500 μl starvation medium containing 50 or 100 μCi/ml [^{35}S]-methionine per well of a 12- or 6-well plate, respectively. Incubate for 20 min at 37°C.

3. Remove medium and wash cells twice with 37°C PBS. Add 1 to 2 ml chase medium and incubate for 15 to 30 min at 37°C.

4. Remove chase medium and wash cells once with 37°C PBS. Add 150 to 300 μl lysis buffer and incubate cells for 10 min on ice. Detach lysed cells with the "wrong" side of a pipet tip. Resuspend each cell lysate separately, transfer it into a 1.5-ml microcentrifuge tube, and store at −20°C.

5. Mix 10 μl of each lysate with 10 μl of 2× SDS-PAGE sample buffer and load onto a precast 10% Tris-glycine polyacrylamide gel (see *CPMB UNIT 10.2*). Electrophorese for ~2.5 hr, 130 V.

6. Fix the gel in fixing solution for 30 min at room temperature. Then replace fixing solution with Amplify solution, and incubate for 30 min at room temperature. Dry in a Bio-Rad gel dryer.

7. Expose the gel on Hyperfilm-MP for 2 hr to overnight (depending on the radioactivity) at room temperature. (Expose at −80°C if using radioactivity-intensifying screens.)

References: Ehrengruber et al., 1999

Contributors: Markus U. Ehrengruber and Kenneth Lundstrom

Table 4.12.1 Troubleshooting Guide for Alphavirus-Mediated Transfection into Neurons

Problem	Possible Cause	Solution
Smearing of RNA bands	RNA degradation	Use RNase-free conditions for in vitro transcription.
	Agarose gel electrophoresis was too long	Run RNA 2–3 cm into the gel.
Low RNA yields	Partially linearized plasmid	Check the linearized plasmid on an agarose gel to verify complete digestion.
	In vitro transcription reaction not optimal	Use reagents as specified in these protocols. To obtain higher RNA yields one may increase the total amount of SP6 RNA polymerase to 100 U.
Low transfection efficiency	Insufficient RNA	Optimize the RNA amount used for the cotransfection.
	Capping reaction (incorporation of cap analogue $m^7G(5')ppp(5')G$) not working	Try different source or new stock of cap.
		Adjust the amount of SP6 RNA polymerase. Only a fraction of the RNA will be capped when too much SP6 RNA polymerase is being used.

continued

Table 4.12.1 Troubleshooting Guide for Alphavirus-Mediated Transfection into Neurons, Continued

Problem	Possible Cause	Solution
	Not using optimal electroporation conditions or amount of DMRIE-C reagent	Optimize transfection conditions using vector RNA expressing GFP or β-galactosidase.
	Cell density too high or too low	Use cells that are ~80% confluent at time of transfection.
Poor cell viability following cotransfection of RNA	Suboptimal electroporation conditions or excessive DMRIE-C reagent	Optimize transfection conditions.
	Cell density too low	Use cells that are ~80% confluent at time of DMRIE-C-mediated transfection.
Low titer of the viral stock	RNA degradation	Use RNase-free conditions for in vitro transcription and cotransfection.
	Insufficient RNA or uneven concentrations of recombinant vector and helper RNA	Optimize packaging reaction by using vector RNA expressing GFP or β-galactosidase. Estimate relative RNA concentrations by agarose gel electrophoresis. Use vector to helper RNA ratios of 1–2 to 1 μg.
	Did not use BHK-21 cells	All packaging procedures have been optimized for BHK-21 cells.
Low recombinant protein expression in infected cells	Did not use an optimal concentration of infectious replicons	Do a dose-response curve with the infectious replicons to determine the amount required for optimal expression.
	Did not activate the packaged SFV replicons	Prior to infection activate the SFV stock with α-chymotrypsin. Use a new preparation of α-chymotrypsin.
	Recombinant protein is unstable	Remove any protein degradation signals and/or use protease inhibitors.
Poor cell viability following infection	Used too much of the viral stock	Optimize infection conditions.
	Cell density is too low	Use cells that are ~80% confluent at the time of infection.
	BHK-21 cell supernatant is toxic for neurons	Transfer the viral particles into neuronal medium before infecting the neurons.
Limited infection of slice cultures	Clogging of the glass micropipet during virus injection	During the injection procedure verify repeatedly that virus solution exits from the pipet tip by lifting the micropipet out of the cutting medium and applying pressure from the syringe. To prevent or remove salt crystal-induced clogging of the tip, lower the micropipet tip into cutting medium.
	Dilution of virus in micropipet by capillary action	Use micropipets with a smaller tip opening.
Slice cultures show many dead, noninfected cells around injection sites	Excessive pressure used for virus injection	Use micropipets with a smaller tip opening; apply less pressure.

Overview of Gene Delivery into Cells Using HSV-1-Based Vectors

Strategies for gene delivery into neurons, either to study the molecular biology of brain function or for gene therapy, must utilize vectors that persist stably in postmitotic cells and that can be targeted both spatially and temporally in the nervous system in vivo. A number of such gene delivery systems are currently under development. This overview describes the considerations involved in the preparation and use of herpes simplex virus type 1 (HSV-1) as a vector for gene transfer into neurons.

HSV-1 possesses multiple features that make it an ideal vector for genetic intervention in the nervous system: it accepts large molecules of exogenous DNA; it infects nondividing cells from a wide range of hosts with high efficiency and persistence; it enables strong expression of foreign genes; it is episomal and thereby does not cause integration effects; and its particles can be concentrated to relatively high titers. HSV-1, especially because of its ability to incorporate large molecules of foreign DNA, is currently one of the best vectors available for functional analysis of genes in the nervous system.

4

BIOLOGY OF HSV-1

Entry Into the Neuron

HSV-1 is a 150-kb linear double-stranded DNA virus that carries >75 genes. It is an enveloped virus comprising an icosahedral-shaped capsid inside a layer of proteins termed the tegument. HSV-1 has been observed to be capable of infecting most differentiated mammalian cell types. Natural HSV-1 infection is initiated when the virus breaches the epithelial cells of the skin or mucous membrane. Initial attachment of the enveloped particle to the plasma membrane is mediated by nonspecific charge interactions between viral envelope glycoproteins (gB or gC or both) and glycosaminoglycan chains of proteoglycans on the cell surface (reviewed by Spear, 1993). Subsequently, the virus specifically recognizes a cellular receptor—at least one of which has been identified as a member of the tumor necrosis factor/nerve growth factor (TNF/NGF) receptor family (Montgomery et al., 1996)—after which multiple viral glycoproteins mediate pH-independent fusion of the virion envelope with the cell membrane (reviewed by Spear, 1993). The virus travels by retrograde axonal transport from the site of its original entry to the nucleus of the neuron, where viral DNA is released via a nuclear pore. At this point, the virus may begin a lytic infection, resulting in viral replication, or may enter a latent state, in which the expression of lytic genes is silenced and a family of latency-associated transcripts is expressed.

The Lytic Cycle

The HSV-1 genome is a linear double-stranded DNA molecule composed of two unique segments—unique long (U_L) and unique short (U_S)—each flanked by a pair of inverted repeat (I_R) segments. During lytic infection with HSV-1, the viral genome is expressed in a temporally regulated sequence (Honess and Roizman, 1974). Immediately after the release of the viral DNA into the nucleus, the five genes encoding the immediate early (IE) proteins—ICP0, ICP4, ICP22, ICP27, and ICP47—are transcribed, an operation that does not require viral protein synthesis. This process is enhanced by at least one tegument protein referred to as VP16 (also known as Vmw65, or αTIF), which interacts with a cellular molecule termed octamer-binding protein one (Oct 1) to bind to a consensus TAATGARAT sequence (where R is A or G) in IE gene promoter enhancers (Mackem and Roizman, 1982a,b; Gaffney et al., 1985). At the same time, a second tegument protein, virion host shutoff (vhs), degrades cellular mRNA and shuts down host-cell protein synthesis (Oroskar and Read, 1989).

Two members of the IE gene product cascade, ICP4 and ICP27, regulate the expression of early (E) and late (L) genes, which are required for viral replication (reviewed by Fink et al., 1996). E gene products control viral DNA synthesis, which proceeds via a rolling circle mechanism that generates head-to-tail concatemers of the HSV-1 genome. Following viral synthesis, the IE gene products ICP4 and ICP27 activate the expression of the L genes, which encode structural proteins of the capsid, tegument, and envelope. Viral DNA is clipped into genome-size pieces, which are packaged into the capsid sequences via the recognition of the packaging "a" sequence in the I_R. The virus particle, consisting of capsid and tegument, buds through a patch of nuclear membrane modified to contain viral glycoproteins, thereby acquiring its envelope. This lytic cycle can destroy the cell within 10 hr in vitro.

HSV-1 genes have been categorized according to whether or not they are required for production of infectious virus in African Green Monkey kidney (Vero) cells. Two of the IE genes, *ICP4* and *ICP27*, are essential. At least twenty structural genes, whose expression is triggered by ICP4 and ICP27, and the seven viral gene products involved in viral DNA replication are also required. Viruses with deletions in essential genes can propagate only in permissive host cells that provide the missing gene product in *trans*. Such deletion mutants provided the framework for the creation of the present-day replication-incompetent HSV-1 vectors.

Establishment of Latency

Although the lytic cycle follows initial infection of most cell types with HSV-1, it is aborted in sensory neurons, where the virus can establish long-lasting latent infections in which no replicating viral DNA is present (reviewed by Latchman, 1990; Roizman and Sears, 1990). During latency, the viral genome is completely silent except for one region encoding the latency-associated transcripts (Croen et al., 1987; Stevens et al., 1987). Importantly, the viral genes encoding the IE proteins are not transcribed in latently infected cells. HSV-1 mutants in which IE gene transcription is blocked readily establish latency in neuronal cells (Katz et al., 1990; Valyi-Nagi et al., 1991), suggesting that the initial establishment of latency itself depends on the failure of IE gene expression following virus entry into the neuronal cell. These data helped lead to the design of replication-incompetent HSV-1 vectors for gene delivery into cells.

DEVELOPMENT OF AMPLICON AND GENOMIC HSV-1 VECTORS

There are two types of replication-deficient HSV vectors: those in which the foreign DNA of interest is cloned into the viral genome itself, and those that are composed of a plasmid (amplicon) carrying minimal HSV sequences that allow it to be packaged into virus particles with the aid of a helper virus. A number of genes within the wild-type HSV genome are dispensable for its growth in cells in vitro. Roizman and his colleagues capitalized on this finding to create recombinant HSV-1 viruses that could be used as vectors for gene transfer into cells (Roizman and Jenkins, 1985). This type of genetically engineered genomic vector, with modifications (Marconi et al., 1996), has been used by a number of investigators and is described in detail by Fink et al. (1996).

The idea of the amplicon vector originated during characterization of defective HSV-1 particles that arose in and interfered with HSV-1 stocks passaged at high multiplicities of infection

(MOI; Spaete and Frenkel, 1982; Stow and McMonagle, 1982). Analysis of the genomes of these defective HSV-1 particles revealed that they retained only a minimal subset of DNA sequences from the wild-type genome. These sequences included an origin of DNA replication and a cleavage/packaging site (Spaete and Frenkel, 1982, 1985; Stow and McMonagle, 1982). It was discovered that incorporation of these two sequences into a plasmid (the amplicon) conferred on the plasmid the ability to be replicated and packaged into virus particles when it was transfected into a cell that was superinfected with wild-type virus (which supplied HSV replication and virion assembly functions in *trans*). The plasmid sequences that were packaged into virus particles consisted predominantly of 150-kb concatemers of the original plasmid (Spaete and Frenkel, 1982, 1985; Stow and McMonagle, 1982; Vlazny et al., 1982).

The chief advantage of this amplicon vector type, which is now packaged with replication-defective helper viruses, is that cloning manipulations are relatively easy due to the small size of the plasmid (5 to 10 kb). One disadvantage is that production of amplicon vectors requires a copropagated HSV helper virus, resulting in viral stocks that are a mixture of helper and vector viruses. Cytotoxic effects of these stocks placed an upper limit on the concentrations of vector that could be used to infect cells. Despite the inability of replication-incompetent helper viruses to progress through the lytic cycle in normal cells, cytopathic effects resulted both from proteins present in the HSV-1 particles and from expression of HSV-1 IE genes (Johnson et al., 1992). The occasional emergence of wild-type HSV-1 revertants during the amplicon packaging process sometimes exacerbated the cytotoxicity of the virus preparations (Geller et al., 1990; Johnson et al., 1992). Recent improvements in the amplicon packaging procedure, which are discussed in the following section, have minimized most of these limitations. The development of a helper virus-free packaging system for the HSV vector (Fraefel et al., 1996) has virtually eliminated any lingering cytotoxicity in the preparations.

ADVANTAGES AND DISADVANTAGES OF PRESENT-DAY AMPLICON VECTORS

Genomic and plasmid defective HSV-1 vectors have been used to manipulate neuronal physiology, both in culture and in the intact organism. These studies have been promising, but they have also revealed limitations of the current HSV vector systems. The enveloped HSV is too fragile to purify on cesium chloride gradients, so it cannot be concentrated to the same degree as encapsulated viruses such as adenovirus. Moreover, as noted above, nonspecific cytopathic effects of the defective vectors have placed an upper limit on the number of viral particles that can be used to infect neurons. Finally, lack of persistence of high expression levels from the viral recombinants has hampered long-term in vivo studies and has restricted the usefulness of the vectors for both experimentation and gene therapy.

Recent improvements in the amplicon packaging procedure have corrected some of the limitations listed above. The most widely used helper virus initially was HSV-1 *tsK*, with a temperature-sensitive single-base mutation in the *ICP4* gene. Revertants of this mutant arose at a finite frequency during the packaging procedure, so lytic virus was present in some preparations (Geller et al., 1990; Johnson et al., 1992). The development of an efficient packaging system using a deletion mutant of *ICP4* as helper virus (Geller et al., 1990) reduced the frequency of revertants. However, occasional lytic virus particles continued to appear, albeit at a greatly reduced frequency, presumably as a result of recombination between the helper virus and the sequences flanking both sides of the *ICP4*-containing fragment present in the permissive host.

Lim et al. (1996) compared three replication-defective HSV-1 mutants—KOS strain 5*dl*1.2, deleted in the *ICP27* (IE 2) gene; strain 17 D30EBA and KOS strain *dl*120, both deleted in the *ICP4* (IE 3) gene—for their usefulness as helper virus for packaging the amplicon vector pHSVlac, which uses the HSV-1 IE 4/5 promoter to regulate expression of the *E. coli lacZ* gene. Historically, *ICP4* mutants have been preferred because they express fewer HSV-1 genes under nonpermissive conditions than do *ICP27* mutants. However, it was determined that use of the IE 2 mutant 5*dl*1.2 yielded higher vector titers than did use of the IE 3 mutants, with no measurable difference in cytotoxicity between the systems. In addition, wild-type lytic virus arose at a much reduced rate when 5*dl*1.2 was used in conjunction with 2-2 cells (10^{-6} in ~5% of the stocks) than when D30EBA was used together with M64A cells (10^{-5} to 10^{-4}). This difference in production of wild-type HSV-1 between the two systems is likely due to the fact that 5*dl*1.2 is a more complete deletion than D30EBA, and that M64A cells have a substantial amount of flanking sequence that is homologous to that in the helper virus.

To achieve a favorable ratio of recombinant vector to helper virus, the stocks derived from transfection of the packaging cells followed by superinfection with helper virus are passaged three times on the permissive host. The recombinant vector is packaged as long concatemers that contain multiple origins of replication, conferring a selective replicative advantage on the vector-containing virus relative to the helper virus. The efficiency of the initial transfection of vector DNA into the packaging line is, therefore, crucial to the success of the packaging. Lim et al. (1996) showed that the transfection of the vector DNA at the start of the packaging procedure was significantly more efficient using LipofectAMINE (Life Technologies) than using calcium phosphate, and thereby achieved a favorable ratio (≥1:1) of vector to helper. The protocol described in *UNIT 4.14* has yielded vector/helper ratios >100.

For a viral vector to have utility for gene therapy, cytopathic effects of the virus must not eclipse the effects of the transgene. Investigators have not always been successful in generating nontoxic HSV-based replication-defective vectors. Such vectors have been reported to be toxic to neurons in vitro (Johnson et al., 1992). Significant necrosis, often accompanied by inflammation and gliosis, has been identified at the injection site with some genetically engineered HSV-1 vectors used for in vivo studies (e.g., Isacson, 1995). However, the achievement of both a high titer of vector (≥10^6/ml) and a more favorable ratio of vector to helper, and the virtual elimination of wild-type virus in the vector preparations (Lim et al., 1996), has greatly reduced the cytotoxicity of present-day defective HSV-1 amplicon vectors (Carlezon et al., 1997; Neve et al., 1997; Bursztajn et al., 1998). The banding of the virus on a sucrose step gradient, followed by a high-speed centrifugation to pellet the virus (see *UNIT 4.14*), has had the dual effect of reducing further the cytotoxicity of the virus preparations (presumably by removing toxic factors present in the crude cell lysates) and enabling the concentration of the vector to reach titers exceeding 10^8/ml.

A troubling problem that has not yet been resolved for any viral vector used in the brain is that of persistence of expression, an important criterion for the use of vectors for gene transfer. Numerous investigators have observed an initial peak in expression of an HSV transgene in vivo or in vitro, followed by loss of the bulk of the expression by 1 to 2 weeks postinfection (e.g., During et al., 1994; Lim et al., 1996). Interestingly, superinfection with helper virus 5*dl*1.2 1 week postinfection rescued expression of a transgene expressed under the control of the IE 4/5 promoter (Lim et al., 1996). It appears that transactivating factors provided by the helper virus reactivated transcription of the transgene. In contrast, Lowenstein et al. (1995) showed that nonpermissive cells infected with only the amplicon vector and not the helper virus exhibited transgene expression under control of the HSV IE 3 promoter, indicating that the IE 3 promoter can be functional in the absence of virus transactivating factors.

Two recent developments suggest that the problem of persistence of expression is not insoluble. Use of a 9.0-kb fragment of the tyrosine hydroxylase promoter to drive reporter gene expression in an HSV-1 amplicon vector resulted in prolonged gene expression in vivo (Jin et al., 1996), suggesting that neuronal, unlike viral, promoters in HSV-1 vectors have the potential to produce stable gene expression. Additionally, the development of hybrid amplicons that incorporate elements that allow autonomous replication of the episome (Wang and Vos, 1996), or that incorporate adeno-associated virus (AAV) elements for genomic integration of the amplicon (Fraefel et al., 1997; Johnston et al., 1997), have resulted in vectors that support long-term gene expression both in vitro and in vivo.

COMPARISON OF HSV-1 WITH OTHER VECTORS FOR GENE TRANSFER INTO NEURONS

In addition to HSV-1, numerous alternative and increasingly user-friendly means of gene transfer into the brain are now available (Table 4.13.1). Adenovirus vectors, like HSV-1 vectors, infect postmitotic cells and can enter a broad range of mammalian cell types (reviewed by Neve and Geller, 1996; also see Choi-Lundberg et al., 1997; Gravel et al., 1997; Haase et al., 1997). They have the additional advantage that they can be concentrated to very high titers ($\geq 10^{10}$/ml). However, the use of adenovirus vectors has been restricted by the host immune response that they elicit (Byrnes et al., 1996; Tripathy et al., 1996; Kajiwara et al., 1997), and by the fact that present vectors can accommodate only a maximum of 6 to 8 kb. Direct

Table 4.13.1 Comparison of Methods for Manipulating Gene Expression in the Brain

Advantages	Disadvantages
Herpesvirus vectors	
Broad host and cell-type range	Occasional cytotoxicity
Episomal (no possibility of insertional activation of host genes)	Lack of persistence of expression
Easy genetic manipulation of amplicon vectors	
Can accommodate up to 15 kb of foreign DNA	
High level of expression of foreign genes	
Helper virus–free stocks possible	
Adenovirus vectors	
Broad mammalian host and cell-type range	Elicits host immune response
Episomal (no possibility of insertional activation of host genes)	Stability of expression unknown
Growth to high titers ($\sim 10^{10}$/ml)	Can accommodate only ~7.5 kb of foreign DNA
High level of expression of foreign genes	Genetic manipulation is unwieldy
Adeno-associated vectors	
Broad mammalian host and cell-type range	Can accommodate only 4.7 kb of foreign DNA
Nonpathogenic	Lack of persistence of expression
Helper virus–free stocks possible	
Lentivirus vectors	
Integrates into host chromosome	Can accommodate only 6-8 kb of foreign DNA
Expression is persistent	Low titers
	Potent human pathogen
DNA-liposome complexes	
Easy to manipulate genetically: unmodified plasmid vectors without eukaryotic replicons can be used	Cytotoxicity of cationic lipids
Virtually no size limitations	Low efficiency of transfection
Probably extrachromosomal	Stability of expression unknown
Antisense oligonucleotides	
Synthesis is straightforward	Effects are acute, necessitating repeated introduction of the nucleic acid
No viral vector intermediate required	Efficiency of introduction into cells is unknown

in vivo transfer of genes into the brain has also been achieved using adeno-associated virus vectors (e.g., Kaplitt et al., 1994; Mandel et al., 1998), lentivirus vectors (e.g., Zuffery et al., 1997; Poeschla et al., 1998), and plasmid-liposome complexes (reviewed by Neve and Geller, 1996; also see Hannas-Djebbara et al., 1997). Neuronal gene expression in vivo can also be manipulated by the introduction of antisense oligonucleotides (reviewed by Neve and Geller, 1996), although this technology remains problematic (Hunter et al., 1995).

At present, HSV vectors can accommodate larger pieces of foreign DNA (on the order of 15 kb) than any other viral vector for gene transfer into the brain. Foreign genes are cloned into easy-to-manipulate amplicon vectors that can be packaged directly into viral particles as head-to-tail repeats in the presence of the helper virus, with no intermediate recombination step required. This enables rapid construction of a large number of recombinant vectors simultaneously, and is particularly useful for mutation analysis and studies of multiple genes. Thus, the HSV-1 vector is a gene delivery system that is coming of age.

References: Fraefel et al., 1996, 1997

Contributors: Rachael L. Neve and Filip Lim

UNIT 4.14

Generation of High-Titer Defective HSV-1 Vectors

4

Because HSV-1-based amplicon vectors (e.g., Fig. 4.14.1) contain only a small proportion of the viral genome, their packaging into virus particles requires the participation of additional HSV-1 genes that are expressed in *trans*. These genes are expressed either by replication-defective mutants of HSV-1 (helper viruses), as described in this unit, or by replication-competent but packaging-defective HSV-1 genomes (*UNIT 4.15*).

NOTE: All incubations are performed in a humidified 37°C, 10% CO_2 incubator unless otherwise specified. All other procedures are performed in a laminar flow hood to maintain sterility.

BASIC PROTOCOL 1

PREPARATION OF HELPER VIRUS STOCKS

Materials

Adherent 2-2 cells (Table 4.14.1) maintained in DMEM/10% FBS
✓ DMEM, supplemented
✓ DBPS
CMF-DPBS (see recipe for DPBS)
0.05% (w/v) trypsin/0.02% (w/v) EDTA (e.g., Sigma)
DMEM/2%FBS; DMEM/10% FBS: DMEM, supplemented, containing 2% or 10% (v/v) FBS
5*dl*1.2 helper virus stock (Table 4.14.1)
✓ Plaque agarose

Laminar flow hood
60- and 100-mm tissue culture dishes
Cell lifters
15-ml polypropylene conical tube with a plug seal
Dry ice/ethanol bath

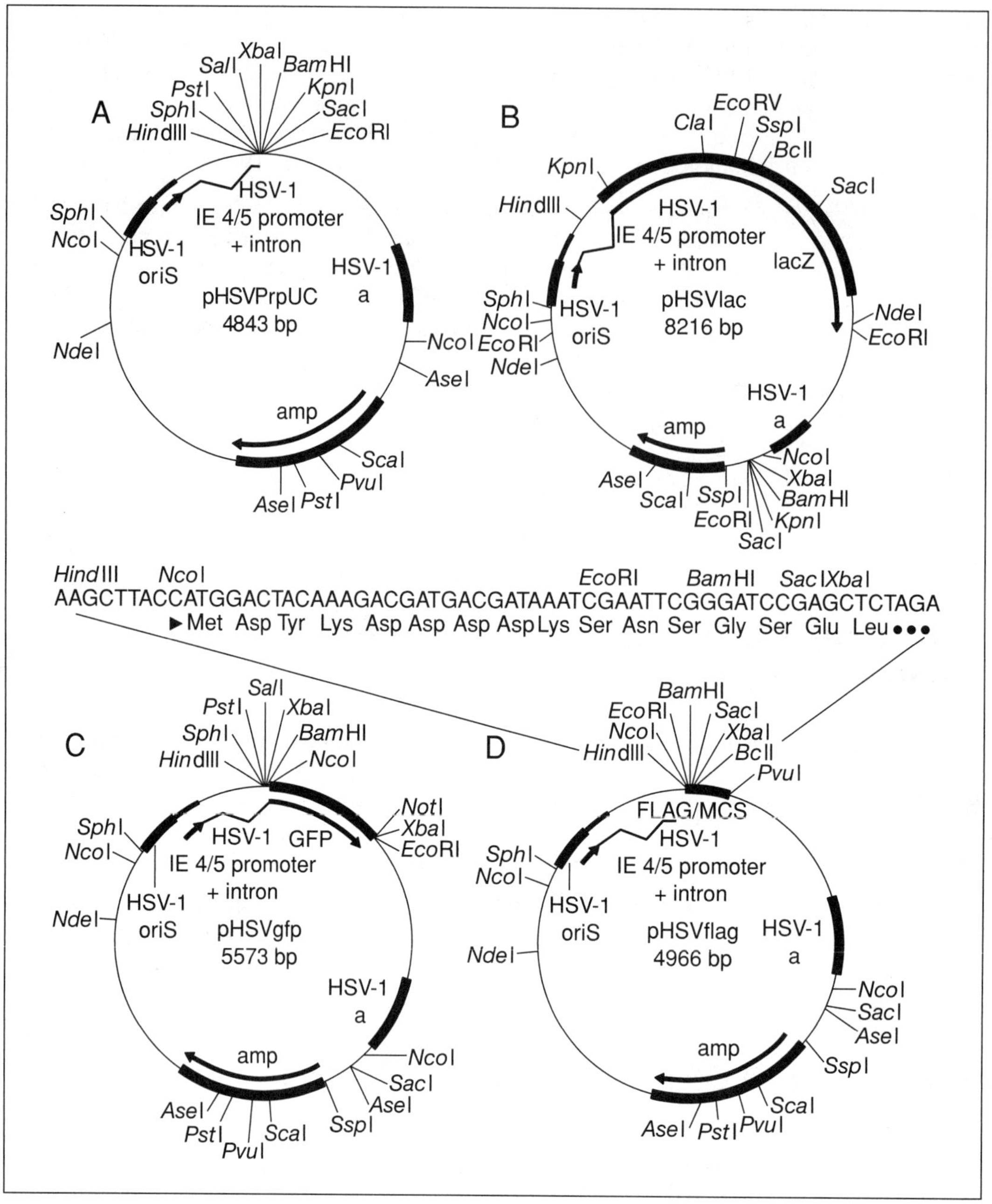

Figure 4.14.1 Maps of several currently available HSV-1 amplicon vectors. (**A**) pHSVPrpUC (Geller et al., 1993) is the basic cloning vector. (**B**) pHSVlac (Geller and Breakefield, 1988) and (**C**) pHSVgfp (Aboody-Guterman et al., 1997) express *E. coli* β-galactosidase and *A. victoria* green fluorescent protein, respectively. (**D**) pHSVflag (Geller et al., 1993) carries the Flag (Prickett et al., 1989) epitope for N-terminal fusions of this tag onto expressed proteins. MCS, multiple cloning site.

15-ml polystyrene conical tubes
Cup-type sonicator
Screw-cap vials for virus storage

1. Passage 2-2 cells by trypsinization by first removing the medium in which the cells are maintained (DMEM/10% FBS), and washing the dish(es) twice with CMF-DPBS. Apply a

Table 4.14.1 Properties and Sources for Virus and Cell Lines

Virus/cell line	Properties[a]	Reference
5*dl*1.2 virus	HSV-1 (KOS strain) with partial deletion in *IE* 2 (*ICP27*)	McCarthy et al. (1989)
2-2 cell line	Vero cells containing *IE 2* (*ICP27*) gene and promoter	Smith et al. (1992)
Vero cells	African green monkey kidney epithelium cell line	ATCC #CCL81
PC12 cell line	Derived from rat pheochromocytoma (adrenal cells with NGF reponsiveness)	ATCC #CRL 172; Greene and Tischler (1976)
d120 virus	HSV-1 (KOS strain) with deletion in *IE 3* (*ICP4*)	DeLuca et al. (1985)
D30EBA virus	HSV-1 (strain 17^+) with deletion in *IE 3* (*ICP4*)	Paterson and Everett (1990)
E5 cells	Vero cells containing *IE 3* (*ICP4*) gene	DeLuca and Schaffer (1987)
RR1 cells	BHK cells containing *IE 3* (*ICP4*) gene	Johnson et al. (1992)

[a]Abbreviations: BHK, baby hamster kidney; NGF, nerve growth factor.

thin layer of 0.05% trypsin/0.02% EDTA and incubate 10 min at 37°C (minimizeexposure to trypsin).

2. Count cells with a hemacytometer (see *APPENDIX 2I*) and seed a 60-mm tissue culture dish with 2×10^6 cells in 5 ml DMEM/10% FBS. Culture overnight in a humidified 37°C incubator with 10% CO_2.

3. Remove 3 ml medium from the culture and add an amount of 5*dl*1.2 helper virus stock that will give well-separated plaques (~50 pfu/plate). Incubate for 1.5 to 6 hr. Remove medium and overlay cells with 3 ml plaque agarose that is no warmer than 40°C. Allow agarose to solidify completely, return the plate to the 37°C incubator, and incubate overnight.

4. The next day, add 2 ml DMEM/2% FBS on top of the agarose. Incubate overnight. Then remove the medium and replace with another 2 ml DMEM/2% FBS.

5. One to three days after infection (step 3), prepare several 60-mm dishes of uninfected 2-2 cells as in steps 1 to 2.

6. The following day (2 to 4 days after infection, when plaques are visible), pick a single plaque from the agarose-overlaid dish, using a pipet tip to stab through the agar and into the cells, and eject the plug into the medium of one dish of uninfected cells. Repeat, sampling from several different plaques, with remaining uninfected dishes. Culture 1 or 2 days (depending upon amount of virus obtained from plaque), until >95% of the cells show cytopathic effects (CPE: cells round up but remain stuck to the plate).

7. Harvest cells by scraping them into the medium with a cell lifter and transferring them, with the medium, to 15-ml polypropylene conical tubes with plug seals (one tube/dish). Proceed to step 8 or store preparation at this stage for up to several days at −70°C.

8. Freeze/thaw three times using a dry ice/ethanol bath and a 37°C water bath (about 10 min for each freezing, 10 to 15 min, each thawing). Minimize the time that cells are thawed at 37°C.

9. Transfer thawed cells from the polypropylene tubes to 15-ml polystyrene conical tubes and sonicate for 4 min in a cup-type sonicator (power setting 6, 50% duty cycle, 1-sec cycles) to disrupt the cells and release virus. Immediately centrifuge 5 min at low speed

(1000 × g in a refrigerated or room temperature benchtop centrifuge) to pellet cell debris but not virus particles.

10. Transfer each supernatant to a fresh screw-cap tube. Measure virus titer (see Support Protocol 1) and store at −70°C to use as seed stock for future viral preparations.

    Seed stocks usually have titers of 0.5–1 × 10^7 pfu/ml and can be stored for several years or more.

11. To amplify seed stock, seed four 100-mm tissue culture dishes with 2 × 10^6 2-2 cells per dish in 10 ml DMEM/10% FBS and incubate 1 day at 37°C.

12. Change medium to 10 ml DMEM/2% FBS and add 50 μl of seed stock virus. Incubate at 37°C, ~1 day (0.5–2 × 10^7cells/dish). Infect the cells at an MOI 0f <0.1

13. Harvest the cells (see steps 7 to 10) when they have rounded up but still remain stuck to the plates. Store in aliquots in screw-cap vials at −70°C.

SUPPORT PROTOCOL 1

TITRATION OF HELPER AND WILD-TYPE VIRUS BY PLAQUE ASSAY

Additional Materials *(also see Basic Protocol 1; see* APPENDIX 1 *for items with ✓)*

Vero cells (for titrating wild-type HSV-1 virus; Table 4.14.1)
Virus stock: helper virus seed stock (see Basic Protocol 1) or wild-type HSV-1 virus at desired dilutions
Plaque-fixing solution: 5% (v/v) methanol/10% (v/v) acetic acid (store at room temperature)
✓ Crystal violet stain
Dissecting microscope

1. Plate 2-2 or Vero cells (for helper or wild-type virus, respectively) in 60-mm tissue culture dishes (2 × 10^6 per dish) in 5 ml DMEM/10% FBS (see Basic Protocol 1, steps 1 to 3) and grow for 1 day in a humidified 37°C, 10% CO_2 incubator. Prepare one dish for each dilution of virus stock to be titered using a series of 10-fold dilutions up to 10^{-7}.

2. Infect cells by removing 3 ml medium from each dish and adding up to 100 μl of the appropriate virus dilution. Allow the virus to adsorb at 37°C for ≥90 min, but <6 hr. Remove medium and overlay cells with 3 ml plaque agarose no warmer than 40°C. Allow the agar to solidify completely, return dishes to the incubator, and incubate overnight. Add 2 ml DMEM/2% FBS on top of the agarose and incubate overnight.

3. The next day, replace medium with another 2 ml DMEM/2% FBS and incubate overnight.

4. Three days after infection, examine the dishes under a microscope for plaques. If they are too small to reveal clear holes in the cell monolayer, leave the dishes for another day before staining.

5. Remove medium and fix cells with 3 ml plaque-fixing solution for ≥15 min.

6. Remove plaque-fixing solution and discard agarose with a quick flick of the wrist (into a disposal container). Add 1 ml crystal violet stain and leave for 5 min. Remove stain, wash plate with 1 ml water, decant, and air dry.

7. Count the number of plaques (visible macroscopically, as holes, but microscopic examination more reliable), and use the dilution factor of the virus stock to calculate the virus plaque-forming titer in pfu/ml.

BASIC PROTOCOL 2

PACKAGING AMPLICON INTO VIRUS PARTICLES

Fig. 4.14.2 illustrates the packaging strategy.

NOTE: The optimum multiplicity of infection (MOI) for superinfection by helper virus is between 0.2 and 0.6. A preliminary transfection and superinfection can be done to determine the optimum amount of helper stock to use for packaging (see Fig. 4.14.3). The transfection and superinfection lysates should be titered for vector on PC12 cells (see Support Protocol 2). Do not continue with preparations that have titers $<10^4$ ivu/ml.

Materials *(see* APPENDIX 1 *for items with ✓)*

Adherent 2-2 cells (Table 4.14.1) maintained in supplemented DMEM/10% FBS
✓ DMEM, supplemented
DMEM/2% FBS; DMEM/10% FBS: DMEM, supplemented, containing 2% or 10% (v/v) FBS), room temperature and prewarmed to 37°C
✓ DPBS, prewarmed to 37°C
CMF-DPBS (see recipe for DPBS)
0.05% (w/v) trypsin/0.02% (w/v) EDTA (e.g., Sigma)
≥50 ng/µl DNA for transfection, purified with a Qiagen column (per manufacturer's instructions) and resuspended in water
Opti-MEM (Life Technologies), prewarmed to 37°C
LipofectAMINE (Life Technologies)

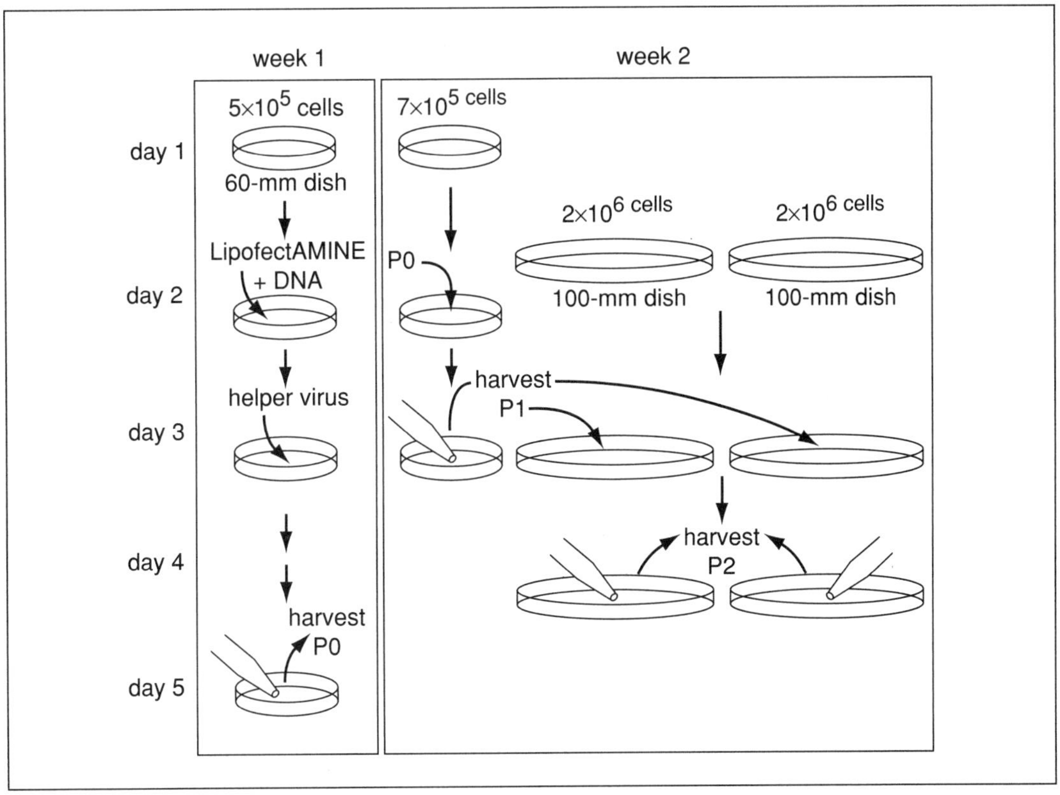

Figure 4.14.2 Overview of the packaging procedure as described for collection of a P2 stock (see Basic Protocol 2, steps 1 to 15 and 16a to 18a). An alternate procedure allows the addition of a P3 passage, followed by sucrose gradient purification of virus (not shown; see Basic Protocol 2, steps 16b to 27b).

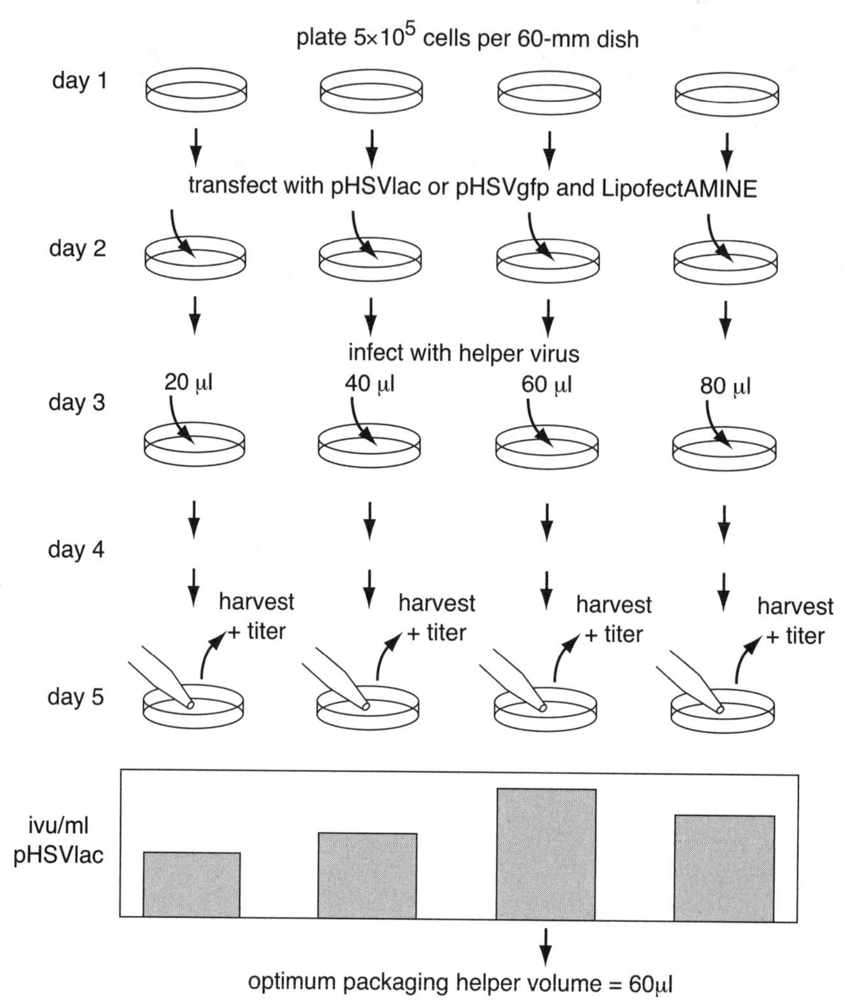

Figure 4.14.3 Illustration of the titration procedure used to determine the optimal amount of a given helper virus preparation for superinfection.

5*dl*1.2 helper virus seed stock (see Basic Protocol 1)
BSA
✓ 10%, 30%, and 60% (w/v) sucrose solutions

Laminar flow hood
60-mm, 100-mm, and 150-mm tissue culture dishes
Cell lifters (optional)
15- and 50-ml polypropylene conical tubes with plug seals
Dry ice/ethanol bath
15- and 50-ml polystyrene conical tubes
Cup-type sonicator
25 × 89–mm and 14 × 89–mm clear ultracentrifuge tubes
Ultracentrifuge and swinging-bucket rotors of appropriate sizes
18-G needles
5-ml syringes
Screw-cap vials for virus storage

1. Maintain 2-2 cells at 37°C in a humidified, 10% CO_2 incubator in DMEM/10% FBS. Passage by trypsinization (see Basic Protocol 1, steps 1 and 2), being careful to keep the cells at subconfluence (50% to 80% confluent)). When a fresh aliquot of cells is thawed, passage at least two times before the cells are plated for transfection.

   If 0.5–1 × 10^6 cells are passaged per dish, it will be 3 to 4 days before they become confluent.

2. One day before transfection, plate 2-2 cells at 5 × 10^5 per 60-mm tissue culture dish in 5 ml DMEM/10% FBS (one dish of cells per DNA construct).

3. Dilute 2 μg of ≥50 ng/μl DNA in 100 μl of prewarmed 37°C Opti-MEM in a sterile 1.5-ml microcentrifuge tube (one tube per dish—i.e., per DNA construct).

4. Dilute 12 μl LipofectAMINE in 100 μl prewarmed Opti-MEM in a second tube (one per dish), and add it to the 100-μl DNA mixture in the first tube. Leave for 20 to 45 min at room temperature to allow liposomes to form.

5. Remove medium from the dishes (step 2), wash once with 2 ml prewarmed Opti-MEM, remove wash, and replace with another 2 ml Opti-MEM.

6. Add 800 μl prewarmed Opti-MEM to each DNA/LipofectAMINE mix. Add this 1-ml mixture to the cells, dropwise and evenly over the whole dish. Incubate for 5 to 7 hr at 37°C.

7. Wash cells three times with 2 ml prewarmed, 37°C DPBS (use CMF-DPBS for the first wash only). Replace DPBS with 3 ml prewarmed DMEM/10% FBS, and incubate cells overnight. Allow the cells to recover ≥20 hr from the time of the last wash.

8. For superinfection, remove medium from dsihes and add 3 ml prewarmedDMEM/2% FBS. Add ~6 × 10^5 pfu of 5*dl*1.2 helper virus seed stock and incubate until 95% of the cells show cytopathic effects (CPE) and have rounded up (~30 to 40 hr).

9. Harvest cells by pipetting the medium onto them until they have all detached from the plate, or by scraping them up with a cell lifter. Transfer cells and medium to 15-ml polypropylene conical tubes with plug seals. Proceed to step 10 or store harvested cells at −70°C (up to several days).

10. Freeze/thaw three times using a dry ice/ethanol bath and a 37°C water bath, being careful to minimize the amount of time that the cells are thawed at 37°C (virus particles are thermolabile). In each cycle, freeze for 5 to 10 min and thaw for 10 to 15 min.

11. Transfer cells from the polypropylene tubes to 15-ml polystyrene conical tubes and sonicate for 2 min in a cup-type sonicator (power setting 6, 50% duty cycle, 1-sec cycles).

12. Centrifuge cells 5 min at low speed (1000 × *g* in a refrigerated or room-temperature benchtop centrifuge) to pellet cell debris but not virus particles. Collect supernatant (PO; see Fig. 4.14.2). Plate PO supernatant on cells for the P1 passage (step 14) or store at −70°C for up to several days.

13. Plate fresh 2-2 cells at 7 × 10^5 cells per 60-mm tissue culture dish in 5 mlDMEM/10% FBS. Incubate overnight.

14. The next day, replace medium with 4 ml DMEM/2% FBS and add the P0 supernatant (~3 ml). Incubate until cells show 95% CPE (~24 hr).

15. Harvest and process cells (P1) as in steps 9 to 12.

    At this point, there are two alternative procedures, depending on the desired concentration and purity of the virus. To obtain virus stocks that contain ~10^7 infectious vector units (ivu)/ml and that are reasonably but not optimally pure, amplify and collect a P2 virus stock (steps 16a

4

to 18a). For a highly concentrated (>10⁸ ivu/ml) virus stock of exceptional purity (e.g., for stereotactic injection), collect and purify a P3 virus stock (steps 16b to 27b).

For P2 virus stock (Fig. 4.14.2)

16a. On same day as step 14, plate fresh 2-2 cells (two dishes per sample) at 2×10^6 cells per 100-mm tissue culture dish in 10 ml DMEM/10% FBS. Incubate overnight.

17a. The next day, replace medium in each dish with 2 ml DMEM/2% FBS, and add 3.5 ml P1 supernatant (from step 15) per dsih. Incubate for ≥90 min.

18a. Replace medium with 2 ml prewarmed Opti-MEM with 1% (w/v) BSA and incubate until the cells show 95% CPE (~24 hr). Harvest and process cells as in steps 9 to 12. Dispense supernatant (P2 virus stock) into 400-μl aliquots in screw-cap vials and store at −70°C for up to 1 year.

For P3 virus stock

16b. On same day as step 14, plate fresh 2-2 cells (two dishes per sample) at 2×10^6 cells per 100-mm tissue culture dish in 10 ml supplemented DMEM/10% FBS. Incubate overnight at 37°C.

17b. The next day, replace medium in each dish with 6.5 ml DMEM/2% FBS and add 3.5 ml P1 supernatant (from step 15) per dish. Incubate until the cells show 95% CPE (~24 hr).

18b. On the same day, plate fresh 2-2 cells (six dishes per sample) at 2.4×10^6 cells per 100-mm dish in 10 ml DMEM/10% FBS. Incubate overnight.

19b. The next day, harvest and process cells from the P1-infected dishes (P2; from step 17b) as in steps 9 to 12, but use 50-ml instead of 15-ml conical tubes.

20b. Replace the medium in each of the newly grown 2-2 dishes (step 18b) with 6 ml DMEM/10% FBS, and add 4.0 ml P2 supernatant per dish. Incubate until the cells show 95% CPE (~24 hr). Harvest and process cells (P3) as in steps 9 to 12, using two 50-ml conical tubes per virus sample (six dishes).

21b. Prepare three sucrose step gradients for each sample in 40-ml (25 × 89–mm) ultracentrifuge tubes. Prepare at room temperature by layering the following sucrose solutions (highest density at the bottom) into each tube:

7 ml 60% sucrose solution
6 ml 30% sucrose solution
3 ml 10% sucrose solution.

22b. Load each crude virus sample (P3 supernatant; step 20b) onto three gradients (20 ml per tube) and centrifuge 1 hr at 125,000 × *g* at 18°C.

23b. Immobilize each tube in a clamp on a ring stand, and place it in front of a black background.

The virus will appear as a sharp, very thin band at the 30%/60% interface, while contaminants will form a diffuse band close to the 10%/30% interface.

24b. Attach an 18-G needle to a 5-ml syringe. Holding the syringe so that the beveled edge of the needle points up, pierce one tube underneath the virus band. Slowly pull the band into the syringe in a volume of 2 ml. Transfer the band to an 11.5-ml (14 × 89–mm) ultracentrifuge tube, and discard the remainder. Repeat for remaining gradients.

25b. Dilute the virus in each tube with 9.5 ml DPBS and gently mix the contents by pipetting up and down. Centrifuge for 75 min at 125,000 × *g* at 18°C. Carefully aspirate the supernatant.

26b. Add 200 μl of 10% sucrose solution to the pellet in each tube, and resuspend by shaking in a rack on a platform shaker at 4°C overnight.

27b. The next day, very briefly triturate the virus, dispense into 40-μl aliquots in screw-cap vials, and store at −70°C (up to one year). To determine virus titer, see Support Protocol 2.

If the virus is triturated without an overnight incubation in the buffer, it is very difficult to resuspend. Moreover, the mechanical disruption necessary to resuspend the virus under these conditions will greatly decrease the viability of the virus. Therefore, slow resuspension overnight at 4°C is recommended.

SUPPORT PROTOCOL 2

TITRATION OF AMPLICON VIRUS BY VECTOR ASSAY

PC12 cells are used for vector assays because they are round and easy to distinguish as single cells when positive for expression. Expression is usually quite high in this cell line, though it will vary with different proteins.

Materials *(see APPENDIX 1 for items with ✓)*

✓ 20 μg/ml poly-D-lysine solution
PC12 cells (Table 4.14.1)
✓ DMEM, supplemented
DMEM/10% HS/5% FBS: DMEM, supplemented, without G418, containing 10% (v/v) horse serum and 5% (v/v) FBS
Amplicon stock to be measured (see Basic Protocol 2)
PBS with and without 10 mM EDTA (see individual recipes)
✓ 4% (w/v) paraformaldehyde solution
✓ TBS[1]
✓ Fe solution (optional)
50 mg/ml 5-bromo-4-chloro-3-indolyl-β-D-galactopyranoside (Xgal) in dimethyl sulfoxide (store in aliquots at −20°C)
Rabbit (or mouse) anti-HSV primary antibody (optional)
TST (optional): TBS containing 1% (v/v) goat serum and 0.1% (v/v) Triton X-100 (store at 4°C)
Alkaline phosphatase (AP)-conjugated anti-rabbit (or anti-mouse) secondary antibody
✓ AP buffer (optional)
✓ AP substrate solution (optional)

Laminar flow hood
24-well tissue culture plates
21-G needle
Dissecting microscope

1. Plate PC12 cells to yield a monolayer that will be confluent within 24 hr.

2. Meanwhile, add 500 μl of 20 μg/ml poly-D-lysine solution to each well of a 24-well tissue culture plate and incubate ≥5 min at room temperature to coat. Aspirate completely before using plate.

3. Harvest PC12 cells by trypsinization (see Basic Protocol 1, steps 1 to 2) and pass through a 21-G needle to dissociate aggregates.

4. Count cells with a hemacytometer (see APPENDIX 21) and seed them in the coated 24-well plate at 3×10^5/well in 500 μl DMEM/10% HS/5% FBS. Prepare four wells for

each amplicon stock to be titered. Incubate overnight in a humidified 37°C, 10% CO_2 incubator.

5. The next day, change medium, and add 0, 1, 2, and 5 μl amplicon stock to the four wells, the uninfected well serving as a negative control for background staining. Incubate at 37°C overnight.

6. The following day, wash cells once with 500 μl PBS and fix with 500 μl of 4% paraformaldehyde solution for ≥15 min at room temperature. Wash cells once with 500 μl PBS (for Xgal) or TBS (not a phosphate-containing buffer for immunocytochemistry).

For Xgal staining

7a. Warm Fe solution (500 μl/well of cells to be stained) and 50 mg/ml Xgal solution in a 37°C water bath. When both solutions are warmed completely to 37°C, add Xgal to the Fe solution at a final concentration of 1 mg/ml.

8a. Remove PBS from the cells and add 500 μl/well Xgal/Fe solution. Incubate 30 min to overnight at 37°C. Stop the reaction by washing twice with 500 μl PBS. If crystals of Xgal are seen after the staining (due to low ambient temperature), remove these precipitates by incubating the sample for 5 min in 1:1 DMSO/H_2O.

9a. Examine cells for specific staining using a tissue culture microscope.

10a. Estimate the percentage of stained cells in one or more of the wells infected for each amplicon stock, and use this information to extrapolate the titer of the stock. For example, if 20% of the cells in a well infected with 2 μl of virus are stained, and if it is assumed that each stained cell represents one infectious unit of virus, then it can be inferred that the 2 μl of virus contained 6×10^4 (20% of 3×10^5 cells plated/well) infectious units. Thus, the titer of the stock would be 3×10^4 ivu/μl, or 3×10^7 ivu/ml.

For immunocytochemical staining

7b. Replace TBS with 500 μl of 1:10,000 rabbit (or mouse) anti-HSV primary antibody in TST. Incubate overnight at 4°C.

8b. Wash twice with 500 μl TBS, then allow to sit for 10 min at room temperature.

9b. Replace TBS with 500 μl of 1:2000 AP-conjugated anti-rabbit (or anti-mouse) secondary antibody in TST for ~1 hr at room temperature.

10b. Wash the cells twice with 500 μl TBS, then twice with 500 μl AP buffer.

11b. Add 500 μl AP substrate solution and incubate in a dimly lit area. Monitor color development with a dissecting microscope.

12b. Stop color development by washing twice with 500 μl PBS. Leave the cells in PBS/10 mM EDTA to prevent further darkening.

13b. Estimate the percentage of stained cells in one or more of the wells infected for each amplicon stock, and use this information to extrapolate the titer of the stock. For example, if 20% of the cells in a well infected with 2 μl of virus are stained, and if it is assumed that each stained cell represents one infectious unit of virus, then it can be inferred that the 2 μl of virus contained 6×10^4 (20% of 3×10^5 cells plated/well) infectious units. Thus, the titer of the stock would be 3×10^4 ivu/μl, or 3×10^7 ivu/ml.

References: Lim et al., 1996

Contributors: Filip Lim and Rachael L. Neve

Gene Delivery Using Helper Virus–Free HSV-1 Amplicon Vectors

Herpes simplex virus type 1 (HSV-1)-based amplicon vectors contain only ~1% of the 152-kbp viral genome. Consequently, replication and packaging of amplicons depend on helper functions that are provided either by replication-defective mutants of HSV-1 (helper viruses; see *UNIT 4.14*) or by replication-competent, but packaging-defective, HSV-1 genomes.

Sets of cosmids that overlap and represent the entire HSV-1 genome can form, via homologous recombination, circular replication-competent viral genomes, which give rise to infectious virus progeny. However, if the DNA cleavage/packaging (*pac*) signals are deleted, reconstituted virus genomes are not packageable, but still provide all the helper functions required for the packaging of cotransfected amplicon DNA. The resulting stocks of packaged amplicon vectors are essentially free of contaminating helper virus (Fig. 4.15.1).

CAUTION: HSV-1 is a human pathogen. Follow biosafety level 2 practices when working with HSV-1 or vectors that are based on HSV-1. Wear safety glasses and gloves at all times.

NOTE: All solutions and equipment coming into contact with cells must be sterile, and proper aseptic technique should be used accordingly. All cell culture incubations are performed in a humidified 37°C, 5% CO_2 incubator unless otherwise stated.

4

BASIC PROTOCOL

PREPARATION OF HELPER VIRUS-FREE AMPLICON STOCKS

The use of an HSV-1 amplicon vector that expresses a reporter gene is highly recommended when establishing this protocol in the laboratory. The gene for green fluorescent protein (GFP) is an ideal reporter. Because GFP fluorescence is independent of substrates or cofactors, transfection and packaging efficiencies can be monitored in living cultures during the entire course of the packaging process.

Materials *(see APPENDIX 1 for items with ✓)*

2-2 cells (Smith et al., 1992)
DMEM (Life Technologies) with 10% and with 6% FBS (DMEM/10% FBS and DMEM/6% FBS)
G418 (Geneticin; Life Technologies)
0.25% (w/v) trypsin/0.02% (w/v) EDTA (Life Technologies)
Opti-MEM I reduced-serum medium (Life Technologies)
*Pac*I-digested cosmid DNA of set C6Δa48Δa (see Support Protocol 1)
HSV-1 amplicon DNA (maxiprep DNA isolated from *E. coli*)
LipofectAMINE reagent (Life Technologies)
10%, 30% and 60% (w/v) sucrose in PBS
✓ PBS

75-cm^2 tissue culture flasks
Humidified 37°C, 5% CO_2 incubator
60-mm-diameter tissue culture dishes
15-ml conical centrifuge tubes
Dry ice/ethanol bath

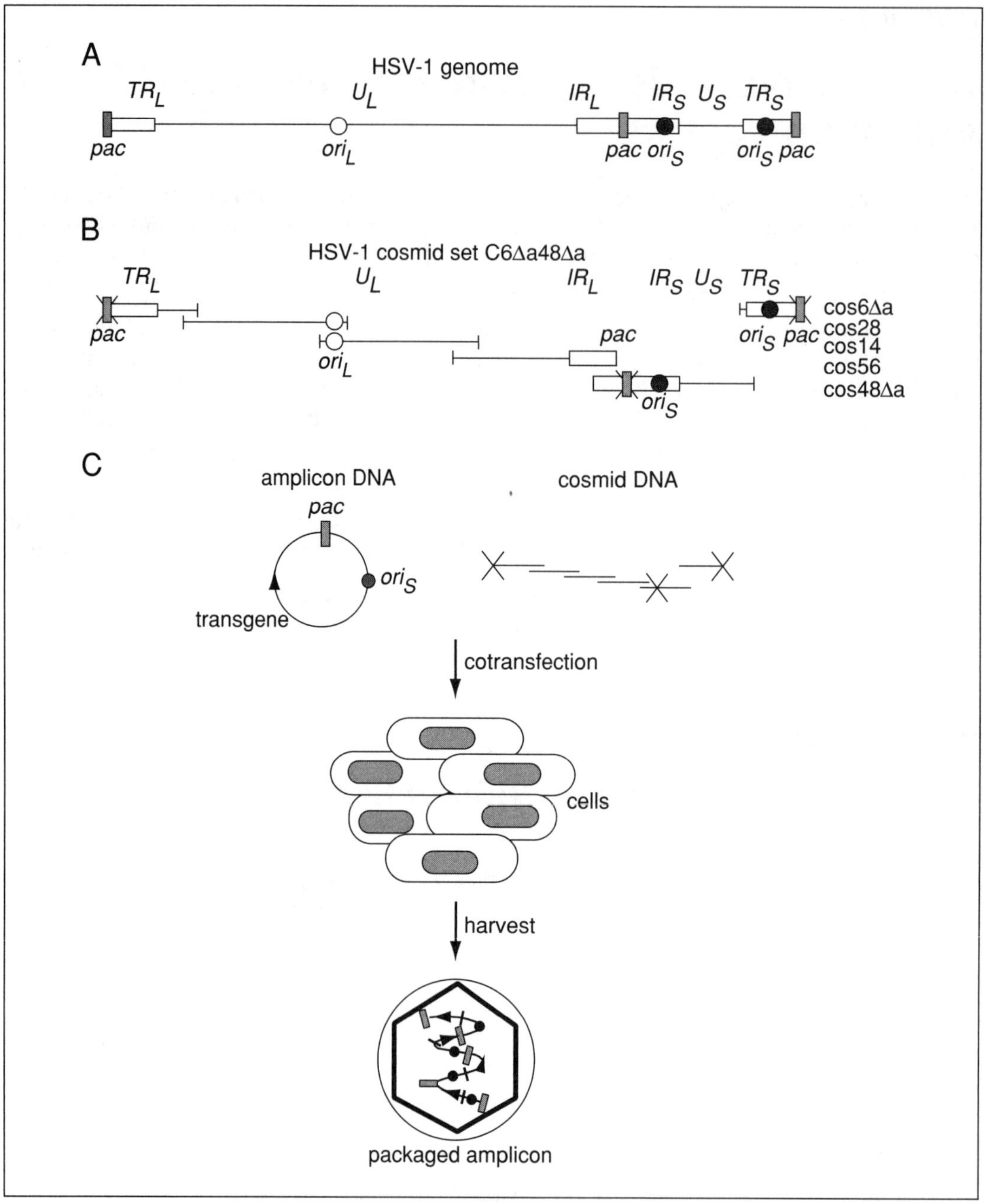

Figure 4.15.1 Helper virus-free packaging of HSV-1 amplicons into HSV-1 particles. (**A**) The HSV-1 genome (152 kbp) is composed of unique long (U_L) and unique short (U_S) segments (horizontal lines), which are flanked by inverted repeats (open rectangles): IR_S, internal repeat of the short segment; TR_S, terminal repeat of the short segment; IR_L, internal repeat of the long segment; TR_L, terminal repeat of the long segment. The origins of DNA replication (ori_S, solid circle; ori_L, open circle) and the DNA cleavage/packaging signals (*pac*, solid rectangles) are also shown. (**B**) Schematic diagram of HSV-1 cosmid set C6Δa48Δa with deleted *pac* signals (X) (Fraefel et al., 1996; Cunningham and Davison, 1993), which includes cos6Δa, cos14, cos28, cos48Δa, and cos56. (**C**) For the helper virus-free packaging of amplicons into HSV-1 particles, cells that are susceptible for HSV-1 replication are cotransfected with amplicon DNA and DNA from cosmid set C6Δa48Δa. In the absence of the *pac* signals, this cosmid set cannot generate a packageable HSV-1 genome, but can provide all the functions required for the replication and packaging of the cotransfected amplicon DNA, which contains a functional *pac* signal. The resulting vector stocks are essentially free of helper virus.

Probe sonicator
0.45-μm polyethersulfone syringe-tip filters (Sarstedt)
20-ml disposable syringes
30-ml clear ultracentrifuge tubes, 25 × 89–mm and 14 × 95–mm
Fixed-angle rotor
Fiber-optic illuminator
Ultracentrifuge with swinging-bucket rotors

1. Maintain 2-2 cells in DMEM/10% FBS containing 500 μg/ml G418 for several weeks before starting a packaging experiment. Propagate the culture twice a week by splitting ~1:5 in fresh medium (20 ml) into a new 75-cm^2 tissue culture flask.

2. On the day before transfection, remove culture medium, wash each dish twice with PBS, add a thin layer of trypsin/EDTA, and incubate 10 min at 37°C to allow cells to detach from dish. Count cells using a hemacytometer (see *APPENDIX 2I*) and plate 1.2 × 10^6 cells per 60-mm-diameter tissue culture dish in 3 ml DMEM/10% FBS to give a confluent monolayer by the time of transfection.

3. For each 60-mm dish to be transfected, place 100 μl Opti-MEM I reduced-serum medium into each of two 15-ml conical tubes. To one tube, add 2 μg of the *Pac*I-digested cosmid DNA mixture (0.4 μg of each of the five clones; see Support Protocol 1) and 0.6 μg amplicon DNA. To the other tube, add 12 μl LipofectAMINE. Combine the contents of the two tubes (the transfection mixture). Mix well without vortexing and incubate 45 min at room temperature.

4. If the cells prepared the previous day (step 2) are confluent, wash once with 2 ml Opti-MEM I. Add 1.1 ml Opti-MEM I to the tube containing the DNA-LipofectAMINE transfection mixture (1.3 ml total volume). Aspirate all medium from the culture, add the transfection mixture, and incubate 5.5 hr.

5. Aspirate the transfection mixture and wash the cells three times with 2 ml Opti-MEM I. After aspirating the last wash, add 3.5 ml DMEM/6% FBS and incubate 2 to 3 days, when at least 50% of the cellls should show cytopathic effect (CPE: the cells round up but remain stuck to the plate).

6. Scrape cells into the medium using a rubber policeman. Transfer the suspension to a 15-ml conical centrifuge tube and perform three freeze-thaw cycles using a dry ice/ethanol bath and a 37°C water bath. Do not leave suspension at 37°C any longer than is necessary to thaw them (they can, however, be kept frozen for extended periods).

7. Place the tube containing the cells into a beaker of ice water. Submerge the tip of the sonicator probe ~0.5 cm into the cell suspension and sonicate 20 sec with 20% output energy.

8. Remove cell debris by centrifuging for 10 min at 1400 × *g*, 4°C; filter the supernatant through a 0.45-μm syringe-tip filter attached to a 20-ml disposable syringe into a new 15-ml conical tube. Remove a sample for titration (see Support Protocol 2), then divide the remaining stock into 1-ml aliquots, freeze them in a dry ice/ethanol bath, and store at −80°C. Alternatively, concentrate (steps 9a and 10a) or purify and concentrate (steps 9b to 12b) the stock before storage.

   Vector stocks from steps 8, 10a, and 12b can be stored at least 6 months at −80°C without a decrease in vector titers. Avoid repeated freezing and thawing.

Concentrate by centrifugation

9a. Transfer the vector solution from step 8 to a 30-ml centrifuge tube and centrifuge, in a fixed-angle rotor, 2 hr at 20,000 × *g*, 4°C.

10a. Resuspend the pellet in a small volume (e.g., 300 μl) of 10% sucrose. Remove a sample of the stock for titration (see Support Protocol 2), then divide into aliquots (e.g., 30 μl) and freeze in a dry ice/ethanol bath. Store at −80°C.

Purify and concentrate using a discontinuous sucrose gradient

9b. Prepare a sucrose gradient, in a clear 25 × 89–mm ultracentrifuge tube by layering the following solutions in the tube:

7 ml 60% sucrose
7 ml 30% sucrose
3 ml 10% sucrose.

10b. Carefully add the vector stock from step 8 (up to 20 ml) on top of the gradient and centrifuge 2 hr at 100,000 × *g*, 4°C, using a swinging-bucket rotor.

11b. Aspirate the 10% and 30% sucrose layers from the top (discard), and collect the virus band (a cloudy band when viewed with a fiberoptic illuminator) at the interface between the 30% and 60% layers. Transfer to a 14 × 95–mm ultracentrifuge tube, add ~15 ml PBS, and pellet virus particles for 1 hr at 100,000 × *g*, 4°C, using a swinging-bucket rotor.

12b. Resuspend the pellet in a small volume (e.g., 300 μl) of 10% sucrose. Divide into aliquots (e.g., 30 μl) and freeze in a dry ice/ethanol bath. Store at −80°C. Before freezing, retain a sample of the stock for titration (see Support Protocol 2).

SUPPORT PROTOCOL 1

PREPARATION OF HSV-1 COSMID DNA FOR TRANSFECTION

Of the variables that affect transfection efficiency, and ultimately titers of packaged amplicon stocks, the quality of the DNA is one of the most critical. Because the cosmids of set C6Δa48Δa contain high-copy colE1 plasmid origins of DNA replication, special care must be taken to prevent random mutations during amplification.

Materials *(see* APPENDIX 1 *for items with ✓)*

E. coli clones of HSV-1 cosmid set C6Δa48Δa, which includes cos6Δa, cos14, cos28, cos48Δa, and cos56 (see Fig. 4.15.1)
✓ SOB medium containing 50 μg/ml ampicillin (SOB/amp)
Dimethyl sulfoxide (DMSO)
Plasmid Maxi Kit (Qiagen), which includes Qiagen-tip 500 columns and buffers P1, P2, P3, QBT, QC, and QF (prewarm buffer QF to 65°C)
Isopropanol
70% (v/v) ethanol
✓ TE buffer, pH 7.5
Restriction endonucleases *Dra*I, *Kpn*I, and *Pac*I
High-molecular-weight DNA standard (Life Technologies)
1-kb DNA ladder (Life Technologies)
Electrophoresis-grade agarose
✓ TAE electrophoresis buffer
1 mg/ml ethidium bromide in H_2O
25:24:1 (v/v/v) phenol/chloroform/isoamyl alcohol
24:1 (v/v) chloroform/isoamyl alcohol
100% ethanol
✓ 3 M sodium acetate, pH 5.5

17 × 100–mm graduated snap-cap tubes (e.g., Falcon 2059), sterile
Centrifuge and fixed-angle rotors
65° and 37°C water baths
250-ml polypropylene centrifuge tubes
30-ml centrifuge tubes

1. For each of the five clones of HSV-1 cosmid set C6Δa48Δa, prepare a 17 × 100–mm sterile snap-cap tube containing 5 ml sterile SOB/amp medium. Inoculate each with a loop of frozen long-term culture of the appropriate cosmid clone. Incubate 8 hr at 37°C in a shaker.
2. Transfer 1 ml of each culture to an individual 2-liter flask containing 300 ml sterile SOB/amp, and incubate another 12 to 16 hr at 37°C, with shaking. Do not exceed the stated incubation times, as this may increase the risk of introducing modifications. For the same reason, avoid colony-purifying the clones.
3. Place 1-ml aliquots into cryogenic storage vials and add 70 μl DMSO to each. Mix well and freeze at −80°C for long-term storage (up to several years).
4. To start cosmid DNA extraction, pellet the remaining bacterial culture in 250-ml polypropylene centrifuge tubes by centrifugation in fixed-angle rotor, 10 min at 4000 × *g*, 4°C. Decant medium and invert each tube on a paper towel for 1 to 2 min to drain all liquid. Resuspend pellets in 15 ml buffer P1 and add 15 ml buffer P2. Mix by inverting each tube four to six times, and incubate 5 min at room temperature.
5. Add 15 ml buffer P3 and mix immediately by inverting tubes six times. Incubate the tubes for 20 min on ice. Invert once more and centrifuge 30 min at 16,000 × *g*, 4°C.
6. Meanwhile, equilibrate five Qiagen-tip 500 columns with 10 ml buffer QBT, and allow the columns to empty by gravity flow. Label each column with the name of the clone and drape with a small piece of a Kimwipe tissue to act as filter to remove cell debris.
7. Carefully filter the supernatants from step 5 through the tissue into the Qiagen-tip 500 columns, and allow the liquid to enter the resin by gravity flow. Wash each column twice with 30 ml buffer QC, then elute DNA, with 15 ml prewarmed (65°C) buffer QF, into a 30-ml centrifuge tube.
8. Precipitate the DNA with 10.5 ml (0.7 vol) isopropanol, and immediately centrifuge 30 min at 20,000 × *g*, 4°C.
9. Carefully—without disturbing the pellets—remove the supernatants from step 8 and mark the locations of the pellets on the outside of the tubes. Wash with chilled 70% ethanol and, if necessary, re-pellet at the same settings as in step 8.
10. Aspirate the supernatants completely, but avoid drying the pellets. Resuspend in 200 μl TE buffer, transfer to microcentrifuge tubes, and store at 4°C. After characterization of the DNA (concentration and restriction enzyme analysis; steps 11 and 12), store aliquots of 10 to 50 μg at −20°C.
11. Determine the absorbance of the DNA solutions from step 10 at 260 nm (A_{260}) and 280 nm (A_{280}) using a UV spectrophotometer.

    A value of 1.0 for A_{260} is equivalent to 50 μg/ml of double-stranded DNA. The ratio between A_{260} and A_{280} provides information about DNA purity. Typically, pure DNA preparations have a A_{260}/A_{280} value of 1.8; do not use DNA preparations with a ratio below this value.

12. In a parallel procedure, digest 2 μg of each cosmid DNA from step 10 with 10 U each of *Dra*I and *Kpn*I restriction endonuclease for 2 hr at 37°C. Separate the fragments overnight by electrophoresis on a 0.4% agarose gel at 40 V in TAE electrophoresis buffer, using

high-molecular-weight DNA and 1-kb DNA ladder as size standards. Stain with ethidium bromide and examine restriction fragment patterns. Treat gel with care.

13. Pool 10 μg of each of the five cosmid DNAs in a microcentrifuge tube and digest with 50 U of *Pac*I restriction endonuclease in a total volume of 100 μl for ≥3 hr at 37°C, using high-molecular-weight DNA and 1-kb DNA ladder as size standards. Confirm completion of the digestion by electrophoresis of 1- to 2-μl aliquots of the reaction mixture on a 0.4% agarose gel.

14. Extract DNA in the reaction mixtures, first with 100 μl (1 vol) of 25:24:1 (v/v/v) phenol/chloroform/isoamyl alcohol, and then with 100 μl (1 vol) of 24:1 (v/v) chloroform/isoamyl alcohol. Precipitate DNA by adding 250 μl (2.5 vol) 100% ethanol and 10 μl (0.1 vol) 3 M sodium acetate. Do not vortex: mix tubes gently by tapping with finger. Then incubate overnight at −20°C to produce a clearly visible DNA precipitate.

IMPORTANT NOTE: Because the DNA will be transfected into mammalian cells (see Basic Protocol), perform the following manipulations under sterile conditions.

15. Microcentrifuge the tubes 10 min at 13,000 rpm, room temperature. Carefully dispose of the supernatant and wash the pellet once with 70% ethanol. Decant the ethanol, allow the pellets to dry for 1 min, and resuspend (with minimal pipetting) in 100 μl TE buffer.

16. Measure DNA concentration as described in step 11, and store 10-μg aliquots at −20°C until needed for cotransfection.

4

SUPPORT PROTOCOL 2

TITRATION OF AMPLICON STOCKS

Materials *(see* APPENDIX 1 *for items with ✓)*

Vero (clone 76; ECACC #85020205), BHK (clone 21; ECACC #85011433), or 293 (ATCC #1573) cells
DMEM (e.g., Life Technologies) supplemented with 10% and 2% FBS (DMEM/10% FBS and DMEM/2% FBS)
✓ PBS
Samples collected from vector stocks (see Basic Protocol, steps 8, 10a, or 12b)
✓ 4% (w/v) paraformaldehyde solution, pH 7.0
✓ X-gal staining solution
✓ GST solution
Appropriate primary and conjugated secondary antibodies

24-well tissue culture plates
Humidified 37°C, 5% CO_2 incubator
Inverted fluorescence microscope
Inverted light microscope

NOTE: The titers are expressed as transducing units per milliliter (t.u./ml). These are relative titers and do not necessarily reflect numbers of infectious vector particles per milliliter. Factors influencing relative transduction efficiencies include: (1) the cells used for titration, (2) the promoter regulating the expression of the transgene, (3) the transgene, and (4) the sensitivity of the detection method.

1. Plate cells (e.g., Vero 76, BHK 21, or 293 cells) at a density of 1.0×10^5 per well of a 24-well tissue culture plate in 0.5 ml DMEM/10% FBS. Incubate overnight.

2. Aspirate the medium and wash each well once with PBS. Remove PBS and add 0.1-μl, 1-μl, or 5-μl samples collected from vector stocks, each sample diluted to 250 μl in DMEM/2% FBS.
3. Incubate 1 to 2 days and remove the inoculum. Fix cells for 20 min at room temperature with 250 μl of 4% paraformaldehyde. Wash the fixed cells three times with PBS, then proceed (depending on the transgene) with a detection protocol such as green fluorescence (step 4a), X-gal staining (steps 4b and 5b), or immunocytochemical staining (steps 4c to 6c). In each case, determine the vector titer (t.u./ml) by multiplying the number of transgene-positive cells by the vector dilution factor.

Detect cells expressing the gene for green fluorescent protein (GFP)

4a. Examine the culture from step 3 (before or after fixation; GFP can be detected in living cells), using an inverted fluorescence microscope. Count green fluorescent cells and determine the vector titer.

Detect cells expressing the E. coli lacZ gene

4b. Add 250 μl X-gal staining solution per well of the 24-well tissue culture plate from step 3, and incubate 4 to 12 hr (depending on the cell type and the promoter regulating expression of the transgene), at 37°C.

5b. Stop the staining reaction by washing the cells three times with PBS. Count blue cells using an inverted light microscope, and determine the vector titer.

Detect transgene-expressing cells by immunocytochemical staining

4c. Add 250 μl GST solution per well of the 24-well tissue culture plate from step 3 (to block nonspecific binding sites and to permeabilize cell membranes) and let stand 30 min at room temperature. Replace the blocking solution with the primary antibody (diluted in GST; the optimal dilution is typically 1/100 to 1/10,000) and incubate overnight at 4°C.

5c. Wash the cells three times with PBS, leaving the solution in the well for 10 min each time. Add secondary antibody (diluted in GST) and incubate at least 4 hr at room temperature.

6c. Wash the cells twice with PBS and develop according to the appropriate visualization protocol. Count transgene-positive cells using an inverted light microscope and determine the vector titer.

References: Fraefel et al., 1998; Pechan et al, 1996

Contributor: Cornel Fraefel

UNIT 4.16

Overview of Nucleic Acid Arrays

Nucleic acid array technology refers to the fabrication and use of arrays containing thousands of nucleic acid samples bound to solid substrates, such as glass microscope slides or silicon wafers. Because the physical area occupied by each sample is usually 50 to 200 μm in diameter, nucleic acid samples representing entire genomes, ranging in size from 3,000 to 32,000 genes, may be efficiently packaged onto a single regular microscope slide in an area easily covered by a coverslip. Such "genomes on a chip" then serve as a target to which fluorescently labeled nucleic acid probes can be applied. Nucleic acid arrays, or microarrays, allow all genes of a given genome to be simultaneously monitored with respect to some experimental condition of interest. This fact has fundamentally changed the manner in which the study of genomics and gene expression can be pursued.

The majority of applications discussed in this overview relate to DNA microarrays fabricated by the mechanical deposition of nucleic acid samples onto glass. Typically, these samples are in the form of PCR products, ranging in size from 100 bps to 9 kb. However, the term "DNA microarray" may apply to several different forms of the technology, each differing in the type of nucleic acid applied and the method of application. For example, Affymetrix sells DNA arrays produced by photolithographic synthesis of individual short oligonucleotides directly on the substrate (Fodor et al., 1991).

WHAT ARE MICROARRAYS GOOD FOR?

Gene Expression Analysis

Undoubtedly the most common use for DNA microarrays is for monitoring gene expression levels. The broad appeal of this approach stems from the fact that it can be applied to virtually any organism, tissue, or cell line from which RNA may be isolated. In a typical experiment, total RNA or mRNA is collected from two or more individuals, cultures, or conditions. The amount of RNA needed for a microarray experiment depends on many factors, such as genome complexity and message content. Most experiments use anywhere from 100 ng to as much as 20 μg of RNA. The next step is the separate conversion of the RNA samples into cDNA by reverse transcription. This is usually accomplished by priming with randomized oligonucleotides or, in the case of organisms that produce polyadenylated messages, an oligo-dT primer. The basic principle behind these manipulations is to convert the RNA from each sample into a form that can be readily distinguished from another RNA sample. This is usually accomplished by labeling the cDNA samples with different fluorescent dyes, either during the reverse transcription process through direct incorporation by reverse transcriptase, or afterwards by chemical conjugation.

The resulting pools of cDNA are mixed together, and when the pool is hybridized to a microarray, the ratio between the intensities observed for two of the fluorophores at any given location in the array is a direct measure of the relative abundance of the corresponding cDNA transcript (see Fig. 4.16.1). By using a single reference sample as the control for a series of experimental samples collected over time, one can compare relative levels of transcript abundance among samples (see Fig. 4.16.2). This in turn allows one to identify gene expression trends. For those who wish to study global regulation of gene expression, the most significant data lies in these trends and patterns. Many instances where this methodology has been successfully put into practice can be found in the literature (DeRisi et al., 1997; Alizadeh et al., 1998; Cho et al., 1998; Chu et al., 1998; Eisen et al., 1998; Spellman et al., 1998; Amundson et al., 1999; Iyer et al., 1999; Perou et al., 1999).

It is certain that the use of microarrays to analyze gene expression will continue to increase. Aside from looking at developmental time courses, mutations, and other genetic modifications, many novel expression experiments will undoubtedly be developed. One recent variation includes the use of expression analysis to reverse-engineer the changes that occurred during a 200-generation yeast evolution experiment (Ferea et al., 1999). Gene expression analysis involving microarrays may also be used as a method to attack problems that were formerly too cumbersome to approach using standard molecular biology techniques. This is especially true for potentially dangerous and difficult-to-culture organisms such as *Mycobacterium tuberculosis* and *Plasmodium falciparum* (Wilson et al., 1999; Hayward et al., 2000). Because only relatively small amounts of RNA are required to determine expression profiles for entire genomes, the data return on the up-front labor investment is much more substantial than with older techniques, thus enlarging the potential scope and depth of experiments. Consider the

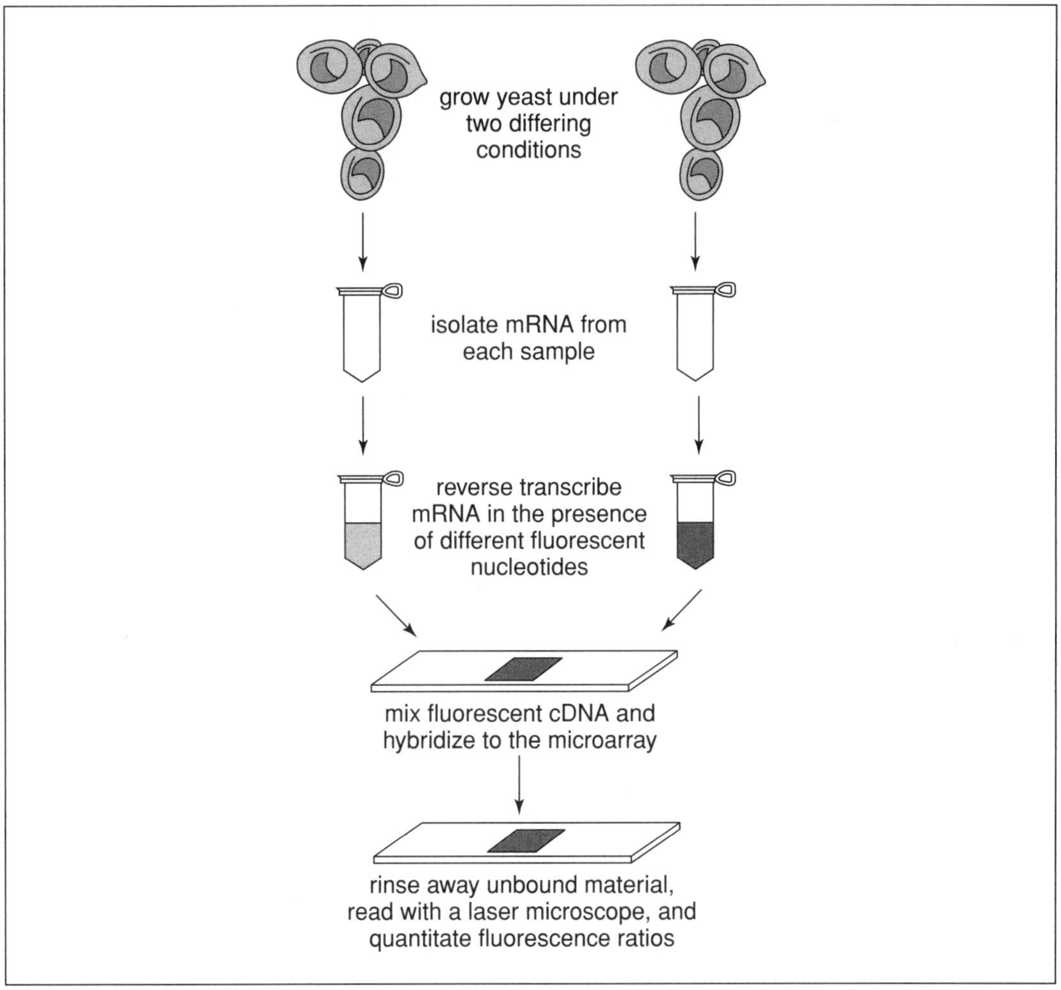

Figure 4.16.1 Scheme for a typical gene expression experiment.

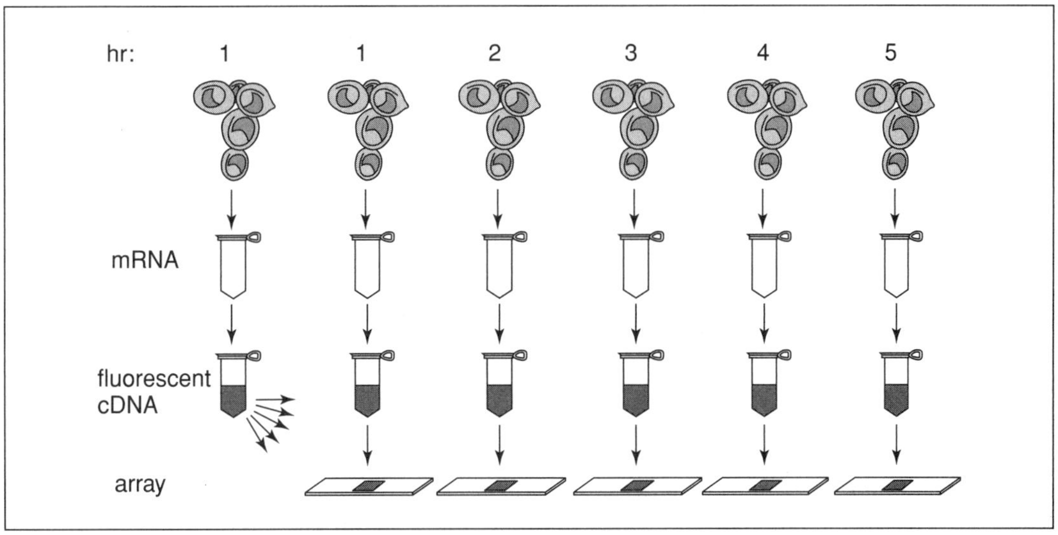

Figure 4.16.2 Monitoring levels of transcript abundance. Time zero is the reference probe for each comparison.

case of *Plasmodium*: genetic crosses are extremely difficult and transformation efficiencies are abysmal. These facts make many of the tools researchers have relied upon for cloning new genes nearly useless. Therefore, determining which genes are induced by various drugs becomes a large undertaking, especially considering that the *Plasmodium* genome has yet to be sequenced completely. However, it is a relatively simple matter to collect RNA samples, so DNA microarrays (utilizing known or random genomic fragments) provide an easy means of making these determinations.

WHAT ELSE ARE NUCLEIC ACID MICROARRAYS GOOD FOR?

Despite all the recent emphasis on gene expression analysis, the number of different uses for DNA microarrays is currently limited only by our imagination. Meaningful data can be derived from any set of biological experiments that result in a differential recovery of nucleic acid material. Aside from gene expression, these uses fall into several basic categories: genetic mapping and genotyping, assessment of genome structure and copy number, polysome analysis, and assays of DNA-protein interaction. Given that there are dozens of variations in experimental design and approach within each of these categories, the following is meant only to pique the imagination and alert readers to the diverse opportunities afforded by microarray technology.

Genotyping and Genetic Mapping

Efficient genotyping of hundreds or thousands of markers for the purpose of mapping multigenic traits may be carried out using DNA microarrays. In one approach, genotyping by array analysis is accomplished by directly detecting hybridization differences caused by single-nucleotide polymorphisms (SNPs). Arrays utilizing short oligo features are well suited for this method due to their low annealing temperatures (Hacia et al., 1996). Another approach, genomic mismatch scanning (GMS), uses mismatch repair enzymes to recognize and selectively degrade hybrid DNA fragments between two individuals that possess SNPs (Nelson et al., 1993). The hybrid DNA is created by denaturing and reannealing fragments from two related individuals. The resulting perfectly matched DNA fragments can then be fluorescently labeled and applied to a microarray, revealing which segments are identical by descent (Cheung et al., 1998; McAllister et al., 1998). Regardless of the method used, genotyping by array analysis has the potential to increase the throughput by several orders of magnitude when compared to traditional gel electrophoresis-based methods.

Comparative Genomic Hybridization

Differences in gene copy number have been associated with various tumorigenic phenotypes. These polymorphisms, which consist of deletions and/or amplifications, can easily be assayed using DNA microarrays, which provide a viable alternative to traditional comparative genomic hybridization (CGH) techniques. The primary limitation of traditional CGH techniques lies in the fact that the resolution to which an amplification or deletion may be mapped is ~20 Mb (Kallioniemi et al., 1992). With array-based CGH methods, the resolution depends only on how many DNA elements are present on the microarray; therefore, it is entirely feasible that all human genes may be represented as discrete elements on future microarrays, taking the resolution of the map to its logical limit. Indeed, with the conclusion of the human genome sequencing project, it should be feasible to detect the deletion or amplification of virtually any and all genes simultaneously for a given human cell population. Several examples already exist in the literature, and advances are sure to follow (Solinas-Toldo et al., 1997; Trent et al., 1997; Pinkel et al., 1998; Behr et al., 1999; Pollack et al., 1999).

Polysome Analysis

Although transcriptional regulation is the primary focus of most microarray experiments, changes in translation can also be readily assayed. This may be accomplished by fluorescently labeling the nucleic acid portion of polysomes, which are typically isolated as low-velocity nucleoprotein fractions separated on sucrose gradients. Higher-molecular-weight material, representing messages bound by several ribosomes, may be differentially labeled with respect to lower-molecular-weight material, those messages with few or no bound ribosomes. The resulting hybridization can then be used to simultaneously assay the degree to which each mRNA is associated with translation machinery. Tracking these associations provides a means of measuring the degree to which the expression of any particular message is likely to be controlled at this level (K. Kuhn and P. Sarnow, pers. comm.). Membrane-bound polysomes may also be collected and analyzed. Because membrane-bound polysomes contain messages undergoing cotranslational secretion, it is then possible to quickly identify these gene products with respect to any given experimental condition, whether it be time, cell type, or environment (Diehn et al., 2000).

DNA-Protein Interaction

The separation of nucleoprotein components for array analysis is not limited to mRNA bound to polysomes. Indeed, any method that creates a stable or covalent cross-link between the nucleic acid and the protein component may be used. Subsequent separation by immunoprecipitation (e.g., *CPNS UNIT 5.8*; also see *CPMB UNIT 10.6* and *APPENDIX 2A* in this manual), filter binding, or other chromatographic procedures may then produce a nucleic acid fraction that can be directly labeled and applied to a microarray representation of the genome. Such genome-wide studies are underway for epitope-tagged DNA-binding proteins and for proteins that cross-link enzymatically to DNA as part of their normal function. Obvious applications of this approach include the study of DNA binding proteins, endonucleases, and chromatin-associated proteins. These methods often produce vanishingly small amounts of nucleic acid source material, often in the sub-nanogram range. For this reason, there is likely to be much development in the area of linear DNA and RNA amplification. Indeed, amplification technology coupled to microarray analysis will allow the study of single-cell expression patterns (*UNIT 5.3*; Eberwine, 1996).

WHAT ABOUT DATA ANALYSIS?

In addition to the many ways microarray experiments may be carried out, there are equally as many ways to analyze the resulting data. Especially for gene expression, it is likely that genome-wide expression analysis will become a field of its own and that an ever-increasing number of innovative approaches will be developed to address reverse engineering of the mechanisms guiding the observed expression patterns. Several methods for visualization and analysis already exist, and many of the necessary tools are readily available on the Internet. The method known as clustering, whereby similar expression patterns are grouped together, has been particularly successful (Eisen et al., 1998; Bassett et al., 1999; Brown and Botstein, 1999; Iyer et al., 1999). Its success relies upon the fact that genes that participate in a common process, pathway, or function often share common regulatory mechanisms and this, in turn, results in similar expression profiles. Therefore, the functions of previously uncharacterized genes may be revealed by other genes that cluster with it over a broad range of experiments.

WHERE CAN I GET MORE INFORMATION?

One way to stay connected to new uses and methods of this rapidly evolving field is to frequent the various noncommercial web sites listed below.

Stanford Genomic Resources

http://www-genome.stanford.edu One-stop shopping for genome-wide expression analysis. This site contains links to several yeast expression databases and companion research articles. Links to several human expression databases are also available here.

Stanford Genome Analysis Group software download area

http://rana.stanford.edu/software This site provides access to several free microarray-analysis tools that run on your PC, written by Mike Eisen.

The MGuide

http://cmgm.stanford.edu/pbrown/mguide Home of a do-it-yourself guide to building your own microarraying robot, maintained by Patrick Brown's lab at Stanford University.

4

National Human Genome Research Institutes (NHGRI) Microarray Project

http://www.nhgri.nih.gov/DIR/LCG/15K/HTML This site features database and data handling information, as well as protocols and research projects.

Genome-Wide Expression Homepage

http://web.wi.mit.edu/young/expression Maintained by Rick Young of the Whitehead Institute. Descriptions of various yeast transcription experiments, protocols, and experiment designs are available.

Large-Scale Gene Expression and Microarray Links and Resources

http://industry.ebi.ac.uk/~alan/MicroArray A "personal not-for-profit web site" containing literally hundreds of links to microarray articles, databases, and companies, maintained by Alan Robinson, a researcher at the EMBL-European Bioinformatics Institute (EBI).

References: Brown and Botstein, 1999; DeRisi et al, 1997

Contributor: Joseph DeRisi

UNIT 4.17

Preparation of mRNA for Expression Monitoring

Expression monitoring is accomplished by hybridizing mRNA, which has been quantitatively amplified and labeled with biotin, to DNA chips that display thousands of oligonucleotides complementary to the mRNAs of interest (see Fig. 4.17.1).

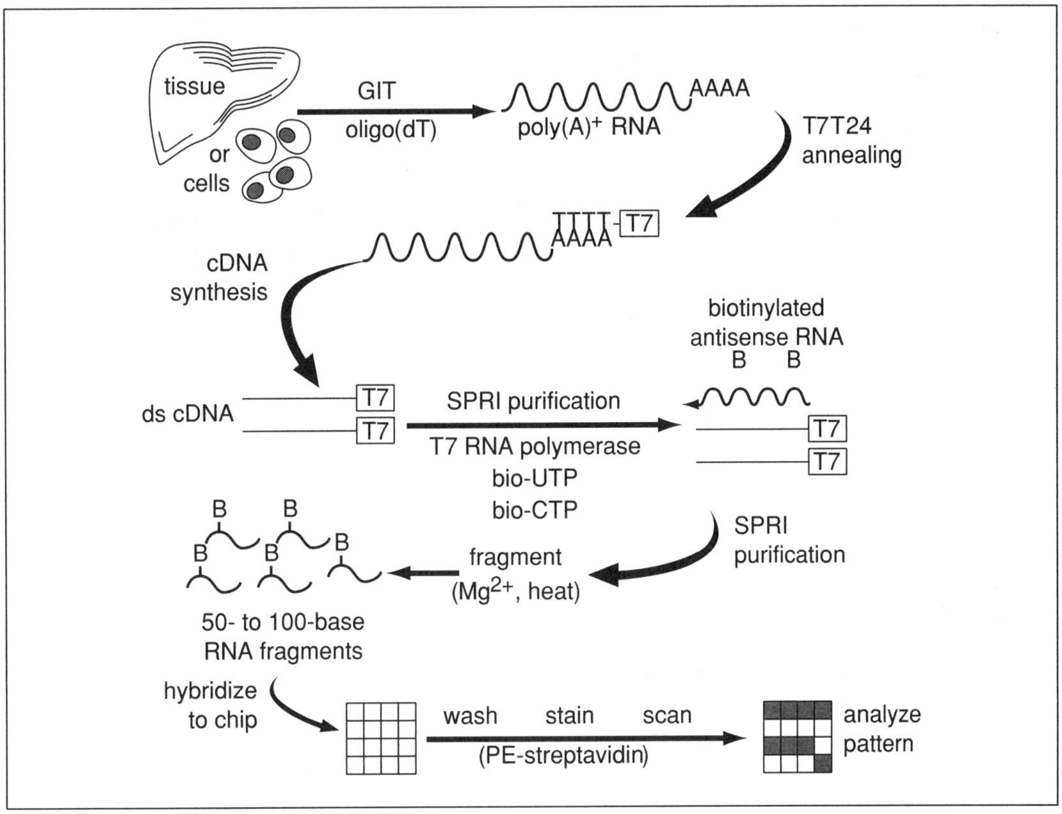

Figure 4.17.1 Chip analysis overview. Abbreviatiosn: GIT, guanidine isothiocyanate, PE, phycoerythrin; SPRI, solid-phase reversible immobilization.

CAUTION: Diethylpyrocarbonate (DEPC) is a suspected carcinogen and should be handled with care.

STRATEGIC PLANNING

Care should be taken at every step to avoid contamination of reagents and materials with RNases (*see* CPMB *UNIT 4.1*). New plasticware can be used without decontamination processes if kept free of dust and handled exclusively with gloved hands. Filtered pipet tips should be used routinely to avoid contamination with the micropipettor. Any reagent that needs to be made in the laboratory, or that is not supplied RNase-free by the manufacturer, should be treated with DEPC.

The amplification procedure is absolutely dependent on clean, intact starting RNA. For cultured cells, the authors use a protocol of lysis in guanidine followed by resin-based purification (such as the Qiagen RNeasy kit).

The mRNA amplification can be performed starting with either poly(A)$^+$ or total RNA. Although poly(A)$^+$ RNA gives higher sensitivity because it largely eliminates mispriming on rRNA during the cDNA reaction, the ability to use total RNA makes it possible to analyze biological systems in which cells or tissues are limited. A key modification in using total RNA is elevation of the first-strand cDNA reaction temperature from 37° to 50°C. Restrict probe selection to the last 600 bases of coding sequence unless the 3′-untranslated region (3′-UT) is >800 bases, in which case some untranslated sequence is also included.

NOTE: A Troubleshooting Guide (see Table 4.17.2 at end of this unit) summarizes problems that may arise during these procedures and possible solutions.

BASIC PROTOCOL

AMPLIFICATION OF mRNA FOR EXPRESSION MONITORING AND HYBRIDIZATION TO OLIGONUCLEOTIDE ARRAY CHIPS

Materials *(see APPENDIX 1 for items with ✓)*

SuperScript cDNA kit (Life Technologies), including:
5× First Strand Buffer
200 U/µl SuperScript II reverse transcriptase
5× Second Strand Buffer
10 mM dNTPs
10 U/µl *E. coli* ligase
2 U/µl *E. coli* RNase H
10 U/µl *E. coli* DNA polymerase
5 U/µl T4 DNA polymerase
T7T24 primer:
5′-GGCCAGTGAATTGTAATACGACTCACTATAGGGAGGCGGTTTTTTTTTTTT TTTTTTTTTTTT-3′ (HPLC purification is recommended)
RNase inhibitor (Life Technologies or Ambion)
✓ RNase-free H_2O (DEPC-treated water; prepare from glass-distilled H_2O)
Sample RNA: poly(A)$^+$ or total RNA
Sense control transcript pool (see Support Protocol 1)
25:24:1 (v/v/v) phenol/chloroform/isoamyl alcohol (molecular biology grade)
7.5 M ammonium acetate
Absolute ethanol
70% (v/v) ethanol in RNase-free H_2O, prechilled to −20°C
10× transcription buffer (Ambion)
✓ 10× rNTP mix
100 mM dithiothreitol (DTT)
10 mM Bio-11-CTP and Bio-11-UTP (Enzo Diagnostics)
2500 U/µl T7 RNA polymerase (Epicentre)
RNeasy mini columns with RLT and RPE buffers and collection tubes (Qiagen)
✓ 5× fragmentation buffer
20× SSPE (Bio-Whittaker)
0.5% (v/v) Triton X-100 (molecular biology grade; Sigma) in RNase-free H_2O
10 mg/ml herring sperm DNA (Promega)
✓ 500 pM Bio948
20× antisense control transcript pool (see Support Protocol 1)
6× SSPET: 6× SSPE containing 0.005% (v/v) Triton X-100
phycoerythrin

Thermocycler (e.g., Perkin-Elmer 9600 PCR machine with heated lid)
0.1- to 10-µl filtered micropipet tips (Continental)
Lyophilizer
Small, thin-walled PCR tubes
GeneChip (Affymetrix)
1- to 200-µl filtered gel-loading micropipet tips (Fisher)
Rotisserie-type rotator (Appropriate Technical Resources)
50°C oven

NOTE: Many buffers and enzymes are supplied with the SuperScript II cDNA kit. Kit enzymes that are limiting may be ordered separately from Life Technologies.

NOTE: All temperature-controlled reactions are performed in an appropriate thermocycler.

1. Set up a linked program on a thermocycler as follows:

10 min	70°C
65 min	37°C (or 50°C for total RNA)
150 min	15.8°C
indefinitely	4°C (hold).

2. Prepare 10 μl first-strand reagent cocktail for each sample RNA by combining the following reagents, using filtered pipet tips:

   4 μl 5× First Strand Buffer
   200 pmol T7T24 primer
   1 μl RNase inhibitor
   1 μl 200 U/μl SuperScript II reverse transcriptase
   RNase-free H_2O to 10 μl.

3. Combine each sample RNA with sense control transcript pool (see Support Protocol 1). Use 5 μl sense control transcript pool per 1 μg sample poly(A)$^+$ RNA or 1 μl control pool per 10 to 20 μg total RNA. Lyophilize to reduce the volume per tube to <10 μl. Adjust volume to 10 μl with RNase-free H_2O. Prewarm first-strand reagent cocktail in hand during this time.

4. Transfer each RNA to a small, thin-walled PCR tube and place in the thermocycler. Begin linked program (step 1).

5. After the 70°C step has been completed, allow 2 min for solution temperature to reach 37°C (or 50°C), and add the hand-warmed first-strand reagent cocktail (step 2; 10 μl per tube).

6. Hold at the appropriate reaction temperature for the remaining 60 min.

7. Prepare 130 μl second-strand reagent cocktail per tube:

   91 μl RNase-free H_2O
   30 μl 5× Second Strand Buffer
   3 μl 10 mM dNTPs
   1 μl 10 U/μl *E. coli* ligase
   1 μl 2 U/μl *E. coli* RNase H
   4 μl 10 U/μl *E. coli* DNA polymerase.

   Store cocktail on ice for up to 90 min.

8. Add 130 μl second-strand reagent cocktail to the tube (total 150 μl) once the thermocycler has shifted to 15.8°C. Mix by pipetting up and down after addition of cocktail. Incubate at 15.8°C for ≥2 hr.

9. Add 2 μl of 5 U/μl T4 DNA polymerase and incubate an additional 5 min. Remove samples and place on ice.

10. Add 150 μl of 25:24:1 phenol/chloroform/isoamyl alcohol, vortex, and centrifuge 5 min at 13,000 × *g*, room temperature.

11. Carefully transfer aqueous top phase to new tube, avoiding contamination with interface. Add 70 μl of 7.5 M ammonium acetate and 0.5 ml absolute ethanol and mix. Centrifuge 20 min at 13,000 × *g*, room temperature.

12. Remove supernatant. Wash pellet with 0.5 ml of cold (−20°C) 70% ethanol. Vortex well. Centrifuge 10 min at 13,000 × *g*, room temperature, remove supernatant, and allow pellet to air dry for several minutes.

13. Resuspend the cDNA pellet in 25 µl RNase-free H_2O and quantitate cDNA (see Support Protocol 2).

14. For each cDNA, set up one in vitro transcription (IVT) reaction as follows:

    100 ng cDNA
    6 µl 10× transcription buffer
    6 µl 10× rNTP mix
    3 µl 100 mM DTT
    2.4 µl 10 mM Bio-11-UTP
    2.4 µl 10 mM Bio-11-CTP
    2 µl RNase inhibitor
    2 µl 2500 U/µl T7 RNA polymerase
    RNase-free H_2O to 60 µl.

    Store cocktail (without cDNA) up to 3 hr on ice. If stored, then warm to room temperature before adding cDNA.

15. Incubate at 37°C for 8 hr to overnight. Store up to 48 hr at −80°C before purification.

16. Bring volume of IVT reaction to 100 µl with RNase-free H_2O, add 350 µl RLT buffer, and mix. Add 250 µl absolute ethanol and mix.

17. To remove unincorporated nucleotides, apply each sample to an RNeasy mini column placed in a collection tube. Microcentrifuge 15 sec at >8000 × *g*, room temperature.

18. Transfer column to a new collection tube. Add 500 µl RPE buffer. Centrifuge 15 sec at >8000 × *g*, room temperature.

19. Discard flowthrough and replace column on the same collection tube. Add 500 µl RPE buffer and microcentrifuge 2 min at maximum speed.

20. Transfer column to a new collection tube. Add 50 µl RNase-free H_2O and centrifuge 1 min at >8000 × *g*. Repeat elution step, pooling the second eluate with the first.

21. Quantitate RNA yield spectrophotometrically at 260 nm (*APPENDIX 2G*).

22. Bring volume of 10 µg IVT RNA to 24 µl with RNase-free H_2O. Add 6 µl of 5× fragmentation buffer. Mix carefully and incubate 35 min at 95°C in the thermocycler. Allow to cool to room temperature. Store up to 1 year at −80°C, and thaw 5 min at 37°C before using for hybridization.

23. For each reaction, prepare 170 µl hybridization master mix as follows:

    51 µl 20× SSPE
    1.7 µl 0.5% Triton X-100
    1.7 µl 10 mg/ml herring sperm DNA
    20 µl 500 pM Bio948
    10 µl 20× antisense control transcript pool
    85.6 µl RNase-free H_2O.

24. Add 170 µl hybridization master mix to 30 µl of each fragmented IVT RNA. Heat to 99°C for 10 min in the thermocycler, then move to 37°C for ≥5 min. Microcentrifuge 5 min at maximum speed.

25. Insert a filtered micropipet tip into the upper septum of a GeneChip to provide a vent. Fill the chip from the bottom septum with the 200 µl hybridization solution, using a

filtered gel-loading micropipet tip. Remove the vent and cover both septa with transparent tape.

26. Incubate in a 40°C oven overnight (16 to 18 hr) on a rotisserie-type rotator running at ~60 rpm.

27. Transfer the chip to a 50°C oven and continue rotating the chip for exactly 1 hr. Then remove chip from the oven and insert a filtered micropipet tip into the upper septum to vent the chip.

28. Using a micropipettor with plunger fully depressed, insert a 200-µl gel-loading pipet tip into the lower septum. Holding the chip vertically, slowly draw out all of the hybridization solution and store it in a microcentrifuge tube at −20°C.

29. Fill the chip with 6× SSPET.

30. Wash and stain the chip with phycoerythrin according to the manufacturer's protocols. Scan the chip as soon as possible.

SUPPORT PROTOCOL 1

IN VITRO TRANSCRIPTION OF CONTROL GENES AND PREPARATION OF TRANSCRIPT POOLS

4

It is essential to include labeled antisense control RNAs of known concentration in each hybridization reaction to normalize chip-to-chip variation and to allow construction of a standard curve for converting hybridization intensity to mRNA frequency. It is also very helpful to add unlabeled sense transcripts to monitor performance of the amplification protocol.

Additional Materials *(also see Basic Protocol)*

Plasmids (Table 4.17.1; ATCC #87482 to 87490)
2500 U/µl T3 RNA polymerase (Enzo Diagnostics)
25 mM 4rNTP mix: 25 mM each rGTP, rCTP, rATP, and UTP (Ultrapure; Pharmacia Biotech) in RNase-free H_2O

1. Prepare linearized plasmid templates according to Table 4.17.1.

2. Purify plasmid DNA by phenol/chloroform extraction and ethanol precipitation (see Basic Protocol, steps 10 to 12). Resuspend, at ~0.1 mg/ml in RNase-free H_2O, and quantitate DNA (see Support Protocol 2).

3. For labeled antisense transcripts, prepare four tubes containing a master mix that includes all components except plasmid DNA:

   6 µl 10× transcription buffer
   6 µl 10× rNTP mix
   3 µl 100 mM DTT
   2.4 µl 10 mM Bio-11-UTP
   2.4 µl 10 mM Bio-11-CTP
   2 µl RNase inhibitor
   2 µl 2500 U/µl T7 RNA polymerase
   RNase-free H_2O to 60 µl, allowing for DNA to be added at step 5.

4. For unlabeled sense transcripts, prepare four tubes of master mix as in step 3, but substitute 25 mM 4rNTP mix for the 10× rNTP mix, eliminate the biotinylated nucleotides, and substitute T3 for T7 RNA polymerase.

Table 4.17.1 Preparation of Plasmid Template Controls

			Sense RNA		Antiense RNA	
Name	ATCC#	Transcript size (kb)	Linearize with	Polymerize with	Linearize with	Polymerize with
pGIBS-LYS[b]	87482	1.0	*Not*I	T3		
pGIBS-PHE[b]	87483	1.3	*Not*I	T3		
pGIBS-THR[b]	87484	2.0	*Not*I	T3		
pGIBS-TRP[b]	87485	2.5	*Not*I	T3		
pGIKS-BioB	87487	1.1			*Xho*I	T7
pGIKS-BioC	87488	0.8			*Xho*I	T7
pGIKS-BioD	87489	0.7			*Xho*I	T7
pGIKS-CRE	87490	1.0			*Xho*I	T7

[a]Abbreviations: BioB, BioC, and BioD are cloned fragments from the *E. coli bioB*, *bioC*, and *BioD* genes, respectively. LYS, PHE, THR, and TRP are fragments from the *Bacillus subtilis lysA, pheA, thrBC*, and *trpEDCF* genes, respectively. CRE is a fragment from the Cre recombinase derived from *E. coli* bacteriophage P1. PGIBS and pGIKS are derived from the Bluescript KS II vector (Stratagene).

[b]pGIBS-LYS, -PHE, -THR, -TRP, and DAP contain a 40-nucleotide synthetic poly(A) tract at the 3′ end of the respective genomic fragments derived from *B. subtilis*. Sense IVT transcripts derived from *Not*I-linearized plasmid templates will contain the artificial poly(A) tail. Plasmids linearized with *Bam*HI prior to T3 IVT will generate sense transcripts without the synthetic poly(A) tract.

4

5. Add 100 ng of each linearized plasmid DNA to a separate tube (the four antisense and four sense plasmids to tubes prepared in steps 3 and 4, respectively; see Table 4.17.1) and incubate at 37°C for 8 hr to overnight.
6. Purify IVT products as described (see Alternate Protocol).
7. Quantitate carefully by absorbance at 260 nm (*APPENDIX 2G*).
8. Dilute unlabeled sense transcripts to 200 nM stock solutions in RNase-free H_2O for long-term storage (up to 1 year) at −80°C.
9. Fragment biotin-labeled antisense transcripts for standard chip controls (BioB, BioC, BioD, CRE) as described (see Basic Protocol, step 22).
10. Dilute fragmented antisense transcripts to 20 nM stock solutions in 6× SSPET with 0.1 mg/ml herring sperm DNA for long-term storage (up to 1 year) at −80°C.
11. Prepare 20× antisense control transcript pool by combining the following transcripts in 6× SSPET with 0.1 mg/ml herring sperm DNA:

    30 pM fragmented BioB transcript
    100 pM fragmented BioC transcript
    500 pM fragmented BioD transcript
    2 nM fragmented CRE transcript.

    Store up to 6 months at −80°C.
12. Prepare sense control transcript pool by combining the following transcripts in RNase-free H_2O:

    10 pM LYS transcript
    30 pM PHE transcript
    90 pM THR transcript
    180 pM TRP transcript.

    Store up to 6 months at −80°C in aliquots of 50 to 100 µl.

ALTERNATE PROTOCOL

SOLID-PHASE REVERSIBLE IMMOBILIZATION (SPRI) PURIFICATION OF cDNA AND IVT PRODUCTS

A magnetic bead-based purification of cDNA and IVT products avoids the use of organic solvents and centrifugation.

Materials *(see* APPENDIX 1 *for items with ✓)*

Carboxy-coated magnetic beads (PerSeptive BioSystems for cDNA purification; Bangs Laboratories for IVT purification)
✓ 0.5 M EDTA
Sample to be purified: cDNA (see Basic Protocol, step 9) or IVT RNA (see Basic Protocol, step 15, or see Support Protocol 1, step 6)
2.5 M NaCl/20% (w/v) PEG 8000 (molecular biology grade; RNase free)
70% (v/v) ethanol in RNase-free H_2O
10 mM Tris acetate, pH 7.8 (RNase free)
Magnetic stand (CPG)

Purify cDNA (equivalent to Basic Protocol steps 10 to 13):

1a. Aliquot 10 µl PerSeptive carboxy-coated magnetic beads per 150-µl cDNA reaction, dispensing the total volume of beads into a single microcentrifuge tube. Place the tube on a magnetic stand and allow the beads to separate to the side of the tube. Carefully remove supernatant with a micropipet.

2a. Add 0.5 M EDTA equal to the starting volume and resuspend the beads by gentle vortexing or agitation. Replace tube on magnetic stand, wait for beads to separate, and remove supernatant. Repeat this washing procedure two more times. Resuspend beads in 0.5 M EDTA equal to the starting volume.

3a. To each tube of cDNA, add 150 µl of 2.5 M NaCl/20% PEG 8000 and 10 µl beads and mix by gentle vortexing or agitation. Incubate 10 min at room temperature.

4a. Place tubes on magnetic stand and allow beads to separate to the side of the tube (~2 min for the original separation, faster for the washes). Draw off the supernatant, then wash the beads twice with 150 µl of 70% ethanol. Remove as much of the final ethanol wash as possible and allow to air dry for 2 min.

5a. Elute RNA by adding 25 µl of 10 mM Tris acetate, pH 7.8, and incubating 5 min at room temperature. Place tube on magnetic stand and save supernatant.

6a. Determine cDNA concentration by PicoGreen fluorescence (see Support Protocol 2).

Purify IVT RNA (equivalent to Basic Protocol steps 16 to 21):

1b. Aliquot 20 µl Bangs Laboratories carboxy-coated magnetic beads per 60-µl IVT reaction, dispensing the total volume of beads into a single microcentrifuge tube. Place the tube on a magnetic stand and allow the beads to separate to the side of the tube. Carefully remove supernatant with a micropipet.

2b. Add 0.5 M EDTA equal to the starting volume and resuspend the beads by gentle vortexing or agitation. Replace tube on magnetic stand, wait for beads to separate, and remove supernatant. Repeat this washing procedure two more times. Resuspend beads in 1.25 M NaCl/10% PEG equal to the starting volume.

3b. To each tube of IVT RNA, add 60 µl of 2.5 M NaCl/20% PEG 8000 and 20 µl beads and mix by gentle vortexing or agitation. Incubate 10 min at room temperature. Place tubes on magnetic stand and allow beads to separate to the side of the tube (~2 min for the original separation, faster for the washes).

4b. Draw off the supernatant, then wash the beads twice with 150 μl of 70% ethanol. Remove as much of the final ethanol wash as possible and allow to air dry for 3 min.

5b. Elute RNA by adding 25 μl of 10 mM Tris acetate, pH 7.8, and incubating 5 min at room temperature. Place tube on magnetic stand and save supernatant.

6b. Determine RNA concentration by absorbance at 260 nm (*APPENDIX 2G*).

SUPPORT PROTOCOL 2

QUANTITATION OF cDNA

For optimal IVT yield, do not exceed 100 ng cDNA in the standard reaction.

Materials

PicoGreen dsDNA Quantitation Kit (Molecular Probes), including:
- 100 ng/μl standard DNA stock solution
- 20× TE buffer
- PicoGreen reagent

cDNA to be quantitated (see Basic Protocol and Alternate Protocol)

Black-walled 96-well plate (Corning)
Fluorimager (Molecular Dynamics, model FSI)

1. Prepare 1 ml of 2 μg/ml diluted standard DNA by diluting 20 μl of 100 ng/μl standard DNA stock solution in 980 μl of 1× TE buffer.
2. Dilute PicoGreen reagent 1:200 (v/v) with 1× TE buffer. Prepare enough diluted reagent so that 100 μl can be placed in each well.
3. Pipet 100, 50, 20, 10, 5, 2, and 0 μl diluted standard DNA into seven wells in a black-walled 96-well plate. Bring each to a total of 100 μl with 1× TE buffer.
4. Pipet 2 μl cDNA to be quantitated into additional wells, as needed.
5. Add 100 μl diluted PicoGreen reagent to each well.
6. Read fluorescence of each well, using a fluorimager.
7. Generate a volume report on each well according to manufacturer's instructions. Use volume number versus ng per well to generate a standard curve.
8. Calculate the concentration of cDNA in test wells.

References: Lockhart et al., 1996; Schena et al., 1995; Wodicka et al., 1997.

Contributors: Michael C. Byrne, Maryann Z. Whitley, and Maximillian T. Follettie

Table 4.17.2 Troubleshooting Guide for Expression Monitoring

Problem	Possible cause	Action
Little or no cDNA yield	Beginning RNA quality was poor	Visualize size range of starting RNA on a gel. There should be a smear extending well above 5 kb, and not much <500 bases.
	RNA or reactants were RNase contaminated	Check RNA by incubating an aliquot at room temperature several hours and running on gel as above. Reactants can be incubated with test RNA and run on gel for visualization of degradation.
		Include RNase inhibitor in cDNA reaction.
	Enzyme inhibitors present in RNA preparation	Check signals for members of the sense control transcript pool. If 5′ end signals are poor and there is no evidence for RNase contamination, reisolate sample RNA using a protocol with a more stringent purification method.
Little or no IVT yield	Too little cDNA template	Quantitate cDNA, use ≥50 ng in IVT reaction. If cDNA yield is low, refer to above section (little or no cDNA yield).
	Inhibitor of polymerase present	Residual phenol/chloroform or dNTPs can inhibit polymerase reaction. Reprecipitate cDNA, being careful to wash pellet with 70% ethanol.
	Wrong polymerase used	Be sure polymerase matches the incorporated promoter
	DTT inactive	Make fresh DTT stock
	Limiting nucleotide concentration	If making unlabeled RNA, be sure to substitute 25 mM 4rNTP solution for the 10 × rNTP mix used with biotin nucleotides.
Blank or low signal on GeneChip	Poor RNA quality	Even when cDNA and IVT yields seem adequate, poor RNA quality can be a cause of poor chip data. Look at housekeeping gene data for ratio of 5′ and 3′ signal. With oligo(dT)-primed cDNA, expect 5′ signal of β-actin and GAPDH[a] to be at least half of the 3′ signal.
	Stringency too high	Check incubator temperature and buffer solutions
	Poor staining of biotin with streptavidin conjugate, or bleaching of phycoerythrin	SA-PE[a] reagents can be variable. Try a different lot, or increase the amount of SA-PE used in the staining solution. Store SA-PE in the dark. Wrap chips in aluminum foil after staining to prevent bleaching before scanning.
	Scanning at wrong wavelength (even edge controls will be very light or absent)	Ensure that scanner is set at 560 nm and rescan

continued

Table 4.17.2 Troubleshooting Guide for Expression Monitoring, continued

Problem	Possible cause	Action
	Hybridization inhibitor in solution	Try using the same hybridization solution on a second chip of the same lot. If signal improves, it may be necessary to incorporate a prehybridization step.
	Scanner needs adjustment	Call for service; try another scanner if available
	Defective lot of GeneChips	Rehybridize saved hybridization solution to a lot of chips that has previously given strong signals. If signal improves, and hybridization inhibitor is not indicated, contact manufacturer.
A few very bright features per gene, the rest very quiet	Failure of fragmentation reaction	Check 5× fragmentation buffer and 94°C temperature control unit. If additional IVT RNA is available, try repeating the fragmentation reaction and check on a gel. Size should be ~20 to 50 bases.
High background values	Stringency too low	Try rewashing the chips at increased temperature and rescanning
	Contaminant	Recommend hybridization solutions be made with glass-distilled water. Some deionization systems can be problematic. If this is suspected, glass-distilled water is commercially available.
Sporadic pattern of round dots all over chip	SA-PE[a] aggregates	Manufacturer recommends the solution be vortexed and then microcentrifuged 2 min prior to taking aliquot
Round area in middle of chip has very low signal compared to rest of chip	Volume of hybridization solution is too low	Need a minimum of 180 µl. If hybridization solution cannot be remade, add 6× SSPET, 0.1 mg/ml herring sperm DNA to the saved hybridization solution to bring the volume to 200 µl and rehybridize to new chip.
	Rotator stopped during hybridization	Rehybridize saved solution to new chip, checking that rotator is functioning properly.

[a]GAPDH, glyceraldehyde 3-phosphate dehydrogenase; SA-PE, streptavidin-phycoerthyrin.

UNIT 4.18

Gene Expression Analysis Using cDNA Microarrays

A Troubleshooting Guide at the end of this unit summarizes some problems that can arise, and possible solutions (see Table 4.18.2). The lower limit of detection of most cDNA array systems is represented by species present at ~1/100,000 to 1/250,000 in a sample. For a comparison of methods for labeling and detection of cDNA on microarry slides, see Table 4.18.1.

STRATEGIC PLANNING

Although there are a number of different protocols for cDNA labeling and detection in microarray format, most protocols using fluorescent dyes will require the following materials:

DEPC-treated H_2O (*see* APPENDIX 1)
DNase- and RNase-free 1.6-ml microcentrifuge tubes

Table 4.18.1 Comparison of cDNA Microarray Labeling and Detection Techniques

Method	RNA per slide (μg)		Comments	Reference
	mRNA	Total RNA		
Klenow direct labeling	1–2	50–100	Reliable, high labeling efficiency, requires the most input RNA	Welford et al., 1998; Geschwind et al., 2001
Reverse transcriptase direct labeling	1–2	25–50	Reproducible, reasonable labeling efficiency, requires less RNA	DeRisi et al., 1997; Brown and Botstein, 1999
TSA	0.1	1–2	Reproducible, high rates of signal amplification, low amount of RNA	Micromax[a]; NEN[a]
PCR amplicons	0.02	1–2	Reproducible, high rates of signal amplification, low amount of RNA, time consuming	Welford et al., 1998; Geschwind et al., 2001
Amino-allyl labeling	2	10–20	Reproducible, requires more input RNA than TSA, less dye-incorporation effect	TIGR[b]
T7 in vitro transcription	$<0.05 \times 10^{-3}$	<1	Reproducible, very low input possible if two rounds of amplification are performed. Lower limit of input RNA is not established, but <0.05 ng poly(A)$^+$ RNA have been used.	*UNIT 5.3*; Luo et al, 1999
DNA dendrimers	0.25	1–5	Simple protocol; company data are promising but not widely adopted	Stears et al., 2000; Genisphere[c]

[a]See *http://www.nen.com.*

[b]See *http://www.tigr.org/tdb/microarray.*

[c]See *http://www.genisphere.com.*

15- and 50-ml polypropylene conical tubes (Becton Dickinson)
22 × 22–mm and 24 × 50–mm microscope cover glass (Fisher Scientific)
Scienceware Coplin staining jar (Fisher Scientific) or other covered slide box for post-hybridization slide incubations and washes
Powder-free plastic gloves
Canned-air cleaning duster
Micropipettors with RNase-free aerosol-resistant tips
Plastic forceps
Low-speed tabletop centrifuge with microtiter plate adapters (e.g., GS-6R, Beckman)
Orbital shaker (Bellco Glass)
Heating block
Hybridization chambers or humidified hybridization oven
Thermal cycler (e.g., PTC-100 Programmable Thermal Controller, MJ Research) or water baths for at least three different temperatures
Laser scanner (see Internet Resources)
Data analysis software program (e.g., BioDiscovery's GeneSight 3.0 or Silicon Genetics' GeneSpring 4.2; see Internet Resources)
Image analysis software (e.g., ARRAY-VIEWER, TIGR; ImaGene 4.2, BioDiscovery; see Internet Resources)

CAUTION: DEPC is a potential carcinogen and should be handled with care.

Consistent humidity is crucial for successful hybridization.

In every microarray-based experiment, one of the most important issues is to minimize background signals. Several simple rules should always be followed:

1. Wear powder-free gloves throughout the procedure.
2. Handle the microarray and cover slips with flat-ended forceps.
3. Pass all buffers and solutions that come in contact with the microarray through a 0.22-μm filter.
4. Keep the microarray wet after adding the hybridization solution, until the slide is scanned.
5. Perform all steps involving handling or use of RNA under RNase-free conditions. Use only RNase-free reagents, plasticware, and pipet tips, as well as DEPC-treated water. Wear gloves.
6. While preparing reactions, always keep the component vials on wet ice. Return the reagents to storage temperatures as soon as possible.
7. Since both Cy3 and Cy5 are photosensitive, minimize their exposure to light. Aliquot Cy-labeled nucleotides or reagents into single-use foil-wrapped tubes and store at −20°C. Perform all reactions in foil-wrapped tubes or containers in a darkenedroom.

4

BASIC PROTOCOL 1

DIRECT LABELING OF cDNA USING KLENOW FRAGMENT

This protocol can easily be used for labeling as little as 0.5 μg poly(A)$^+$ RNA (Luo et al., 1999), but the optimal range is from 1 to 3 μg poly(A)$^+$ RNA, or ∼50 μg total RNA.

Materials *(see APPENDIX 1 for items with ✓)*

Total or poly(A)$^+$ RNAin TE buffer
0.5 μg/μl oligo(dT)$_{12-18}$
SuperScript II cDNA kit (Invitrogen Life Technologies)
- 5× first-strand buffer
- 0.1 M dithiothreitol (DTT)
- 10 mM 4dNTP mix (10 mM each dATP, dCTP, dGTP, and dTTP)
- 200 U/μl SuperScript II reverse transcriptase
- 5× second-strand buffer
- 10 U/μl *E. coli* DNA polymerase I
- 2 U/μl RNase H
- 10 U/μl *E. coli* DNA ligase
- 5× random primers (hexamers)

10 U/μl RNasin (Promega)
15 mM β-Nicotinamide adenine dinucleotide (β-NAD; Sigma)
25:24:1 (v/v/v) phenol/chloroform/isoamyl alcohol (molecular biology grade)
Chloroform
10 M ammonium acetate
1 μg/μl glycogen carrier (Boehringer Mannheim)
100% ethanol (molecular-biology grade), −20° (Sigma)
70% (v/v) ethanol diluted with DEPC-treated H_2O, ice cold
5× dCTP buffer
0.25 mM dCTP
2.5 mM 3dNTP mix (2.5 mM each dATP, dGTP, and dTTP)
1 mM cyanine-3- and cyanine-5-dCTP (Cy3- and Cy5-dCTP; Amersham Pharmacia Biotech)

5 U/μl Klenow fragment (exo$^-$) (Stratagene)
1 μg/μl mouse Cot-1 DNA
✓ 3 M sodium acetate, pH 5.2
✓ Hybridization buffer
✓ TE buffer
✓ 2× SSC
0.2× and 2× SSC/0.1% SDS
Microarray (commercial or custom made; see Internet Resources)

NOTE: Unless otherwise mentioned, all reagents are from Invitrogen Life Technologies.

Day 1

1. For each of the two targets to be compared, start synthesis of first DNA strand by preparing an RNA/oligonucleotide mix (total 34 μl) comprising the following components:

    2.5 μl 2 μg poly(A)$^+$ RNA or 50 μg total RNA in TE buffer (sample)
    2.5 μl 0.5 μg/μl oligo(dT)$_{12-18}$
    29 μl DEPC-treated H_2O.

    Incubate RNA/oligonucleotide mixes 10 min at 70°C to denature secondary RNA structures. Cool 5 min at room temperature to anneal primers to the RNA template, and microcentrifuge 10 sec at 1000 × *g*, room temperature.

2. Add the following components in the order indicated:

    10 μl 5× first-strand buffer
    0.5 μl 10 U/μl RNasin
    0.5 μl 0.1 M DTT
    2.5 μl 10 mM 4dNTP mix
    2.5 μl 200 U/μl SuperScript II reverse transcriptase.

    Incubate 1 hr at 42°C.

3. Place above mixture on ice and, to synthesize second strand, add the following components in the order indicated:

    20 μl 5× second-strand buffer
    1 μl 15 mM β-NAD
    0.5 μl 0.1 M DTT
    2.5 μl 10 U/μl *E. coli* DNA polymerase I
    0.6 μl 2 U/μl RNase H
    0.5 μl 10 U/μl *E. coli* DNA ligase
    24.9 μl DEPC-treated H_2O to final volume of 100 μl.

    Incubate 3 hr at 16°C.

4. Add 100 μl molecular-biology-grade 25:24:1 (v/v/v) phenol/chloroform/isoamyl alcohol and vortex 20 sec. Separate the aqueous and organic phases by microcentrifuging 2 min at top speed (12,000 to 14,000 × *g*), 4°C. Carefully collect the top (aqueous) phase using a micropipettor and transfer to a new microcentrifuge tube.

5. Add 1 vol chloroform to aqueous phase, vortex well, and repeat step 4.

6. Add the following reagents, mixing well after each addition:

    25 μl 10 M ammonium acetate
    1 μl 1 μg/μl glycogen carrier
    250 μl 100% molecular-biology-grade ethanol, −20°C.

Incubate at least 1 hr (usually 2 to 3 hr) at −20°C. Proceed to step 7 or store overnight at −20°C.

7. Precipitate cDNA by microcentrifuging at least 20 to 30 min at top speed, 4°C. Carefully decant supernatant. Wash pellet with 250 μl ice-cold 70% ethanol diluted with DEPC-treated water. Microcentrifuge 15 min at top speed, aspirate ethanol, and air dry pellet. If using a vacuum dryer, do not overdry.

8. Dissolve pellet (double-stranded cDNA) in 30 μl DEPC-treated H_2O and measure the concentration by reading the OD_{260} (*APPENDIX 2G*). Store unused cDNA up to 3 months at −20°C.

9. Combine the following components in a 1.6-ml microcentrifuge tube in the order indicated:

   23.4 μl (1 to 2 μg) double-stranded cDNA (step 9)
   10 μl 5× random primers (hexamers).

   Mix and incubate 5 min at 95°C to denature DNA, then chill on ice and microcentrifuge to collect solution in the bottom of the tube.

10. Using Cy3-dCTP to label one sample and Cy5-dCTP to label the other, add the following components in a darkened room (50 μl total volume per synthesis):

    10 μl 5× dCTP buffer
    2 μl 0.25 mM dCTP
    1.6 μl 2.5 mM 3dNTP mix
    1 μl 1 mM Cy3- *or* Cy5-dCTP
    2 μl 5 U/μl Klenow fragment (exo^-).

    Briefly vortex and incubate overnight in a 37°C water bath in the dark.

Day 2

11. Mix the Cy3- and Cy-5-dCTP-labeled cDNAs from step 10. Add the following components in the order indicated:

    10 μl 3 M sodium acetate, pH 5.2
    250 μl 100% ethanol.

    Incubate at least 1 hr (or up to overnight) at −20°C.

12. Microcentrifuge the cDNA for 30 min at top speed, 4°C. Decant supernatant. Wash pellet with ice-cold 70% ethanol. Microcentrifuge 15 min at top speed, 4°C. Aspirate ethanol and air dry pellet. Do not overdry.

13. Resuspend labeled cDNA in ~20 μl hybridization buffer, adjusting volume according to the array surface area (for a 24 × 50–mm microscope cover glass, use 30 μl probe solution). Let the pellet dissolve at least 15 to 20 min. Store up to 1 week at −70°C.

14. Denature labeled cDNA 2 to 5 min in a pre-equilibrated 95°C heating block. Transfer to a pre-equilibrated humidified hybridization oven or water bath and prehybridize the cDNA 30 to 60 min at 65°C.

15. If using a hybridization chamber, add 60 μl of 2× SSC into the chamber well, and place the microarray slide in the chamber over the well.

16. Carefully apply the cDNA suspension in one drop onto the hybridization area of the slide. Ensure that the solution covers the entire array surface (the region containing gridded

cDNA clones). Be careful not to touch the array surface with the micropipet tip when applying the hybridization mixture. To prevent bubbles, make sure there is no debris or dust on the hybridization surface before applying cover slip or hybridization mixture—e.g., by dusting the slide with a compressed air mixture.

17. Very carefully place the coverslip over the hybridization mixture using fingers or forceps. Start making contact between the edge of the cover slip and the hybridization solution on the slide, then slowly lay the coverslip down onto the hybridization area of the slide, to avoid bubbles and provide an even distribution of hybridization mixture across the array. Alternatively, gently overlay a coverslip onto the array hybridization area and apply the cDNA suspension (hybridization mixture) to the edge of the cover slip. The mixture is drawn underneath the cover slip by capillary action.

18a. *For standard hybridization chamber:* Seal the chamber tightly and incubate in a 65°C water bath overnight (not more than 16 hr). Be sure the array is positioned horizontally during transfer and incubation.

18b. *For Boekel hybridization oven or equivalent*: Carefully place the cover-slipped slide inside the oven according to manufacturer's instructions and seal the oven. Incubate at 65°C overnight (not more than 16 hr).

Day 3

19. Dip the slide with the coverslip in 2× SSC/0.1% SDS and let the cover slip detach into the solution.

20. Wash twice for 5 min each in Coplin jars containing 2× SSC/0.1% SDS buffer, room temperature, on an orbital shaker at low speed. Repeat the two washes, using 0.2× SSC/0.1%SDS.

21. To dry the slide, place it carefully in the grooves of a shallow covered slide box and centrifuge in a low-speed tabletop centrifuge with microtiter plate adaptor for 5 min at 500 × *g*, room temperature.

22. Scan the slide.

4

ALTERNATE PROTOCOL 1

DIRECT LABELING USING REVERSE TRANSCRIPTASE

This method is widely used and can be applied when smaller amounts of starting material limit the use of Klenow direct labeling.

Additional Materials *(also see Basic Protocol 1; see APPENDIX 1 for items with ✓)*

100 mM 3dNTP mix (100 mM each dATP, dGTP, and dTTP; store up to several months at −20°C)
100 mM dCTP
0.5 μg/μl oligo$(dT)_{18-20}$
1 mM cyanine-3- and cyanine-5-dCTP (Cy3-dCTP and Cy5-dCTP; Amersham Pharmacia Biotech)
✓ 20 mM EDTA, pH 8.0
0.5 M NaOH
0.5 M HCl
10 M ammonium acetate
100% isopropanol

1. Prepare 300 μl reverse transcription labeling mix by combining the following components in a 1.6-ml microcentrifuge tube:

   120 μl 5× first-strand buffer
   60 μl 5 mM DTT
   3 μl 100 mM 3dNTP mix
   0.6 μl 100 mM dCTP
   116.4 μl DEPC-treated H_2O.

   Store unused solution up to several weeks at −20°C.

2. For each of the two targets to be compared, add the following in a 1.6-ml microcentrifuge tube:

   1 to 2 μg poly$(A)^+$ RNA or 25 to 50 μg total RNA
   2 μl 0.5 μg/μl oligo$(dT)_{18-20}$
   DEPC-treated H_2O to 10 μl.

3. Incubate RNA/oligonucleotide mixture 10 min at 65°C to denature secondary RNA structures. Cool to room temperature 5 min to anneal the primers to the RNA template. Microcentrifuge 30 sec at 1000 × *g*, room temperature.

4. Using Cy3-dCTP to label one sample and Cy5-dCTP to label the other, add the following components to the annealed RNA/oligonucleotide mixture in the order indicated:

   15 μl reverse transcription labeling mix (step 1)
   3 μl 1 mM Cy3- *or* Cy5-dCTP
   2 μl 200 U/μl SuperScript II reverse transcriptase.

   Vortex briefly and incubate at least 3 hr at 42°C.

5. Place the tubes on ice 5 min, then add 1.5 μl of 20 mM EDTA, pH 8.0, and 1.5 μl of 0.5 M NaOH to stop cDNA synthesis and hydrolyze the RNA. Incubate 30 min at 65°C.

6. Place tube on ice 5 min and neutralize the solution by adding 1.5 μl of 0.5 M HCl.

7. Combine the Cy3 and Cy5-dCTP-labeled cDNAs (step 6) and add the following reagents, mixing well after each addition:

   2.5 μl 10 M ammonium acetate
   25 μl 100% isopropanol.

   Incubate at least 1 hr (usually 2 to 3 hr, or overnight, if preferred) at −20°C.

8. Precipitate cDNA and use for hybridization as described (see Basic Protocol 1, steps 12 to 17).

BASIC PROTOCOL 2

INDIRECT LABELING AND DETECTION OF cDNA USING TYRAMIDE SIGNAL AMPLIFICATION (TSA)

This process is more tedious than direct methods, but allows the use of 50 to 100 times less starting RNA. Nonlinearity is introduced and dye effects are magnified by signal amplification, so it is best to perform array hybridizations in duplicate, switching the dyes.

NOTE: NEN's TSA labeling and detection kit contains all the necessary reagents to check the quality of cDNA synthesis and the efficiency of incorporation of labeled nucleotide prior to the hybridization step. The protocol uses a semi-quantitative membrane-based colorimetric assay. Considering the costs of reagents and slides, control of cDNA quality is desirable for every microarray protocol.

Materials *(also see Basic Protocol 1 and Strategic Planning; see* APPENDIX 1 *for items with ✓)*

Total or poly(A)$^+$ RNA
✓ DEPC-treated H_2O
MICROMAX TSA Labeling and Detection Kit (NEN Life Science Products)
  Reaction mix concentrate
  Biotin- and fluorescein-labeled dCTP (B- and Fl-dCTP)
  Reaction buffer
  AMV reverse transcriptase/RNase inhibitor mix
  Q buffer
  Biotin- and fluorescein-labeled control cDNA
  ✓ 1× TE buffer
  ✓ 2× SSC
  Streptavidin-horseradish peroxidase conjugate (Streptavidin-HRP)
  Anti-Fl-HRP conjugate
  4CN Plus reagent
  Diluent working solution: 1 ml BB diluent in 9 ml H_2O; make fresh
  Cyanine-3- and -5-tyramide (Cy3- and Cy5-tyramide)
  Amplification diluent
  HRP inactivation reagent
  Nylon membrane
✓ 0.5 M EDTA, pH 8.0
1 N NaOH, fresh
✓ 1 M Tris·Cl, pH 7.5 (Invitrogen Life Technologies)
5 M ammonium acetate
100% isopropanol
70% (v/v) ethanol diluted with DEPC-treated water, ice cold
✓ TNB-G blocking buffer
✓ TNT buffer
0.5× and 0.06× SSC/0.01% (w/v) SDS
✓ 0.06× SSC
Dimethyl sulfoxide (DMSO)
✓ 3 M sodium acetate, pH 5.2

UV cross-linking apparatus (e.g., GS GeneLinker UV Chamber, Bio-Rad)
Large 24 × 50–mm commercial or custom made microarray (see Critical Parameters and Internet Resources)

Day 1

1. Using B-dCTP for one target and Fl-dCTP for the other, combine the following in separate 1.6-ml microcentrifuge tubes:

   0.5 to 10 μg total RNA or 100 ng poly(A)$^+$ RNA
   1 μl reaction mix concentrate
   1 μl B- *or* Fl-dCTP
   DEPC-treated H_2O to 20 μl.

   Incubate the tubes 10 min at 65°C to denature secondary RNA structures. Cool to room temperature 5 min to anneal primers to the RNA template.

2. Place each tube in a 42°C water bath. Add 2.5 μl reaction buffer and 2 μl AMV reverse transcriptase/RNase inhibitor mix. Incubate at least 3 hr at 42°C.

3. Place the tube on ice 5 min and add 2.5 μl of 0.5 M EDTA, pH 8.0, and 2.5 μl fresh 1 N NaOH to stop the reaction and hydrolyze the RNA. Mix components well by inverting the tube and incubate no longer than 30 min at 65°C.

4. Place the tube on ice 5 min and neutralize the solution by adding 6.5 μl of 1 M Tris·Cl, pH 7.5.

5. To precipitate labeled cDNA (thereby eliminated unincorporated labeled nucleotides), add 2.7 μl of 5 M ammonium acetate and 31 μl of 100% isopropanol. Vortex, and incubate 1 hr at 4°C. Microcentrifuge 30 min at top speed, 4°C, and carefully decant supernatant.

6. Wash pellet twice with 200 μl ice-cold 70% ethanol diluted with DEPC-treated water. Microcentrifuge 25 min at top speed, 4°C. Aspirate ethanol and dry the pellet under vacuum; do not overdry.

7. Resuspend in a suitable volume of Q buffer. For a large 24 × 50–mm commercial or custom microarray, resuspend each labeled pellet in 15 μl Q buffer for a total hybridization volume of 30 μl. Use 1 μl (100 ng) for cDNA analysis (steps 8 to 15). Store the remainder at −20°C in the dark for up to one week.

8. Make separate serial dilutions of each labeled control DNA by adding:

   2 μl B- or Fl-labeled control cDNA to 18 μl 1× TE buffer (1:10)
   5 μl 1:10 dilution to 5 μl 1× TE (1:20)
   5 μl 1:20 dilution to 5 μl 1× TE (1:40)
   5 μl 1:40 dilution to 5 μl 1× TE (1:80).

9. Make serial dilutions of each experimental cDNA by adding:

   1 μl experimental cDNA (step 7) to 9 μl 1× TE (1:10)
   5 μl 1:10 dilution to 5 μl 1× TE (1:20)
   5 μl 1:20 dilution to 5 μl 1× TE (1:40).

10. Apply 1 μl of each control and test dilution in duplicate on a nylon membrane. Air dry.

11. Wet the membrane in 2× SSC and cross-link using 1.2×10^5 μJ from a UV cross-linking apparatus in order to covalently attach cDNA to the support.

12. Block the membrane 30 min in 3.5 ml TNB-G blocking buffer.

13. Dilute 20 μl streptavidin-HRP and 20 μl anti-Fl-HRP in 3.5 ml TNB-G blocking buffer and mix. Remove blocking buffer from membrane, add diluted HRP conjugate solution, and incubate 30 min at room temperature. Wash four times for 5 min each with TNT buffer and air dry the filter.

14. Mix 0.2 ml 4CN Plus reagent and 10 ml diluent working solution and add to the membrane. Incubate 30 min in the dark and observe spots.

    Distinct signals should be visible from all three serial dilutions. The 1:40 experimental sample dilutions should be of the same intensity as the 1:80 control dilutions. If all spots are visible, proceed with hybridization. If two or fewer experimental spots are visible, repeat the dot-blot analysis. If similar results are obtained, perform another cDNA synthesis reaction.

15. Combine equal amounts of experimental B- and Fl-labeled cDNA samples, mix well, and denature secondary structures by heating 2 min at 90°C. Immediately add to the array, apply the cover slip, and perform hybridization (see Basic Protocol 1, steps 15 to 18).

Day 2

16. Vertically immerse the slide in 0.5× SSC/0.01% SDS and let the cover slip fall off. Place the slide in a new Coplin jar with the same buffer and wash the microarray 5 min with gentle agitation on an orbital shaker. (Minimize drying during all washing steps.)

17. Wash 5 min in 0.06× SSC/0.01% SDS and decant. Wash 2 min in 0.06× SSC (without SDS) and decant.

18. Incubate slide 10 min with 600 µl TNB-G blocking buffer. Wash 1 min with TNT buffer in a Coplin jar.

    Perform all incubation steps at room temperature in a covered container with wet blotting paper underneath the slide without agitation. It is important to keep the slide surface from drying out during incubation.

19. Dilute 4 µl anti-Fl-HRP conjugate in 400 µl TNB-G blocking buffer. Decant TNT from slide, apply diluted conjugate to hybridization area, completely covering the area, and incubate 10 min at room temperature, in covered container out of direct light.

20. Remove solution by gently tilting the slide, and dry the sides of the slide by capillary action with a dust-free Kimwipe. Do not touch the hybridization area.

21. Wash three times (all washes with agitation at low speed), 1 min each, in Coplin jars containing TNT buffer, then decant.

22. Dissolve 1 µl Cy3-tyramide diluted in 20 µl DMSO into 500 µl amplification diluent. Apply to hybridization area and incubate 10 min. Wash three times, 5 min each, in a Coplin jar containing TNT buffer, then decant.

23. Add 10 µl of 3 M sodium acetate, pH 5.2, to 290 µl HRP inactivation reagent, apply to hybridization area, and incubate 10 min. Wash three times, 1 min each, in Coplin jars containing TNT buffer, then decant.

24. Perform biotin detection by repeating steps 19 to 20 but using anti-FL-HRP conjugate and Cy5-tyramide.

25. Wash 1 min in a Coplin jar containing 0.06× SSC, then decant.

26. Centrifuge the slide 5 min at 500 × *g*, room temperature.

27. Scan the slide.

ALTERNATE PROTOCOL 2

INDIRECT LABELING AND DETECTION OF cDNA USING PCR AMPLIFICATION

Additional Materials *(also see Basic Protocol 1)*

10 U/µl *Dpn*II and 10× buffer (NEB)
T4 DNA ligase and 10× buffer (NEB)
10 M ammonium acetate
10 mM ATP
2 µg/µl custom-synthesized oligonucleotide adaptors R1 and R2 (*APPENDIX 2A*):
R1: 5′-AGCACTCTCCAGCCTCTCACCGCA-3′
R2: 5′-GATCTGCGGTGAGAGGCTGGAGAGTGCT-3′
10× PCR reaction buffer (Qiagen)
25 mM $MgCl_2$ (Qiagen)
10 mM 4dNTP mix (10 mM each dATP, dCTP, dGTP, and dTTP)
5 U/µl *Taq* DNA polymerase (Qiagen) *or* Ampli *Taq* DNA polymerase (Perkin-Elmer)
Small thin-walled PCR tubes
Thermal cycler (e.g., Perkin-Elmer 9600 PCR machine with heated lid)

Day 1

1. Synthesize double-stranded cDNA as described (see Basic Protocol 1, steps 1 to 8).

Day 2

2. In a 1.6-ml microcentrifuge tube, combine the following reagents for DNA digestion (also see *CPNS APPENDIX 1M*):

   10 µl (2 µg) double-stranded cDNA (step 1)
   10 µl 10× *Dpn*II buffer
   78 µl DEPC-treated H_2O
   2 µl 10 U/µl *Dpn*II.

   Incubate the reaction mixture 2 hr at 37°C.

3. Confirm cDNA digestion by electrophoresing 2 µl of 0.2 µg/µl undigested cDNA (step 1) and 20 µl of 0.02 µg/µl digested cDNA (step 2) in parallel on a 1% agarose gel using 1× TAE electrophoresis buffer (*CPNS APPENDIX 1N*). If digestion is complete, proceed to step 4.

4. Add 1 vol 25:24:1 (v/v/v) phenol/chloroform/isoamyl alcohol to the digested cDNA, and vortex 20 sec. Separate the aqueous and organic phases by microcentrifuging 2 min at top speed, 4°C. Carefully collect the top (aqueous) phase and transfer to a new microcentrifuge tube. Add 1 vol chloroform and repeat.

5. Add the following reagents, mixing well after each addition:

   25 µl 10 M ammonium acetate
   1 µl 1 µg/µl glycogen carrier
   250 µl 100% ethanol, −20°C.

   Incubate at least 1 hr (usually 2 to 3 hr) at −20°C. Proceed to step 6 or store samples overnight at −20°C.

6. Precipitate *Dpn*II-digested cDNA by microcentrifuging at least 30 min at top speed, 4°C. Remove the supernatant. Wash the pellet with ice-cold 70% ethanol. Microcentrifuge 15 min, aspirate ethanol, and air dry pellet. Do not overdry. Resuspend *Dpn*II-digested cDNA in 37 µl water.

7. Combine the following components:

   37 µl *Dpn*II-digested cDNA
   6 µl 10× ligase buffer
   6 µl 10 mM ATP
   4 µl 2 µg/µl R1 adaptor
   4 µl 2 µg/µl R2 adaptor.

   Adaptors consist of two oligonucleotides: a 24-mer (R1) and complementary 28-mer (R2). Annealing of R1 and R2 oligonucleotides creates an overhang, which is complementary to the DpnII restriction site.

   Preheat the reaction mixture in a heating block 2 min at 55°C. Anneal the adaptors by cooling the heating block to a 10°C in a cold room and incubating 1 hr.

8. Add 3 µl of 400 U/µl T4 DNA ligase to the reaction mixture and incubate 12 to 18 hr (overnight) at 16°C. After ligation, bring the final volume to 100 µl with molecular-biology-grade water.

Day 3

9. Add the following components to a thin-walled PCR tube in the order indicated:

   5 μl 10× PCR reaction buffer
   8 μl 25 mM $MgCl_2$
   1.25 μl 10 mM 4dNTP mix
   0.25 μl 5 U/μl *Taq* or Ampli *Taq* DNA polymerase
   34.25 μl DEPC-treated H_2O
   0.25 μl 2 μg/μl R1 adaptor
   1 μl cDNA/R1/R2 ligation mixture (step 8).

   Alternatively, when running several reactions at once, make a master mix containing all common components, minus the cDNA/R1/R2 ligation mixture. Use the same amount of master mix in each reaction to minimize variation between the PCR reactions.

10. Heat a thermal cycler to 72°C, insert the tubes, and start the following program:

Initial step:	5 min	72°C	(extension)
30 cycles (initial):	1 min	95°C	(denaturation)
	3 min	72°C	(annealing and extension)
Final step:	indefinite	4°C	(hold).

   In the initial experiment, withdraw 5-μl aliquots from the PCR reaction every two cycles starting at cycle 12 (i.e., cycle 12, 14, 16 . . . 30) and analyze on an agarose gel as described (step 11).

11. In the initial experiment (not necessary for subsequent experiments), electrophorese the PCR products (amplicons) on a 1.2% agarose gel using 1× TAE electrophoresis buffer (*CPNS APPENDIX 1N*). Analyze overall intensity to determine the plateau of product formation. Adjust the number of PCR cycles in subsequent experiments so that the total number is two less than the plateau phase to optimize the likelihood of obtaining an unbiased cDNA sample.

12. Extract cDNA as in step 4.

13. Add the following reagents, vortex, and incubate the sample 2 hr at −20°C.

    1/10 vol 3 M sodium acetate, pH 5.2
    1 μl 1 μg/μl glycogen carrier
    2.5 vols 100% ethanol.

14. Precipitate amplicons by microcentrifuging 30 min at top speed, 4°C. Remove the supernatant. Wash the pellet with ice-cold 70% ethanol, microcentrifuge briefly, and vacuum dry.

15. Resuspend amplicons in 250 μl water and measure the concentration by reading the OD_{260} (*APPENDIX 2G*). Use amplicons for labeling and hybridization. Store unused amplicons at −20°C in the dark.

Internet Resources

General information and protocols

http://www.nhgri.nih.gov/DDIR/LCG/15K/HTML/protocol.html

RNA isolation and cDNA probe labeling protocols from the National Human Genome Research Institute.

4

http://cmgm.stanford.edu/pbrown/protocols/index.html
Experimental protocols available from Stanford University.

http://www.microarrays.org
The microarrays.org main page. A public source for microarray protocols and software. Maintained by the DeRisi Laboratory, University of California at San Francisco.

http://genomics.med.upenn.edu/vcheung/protocols.htm
Protocols available from the University of Pennsylvania.

http://www.tigr.org/tdb/microarray
The Institute for Genomic Research (TIGR) microarray resources.

http://ihome.cuhk.edu.hk/~b400559/array.html
Y.F. Leung's Functional Genomics Web site.

http://arrayit.com/DNA-Microarray-Protocols
The TeleChem International/arrayit.com home page.

Microarray image and data analysis software resources
http://www.tigr.org/softlab
ARRAY-VIEWER software from TIGR. Also see above.

http://ep.ebi.ac.uk
Expression Profiler available from the EMBL, European Bioinformatics Institute (EBI).

4

http://www.cbil.upenn.edu/PaGE
Patterns from Gene Expression (PaGE) software from the University of Pennsylvania.

http://www.microarrays.org/software.html
SCAN-ALYZE available from microarray.org, maintained by Lawrence Berkeley National Laboratory.

http://rana.lbl.gov/EisenSoftware.htm
Eisen laboratory software page at Stanford University.

http://www.rosettabio.com/products/resolver/default.htm
The Rosetta Resolver System product page.

http://imaging.brocku.ca/products/Genomics_Software.asp
Genomics software from Imaging Research.

http://www.biodiscovery.com
ImaGene and GeneSight software from BioDiscovery.

http://www.sigenetics.com
GeneSpring available from Silicon Genetics.

http://www.spotfire.com
Spotfire DecisionSite for Functional Genomics.

Data handling and interpretation
http://www.microarrays.ca
Microarray Center at University Health Network, Ontario Cancer Institute.

http://lgsun.grc.nia.nih.gov
NIA/NIH Mouse Genomics Home Page.

http://www.genetics.ucla.edu/horvathlab/adag.htm
Home page of the Array Data Analysis Group at the University of California, Los Angeles.

http://www.cbil.upenn.edu/PaGE
Penn Center for Bioinformatics at the University of Pennsylvania.

http://www.mged.org
Home page of the Microarray Gene Expression Database (MGED) Group.

http://ep.ebi.ac.uk
The European Bioinformatics Institute (EMBL) Web site.

Partial list of academic microarray core facilities
http://public.bcm.tmc.edu/mcfweb
The Microarray Core Facility at Baylor College of Medicine, Houston, Tex.

http://www.genetics.ucla.edu/microarray
Microarray at UCLA hosted by the University of California at Los Angeles School of Medicine.

http://genome.genetics.duke.edu
The Duke Genome Core Facility at the Comprehensive Cancer Center, Duke University, Durham, North Carolina

http://cgr.harvard.edu/services/services.html
The Harvard Center for Genomic Research (CGR), Molecular and Cellular Department, Cambridge, Mass.

http://minihelix.mit.edu
The Bioinformatics and MicroArraying Center, Massachusetts Institute of Technology (MIT) Center for Cancer, Department of Biology and Biological Engineering, Cambridge, Massachusetts

http://sgio2.biotec.psu.edu
Penn State Microarray Facility sponsored by the Life Sciences Consortium.

http://www.rockefeller.edu/genearray
The Rockefeller University Gene Array Resource Center, New York.

http://genome-www4.stanford.edu/cgi-bin/sfgf/home.pl
Stanford University School of Medicine Functional Genomics Facility, Stanford, Calif.

http://info.med.yale.edu/wmkeck/dna_arrays.htm
The HHMI Biopolymer/Keck Foundation Biotechnology Resource Laboratory at the Yale University School of Medicine, New Haven, Conn.

Public microarray databases
http://www.ebi.ac.uk/arrayexpress
The ArrayExpress Database at EMBL-EBI.

http://pevsnerlab.kennedykrieger.org/dragon.htm
The Pevsner Laboratory Dragon Database at the Kennedy Krieger Institute.

http://www.ncbi.nlm.nih.gov/geo
Gene Expression Omnibus at NCBI.

http://www.ncgr.org/genex
GeneX at the National Center for Genome Resources hosted by the University of California at Irvine.

http://genome-www5.stanford.edu/MicroArray/SMD
The Stanford Microarray Database.

http://info.med.yale.edu/microarray
The Yale Microarray Database (YMD).

Partial list of commercial sources of microarrays
http://www.affymetrix.com
The Affymetrix home page.

http://we.home.agilent.com
The Agilent Technologies home page.

http://www.clontech.com
Home page of the Clontech division of Becton Dickenson.

http://www.incyte.com
The Incyte Genomics Web site.

http://www.invitrogen.com
The Invitrogen home page.

http://www.motorola.com/lifesciences
The Motorola Life Sciences Web site.

http://www.operon.com
The Qiagen Operon Web site.

http://lifesciences.perkinelmer.com
The PerkinElmer Life Sciences Web page.

4

References: DeRisi et al., 1997; Geschwind, 2001; Luo et al., 1999; Mirnics et al, 2000; Pomeroy et al., 2002; Sandberg et al., 2000; Schena et al., 1995; Yue et al., 2001

Contributors: Stanislav L. Karsten and Daniel H. Geschwind

Table 4.18.2 Troubleshooting Guide for cDNA Microarray Analysis

Problem	Possible cause	Solution
Poor cDNA yield	Poor RNA quality	Analyze an aliquot of starting RNA on a 1% denaturing agarose gel (*UNIT 5.17* and *APPENDIX 1N*). Clear and distinct rRNA bands should be observed following ethidium bromide staining with a 2:1 ratio of 28S:18S species. If not, RNA may be degraded. Isolate new RNA.
	Low amount of RNA in the reaction	Measure concentration of RNA again and make corresponding adjustments.
	RNA was not properly dissolved	Incubate RNA 20 min at 42°C with occasional mixing and measure RNA concentration again. Adjust amounts according to new readings.
	RNA was contaminated with RNases	Incubate an aliquot 1 hr at room temperature and analyze on a denaturing agarose gel. If degradation occurred, isolate new RNA and repeat the reaction in a nuclease-free environment using RNase inhibitor, fresh reagents, and buffers.
	Enzyme is partially inactivated	Repeat the reaction with fresh enzyme.

continued

Table 4.18.2 Troubleshooting Guide for cDNA Microarray Analysis, continued

Problem	Possible cause	Solution
Poor cDNA labeling	Low amount of cDNA in the reaction	Quantify cDNA and adjust the volumes according to the protocol.
	Modified nucleotides are partially degraded (old dye)	Repeat the reaction with new nucleotides.
Air bubble under the cover slip	Surface of cover slip was not clean	Try to remove the bubble by very gently tapping on top of the cover slip with a micropipet tip. Do not move the cover slip. Most often the bubbles tend to be released during overnight incubation at 65°C.
Hybridization buffer dries out during overnight incubation	Low level of humidity	Increase the volume of 2× SSC and make sure that the chamber remains humid overnight. Start the entire process over with a new microarray.
		Use another hybridization oven or a water bath. Use a new microarray.
Cover slip slides off the array during hybridization	Hybridization oven is not properly installed	Make sure that the hybridization oven is level (i.e., horizontal).
	Too much hybridization solution; the cover slip is floating	Use less solution
No signal after hybridization	Insufficient probe denaturation prior to hybridization	Use new probes and slide
	cDNA synthesis or labeling did not work	Refer to Commentary sections about poor cDNA synthesis and labeling
	Hybridization probe degraded	All probe handling should be performed in a nuclease-free environment
Low signal/high noise	Poor starting RNA quality	Isolate new RNA and check its quality on a denaturing gel
	Poor probe	Make sure cDNA synthesis and labeling reactions worked properly. Refer to Commentary sections on poor cDNA synthesis and labeling.
	Buffers contain impurities	To identify which buffer or step is problematic, scan slide at each step of the process. Filter all buffers and start the entire process over with a new microarray.
	Defect in the array	The array was not properly blocked post-arraying or slide coating was flawed pre-arraying. Prescan slides at each step (e.g., prior to arraying or hybridization) in order to measure inherent background. May need to rearray or obtain new slides.
High background/low signal near array corners	Partial drying of probe during hybridization	Insufficient amount of moisture in the hybridization vessel. Increase the volume of 2× SSC and ensure that the chamber remains humid overnight.

4

Overview of Gene Targeting by Homologous Recombination

The analysis of mutant organisms and cell lines has been important in determining the function of specific proteins. Until recently, mutants were produced by mutagenesis followed by selection for a particular phenotypic change. Recent technological advances in gene targeting by homologous recombination in mammalian systems enable the production of mutants in any desired gene (Mansour, 1990; Robertson, 1991; Zimmer, 1992). This technology can be used to produce mutant mouse strains and mutant cell lines. In most cases, both alleles of a gene in a diploid mammalian cell must be inactivated to produce a discernible phenotypic change in a mutant. The conversion from heterozygosity to homozygosity is accomplished by breeding in the case of mouse strains and by direct selection in cell lines.

Bacteriophage recombinases such as Cre and its recognition sequence, *loxP*, have also allowed spatial control of knockouts. Another recombinase system, the yeast Flp/FRT system, can also be used (Fiering et al., 1993, 1999). The control can function along actual spatial coordinates when a viral gene transfer system is used, or in a cell type- or tissue-specific fashion when restricted promoters are employed. Adding temporal regulation of Cre, such as that achievable with the tetracycline-regulatable expression system (see *APPENDIX 2A; CPMB UNIT 16.21*), allows temporal control as well.

To produce a mutant mouse strain by homologous recombination, two major elements are needed: an embryonic stem (ES) cell line capable of contributing to the germ line, and a targeting construct containing target-gene sequences with the desired mutation. Maintaining ES cells in their undifferentiated state is a major task during gene targeting (*APPENDIX 2A; CPMB UNIT 23.3*). This usually is accomplished by growing cells on a layer of feeder cells (*APPENDIX 2A; CPMB UNIT 23.2*). The targeting construct is then transfected into cultured ES cells (*CPNS UNIT 4.30*). ES cell lines are derived from the inner cell mass of a blastocyst-stage embryo. Homologous recombination occurs in a small number of the transfected cells, resulting in introduction of the mutation present in the targeting construct into the target gene. Once identified, mutant ES cell clones can be microinjected into a normal blastocyst in order to produce a chimeric mouse. Because many ES cell lines retain the ability to differentiate into every cell type present in the mouse, the chimera can have tissues, including the germ line, with contribution from both the normal blastocyst and the mutant ES cells. Breeding germ-line chimeras yields animals that are heterozygous for the mutation introduced into the ES cell, and that can be interbred to produce homozygous mutant mice.

Homologous recombination can also be used to produce homozygous mutant cell lines. Previously, inactivation of both alleles of a gene required two rounds of homologous recombination and selection (te Riele et al., 1990; Cruz et al., 1991; Mortensen et al., 1991). Now, however, inactivation of both alleles of many genes requires only a single round of homologous recombination using a single targeting construct (Mortensen et al., 1992). The homozygous mutant cells can then be analyzed for phenotypic changes to determine the function of the gene.

ANATOMY OF TARGETING CONSTRUCTS

Two basic configurations of constructs are used for homologous recombination: insertion constructs and replacement constructs (Fig. 4.19.1). Each can be used for different purposes in specific situations, as discussed below. The insertion construct contains a region of homology to

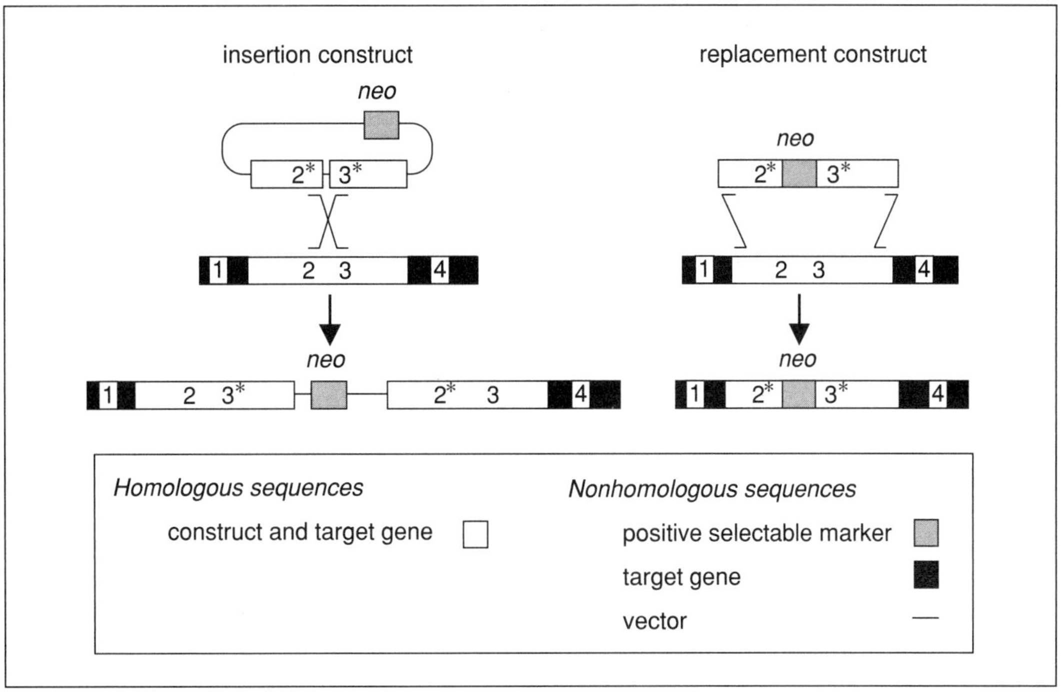

Figure 4.19.1 Two configurations of constructs used for homologous recombination. Numbers indicate target-gene sequences in the genome. An asterisk indicates homologous target-gene sequences in the construct. Replacement constructs substitute their sequences (2*, *neo*, and 3*) for the endogenous target-gene sequences (2 and 3). Insertion constructs add their sequences (2*, *neo*, and 3*) to the endogenous target gene, resulting in tandem duplication and disruption of the normal gene structure.

the target gene cloned as a single continuous sequence, and is linearized by cleavage of a unique restriction site within the region of homology. Homologous recombination introduces the insertion construct sequences into the homologous site of the target gene, interrupting normal target-gene structure by adding sequences. As a result, the normal gene can be regenerated from the mutated target gene by an intrachromosomal recombination event.

The replacement construct is the second, more commonly used construct. It contains two regions of homology to the target gene located on either side of a mutation (usually a positive selectable marker; see below). Homologous recombination proceeds by a double cross-over event that replaces the target-gene sequences with the replacement-construct sequences. Because no duplication of sequences occurs, the normal gene cannot be regenerated.

METHODS OF ENRICHMENT FOR HOMOLOGOUS RECOMBINANTS

Positive Selection by Drug-Resistance Gene

Nearly all constructs used for homologous recombination rely on the positive selection of a drug-resistance gene (e.g., neomycin or neo) that is also used to interrupt and mutate the target gene. When either insertion or replacement constructs are linearized, the drug-resistance gene is flanked by two regions of homology to the target gene. Selection of the cells using drugs (e.g., G418) eliminates the great majority of transformants that have not stably incorporated the construct (*APPENDIX 2A; CPMB UNIT 9.5*). However, in many of the surviving clones the construct has been incorporated into the genome not by homologous recombination, but through random integration. Therefore, methods to enrich for homologous recombinant clones have been developed.

Positive-Negative Selection

The most commonly used method for eliminating cells in which the construct integrated into the genome randomly, thus further enriching for homologous recombinants, is known as positive-negative selection. It is only applicable to replacement constructs (Fig. 4.19.2; Mansour et al., 1988). In these constructs, a negative selectable marker (e.g., herpes simplex virus thymidine kinase, HSV-TK) is included outside the region of homology to the target gene. In the presence of the TK gene, the cells are sensitive to acyclovir and its analogs (e.g., gancyclovir, GANC). The HSV-TK enzyme activates these drugs, resulting in their incorporation into growing DNA, causing chain termination and cell death. During homologous

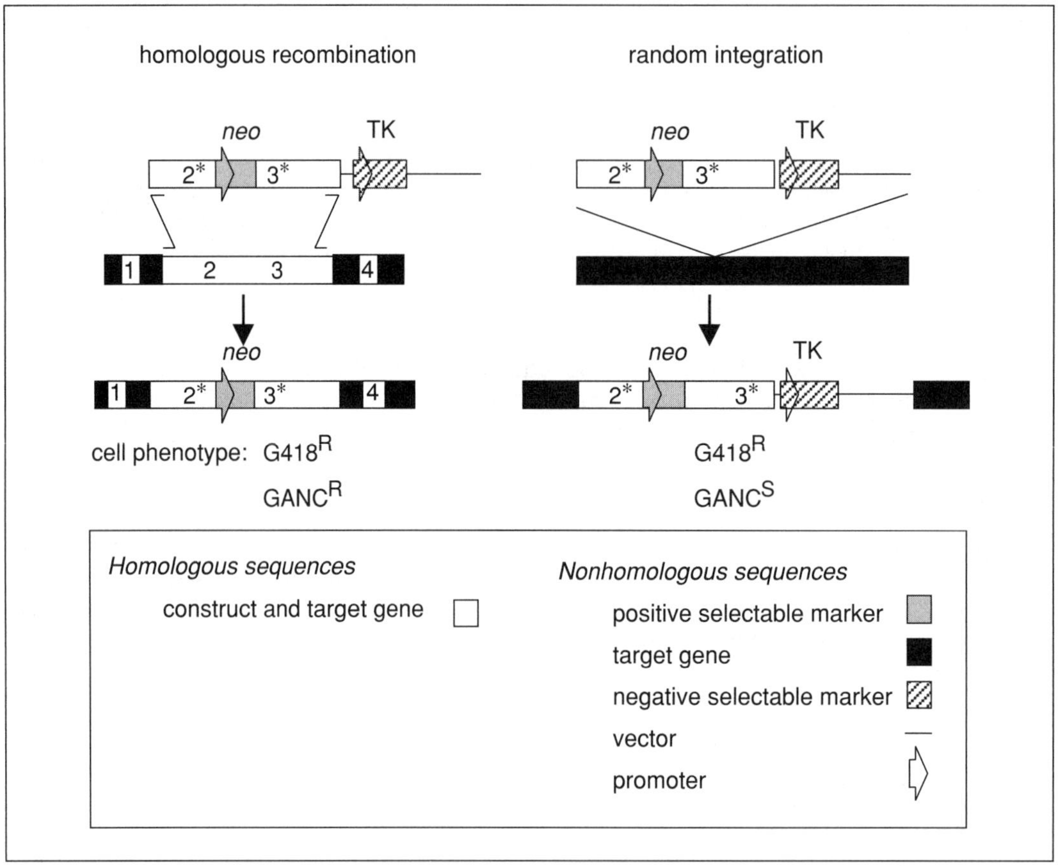

Figure 4.19.2 Enrichment for homologous recombinants by positive-negative selection using the TK gene. Homologous recombination involving cross-overs on either side of the *neo* gene results in loss of the TK gene. Random integration tends to preserve the TK gene, which can be selected against because any cell expressing the gene will be killed by gancyclovir (GANC). Although both homologous recombinants and clones in which the construct integrated randomly are G418-resistant, only homologous recombinants are gancyclovir-resistant. The construct is shown linearized so that the plasmid vector sequences remain attached to the TK gene. This configuration helps preserve the integrity of the TK gene. Superscript R, resistance; superscript S, sensitivity.

recombination, sequences outside the regions of homology to the target gene are lost due to crossing over. In contrast, during random integration all sequences in the construct tend to be retained because recombination usually occurs at the ends of the construct. The presence of the TK gene can be selected against by growing the cells in gancyclovir; the homologous recombinants will be G418-resistant and gancyclovir-resistant, whereas clones in which the construct integrated randomly will be G418-resistant and gancyclovir-sensitive. In some cases, TK is inactivated without homologous recombination; thus, the gancyclovir-resistant clones must be screened to identify the true homologous recombinants. Other markers that are lethal to cells have also been used instead of TK and gancyclovir (e.g., diphtheria toxin; Yagi et al., 1990).

Endogenous Promoters

Constructs that rely on an endogenous promoter to express the positive selectable marker can also cause enrichment of homologous recombinants (Fig. 4.19.3), but they can only be used

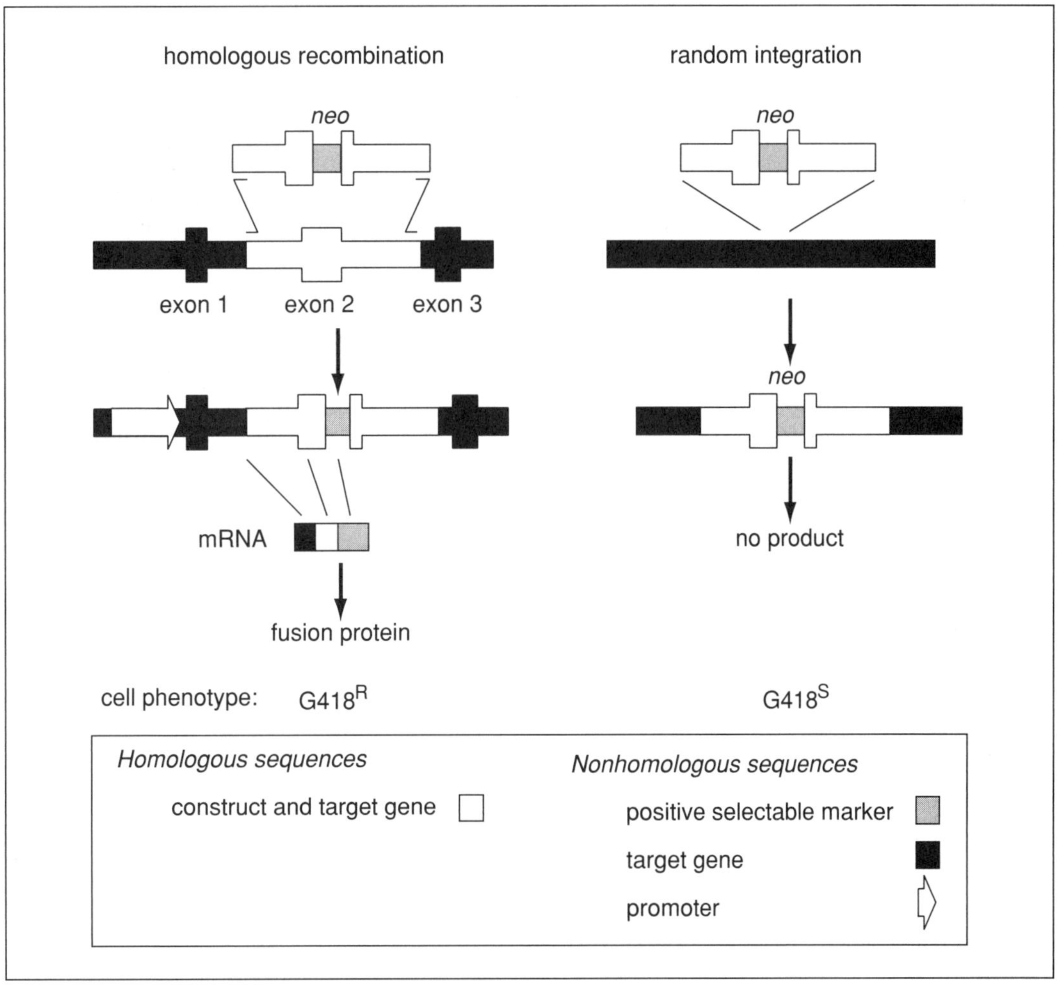

Figure 4.19.3 Enrichment for homologous recombinants using a positive selectable marker *(neo)* lacking a promoter. The construct is designed so that homologous recombination will provide an endogenous promoter leading to *neo* expression (and therefore G418 resistance), whereas random integration will most likely not provide a promoter, thus precluding *neo* expression.

if the gene of interest is expressed in the cell line. Such constructs contain the coding region of a selectable marker (e.g., *neo*) but lack a promoter for the marker. The coding sequence for the marker usually interrupts, and is in frame with, an exon of the target gene. Thus, when homologous recombination occurs, a fusion protein is produced, driven by the endogenous target-gene promoter. In contrast, when random integration occurs, the selectable-marker protein is not usually produced. Therefore, homologous recombinants are G418-resistant, whereas cells in which the construct integrated randomly are G418-sensitive. Constructs containing a promoterless selectable marker can be constructed in either replacement or insertion structure and can result in dramatic enrichment for homologous recombinants.

TYPES OF MUTATIONS

Gene Inactivation

Homologous recombination has most often been used to completely inactivate a gene (commonly termed "knockout"). Usually, an exon encoding an important region of the protein (or an exon 5′ to that region) is interrupted by a positive selectable marker (e.g., *neo*), preventing the production of normal mRNA from the target gene and resulting in inactivation of the gene.

A gene may also be inactivated by creating a deletion in part of a gene, or by deleting the entire gene. By using a construct with two regions of homology to the target gene that are far apart in the genome, the sequences intervening the two regions can be deleted. Up to 15 kb have been deleted in this way; thus, many genes could be completely eliminated (Mombaerts et al., 1991). Gene inactivations may also be controlled using the Cre/*loxP* recombinase system either spatially, as in cell type- or tissue-specific knockout, or temporally, through control of the activity or expression of the recombinase (see Cre/*loxP* System, below).

Mutations can be introduced that have multiple purposes. Homologous recombination has been used to introduce a replacement construct containing the coding sequence of β-galactosidase in frame with the 5′ end of the target gene. Downstream of the *lacZ* gene is a positive selectable marker driven by a heterologous promoter (Fig. 4.19.4). This construct not only disrupts target-gene function but also expresses a fusion protein with β-galactosidase activity, and thus can be used to monitor the activity of the endogenous gene's promoter in various tissues during development (Mansour et al., 1990).

Subtle Gene Mutations

Homologous recombination can also be used to introduce subtle mutations in a gene. One method is analogous to the transplacement, or allele replacement, method used with yeast. It is called "hit and run" (see *APPENDIX 2A; CPMB UNIT 13.10*) because duplications are introduced into the target gene and then removed. An insertion construct containing both positive and negative selectable markers (e.g., *neo* and TK) is used to introduce a duplication that contains a subtle mutation, such as a point mutation, into the target gene sequence (Fig. 4.19.5). After selection for integration of the construct via the positive selectable marker (e.g., G418), homologous recombinants are identified by screening. A homologous recombinant clone is cultured and then the presence of the negative selectable marker is selected against (e.g., selection against TK using gancyclovir). This leads to selection for an intrachromosomal recombination that eliminates the target-gene duplications and the selectable markers but leaves the mutant target-gene sequences substituting for the normal target-gene sequences. Surviving clones are screened for the correct intrachromosomal rearrangements. A second method of introducing

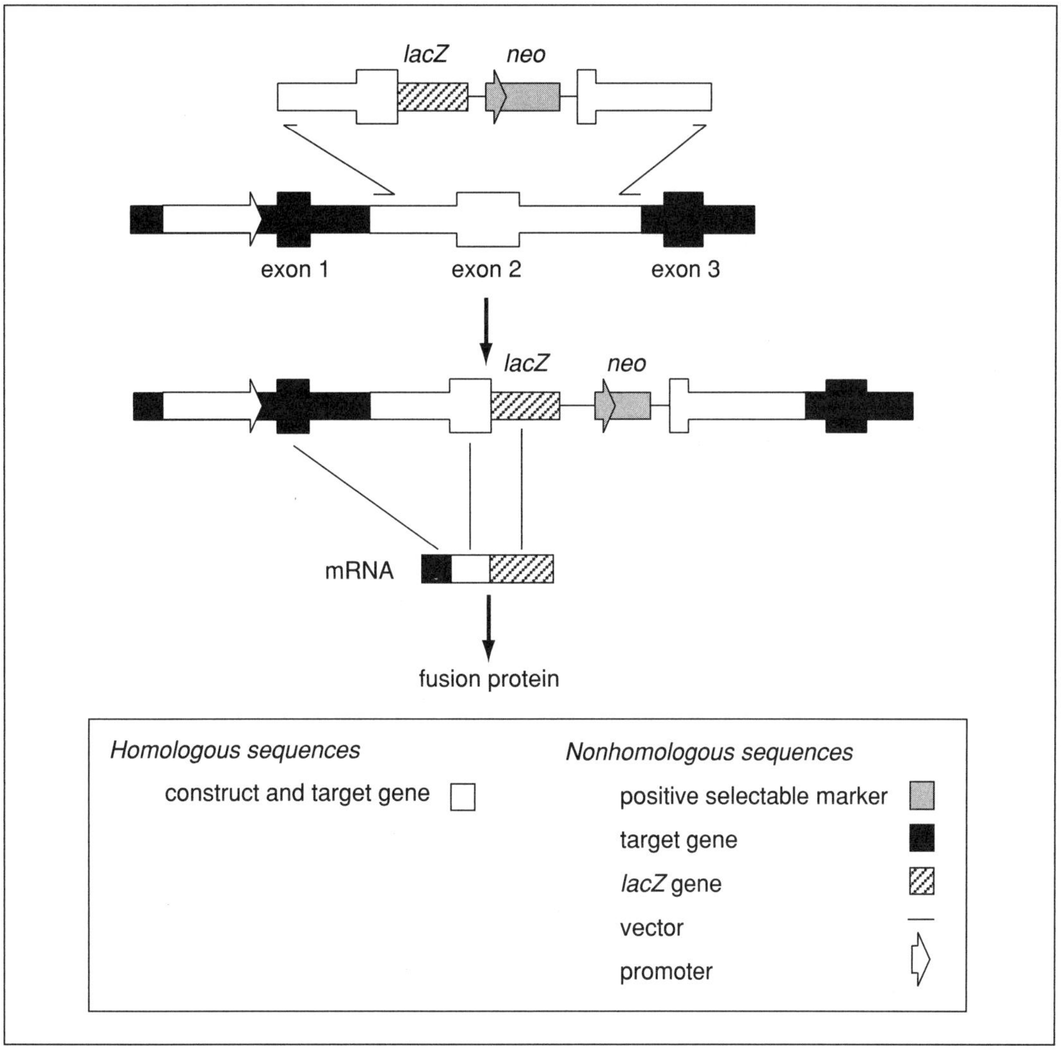

Figure 4.19.4 *LacZ* reporter construct for gene targeting. This construct has two purposes: to disrupt the target gene, and to express the *lacZ* gene as a marker of activity of the endogenous target gene's promoter.

subtle mutations into a gene is to insert the mutation by homologous recombination and then use the Cre/*loxP* system to remove the selectable marker.

CRE/*loxP* SYSTEM

The Cre/*loxP* system is derived from the bacteriophage P1. The recombinase Cre acts on the DNA site *loxP*. If there are two *loxP* sites in the same orientation near each other, Cre can act to loop out the sequence between the two sites, leaving a single *loxP* site in the original DNA and a second *loxP* in a circular piece of DNA containing the intervening sequence. Therefore, a properly designed targeting construct containing *loxP* sites can be used for introducing subtle mutations or for a temporally or spatially controlled knockout (for a review of the control of transgenes, see Sauer, 1993). Other recombinase systems, such as the Flp/FRT system, can be similarly useful (Fiering et al., 1995; Vooijs et al., 1998).

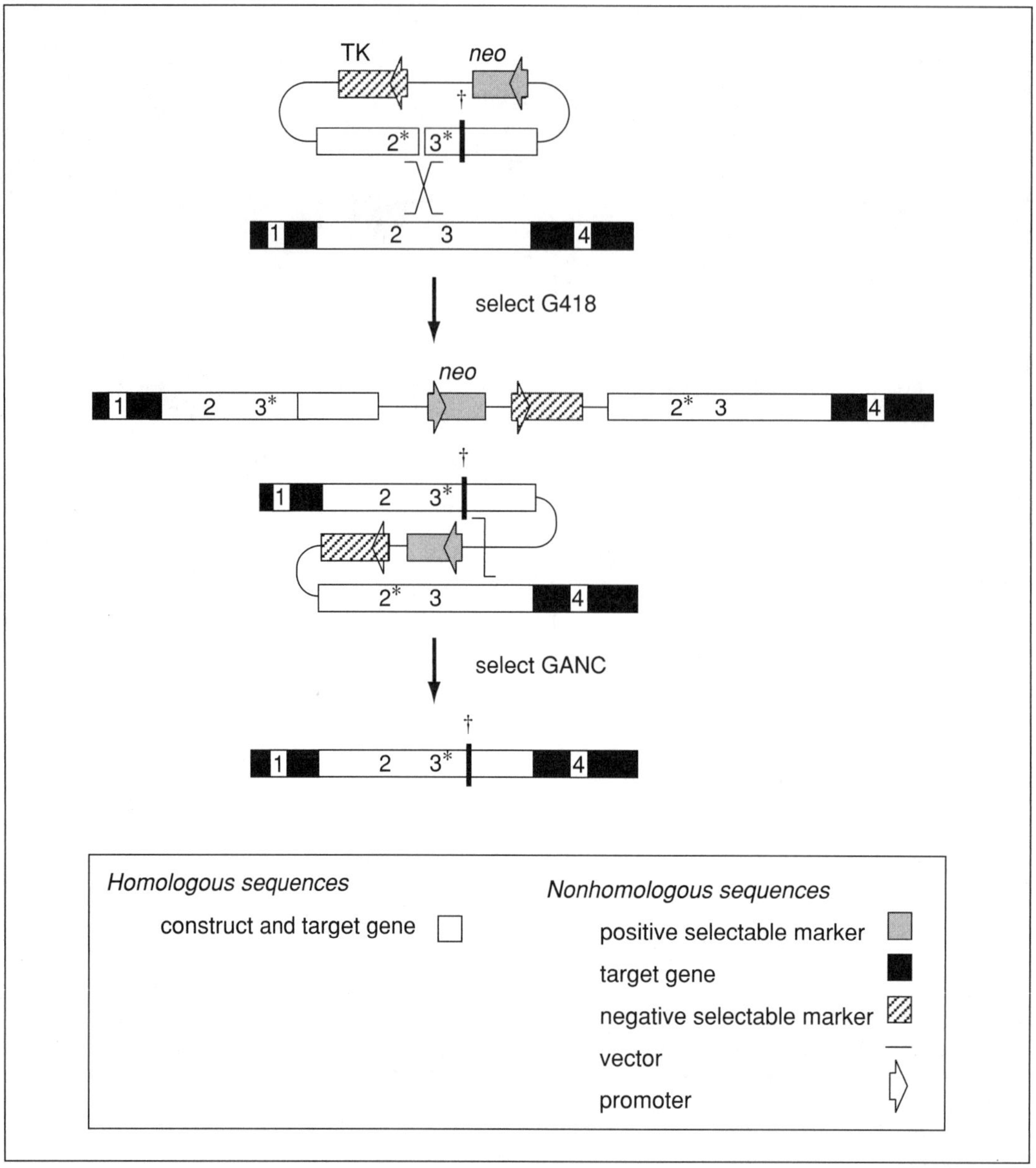

Figure 4.19.5 Allele replacement "hit and run." Cells are cultured in G418 to select for the integration of *neo*, then homologous recombinants are identified by screening and are cultured in gancyclovir (GANC) to select against the presence of the TK gene. This strategy may yield a reconstituted gene containing the subtle mutation present in the construct (indicated by the dark bar and †). Because intrachromosomal recombination may result in the loss of the subtle mutation, its presence must be verified (e.g., by a change in restriction site).

Removing the Positive Selectable Marker

Although many gene inactivation approaches involving homologous recombination still use constructs that leave the positive selectable marker in the genomic DNA, it has become increasingly clear that this can cause a number of unanticipated effects. For example, the presence of the *neo* gene, often with its own promoter, can alter the expression of neighboring loci (Olson et al., 1996; Pham et al., 1996). This can be a particular problem in gene clusters where neighboring genes are in the same family, since the genes affected may have similar or identical functions. As a result, slight differences in targeting constructs have led to marked differences in phenotype.

If the targeting construct includes *loxP* sites flanking the *neo* gene, then *neo* can be removed after targeting by transient expression of the Cre recombinase (as discussed in Fig. 4.19.3). This will leave the small *loxP* site in the genomic DNA, but the construct can be engineered so that this is in an innocuous location, such as an intron. Although theoretically even a *loxP* site could cause alterations in the expression of neighboring genes, no such cases have yet been reported. The efficiency of Cre recombination from transient expression reported in the literature varies widely, from ~2% to ~15% (Sauer and Henderson, 1989; Abuin and Bradley, 1996). This rate should be distinguished from the efficiency of Cre recombination in vivo, where the expression of Cre is derived from sequences integrated into the genome and therefore will show longer-lasting expression in nearly all cases.

Introduction of Subtle Mutations Using Cre/*loxP*

The strategy described in the previous section involves introducing subtle mutations by first duplicating sequences and then screening for intrachromosomal recombination that removes the redundant sequences and leaves the mutation (see Fig. 4.19.5). A limitation to this approach is that the second homologous recombination event occurs only infrequently. A more efficient method is to use a replacement construct containing the subtle mutation and then remove the positive selectable marker, which is flanked by *loxP* sites, using the Cre recombinase system (Fig. 4.19.6). This is identical in effect to removing the *neo* locus after gene inactivation, except that instead of an inactive gene, the replaced sequences contain a subtly mutated version.

Spatial Control of Knockout

Spatially controlled targeted gene inactivations can be performed in two ways. The most common makes use of cell type–specific promoters (sometimes misleadingly called tissue-specific promoters). This approach begins with the creation of a transgenic animal that expresses Cre in only some cells, using a cell type–restricted promoter. A second transgenic animal line is then created by homologous recombination that contains *loxP* sites flanking a portion of the gene that is critical for activity (Fig. 4.19.7). Initially there are three *loxP* sites flanking this important gene region and the selectable marker. After homologous recombination has been verified, Cre is transiently expressed, and loops out regions of DNA between pairs of *loxP* sites. The resultant colonies are screened for the desired recombination (loss of the selectable marker but retention of all regions of the gene). Depending on the frequency of recombination at the site, it may be useful to use a construct that contains a negative selectable marker (such as cytosine deaminase in the example shown in Fig. 4.19.2) between the *loxP* sites along with the positive selectable marker. In this way cells that have lost the markers can be selected.

The targeted line will have normal expression of the targeted gene, since its only modification is the presence of *loxP* sites in innocuous sites (e.g., introns). When the two lines are bred together, the Cre recombinase will loop out the DNA—inactivating the gene—only in those cells where it is expressed. In this way, tissue-specific knockouts of a number of genes have been generated (Gu et al., 1994; Agah et al., 1997). The method also has the advantage that, once a transgenic line is generated with the desired restricted expression of Cre, the approach can be applied to a number of targeted lines. In addition, it is not necessary to make separate constructs for a restricted and a complete knockout, since Cre-expressing lines have been made that will produce rearrangement in all tissues when bred to the targeted line (Schwenk et al., 1995).

Another way of spatially controlling knockout is to use an expression system for Cre that can be applied to absolute location. In some cases, no restricted expression pattern is known

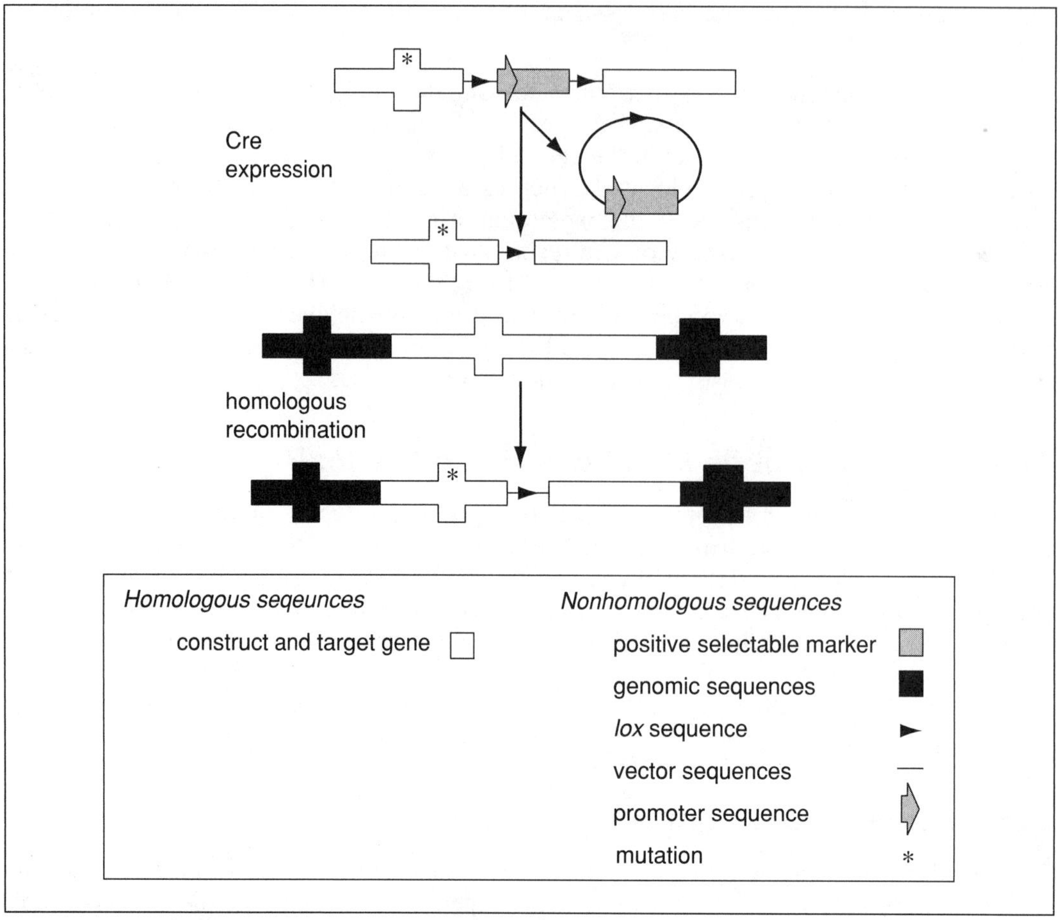

Figure 4.19.6 Using the Cre/*loxP* system to introduce subtle mutations. The subtle mutation is introduced along with the selectable marker in the targeting vector. The selectable marker is then removed by transient expression of Cre, which leaves only the small *loxP* site in the genome in a silent location.

for a gene that matches the desired spatial alteration; in others, the site may be particularly amenable to viral manipulation (as with an epithelial or endothelial surface) or accessible by direct injection (such as sterotactic injection of the central nervous system). By using a viral vector to express the Cre protein, it is possible to obtain knockouts that are spatially limited by the viral infection. This strategy has been applied to a number of tissues including the brain, liver, colon, and heart (Rohlmann et al., 1996; Wang et al., 1996; Agah et al., 1997; Shibata et al., 1997; van der Neut, 1997).

Temporal Control of Knockout

In many cases the phenotype of interest is in the adult animal but, because the gene is necessary for development, no adult animals are obtained. Delaying the expression of Cre activity until the animal is an adult would allow normal development, and then the knockout could be created in the adult (Rajewsky et al., 1996). This can be accomplished by using a conditional expression system (e.g., the tet-on, tet-off, or ecdysone systems; see St-Onge et al., 1996) or other inducible system (such as an interferon-inducible promoter; Kuhn et al., 1995) to express Cre at the proper time. This would, however, require the construction of animals containing three transgenes. Another approach that has been used is the creation of a fusion protein with either a modified estrogen receptor (Feil et al., 1996, 1997; Zhang et al., 1996; Brocard et al.,

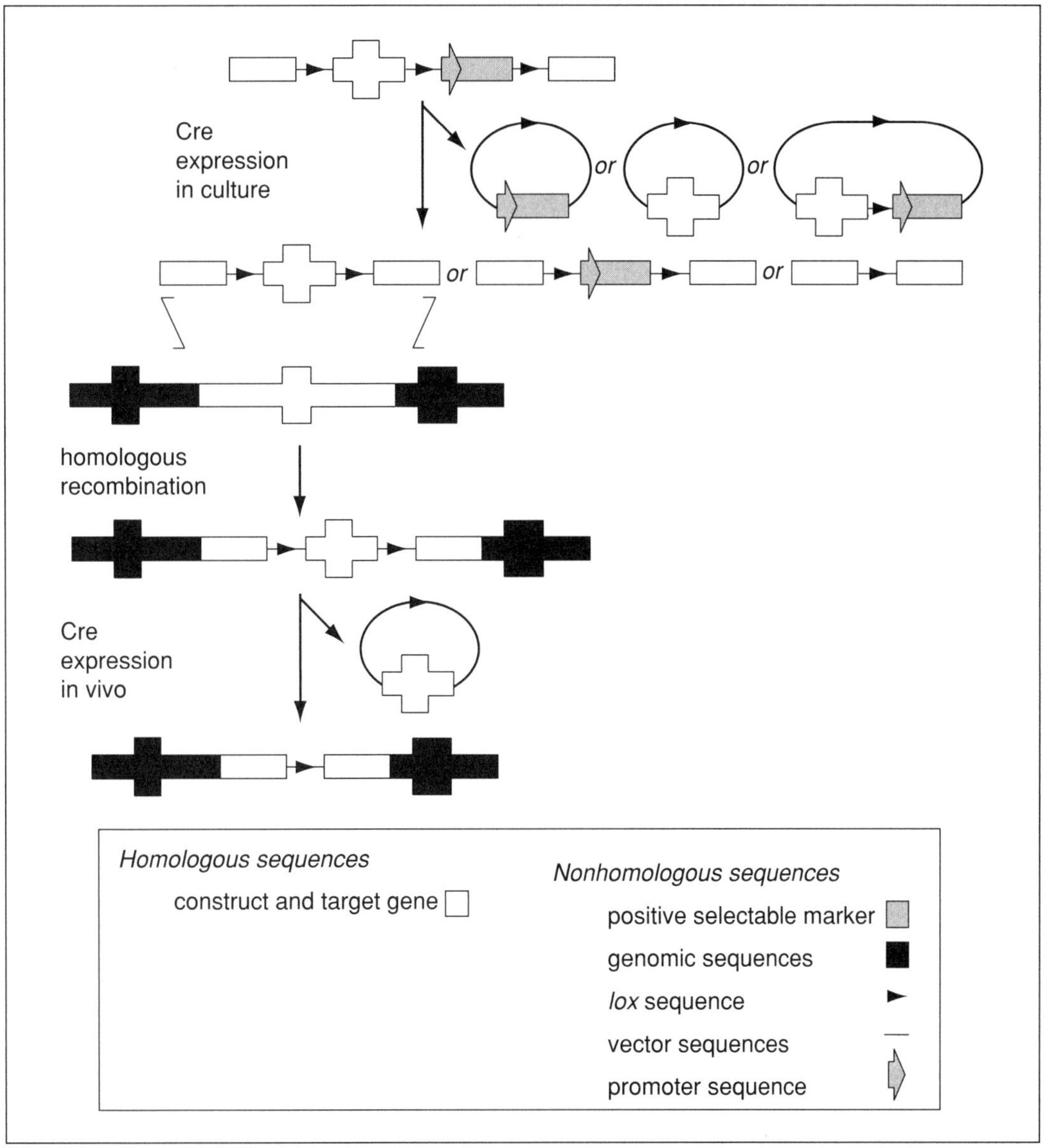

Figure 4.19.7 Conditional gene targeting using the Cre/*loxP* system. The targeting vector contains three *loxP* sites that flank the regions of the gene to be removed and the positive selectable marker *neo*. After homologous recombination is obtained, the selectable marker is excised from the gene by transient expression of Cre. The correct recombination is identified by screening by Southern analysis or PCR. The mutant ES cells are then used to produce a transgenic animal, which is finally bred to an animal line expressing Cre under either temporal or spatial control. Cre can also be ubiquitously expressed to obtain a knockout in all tissues, including the germ line.

1997) or a modified glucocorticoid receptor (Brocard et al., 1998). These fusion proteins are inactive for recombination until the appropriate ligand is added, allowing temporal control in an animal with only transgenes. The Flp/FRT recombinase system can be used in an analogous way. Combination of the two systems can allow the production of complex schemes for gene mutation.

Contributor: Richard Mortensen

CHAPTER 5

Molecular Neuroscience

Chapter 5 presents a selection of various molecular and biochemical techniques with important applications to neuroscience research. Initially, three major areas of technology have been addressed. These include the analysis and manipulation of RNA, the production and use of antisera/antibodies, and the purification and characterization of proteins. Broader coverage of each of these areas can be found in other Current Protocols volumes, including *Current Protocols in Molecular Biology* (CPMB; Ausubel et al., 2006), *Current Protocols in Immunology* (CPI; Coligan et al., 2006a), and *Current Protocols in Protein Science* (CPPS; Coligan et al., 2006b), from which certain of the units that follow have been adapted. The goal here is to provide a selection of basic as well as novel techniques that will be of particular use to neuroscientists. In some cases, the methods may have specific neuroscience applications; in others (e.g., *UNIT 5.3*) they may be unique to research on neural systems.

The first several units in this chapter deal with a variety of techniques for analyzing and manipulating RNA. The presence of a particular RNA can be analyzed by northern or slot blot hybridization (*UNIT 5.1*). This is a classic, but still widely used and important method to detect and characterize RNA in tissues and cells. The ribonuclease (RNase) protection assay, an extremely sensitive procedure for detecting and quantitating mRNA species in complex mixtures of total cellular RNA, is presented in *UNIT 5.2*. The approach used is the polymerase chain reaction (PCR)-based method of differential display, an alternative to classical techniques (such as the construction of subtractive hybridization cDNA libraries) for detecting differentially expressed genes. Next, *UNIT 5.3* presents a highly novel and useful technique for analyzing mRNA species within single neurons. Given the cellular complexity of the nervous system, in which morphologically identical cells may be functionally different, this single-cell analysis approach for mRNA detection is likely to become increasingly popular.

The remaining units in the chapter present a variety of immunological techniques, mostly dealing with the production of polyclonal antisera. *UNIT 5.5* describes a method for preparing antisera against small peptides. This antipeptide approach is particularly useful when it is necessary to produce antiserum against a protein whose sequence is known but that is not available in sufficient quantity to use for immunization. In most instances, the antiserum directed against the peptide derived from the protein is able to recognize the native protein itself. If a clone encoding a protein is available, on the other hand, a highly effective method for producing antisera against that protein is through the use of fusion proteins, as detailed in *UNIT 5.6*. Immunoblotting and immunodetection are described in *UNIT 5.7*. The novel immunological technique of phage display is introduced in *UNIT 5.8*. This method is used to study protein-protein interactions and receptor and antibody sites, and to select antibodies against a wide range of different antigens. Actual protocols describing the use of phage display in neurobiology are presented in *UNIT 5.9*.

Techniques for the detection and analysis of protein phosphorylation in tissues and cells are presented in *UNIT 5.10*. Methods for protein solubilization are reviewed in *UNIT 5.11*, while *UNIT 5.12* describes immunoprecipitation and *UNIT 5.13* describes coimmunoprecipitation. Analysis of protein-protein interactions by fluorescence resonance energy transfer (FRET) is covered in *UNIT 5.14*.

UNITS 5.15 & 5.16 discuss the use of bacterial artificial chromosome (BAC) technology in neurobiology research. BACs provide an easy approach to manipulate large segments of DNA and are highly useful for the creation of transgenic mice. Due to the large size of BACs, the gene under investigation is likely to contain all of the regulatory elements to confer accurate (cell type-specific) transgene expression in vivo. *UNIT 5.15* provides an overview of BAC technology and the applications and uses of BAC transgenic mice in neuroscience research. *UNIT 5.16* presents experimental protocols for manipulating BACs and their preparation for pronuclear microinjections for the creation of transgenic mice.

References: Ausubel et al., 2006; Coligan et al., 2006a; Coligan et al., 2006b

Contributor: David Sibley

UNIT 5.1

Analysis of RNA by Northern and Slot Blot Hybridization

NOTE: To inhibit RNase activity, all solutions for northern blotting should be prepared using sterile deionized water that has been treated with DEPC (*APPENDIX 1*). For full details on establishment of an RNase-free environment, see Wilkinson (1991).

BASIC PROTOCOL

NORTHERN HYBRIDIZATION OF RNA FRACTIONATED BY AGAROSE-FORMALDEHYDE GEL ELECTROPHORESIS

5

Materials *(see APPENDIX 1 for items with ✓)*

✓ 10× MOPS running buffer
12.3 M (37%) formaldehyde, pH >4.0
RNA sample: total cellular RNA or poly(A)$^+$ RNA
Formamide
✓ Formaldehyde loading buffer
Staining solutions (select one):
 0.5 M ammonium acetate (*APPENDIX 1*) with and without 0.5 μg/ml ethidium bromide
 1.1 M formaldehyde/10 mM sodium phosphate (pH 7.0; *APPENDIX 1*) with and without 10 μg/ml acridine orange
0.05 M NaOH/1.5 M NaCl
✓ 0.5 M Tris·Cl (pH 7.4)/1.5 M NaCl
✓ 20× SSC
0.03% (w/v) methylene blue in 0.3 M sodium acetate, pH 5.2 (optional)
DNA suitable for use as probe template
✓ Formamide prehybridization/hybridization (FPH) solution
2× SSC, 0.2× SSC, and 0.1× SSC in 0.1% (w/v) SDS

✓ RNase-free glass dishes (see DEPC treatment)
UV transilluminator, calibrated
Nylon or nitrocellulose membrane
3MM Whatman paper
Vacuum oven
UV-transparent plastic wrap (e.g., Saran Wrap or other polyvinylidene wrap)

1. Prepare a 1% agarose gel that is 2 to 6 mm thick and has wells for 60-μl samples by dissolving 1.0 g agarose in 72 ml water. Cool to 60°C and add 10 ml of 10× MOPS running buffer and 18 ml of 12.3 M formaldehyde. Place solidified gel in tank and add 1× MOPS running buffer to a depth of ~1 mm over the gel.

   CAUTION: *Formaldehyde is toxic through skin contact and inhalation of vapors. All operations involving formaldehyde should be carried out in a fume hood.*

2. Adjust volume of each RNA sample (0.5 to 10 μg for each duplicate lane) to 11 μl with water, then add:

   5 μl 10× MOPS running buffer
   9 μl 12.3 M formaldehyde
   25 μl formamide.

   Mix by vortexing, microcentrifuge briefly (5 to 10 sec), and incubate 15 min at 55°C.

3. Add 10 μl formaldehyde loading buffer, vortex, microcentrifuge briefly, and load onto gel so that one set of all samples is followed by a duplicate set. Run gel at 5 V/cm until the bromphenol blue dye has migrated one-half to two-thirds the length of the gel (~3 hr).

   Molecular weight markers are not usually needed on total RNA gels as rRNA molecules are stained as sharp bands and can be used as internal markers. If poly(A)$^+$ RNA is being fractionated, commercial RNAs (e.g., 0.24- to 9.5-kb RNA ladder; Life Technologies) can be used.

4a. *To stain with ethidium bromide:* Remove gel, cut off lanes to be stained, and wash twice, 20 min each, in an RNase-free glass dish containing 0.5 M ammonium acetate. Replace solution with 0.5 μg/ml ethidium bromide in 0.5 M ammonium acetate and stain 40 min. If necessary, destain up to 1 hr in 0.5 M ammonium acetate.

4b. *To stain with acridine orange:* Remove gel, cut off lanes to be stained, and stain 2 min in 1.1 M formaldehyde/10 mM sodium phosphate containing 10 μg/ml acridine orange. If necessary, destain 20 min in the same buffer without acridine orange.

5. Visualize RNA on a UV transilluminator and photograph with a ruler laid alongside the gel so that band positions can later be identified on the membrane.

6. Place unstained portion of gel in an RNase-free glass dish and rinse with several changes of deionized water. Treat as follows:

   30 min in ~10 gel vol 0.05 M NaOH/1.5 M NaCl (partial hydrolysis)
   20 min in 10 gel vol 0.5 M Tris·Cl (pH 7.4)/1.5 M NaCl (neutralization)
   45 min in 10 gel vol 20× SSC.

7. Set up transfer stack as for Southern blotting and transfer overnight.

8. Recover membrane and flattened gel together. Mark in pencil the position of the wells on the membrane, and ensure that the up-down and back-front orientations are recognizable (e.g., by cutting one corner of membrane). Rinse membrane in 2× SSC and then allow to dry on a sheet of Whatman 3MM paper.

9. Place membrane between two sheets of Whatman 3MM filter paper and cross-link by baking 2 hr at 80°C under vacuum. Alternatively, for nylon only, wrap dry membrane in UV-transparent plastic wrap, place RNA-side-down on a UV transilluminator (254 nm), and irradiate for the appropriate length of time.

10. If desired, check transfer efficiency by staining the gel in ethidium bromide or acridine orange (step 4) *or* (for nylon only) by staining the membrane 45 sec in 0.03% (w/v) methylene blue in 0.3 M sodium acetate, pH 5.2, and destaining 2 min in water.

11. Prepare RNA or DNA probe, ideally 100 to 1000 bp in length, labeled to a specific activity of $>10^8$ dpm/μg. Remove unincorporated nucleotides.

12. Wet blot in 6× SSC, place RNA-side-up in a hybridization tube or heat-sealable plastic bag, and add ~1 ml FPH solution per 10 cm^2 membrane. Incubate with rotation 3 hr at 42°C (for DNA probe) or 60°C (for RNA probe). For nylon, reduce time to 15 min.

13. If the probe is double-stranded, denature by heating 10 min in a 100°C water bath or incubator. Transfer to ice. Add probe to the tube or bag at 10 ng probe/ml if the specific activity is 10^8 dpm/μg or 2 ng/ml if the specific activity is 10^9 dpm/μg. Incubate overnight.

14. Pour off hybridization solution and wash in:

    1 vol 2× SSC/0.1% SDS, twice for 5 min each
    1 vol 0.2× SSC/0.1% SDS, twice for 5 min each (low-stringency wash)
    prewarmed (42°C) 0.2× SSC/0.1% SDS, twice for 15 min each (moderate-stringency wash; optional)
    prewarmed (68°C) 0.1× SSC/0.1% SDS, twice for 15 min each (high-stringency wash; optional).

15. Rinse in 2× SSC at room temperature, blot excess liquid, cover in UV-transparent plastic wrap, and perform autoradiography.

ALTERNATE PROTOCOL 1

NORTHERN HYBRIDIZATION OF RNA DENATURED BY GLYOXAL/DMSO TREATMENT

Sharper bands are produced when RNA is denatured with glyoxal and DMSO and then run on an agarose gel made with phosphate buffer.

Additional Materials *(also see Basic Protocol; see* APPENDIX 1 *for items with ✓)*

✓ 10 and 100 mM sodium phosphate, pH 7.0
Dimethyl sulfoxide (DMSO)
✓ 6 M glyoxal, deionized
✓ Glyoxal loading buffer
✓ 20 mM Tris·Cl, pH 8.0
Apparatus for recirculating running buffer during electrophoresis

1. Prepare a 1.0% agarose gel that is 2 to 6 mm thick and has wells for 60-μl samples by dissolving 1.0 g agarose in 100 ml of 10 mM sodium phosphate, pH 7.0. Place solidified gel in tank and add 10 mM sodium phosphate, pH 7.0, to a depth of ~1 mm over the gel.

2. Adjust volume of each RNA sample (0.5 to 10 μg for each duplicate lane) to 11 μl with water, then add:

    4.5 μl 100 mM sodium phosphate, pH 7.0
    22.5 μl DMSO
    6.6 μl 6 M glyoxal.

    Mix by vortexing, microcentrifuge briefly (5 to 10 sec), and incubate 1 hr at 50°C.

3. Cool samples on ice, add 12 μl glyoxal loading buffer, and load onto gel so that one set of all samples is followed by a duplicate set. Run gel at 4 V/cm with constant recirculation of running buffer until bromphenol blue dye has migrated one-half to two-thirds the length of the gel (~3 hr).

    Recirculation prevents formation of a H^+ gradient in the buffer.

4. Transfer RNA from one set of lanes to membrane (see Basic Protocol, steps 6 to 10).
5. Once transfer is started, stain the other portion of the gel with ethidium bromide (see Basic Protocol, steps 4a and 5).
6. Perform hybridization (see Basic Protocol, steps 11 to 15), but immediately before hybridization, soak membrane 5 min in 20 mM Tris·Cl, pH 8.0, at 65°C to remove glyoxal.

ALTERNATE PROTOCOL 2

NORTHERN HYBRIDIZATION OF UNFRACTIONATED RNA IMMOBILIZED BY SLOT BLOTTING

Immobilized unfractionated RNA can be used in hybridization analysis to determine the relative abundance of target mRNA sequences. Total cellular or poly(A)$^+$ RNA can be used.

Additional Materials *(also see Basic Protocol; see* APPENDIX 1 *for items with ✓)*

0.1 M NaOH
✓ Denaturing solution
✓ 100 mM sodium phosphate, pH 7.0
Dimethyl sulfoxide (DMSO)
✓ 6 M glyoxal, deionized
3% (w/v) gelatin (optional)
✓ 20 mM Tris·Cl, pH 8.0
Manifold apparatus with a filtration template for slot blots (e.g., Bio-Rad Bio-Dot SF, Schleicher and Schuell Minifold II)

1. Clean manifold with 0.1 M NaOH and rinse with distilled water.
2. Cut a piece of nylon or nitrocellulose membrane to the size of the manifold. Pour 10× SSC (nylon) or 20× SSC (nitrocellulose) into a glass dish, place membrane on top of liquid, and allow to submerge. Leave 10 min.
3. Place membrane in the manifold and assemble according to manufacturer's instructions. Fill each slot with 10× SSC. Ensure there are no air leaks in the assembly.

4a. *For denaturation with formaldehyde:* Add 3 vol denaturing solution to RNA sample (up to 20 μg per slot), incubate 15 min at 65°C, and place on ice.

Total cellular RNA or poly(A)$^+$ RNA can be used, although the latter is preferable.

4b. *For denaturation with DMSO/glyoxal:* Mix RNA (up to 20 μg per slot) with water to make 11 μl, and then add:

4.5 μl 100 mM sodium phosphate, pH 7.0
22.5 μl DMSO
6.6 μl 6 M glyoxal.

Mix by vortexing, microcentrifuge briefly, and incubate 1 hr at 50°C.

5. Add 2 vol ice-cold 20× SSC to each sample.
6. Switch on the manifold suction and allow 10× SSC (step 3) to filter through. Adjust so that 500 μl buffer passes through in ~5 min and leave suction on.
7. Block slots that are not being used with masking tape or by applying 500 μl of 3% (w/v) gelatin to each onc. Alternatively, apply 10× SSC instead of sample to unused slots.

5

8. Load each sample to a slot and allow to filter through, being careful not to touch the membrane with the pipet tip. Wash twice with 1 ml of 10× SSC.
9. Place membrane on a sheet of Whatman 3MM paper, and allow to dry. Immobilize the RNA (see Basic Protocol, step 9).
10. Carry out hybridization analysis (see Basic Protocol, steps 11 to 15). If glyoxal/DMSO denaturation was used, immediately before hybridization, soak membrane 5 min in 20 mM Tris·Cl, pH 8.0, at 65°C to remove glyoxal.

SUPPORT PROTOCOL

REMOVAL OF PROBES FROM NORTHERN BLOTS

Radioactive and chemiluminescent hybridization probes can be removed from northern blots on nylon membranes without damage to the membrane or loss of the transferred RNA.

Materials *(see APPENDIX 1 for items with ✓)*

RNA blot containing probe (see Basic Protocol or Alternate Protocol 1 or 2)
✓ Stripping solution

1. Place membrane in a hybridization bag or open container with stripping solution without formamide (enough to cover membrane). Place bag in 80°C water for 5 min. Pour out solution and repeat wash three to four times.
2. Monitor radioactivity. If further stripping is needed, place membrane in a bag with fresh stripping solution and place bag in boiling water for 5 min. Pour out solution and repeat three to four times.
3. Monitor radioactivity. If further stripping is needed, place membrane in a bag with fresh stripping solution containing formamide. Place bag in 65°C water for 5 min. Pour out solution, and repeat wash three times with formamide and one time without formamide.
4. Place membrane on filter paper to remove excess solution. Wrap membrane in plastic wrap and perform autoradiography or chemiluminescent detection to verify probe removal.

Reference: Thomas, 1980

Contributor: Terry Brown

UNIT 5.2

RNA Analysis by Nuclease Protection

Commercial kits are available that contain preoptimized reagents for RNase protection assays (available from Ambion, Roche Applied Sciences, and Pharmingen), and some of these products have streamlined the procedure by eliminating the protease digestion and organic extraction steps (e.g., Ambion RPA III kit). Additional products facilitate the method by permitting direct analysis of mRNA in cell or tissue lysates, without prior purification of the RNA fraction (Ambion Direct Protect kit), or by increasing the hybridization rate between probe and target by several orders of magnitude to eliminate the overnight hybridization step (Ambion HybSpeed RPA kit). Some commercial products are designed to detect and quantify multiple target mRNAs in the same hybridization reaction by providing cocktails of probe templates that can be used for multiplex RPAs (Pharmingen RiboQuant system and Ambion MultiProbe qRPA templates). A commercial kit designed for nonisotopic RPAs is also available (Pierce SuperSignal RPAIII chemiluminescent detection system). However, nonisotopic detection requires the additional step of transferring the protected fragments onto a solid support.

NOTE: Use DEPC-treated water (*APPENDIX 1*) to prepare all solutions.

BASIC PROTOCOL 1

RNase PROTECTION ASSAY

The sequence to be transcribed should be oriented tail-to-tail with respect to the promoter (i.e., with the carboxy-terminal end of the coding region facing the phage promoter) to make probes for detecting mRNA. The sensitivity of the assay is determined by the specific activity of the in vitro transcript, which depends on the ratio of radiolabeled to unlabeled limiting nucleotide in the reaction. For maximum sensitivity, all "cold" (nonradioactive) limiting nucleotide should be omitted.

Materials *(see APPENDIX 1 for items with ✓)*

Radiolabeled RNA probe (see Support Protocol 1) in probe elution buffer or hybridization buffer (*APPENDIX 1*)
RNA sample: aqueous or in hybridization buffer (*APPENDIX 1*; storage in high concentrations of formamide enhances stability)
RNase-free sheared total yeast RNA (see Support Protocol 2) or tRNA that does not contain the target sequence
5 M ammonium acetate
100% ethanol or isopropanol
✓ Hybridization buffer
✓ RNase A/RNase T1 stock solution
✓ RNase digestion buffer
✓ Proteinase K/SDS solution
1 mg/ml carrier nucleic acid or glycogen
25:24:1 (v/v/v) phenol/chloroform/isoamyl alcohol
✓ Gel loading buffer
Denaturing 5% (19:1 acrylamide/bisacrylamide) polyacrylamide gel

RNase-free microcentrifuge tubes (e.g., Ambion)
Pasteur pipets with drawn-out tips *or* fine-gauge needles
Heating block
42° to 45°C incubator or water bath
Chromatography paper (Whatman)
Plastic wrap (if using ^{32}P-labeled probe)
X-ray film (Kodak XRP)

1. Mix radiolabeled probe (usually at 2–8 × 10^4 cpm) with desired amount of sample RNA (usually 1 to 20 μg total cellular RNA) in an RNase-free microcentrifuge tube. For each probe used, include two control reactions containing the probe in combination with 10 μg yeast RNA or tRNA that does not contain the target sequence.

   For quantitative results, it is important that the probe be present in about five-fold molar excess over the target RNA in the hybridization reaction. For detecting very low-abundance mRNAs, poly(A)$^+$ purification of the mRNA fraction may be needed. Kits for mRNA poly(A)$^+$ purification are made by Ambion and Boehringer Mannheim.

   The RNA used in the control reactions should not be completely intact, but the average size should be ~300 nucleotides (see Support Protocol 2).

5

2. Add 5 M ammonium acetate to all reactions to a final concentration of 0.5 M (1:9 5 M ammonium acetate/reaction mixture). Mix well. Add 2 vol ethanol or 1 vol isopropanol to all reactions. Mix well. Chill reactions at least 15 min at −20°C. Microcentrifuge 15 min at maximum speed, 4°C, to obtain the probe and DNA pellet.

3. Carefully remove the supernatant from each tube by gentle aspiration. Microcentrifuge tubes again for a few seconds at maximum speed and remove residual fluid with a drawn-out Pasteur pipet or fine-gauge needle.

4. Resuspend pellet in 20 μl hybridization buffer. Vortex vigorously to resuspend, then microcentrifuge briefly at maximum speed to collect all fluid at the bottom of the tubes. Heat reactions 3 to 4 min at 95°C in a heating block, then vortex again and microcentrifuge again briefly at maximum speed. Incubate reactions, tightly capped, overnight in a 42° to 45°C incubator or heating block to permit hybridization of probe and target RNA. Microcentrifuge briefly to collect all liquid.

5. Prepare a 1:100 working dilution of the RNase A/RNase T1 stock solution in RNase digestion buffer (200 μl of diluted RNaseA/RNase T1 solution is needed for each reaction).

   For probes very rich in A and/or U, better results will be obtained by using RNase T1 alone in the digestion step, rather than in combination with RNase A. In an initial experiment using a new probe, include at least one reaction using RNase T1 alone. To digest excess probe with RNase T1 alone, prepare the RNase stock solution using RNase T1 at 10,000 U/ml; when ready to perform the digestion, dilute the enzyme (1:100) in RNase digestion buffer to a final concentration of ~100 U/ml.

6. Remove reactions from the incubator or heating block after the appropriate digestion time and microcentrifuge briefly at maximum speed if any condensation is visible. Add 200 μl of the diluted RNase solution prepared in step 5 to each experimental tube and to one of the yeast RNA control tubes. Add 200 μl of RNase digestion buffer (without RNase) to the other yeast RNA control tube (to obtain a mock-digested control). Mix thoroughly and microcentrifuge briefly at maximum speed. Incubate reactions 30 min at 37°C.

7. Add 20 μl proteinase K/SDS solution to each reaction. Mix thoroughly, then microcentrifuge briefly at maximum speed. Incubate reactions 15 min at 37°C.

8. Add sufficient 1 mg/ml carrier nucleic acid or glycogen to each reaction for a final concentration of 5 μg, then add 250 μl of 25:24:1 phenol/chloroform/isoamyl alcohol. Vortex vigorously for at least 10 sec, then microcentrifuge at least 1 min at maximum speed to separate the aqueous and organic phases.

9. Transfer each aqueous (top) phase to a fresh tube and add 625 μl ethanol. Mix thoroughly, then incubate at least 15 min at −20°C.

10. Microcentrifuge reactions 15 min at maximum speed, 4°C. Carefully remove the supernatant from each reaction, then microcentrifuge again for a few seconds at maximum speed and remove the residual fluid with a drawn-out Pasteur pipet or fine-gauge needle.

11. Resuspend each pellet in ~8 μl of gel loading buffer with vigorous vortexing. Heat 3 to 4 min at 95°C to completely solubilize and denature the RNA, then vortex and microcentrifuge the tubes briefly at maximum speed.

12. Load resuspended DNA pellets on a 5% (19:1 acrylamide/bisacrylamide) denaturing polyacrylamide gel and electrophorese at ~250 V for a length of time appropriate to resolve the protected fragments (usually ~45 to 60 min).

13. Place gel on chromatography paper, mark the origins and the orientation of lanes, and cover with plastic wrap.

14. Autoradiograph by exposing gel to single-side-coated X-ray film at −20°C or −80°C for an appropriate period of time, usually overnight, with an intensifying screen for rare targets or without a screen for abundant targets.

ALTERNATE PROTOCOL

SMALL-VOLUME RNase PROTECTION ASSAY

When the probe and sample RNA are in a small volume ($\leq$20 µl), preferably dissolved in hybridization buffer, the following short-cut procedure may be used.

Additional Materials *(also see Basic Protocol 1)*

RNA sample in volume $\leq$20 µl (in hybridization buffer; see *APPENDIX 1*)
Radiolabeled RNA probe (see Support Protocol 1) in volume $\leq$20 µl (in hybridization buffer; see *APPENDIX 1*)

1. Mix radiolabeled probe (usually ~2–8 × 10^4 cpm) with desired amount of sample RNA (usually 1 to 20 µg total cellular RNA) in an RNase-free microcentrifuge tube. For each probe used, include two control reactions containing the probe in combination with 10 µg yeast RNA.
2. Add hybridization buffer, if necessary, to give a total of $\geq$20 µl (but not >35 µl) of hybridization buffer in each reaction and to adjust all reactions to the same final volume. Mix thoroughly.
3. Heat reactions 3 to 4 min at 95°C, then incubate overnight at 42° to 45°C.
4. Digest reactions with RNase, then isolate and detect protected fragments (see Basic Protocol 1, steps 5 to 14).

5

SUPPORT PROTOCOL 1

SYNTHESIS AND GEL PURIFICATION OF FULL-LENGTH RNA PROBE

Recommended probe size is 200 to 500 bases; shorter probes may be used for abundant targets. Although this procedure is for preparing ^{32}P-labeled probes, ^{35}S-labeled nucleotides may also be used; however, in that case the assay will be less sensitive by a factor of 10- to 100-fold. The radiolabeled nucleotide is at a limiting concentration in the reaction.

Materials *(see APPENDIX 1 for items with ✓)*

✓ DEPC-treated H_2O
✓ 10× transcription buffer
✓ 3NTP mix
Dilute solution of limiting nucleotide: e.g., 50 µM cold UTP or CTP
10 to 20 mCi/ml [α-^{32}P]UTP or [α-^{32}P]CTP (400 to 800 Ci/mmol) in aqueous buffer (not in ethanol)
10 to 50 U/µl placental ribonuclease inhibitor (Ambion or Boehringer Mannheim)
Template DNA: 0.5 µg/µl linearized plasmid or 50 ng/µl amplified PCR product
10 to 20 U/µl bacteriophage RNA polymerase (T7, T3, or SP6) appropriate to promoter used in template DNA (keep on ice before use)
1 to 2 U/µl RNase-free DNase I
Gel loading buffer
5% denaturing polyacrylamide gel

✓ Probe elution buffer
5 M ammonium acetate
Nucleic acid precipitation aid: 5 mg/ml RNase-free sheared yeast RNA (see Support Protocol 2) *or* 5 mg/ml glycogen
100% isopropanol (ACS grade)
✓ Hybridization buffer

RNase-free microcentrifuge tubes
Heating block
Scalpel
Forceps
Pasteur pipet with drawn-out tip *or* fine-gauge needle

1. Add the following to an RNase-free microcentrifuge tube at room temperature in the order indicated, with gentle mixing:

   DEPC-treated H_2O to make final volume of 20 μl
   2 μl 10× transcription buffer
   3 μl 3NTP mix
   1 to 5 μl dilute cold limiting nucleotide (optional; see Notes 1 and 2, below)
   1 to 5 μl 10 to 20 mCi/ml [α-^{32}P]CTP or [α-^{32}P]UTP (see Notes 1 and 2, below)
   1 μl 10 to 50 U/μl placental ribonuclease inhibitor
   2 μl template DNA (0.5 μg plasmid or 50 to 100 ng of PCR template; see Note 3, below)
   0.5 to 1 μl 10 to 20 U/μl T7, SP6, or T3 bacteriophage RNA polymerase (10 U total).

   Incubate reaction mixture 1 hr at 37°C.

   NOTE 1: *Either of the pyrimidine ribonucleotides (UTP or CTP) is chosen as the radiolabeled nucleotide; the purines may also be used, although GTP is less stable and ATP requires a higher concentration for efficient incorporation. A radiolabeled nucleotide of medium specific activity (~800 Ci/mmol), rather than high specific activity (3000 Ci/mmol), is recommended.*

   NOTE 2: *For maximum sensitivity, probes should be synthesized at the highest practical specific activity, which is done by omitting all unlabeled ("cold") limiting nucleotide. For synthesizing long transcripts (greater than ~500 bases), it is necessary to add cold limiting nucleotide to keep the limiting nucleotide concentration high enough to produce significant amounts of full-length transcript. For detecting low-abundance targets, gel purification of the full-length transcript is a better remedy for premature transcription than increasing the cold limiting nucleotide concentration. To synthesize probes of medium specific activity (~3 × 10^8 cpm/μg), add cold limiting nucleotide to a final concentration of ~3 μM. When using two or more probes whose target RNAs differ widely in abundance, adjust their specific activities by using different amounts of cold limiting nucleotide in the transcription reactions in order to reduce the differences in intensity of signal between rare and abundant targets.*

   NOTE 3: *Template DNA is added as one of the last components to prevent precipitation of spermidine from the transcription buffer. Template DNA may consist of sequences cloned in transcription vectors (plasmids with T7, T3, or SP6 bacteriophage promoters), or of PCR products amplified with a bacteriophage promoter added to the reverse PCR amplimer. Linearize plasmid templates with an appropriate restriction enzyme (with a site some distance from the cloning site) on the amino-terminal side of the insert to make transcripts complementary to the mRNA.*

2. Add 1 μl RNase-free DNase I and mix gently. Incubate reaction mixture 15 min at 37°C.

To gel-purify probe

3a. Add 20 μl gel loading buffer, then heat reaction 3 min in a 95°C heating block.

4a. Run all or part of reaction on a 5% preparative denaturing polyacrylamide gel.

5a. After electrophoresis, cover the gel with plastic wrap and expose it to X-ray film for several minutes.

6a. Localize the full-length transcript on the autoradiogram (usually this is the slowest-migrating and most intense band) and excise the band from the gel with a scalpel blade. With clean forceps, transfer to a microcentrifuge tube containing 300 to 500 µl probe elution buffer at 37°C. Incubate several hours to overnight at 37°C.

If using probe without gel purification

3b. After DNase treatment, remove a 1-µl aliquot of the reaction and determine the percentage of radiolabel that was incorporated into transcript—e.g., by TCA precipitation.

4b. Add to the reaction (from step 2):

250 µl DEPC-treated H_2O
30 µl 5 M ammonium acetate
2 µl 5 mg/ml RNase-free sheared yeast RNA (10 µg total) or 10 µg glycogen (to act as carrier; optional)
300 µl 100% isopropanol.

Mix thoroughly.

5b. Chill reaction 15 min at −20°C, then microcentrifuge 15 min at maximum speed, 4°C. Carefully remove the supernatant with gentle aspiration, then microcentrifuge again for a few seconds and remove the residual fluid with a drawn-out Pasteur pipet or fine-gauge needle.

6b. Resuspend the pellet with vigorous vortexing in 100 µl hybridization buffer. Heat 3 min at 95°C, then vortex and microcentrifuge briefly at maximum speed to collect all fluid at the bottom of the tube. Store probe at −20°C until use.

5

SUPPORT PROTOCOL 2

PREPARATION OF RNase-FREE SHEARED YEAST RNA

Use yeast RNA that lacks the target sequence as a negative control in nuclease protection assays. Add yeast RNA to experimental samples to attain the same amount of total RNA in all the samples.

Materials *(see* APPENDIX 1 *for items with ✓)*

Total yeast RNA (from *Torula* yeast; Sigma)
Sodium dodecyl sulfate (SDS)
Proteinase K
5 M ammonium acetate
100% ethanol
✓ 0.1 mM EDTA in DEPC-treated H_2O
55°C water bath

1. Dissolve total yeast RNA at a final concentration of 10 mg/ml in distilled water. Incubate at 55°C for several hours to increase solubilization.

2. Add SDS to 0.1% (w/v) final concentration and proteinase K to 0.2 mg/ml final concentration and mix well. Incubate ≥1 hr at 55°C.

3. Extract twice with 25:24:1 phenol/chloroform/isoamyl alcohol and once with 24:1 chloroform/isoamyl alcohol. Add 1/10 vol of 5 M ammonium acetate (0.5 M final) and

precipitate the RNA with 2 vol ethanol. Resuspend the RNA in DEPC-treated water containing 0.1 mM EDTA. Store indefinitely at −20°C.

RNA concentration is determined by UV absorbance; an A_{260} of 1.0 corresponds to a concentration of 35 μg/ml. Yeast RNA prepared according to this protocol will typically have an average size of ~300 bases.

SUPPORT PROTOCOL 3

ABSOLUTE QUANTITATION OF mRNA

Absolute quantitation of target RNA is based on construction of a standard curve using known amounts of in vitro–synthesized sense-strand RNA hybridized with an excess of ^{32}P-labeled antisense probe. The intensity of the signal in experimental samples is then compared with the intensity generated by known amounts of the synthetic target in the standard curve and used to determine the absolute amount of target RNA. The sense-strand transcripts used to make the standard curve can be made by linearizing the DNA template used for antisense-probe synthesis on the other side of the probe insert and synthesizing a DNA or RNA probe from the opposite strand. Quantitation of the sense-strand transcript is most easily accomplished by including a tracer amount of ^{32}P-labeled nucleotide in the presence of a high concentration of all four unlabeled nucleotides during probe synthesis.

Alternatively, a low-specific-activity sense-strand transcript can be synthesized and quantitated by using a ^{3}H-labeled ribonucleotide as a tracer. The specific activity of the sense-strand probe can be calculated to determine the amount of labeled sense strand to use to construct the standard curve (typically this will range from picogram to nanogram amounts, and the actual cpm of sense strand used will be negligible). If the protected fragment in the sample RNA is the same size as the sense strand transcript used to generate the standard curve, equivalent intensities of these protected fragments indicate equimolar amounts of protecting RNA (mRNA or sense-strand). If the sizes of the protected fragments differ, the molar amounts and band intensities will differ in proportion to the sizes of the protected fragments.

The assay can be quantitated by densitometric scanning of the autoradiogram, or by excising and counting the regions of the gel that contain the protected fragments. Alternatively, the standard curve can be constructed by TCA precipitation and counting of the protected species.

BASIC PROTOCOL 2

PROTECTION OF mRNA FROM S1 NUCLEASE DIGESTION USING SINGLE-STRANDED DNA OR RNA PROBES

The probes may consist of internally labeled single-stranded RNA (synthesized by in vitro transcription), single-stranded DNA (made by primer extension), or end-labeled oligonucleotides.

Materials *(see* APPENDIX 1 *for items with ✓)*

- Sample RNA
- Labeled probe (see Support Protocol 1, 4, 5, or 6)
- RNase-free total yeast RNA (see Support Protocol 2)
- 5 M ammonium acetate
- 100% ethanol, ice-cold
- ✓ Hybridization buffer

✓ 2× S1 nuclease digestion buffer
✓ DEPC-treated H_2O
250 to 500 U/μl S1 nuclease (Boehringer Mannheim or Pharmacia Biotech)
✓ S1 nuclease stop buffer
Gel loading buffer

RNase-free microcentrifuge tubes (e.g., Ambion)
Heating block
42° to 45°C incubator or water bath
Pasteur pipets with drawn-out tips
Chromatography paper (Whatman)
Plastic wrap
X-ray film (Kodak XRP)

1. Mix 1 to 10 μg (up to 50 μg) total or poly(A)$^+$ RNA sample RNA and 1–4 × 10^4 cpm labeled probe (2 to 8 pg) in an RNase-free microcentrifuge tube. For each probe used, include two control reactions containing the probe in combination with RNase-free total yeast RNA at the same concentration as sample RNA.

2. Add 1/10 vol of 5 M ammonium acetate to each tube, mix, then add 2.5 vol ice-cold 100% ethanol. Mix, then chill 15 to 30 min at −20°C. Microcentrifuge 15 min at maximum speed, 4°C, and remove supernatant.

3. Dissolve the pellets in 10 μl hybridization buffer. Vortex each tube ~1 min, then microcentrifuge briefly at maximum speed to collect the liquid at the bottom of the tube.

4. Incubate tubes 3 to 4 min at 90° ± 5°C to denature the RNA and aid in its solubilization, then vortex and microcentrifuge briefly at maximum speed.

5. Incubate tubes overnight in a 42° to 45°C cabinet-type incubator, water bath, or heating block, to allow hybridization of probe and complementary mRNA.

6. Thaw the 2× S1 nuclease digestion buffer and remove an appropriate volume (100 μl multiplied by the number of assay tubes). Dilute to 1× using DEPC-treated water. Add 50 to 500 U S1 nuclease per 200 μl of the 1× S1 nuclease digestion buffer. Mix contents by gentle flicking (do not vortex) to assure even distribution of S1 nuclease in the digestion buffer. Keep >200 μl of 1× S1 nuclease digestion buffer without enzyme for control.

7. Remove the hybridization reactions from the incubator and microcentrifuge briefly at maximum speed if any condensation is present on the sides or top of the tube.

8. To each tube containing sample RNA, and to one of the yeast RNA control tubes, add 200 μl of the 1× S1 nuclease digestion buffer containing S1 nuclease. To the other yeast RNA control tube, add 200 μl of 1× S1 nuclease digestion buffer without S1 nuclease. Mix gently.

9. Microcentrifuge tubes briefly, then incubate 30 to 60 min at 37°C to digest unhybridized single-stranded probe and RNA.

10. Add 40 μl S1 nuclease stop buffer to each tube. Vortex, then microcentrifuge tubes briefly at maximum speed.

11. Add 625 μl ice-cold 100% ethanol and vortex briefly. Transfer tubes to −20°C freezer for 15 to 30 min. Microcentrifuge reactions 15 min at maximum speed, 4°C.

12. Remove all supernatant from each tube, taking care not to dislodge the pellet. Remove the last traces of supernatant by microcentrifuging tubes ~5 sec at maximum speed, room temperature, then withdrawing the residual fluid with a drawn-out Pasteur pipet.

13. Prepare an 8 M urea/1× TBE buffer denaturing 5% polyacrylamide gel.

14. Resuspend the pellets in ~8 µl of gel loading buffer. Vortex tubes 1 to 2 min, then microcentrifuge briefly at maximum speed.

15. Heat tubes 3 to 4 min in a heating block at 90° ± 5°C to completely solubilize and denature the probe/mRNA duplex.

16. Load each sample on the gel (step 13) and run at ~150 to 300 V (constant voltage) for a length of time appropriate for resolving the protected fragment(s).

17. Place gel on chromatography paper, mark the origins and the orientation of lanes, and cover with plastic wrap. Autoradiograph by exposing to single-side-coated X-ray film overnight to several days at −80°C with or without intensifying screens.

SUPPORT PROTOCOL 4

SYNTHESIS OF DNA PROBES BY PRIMER EXTENSION OF DOUBLE-STRANDED PLASMID OR PCR PRODUCT USING KLENOW FRAGMENT

For probes made by primer extension from PCR products, the primers should be designed to anneal internally to produce probes that are shorter than the PCR template; this will facilitate separation of the labeled probe from the template after synthesis. Figure 5.2.1 illustrates the use of this protocol using a plasmid (pT7/T3; panel A) or a PCR product (panel B) as the template.

Materials *(see* APPENDIX 1 *for items with* ✓*)*

Template DNA (linearized plasmid or purified PCR product)
Primer oligonucleotide (typically 20 bases in length)
✓ DEPC-treated H_2O
Liquid nitrogen or dry ice/methanol bath
✓ 10× primer-extension buffer
10 mCi/ml [α-^{32}P]dATP or [α-^{32}P]dCTP (3000 Ci/mmol)
✓ 10× 3dNTP mix (appropriate to radiolabeled dNTP used)
50 mM DTT
5 U/µl Klenow fragment of *E. coli* DNA polymerase I
✓ 25 mM EDTA, pH 8.0
25:24:1 (v/v/v) phenol/chloroform/isoamyl alcohol
5 M ammonium acetate
100% ethanol, ice-cold
Gel loading buffer
0.5-ml RNase-free microcentrifuge tubes
95° to 100°C heating block or water bath

1. Combine template DNA (1.0 µg linearized plasmid or 0.1 µg purified PCR product) in a 0.5-ml RNase-free microcentrifuge tube with a 5- to 50-fold molar excess of the specific primer oligonucleotide. Bring the volume to 10 µl with DEPC-treated water, cap the tube, and incubate 5 to 10 min in a 95° to 100°C heating block or water bath to insure complete denaturation.

2. Quickly freeze reaction mixture in liquid nitrogen or a dry ice/methanol bath to prevent reannealing of the template DNA strands. Place tube on ice, allow reaction to thaw, then add the following components to the tube on ice (final reaction volume ~30 µl):

   3.0 µl 10× primer-extension buffer
   10 µl 10 mCi/ml [α-^{32}P]dATP or [α^{32}P]dCTP (3000 to 6000 Ci/mmol)

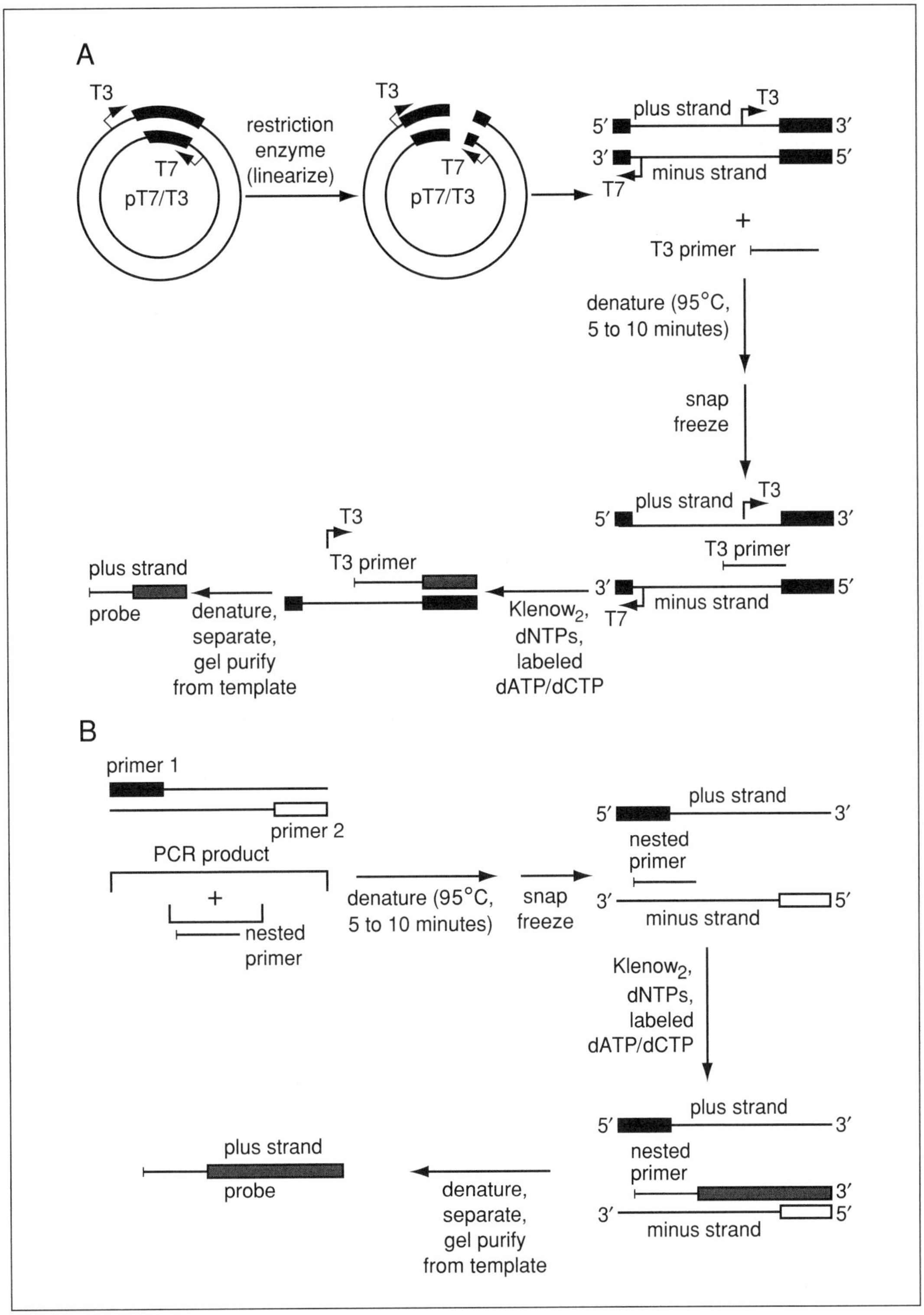

Figure 5.2.1 Synthesis of single-stranded (ss) DNA probes by primer extension. (**A**) Primer extension from DNA sequences cloned into pT7/T3. The plasmid is first linearized with a restriction enzyme at the opposite end of the insert from that used for priming. A specific primer oligonucleotide (at the 5′ side of the insert) designed to hybridize to a specific sequence in the plasmid is added, then the DNA is heat denatured. The reaction is snap frozen, thawed, and put on ice, then incubated with a master mix containing salts, dNTPs, radiolabel, and Klenow fragment. The resulting ssDNA probe is then gel-purified. (**B**) Primer extension from double-stranded PCR products. A specific nested primer oligonucleotide designed to hybridize to a specific sequence on the PCR product is added, then the resulting ssDNA probe is generated and purified as described above.

3.0 μl 10× 3dNTP mix
1.0 μl 50 mM DTT
1.0 μl 5 U/μl Klenow fragment of DNA polymerase I
2.0 μl DEPC-treated H_2O.

3. Mix contents of the tube by gentle flicking and microcentrifuge briefly at maximum speed. Incubate 30 min at 37°C.
4. Add 75 μl of 25 mM EDTA, mix, and put on ice.
5. Add 1 vol of 25:24:1 phenol/chloroform/isoamyl alcohol to the reaction, mix, then microcentrifuge 3 min at maximum speed. Transfer top phase to a new tube, add 1/10 vol of 5 M ammonium acetate, mix, and add 2.5 vol ice-cold 100% ethanol. Incubate 15 to 30 min at −20°C, then microcentrifuge 15 min at maximum speed. Remove supernatant and resuspend the probe in 10 μl gel loading buffer.
6. Gel purify the probe to remove the template DNA (see Support Protocol 1, steps 4a to 6a).
7. Precipitate the probe out of the probe elution buffer (see Support Protocol 1, step 6a) by adding 1/10 vol of 5 M ammonium acetate (0.5 M final) and 2.5 vol 100% ethanol to remove EDTA.

   EDTA will reduce the effective concentration of zinc, which is essential for S1 nuclease activity.

8. Resuspend the probe in DEPC-treated water and store at −20° or −80°C.

SUPPORT PROTOCOL 5

SYNTHESIS OF DNA PROBES BY PRIMER EXTENSION OF DOUBLE-STRANDED PLASMID OR PCR PRODUCT IN A THERMAL CYCLER USING THERMOSTABLE POLYMERASE

Materials *(see* APPENDIX 1 *for items with ✓)*

Template DNA (linearized double-stranded plasmid or PCR product)
✓ 10× PCR buffer
5 μM primer DNA
✓ 10× 3dNTP mix (appropriate to labeled dNTP used)
10 mCi/ml [α-^{32}P]dATP or [α-^{32}P]dCTP (3000 Ci/mmol)
2 U/μl *Taq* or *Tth* DNA polymerase
✓ DEPC-treated H_2O
Mineral oil
✓ 25 mM EDTA, pH 8.0
25:24:1 (v/v/v) phenol/chloroform/isoamyl alcohol
5 M ammonium acetate
100% ethanol, ice-cold
Gel loading buffer

RNase-free microcentrifuge tubes
Thermal cycler
Pasteur pipets with drawn-out tips

1. Assemble the following components in an RNase-free microcentrifuge tube on ice:

   Template DNA: 1.0 μg linearized double-stranded plasmid or 0.1 μg PCR product
   8.0 μl 10× PCR buffer

5.0 μl 5 μM primer DNA
8.0 μl 10× 3dNTP mix
10.0 μl 10 mCi/ml [α-^{32}P]dATP or [α-^{32}P]dCTP
2.5 μl 2 U/μl *Taq* or *Tth* polymerase
DEPC-treated H_2O to 80 μl.

2. Mix gently (do not vortex), then microcentrifuge briefly at maximum speed. Layer one drop of mineral oil over reaction.

3. Place reaction in thermal cycler and run the following program:

Initial step:	60 sec	95°C	(denaturation)
30 cycles:	60 sec	50°C	(annealing)
	90 sec	72°C	(extension)
Final step:	indefinitely	4°C	(soak).

4. When cycling is complete, either remove mineral oil from top of reaction mix or remove reaction mix from under mineral oil using a drawn-out Pasteur pipet.

5. Add 120 μl of 25 mM EDTA, mix, and put on ice.

6. Add 1 vol of 25:24:1 phenol/chloroform/isoamyl alcohol to the reaction, mix, then microcentrifuge 3 min at maximum speed. Transfer top phase to a new tube, add 1/10 vol of 5 M ammonium acetate, mix, and add 2.5 vol ice-cold 100% ethanol. Incubate 15 to 30 min at −20°C, then microcentrifuge 15 min at maximum speed. Remove supernatant and resuspend the probe in 10 μl gel loading buffer.

7. Gel purify the probe to remove the template DNA (see Support Protocol 1, steps 4a to 6a).

8. Precipitate the probe out of the probe elution buffer (see Support Protocol 1, step 6a) by adding 1/10 vol of 5 M ammonium acetate (0.5 M final) and 2.5 vol of 100% ethanol to remove EDTA.

   EDTA will reduce the effective concentration of zinc, which is essential for S1 nuclease activity.

9. Resuspend the probe in DEPC-treated water and store at −20° or −80°C.

5

SUPPORT PROTOCOL 6

SYNTHESIS OF DNA PROBES BY T4 POLYNUCLEOTIDE KINASE END LABELING OF OLIGONUCLEOTIDES

Materials *(see* APPENDIX 1 *for items with ✓)*

0.1 to 10 pmol/μl oligonucleotide to be labeled
150 mCi/ml [γ-^{32}P]ATP (7000 Ci/mmol)
✓ 10× T4 polynucleotide kinase buffer
10 U/μl T4 polynucleotide kinase
Gel loading buffer
20% denaturing polyacrylamide gel
5 M ammonium acetate
100% ethanol

RNase-free microcentrifuge tubes (e.g., Ambion)
95°C heating block

1. Assemble the following components in an RNase-free microcentrifuge tube on ice:

    0.1 to 10 pmol oligonucleotide to be labeled
    1 μl 150 mCi/ml [γ-^{32}P]ATP (25 pmol total)
    1 μl 10× T4 polynucleotide kinase buffer
    1 μl 10 U/μl T4 polynucleotide kinase
    DEPC-treated H_2O to 10 μl.

    Careful design of the oligo (length and base composition) is critical for obtaining nuclease-stable duplexes. Using a software ackage, select a region of the target mRNA where the sequence contains an even distribution of the four bases. The annealing site should be an internal site of 20 to 50 bases, located at least 20 bases from the 3′ or 5′ ends unless the probe is intended for use in end-mapping experiments. Because the label is on the 5′ end, this end must be 100% homologous to the target mRNA. Ideally, the 5′ end of the oligonucleotide probe should begin with at least two GC residues to minimize mRNA duplexing and loss of the 5′ end label during S1 digestion. By adding several noncomplementary bases to the 3′ end of the oligonucleotide probe, a size shift will be produced between the protected fragment and the full-length probe.

2. Mix the reaction by gentle flicking, then microcentrifuge briefly at maximum speed and incubate 30 min at 37°C.

3. Incubate 2 min at 95°C to inactivate the enzyme.

4. To determine the efficiency of the reaction, calculate the mass amount of labeled nucleotide incorporated into the probe after removal of unincorporated label by column chromatography or ethanol precipitation.

5. Add an equal volume of gel loading buffer and heat 3 min at 95°C.

6. Gel purify the probe on a 20% denaturing polyacrylamide gel (see Support Protocol 1, steps 4a to 6a).

7. Precipitate the probe out of the probe elution buffer (see Support Protocol 1, step 6a) by adding 1/10 vol of 5 M ammonium acetate (0.5 M final) and 2.5 vol 100% ethanol. Incubate 15 to 30 min at −20°C, then microcentrifuge 15 min at maximum speed to remove EDTA.

    EDTA will reduce the effective concentration of zinc, which is essential for S1 nuclease activity.

8. Remove supernatant and resuspend the probe in DEPC-treated water and store at −20° or −80°C.

SUPPORT PROTOCOL 7

5′ END MAPPING OF mRNA TRANSCRIPTION START SITES

Probes synthesized for mapping transcription start sites should contain both nontranscribed 5′ genomic sequences and exon sequences. The portion of the probe corresponding to nontranscribed sequences will be digested by the S1 nuclease and the size difference between the full-length probe and the protected fragment reflects the position of the 5′ end of the mRNA target. The transcription start site is mapped by comparing the size of the protected fragment to the size of the undigested probe (Fig. 5.2.2). For example, if the protected fragment is 75 nucleotides shorter than the full-length probe, the transcription start site would lie 75 nucleotides upstream from the 3′ end of the probe. For exact determination of the size of the protected fragment (which is necessary to map the transcription start site to the single-nucleotide level), the protected fragment may be analyzed on a sequencing gel in conjunction with a DNA "sequencing ladder" reaction. If such a high degree of resolution is not required, adequate molecular size markers may be prepared by end labeling DNA fragments generated

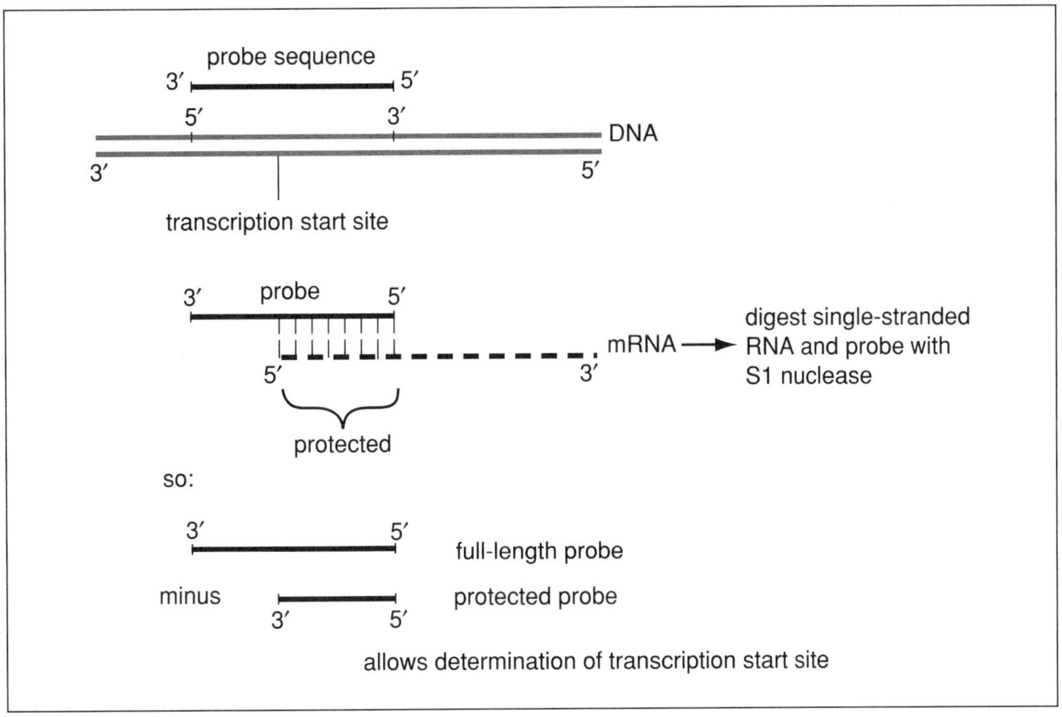

Figure 5.2.2 5′ end mapping of mRNA transcription start sites.

by digestion of a plasmid with a frequent-cutting restriction enzyme (e.g., pUC19 digested with *Sau*3A).

Contributors: Marianna Goldrick and Donald Kessler

5

UNIT 5.3

Analysis of mRNA Populations from Single Live and Fixed Cells of the Central Nervous System (CNS)

NOTE: When aspirating cell contents and preparing for single-stranded cDNA synthesis, all reagents and materials should be RNase-free and kept on ice.

BASIC PROTOCOL

SINGLE-CELL mRNA AMPLIFICATION FROM LIVE CELLS

The use of an oligo(dT) primer to amplify aRNA (amplified antisense RNA) is detailed in Figure 5.3.1.

Materials *(see* APPENDIX 1 *for items with ✓)*

✓ Electrode solution with T7-oligo$(dT)_{24}$ primer
Cells
✓ 10× electrode buffer, adjusted to pH 8.3 with 2 M NaOH

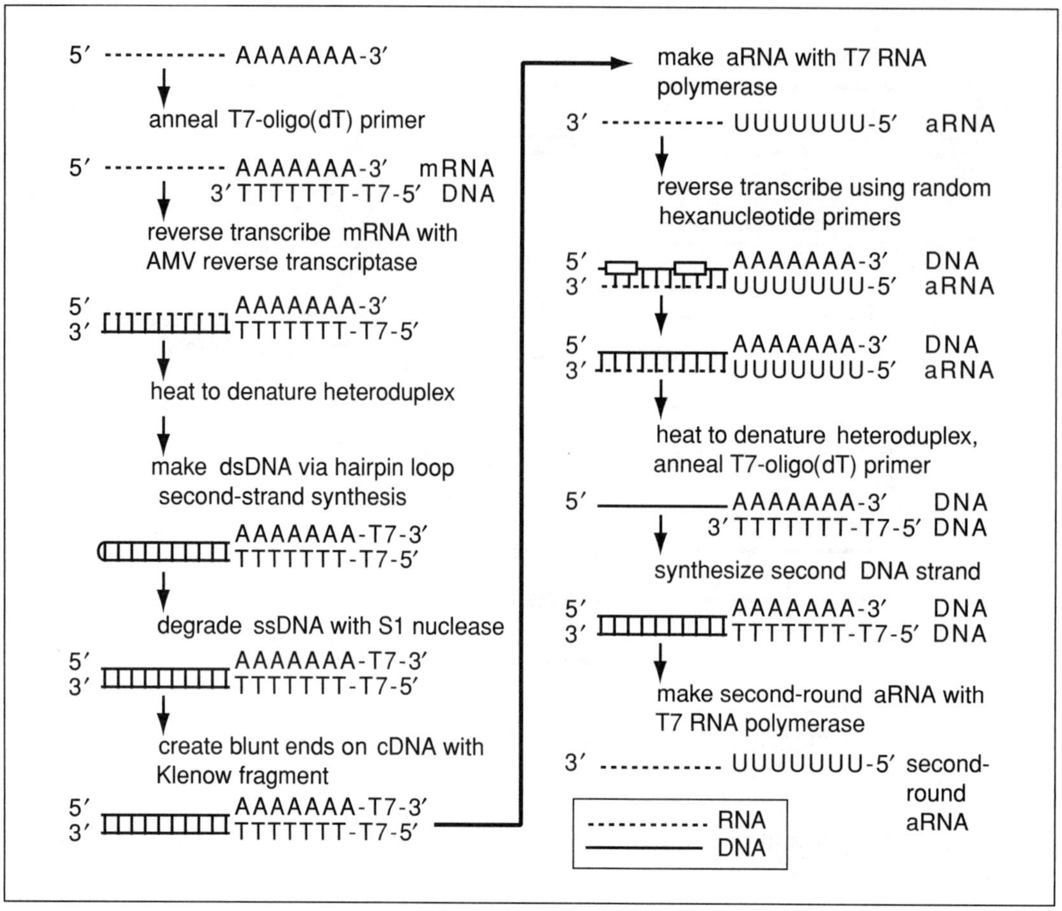

Figure 5.3.1 aRNA amplification schematic.

2.5 mM 4dNTP mix
20 U/μl AMV-RT (Seikagaku America)
5 M NaCl
100% ethanol
E. coli tRNA
✓ 10× second-strand buffer
1 U/μl T4 DNA polymerase
1 U/μl Klenow fragment of *E. coli* DNA polymerase
✓ 10× S1 buffer
1 U/μl S1 nuclease (Boehringer Mannheim) in 1× S1 buffer
1:1 (v/v) buffered phenol/chloroform
✓ 10× Klenow filling in (KFI) buffer
10 mM each dATP, dCTP, dGTP, and dTTP
✓ TE buffer, RNase-free (e.g., made with DEPC-treated water)
✓ 10× RNA amplification buffer
100 mM DTT
2.5 mM 4NTP mix
2.5 mM 3NTP mix (A, G, U)
0.1 mM CTP
10 mCi/ml [α-^{32}P]CTP (800 Ci/mmol)
0.1 U/μl RNasin (Promega) or equivalent RNase inhibitor
2000 U/μl T7 RNA polymerase (Epicentre Technologies)

3 M sodium acetate, pH 6.5
Random hexanucleotide primers
✓ 10× RT buffer
100 ng/μl T7-oligo(dT)$_{24}$ (for T7 sequence, see recipe for electrode buffer with primer)
✓ DEPC-treated water
Formaldehyde
✓ 10× MOPS
Formamide
✓ Gel loading dye
10% (w/v) TCA

Electrode apparatus
1-ml sterile syringe
14°, 37° to 42°, 75°, 85°, and 95°C heating blocks or water baths
Dry ice/ethanol bath
0.025-μm Millipore filter

1. Backfill electrode with 50 μl of electrode solution with T7-oligo(dT)$_{24}$ immediately before use, using 1-ml sterile syringe that has been heated and pulled to form a drawn tip. Keep syringe on ice between fills. Penetrate cell and either allow electrode contents to diffuse within the cell or pressure-inject some of it into the cell. Wait 1 to 20 min before withdrawing cytoplasm into electrode tip.

2. Make up the reaction solution in a sterile 0.5-ml microcentrifuge tube (20 μl final volume):

   15.5 μl H_2O
   2 μl 10× electrode buffer, pH 8.3
   2 μl 2.5 mM 4dNTP mix
   0.5 μl 20 U/μl AMV-RT.

   Introduce mRNA-containing sample according to step 3a or 3b.

3a. *Preferred method:* Remove holder (with electrode still attached) and blow contents into the microcentrifuge tube containing the reaction solution, rinsing the electrode into the solution by alternatively sucking and blowing.

3b. *Alternative method:* Remove electrode from holder and break tip into the microcentrifuge tube containing the reaction solution. Microcentrifuge 2 min at maximum speed, 4°C, to empty contents.

4. Incubate 60 to 90 min at 37° to 42°C.

5. Add 2 μl of 5 M NaCl, 40 μl of 100% ethanol, and 1 μg *E. coli* tRNA to each tube to aid precipitation. Mix gently; place at −70°C for 30 min. Microcentrifuge 15 min at maximum speed, 4°C. Remove supernatant (the pellet should be obvious) and add 20 μl water and suspend pellet.

6. Heat 3 min at 95°C to separate RNA/DNA duplexes. Quick-cool on ice; microcentrifuge briefly to collect drops.

7. Add the following to the tube containing single-stranded cDNA in 20 μl:

   10 μl H_2O
   4 μl 10× second-strand buffer
   4 μl 2.5 mM 4dNTP mix
   1 μl 1 U/μl T4 DNA polymerase
   1 μl 1 U/μl Klenow fragment.

   Mix gently; incubate ≥2 hr at 14°C.

8. Add 319 μl water and mix gently. Add 40 μl of 10× S1 buffer and 1 μl S1 nuclease (1 U/μl in 1× S1 buffer) to degrade single-stranded DNA. Mix and incubate 5 min at 37°C.

9. Add 1 vol (400 μl) 1:1 (v/v) phenol/chloroform. Vortex 30 sec and microcentrifuge 60 sec at maximum speed, room temperature. Transfer supernatant to new tube. Add 1 ml of 100% ethanol. Put in dry ice/ethanol bath for 2 hr. Microcentrifuge 15 min at maximum speed, 4°C. If a large salt pellet is produced, repeat ethanol precipitation.

10. Discard supernatant (unless no pellet is visible) and air dry. Suspend pellet in 20 μl water. Add the following to the suspended DNA:

    2.5 μl 10× KFI buffer
    0.5 μl 10 mM dATP
    0.5 μl 10 mM dCTP
    0.5 μl 10 mM dGTP
    0.5 μl 10 mM dTTP
    0.5 μl Klenow fragment (1 U/μl).

    Mix and incubate 15 min at 37°C.

11. Add 25 μl of 1:1 phenol/chloroform. Vortex 30 sec and microcentrifuge 60 sec at maximum speed, room temperature. Transfer supernatant to new tube. Add 2.5 μl of 5 M NaCl and 75 μl of 100% ethanol. Mix and place in dry ice/ethanol bath for 2 hr.

12. Microcentrifuge 15 min at maximum speed, 4°C. Discard supernatant (unless no pellet is visible) and air dry. Suspend pellet in 20 μl RNase-free TE buffer.

13. Drop-dialyze cDNA sample 4 hr to overnight in two batches by dropping 10-μl on top of a 0.025-μm Millipore filter against 50 ml of RNase-free TE or water in a sterile 50-ml conical tube. When finished, transfer drop to a sterile 0.5-ml microcentrifuge tube, rinsing spot on filter with 10 μl TE and adding rinse to tube.

14. Assemble the components for aRNA amplification in a 0.5-ml microcentrifuge tube (see Table 5.3.1; 20 μl total volume). Mix gently and incubate 4 hr at 37°C.

15. Add 20 μl of 1:1 phenol/chloroform and 2 μl of 3 M sodium acetate, pH 6.5. Vortex 30 sec and microcentrifuge 60 sec at maximum speed, room temperature. Transfer supernatant to new tube. Add 70 μl of 100% ethanol. Mix and put in dry ice/ethanol bath for 2 hr. Microcentrifuge 15 min at 4°C. Discard supernatant (unless no pellet is visible) and air dry. Suspend pellet in 20 μl RNase-free water. Store indefinitely at −70°C.

Table 5.3.1 aRNA Amplification Mixture

Component	Unlabeled reaction	Radiolabeling reaction
H_2O	11.5 μl	7.5 μl
Double-stranded cDNA (after drop dialysis)	2 μl	2 μl
10 RNA amplification buffer	2 μl	2 μl
100 mM dithiothreitol (DTT)	1 μl	1 μl
2.5 mM 4NTP mix	2 μl	—
2.5 mM 3NTP mix (A, G, U)	—	2 μl
0.1 mM CTP	—	2.5 μl
1 mCi/100 μl [α-^{32}P]CTP (800 Ci/mmol)	—	1.5 μl
0.1 U/μl RNasin	0.5 μl	0.5 μl
2000 U/μl T7 RNA polymerase	1 μl	1 μl

5

16. Denature 2 to 20 µl aRNA by heating 3 min at 75°C. Add the following:

    x µl H_2O (RNase-free; amount depends on volume of aRNA and amount of primers)
    10 to 100 ng random hexanucleotide primers (depends on yield of first-round aRNA)
    3 µl 10× RT buffer
    3 µl 100 mM DTT
    3 µl 2.5 mM 4dNTP mix
    0.5 µl RNasin (0.1 U/µl)
    2 µl AMV-RT (20 U/µl).

    Mix gently; incubate for 1 hr at 37°C.

17. Add 30 µl of 1:1 phenol/chloroform and 3 µl of 3 M sodium acetate. Vortex 30 sec and microcentrifuge 60 sec at maximum speed, room temperature. Transfer supernatant to new tube. Add 65 µl ethanol, mix, and place 30 min at −70°C. Microcentrifuge 15 min at maximum speed, 4°C. Discard supernatant, air dry pellet, and suspend in 10 µl water.

18. Denature aRNA/cDNA heteroduplexes by heating 3 min at 95°C. Quick-cool on ice, microcentrifuge briefly to collect drops, and add:

    3 µl H_2O
    2 µl 10× KFI buffer
    2 µl 4 dNTP mix (2.5 mM each)
    1 µl T7-oligo$(dT)_{24}$ (100 ng/µl)
    1 µl T4 DNA polymerase (1 U/µl)
    1 µl Klenow fragment (1 U/µl).

    Incubate ≥2 hr at 14°C.

19. Add 20 µl of 1:1 phenol/chloroform and 2 µl of 3 M sodium acetate. Vortex 30 sec and microcentrifuge 60 sec at maximum speed, room temperature. Transfer supernatant to new tube, add 45 µl of 100% ethanol, and place at −70°C for 30 min. Microcentrifuge 15 min at maximum speed, 4°C. Discard supernatant, air dry, and suspend pellet in 20 µl TE.

20. Drop-dialyze against 50 ml TE ≥4 hr as described in step 13.

21. Amplify aRNA as described in step 14 (also see Table 5.3.1).

22. Add 40 µl of 1:1 phenol/chloroform and 2.5 µl of 3 M sodium acetate. Vortex 30 sec and microcentrifuge 60 sec at maximum speed, room temperature. Transfer supernatant to new tube, add 70 µl of 100% ethanol, and mix. Put in dry ice/ethanol bath for 2 hr. Microcentrifuge 15 min at maximum speed, 4°C. Discard supernatant (unless no pellet is visible) and air dry. Suspend pellet in 20 µl DEPC-treated water. Store at −70°C.

23. After aRNA synthesis, prepare an aliquot of aRNA for electrophoresis on a 1% agarose denaturing gel by combining:

    1 µl aRNA reaction mix
    3.5 µl DEPC-treated H_2O
    3.5 µl formaldehyde
    2 µl 10× MOPS
    10 µl formamide.

    Heat-denature 5 min at 85°C.

24. Add 2 µl gel loading dye. Load the denaturing gel and electrophorese until the bromphenol blue dye has migrated 75% of the way down the gel. After electrophoresis, fix the gel in 10% (w/v) TCA (three 15-min washes at room temperature).

ALTERNATE PROTOCOL

SINGLE-CELL mRNA AMPLIFICATION FROM IMMUNOSTAINED FIXED CELLS

Brain or other neural tissue to be used for single-cell mRNA analysis must first be processed for immunohistochemistry (IHC; *UNIT 1.2*). Water used in making solutions should be treated with diethylpyrocarbonate (DEPC) to remove RNases.

Additional Materials *(also see Basic Protocol; see APPENDIX 1 for items with ✓)*

Neural tissue
4% (w/v) paraformaldehyde *or* 70% ethanol/150 mM NaCl
✓ 0.5×, 2×, and 5× SSC
✓ IST reaction buffer
0.5 U/μl avian myeloblastosis virus reverse transcriptase (AMV-RT; Seikagaku America)
✓ cDNA synthesis buffer
✓ Microelectrode solution

Paraffin
Rubber cement
Poly-L-lysine-coated glass slides
Dissecting microscope

1. Immerse neural tissue in 4% (w/v) paraformaldehyde or 70% ethanol/150 mM NaCl overnight to fix the tissue.

2. Embed tissue blocks in paraffin and section at 7 μm. Mount sections onto poly-L-lysine-coated glass slides (see *UNIT 1.1* for sectioning and mounting instructions).

3. Label neurons using appropriate method (*UNIT 1.2*).

4. Following the labeling immunoreaction, rinse sections by dipping slides in DEPC-treated water.

   For archival material typically found in pathology laboratories that is fixed with neutral buffered formalin or 4% paraformaldehyde, the aRNA yield may be enhanced by treating the tissue with 0.2 N HCl for 20 min, rinsing in phosphate-buffered saline (PBS, pH 7.4; APPENDIX 2A), and then digesting 30 min with proteinase K (50 μg/ml).

5. Construct a 1-cm^2 rubber cement reaction well on the mounted sections by applying cement by syringe to circumscribe the section. Allow 5 min for drying.

6. Initiate IST by hybridizing the T7-oligo$(dT)_{24}$ primer to poly$(A)^+$ mRNA overnight at room temperature directly on the tissue section, using ~50 μl reaction buffer. Include 50% formamide in the reaction mixture to reduce the effective hybridization temperature (T_m) and use 5× SSC as a low-stringency buffer.

7. The next morning, wash off excess T7-oligo(dT) first with 2× SSC and then with IST reaction buffer.

8. Perform cDNA synthesis directly on the section with 0.5 U/μl AMV-RT in ~50 μl cDNA synthesis buffer; incubate reaction mixture 90 min at 37°C.

9. Wash sections 8 to 12 hr in 0.5× SSC at room temperature Use these for single-cell dissections.

10. Following IHC and IST, dissect individual immunolabeled cells from the surrounding neuropil by viewing with a dissecting microscope and using a recording microelectrode and micromanipulator. Gently aspirate dissected cell into microelectrode under a thin layer (~500 µl) of DEPC-treated water. To assure cDNA synthesis in the cell, fill microelectrode with microelectrode solution. Let cDNA synthesis proceed 90 min at 40°C.

11. Extract cDNA with 1:1 (v/v) phenol/chloroform and precipitate with 100% ethanol (see Basic Protocol, steps 9 to 10).

12. Synthesize double-stranded template cDNA as described starting at Basic Protocol, step 6. From this point on, follow Basic Protocol to complete the synthesis of aRNA from the fixed single cell.

SUPPORT PROTOCOL 1

REVERSE NORTHERN ANALYSIS OF mRNA

Initially, cloned plasmid cDNAs that have been linearized are adhered to nylon membranes (Hybond N+, Amersham). The cDNAs should be selected to reflect a sample of mRNAs expected to be present in individual cells so that a molecular fingerprint of expressed sequences can be generated. Blots are probed with ^{32}P-labeled aRNA from whole sections or individual cells (see Basic or Alternate Protocol). Prehybridization (8 hr) and hybridization (24 hr) steps are performed in 6× SSPE buffer containing 5× Denhardt's solution, 50% (v/v) formamide, 0.1% (w/v) SDS, and salmon sperm DNA (200 µg/ml) at 42°C. Blots are washed in 2× SSC and then apposed to film for 24 to 48 hr. Detection of radiolabeled aRNA can be performed by scanning densitometry of the autoradiographs (e.g., with ImageQuant 3.3 software from Molecular Dynamics).

5

SUPPORT PROTOCOL 2

ACRIDINE ORANGE LABELING

Perform all procedures at room temperature, and protect dye-containing solutions from light.

Materials *(see* APPENDIX 1 *for items with ✓)*

Fixed tissue sections (see Alternate Protocol)
✓ Citric acid/sodium phosphate buffer (CPBS buffer)
✓ Acridine orange staining solution
Fluorescence microscope (see *UNIT 2.1*)

1. Wash nonimmunostained section in CPBS.

2. Stain section 15 to 20 min with acridine orange staining solution.

3. Rinse section 1 min in CPBS.

4. View stained section using fluorescence microscopy with excitation at 470 nm, detecting emission at 630 nm.

References: Eberwine et al., 1992; Eberwine et al., 1995

Contributors: James Eberwine and Peter Crino

UNIT 5.4

Overview of RNA Interference and Related Processes

HISTORY

The history of RNA interference (RNAi) has unfolded rapidly since 1997 with a series of discoveries from plants, fungi, and animals. As the underlying mechanism was revealed, it became apparent that RNAi manifests a novel system of genetic regulation that was only hinted at by previous data.

From the initial findings, it was not clear whether the antisense and cosuppression mechanisms were related, although both involved a block at the post-transcriptional level (Kooter and Mol, 1993At about the same time, there was a similar controversy in the *Caenorhabditis elegans* field. Gene expression could be specifically suppressed by direct injection of antisense RNA; however, the process was inefficient because large amounts of antisense RNA had to be injected (Fire et al., 1991). Additionally, there was a remarkable resonance of the plant "antisense versus cosuppression" mystery, because injected sense RNA was as potent as antisense RNA (Guo and Kemphues, 1995).

An inspired analysis of the *C. elegans* phenomenon provided the key to understanding these unexpected findings. It revealed that RNAi in injected *C. elegans* was mediated by a small amount of double-stranded (ds) RNA that contaminated the sense and antisense RNA preparations (Montgomery et al., 1998). Subsequently, it was established in plants that if dsRNA was produced, suppression of endogenous gene expression was more efficient than with sense or antisense transgenes (Waterhouse et al., 1998; Chuang and Meyerowitz, 2000). Inverted repeat transgenes were particularly efficient, but simultaneous expression of sense and antisense RNA was also effective (Waterhouse et al., 1998).

5

A second important discovery followed from the search for the specificity determinant of cosuppression in plants. It was reasoned that, as cosuppression is nucleotide sequence-specific and acts at the RNA level, there should be antisense RNA corresponding to the target species. Presumably this antisense RNA would guide the degradation of the cosuppressed RNA. An initial unsuccessful search for this antisense RNA focused on RNA of >100 nucleotides length. However, when the hunt was redirected to small RNAs, an antisense species of ~25 nucleotides length was discovered (Hamilton and Baulcombe, 1999). These short RNAs also corresponded to the sense strand of the cosuppression target and it seemed likely that they were derived from a dsRNA precursor. This link with RNAi was subsequently confirmed when short RNAs were associated with RNAi in *Drosophila melanogaster* (Hammond et al., 2000). The *D. melanogaster* work also confirmed the prediction that the short RNAs guide a ribonuclease complex (RISC) to its target RNA.

The discovery that short RNAs play a key role in RNAi-mediated suppression of gene expression precipitated an avalanche of discoveries that are relevant to the application of RNAi and cosuppression, as well as to understanding of the natural roles of these processes. For example, the short RNAs have been characterized in detail and are now known to exist in a double-stranded form, with two-nucleotide overhangs at each 3′ end, and are known as siRNAs (Elbashir et al., 2001a). Several different proteins have been identified that are associated with siRNAs in ribonucleoproteins (Caudy et al., 2002; Mourelatos et al., 2002). At least some of these proteins are part of RISC (Hammond et al., 2001). It is also now understood how dsRNA is processed into the siRNAs. The processing enzyme is known as Dicer and is a member of the RNase III family with dsRNA binding regions and a conserved PAZ domain shared with members of the piwi, argonaute, and zwille family (Bernstein et al., 2001), from which it takes its name. ATP is required at several stages in the processing of dsRNA and assembly of RISC (Nykanen et al., 2001). In some instances an RNA-dependent RNA polymerase is also involved (Cogoni and Macino, 1999; Dalmay et al., 2000; Mourrain et al., 2000). It converts a single-stranded RNA into a double-stranded siRNA precursor.

NATURAL ROLES OF RNAi AND RELATED PROCESSES

In plants, a natural role of RNAi and cosuppression is as an antiviral defense system (Voinnet, 2001). Viruses produce a dsRNA replication intermediate that is the substrate for Dicer and that leads to production of siRNA. The accumulation of siRNA guides RISC to the single-stranded viral RNA so that a feedback system is established in which there is an equilibrium between the level of RISC and the rate of virus accumulation. Accordingly, viruses produce suppressor proteins of RNAi as a counter-defense system (Voinnet, 2001). It is thought that if a suppressor is weak then virus accumulation would be transient, whereas a strong suppressor would permit prolonged accumulation to a high level. At present it is not known whether the mechanism has an antiviral role in vertebrates; however, the recent findings that siRNAs have antiviral activity in mammalian cells are certainly consistent with that possibility (Novina et al., 2002).

The RNAi mechanism may also provide protection against transposons or other genome perturbations. Thus, in plants and protozoans there are endogenous short RNAs from retroelements (Hamilton et al., 2002), while in *C. elegans* mutations that affect RNAi may also cause genome instability (Ketting et al., 1999). Presumably, retroelement double-stranded RNA is processed by Dicer into short RNAs that prevent expression and mobilization of the corresponding genomic elements.

Many of the proteins required for RNAi are conserved in animals, plants, and fungi (Zamore, 2001). It is therefore likely that the mechanism evolved in primitive eukaryotes as a defense system against viruses and selfish DNA, as in modern plants. However, it seems that the RNAi mechanisms have been recruited during evolution of modern eukaryotes into other aspects of genetic regulation. It is not known at present whether RISC is involved in these effects at the genome level. The alternative possibility is that a separate complex recruits short RNAs and targets DNA rather than RNA.

Additional roles of RNAi-related mechanisms are implied from the identification of microRNAs (miRNAs) in plants and animals (Lagos-Quintana et al., 2001; Lau et al., 2001; Lee and Ambros, 2001; Rheinhart et al., 2002; Rhoades et al., 2002). These miRNAs are similar to siRNAs in that they are produced by Dicer-like enzymes and are a similar length. They may also be recruited into RISC and mediate targeted RNA degradation or translational arrest. However, unlike siRNAs that are from transposons, transgenes, or viruses, the miRNAs are transcribed from endogenous DNA. A typical dsRNA precursor of miRNA is the transcript of inverted-repeat DNA in regions between conventional genes.

The likely role of miRNAs is in the control of mRNA translation or stability. Recently, an miRNA in *Arabidopsis* (*miR171*) has been identified that apparently regulates site-specific cleavage of a transcription factor mRNA (Llave et al., 2002a).

For most animal miRNAs, the putative miRNA targets cannot be easily identified because the mismatch is greater than in plants. A strategy for miRNA target identification is a major outstanding challenge.

STRATEGIES FOR RNAi AND COSUPPRESSION IN PLANTS AND ANIMALS

RNAi and cosuppression are promising new approaches to genetic analysis and manipulation. It should be possible, for example, to infer the function of an uncharacterized gene, to correct dominant genetic defects, and to develop antiviral strategies using this approach. RNAi and related procedures may have several advantages over alternative strategies or may provide solutions to currently intractable problems. For example, if a gene of interest is a member of a multigene family, it may not be possible to infer function from standard genetic knockout strategies. The expression of functionally redundant gene family members may compensate for the mutant gene. However, because the RNAi-based procedures are nucleotide-sequence rather than genetic-locus specific, they allow suppression of many similar family members with a single manipulation.

Despite the relative novelty of RNAi and cosuppression technologies, there are already many different procedures that can be used, depending on the organism and the function of the target RNA. The most cumbersome of these involve transformation of the organism to produce a dsRNA precursor of an siRNA. This approach has been used in plants and animals (Wesley et al., 2001; Giordano et al., 2002; Paddison et al., 2002; Sui et al., 2002) and has the advantage that the phenotype is stable through several generations.

More direct approaches do not require stable transformation and are particularly amenable to high-throughput applications, including genome-wide surveys of gene function. It is possible, for example, to transfect with dsRNA or siRNA targeted at a gene of interest (Elbashir et al., 2001b; Harborth et al., 2001). If chemically synthesized siRNAs are used, this approach is relatively expensive. However siRNA produced in vitro by enzymatic cleavage of dsRNA may be a less expensive alternative (Yang et al., 2002).

5

Direct introduction of dsRNA or siRNA is effective in cultured cells. Surprisingly, however, direct introduction of siRNA or dsRNA to intact multicellular animals can also result in RNAi. With *C. elegans*, for example, it is necessary only to feed animals on *E. coli* that is transformed to produce the dsRNA (Timmons and Fire, 1998). Thus, by producing a set of *E. coli* strains with dsRNA targeted against each gene, it was possible to generate a resource that will greatly facilitate characterization of the *C. elegans* genome (Kamath et al., 2003). Short dsRNA or siRNA could also be used in adult mice to target transgene expression in the liver (McCaffrey et al., 2002).

In plants, RNAi can be achieved in transgenic plants expressing the dsRNA of the target. An alternative strategy involves infecting plants with virus vectors carrying endogenous gene fragments. The corresponding host RNA is suppressed in the infected plant and the symptoms resemble a loss of function mutant phenotype. This approach has been refined on the model plant *Nicotiana benthamiana* and has been used to investigate genes involved in cell wall biosynthesis and disease resistance (Burton et al., 2000; Peart et al., 2002a,b). The potential of virus-mediated RNAi has not been fully explored in animals or animal cells.

DETECTION AND CHARACTERIZATION OF siRNAs AND miRNAs

A direct procedure for detection of siRNA and miRNA is northern blotting. The procedures are similar to the standard protocol for detection of mRNA, but are modified to account for the small size of these RNAs. Thus, the RNA is fractionated by electrophoresis on a polyacrylamide gel instead of agarose and electroblotted onto a nylon membrane. The stringency of the hybridization is also reduced. Probes can be labeled cDNA, cRNA, or oligonucleotides (Hamilton and Baulcombe, 1999).

Depending on its abundance, the short RNA may be detectable in total RNA preparations. However, for rare species it is necessary to enrich for short RNA by selective precipitation or enrichment for short RNA protein complexes (Mourelatos et al., 2002). Alternatively, for rare short RNAs, RNase protection can be used instead of northern blotting. However the cost of the increased sensitivity is the loss of size resolution—i.e., rather than appearing as a discrete-sized species, the RNase protection reveals the short RNA as a smear (Sijen et al., 2001).

There are several procedures for cloning of cDNA corresponding to siRNA or miRNA (Djikeng et al., 2001; Lagos-Quintana et al., 2001; Llave et al., 2002b). These procedures all involve addition of adaptors to the termini of the short RNA using RNA ligase followed by reverse transcription and PCR amplification. The various protocols employ different strategies to avoid circularization or concatemerization of the reaction intermediates. However, some of the protocols do allow for concatemerization of the PCR products prior to cloning so that a single sequence reaction will be informative about several short RNA species.

Contributor: David Baulcombe

UNIT 5.5

Production of Antipeptide Antisera

A flowchart for the preparation of antipeptide antibodies (antibodies reactive with a synthetic peptide) is shown in Figure 5.5.1.

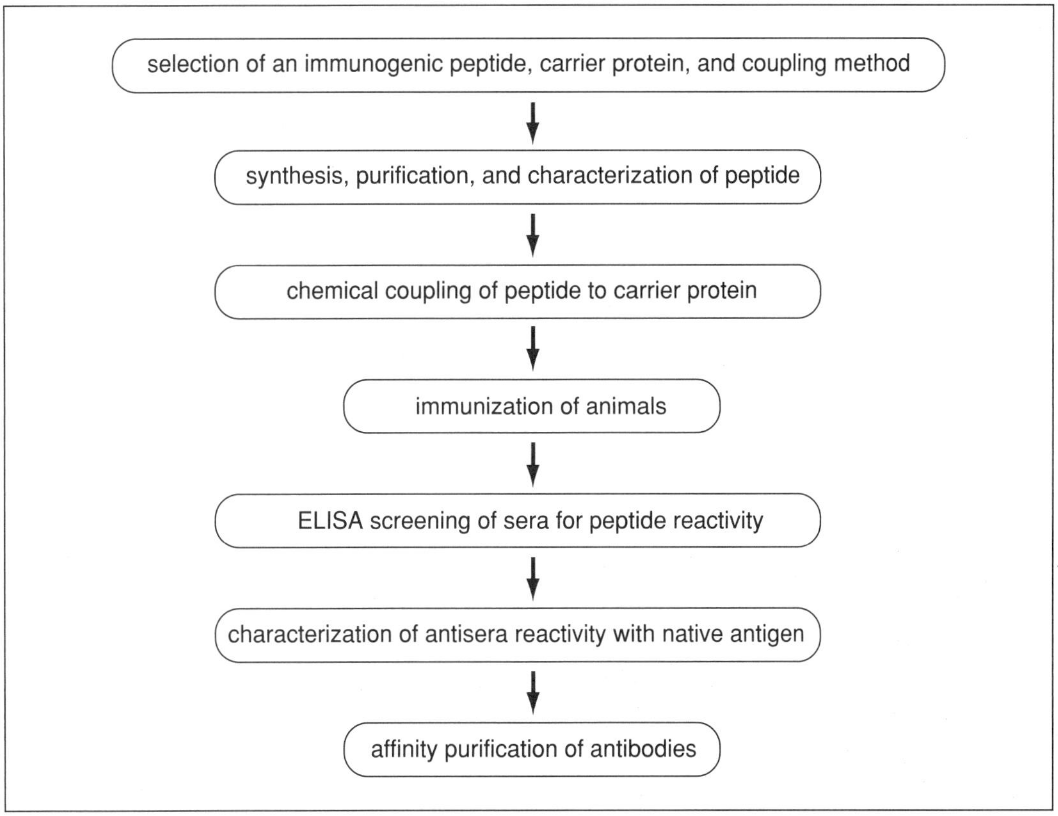

Figure 5.5.1 Flow chart for preparation and analysis of antipeptide antibodies.

Selection of an immunogenic peptide is the most critical step in obtaining an antibody that reacts with the native antigen. In practice, a 10- to 15-residue peptide sequence inferred from a cDNA sequence or from an N- or C-terminal amino acid sequence is selected. Alternatively, nonterminal sequences that are likely to be antigenic—exposed on the surface, located in flexible regions of the protein, or found in reverse turns or loop structures—are chosen. Sequences should be avoided that are likely to be identical or highly homologous to those in the animal to be immunized (usually rabbits). After synthesis and purification, the peptide is cross-linked to a carrier protein such as keyhole limpet hemocyanin (KLH).

Selection of a peptide from an internal part of the protein sequence can be aided by the use of algorithms to predict those regions most likely to be exposed on the surface of the protein. The only information needed is the primary amino acid sequence.

Using the first algorithmic method, a hydrophilicity value is assigned for each overlapping six-amino-acid segment of the protein sequence based on the average of the hydrophilicity values (Table 5.5.1) of the amino acids in that segment. The highest point of average local hydrophilicity is usually located in or near an antigenic determinant. A computer program written in Basic is available for analysis (Hopp and Woods, 1983).

An alternative algorithm evaluates the hydrophobic and hydrophilic tendencies of a polypeptide chain based on water vapor free-energy transfers and the interior versus exterior distributions of amino acid side chains. Values for each amino acid are listed in the second column of Table 5.5.1. A computer program for calculating this hydropathicity profile has been written

Table 5.5.1 Hydrophobic/Hydrophilic Index of Amino Acids

Amino acid	Hydrophilicity value[a]	Hydropathy index[b]
Arginine (R)	3.0	−4.5
Aspartic acid (D)	3.0	−3.5
Glutamic acid (E)	3.0	−3.5
Lysine (K)	3.0	−3.9
Serine (S)	0.3	−0.8
Asparagine (N)	0.2	−3.5
Glutamine (Q)	0.2	−3.5
Glycine (G)	0.0	−0.4
Proline (P)	0.0	−1.6
Threonine (T)	−0.4	−0.7
Alanine (A)	−0.5	1.8
Histidine (H)	−0.5	−3.2
Cysteine (C)	−1.0	2.5
Methionine (M)	−1.3	1.9
Valine (V)	−1.5	4.2
Isoleucine (I)	−1.8	4.5
Leucine (L)	−1.8	3.8
Tyrosine (Y)	−2.3	−1.3
Phenylalanine (F)	−2.5	2.8
Tryptophan (W)	−3.4	−0.9

[a]Hopp and Woods (1981).
[b]Kyte and Doolittle (1982).

by Kyte and Doolittle (1982). This profile is useful for determining exterior and interior regions of a native protein, as well as for locating signal sequences and transmembrane sequences.

A third algorithm is an empirical method that relies on a library of known structures to determine the frequency with which each amino acid occurs in the various conformational states (i.e., α-helix, β-sheet, β-turn, or all other structural forms; Chou and Fasman, 1974). Using these frequencies, predictions can be made about secondary structure for a given sequence. For making antipeptide sera, regions that are predicted to form turns or loops, or extended sequences (20 to 25 residues) that have a very high probability for formation of an α-helix, are useful. A computer program for performing this analysis can be found in Corrigan and Huang (1982).

Avoid choosing peptides containing predicted polysaccharide attachment sites, most notably the sequence Asn-X-Ser or Asn-X-Thr, which predict the presence of an Asn-linked polysaccharide. It is likely that the presence of polysaccharide moieties at such sites in the native protein would interfere with antibody accessibility.

The above-mentioned computer programs, plus programs to predict flexibility, location of transmembrane regions, Asn-linked glycosylation sites, and sites of signal sequence cleavage, are all contained in a package called PC Gene produced by IntelliGenetics.

Generally, peptides with a length of 10 to 15 residues are used to make antipeptide sera that react with the native protein.

Thus, in summary, a reasonable order of suggestions for choosing peptide sequences for making antipeptide sera would be:

1. If possible, use more than one peptide.
2. Use the C-terminal sequence (7 to 15 residues) if it is hydrophilic and if a suitable coupling group is available or can be added.
3. Use the N-terminal sequence (7 to 15 residues) if it is hydrophilic and if a suitable coupling group is available or can be added.
4. Use internal hydrophilic regions, perhaps using longer peptides (15 to 20 residues).

Synthetic peptides used for the production of antisera should at least be purified by gel-filtration, if not by RP-HPLC, to the extent that any cleavage by-products and remaining chemical scavengers employed during synthesis are removed. If pure antipeptide antibodies are required for the experiment at hand (e.g., as needed for immunohistochemical staining; see *UNIT 1.2*) because of the high background inherent to rabbit antisera, these can be obtained by immunoaffinity chromatography.

A minimal assessment of peptide purity should include an analytical RP-HPLC analysis, with the number of peaks visible at A_{215} being a rough gauge of purity. The assessment of purity should also include some form of compositional analyses done by a mass spectrometric mass determination of the crude synthetic peptide or its component peaks resolved by RP-HPLC.

Other features of the peptide must be considered in order for it to mimic the native antigen as closely as possible. The N-terminal amino group can be modified by acetylation of the peptide α-amino group during synthesis, and the C-terminal carboxyl group can be modified with a C-terminal amide during peptide synthesis to more closely mimic native structure.

A carrier protein should be a good immunogen and have a sufficient number of amino acid residues with reactive side-chains (see Table 5.5.2) for coupling to the synthetic peptide. Table 5.5.2 lists the most common protein carriers and their relevant properties.

In addition to the choice of peptide, a method for coupling the peptide to a protein carrier must also be selected. The chosen coupling method should link the peptide to the carrier via either the

Table 5.5.2 Principal Carriers Used for Coupling Peptides[a]

Carrier[b]	M_r (kDa)	Number of groups/molecule			
		ε-NH_2	-SH	Phenol	Imidazole
BSA	67	59[c]	1	19	17
Ovalbumin	43	20	4	10	7
Myoglobin	17	19	0	3	12
Tetanus toxoid	150	106	10	81	14
KLH	>2000	6.9[d]	1.7[d]	7.0[d]	8.7[d]

[a]Adapted from Van Regenmortel et al. (1988).

[b]Abbreviations; BSA, bovine serum albumin; KLH, keyhole limpet hemocyanin.

[c]Only 30 to 35 of the 59 Lys residues of BSA are accessible.

[d]For KLH, the amino acid groups are expressed in grams of amino acid containing this functional group per 100 g.

C- or N-terminal residue. Peptides corresponding to the amino terminus of proteins should be coupled through their carboxyl-terminal amino acid residue, whereas peptides corresponding to the carboxyl terminus of proteins should be coupled through their amino-terminal amino acid residue.

One coupling procedure that has proved to be particularly effective employs *m*-maleimidobenzoyl-*N*-hydroxysuccinimide ester (MBS) as the coupling reagent. If coupling with MBS is not desirable for some reason (such as the presence of non-terminal Cys residues that are not disulfide-linked in the native protein), coupling can then be accomplished through a C- or N-terminal Tyr using *bis*-diazotized benzidine (BDB), through the C or N terminus with 1-ethyl-3-(3-dimethylaminopropyl)-carbodiimide (EDCI), or through the N-terminal α-amino group with glutaraldehyde. BDB coupling is not advisable if there are internal Tyr residues in the peptide. EDCI coupling is not advisable when internal Glu, Asp, or Lys residues are present. Glutaraldehyde coupling may not be appropriate if there are internal Lys residues in the peptide. Table 5.5.3 lists the principal reagents used for peptide-protein conjugation and the functional groups involved.

For example, if the carboxy-terminal sequence of a protein is

1 2 3 4 5 6 7 8 9 10
S Y G R N Q A E C Q – COOH

then coupling via MBS by adding a Cys residue to the N terminus may not be appropriate because of the Cys at position 9 (see Table 5.5.1 for single-letter codes). In this case, it may

Table 5.5.3 Principal Reagents Used for Peptide Protein Conjugation[a]

Coupling agent	Modified amino acid	
	Primary reaction	Secondary reaction
Glutaraldehyde	ε-NH_2Lys, α-NH_2, SH-Cys	Tyr, His
Bis-imido esters	α-NH_2, ε-NH_2Lys	Negligible
BDB	Tyr, SH-Cys, His, ε-NH_2Lys	Trp, Arg
Carbodiimides (EDCI)	α-NH_2, ε-NH_2Lys, α-COOH, Glu, Asp	Tyr, Cys
MBS	Cys-SH	Not observed

[a]Adapted from Van Regenmortel et al. (1988). Abbreviations: BDB, *bis*-diazotized benzidine; EDCI, 1-ethyl-3-(3-dimethylaminopropyl)-carbodiimide; MBS, *m*-maleimidobenzoyl-*N*-hydroxysuccinimide.

be preferable to couple the peptide via the N terminus using glutaraldehyde. However, if the Cys at position 9 is known to be part of a disulfide loop in the native protein, it may be better to couple with MBS through the naturally occurring Cys at position 9.

NOTE: Tris buffers should not be used in the protocols.

BASIC PROTOCOL 1

COUPLING OF SYNTHETIC PEPTIDE TO CARRIER PROTEIN USING MBS

MBS is the best known heterobifunctional reagent; at neutral pH it cross-links thiol groups with amino groups. The linkage proceeds via two separate reactions, thus avoiding bonds between identical molecules (Fig. 5.5.2).

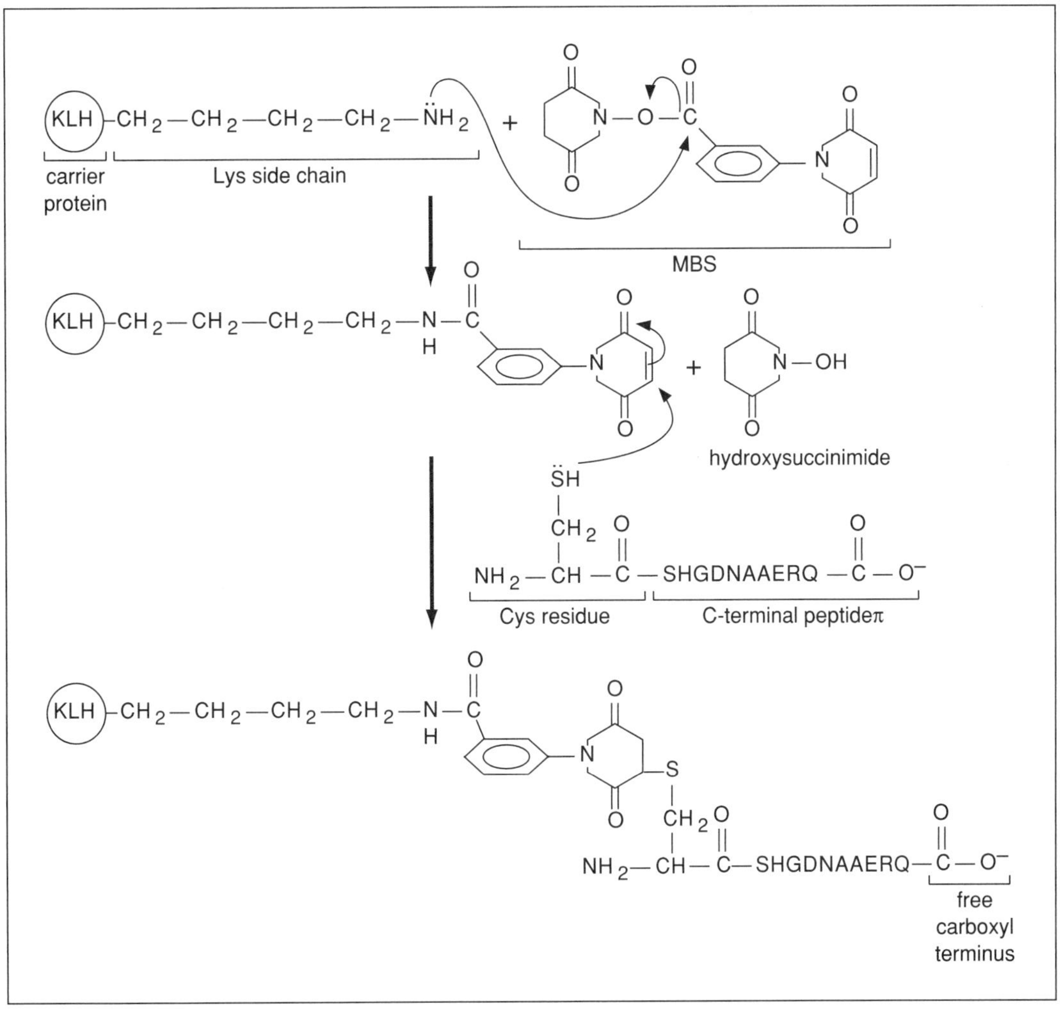

Figure 5.5.2 Coupling a C-terminal peptide to KLH carrier protein using MBS. The coupling can also be done using an N-terminal peptide if the Cys residue is present at the C terminus (instead of at the N terminus). KLH, keyhole limpet hemocyanin; MBS, *m*-maleimidobenzoyl-*N*-hydroxysuccinimide ester.

Materials *(see* APPENDIX 1 *for items with ✓)*

Keyhole limpet hemocyanin (KLH; Calbiochem)
✓ 0.01 M potassium phosphate buffer, pH 7.0
15 mg/ml *m*-maleimidobenzoyl-*N*-hydroxysuccinimide ester (Pierce) in dimethylformamide (MBS/DMF), prepared within 1 hr of use
✓ 0.05 M potassium phosphate buffer, pH 6.0
Cys-containing synthetic peptide
✓ PBS
0.1 M HCl
0.1 M NaOH

Dialysis tubing (10,000 MWCO; Spectrapor; Fisher)
10 × 75–mm (3-ml) and 15-ml glass test tubes
PD-10 column, prepacked (Pharmacia Biotech)

1. Dissolve 5 mg KLH in ~0.5 ml of 0.01 M phosphate buffer, pH 7.0. Place in dialysis tubing and dialyze against 4 liters of 0.01 M phosphate buffer, pH 7.0, overnight at 4°C. Transfer dialyzed solution into a 10 × 75–mm glass test tube. Add 70 μl MBS/DMF to dialyzed KLH solution. Stir gently with magnetic stirbar 30 min at room temperature.

2. Pre-equilibrate a PD-10 column (see manufacturer's instructions) by washing with 50 ml of 0.05 M phosphate buffer, pH 6.0, and load the KLH reaction mixture (from step 1). Elute the column with 0.05 M phosphate buffer, pH 6.0, and collect about twenty 0.5-ml fractions. Read the A_{280} of the fractions. Pool the MB/KLH conjugate fractions and place them in a 15-ml test tube.

   The first peak represents MB/KLH conjugate and the second peak represents free MBS. The MB/KLH peak is readily identified by its cloudiness.

3. Dissolve 5 mg synthetic peptide in 1 ml PBS. If the peptide is not soluble in PBS, dissolve it in 6 M guanidine·HCl/0.01 M phosphate buffer, pH 7.0. Add the peptide solution to the MB/KLH conjugate (from step 2). Check pH and adjust to 7.3 with 0.1 M HCl or 0.1 M NaOH. Stir 3 hr at room temperature with magnetic stirbar.

4. Dialyze the mixture against 4 liters of water overnight at 4°C. Replace with fresh water and dialyze for ≥4 hr at 4°C. Save an aliquot for step 5, maintaining conjugate at 4°C. Lyophilize the remainder and store at −20°C.

5. Calculate the level of coupling (see Support Protocol 2 or use Ellman's reagent) by obtaining the amino acid composition of a small amount of the conjugate.

6. Immunize rabbits to obtain antiserum (see Support Protocol 3).

ALTERNATE PROTOCOL 1

COUPLING OF SYNTHETIC PEPTIDE TO CARRIER PROTEIN USING GLUTARALDEHYDE

Because glutaraldehyde cross-links peptide and carrier molecules through their amino groups, peptides having Lys residues at positions other than the amino terminus are best avoided (Fig. 5.5.3).

Additional Materials *(also see Basic Protocol 1; see* APPENDIX 1 *for items with ✓)*

✓ Borate buffers, pH 10 and pH 8.5
Synthetic peptide

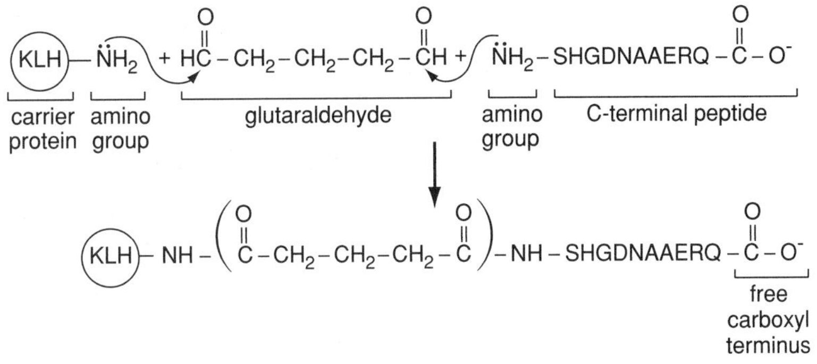

Figure 5.5.3 Coupling a C-terminal peptide to KLH carrier protein using glutaraldehyde. KLH, keyhole limpet hemocyanin. This is one of several mechanisms postulated for this reaction; the exact mechanism is not known.

✓ 0.3% (v/v) glutaraldehyde solution
1 M glycine

1. To a 15-ml glass test tube, add 10 mg KLH and 2 ml borate buffer, pH 10. Dissolve by gentle mixing on a stir plate, then add 10 μmol (∼10 to 15 mg) synthetic peptide. Slowly add 1 ml of freshly prepared 0.3% glutaraldehyde solution while stirring at room temperature. Allow to react 2 hr (the solution will turn yellow). Add 0.25 ml of 1 M glycine to block unreacted glutaraldehyde. Allow to react for 30 min.

2. Dialyze the peptide/protein conjugate (see Basic Protocol 1, step 4), substituting borate buffer, pH 8.5, for water. Calculate the level of coupling and immunize the rabbits (see Basic Protocol 1, steps 5 and 6).

5

ALTERNATE PROTOCOL 2

COUPLING OF SYNTHETIC PEPTIDE TO CARRIER PROTEIN USING EDCI

NOTE: Buffers containing phosphate, carboxylates, or amino groups (e.g., Tris) will interfere with this reaction and should be avoided.

Additional Materials *(also see Basic Protocol 1)*

1-ethyl-3-(3-dimethylaminopropyl)-carbodiimide·HCl (EDCI), freshly prepared (Sigma or Pierce)
12 × 75–mm glass test tubes (∼5-ml capacity)

1. Dissolve 10 mg synthetic peptide in 1 ml water and add to a 12 × 75–mm glass test tube. Stir gently using magnetic stirbar. Continue stirring and add 40 mg EDCI. Adjust pH to 4.5 with 0.1 M HCl and allow to react 5 to 10 min at room temperature.

2. Dissolve 10 mg KLH in 0.5 ml water and add to solution from step 1. Allow to react 2 hr at room temperature.

3. Dialyze the peptide/protein conjugate, calculate the level of coupling, and immunize the rabbit (see Basic Protocol 1, steps 4 to 6).

ALTERNATE PROTOCOL 3

COUPLING OF SYNTHETIC PEPTIDE TO CARRIER PROTEIN USING BDB

CAUTION: Benzedine and its salts and BDB are considered to be carcinogens. Use appropriate precautions when handling.

BDB primarily couples Tyr side chains in peptides to Tyr side chains in carrier proteins. However, interactions with the side chains of His, Cys, and Lys readily occur.

Additional Materials *(also see Basic Protocol 1; see* APPENDIX 1 *for items with ✓)*

Benzidine·HCl (Sigma)
0.2 M HCl
$NaNO_2$
✓ Borate buffer, pH 9.0

15-ml test tube
50-ml beaker

1. In a 15-ml test tube, dissolve 5 mg benzidine·HCl in 1 ml of 0.2 M HCl. Add 3.5 mg $NaNO_2$ at 4°C with constant agitation and stir 1 hr at 4°C in the dark. Store up to 1 year at –70°C.
2. In a 50-ml beaker, dissolve 5 mg KLH in 10 ml borate buffer, pH 9.0. Add 2 mg synthetic peptide and cool on ice.
3. Add 0.1 ml BDB solution (from step 1) dropwise to the peptide/protein solution (the solution will turn dark brown). Stir continuously 2 hr at 4°C, adjusting pH to 9.0 as necessary with 0.1 M NaOH (the solution will turn yellow with time).
4. Dialyze the peptide/protein conjugate, calculate the level of coupling, and immunize the rabbit (see Basic Protocol 1, steps 4 to 6) with homogenized suspensions.

SUPPORT PROTOCOL 1

DETECTION OF FREE SULFHYDRYL GROUPS IN PEPTIDES

This reaction should be carried out if Basic Protocol 1 for coupling with MBS is used.

Materials *(see* APPENDIX 1 *for items with ✓)*

✓ 0.05 M potassium phosphate buffer, pH 8.0
✓ DTNB solution, freshly prepared
3 mM 2-mercaptoethanol (2-ME), freshly prepared in water
5 mg/ml synthetic peptide in 0.05 M potassium phosphate buffer, pH 8.0
10 × 75–mm glass test tubes

1. Add 0.5 ml of 0.05 M phosphate buffer, pH 8.0, to each of three 12 × 75–mm glass test tubes labeled *blank*, *control*, and *sample*. Add 3.3 µl DTNB solution to each tube.
2. Add 10 µl water to the blank tube, 10 µl of 3 mM 2-ME to the control tube, and 10 µl of 5 mg/ml synthetic peptide, pH 8.0, to the sample tube. Vortex the tubes. Read A_{412} of the 2-ME control (A_{412} = 0.74) and sample tubes, using the blank tube as reference.

The sample will also have an A_{412} of 0.74 if it is 3 mM [molecular weight = (5 mg/ml)/ 3 mM = 1667] and if there is only one unoxidized Cys residue in the peptide. It is more likely that the molecular weight of the peptide will be higher or lower than 1667, causing the observed absorbance to be proportionately lower or higher. If some Cys residues have been oxidized prior to the DTNB reaction, the absorbance will also be lower.

3. Calculate the percentage of oxidized Cys residues as follows:
   a. theoretical absorbance (A_{th}) = [1667/(peptide molecular weight)] × 0.74.
   b. percentage of oxidized Cys = [1 − (A_{412} peptide/A_{th} × 100)].

SUPPORT PROTOCOL 2

CALCULATION OF THE MOLAR RATIO OF PEPTIDE TO CARRIER PROTEIN

By performing the calculations presented in this protocol, the molecules of peptide in the conjugate per molecule of carrier protein in the conjugate can be determined.

1. Obtain the amino acid composition of the carrier protein, the peptide, and the peptide/carrier conjugate. Amino acid compositional analysis (of these hydrolysates) is usually available at sources that provide automated peptide synthesis. Be sure that the conjugate is free of unconjugated peptide (i.e., well dialyzed).

2. Determine a scaling factor (SF) that relates the moles of protein in the unconjugated carrier protein to the moles of protein in the peptide/carrier conjugate. This is done by comparing the molar ratio of ≥3 amino acids present in the carrier protein and peptide/carrier conjugate but not present in the peptide. For example, if the peptide TGLRDSC (Table 5.5.4) is coupled to a carrier protein, choose A, K, and I. The calculation is done as:

$$\begin{aligned} SF = [&(\text{pmol A in conjugate/pmol A in carrier}) \\ &+ (\text{pmol K in conjugate/pmol K in carrier}) \\ &+ (\text{pmol I in conjugate/pmol I in carrier})]/3 \end{aligned}$$

   For these amino acids, the carrier protein yields are as follows: A = 103 pmol, K = 65 pmol, and I = 65 pmol. For these same amino acids, the peptide/carrier conjugate yields: A = 206 pmol, K = 125 pmol, and I = 135 pmol.

   From these values, the relative amount of carrier protein in the conjugate versus the unconjugated carrier protein (SF) can be calculated as follows: (206/103 + 125/65 + 135/65)/3 = 2.0, indicating that there is twice as much carrier protein in the peptide/carrier conjugate hydrolysate as in the carrier-protein hydrolysate.

3. Calculate the moles of peptide present in the conjugate by subtracting the moles of amino acid present in the carrier from the moles of amino acid present in the conjugate. Choose ≥3 amino acids present in the peptide. The relative amount (SF) of protein present in the carrier protein versus the amount in the conjugate as calculated in step 2 must also be considered as follows:

$$\begin{aligned} \text{pmol peptide in conjugate} = \{&[\text{pmol G in conjugate} - (SF \times \text{pmol G in carrier})] \\ &+ [\text{pmol L in conjugate} - (SF \times \text{pmol L in carrier})] \\ &+ [\text{pmol R in conjugate} - (SF \times \text{pmol R in carrier})]\}/3 \end{aligned}$$

5

Table 5.5.4 Sample Calculation of the Extent of Coupling of the Peptide TGLRDSC to Carrier Protein[a]

Amino acid	Composition of carrier protein	Composition of peptide/carrier conjugate	Amount of carrier protein amino acids in conjugate	Amount of peptide amino acids in conjugate
D	80	185	160	25
E	110	222	220	—
G	95	215	190	25
S	65	150	130	20
T	70	163	140	23
H	10	19	20	—
P	25	51	50	—
A	103	206	206	—
M	5	11	10	—
V	60	118	120	—
F	22	45	44	—
L	55	133	110	23
I	65	135	130	—
C	7	22	14	8
Y	13	25	26	—
K	65	125	130	—
R	75	177	150	27
Total pmol amino acid	925	2002	1850	151

[a]For use in calculations of peptide coupling, the molecular weight is taken at 100 kDa. All numbers are picomoles.

Therefore, the amount of peptide in the conjugate hydrolysate for the example shown in Table 5.5.4, calculated using the amino acids G, L, and R, is $\{[215 - (2 \times 95)] + [133 - (2 \times 55)] + [177 - (2 \times 75)]\}/3 = 25$ pmol.

4. Calculate the number of moles of protein in the conjugate hydrolysate as follows:

$$\text{pmol carrier protein in conjugate} = \frac{\text{(total pmol carrier protein amino acids)}}{\text{molecular weight of carrier protein}} \times 110$$

where total pmol carrier protein amino acids = SF × (total amino acid composition of carrier in pmol) and 110 is the average molecular weight of an amino acid.

In this example, there are 1850 pmol of carrier protein amino acids in the conjugate; therefore, 1850 pmol × (110/100,000) = 2.04 pmol carrier protein in conjugate.

5. Determine the ratio of peptide to carrier protein as follows:

molecules peptide in conjugate/molecules carrier protein in conjugate

= pmol peptide in conjugate/pmol carrier protein in conjugate.

Using the values calculated in steps 3 and 4, the result is: 25 pmol peptide in conjugate/ 2.04 pmol carrier protein in conjugate = 12.2 molecules peptide in conjugate per molecule carrier protein in conjugate.

SUPPORT PROTOCOL 3

IMMUNIZATION SCHEDULE FOR PRODUCING ANTIPEPTIDE SERA IN RABBITS

Materials *(see* APPENDIX 1 *for items with* ✓*)*

- Rabbits
- Peptide/carrier conjugate (see Basic Protocol 1 and Alternate Protocols 1, 2, and 3)
- ✓ PBS
- Complete Freunds adjuvant (Sigma)
- Incomplete Freunds adjuvant (Sigma)

1. Obtain 2 to 5 ml preimmune serum from each rabbit (at least two rabbits per peptide).
2. Dissolve or suspend 0.7 to 1 mg peptide/carrier conjugate in 1 ml PBS and mix 1:1 with complete Freunds adjuvant. Immunize each rabbit subcutaneously with this mixture.
3. Boost each rabbit with 0.3 to 0.5 mg peptide/carrier conjugate, dissolved or suspended in 1 ml PBS, at 2 weeks and again at 4 weeks.
4. Dissolve or suspend 0.7 to 1 mg peptide/carrier conjugate in 1 ml PBS and mix 1:1 with incomplete Freunds adjuvant. Boost each rabbit with this mixture at 11 weeks.
5. Bleed rabbits at 7 and 12 weeks (the best sera are obtained at 12 weeks) after initial immunization to obtain antipeptide serum. To determine level of antipeptide antibodies in the serum, see Support Protocol 4.
6. If purified antipeptide antibodies are required, set up a peptide affinity column (see Support Protocol 5).

SUPPORT PROTOCOL 4

INDIRECT ELISA TO DETERMINE ANTIPEPTIDE ANTIBODY TITER

Materials *(see* APPENDIX 1 *for items with* ✓*)*

- 0.2 to 2.5 μM synthetic peptide in carbonate buffer (see *APPENDIX 1* for buffer)
- ✓ PBS containing 0.05% (v/v) Tween 20 (PBS/Tween; Sigma)
- 10 mg/ml BSA in PBS/Tween
- Antipeptide antiserum (see Support Protocol 2)
- Goat anti-rabbit globulin conjugated to alkaline phosphatase (Sigma)
- ✓ Enzyme substrate: 1 mg/ml *p*-nitrophenyl/phosphate (NPP; Sigma) in 0.1 M diethanolamine buffer, pH 9.8

- 96-well microtiter plate
- Microtiter plate reader: spectrophotometer with 405-nm filter or spectrofluorometer (Dynatech) with 365-nm excitation filter and 450-nm emission filter

1. Coat the wells of a 96-well microtiter plate with 100 to 300 μl of 0.2 to 2.5 μM synthetic peptide by incubating overnight at 4°C or 2 to 6 hr at room temperature.

   If the peptide does not adsorb, try other buffers in the pH 4 to 8 range (Geerligs et al., 1988). For peptides of <15 residues, it may be necessary to coat the plate with the peptide conjugated to a carrier protein different from that used to obtain the antisera. The concentration of peptide in the conjugate solution should still be ~1 μM.

2. Discard the uncoated synthetic peptide. Wash the coated wells *at least* three times with PBS/Tween by filling wells and flicking the plate.
3. Block remaining active sites by incubating plate with 300 μl/well of 10 mg/ml BSA in PBS/Tween for 1 hr at 37°C. Use ovalbumin if BSA is used as the carrier protein for immunization. Wash the wells *at least* three times with PBS/Tween as in step 2.
4. Prepare two-fold serial dilutions of 300 μl of antipeptide antiserum with PBS/Tween ranging from 1:100 to 1:12,800. Add to wells and incubate 2 hr at 37°C. Wash the wells *at least* three times with PBS/Tween as in step 2.
5. Dilute goat anti-rabbit globulin conjugated to alkaline phosphatase 1:500 to 1:8000 in PBS/Tween. Add diluted conjugate to wells and incubate 1 to 3 hr at 37°C. Wash the wells *at least* three times with PBS/Tween as in step 2.
6. Add 50 μl enzyme substrate and incubate 1 to 3 hr at 37°C.
7. Read the absorbance at 405 nm with a microtiter plate reader.

SUPPORT PROTOCOL 5

PREPARATION OF PEPTIDE AFFINITY COLUMN

Materials

Synthetic peptide
CNBr-activated Sepharose-4B (Pharmacia)
1 × 10–cm column (e.g., Bio-Rad Econo Column)

1. Couple 10 mg synthetic peptide to 1 g CNBr-activated Sepharose-4B (3.5 ml when swelled) according to the manufacturer's instructions. Pour the coupled mixture into a 1 × 10–cm column and wash the resin thoroughly with 100 bed volume of PBS.
2. Purify antipeptide antibodies. Elute bound antibody with acid, base, chaotropic agents, or specific peptide. Then remove from the eluted antibody by extensive dialysis against PBS.

5

BASIC PROTOCOL 2

USE OF MULTIPLE ANTIGEN PEPTIDE (MAP) SYSTEMS

The multiple antigen peptide (MAP) system was designed as a novel approach to prepare peptide immunogens that overcome the limitations of peptide-carrier conjugates—the small portion of peptide antigen represented in the whole conjugate; chemical ambiguity in antigen composition and structure; production of irrelevant epitopes and antibodies; and carrier toxicity and carrier-induced epitope suppression. The MAP system produces an immunogen that is carrier-free, chemically defined, and homogeneous. MAP systems have been used successfully to produce both polyclonal and monoclonal antibodies that specifically recognize the native proteins. They have also been used to produce sera that have a significantly higher titer of antibodies than sera with antibodies against the same peptides conjugated to keyhole limpet hemocyanin (KLH) as a carrier protein (Tam, 1988).

A MAP has three important structural features: an amino acid bound to a resin, an inner core matrix consisting of layers of lysine, and a surface layer of peptide antigens attached to the

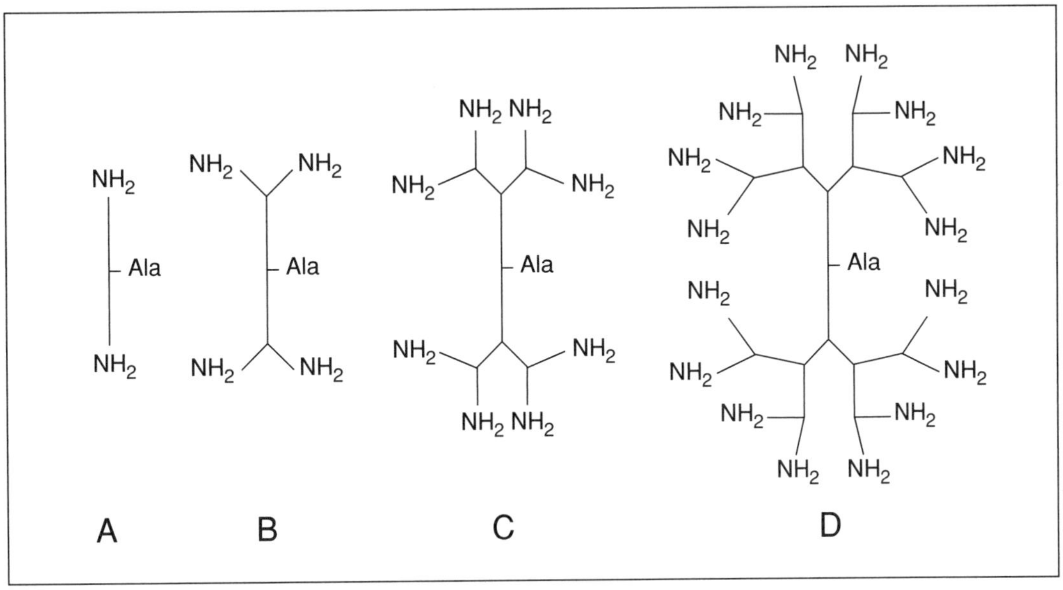

Figure 5.5.4 Schematic drawing of the MAP core matrix. (**A**) Divalent; (**B**) tetravalent; (**C**) octavalent; and (**D**) hexadecavalent matrix core.

lysine inner core matrix (Fig. 5.5.5). When the MAP is cleaved from the resin, the resulting immunogen is a chemically defined peptide.

MAP peptides can be completely prepared in a peptide synthesis laboratory by the direct approach, in which the core resin and antigenic peptide are synthesized in one contiguous synthesis (Fig. 5.5.4), or indirectly, in which the peptide and the core matrix are synthesized separately and conjugated by one of several possible condensation methods (Fig. 5.5.5). Alternatively, the core matrix resin can be purchased from commercial sources.

As in conventional procedures using carrier proteins, the crucial point in the preparation of reactive antipeptide sera and antipeptide antibodies using the MAP approach is selection of the peptide antigen sequence. In general, C-terminal, N-terminal, or hydrophilic internal peptide sequences are selected. C- and N-terminal portions are usually more immunogenic because they are more flexible and are often exposed on the surface of proteins. Similarly, hydrophilic sequences are more likely to be exposed on external surfaces and are therefore generally more immunogenic.

All the responses to MAP systems have been identified as T cell dependent, suggesting that the B cell epitope may also act as a T cell epitope in the MAP format. The following points should be considered in selecting peptides for the MAP approach. Peptides should be 14 residues or longer so that it is more likely that a helper T epitope is included. The B cell epitope should be located at or near the amino-terminal (distal) portion of the MAP. If a short peptide antigen is used, a helper T epitope should be incorporated. If inbred mice are used for immunization, a known helper T epitope should be incorporated. If monoclonal antibodies are being produced, several strains of mice should be used. To ensure obtaining large amounts of reactive polyclonal sera, larger outbred animals such as rabbits, guinea pigs, or goats should be used.

References: Van Regenmortel et al., 1988

Contributors: John E. Coligan, James P. Tam, and Jun Shao

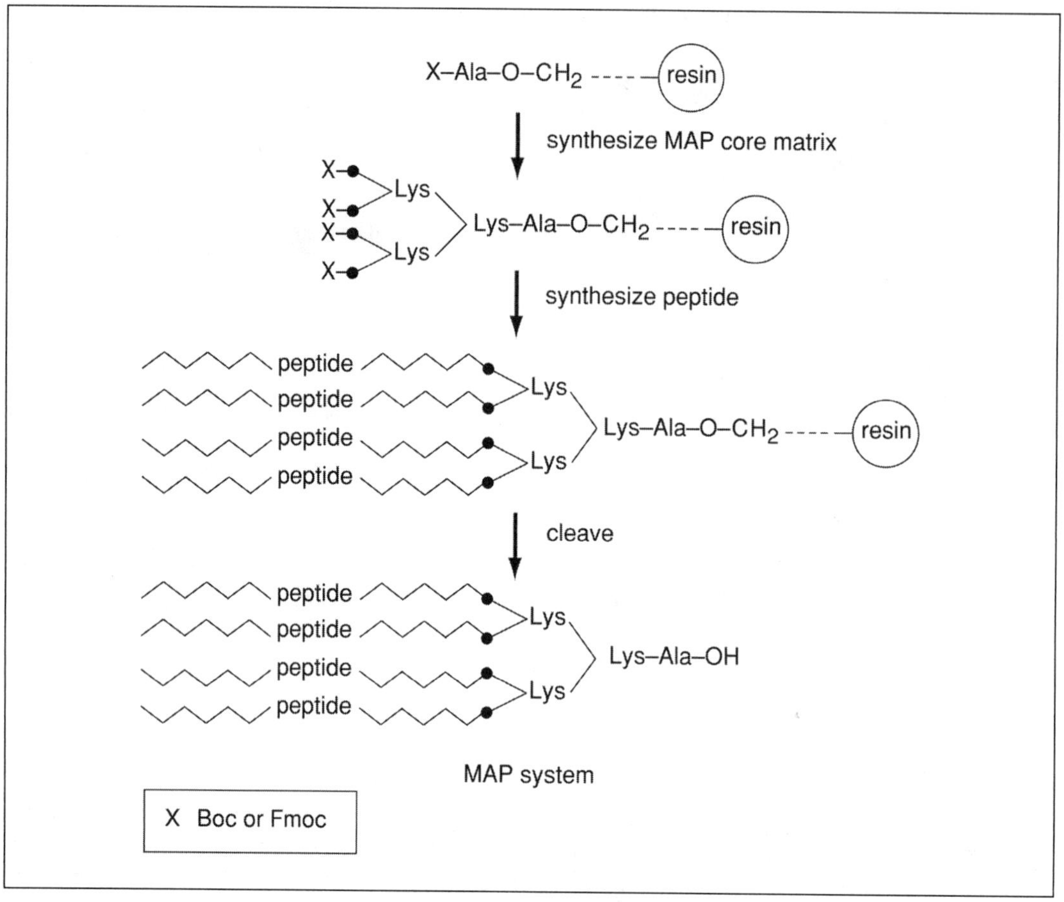

Figure 5.5.5 Direct approach for preparing MAP system on a tetrabranched core matrix.

5

UNIT 5.6

Production of Antisera Using Fusion Proteins

NOTE: All reagents and equipment coming in contact with live cells must be sterile, and proper aseptic technique should be followed accordingly. High-quality water (e.g., tissue culture grade) should be used in all solutions.

BASIC PROTOCOL 1

CONSTRUCTION AND PURIFICATION OF RECOMBINANT FUSION-PROTEIN EXPRESSION PLASMID

Materials *(see* APPENDIX 1 *for items with* ✓*)*

Template cDNA encoding region of interest for antibody production
pGEX-2T vector (Pharmacia Biotech)
*Bam*HI and *Eco*RI (or other chosen restriction enzymes) and appropriate restriction endonuclease buffers (see *APPENDIX 1* for restriction endonuclease buffers)
T4 DNA ligase
E. coli strain RR1 (Life Technologies)

✓ 2YT plates containing 50 μg/ml ampicillin
✓ 2YT medium containing 50 μg/ml ampicillin

1. Design and synthesize oligonucleotide primers, 15 to 20 nucleotides in length, complementary to the sense and antisense strands of the cDNA sequence to be amplified. Incorporate restriction sites for *Bam*HI and *Eco*RI into the primers (to facilitate subcloning into the pGEX-2T vector) and add ~6 base pairs 5′ of the restriction site for efficient cutting.

2. Amplify the cDNA template by PCR using an appropriate annealing temperature based on the sequence of the oligonucleotide primers and the cDNA and the number of matching base pairs, calculated according to the formula:

$$\text{annealing temperature} = [(4°\text{C} \times \text{no. of GC base pairs}) + (2°\text{C} \times \text{no. of AT base pairs})] - [5°\text{C} \times \text{no. of mismatch base pairs}].$$

   Calculate annealing temperature for both the 5′ and 3′ oligonucleotide primer and use the lower of the annealing temperatures for amplification. Confirm the approximate length of the PCR product on an agarose gel.

3. Digest the PCR product and the pGEX-2T vector with *Bam*HI and *Eco*RI according to the conditions for each enzyme, either sequentially or as a double digest. Confirm efficient digestion on an agarose gel.

4. Purify the digested PCR product and vector on a low-gelling/melting temperature agarose gel and extract from the gel using Elutip column chromatography or glass beads. Run a standard agarose gel to estimate the quantity of purified fragment and vector.

5. Ligate the fragment (e.g., five-fold molar excess) into vector (e.g., 50 ng) with T4 DNA ligase using appropriate quantities of each (according to the supplier's guidelines) to produce recombinant plasmid.

6. Transform competent *E. coli* RR1 with the recombinant plasmid, plate on 2YT agar plates containing 50 μg/ml ampicillin, and grow overnight at 37°C. Pick individual colonies into 2 ml of 2YT medium containing 50 μg/ml ampicillin. Incubate overnight with shaking at 37°C.

7. Purify plasmid DNA from transformants by alkaline lysis miniprep and screen miniprep DNA by restriction-digest analysis.

8. Prepare a large-scale *E. coli* lysate. Purify microgram amounts of restriction-digest selected DNA (containing the insert of interest) from the lysate by column chromatography using a Qiagen column. Store purified recombinant pGEX plasmid indefinitely at −20°C.

9. Verify the recombinant plasmid by double-stranded dideoxy sequencing.

BASIC PROTOCOL 2

EXPRESSION AND PURIFICATION OF SOLUBLE FUSION PROTEIN

To avoid repeating this protocol several times, two 800-ml cultures can be grown and separated into four 400-ml cultures prior to harvesting the cells. To do this, simply double the amount of 2YT broth in steps 2 and 3 and the IPTG. The washed pellets can be stored frozen at −80°C and used to purify fusion protein in the future, picking up the purification process.

A separate aliquot of competent BL21 cells should be transformed with the parent pGEX-2T vector. One 400-ml culture should be processed through step 11 and the supernatant used to

5

make the BL21/GST lysate affinity column for antibody purification (see Support Protocol 1). Another 400-ml culture can be processed to the end of this protocol to provide purified GST.

Materials *(see* APPENDIX 1 *for items with ✓)*

E. coli strain BL21(DE3) (Novagen)
Recombinant pGEX-2T fusion-protein expression plasmid (see Basic Protocol 1)
pGEX-2T vector (Pharmacia Biotech)
✓ 2YT plates containing 50 μg/ml ampicillin and 150 μg/ml chloramphenicol
✓ 2YT medium containing 50 μg/ml ampicillin and 150 μg/ml chloramphenicol
1 M isopropyl thio-β-D-galactosidase (IPTG; store in aliquots indefinitely at −20°C)
✓ Buffer A, ice-cold
20 mg/ml lysozyme in buffer A (prepare fresh)
Protease inhibitor stock solutions (store up to 1 year at −20°C):
  1 mg/ml aprotinin in 0.01 M HEPES, pH 8.0
  1 mg/ml leupeptin in H_2O
  1 mg/ml pepstatin A in ethanol
Dry ice/ethanol slurry
10% (v/v) Triton X-100
DNase I
1 M $MgSO_4$
✓ TE buffer, pH 8.0, ice-cold
✓ Buffer C, ice-cold
✓ Glutathione-agarose bead slurry
50 mM reduced glutathione in buffer C, adjusted pH to ~7.5 with NaOH (prepare fresh)
✓ PBS, pH 7.4

250-ml flasks
1-liter baffled flask(s)
Refrigerated centrifuge and centrifuge buckets
50-ml conical polypropylene centrifuge tubes
Spectra/Por dialysis tubing, MWCO 6000 to 8000 (Spectrum)

1. Transform separate aliquots of competent *E. coli* BL21 with the recombinant plasmid prepared in Basic Protocol 1 and with the parent pGEX-2T vector (for production of BL21/GST lysate; see Support Protocol 1). Plate on 2YT agar plates containing 50 μg/ml ampicillin and 150 μg/ml chloramphenicol. Grow overnight at 37°C.

2. Pick a small isolated colony and transfer to a 250-ml flask containing 40 ml of 2YT broth containing 50 μg/ml ampicillin and 150 μg/ml chloramphenicol. Grow overnight on a shaker at 37°C.

3. Inoculate 400 ml of fresh 2YT broth containing 50 μl/ml ampicillin and 150 μg/ml chloramphenicol with the 40-ml overnight culture in a 1-liter baffled flask and grow on a shaker ~1 hr at 37°C until the A_{600} is 0.4 to 0.5. Set aside a 100-μl aliquot of this culture (preinduction) for later SDS-PAGE analysis.

4. Induce the bacteria to produce the fusion protein by adding 400 μl of 1 M IPTG per 400 ml of broth (1 mM IPTG final). Grow 4 hr on shaker at 37°C. Set aside a 10-μl aliquot of this culture (post-induction) for SDS-PAGE analysis.

5. Check protein production by analyzing the pre- and post-induction samples via SDS-PAGE and staining the gel with Coomassie blue.

6. Transfer the induced culture from step 4 to as large a vessel as the available refrigerated centrifuge will accommodate. Harvest cells by centrifuging 15 min at 8000 × *g*, 4°C.

7. Decant the supernatant and resuspend the pellet in 10 ml ice-cold buffer A. Transfer to a 50-ml polypropylene conical screw-cap centrifuge tube. Centrifuge 10 min at 5000 × *g*, 4°C. Decant supernatant. Store washed pellets at −80°C for future purification.

8. Resuspend pellet in 10 ml ice-cold buffer A. Add 2 ml of fresh 20 mg/ml lysozyme in buffer A and incubate 1 hr on ice to digest cell walls. After this incubation, add 20 μl/ml of total resuspended pellet volume of the 1 mg/ml stock solutions of the protease inhibitors aprotinin, leupeptin, and pepstatin A.

9. Freeze/thaw the pellet suspension twice on a dry ice/ethanol slurry. Add 1/10 vol of 10% Triton X-100, then freeze/thaw two more times.

10. Add 20 μg DNase I per ml of total resuspended pellet volume, then add 3 ml of 1 M $MgSO_4$. Incubate 30 min to 1 hr at room temperature with occasional gentle shaking until viscosity decreases.

11. Centrifuge 10 min at 3400 × *g*, 4°C. Decant the supernatant into a new 50-ml tube and keep on ice. Resuspend the pellet fraction in a volume of TE buffer, pH 8.0, equal to that of the supernatant. Test 5-μl aliquots of both fractions for the presence of fusion protein via SDS-PAGE and Coomassie blue staining.

12. Add 15 ml ice-cold buffer C containing 1 μg/ml protease inhibitors (added from the 1 mg/ml stocks) to the supernatant fraction, then add 4 ml glutathione-agarose bead slurry. Incubate 1 to 2 hr (to overnight) at 4°C with shaking.

13. Centrifuge 5 min at 800 × *g*, 4°C. Remove supernatant and check for unadsorbed fusion protein by SDS-PAGE/Coomassie staining. Wash the beads three times, each time by filling the tube with ice-cold buffer C, inverting several times to disperse the beads, then centrifuging 5 min at 800 × *g*, 4°C, and carefully aspirating the supernatant.

14. Elute protein six times, each time by adding 2 ml of 50 mM reduced glutathione in buffer C to the beads, mixing gently, then centrifuging 5 min at 800 × *g*, 4°C, and carefully collecting the supernatant.

    To track elution, the concentration of fusion protein in each elution can be estimated by optical density. An A_{280} of 1.0 is approximately equivalent to 1 mg/ml of fusion protein.

15. Combine elutions and centrifuge 5 min at 5000 × *g*, 4°C, then transfer the supernatant to remove any remaining beads.

    The glutathione-agarose beads can be recycled by washing them in 3 M NaCl overnight followed by several rinses/washes in buffer C as described in step 13. Store in buffer C at 4°C.

16. Dialyze combined elutions against PBS using dialysis tubing of MWCO 6000 to 8000. Change the dialysis buffer three times (approximately every 6 hr) to completely remove glutathione. Estimate the final concentration of fusion protein by spectrophotometry (A_{280}), SDS-PAGE/Coomassie staining, or a protein assay. Store protein solution at −80°C.

BASIC PROTOCOL 3

IMMUNIZATION USING FUSION PROTEINS

This process may also be performed by an off-site commercial laboratory.

Materials

New Zealand white rabbits (~2.5 kg)
Purified fusion protein (see Basic Protocol 2)
Complete and incomplete Freund's adjuvant (CFA and IFA; e.g., Pierce)

1. Take a blood sample from the rabbit prior to immunization, prepare serum (*UNIT 5.6*) and freeze at −80°C for use as control when testing specificity of antiserum.

2. Prepare aliquots of fusion protein sufficient for all immunizations (200- to 500-μg for primary immunizations and 100- to 250-μg for booster immunizations). Freeze until needed.

3. Just prior to immunization, emulsify a 200- to 500-μg aliquot of soluble fusion protein with an equal volume of complete Freund's adjuvant (CFA).

4. Perform primary immunization by injection of fusion protein/adjuvant.

5. Two weeks after primary immunization, emulsify an aliquot containing 100 to 250 μg of fusion protein (i.e., half the amount used for the primary immunization) with an equal volume of incomplete Freund's adjuvant (IFA). Perform booster immunization by injection of this fusion protein/IFA mixture. Perform additional booster immunizations monthly.

6. Collect 15- to 25-ml bleeds 10 and 24 days after booster immunization (or according to other established animal care and handling guidelines).

7. Prepare serum by clotting blood overnight at 4°C and clarifying by centrifugation . Store aliquots at −80°C.

8. Test the various bleeds to determine that they contain antibodies to the region of interest by western blotting or immunocytochemistry (*UNIT 1.2*).

BASIC PROTOCOL 4

AFFINITY PURIFICATION OF ANTISERA

NOTE: Affinity purification should be performed at 4°C, either in a refrigerator or a cold room, to prevent degradation of antibodies.

Materials *(see APPENDIX 1 for items with ✓)*

Crude, clarified antiserum (see Basic Protocol 3)
Heparin
10-ml BL21/GST affinity column (see Support Protocol 1)
2-ml fusion-protein affinity column (see Support Protocol 1)
✓ 10 mM, 0.1 M, and 1 M Tris·Cl, pH 8.0, ice-cold
0.1 M glycine, pH 3.0, ice-cold
Glycerol
20% (w/v) sodium azide (1000× stock)

15- and 50-ml tubes
Quartz spectrophotometer cuvette
UV spectrophotometer

1. Add 250 U heparin (50 U/ml) to a 5-ml aliquot of serum to be purified to prevent endogenous thrombin activity.

2. Pass heparinized serum over BL21/GST lysate column and collect effluent in a 50-ml tube. Wash column with 10 bed volumes of 10 mM Tris·Cl, pH 8.0, and combine washings with serum. Cap tube and set aside at 4°C.

3. Strip BL21/GST lysate column by washing with 10 bed volumes of 0.1 M glycine, pH 3.0. Collect the first few drops of glycine off the column in a quartz cuvette and measure A_{280} to monitor the removal of GST and bacterial antibodies from the serum.

4. Neutralize column with ~10 bed volumes of 0.1 M Tris·Cl, pH 8.0, until column eluate is pH 8.0 as determined with pH paper.
5. Repeat steps 2 to 4 as necessary (two to three times) until the A_{280} of the eluant is near zero, indicating adequate removal of antibodies to GST and bacterial proteins.

   The volume of solution increases with the washes and the entire amount must be put back through the column. The columns used come with funnels that fit into the top, so ~30 ml of liquid can be added at once and allowed to drain through the much smaller column. It is necessary to wait and refill the funnel repeatedly until all of the liquid has passed through the column, being careful not to let the column run dry. This procedure is also necessary with step 6, below. This entire protocol can nevertheless usually be completed in one long day.

6. Pass collected serum/washes from BL21/GST column over fusion-protein affinity column to select antibodies specific for the fusion protein. Wash column with 20 ml of 10 mM Tris·Cl, pH 8.0, to remove unbound protein. Discard wash solution.
7. Place 50 µl of 1 M Tris·Cl, pH 8.0, into each of five microcentrifuge tubes. Elute antibodies from fusion-protein column with five 500-µl aliquots of 0.1 M glycine and collect each 500 µl of eluate in a separate tube. Vortex and measure antibody concentration by A_{280}.
8. Collect a tail elution by passing an additional 7.5 ml of 0.1 M glycine over the column and collecting the eluate in a 15-ml tube containing 750 µl of 1 M Tris·Cl, pH 8.0. Neutralize the fusion-protein column with excess 0.1 M Tris·Cl, pH 8.0.
9. Pool the fractions containing the peak of the antibody-elution curve, add an equal volume of glycerol, and store in aliquots at a concentration ≥0.1 mg/ml in 50% glycerol at −20°C. Treat aliquots in use with 1/1000 vol of 20% sodium azide and store at 4°C.
10. Test the specificity of the antibodies using western blot analysis and/or immunocytochemistry (see *UNIT 1.2*).

5

SUPPORT PROTOCOL 1

MAKING AFFINITY COLUMNS FOR PURIFICATION OF ANTI-FUSION PROTEIN ANTISERA

Materials *(see APPENDIX 1 for items with ✓)*

Affinity resin (Affi-Gel 10 and 15; Bio-Rad)
BL21/GST lysate (see Basic Protocol 2)
Purified fusion protein (see Basic Protocol 2)
0.1 M glycine, pH 3.0
✓ 0.1 M Tris·Cl, pH 8.0
0.1 M triethylamine, pH 11.0

Tabletop centrifuge with bucket rotors, 4°C
Shaker
2-ml (for fusion-protein) and 10-ml (for BL21/GST) double-frit chromatography columns (Pierce)

1. To determine the best affinity resin to use, separately incubate a small amount of fusion protein 1 hr at 4°C with 100 µl of resuspended Affi-Gel 10 and Affi-Gel 15. Run the supernatants before and after incubation with each resin on an SDS-PAGE gel, stain with Coomassie blue, and compare the concentrations of fusion protein by the intensity of staining. Use the resin that gives the highest concentration of fusion protein (i.e., the

lowest concentration of protein in the post-incubation sample). Repeat this determination with a small amount of BL21/GST lysate.

2. Widen the end of a 1-ml pipet tip by cutting with a razor blade. Shake the bottle of affinity-resin slurry to resuspend, then use the cut-off pipet tip to transfer 1 ml of the slurry to a 15-ml tube for the fusion-protein affinity column (0.5-ml bed volume) or 16 ml of the slurry for the BL21/GST affinity column (8-ml bed volume). Centrifuge beads at 800 × *g*, 4°C, aspirate supernatant (being careful not to remove any beads), fill tube with ice-cold deionized water, then repeat this washing procedure two more times.

3a. *To prepare the fusion-protein affinity column:* Add up to 25 mg of fusion protein to the washed beads. Set aside a 2-μg sample of fusion protein for use as the pre-bead control in step 5.

3b. *To prepare the BL21/GST affinity column:* Add the BL21/GST lysate from one 400-ml culture to the washed beads. Set aside a 10-μl sample of BL21/GST lysate for use as the pre-bead control in step 5.

4. Fill any remaining volume of the 15-ml tube with water and incubate with gentle shaking overnight at 4°C.

5. Centrifuge beads at 800 × *g*, 4°C, and decant the supernatant (being careful not to remove any beads) and save. Take a post-bead sample of the supernatant equal in volume to the pre-bead sample. Analyze pre- and post-bead samples by SDS-PAGE and Coomassie staining to confirm efficient binding of the protein to the resin.

6. Wash beads sequentially with

   0.1 M glycine (pH 3.0)
   0.1 M Tris·Cl (pH 8.0)
   0.1 M triethylamine (pH 11.0)
   0.1 M Tris·Cl (pH 8.0)

   each time by adding the appropriate solution, centrifuging at 800 × *g*, aspirating the supernatant, then adding the next solution, centrifuging again, and aspirating the supernatant.

7. Fill the 10-ml column (for BL21/GST column) or the 2-ml column (for fusion-protein affinity column) with water. After ensuring that fluid passes through the column, cap the end while the water level is above the neck of the column. Float the first frit on water, then slowly push it to the bottom of the column with the wide end of a Pasteur pipet.

8. Widen the end of a 1-ml pipet tip by cutting with a razor blade. Use this tip to transfer 12 to 16 ml of BL21/GST beads to the 10-ml column or 1 to 2 ml of the fusion-protein beads to the 2-ml column, washing the tip several times with 0.1 M Tris·Cl, pH 8.0, so as to lose as few beads as possible. Allow the beads to settle.

9. Fill the column with water or 0.1 M Tris·Cl, pH 8.0, and cap the bottom while the liquid volume is still above the neck. Slowly put the second frit in place, leaving a small space between the bead bed and this frit.

10. Wash the column with 30 ml of 0.1 M Tris·Cl, pH 8.0. Measure the A_{280} of the eluant to confirm that it is zero. Add sufficient 0.1 M Tris·Cl, pH 8.0, to cover the second frit, cap both ends of the column, and store at 4°C.

SUPPORT PROTOCOL 2

PREPARATION OF INSOLUBLE FUSION PROTEINS FOR USE IN IMMUNIZATION

Additional Materials *(also see Basic Protocol 2)*

10% (w/v) SDS
Heating block

1. Estimate the amount of fusion protein present per microliter of the resuspended pellet (see Basic Protocol 2, step 11) by SDS-PAGE and Coomassie staining.
2. Pipet an aliquot of resuspended pellet that will be enough for all immunizations. Centrifuge 10 min at 3400 × *g*, 4°C, and aspirate supernatant.
3. Resuspend the pellet in an equal volume of ice-cold buffer C. Centrifuge 10 min at 3400 × *g*, 4°C, and aspirate the supernatant.
4. Add 2 μl of 10% SDS for every 100 μg of protein. Boil for 2 min in a heating block. If the pellet material does not appear solubilized, add more SDS; however, use the minimum amount of SDS required to solubilize the particulate matter because SDS interferes slightly with the binding of proteins to the affinity resin. More importantly, however, SDS can be toxic to the animal.
5. Dilute the pellet/SDS solution 10-fold with PBS. Prepare aliquots sufficient for all immunizations and freeze until required.

SUPPORT PROTOCOL 3

PREPARATION OF AFFINITY COLUMNS USING INSOLUBLE FUSION PROTEINS

5

The procedure is in general the same as for soluble fusion proteins (see Basic Protocol 4), except that the pellet fraction from the lysis of *E. coli*, which contains insoluble protein of interest, must be solubilized with SDS.

1a. *To make a fusion-protein affinity column from an insoluble fusion protein:* Retain the 400-ml culture pellet fraction (see Basic Protocol 2, step 11) and solubilize using SDS (see Support Protocol 2), then bind to affinity resin (see Support Protocol 1).

2a. *To make a BL21/GST affinity column from an insoluble fusion protein:* Grow a 400-ml culture of *E. coli* BL21 transformed with pGEX parent vector (see Basic Protocol 2). Follow Basic Protocol 2 until step 11, keeping both the supernatant and pellet fractions. Solubilize the pellet fraction with SDS (see Support Protocol 2). Bind the supernatant and solubilized pellet fraction to two separate 4-ml aliquots of affinity resin (see Support Protocol 1) for making affinity columns.

References: Harlow and Lane, 1988; Sambrook et al., 1989

Contributors: Michelle Gilmor, Craig Heilman, Norman Nash, and Allan Levey

UNIT 5.7

Immunoblotting and Immunodetection

Immunoblotting (often referred to as western blotting) is used to identify specific antigens recognized by polyclonal or monoclonal antibodies.

NOTE: Deionized, distilled water should be used throughout this unit.

NOTE: Use powder-free gloves when manipulating filter papers, gels, and membranes. Oil from hands blocks protein transfer.

BASIC PROTOCOL 1

PROTEIN BLOTTING WITH TANK TRANSFER SYSTEMS

Materials *(see* APPENDIX 1 *for items with ✓)*

Samples for analysis
Protein molecular weight standards, prestained (Sigma or Bio-Rad) or biotinylated (Vector Laboratories or Sigma)
✓ Transfer buffer
Methanol (optional)

Electroblotting apparatus (EC Apparatus, Bio-Rad, or Hoefer)
Scotch-Brite pads (3M) or equivalent sponge
Whatman 3MM filter paper or equivalent
Transfer membrane: 0.45-µm nitrocellulose (Millipore or Schleicher & Schuell), PVDF (Millipore Immobilon P), neutral nylon (Pall Biodyne A), or positively charged nylon (Pall Biodyne B; Bio-Rad Zetabind)
Power supply
Indelible pen (e.g., Paper-Mate) or soft lead pencil (optional)

1. Prepare antigenic samples and separate proteins using small or standard-sized one- or two-dimensional gels. Include prestained or biotinylated protein molecular weight standards in one or more gel lanes.

2. Disassemble gel sandwich and remove stacking gel. Equilibrate separating gel 30 min at room temperature in the appropriate transfer buffer for the membrane to be used.

3. Fill a tray large enough to hold the plastic transfer cassette with transfer buffer so that the cassette will be covered. Assemble the transfer sandwich under the buffer to minimize trapping of air bubbles. On the bottom half of plastic transfer cassette, place a Scotch-Brite pad or sponge, followed by a sheet of filter paper cut to same size as gel and prewet with transfer buffer (Fig. 5.7.1).

4. Place gel on top of filter paper. Remove any air bubbles between gel and filter paper by gently rolling a test tube or glass rod over surface of gel.

   The side of the gel touching the paper arbitrarily becomes the cathode side (i.e., toward the negative electrode when placed in the tank).

5. Cut transfer membrane to same size as gel plus 1 to 2 mm on each edge. Place nitrocellulose or nylon membrane slowly into distilled water, with one edge at a 45° angle, to wet the entire surface and avoid bubbles. For PVDF membranes, immerse 1 to 2 sec in 100% methanol. Do not let membranes dry out at any time.

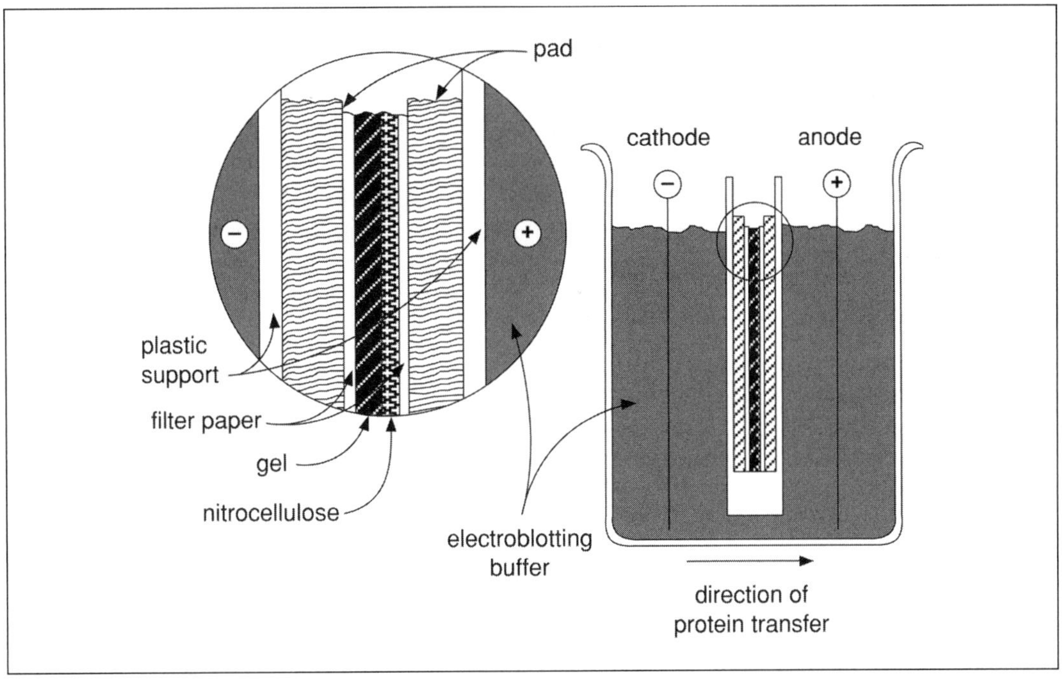

Figure 5.7.1 Immunoblotting with a tank blotting unit. A Scotch-Brite pad is placed on one side of the plastic cassette. This is overlaid with a buffer-saturated sheet of filter paper, the gel, and then a sheet of membranke pre-cut to the size of the gel plus 1 to 2 mm on each edge. This is overlaid with another sheet of filter paper, another Scotch-Brite pad, and the other half of the plastic cassette. The entire assembly is placed in a tank containing transfer buffer. For transfer of negatively charged protein, the membrane is positioned on the anode side of the gel. For transfer of positively charged protein, the membrane is placed on the cathode side of the gel. Charged proteins are transferred electrophoretically from the gel onto the membrane. Transfer is achieved by applying a voltage of 100 V for 1 to 2 hr (with cooling) or 14 V overnight.

6. Equilibrate membrane 10 to 15 min in transfer buffer. Moisten surface of gel with transfer buffer and place membrane directly on the top (i.e., anode side) of the gel. Do not reposition the membrane. Remove all air bubbles as in step 4.
7. Wet another piece of filter paper, place on anode side of membrane, and remove all air bubbles. Place another Scotch-Brite pad or sponge on top of this filter paper, and complete assembly by locking the top half of the transfer cassette into place (Fig. 5.7.1).
8. Fill tank with transfer buffer and place transfer cassette containing sandwich into electroblotting apparatus in correct orientation—i.e., so that the membrane faces the anode (positively charged) side of the tank (Fig. 5.7.1). Adjust level of transfer buffer so that it covers the electrode panels but does not touch the base of the banana plug.
9. Connect leads of power supply to corresponding anode and cathode sides of electroblotting apparatus. Electrophoretically transfer proteins from gel to membrane for 30 to 60 min at 100 V with cooling (10° to 20°C), or overnight at 14 V (constant voltage) in a cold room.

Transfer time depends on gel thickness, percent acrylamide, and protein size. Cooling is required for transfers >1 hr at high power. Heat exchanger cooling cores using a circulating water bath are placed into the transfer unit for cooling.

10. Turn off power supply and disassemble apparatus. Remove membrane and note orientation by cutting a corner or marking with a soft lead pencil or Paper-Mate pen.

    Membranes can be dried and stored in resealable plastic bags at 4°C for 1 year or longer. Before use, dried PVDF membranes must be wet in 100% methanol and rinsed in distilled water to remove the methanol.

11. Stain gel for total protein with Coomassie blue to verify transfer efficiency. If desired, stain membrane reversibly with Ponceau S (see Support Protocol 1) or irreversibly with Coomassie blue, India ink, naphthol blue, or colloidal gold.

    These staining procedures are incompatible with nylon membranes.

12. Proceed with immunoprobing and visualization of proteins (see Basic Protocols 2 and 3 and Alternate Protocols 3 and 4).

ALTERNATE PROTOCOL 1

PROTEIN BLOTTING WITH SEMIDRY SYSTEMS

Additional Materials *(also see Basic Protocol 1)*

Semidry transfer unit (Hoefer, Bio-Rad, or Sartorius)

1. Prepare gel (see Basic Protocol 1, step 1).
2. Wet and equilibrate transfer membrane (see Basic Protocol 1, steps 5 and 6).
3. Disassemble gel sandwich and remove stacking gel.
4. Assemble the transfer stack on top of the anode (Fig. 5.7.2):

    Mylar mask (optional for some equipment)
    Three sheets of Whatman 3MM filter paper saturated with transfer buffer
    Equilibrated transfer membrane
    Gel
    Three more sheets of filter paper.

    Roll out bubbles as each component is added to the stack.

    Multiple gels can be transferred by simply placing a sheet of porous cellophane (Amersham Pharmacia Biotech) or dialysis membrane (Bio-Rad or Sartorius) equilibrated with transfer buffer between the transfer stacks (Fig. 5.7.2).

5. Place top electrode (cathode) onto transfer stack.
6. Carefully connect high-voltage leads to the power supply and apply constant current to initiate protein transfer. Transfers of 1 hr are generally sufficient.

    In general, do not exceed 0.8 mA/cm^2 of gel area. For a typical minigel (8 × 10 cm) and standard-sized gel (14 × 14 cm) this means ~60 and 200 mA, respectively.

    Monitor the temperature of the transfer unit directly above the gel by touch. The unit should not exceed 45°C. If the unit is warm, lower the current.

7. Turn off power supply and disassemble unit. Remove membrane from transfer stack, mark orientation, and proceed with staining and immunoprobing (see Basic Protocol 1, steps 10 to 12).

5

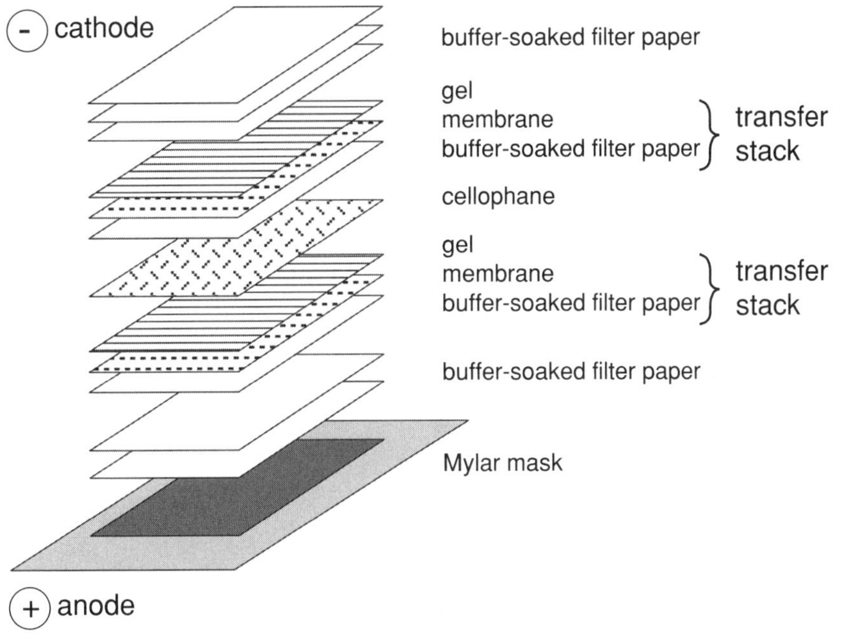

Figure 5.7.2 Immunoblotting with a semidry transfer unit. Generally, the lower electrode is the anode, and one gel is transferred at a time. A Mylar mask (optional in some units) is placed on the anode, followed by three sheets of buffer-soaked filter paper, the membrane, the gel, and, finally, three more sheets of buffer-soaked filter paper. To transfer multiple gels, transfer stacks are separated with a sheet of porous cellophane. For transfer of negatively charged protein, the membrane is positioned on the anode side of the gel. For transfer of positively charged protein, the membrane is placed on the cathode side of the gel. Transfer is achieved by applying a maximum current of 0.8 mA/cm^2 gel area. For a typical minigel (8 × 10 cm) and standard-sized gel (14 × 14 cm), this means 60 and 200 mA, respectively.

5

ALTERNATE PROTOCOL 2

BLOTTING OF STAINED GELS

Gels stained with Coomassie blue can be effectively immunoblotted by the following procedure.

Materials *(also see Basic Protocol 1)*

- Destained gel containing proteins of interest
- 25 mM Tris base/192 mM glycine/1% SDS
- 25 mM Tris base/192 mM glycine/0.1% SDS

1. Soak destained gel containing proteins of interest in distilled water for 15 min. Equilibrate with 25 mM Tris base/192 mM glycine/1% SDS for 1 hr with gentle agitation.

2. Transfer to 25 mM Tris base/192 mM glycine/0.1% SDS and equilibrate 30 min with gentle agitation.

   To increase transfer efficiency of larger proteins, the gel should be transferred to the above solution with 6 M urea for an additional 30 min.

3. Proceed with transfer (see Basic Protocol 1, steps 3 to 10).

   For optimal transfer and binding, the transfer buffer should contain SDS.

4. Soak membrane 10 to 30 min in 45% methanol (nitrocellulose) or 100% methanol (nylon or PVDF) to remove bound Coomassie blue.

   This step is not needed if using chemiluminescent reactions or radiolabeled protein A for immunodevelopment. Destaining of the nitrocellulose membrane is enhanced by adding a small ball of laboratory tissue to the methanol to absorb the Coomassie blue.

5. Proceed with immunoprobing and visualization of proteins (see Basic Protocols 2 and 3 and Alternate Protocols 3 and 4).

SUPPORT PROTOCOL 1

REVERSIBLE STAINING OF TRANSFERRED PROTEINS

Additional Materials *(also see Basic Protocol 1; see* APPENDIX 1 *for items with ✓)*

Nitrocellulose or PVDF membrane with transferred protein (see Basic Protocol 1 or Alternate Protocol 1)
✓ Ponceau S solution

1. Place nitrocellulose or PVDF membrane with transferred protein in Ponceau S solution 5 min at room temperature.

2. Destain 2 min in water. Photograph membrane and mark molecular-weight-standard band locations with indelible ink.

3. Completely destain membrane by soaking an additional 10 min in water.

5

BASIC PROTOCOL 2

IMMUNOPROBING WITH DIRECTLY CONJUGATED SECONDARY ANTIBODY

Materials *(see* APPENDIX 1 *for items with ✓)*

Membrane with transferred proteins (see Basic Protocol 1 or Alternate Protocol 1 or 2)
✓ Blocking buffer appropriate for membrane and detection protocol
Primary antibody specific for protein of interest
✓ TTBS (nitrocellulose or PVDF) or TBS (nylon)
Secondary antibody: anti-Ig conjugated with horseradish peroxidase (HRP) or alkaline phosphatase (AP), diluted per manufacturer (Cappel, Vector Labs, Kirkegaard & Perry, Sigma) and stored frozen in 25-μl aliquots
Heat-sealable plastic bag
Powder-free gloves

1. Place membrane with transferred proteins in a heat-sealable plastic bag with 5 ml blocking buffer. Seal bag and incubate 30 to 60 min at room temperature with agitation on an orbital shaker or rocking platform.

   If membrane is to be stripped and reprobed (see Support Protocol 2), blocking buffer must contain casein (for AP systems) or nonfat dry milk.

   Multiple membranes can be reacted simultaneously in a plastic tray. Increase volumes to 25 to 50 ml.

2. Dilute primary antibody in 5 ml blocking buffer.

   Primary antibody dilution is determined empirically but is typically 1/100 to 1/1000 for a polyclonal antibody, 1/10 to 1/100 for hybridoma supernatants, and ≥1/1000 for murine ascites fluid containing monoclonal antibodies. Ten- to one-hundred-fold higher dilutions can be used with AP-based detection systems. Both primary and secondary antibody solutions can be used at least twice, but long-term storage (i.e., >2 days at 4°C) is not recommended.

3. Open bag and pour out blocking buffer. Replace with diluted primary antibody and incubate 30 to 60 min at room temperature with constant agitation.

4. Remove membrane from plastic bag with gloved hand. Place in plastic box and wash four times, 10 to 15 min each, by agitating with 200 ml TTBS (nitrocellulose or PVDF) or TBS (nylon).

5. Dilute secondary antibody in blocking buffer per supplier (typically 1/200 to 1/2000).

6. Place membrane in a new heat-sealable plastic bag, add diluted secondary antibody, and incubate 30 to 60 min at room temperature with constant agitation.

7. Remove membrane from bag and wash as in step 4. Develop according to appropriate visualization protocol (see Basic Protocol 3 or Alternate Protocol 4).

ALTERNATE PROTOCOL 3

IMMUNOPROBING WITH AVIDIN-BIOTIN COUPLING TO SECONDARY ANTIBODY

Additional Materials *(also see Basic Protocol 2)*

Vectastain ABC (HRP) or ABC-AP (AP) kit (Vector Laboratories) containing: reagent A (avidin), reagent B (biotinylated HRP or AP), and biotinylated secondary antibody (request membrane immunodetection protocols when ordering)

5

1. Place membrane with transferred proteins in a heat-sealable plastic bag with 5 ml blocking buffer. Seal bag and incubate with agitation on an orbital shaker or rocking platform. For nitrocellulose or PVDF, incubate 30 to 60 min at room temperature. For nylon, incubate ≥2 hr at 37°C.

   TTBS is well suited for avidin-biotin systems. For nylon, protein-binding agents are recommended. Because nonfat dry milk contains residual biotin, it must be used in the blocking step only. If membrane is to be stripped and reprobed (see Support Protocol 2), blocking buffer must contain casein (for AP systems) or nonfat dry milk.

   Multiple membranes can be reacted simultaneously in a plastic tray. Increase volumes to 25 to 50 ml.

2. Prepare primary antibody in 5 ml TTBS (nitrocellulose or PVDF) or TBS (nylon).

   Dilutions generally range from 1/100 to 1/100,000. To determine the appropriate dilution, separate antigens on a preparative gel (i.e., a single large sample well) and immunoblot the entire gel. Cut 2- to 4-mm strips by hand or with a membrane cutter (Schleicher and Schuell, Inotech) and incubate individual strips in serial dilutions of primary antibody. The correct dilution should give low background and high specificity.

3. Open bag, remove blocking buffer, and replace with diluted primary antibody solution. Incubate 30 min at room temperature with gentle rocking.

4. Remove membrane from plastic bag and place in a plastic box. Wash membrane three times over a 15-min span in TTBS (nitrocellulose or PVDF) or TBS (nylon). Add enough buffer to fully cover the membrane (e.g., 25 to 50 ml/membrane).

5. Dilute two drops biotinylated secondary antibody in 50 to 100 ml TTBS (nitrocellulose or PVDF) or TBS (nylon).

6. Transfer membrane to fresh plastic bag containing secondary antibody solution. Incubate 30 min at room temperature with slow rocking, then wash as in step 4.

7. While membrane is being incubated with secondary antibody, prepare avidin-biotin-HRP or -AP complex. Mix two drops Vectastain reagent A and two drops reagent B into 10 ml TTBS (nitrocellulose or PVDF) or TBS (nylon). Incubate 30 min at room temperature, then further dilute to 50 ml with TTBS or TBS.

   Casein, nonfat dry milk, serum, and some grades of BSA may interfere with the formation of the avidin-biotin complex and should not be used in the presence of avidin or biotin reagents.

8. Transfer washed membrane (step 6) to avidin-biotin-enzyme solution. Incubate 30 min at room temperature with slow rocking, then wash over a 30-min span as in step 4.

9. Develop according to the appropriate visualization protocol (see Basic Protocol 3 or Alternate Protocol 4).

BASIC PROTOCOL 3

VISUALIZATION WITH CHROMOGENIC SUBSTRATES

Materials *(see APPENDIX 1 for items with ✓)*

Membrane with transferred proteins and probed with antibody-enzyme complex (see Basic Protocol 2 or Alternate Protocol 3)
✓ TBS
Chromogenic visualization solution (Table 5.7.1)

1. Wash membrane 15 min at room temperature in 50 ml TBS.

2. Place membrane into chromogenic visualization solution until bands appear (10 to 30 min).

3. Terminate reaction by washing membrane in distilled water. Air dry and photograph for a permanent record.

ALTERNATE PROTOCOL 4

VISUALIZATION WITH LUMINESCENT SUBSTRATES

Additional Materials *(also see Basic Protocol 3; see APPENDIX 1 for items with ✓)*

✓ Substrate buffer (select one): 50 mM Tris·Cl, pH 7.5 (HRP) or dioxetane phosphate substrate buffer (AP)
5% (v/v) Nitro-Block (AP only; Tropix)
Luminescent visualization solution (Table 5.7.1)
Clear plastic wrap

1. Equilibrate membrane in two 15-min washes with 50 ml substrate buffer.

2. *For AP reactions using nitrocellulose or PVDF membranes:* Incubate 5 min in 50 ml freshly prepared 5% (v/v) Nitro-Block in dioxetane phosphate substrate buffer, followed by 5 min in 50 ml substrate buffer alone.

Table 5.7.1 Chromogenic and Luminescent Visualization Systems[a]

System	Reagent[b]	Reaction/Detection	Comments
Chromogenic			
HRP-based	4CN	Oxidized products form purple precipitate	Not very sensitive (Tween 20 inhibits reaction); fades rapidly upon exposure to light
	DAB/$NiCl_2$[c]	Forms dark brown precipitate	More sensitive than 4CN but potentially carcinogenic; resulting membrane easily scanned
	TMB	Forms dark purple stain	More stable, less toxic than DAB/$NiCl_2$; may be somewhat more sensitive[d]; can be used with all membrane types; kits available from Kirkegaard & Perry, TSI, Moss, and Vector Labs
AP-based	BCIP/NBT	BCIP hydrolysis produces indigo precipitate after oxidation with NBT; reduced NBT precipitates; dark blue-gray stain results	More sensitive and reliable than other AP-precipitating substrates; note that phosphate inhibits AP activity
Luminescent			
HRP-based	Luminol/H_2O_2/ *p*-iodophenol	Oxidized luminol substrate gives off blue light; *p*-iodophenol increases light output	Very convenient, sensitive system; reaction detected within a few seconds to 1 hr
AP-based	Substituted 1,2-dioxetane-phosphates (e.g., AMPPD, CSPD, Lumigen-PPD, Lumi-Phos 530[e])	Dephosphorylated substrate gives off light	Protocol described gives reasonable sensitivity on all membrane types; consult instructions of reagent manufacturer for maximum sensitivity and minimum background

[a]Abbreviations: AMPPD or Lumigen-PPD, disodium 3-(4-methoxyspiro{1,2-dioxetane-3,2′-tricyclo[3.3.1.13,7]decan}-4-yl)phenyl phosphate; AP, alkaline phosphatase; BCIP, 5-bromo-4-chloro-3-indolyl phosphate; 4CN, 4-chloro-1-napthol; CSPD, AMPPD with substituted chlorine moiety on adamantine ring; DAB, 3,3′-diaminobenzidine; HRP, horseradish peroxidase; NBT, nitroblue tetrazolium; TMB, 3,3′,5,5′-tetramethylbenzidine.

[b]Recipes and suppliers are listed in *APPENDIX 1* and *APPENDIX 4* except for TMB, for which use of a kit is recommended.

[c]DAB can be used without the nickel enhancement, but it is much less sensitive.

[d]McKimm-Breschkin (1990) reported that if nitrocellulose filters are first treated with 1% dextran sulfate for 10 min in 10 mM citrate-EDTA (pH 5.0), TMB precipitates onto the membrane with a sensitivity much greater than 4CN or DAB, and equal to or better than that of BCIP/NBT.

[e]Lumi-Phos 530 contains dioxetane phosphate, $MgCl_2$, CTAB, and fluorescent enhancer in a pH 9.6 buffer.

Nitro-Block enhances light output from the dioxetane substrate in reactions using AMPPD, CSPD, or Lumigen-PPD concentrate. It is required for nitrocellulose and recommended for PVDF membranes. It is not needed for Lumi-Phos 530, AP reactions on nylon membranes, or HRP-based reactions on any type of membrane. Lumi-Phos 530 is not recommended for nitrocellulose membranes.

3. Transfer membrane to 50 ml luminescent visualization solution. Soak 30 sec (HRP) to 5 min (AP).

4. Remove membrane, drain, and place face down on a sheet of clear plastic wrap. Fold wrap back onto membrane to form a liquid-tight enclosure.
5. In a darkroom, place membrane face down onto film. Do not reposition. Expose film for a few seconds to several hours.
6. If desired, wash membrane in two 15-min washes of 50 ml TBS and process for chromogenic development (see Basic Protocol 3).

SUPPORT PROTOCOL 2

STRIPPING AND REUSING MEMBRANES

This stripping procedure works with blotted membranes from one- and two-dimensional gels as well as with proteins blotted from previously stained gels. Reprobing PVDF membranes that have been developed with chemiluminescent reagents is simple and straightforward. Although repeated probing can lead to loss of signal, up to five reprobings are generally feasible. The blot should have been blocked with 5% nonfat dry milk prior to treatment.

Materials

0.2 M NaOH

1. Wash blot 5 min in distilled water.

    In order to effectively reprobe the membranes, casein (for AP systems) or nonfat dry milk must be used as the blocking agent. Chromogenic development leaves a permanent stain on the membrane that is difficult to remove, and should not be used when reprobing. The stain can interfere with subsequent analysis if reactive bands from sequential immunostainings are close together.

2. Transfer to 0.2 M NaOH and wash 5 min.
3. Wash blot 5 min in distilled water.
4. Proceed with immunoprobing procedure (see Basic Protocol 2 and Alternate Protocol 3).

    Casein or nonfat dry milk is recommended as blocking agent when reprobing membranes.

References: Bejurrum and Schafer-Nielsen, 1986; Gillespie and Hudspeth, 1991; Harlow and Lane, 1988; Schneppenheim et al., 1991; Suck and Krupinska, 1996

Contributors: Sean Gallagher, Scott E. Winston, Steven A. Fuller, and John G.R. Hurrell

UNIT 5.8

Phage Display in Neurobiology

Phage display is a technique that involves the coupling of phenotype to genotype in a selectable format. It has been extensively used in molecular biology to study protein-protein interactions, receptor and antibody binding sites, and immune responses; to modify protein properties; and to select antibodies against a wide range of different antigens. In the format most often used, a polypeptide is displayed on the surface of a filamentous phage by genetic fusion to one of the coat proteins, creating a chimeric coat protein. As the gene encoding the chimeric coat protein is packaged within the phage, selection of the phage on the basis of the binding properties of the polypeptide displayed on the surface simultaneously results in the isolation of the gene encoding the polypeptide (Fig. 5.8.1).

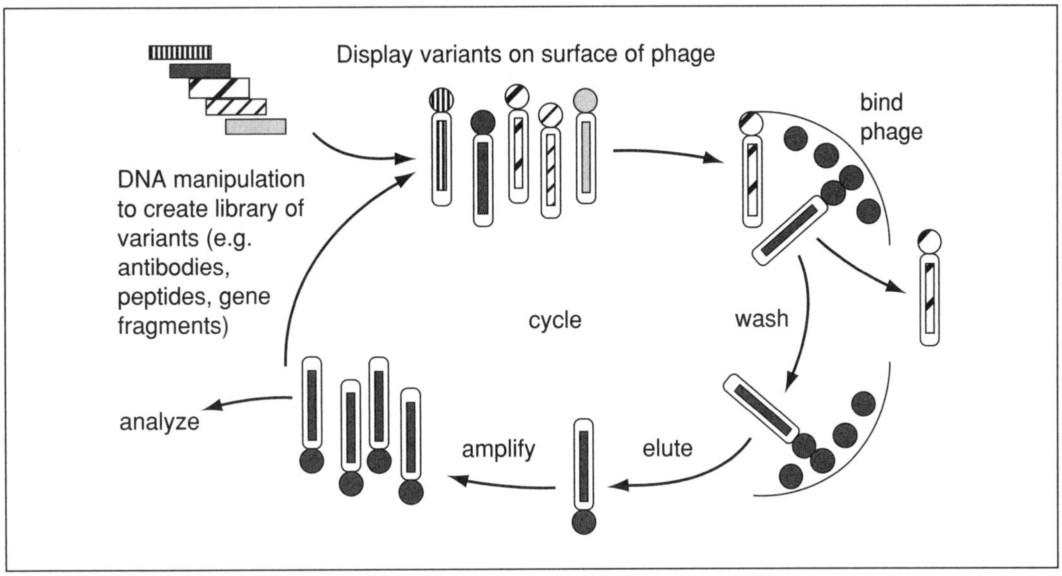

Figure 5.8.1 The phage display cycle. The phage display cycle starts with the creation of diversity at the DNA level. This is translated into phenotypic diversity by the display of different polypeptides on the surface of the phage. The application of selection pressure, manifested by cycles of binding to a ligand (here depicted in solid phase), washing, and elution allow the selection of phage-displaying polypeptides that bind to the ligand.

FILAMENTOUS PHAGE BIOLOGY

The Ff group of filamentous phage (M13, fd and f1) is made up of single-stranded DNA (ssDNA) phages that infect bacteria via the F pilus. After phage has bound to the tip of the pilus, the pilus undergoes retraction, and the phage is brought to the bacterial surface. The phage ssDNA enters the bacteria and is converted to a double-stranded form; this acts as the template for the creation of single-stranded progeny, which are packaged into phage particles. Filamentous phage infection does not kill the host but slows down its growth; ~100 phages are secreted for each cell division.

The phage genome encodes eleven proteins. Of these, five are phage coat proteins and six are involved in phage assembly (Fig. 5.8.2). The major coat protein is p8, present in 2700 copies. p3 and p6 are found at the end of the phage that enters the bacteria first and leaves it last, and p7 and p9 are found at the other end. p3, the protein responsible for binding to the F pilus, is found in three to five copies and possesses a number of functional domains. The structure of p3 has recently been determined and provides some insight into the infection mechanism (Lubkowski et al., 1998). Both p8 and p3 have leader sequences that direct synthesis of the protein to the periplasmic space and that are subsequently removed.

PHAGE DISPLAY

If foreign DNA encoding a polypeptide A is cloned downstream of the gene 3 or gene 8 leader sequence, it will be translated and exposed at the N terminus of the mature p3 or p8 protein. Depending on the format of the vector, the resulting polypeptide A can be expressed so that it does not impede the normal function of either p8 or p3. If an antibody recognizing A (anti-A) is fixed to a solid support, phages displaying A can be selected from a background of billions of other polypeptide-displaying phage that are unable to bind A (Fig. 5.8.1). The collection of

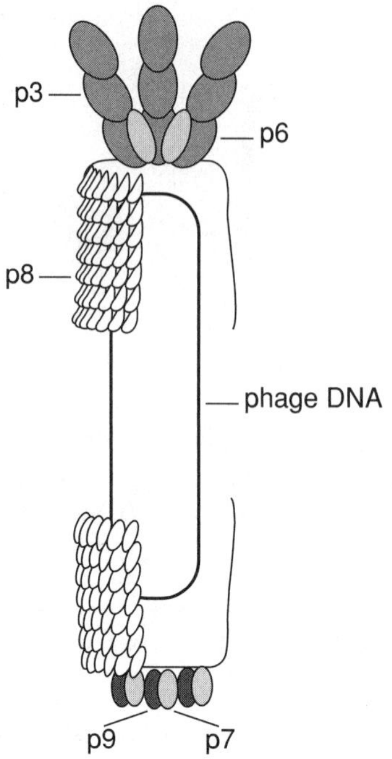

Figure 5.8.2 The structure of filamentous phage. Three copies of p3, p6, p7, and p9 are shown; although the true number is thought to be 3 to 5. There are 2700 copies of p8, which is the major coat protein.

5

billions of phage displaying different polypeptides on their surfaces is termed a *phage library*, which provides a valuable resource and from which useful polypeptides can be isolated.

The fact that the gene encoding A is found within the phage displaying A, means that cloning of the gene for A occurs simultaneously with the selection of phage displaying A. This permits the continuation of the phage display cycle (Fig. 5.8.1) until a population that is homogenous in its binding properties is obtained (i.e., all bind anti-A). The cloning of the gene encoding A also permits the subsequent recombinant manipulation or expression of A for further experiments.

This is the simplest form of selection; phage display has evolved considerably since the mid-1990s, and complex selection schemes involving the selection of ligands able to activate receptors or induce different biological states have been either performed or proposed. Selection systems using complex selectors, such as cell surfaces or cell extracts, have also been carried out, as well as subtractive selection schemes in which the isolation of ligands recognizing specific (but not closely related) proteins or tissues is performed.

Phage display was first carried out using peptide libraries (Smith, 1985) with the goal of identifying monoclonal antibody epitopes. Since then, the number of proteins displayed on phages has expanded enormously (Table 5.8.1), and its present application can be essentially reduced to three general categories:

1. Peptide display: used to identify epitopes that bind to antibodies or other proteins,
2. Antibody display: used to isolate antibodies that recognize different antigens or to study an immune response, and
3. Protein display: used to modify the properties of a particular protein.

A number of vector systems have been adopted for phage display (Fig. 5.8.3), and each has specific advantages and disadvantages.

Table 5.8.1 Proteins Displayed on Filamentous Phage

Protein displayed	Vector type	Reference
cDNA libraries (i.e., random clones)	pJuFo	Crameri and Suter, 1993; Smith and Petrenko, 1997
Genomic libraries (*Staphylococcus* and *Streptococcus* genomes)	3 + 3, 8 + 8	Smith and Petrenko, 1997; Jacobsson and Frykberg, 1998
Protein fragments		
β-galactosidase	3	Smith and Petrenko, 1997
Bluetongue virus VP5 and NS1 proteins	3	Smith and Petrenko, 1997
p53	3	Petersen et al., 1995
RNA polymerase II	3	Petersen et al., 1995
Cytokeratin 19	3	Petersen et al., 1995
PM/Scl	3	Bluthner et al., 1996
ZAG and MAG from streptococci	8 + 8	Jacobsson et al., 1997
myb	3 + 3	Kiewitz and Wolfes, 1997
PAI-1	3 + 3	van Meijer et al., 1996
Small constrained peptide domains		
Hybrid *rop* protein constrained peptide library	3	Santiago Vispo et al., 1993
Cytochrome *b*562		Smith and Petrenko, 1997
Tendamistat constrained peptide library	33	Smith and Petrenko, 1997
Protein A (E, D, A and B domains or B domain alone)	3 + 3	Smith and Petrenko, 1997
Knottins	3 + 3	Smith et al., 1998
Protease inhibitors		
Alzheimer's amyloid β-protein precursor Kunitz domain	3 + 3	Smith and Petrenko, 1997
Human PAI-1	3 + 3	Smith and Petrenko, 1997
Kunitz domain libraries, BPTI	3, 88	Roberts et al., 1992
Ecotin	3 + 3	Smith and Petrenko, 1997
Cystatin	3 + 3	Smith and Petrenko, 1997
Proteases		
Prostate-specific antigen	3 + 3	Smith and Petrenko, 1997
Trypsin	3 + 3, 33, 8 + 8, 88	Smith and Petrenko, 1997
Enzymes		
Glutathione transferase A1-1	3 + 3	Smith and Petrenko, 1997
Staphylococcal nuclease	3, 3 + 3	Smith and Petrenko, 1997
Alkaline phosphatase	pJuFo, 3, 3 + 3	Crameri and Suter, 1993
β-lactamase	3	Smith and Petrenko, 1997
Lysozyme	3 + 3	Smith and Petrenko, 1997
Inactive phospholipase A2	pJuFo	Crameri and Suter, 1993
Cell surface receptor fragments		
CD4, extracellular and individual domains	3, 3 + 3	Chiswell and McCafferty, 1992; Smith and Petrenko, 1997
FceR1 α-chain, extracellular and individual domains	3, 3 + 3	Smith and Petrenko, 1997
PDGF receptor, extracellular domain	3	Chiswell and McCafferty, 1992
Hormones, interleukins and bioactive peptides		
Human growth hormone	3 + 3	Lowman and Wells, 1993
C5a	pJuFo	Hennecke et al., 1997
Thymosin \ 'df4		Rossenu et al., 1997

continued

Table 5.8.2 Proteins Displayed on Filamentous Phage, continued

Protein displayed	Vector type	Reference
Transforming growth factor α	3	Tang et al., 1997
Tumor necrosis factor	3 + 3	Clackson and Wells, 1994
IL-2	3 + 3	Vispo et al., 1997
IL-3	3, 3 + 3	Smith and Petrenko, 1997
IL-6	3 + 3	Smith and Petrenko, 1997
IL-8	3 + 3	Clackson and Wells, 1994
Insulin-like growth factor (IGF) binding protein	3 + 3	Lucic et al., 1998
Heregulin β domain	3 + 3	Ballinger et al., 1998
EGF	3	Souriau et al., 1997
Ciliary neurotrophic factor	3 + 3	Smith and Petrenko, 1997
Atrial natriuretic peptide and derivatives	3 + 3	Smith and Petrenko, 1997
Bone morphogenetic protein (2A)		Liu et al., 1996b
Antibodies and derivatives		
Minibody (61 amino acids, 4 β sheets)	3, 3 + 3	Martin et al., 1996
Fab	3, 3 + 3, 8 + 8	Winter et al., 1994
scFv	3, 3 + 3, 8 + 8	Winter et al., 1994
CH3	3 + 3	Atwell et al., 1997
VH domains	3 + 3	Davies and Riechmann, 1996
Dromedary heavy-chain antibodies	3 + 3	Lauwereys et al., 1998
Toxins		
Ricin β-chain, complete or domain 2	3	Smith and Petrenko, 1997
Fungal ribotoxin reAsp f I/a	3 + 3	Crameri and Suter, 1993
B. thuringiensis CryIA(a) toxin	3 + 3	Marzari et al., 1997
Nucleic acid binding proteins		
HIV *tat*	3 + 3	Hoffmann and Willbold, 1997
U1A protein	3 + 3	Smith and Petrenko, 1997
Zinc finger proteins, domains and libraries	3, 3 + 3	Smith and Petrenko, 1997
Miscellaneous		
T cell receptor α chain	3	Smith and Petrenko, 1997
Peptostreptococcal protein L		Gu et al., 1995
Pseudomonas aeruginosa protein F	3	Kermani et al., 1995

VECTOR SYSTEMS USED FOR PHAGE DISPLAY

Three broad classes of vector systems have been described (Fig. 5.8.3). In the first—type 3 or type 8 vectors—phage DNA is directly modified, and all phage proteins should be recombinant; thus such vectors are termed *polyvalent.* Although p3 display has been used for many peptide libraries and a few proteins, p8 display in this system is limited to peptides smaller than six amino acids, because longer ones cause a severe display bias.

It has proved difficult to work with phage vectors because of the transfection rates and instability of the phage genome to inserts; there is a tendency for the phage to delete extraneous DNA. For this reason, the 3 + 3 vectors are generally preferred for protein display. These are phagemid vectors that contain an Ff origin of replication (which ensures packaging into viral particles) and a copy of g3, which is used to display the recombinant polypeptide. These vectors can be propagated as plasmids. When display is required, bacteria containing the phagemid are infected with helper phages, which provide all the necessary phage proteins for the creation of selectable phagemid particles. The origin of replication in the phagemid is packaged in

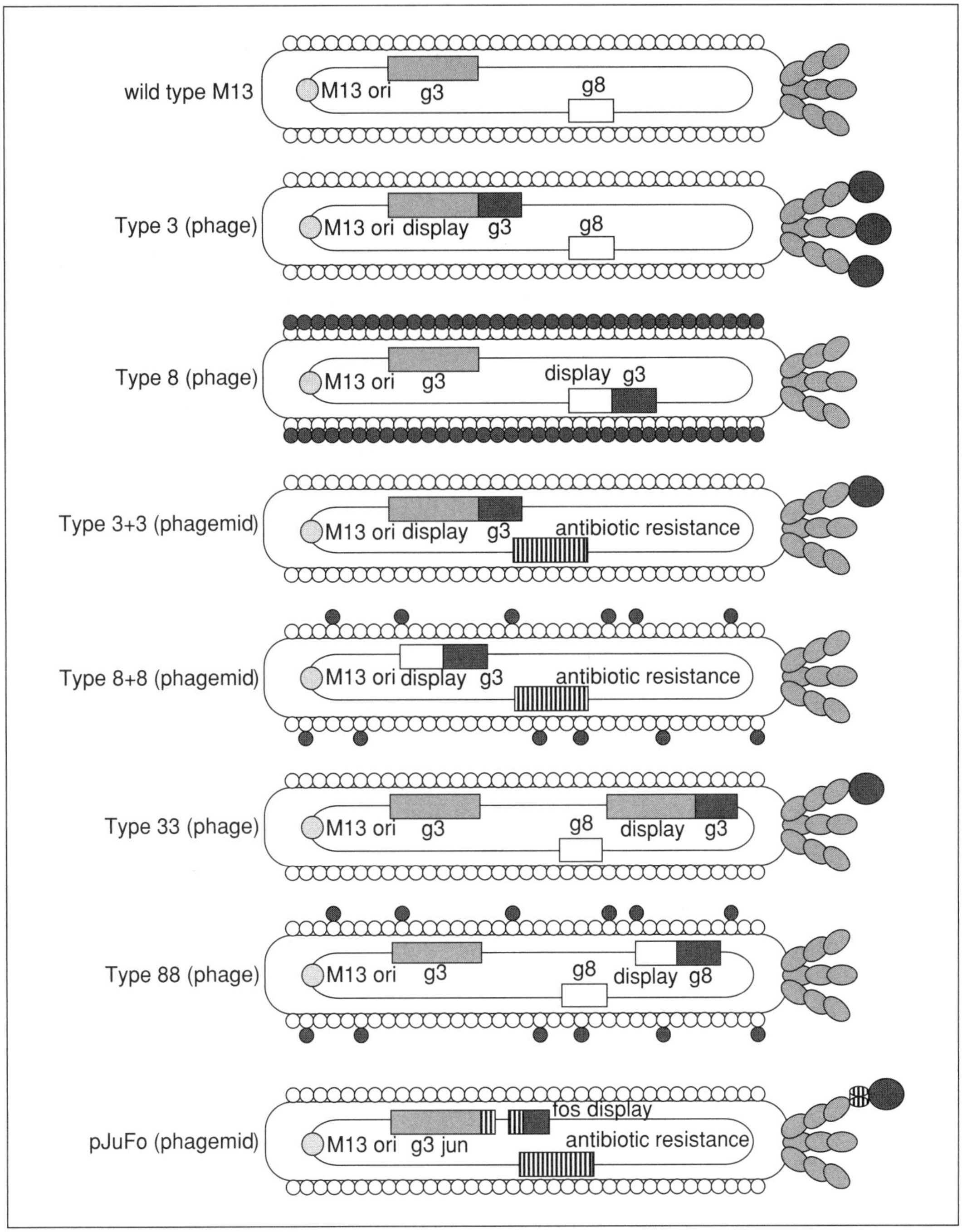

Figure 5.8.3 Different phage display vectors. The type 3 and type 8 vectors are phage vectors that have only one copy of gene 3 or 8, which is modified. As a result, all phages will display recombinant protein. The type 3 + 3 and 8 + 8 vectors are phagemid vectors that contain recombinant display genes. Wild type p3 or p8 is provided by the helper phage; as a result not all p3 or p8 is recombinant. The type 33 or 88 vectors are phage vectors that have one wild type copy of p3 or p8 and one recombinant copy; as a result not all p3 or p8 are recombinant. pJuFo is a phagemid vector that has a jun peptide at the N terminus of p3. This interacts with a fos peptide, which displays the recombinant protein. The interaction between jun and fos is rendered covalent by the presence of two disulfide bonds, one at each end. Modified from Smith (1993).

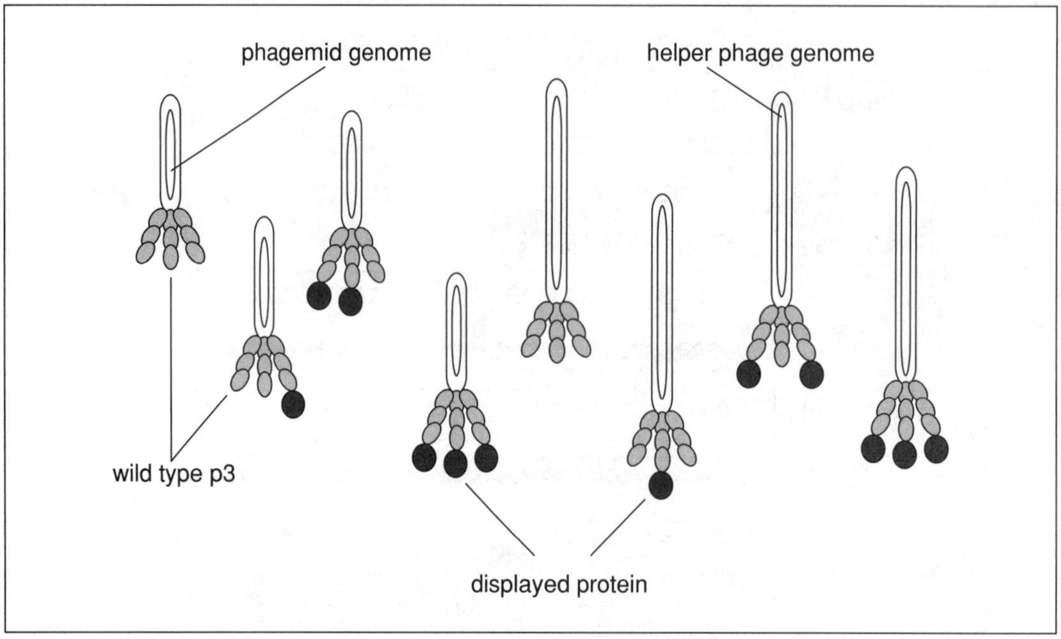

Figure 5.8.4 Phagemid heterogeneity. The presence of both wild type p3/8 and recombinant p3/8 coupled with the presence of genomes derived from both the phagemid display vector and the helper phage result in phenotypic and genotypic heterogeneity in type 3 + 3, 8 + 8, and pJuFo vectors. The phages illustrated display proteins only on p3, although the same applies to p8.

5

preference to that of the helper phage, and a proportion (1% to 10%; Clackson and Wells, 1994) of the phage particles will display recombinant p3. Of these, most will display single copies, the so-called monovalent display. The remaining p3 molecules are derived from the helper phage. As a result, such vector systems are both phenotypically (wild type and recombinant p3 is present) and genotypically (both helper and phagemid genomes are present) heterogenous (Fig. 5.8.4). A similar 8 + 8 system has been used for the display of some peptide libraries.

The 33 and 88 systems are intermediate, because they are phage based and contain both wild type g3 (or g8) in association with recombinant g3 (or g8) cloned within the phage genome. In practice, however, only the 88 system has been used and only for some peptide libraries (Bonnycastle et al., 1996). Phage produced using this system are phenotypically heterogenous but genotypically homogenous. They suffer from some of the same problems as the phage vectors and have not yet been used for protein display.

In general, monovalent display is more efficient at discriminating polypeptides of different affinities, whereas polyvalent display is more efficient at isolating any binding polypeptides, with avidity effects playing an important role.

An alternative system designed for the display of cDNA, pJuFo (Fig. 5.8.3; Crameri and Suter, 1993), involves fusions of translated cDNA to the C terminus of a fos domain. This associates with a jun domain expressed at the N terminus of p3. The use of the C terminal fusion has the advantage that only one fusion in three will be out of frame, as opposed to one in nine if random directional display at the N terminus of p3 is attempted (Fig. 5.8.5).

Ff p3 is the display protein that has been used most frequently for proteins. In fact, the permissivity of this protein (Table 5.8.1) is remarkable. p8 has also been used to display proteins (Kretzschmar and Geiser, 1995), although direct comparison of the quantity of a recombinant protein (single-chain variable fragment; scFv) expressed on either p3 or p8 in a

Random cloning of blunt fragments at N terminus

```
1/2 correct    ATG GTT CAG CCC CGT ATC CGT
orientation    ACG GAT ACG GGG CTG AAC CAT

               ATG GTT CAG CCC CGT ATC CGT
1/3 start       TG GTT CAG CCC CGT ATC CGT
correctly        G GTT CAG CCC CGT ATC CGT

               ATG GTT CAG CCC CGT ATC CGT
1/3 finish     ATG GTT CAG CCC CGT ATC CG
correctly      ATG GTT CAG CCC CGT ATC C

1/18 completely
correct        ATG GTT CAG CCC CGT ATC CGT
```

Random cloning of blunt fragments at C terminus

```
1/2 correct    ATG GTT CAG CCC CGT ATC CGT
orientation    ACG GAT ACG GGG CTG AAC CAT

               ATG GTT CAG CCC CGT ATC CGT
1/3 start       TG GTT CAG CCC CGT ATC CGT
correctly        G GTT CAG CCC CGT ATC CGT

1/6 completely
correct        ATG GTT CAG CCC CGT ATC CGT
```

Figure 5.8.5 The problems of cloning random DNA fragments in frame with proteins (e.g., p3/8) at either the N or the C terminus are illustrated. More clones will have functional proteins in frame if cloned at the C terminus, because there is no need for DNA clones to terminate correctly. If directional cloning is used, the probability of clones being in frame improves 2-fold; i.e., one in nine for N terminal fusions and one in three for C terminal fusions.

phagemid format showed that (Ff contains 2700 copies of p8 and only three to five copies of p3) there was more scFv expressed in the p3 format, indicating that p8 is a rather sensitive site for proteins of this size (Kretzschmar and Geiser, 1995). This is confirmed by experiments with p8 that have attempted to define the length limits of insertion within gp8. It was found that in a phage format almost all hexapeptides are tolerated whatever their sequence, whereas only 40% of phages carrying an octapeptide can form infective particles; this fraction drops to 20% and 1% for insertions of 10 or 16 amino acids long, respectively (Iannolo et al., 1995).

The use of the phagemid format permits the display of a greater diversity of peptides, although the levels of display can vary greatly. Display within p3 appears to be limitless, judging by the large numbers of proteins that have been expressed at its N terminus; but some constraints have been found for peptides with positive charges close to the signal sequence. These appear to interfere with correct insertion of p3 into the *Escherichia coli* inner membrane, thus blocking assembly and extrusion of the phage. In model systems, these can be overcome by suppressor mutations in the prlA (secY) component of the protein export apparatus; however, no systematic study of the use of these bacteria for display purposes has been made. Proteins containing long hydrophobic regions are also poorly displayed, likely a result of problems in translocation across the membrane; and in some cases, proteins are poorly displayed for no obvious reason.

USE OF PHAGE DISPLAY LIBRARIES TO MAP PROTEIN BINDING SITES

Peptide Display Libraries

Antibody epitopes

Peptide libraries have been made in both p3 and p8; and the general feeling is that the multivalent p8 display is probably better for initial screens while the monovalent p3 display is better for subsequent affinity maturation. The first peptide libraries were linear, but constrained by disulfide bonds, and beta-turn libraries have also been made to simulate different epitope targets. The importance of this is shown by the fact that not all libraries yield products that bind to a particular antibody target (Bonnycastle et al., 1996).

Peptide libraries were first used to determine the epitopes recognized by monoclonal antibodies by fixing the monoclonal anti-A to a solid support and interacting with the peptide library. After a number of cycles of binding, washing, and elution, the sequences of selected peptides are determined by sequencing the phage(mid) DNA. Usually, one or more motifs are obtained, although this is not always the case. Such motifs may be similar to a sequence found within A, which—pending further experimental confirmation using synthetic peptides—is considered good evidence that anti-A binds A at that site (Yayon et al., 1993). If the identity of A is unknown, screening the sequences of selected peptides against protein databases may allow the identification of the protein recognized, although the degeneracy of the motif usually obtained can make this difficult, and experimental confirmation is always required.

When the motif has no homologue within A, it is termed a *mimotope* and is thought to mimic a conformational or discontinuous epitope. This probably occurs by displaying the epitopic amino acids found close to one another on the surface of A (but distant in primary sequence), in a linear fashion, so that they mimic the spatial arrangement found on the surface of A.

5

When the antibody used for selection inhibits the binding of a natural ligand to its receptor by binding to either receptor or ligand, identification of the isolated antigen peptide sequences may provide information on the receptor binding site for the natural ligand. This is based on the facts that blocking antibodies will often bind to one of the interacting surfaces of a receptor ligand pair (thus acting as a template for that surface) and that the sequences of peptides binding to such antibodies (templates) may identify the respective parts of the receptor or ligand recognized by the antibody. Some sequences, however, will bear no resemblance to the receptor-ligand sequences, and examination of surface maps may be more useful. For example, success has been achieved with two monoclonal antibodies that bind basic fibroblast growth factor (bFGF) and inhibit binding to its receptor. Two consensus sequences derived from the isolated peptides identified two continuous protein sequences of bFGF. In addition, synthetic peptides corresponding to either the phage epitopes or the similar bFGF sequences specifically inhibited binding of bFGF to the antibodies and its high-affinity receptor and inhibited the biological effect of bFGF, strongly suggesting that these sequences are functionally involved in receptor binding and may constitute part of the receptor-binding determinants on bFGF (Yayon et al., 1993).

Peptide libraries have also been used to identify antigenic epitopes present in the sera of patients suffering from infectious diseases. Although conceptually similar to the use of monoclonal antibodies, it is practically more difficult, because individual patient sera contain many antibodies able to bind epitopes that have no relevance to the disease under study. For this reason, selection on sera from a number of different patients with the same disease is followed by depletion or negative screening using sera from normal or diseased controls, leading to the identification of disease *phagotopes* (Cortese et al., 1994). Such phagotopes can often induce resistance to infectious disease agents when used as immunogens and can provide important information about the immune response. The use of this approach with immunoglobulins from the cerebrospinal fluid (CSF) of patients with multiple sclerosis (MS) revealed that, although individual patients were each able to select consensus phagotopes, no MS-specific phagotope could be selected and that normal individuals also had antibodies that recognized the selected phagotopes, indicating that these phagotopes hold no diagnostic value for MS (Cortese et al., 1998).

Peptide mimics of DNA (Sibille et al., 1997) and polysaccharides (Harris et al., 1997) have also been isolated using antigen-specific antibodies, indicating that it is likely that peptides can mimic many different antigenic structures. It is interesting that immunization with the peptide mimics can often induce an immune response against the nonpeptide antigen.

Protein binding sites

Peptide libraries can be used to determine binding sites of proteins other than antibodies using the same methodology. The binding specificity of immunoglobulin heavy chain binding protein (BiP), an endoplasmic reticulum chaperone, was found to consist of a subset of aromatic and hydrophobic amino acids in alternating positions. This suggests that peptides in unfolded target proteins bind BiP in an extended conformation, with alternate side chains pointing into a cleft on the surface of BiP (Blond-Elguindi et al., 1993). Furthermore, synthetic peptides corresponding to those selected were able to bind to BiP and stimulate its ATPase activity, further confirming the importance of this result.

In another series of experiments, peptides were selected on α-bungarotoxin, a toxin that inactivates the nicotinic acetylcholine receptor (AChR) by binding to it and competing with acetylcholine (Balass et al., 1997). The selected motif (YYXSSL) bore similarities to a region of the torpedo muscle AChR, which is inactivated by this toxin. One of the selected peptides and a receptor-derived peptide taken from the region that resembled the motif were able to inhibit binding of the toxin to the receptor. The biological significance of this result is underscored by the fact that a synthetic peptide derived from the identical region of a receptor that does not bind the toxin and that differs by only a single amino acid was unable to inhibit toxin binding, suggesting that the selected peptide had identified the toxin-binding site.

An alternative to using whole proteins is to synthesize individual domains known to be involved in protein-protein interaction. In fact, much of the work that led to the conclusion that different members of interaction domain families have slightly different specificities stemmed from work using phage peptide libraries and was confirmed on isolated proteins. The binding specificities of *src* homology 2 (SH2; Songyang, 1994), SH3 (Sparks et al., 1994), and *eps* homology (EH; Salcini et al., 1997) domains were determined in this way.

It is often difficult to obtain pure protein preparations with which to perform peptide phage selections, especially proteins that maintain their native structure. In some cases, this has been overcome by expressing receptors (e.g., erythropoietin or thymopoeitin) on the surface of cells and using them for selection. An alternative is to express proteins on the surface of bacteria, as has been done for β-adrenoceptors (Chapot et al., 1990), and to use these as living columns to purify binding phage (Bradbury et al., 1993).

The use of receptors on cell surfaces to select peptides has allowed the selection of functional peptides that are far smaller than the original cytokine. In one case, a dimerized 14-amino-acid peptide selected on the thrombopoietin receptor was as potent as the full-length 322-amino-acid cytokine in cell-based assays (Cwirla et al., 1997), and a 14-amino-acid dimerized peptide selected on the erythropoietin receptor could stimulate erythropoiesis in mice (Smith and Petrenko, 1997).

In one application of phage peptide libraries, peptides binding to the endothelium of specific tissues were isolated by injecting a library into a mouse and eluting the endothelium-bound peptides from homogenized extracts of those tissues (Pasqualini and Ruoslahti, 1996). In this way, peptides with specificities for specific tissues 3 to 33 times greater than unselected peptides were isolated.

Post-translational modification sites

A recent development in the use of peptide libraries has been the determination of enzyme specificities. This is performed by treating the peptide library with the enzyme under study and isolating peptide phage that have been appropriately modified. By using antibodies that recognize phosphorylated tyrosines, the sequence specificity of the p55 fyn (Dente et al., 1997), c-Src, Blk, Lyn, and Syk (Schmitz et al., 1996) kinases were determined. In general, all of these kinases had strong preferences for amino acids in the -1 and $+1$ position with respect to the phosphorylated tyrosine. It is interesting that the specificity is only relative, because if incubated with the kinase for long enough, all tyrosines become phosphorylated. If appropriate antibodies were available, there would be no reason why other post-translational modifications could not be studied in the same way.

Protease specificities

A similar technique has been used to study protease specificities. Phages displaying a binding domain (e.g., growth hormone; Matthews and Wells, 1993) followed by a random protease substrate sequence are bound to an affinity matrix (e.g., growth hormone receptor). After extensive washing, phages are treated with protease. Phages with good protease substrates are released, and those with substrates that resist proteolysis remain bound. By sequencing the DNA encoding the peptide sequences, the investigator can determine protease specificity. In this way, the specificities of subtilisin, factor Xa (Matthews and Wells, 1993), and furin have been determined.

Gene-Fragment Libraries

The perfect peptide library would cover the universe of possible epitopes. The limitations of transfection efficiency means that, for practical purposes, the complete diversity of no more than seven amino acids can be fully explored (requiring a diversity of $\sim 10^9$). Libraries based on peptide sequences longer than this will, by necessity, be incomplete. For example, a 12-mer library containing 10^9 different members will represent only 0.000025% of the potential dodecapeptide diversity. A solution to this problem may be the use of biological sequences rather than synthetic ones. When the cDNA for A is cloned, more nearly accurate results can be obtained by selecting from small gene (cDNA) specific libraries made by displaying variably sized random fragments of A. Although 1 in 18 clones will be nonfunctional owing to the difficulty of cloning in frame with p3/8 (Fig. 5.8.5), the relatively small sequence space that needs to be covered allows the selection of both continuous and discontinuous epitopes from libraries $<10^6$ (Wang et al., 1995). This has been done using either monoclonal or polyclonal antibodies for a number of different genes, including β-galactosidase, p53, RNA polymerase II, cytokeratin 19, PM/Scl, bluetongue virus NS1 and VP5, and Zn-α_2-glycoprotein (ZAG) and myelin-associated glycoprotein (MAG) from streptococci. In one study in which general synthetic peptide libraries were compared to gene-specific libraries for the identification of a number of antibody epitopes, it was found that epitopes were more easily and accurately identified using gene-fragment libraries (Fack et al., 1997). The disadvantage is that new libraries need to be made for each experiment.

Gene-fragment libraries have also been used to identify protein domains involved in protein-protein interactions as an alternative to the yeast two-hybrid system. In one case, the domain of *myb* involved in the interaction with its cofactor, CREB-binding protein (CBP) was identified (Kiewitz and Wolfes, 1997); and in another, the interaction sites of plasminogen activator inhibitor 1 (PAI-1) with thrombin-vanilloid receptor 1 (VR1) were found (van Meijer et al., 1996).

DISPLAYING cDNA AND GENOMIC LIBRARIES

Gene-fragment libraries can be used only when the gene product recognized by the selector has been identified and cloned. As described, synthetic peptide display libraries can sometimes give information about the gene product recognized, but this is not usually the case. The display of cDNA or genomic fragments should overcome this problem, but implementing such display libraries has been difficult. There are two main obstacles. The first is that when cloning random fragments into a protein such as p3, for which only the N terminus can be used, only 1 in 18 clones will be in frame (Fig. 5.8.5). The second is that for polypeptides to be efficiently displayed on filamentous phage, they must cross the bacterial inner membrane. This tends to be easier for secreted proteins and the extracellular domains of membrane proteins, although a number of cytoplasmic protein domains have also been displayed (Table 5.8.1); but the rules that govern protein export, and hence display, are not completely understood. The cloning of fragments at the C terminus of p6, a filamentous phage protein that has a free C terminus (Jespers et al., 1995), or a fos polypeptide, which interacts with a jun fragment at the N terminus of p3 (the pJuFo system; Fig. 5.8.3; Crameri and Suter, 1993), should help the cloning problem by increasing the number of functional clones to 1 in 6 for random fragments. The latter has been used to identify allergens recognized by IgE from allergic patients, but there have been few other published studies that used these systems. Display at the C terminus of proteins from intracellular phages, such as λ, T4, P22, and P4 have been proposed as alternatives.

If a genomic-fragment library is sufficiently large, representative fragments of all genes (except those that span exons) should be displayed, whereas cDNA libraries are limited to genes expressed in the tissue used to prepare the library. Furthermore, genomic display would not be expected to suffer from the bias that is inherent in cDNA libraries owing to variations in levels of gene expression. The display of genomic fragments to date has been carried out only with bacterial genomes, which lack introns. Fibronectin, fibrinogen, and IgG-binding open reading frames have been selected from *Staphylococcus aureus,* and α_2-macroglobulin, IgG, and serum albumin binding domains have been selected from the *Streptococcus* genome (Jacobsson and Frykberg, 1998, and references therein).

5

SELECTING ANTIBODIES BY PHAGE DISPLAY

One of the most successful applications of phage display has been the isolation of monoclonal antibodies using either purified antigens or antigens in complex mixtures, such as cellular surfaces (see Winter et al., 1994, for a review).

Antibody phage libraries can be classified as either naive or immunized. Naive libraries are derived from natural, unimmunized, human rearranged V genes or synthetic human V genes in which the diversity is provided by a synthetic binding domain introduced by polymerase chain reaction (PCR) amplification of cloned V genes. These libraries should not contain any inherent bias in the antibodies present, and it has been found that antibodies can be isolated against virtually any antigen from a good naive library. Immunized libraries, on the other hand, are created from V genes from immunized humans—e.g., human immunodeficiency virus (HIV), hepatitis B, rhesus D, and human tumor tissue—or mice and contain antibodies biased toward a certain specificity. As a result, antibodies of higher affinity can be isolated against the immunogen than can be selected from naive libraries of the same size and the library can be smaller (10^6 to 10^7). Immunized libraries can also be used to select antibodies against antigens that were not used in the immunization, although affinities tend to be lower than those obtained from naive libraries.

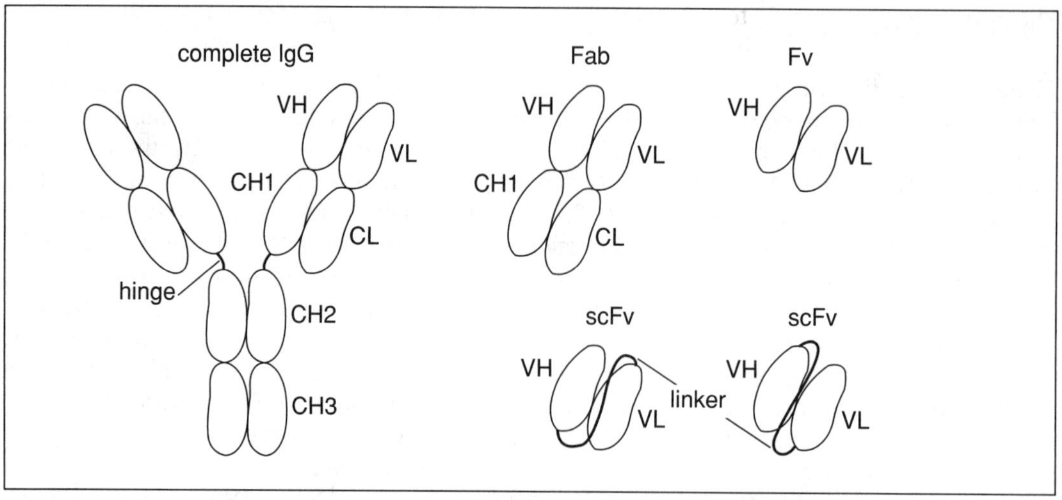

Figure 5.8.6 An antibody with derived polypeptides is shown. V regions are variable and involved in antigen binding; C regions provide in vivo immunological effector functions to the antibody.

Antibody-binding domains are localized to the V regions Fig. 5.8.6); and the C regions play no role. Thus the minimum scFv and Fab binding domains have been used for phage display.

In general for naive libraries, the affinity of the antibodies selected is proportional to the size of the library, ranging from 10^{-6} to 10^{-7} for the smaller libraries to 10^{-9} for the larger ones.

One great advantage of using phage display to isolate antibodies is that antibodies can be isolated against proteins that either are nonimmunogenic or cannot be produced by hybridoma cells for biological reasons (e.g., monoclonal antibodies that recognize BiP are retained within the endoplasmic reticulum).

5

IMPROVING AFFINITY

The affinity of antibodies, derived either from hybridomas or from primary phage antibody libraries, as described earlier, can be improved using a number of mutation techniques—targeted or random.

Error-prone PCR relies on the natural error rate of *Taq* DNA polymerase (which lacks a proofreading function) to incorporate mutations into a PCR-amplified segment of DNA. This has been used with chemical mutagenesis to increase the affinity of an anticarbohydrate scFv by 10-fold (Deng et al., 1994). This is the same order of affinity improvement obtained by using mutator strains (30-fold) and chain shuffling (30-fold; Schier et al., 1996a), in which the VH gene from the best phage antibody isolated is kept constant and displayed in association with a library of light chains; the procedure is then reversed with a library of VH genes for the best new light-chain V (or vice versa). This technique can, however, lead to a drift in the epitope recognition.

DNA shuffling, or sexual PCR (Stemmer, 1994), allows PCR errors from different amplified molecules to combine at random. This has been applied to antibodies and was successfully used to evolve an antibody that lacks the conserved cysteine residues. The same technique has been used to shuffle complementarity determining regions (CDRs) among different antibodies, creating a relatively large library of scFvs from a small collection of starting V regions.

The creation of small, targeted libraries in which mutations are made only in residues known or suspected to be involved in antigen binding (e.g., in the CDRs or in known contact residues) has shown the most spectacular improvements in affinity; the affinity of an antibody that recognizes c-*erb B2* increased 1230 times to 13 pmol (Schier et al., 1996b). Slightly lower affinity increases have been reported for two anti-gp120 antibodies. In one case, the increase in affinity was accompanied by a broadening of the HIV isolates recognized.

USING PHAGE ANTIBODIES AS A DISCOVERY TOOL

When antibodies directed against cell surfaces have been isolated using naive phage antibody libraries, the antigens recognized have often been found to be novel. This is likely owing to the lack of immunological constraints with naive phage antibody libraries compared to the creation of traditional hybridomas in which immunological tolerance plays a role. Similar problems with tolerance also apply to phage antibody libraries made from immunized sources. This problem notwithstanding promising results were obtained from a library made from a mouse immunized with isolated leech R neurons and screened against single R neurons. When used for immunocytochemistry with leech CNS ganglia, a number of binding patterns were obtained; 30% of antibodies stained selective subsets of neurons, including (but not specific for) the R neuron (Blow and Laskey, 1988).

DISEASE-SPECIFIC PHAGE ANTIBODY LIBRARIES

In addition to libraries from naive and immunized sources, phage antibody libraries have also been made from patients suffering from diseases in which antibodies are known to play a role. This work has been most extensively carried out with thyroid disease (McIntosh et al., 1997); the antibodies were selected from such libraries with similar specificities to those found in patients' serum and with the high affinities characteristic of immune libraries. Similar experiments have been done with systemic lupus erythematosus, Sjogren's syndrome, paraneoplastic encephalomyelitis, and myasthenia gravis (MG). The antibodies isolated from patients with MG recognized the AChR in or close to the major immunogenic region, a finding confirmed by the fact that two of the selected antibodies were able to inhibit both donor serum anti-AChR antibodies and those in unrelated MG patients. The antibodies showed evidence of significant somatic mutation, supporting the idea that the anti-AChR response in patients is driven by antigen. Monomeric Fabs were isolated from the library, and in in vitro antigenic modulation studies were able to induce AChR loss only when cross-linked by an anti-Fab antibody. The monovalent Fabs were able to protect against AChR loss by antigenic modulation induced by MG serum antibodies, suggesting a potential therapeutic role for these recombinant Fabs in patients with a myasthenic crisis (Graus et al., 1997).

DISPLAYING OTHER PROTEINS

The permissiveness of p3 to the display of foreign proteins is remarkable (Table 5.8.1). More than 50 proteins have been cited in the literature, including secreted proteins, the extracellular domains of membrane proteins, intracellular proteins, enzymes, toxins, and a number of small scaffolds derived from many sources. Any protein that can be secreted into the periplasmic space of *E. coli* can probably be displayed on phage, and it may be possible to widen the spectrum of proteins that can be displayed by using *prl* mutants. The main reason to display a protein on phage is to select mutants of that protein that have different properties, usually an increase in affinity or biological activity, although decreases in affinity that allow the identification of interaction residues can also be selected (Jespers et al., 1997). This process requires a number of steps, including the creation of the library, selection, and screening.

Although all the techniques described in this unit can be used to create the library of mutants, targeted mutations have been used most frequently and with the greatest success: The most striking example is the selection of a basic pancreatic trypsin inhibitor (BPTI) derivative that binds human neutrophil elastase with an affinity 3.6×10^6 times greater (final affinity 1 pM) than the wild type, from a library in which five residues were mutated (Roberts et al., 1992). Similarly, an ecotin variant with an affinity for urokinase-type plasminogen activator 2800 times higher than the wild type was selected from a targeted library in which only two residues were mutated. Using the same technique, specific inhibitors of either kallikrein or tissue factor VIIa, with a K_i as low as 15 pM, were selected from a library of amyloid protein precursor (APP) Kunitz domains. Plasminogen activator inhibitor libraries have been also made using error-prone PCR.

Protease inhibitors are not the only proteins that have had their affinities increased using phage display: The mutation of human growth hormone at 20 different sites in 5 different libraries (4 sites per library) selected on the receptor yielded a derivative able to bind its receptor with an affinity 400 times greater than the wild type (Lowman and Wells, 1993), whereas the affinity of human interleukin 6 (IL-6) for its receptor has been increased five-fold. Increased receptor binding, however, does not necessarily translate into increased biological activity. This may be overcome by exploiting the fact that displayed proteins are often functional, even if displayed on phage; i.e., phage-displaying proteins can bind and activate cell surface receptors at similar molar ratios to soluble protein. This has been reported for epidermal growth factor (EGF; Souriau et al., 1997), IL-2 (Buchli et al., 1997), α-melanocyte-stimulating hormone (αMSH; Szardenings et al., 1997), and IL-3 (Merlin et al., 1997), allowing for the development of selection strategies that target only binding with biological activity.

Displaying a protein on phage is also an easy way to make sufficient quantities of the product of a cloned gene to immunize mice or rabbits, because the attached phage acts as an excellent adjuvant.

SPECIFIC PHAGE DISPLAY APPLICATIONS IN NEUROLOGY

The potential for the use of phage display in neurobiology is enormous. As described, phage display is especially useful in the study of the binding of ligands to receptors, a field that is particularly rich in neurobiology. At the simplest level, it is possible to derive monoclonal antibodies either to the ligands (such as neurotrophins or neurotransmitters) or to their receptors by screening naive antibody libraries with purified antigen. And it is possible to use phage peptide libraries to identify peptides that bind to a ligand, to its receptor, or to an antibody that recognizes both. The peptides obtained can reveal information about the interacting surfaces; but even if they do not, they may prove to be useful pharmacological reagents.

There are likely to be many undiscovered surface molecules involved in signal transduction, neuronal differentiation and growth, synapse formation, and plasticity that may be characterized by selecting phage antibody libraries on neurons, neuronal cell lines, and neuronal cell extracts. The development of subtractive selection methods should allow the isolation of ligands specific for a particular differentiation or developmental state.

Contributor: Andrew Bradbury

Using Phage Display in Neurobiology

See Table 5.9.1 at the end of this unit for a troubleshooting guide to phage display protocols.

BASIC PROTOCOL 1

CONCENTRATION OF PHAGE OR PHAGEMID PARTICLES BY PEG PRECIPITATION

When precipitating phage from large volumes of supernatant, it is important not to reduce the volume too much in one step (i.e., >50 fold), as this can result in phage remaining precipitated due to the residual PEG present.

Materials *(see APPENDIX 1 for items with ✓)*

Culture of *E. coli* bearing the phage or phagemid—e.g., DH5αF′ (Life Technologies), TG1 (Stratagene): grown in 2× TY medium (*APPENDIX 1*) overnight at 30°C
20% (w/v) PEG 6000/2.5 M NaCl
✓ 10 mM Tris·Cl, pH 7.4/0.1 mM EDTA or 1× PBS
0.45-μm filter

1. Centrifuge the culture of *E. coli* bearing the phage(mid) (e.g., DH5αF′, TG1) 15 min at 10,000 × *g*, 4°C. Retain the supernatant.
2. Add 0.2 (i.e., 1/5) vol of 20% (w/v) PEG/2.5 M NaCl (e.g., 20 ml to 100 ml) and mix well. Incubate 30 to 60 min on ice until the precipitated solution becomes hazy.
3. Centrifuge 15 min at 4000 × *g*, 4°C. Discard the supernatant and save the white pellet. If the pellet is brown (indicative of bacterial contamination), continue with step 4, then repeat steps 2 to 4.
4. Resuspend the phage pellet in 10 mM Tris·Cl, pH 7.4/0.1 mM EDTA or 1× PBS to 1/10 the original volume. Centrifuge the resuspended phage 15 min at 10,000 × *g*, 4°C, to remove bacterial debris. Save the supernatant (phage) for months at 4°C.

5a. *For double PEG precipitation (optional):* Repeat steps 2 to 4 for added purity (double PEG precipitation), especially if the first PEG precipitate is brown (see step 3). Store up to a few days at 4°C or up to 1 year at −20°C.

5b. *For filtration (optional):* Filter the supernatant through a 0.45-μm filter (up to 80% loss of phage). Store up to a few weeks at 4°C.

SUPPORT PROTOCOL 1

PREPARING HELPER PHAGE

In general, helper phage can be purchased from either Pharmacia (i.e., M13-KO7) or Stratagene (i.e., VCS-M13). It is advisable to produce large amounts (i.e., 100 to 500 ml bacterial culture) of helper phage.

When infecting F′ bacteria with phage or phagemid, it is important to ensure that the bacteria are expressing the pilus. This is best done by not allowing the bacteria to overgrow (i.e., keep OD_{600} <0.6) and maintaining them at 37°C before infection, as the pilus is lost after 2 to 3 min at room temperature.

Materials *(see* APPENDIX 1 *for items with ✓)*

- 2×10^{13} pfu/ml helper phage M13K07 (Pharmacia) or VCSM13 (Stratagene)
- ✓ TE or PBS
- Male *E. coli*—e.g., DH5αF′ (Life Technologies Inc.), TG1 (Stratagene)
- ✓ 2× TY medium
- 2× TY top agar: add 6 g bacto agar/liter 2× TY medium; autoclave and allow to cool to 45°C
- ✓ 2× TY plates
- 5 mg/ml (1000×) kanamycin in water: filter sterilize and store in aliquots up to 1 year at −20°C
- 60% (w/v) molecular-biology-grade glycerol, sterile

- Toothpicks, sterile
- 500-ml culture flask

1. Dilute 2×10^{13} pfu/ml helper phage M12K07 or VCSM13 stock to $2–20 \times 10^3$ pfu/ml with TE or PBS.

2. Grow 10 ml *E. coli* to an OD_{600} of 0.5 (a few hours) in 2× TY medium at 37°C in a shaker incubator at 270 rpm.

3. In a culture tube, infect 50 µl of these bacteria with 50 µl of $2–20 \times 10^3$ pfu/ml M13K07 or VCSM13. Incubate 20 to 30 min at 37°C without shaking.

4. Add 100 to 400 µl uninfected *E. coli* from a culture grown overnight or to an OD_{600} of 0.5, and 4 ml top agar, 45°C, to the bacteria-helper phage mixture. Pour directly onto 2× TY plates. Culture overnight at 37°C. Store up to 1 week at 4°C.

5. Grow an uninfected culture of *E. coli* overnight at 37°C. Pick a small plaque of helper phage (from step 4) using a sterile toothpick and inoculate into 3 or 4 ml of 2× TY medium. Add 40 µl (i.e., 1/100 dilution) of the uninfected overnight *E. coli* culture. Incubate ~2 hr at 37°C.

6. Dilute into a 500-ml culture flask containing 100 ml of 2× TY medium, 37°C. Incubate 1 hr at 37°C with shaking. Add 5 mg/ml kanamycin in water to a final concentration of 25 µg/ml. Incubate 8 hr at 37°C. Centrifuge 15 min at 10,000 × *g*, 4°C.

7. PEG precipitate the supernatant once (see Basic Protocol 1).

8. Determine the phage titer in plaque forming units (pfu) per milliliter using top agar as described (see Support Protocol 2; optimally 1×10^{13} to 1×10^{14} pfu/ml). Dilute the helper phage to 2×10^{13} pfu/ml in TE or PBS. Before storage, verify the capacity of the phage to rescue other phagemids (see Basic Protocol 2) as deletion variants can occur. Store in aliquots up to a few weeks at 4°C or add 60% (w/v) molecular-biology-grade glycerol to 15% and store up to a few years at −70°C.

BASIC PROTOCOL 2

RESCUING PHAGE/PHAGEMID PARTICLES FROM LIBRARIES

Growth of phagemid libraries requires the use of helper phage (see Support Protocol 1), which provides all the other proteins needed to produce the phage particles (Fig. 5.9.1). In general, to maintain the diversity of a library, the starting culture should contain ten times more clones than the original library. Different clones have very different effects on bacterial growth rates, and as a result library amplification should be minimized to prevent bias towards the least toxic clones.

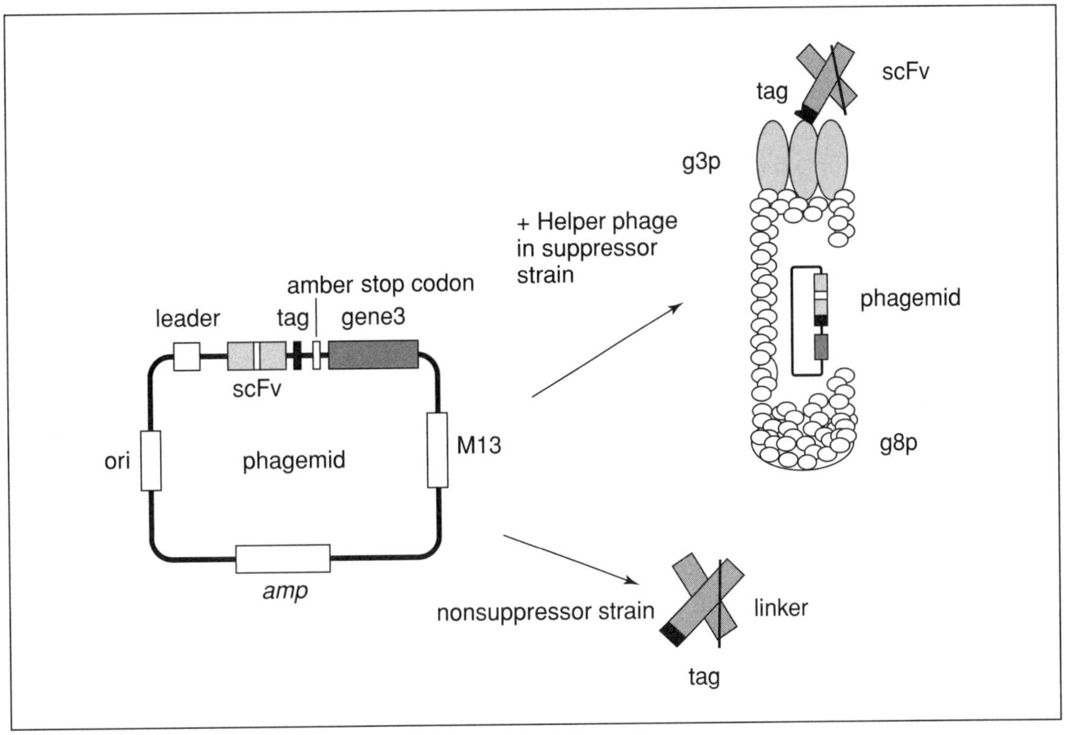

Figure 5.9.1 Schematic of a phagemid display vector.

Materials

Phage library:
 Peptide libraries (Mobitec or New England Biolabs) or
 Antibody libraries (obtain under collaborative agreements)
2× TYAG: 2× TY medium (*APPENDIX 1*) containing 100 μg/ml ampicillin and 2% (w/v) glucose: store up to 1 month at room temperature
Helper phage (see Support Protocol 1)
2× TYAK: 2× TY medium (*APPENDIX 1*) containing 100 μg/ml ampicillin and 25 μg/ml kanamycin: store up to 1 month at room temperature

30°C shaker incubator
0.25- or 0.45-μm filter

1. Measure the OD_{600} of a diluted aliquot of the phage library (i.e., peptide or antibody libraries) and calculate the number of bacteria per milliliter to determine the inoculum.

   An OD_{600} of 1.00 corresponds to ~2–8 × 10^8 bacteria (this should be calibrated for each bacterial strain used). Usually the OD will be around 100, or 3 × 10^{10} cells/ml stock (assuming an OD_{600} of 1.00 corresponds to 3 × 10^8 bacteria). For most rescues, the inoculum is therefore 30 to 300 μl of the glycerol stock (or concentrated solution of bacteria scraped from plate), which will contain 1 × 10^9 to 1 × 10^{10} bacteria. The inoculum size should be ten times the library size in number of bacteria at the start, but should not exceed an OD_{600} of 0.05.

2. Place the inoculum in an appropriate volume of 2× TYAG in a sterile culture flask 5 to 10 times bigger than the culture volume. Incubate 1.5 to 2.5 hr at 37°C with shaking at 270 rpm to an OD_{600} of 0.5. Because not all repertoires will grow at equal rates, check the OD_{600} regularly to ensure that the cells do not overgrow.

   After this incubation, the cells will be in the mid-log phase of growth and will express the F-pilus for infection.

3. When an OD_{600} of 0.5 is reached (i.e., 1–4 × 10^8 bacteria/ml), add a 20-fold excess of helper phage (for phagemid-based libraries). Place culture in a 37°C waterbath for 30 min with occasional agitation.

4. After the infection event, centrifuge the cells 10 min at 4000 × *g*, room temperature. Remove the supernatant. Resuspend the bacterial pellet in a volume of 2× TYAK five times greater than the initial culture volume. Incubate overnight in a 30°C (for scFv antibody libraries) or 37°C (for peptide libraries) shaker incubator set at 270 rpm, using enough culture flasks to ensure that the flask volume is five to ten times greater than the culture volume.

5. Pellet the bacteria by centrifuging the culture 20 min at 10,000 × *g*, room temperature. PEG-precipitate the phage or phagemid particles from the supernatant (see Basic Protocol 1). If there are many contaminating bacteria visible, repeat the PEG precipitation.

6. (*Optional*) To further purify the phage for long-term storage at 4°C, filter sterilize them through a 0.25- or 0.45-μm sterile filter.

   The standard yield is ~2–10 × 10^{12} phages from a 25-ml culture.

BASIC PROTOCOL 3

SELECTION OF PHAGE ANTIBODIES TO AN ANTIGEN IMMOBILIZED INDIRECTLY OR DIRECTLY ON (PLASTIC) SURFACES

To perform two to four rounds of selection and screening, 500 μg of antigen is sufficient; however, when there is only a limited amount of antigen available, it may be re-used, PBS as the coating buffer, no more than three to four times for selection.

Materials *(see* APPENDIX 1 *for items with ✓)*

✓ 1 to 100 μg/ml antigen in carbonate buffer, pH 9.6, or PBS
✓ PBS
MPBS: 2% and 4% (w/v) nonfat milk powder in PBS
1–5 × 10^{12} PEG-concentrated phage/ml (see Basic Protocol 1)
0.1% (w/v) Tween 20 in PBS
100 mM TEA, pH 12: dissolve 140 μl triethylamine in 10 ml H_2O, prepare fresh
✓ 1 M Tris·Cl, pH 7.4

75 × 12–mm immunotube (Nunc)
Rotator

1. Add 2 ml of 1 to 100 μg/ml antigen in carbonate buffer, pH 9.6, or PBS (for antigens that are base-sensitive) to a 75 × 12–mm Nunc immunotube. Incubate overnight at 4°C or 1 hr at 37°C. Decant the antigen solution and wash the tubes three times with PBS.

2. Preblock the immunotube by adding 2% (w/v) MPBS. Seal the tube with Parafilm and incubate 30 min to 2 hr at room temperature.

3. While the immunotube is incubating, preblock the phage mix by adding 500 μl of 1–5 × 10^{12} PEG-concentrated phage/ml to 1 ml of 4% (w/v) MPBS and 500 μl PBS. Incubate 30 min to 1 hr at room temperature.

4. Decant the 2% (w/v) MPBS from the immunotube and wash twice with 0.1% (w/v) Tween 20 in PBS and twice with PBS.

5. Pipet the phage-mix (from step 3) into the immunotube and cover with Parafilm. Incubate 30 min at room temperature, mixing gently on a rotator. Incubate an additional 1.5 hr without agitation. Decant the phage mix and wash the immunotube 20 times with Tween 20 in PBS and then 20 times with PBS. Decant the final PBS wash, add 1 ml freshly prepared 100 mM triethylamine (TEA), pH 12, and cover the tube with fresh Parafilm (this prevents cross-contamination). Incubate 10 min on a rotator at room temperature.
6. Transfer phage mix to a microcentrifuge tube containing 0.5 ml of 1 M Tris·Cl, pH 7.4, and mix by inversion. Transfer phage mix to ice or store up to 1 month at 4°C.
7. Titrate the phage in TG1 or DH5αF cells (see Support Protocol 2) to determine the output.
8. Rescue phage or phagemid as described (see Basic Protocol 2) and harvest the phage.
9. Start new round of selection (up to 3 to 4 rounds), starting from step 1 of this protocol.

   In subsequent rounds of selection, the number of phage added can be lower (e.g., 1×10^{10} to 1×10^{11}), and washing can be more stringent (e.g., by performing longer washes, plus a prewash for 10 to 30 min with Tween in PBS before the washes described in step 5).

ALTERNATE PROTOCOL

SELECTION OF PHAGE ANTIBODIES USING BIOTINYLATED ANTIGEN AND STREPTAVIDIN-PARAMAGNETIC BEADS

The biotinylated antigen is incubated with the phage antibody library after both have been appropriately blocked (Fig. 5.9.2).

Additional Materials *(also see Basic Protocol 3; see* APPENDIX 1 *for items with* ✓*)*

Phage library:
- Peptide libraries (Mobitec or New England Biolabs)
- Antibody libraries (cannot be purchased, but can often be obtained under collaborative agreements)

5

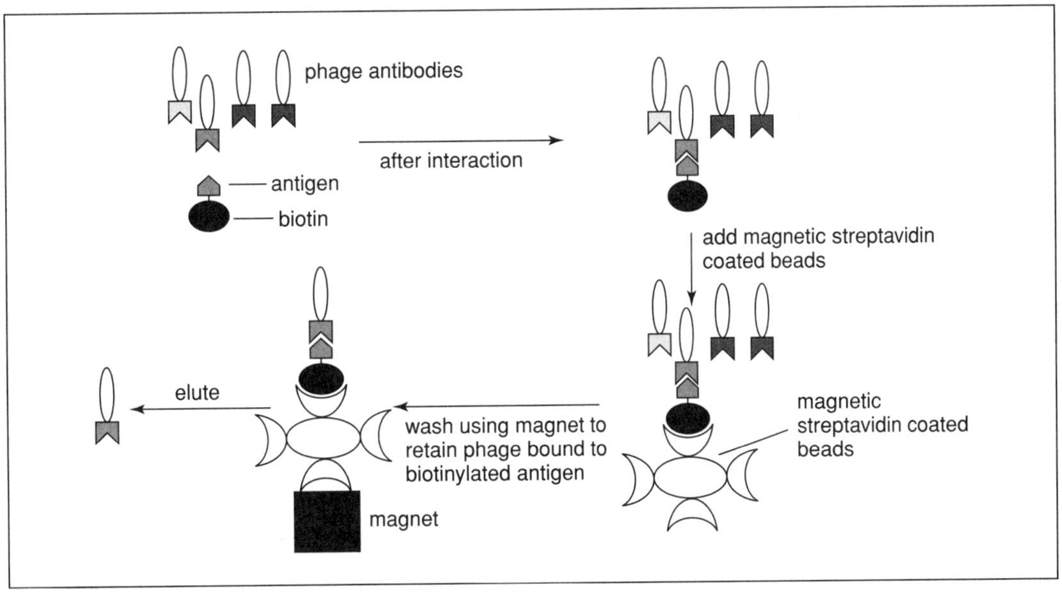

Figure 5.9.2 Biotin selection.

Streptavidin paramagnetic beads (Dynal)
100 to 500 nM biotinylated antigen: label using a commercial kit (e.g., Pierce); storage conditions are variable and antigen-dependent
Dimethyl sulfoxide (DMSO; optional)
2% (w/v) nonfat milk powder/0.1% (w/v) Tween 20 in PBS
✓ PBS
1 mM DTT

End-over-end rotator
Magnetic rack (Dynal)

1. Mix equal volumes (≥1 ml total) of phage library (i.e., peptide or antibody library) and 4% MPBS in a 1.5-ml microcentrifuge tube per antigen selection to preblock the library.

2. Wash 100 µl streptavidin paramagnetic beads once in 2% MPBS, and add to the microcentrifuge tube from step 1. Equilibrate 60 min at room temperature using an end-over-end rotator at slow speed to select for streptavidin-binding phage.

3. While preincubating the phage, place 100 to 200 µl fresh streptavidin paramagnetic beads per selection in 1.5-ml microcentrifuge tubes. Draw to one side using a magnetic rack. Remove buffer and resuspend beads in 2% MPBS solution. Equilibrate 1 to 2 hr at room temperature on an end-over-end rotator.

4. Remove the streptavidin-binding phage (from step 2) by drawing the beads to one side using a magnetic rack and then use a pipet to transfer the supernatant which contains nonstreptavidin-binding phage to a new tube. Add 100 to 500 nM biotinylated antigen to this equilibrated phage mixture. If the antigen is hydrophobic, add dimethyl sulfoxide (DMSO) to a final concentration of 5% (v/v). Incubate 30 min to 1 hr at room temperature on an end-over-end rotator.

5. Draw equilibrated beads from step 3 to one side with a magnetic rack and remove buffer. Resuspend streptavidin-magnetic beads in 250 µl of 2% (w/v) nonfat milk powder/0.1% (w/v) Tween 20 in PBS. Add 12.5 µl DMSO for hydrophobic antigens.

6. Add streptavidin-magnetic beads to the phage-antigen mix from step 4. Incubate 15 min at room temperature on an end-over-end rotator. Place tubes in a magnetic rack and wait until all beads are bound to the magnetic site (20 sec). Tip the rack upside down and back again to wash beads from the cap.

7. Leave tubes in the rack for 1 min. Aspirate the supernatants carefully, leaving the beads on the side of the tube. Wash the beads carefully six times with milk/Tween 20 in PBS. Add 1 ml milk/Tween 20 in PBS, resuspend the beads, and pipet into a new microcentrifuge tube. Repeat once. Wash the beads two times with 1 ml PBS. Add 1 ml PBS, resuspend the beads, and pipet to a new tube.

8a. *To elute with TEA:* Elute phage from beads with 1 ml of 100 mM TEA (10 min maximum incubation), pH 12. Immediately transfer the solution to a microcentrifuge tube containing 0.1 ml 1 M Tris·Cl, pH 7.4, and mix by inversion to neutralize the phage eluate.

8b. *To elute with DTT:* If using -SS-biotin labeled protein/peptide, elute the phage by adding 200 µl of 1 mM DTT and then incubate the tube on a rotator 5 min at room temperature. Transfer solution to a new tube.

9. Titrate in TG1 cells (see Support Protocol 2).

10. Reinfect the selected phages into TG1 cells for another round of selection, repeating steps 1 to 9. Store the remaining beads or phage eluate at 4°C as a backup for up to 1 week.

SUPPORT PROTOCOL 2

RESCUE OF PHAGE/PHAGEMID BY INFECTION OF *E. COLI*

CAUTION: As antibody-p3 fusion proteins can be toxic, it is recommended that plates contain 1% to 2% (w/v) glucose to inhibit expression of the fusion protein.

Materials *(see* APPENDIX 1 *for items with ✓)*

✓ 2× TY medium
E. coli (Life Technologies)—i.e., single colony of TG1 on minimal agar or DH5αF′ in 2× TY medium
Phage/phagemid solution (see Basic Protocol 2 or Alternate Protocol)
✓ 90-mm round and 14-cm round or 243 × 243–mm square 2× TYAG plates containing appropriate antibiotics, dry
2× TYAG medium: 2× TY medium (*APPENDIX 1*) containing 100 µg/ml ampicillin and 2% (w/v) glucose: store up to 1 month at room temperature
60% (w/v) molecular-biology-grade glycerol

Microtiter plates (optional)
30°C incubator
15- and 50-ml conical tubes
Spreader, sterile
Nunc Cryotube

1. Inoculate 25 ml of 2× TY medium with *E. coli*. Incubate ≥12 hr at 37°C with shaking at 270 rpm. Use 500 µl of this culture to inoculate 50 ml of 2× TY medium about 1 hr before elution of phage (see Basic Protocol 2 and Alternate Protocol). Incubate ~1.5 hr at 37°C with shaking until an OD_{600} of 0.5 is reached. Use the cells immediately (preferably) or store them on ice, as standing at room temperature will cause rapid loss of pili. Store stock cultures up to 1 month at 4°C.

To perform titrations using microcentrifuge tubes

2a. Using new pipet tips for each tube, make serial 10- to 100-fold dilutions (500 µl final volume) of phage/phagemid in microcentrifuge tubes containing 2× TY medium:

Stocks (1 × 10^{12} to 1 × 10^{13} pfu/ml): make dilutions to 10^{-10}
Eluates (1 × 10^{6} to 1 × 10^{8} pfu/ml at round 1 or 2 of selections): make dilutions to 10^{-4}
Eluates (1 × 10^{9} pfu/ml at round 3 or 4 of selections): make dilutions to 10^{-6}

For more accurate titrations, make 10-fold rather than 100-fold dilutions around the relevant concentrations.

3a. Add 500 µl exponentially growing *E. coli* (step 2) to the dilutions. Retain uninfected *E. coli* for use as a control. If a long series of titrations are to be plated and compared, place the tubes on ice or use microtiter plates (step 2b).

4a. Plate 100 µl of each of the following dilutions (vector plus *E. coli*) onto dry 90-mm 2× TYAG plates containing the appropriate antibiotic and incubate overnight at 30°C:

Phage stocks (expected 1 × 10^{12} to 1 × 10^{13}): plate the 10^{-8} to 10^{-10} dilutions
Round 1 to 2 phage eluates (expected 1 × 10^{6} to 1 × 10^{8}): plate the 10^{-2} to 10^{-4} dilutions
Round 3 to 4 phage eluates (expected 1 × 10^{9}): plate the 10^{-4} to 10^{-6} dilutions.

Store the remainder of the liquid samples up to 1 day at 4°C, to prevent further growth of the *E. coli.*

To perform titration using microtiter plates:

2b. To reduce the number of culture plates used, titrate in microtiter plates, diluting 11 µl to 100 µl of 2× TY medium from well to well (this represents a 10-fold dilution for each well) to the concentrations listed above (step 2a).

3b. Warm microtiter plates and bacteria to 37°C for 5 min. Add 100 µl exponentially growing *E. coli* (step 2) to each 100-µl aliquot of diluted phage/phagemid. Incubate without shaking 30 min at 37°C. Do not allow bacteria to cool.

4b. Add a 5-µl drop (i.e., spot) of diluted phagemid from each well onto a well dried 90-mm 2× TYAG agar plate, or for phage, use a plate with the appropriate antibiotic.

5. Prepare control plates containing noninfected *E. coli*, grown on a 2× TYAG plate to check plates, *E. coli* culture, and sterile technique. Incubate overnight at 30°C with the titration plates. The next day, count the number of colonies and calculate the vector titer as follows:

   Vector titer (vector/ml) = (no. colonies)/(total dilution × fraction of 1 ml plated)

6a. *Amplify library for panning:* In a 50-ml conical tube, mix 5 ml exponentially growing *E. coli* (see steps 1 and 2) with 4 ml of 2× TY medium and 1 ml phage/phagemid. If TEA has been used for elution, dilute the eluate at least ten-fold for toxicity reasons and scale volumes up if required. Alternatively, elute the phage by adding *E. coli* directly to the tube, although this tends to select for lower-affinity antibodies.

6b. *Amplify library for biotin selections:* Mix 1 ml exponentially growing *E. coli* with 100 to 200 µl phage/phagemid.

7. Incubate 30 min in a 37°C waterbath.

8. (*Optional*) If culture volumes are greater than 2 ml, reduce by centrifuging 10 min at 4000 × *g*, room temperature. Remove the supernatant and resuspend the pellet in a small volume of 2× TYAG medium.

9. For first round selections, bring the final volume to 2 ml and plate the sample on four 14-cm 2× TYAG plates (500 µl per plate) or a single 243 × 243–mm square plate using all eluted vector. Incubate overnight in a 30°C incubator, as growth at higher temperatures may lead to loss of some antibody clones.

10. After overnight growth, add 2 ml of 2× TYAG medium to each 14-cm plate or 5 ml for larger plates, and scrape the bacteria off using a sterile spreader. Alternatively, add 6 to 8 ml of 2× TYAG medium to each 14-cm petri dish and shake gently for 5 to 10 min on a rotating shaker.

11. Transfer the cells to a 15-ml conical tube. Make a homogeneous suspension by pipetting up and down with a sterile pipet or by vortexing to disperse clumps. Wash each plate with an additional 2.5 ml of 2× TYAG medium, scrape the remaining *E. coli* loose, and add to the other cells.

12. After complete resuspension, collect the bacteria by centrifuging 10 min at 4000 × *g*, room temperature. Remove most of the supernatant with a pipet, resuspend cells with gentle pipetting, and transfer a 1-ml concentrated sample to a Nunc Cryotube or micro-centrifuge tube. Add 60% molecular-biology-grade glycerol to 15% final concentration and immediately place at −70°C for storage up to 2 years or more.

SUPPORT PROTOCOL 3

GROWING PHAGE CLONES IN MICROTITER PLATES FOR ELISA TESTING

Materials

2× TYAG medium (see Basic Protocol 2)
Bacterial colonies (i.e., antibody libraries and controls):
- Antibody library of interest—e.g., 2× TYAG plates (see recipe) containing output of selection round with evenly spaced colonies (see Support Protocol 2)
- Uninfected *E. coli* (i.e., phage not produced)
- Phage not displaying antibody fragment
- Phage displaying nonrelevant antibody fragment
- Phage displaying antibody recognizing antigen of interest (positive control; optional)

60% (w/v) molecular-biology-grade glycerol, sterile
2×10^{13} pfu/ml helper phage (e.g., NEB, Stratagene; also see Support Protocol 1) in 2× TY medium (*APPENDIX 1*)
2× TYAK medium without glucose (see Basic Protocol 2)

96-well flat bottomed ELISA plates (e.g., Nunc maxisorb)
Toothpicks, sterile
30°C shaker incubator and microtiter plate holder or enzyme plastic box (Boehringer) cushioned with foam
Adhesive tape
96-well transfer device (e.g., Boekel replicator)
96-well round-bottom plate for bacterial culture (e.g., Nunc or Costar)
Tabletop centrifuge with microtiter plate carrier
Multichannel pipettor or suction device

1. Add 100 µl of 2× TYAG medium into each well of a 96-well flat-bottomed ELISA (i.e., master) plate. Gently touch the top of a bacterial colony from the antibody library of interest with a sterile toothpick and swirl it around in a single well to inoculate the culture. Include control wells in which:

   no phage is produced (uninfected bacteria)
   no antibody fragment-displaying phages are produced
   phages displaying nonrelevant antibody fragments are produced
   phages displaying antibody recognize the antigen of interest (i.e., positive controls, if possible).

   Incubate overnight in a 30°C shaker incubator at 270 rpm in either a holder specifically designed for microtiter plates (e.g., as supplied by the incubator manufacturer) or enzyme plastic box cushioned with foam and tightly secured to the shaker with adhesive tape. Place the box as far as possible from the ventilator to avoid evaporation.

2. The next day, use a 96-well sterile transfer device or micropipettor to place 2 µl per well from the inoculated plate into a 96-well round-bottom (i.e., induction) plate for bacterial culture containing 150 µl of 2× TYAG medium per well. Grow to an OD_{600} of 0.5 (~2.5 hr at 37°C with shaking. To the wells of the master plate, add 50 µl of 60% molecular-biology-grade glycerol per well. Mix thoroughly and store up to 2 years or more at −70°C.

3. To each well of the induction plate (step 2), add 50 μl of 2× TYAG containing 2×10^9 pfu helper phage (e.g., dilute 10 μl of a 2×10^{13} pfu/ml stock into 1 ml 2× TYAG of to make enough for a 96-well plate; use a ratio of phage to bacterium ~20:1). Incubate 30 min at 37°C. After incubation, centrifuge in a tabletop centrifuge with microtiter plate carrier 10 min at 1700 rpm, room temperature. Remove the supernatant with a multichannel pipettor or suction device.

4. Resuspend the bacterial pellets in 150 μl of 2× TYAK medium by pipetting. Incubate overnight at 30°C with shaking.

5. The next day, centrifuge for 10 min at 1700 rpm, room temperature. Retain the supernatant.

BASIC PROTOCOL 4

PHAGE ELISA

Phage selected by phage display technology can be tested for antigen binding activity by using ELISA plates in which the wells are coated with antigens. To increase the accuracy of the test and reduce the background, 0.1% (w/v) Tween 20 can be included in all washing, incubation, and blocking steps.

Materials *(see* APPENDIX 1 *for items with ✓)*

✓ 1 to 100 μg/ml protein antigen in carbonate buffer or PBS
✓ PBS, pH 7.4 to 7.6
MPBS: 2% and 4% (w/v) nonfat milk powder in PBS
0.1% (w/v) Tween 20 in PBS
Culture supernatant containing phage antibody (see Support Protocol 3)
Positive and negative controls (optional)
Horseradish peroxidase (HRP)-conjugated mouse anti-phage mAb, (e.g., Pharmacia): dilute according to manufacturer's instructions
10-mg 2,2′-azinobis(3-ethylbenzthiazolinesulfonic acid) tablet (ABTS; Sigma)
✓ 50 mM citrate buffer, pH 4.5
30% (v/v) hydrogen peroxide
0.02% (w/v) sodium azide or 3.2 mg/ml sodium fluoride (ABTS developing)
✓ TMB working solution
2 N H_2SO_4 (TMB developing)

96-well ELISA plate, e.g., Nunc Maxisorb
ELISA plate reader (optional)

1. Add 100 μl of 1 to 100 μg/ml protein antigen in carbonate buffer or PBS to each well of a 96-well ELISA plate. Incubate overnight at 4°C or 1 hr at 37°C to coat. Discard the antigen solution and rinse the wells twice with PBS, pH 7.4 to 7.6. Wash by submerging the plate into buffer and removing the air bubbles in the wells by agitation.

2. Block with 120 μl of 2% MPBS per well ≥30 min at room temperature. Wash wells three times with 0.1% (v/v) Tween 20 in PBS, and twice with PBS. Add 50 μl of 4% MPBS to each well.

3. Add 50 μl culture supernatant containing the phage antibodies to the appropriate wells. Include positive and negative controls if possible. Mix by pipetting up and down. Incubate ~1.5 hr at room temperature with occasional mild shaking.

   Positive controls include, if possible, supernatant of phage antibodies which are known to bind the antigen coated on the plate. Such phages should be grown in a separate well of the microtiter plate, rather than using a previously prepared stock, as this will control bacterial growth and phagemid rescue (see Support Protocol 3).

Although there are usually negative phagemids in any phage ELISA, it is useful to include a phagemid clone that is known to be negative. This can be empty vector, or better still, a phage antibody that binds to another antigen. This provides the background level of phage binding. A non-inoculated well should also be included to control for the presence of contamination in the medium used.

4. Discard solution, and wash wells three times with Tween in PBS and three times with PBS. Decant final wash.

5. Add 100 µl HRP-conjugated mouse anti-phage mAb prepared at the dilution indicated by the manufacturer (e.g., the Pharmacia antibody is usually used at 1/5000). Incubate 1 hr at room temperature without shaking. Decant the HRP-conjugated mouse anti-phage mAb and wash wells three times with Tween in PBS and three more times with PBS.

To develop with ABTS

6a. Dissolve one 10-mg 2,2′-azinobis(3-ethylbenzthiazolinesulfonic acid) tablet (ABTS) in 20 ml of 50 mM citrate buffer, pH 4.5. Add 10 µl of 30% hydrogen peroxide to the substrate solution immediately before dispensing 100 µl to each well. Incubate 10 to 30 min (or longer if necessary) at room temperature. Time the development period if comparing data with data from other plates.

7a. Quench the peroxidase reaction by adding 50 to 100 µl of either 0.02% sodium azide or 3.2 mg/ml sodium fluoride. Proceed to step 8.

To develop with TMB

6b. Just before use, add 1 ml 30% hydrogen peroxide to 10 ml TMB working solution. Immediately dispense 100 µl to each well. Incubate 10 to 30 min (or longer if necessary) at room temperature in the dark. Time the development period if comparing data with data from other plates. Proceed to step 8.

8. Measure the absorbance at 405 (ABTS) or 450 nm (TMB) using an ELISA plate reader.

In general negative antibodies will produce signals <0.1, while positive antibodies will yield signals >0.25, and sometimes as high as 3.0.

5

SUPPORT PROTOCOL 4

GROWING SOLUBLE FRAGMENTS IN MICROTITER PLATES

Materials *(see* APPENDIX 1 *for items with ✓)*

✓ 2× TYAG plates containing output of selection round with evenly spaced colonies (e.g., titration plates; see Support Protocol 2)
Bacterial colonies with phage producing no antibody fragments and producing nonrelevant antibody fragments (controls)
2× TYAG medium (see Basic Protocol 2)
60% molecular biology grade glycerol, sterile
✓ 2× TY medium containing 100 µg/ml ampicillin and 3 mM IPTG (diluted from a 1 M stock)

96-well flat-bottomed microtiter plates (e.g., Nunc or Costar)
Toothpicks, sterile
96-well transfer device (e.g., Boekel replicator)
Tabletop centrifuge

1. Place 150 µl of 2× TYAG medium into each well of a 96-well flat-bottomed microtiter plate. Gently touch the top of a colony from a 2× TYAG plate containing the output

of a selection round with a sterile toothpick and swirl it around in an individual well to inoculate the culture. Repeat for all wells to be used, including those in which no antibody fragments are produced, and nonrelevant antibodies are produced.

2. Incubate overnight in a 30°C shaker incubator at 270 rpm in either a holder specifically designed for microtiter plates (e.g., as supplied by the incubator manufacturer) or a plastic box cushioned with foam and tightly secured to the shaker with adhesive tape. Place the box as far as possible from the ventilator to avoid evaporation.

3. The next day, use a 96-well sterile transfer device or micropipettor to transfer 2 μl per well from this plate to a 96-well induction plate containing 100 μl of 2× TY medium containing 100 μg/ml ampicillin and 0.1% (w/v) glucose per well. Grow at 37°C with shaking until the OD_{600} is ~0.9 (i.e., ~2 to 3 hr), or alternatively for 3 to 4 hr at 30°C. To the wells of the master plate, add 50 μl of 60% (w/v) sterile glycerol per well (15% final). Mix and store at −70°C after use for up to 2 years or more.

4. Add 50 μl of 2× TY medium containing 100 μg/ml ampicillin and 3 mM IPTG (final concentration 1 mM IPTG) to induce production of the antibody fragment. Continue shaking at 30°C for an additional 16 to 24 hr to produce soluble scFv/Fab.

5. The next day, centrifuge in a tabletop centrifuge 10 min at 1700 rpm, 4°C. Use 2 to 50 μl supernatant in ELISA (see Basic Protocol 5).

BASIC PROTOCOL 5

SOLUBLE FRAGMENT ELISA IN MICROTITER PLATES

Materials *(see* APPENDIX 1 *for items with ✓)*

✓ 1 to 100 μg/ml antigen in carbonate buffer or PBS
✓ PBS
MPBS: 2% and 4% (w/v) nonfat milk powder in PBS
0.1% (w/v) Tween 20 in PBS
Supernatants containing soluble antibody fragment (scFv), antibody which binds the coating antigen (positive control; optional), and nonbinding soluble antibody fragment (negative control; see Support Protocol 4 for all supernatants)
1.6 μg/ml mAb 9E10 (ATCC# CRL-1729) in 2% MPBS
Horseradish peroxidase (HRP)-conjugated anti-mouse secondary antibody: use according to manufacturer's specifications (e.g., Dako D260 is used at 1/1000 in 2% MPBS) *or*
Alkaline phosphatase (AP)-conjugated anti-mouse secondary antibody: use according to manufacturer's specifications
0.9% (w/v) NaCl (detection with alkaline phosphatase; optional)
10-mg ABTS substrate tablet
✓ 50 mM citrate buffer, pH 4.5
30% (v/v) hydrogen peroxide
✓ TMB solution, fresh
20-mg *p*-nitrophenyl phosphate (pNPP) tablet
✓ 20 ml pNPP buffer
0.02% (w/v) sodium azide
3.2 mg/ml sodium fluoride/ml in H_2O
2 N H_2SO_4

96-well ELISA plate (e.g., Nunc Maxisorb)
Buffer tank

1. Coat a 96-well ELISA plate with 100 μl per well of 1 to 100 μg/ml protein antigen in carbonate buffer or PBS. Incubate plate overnight at 4°C or 1 hr at 37°C.

2. If the antigen solution is to be reused, remove it and rinse the wells twice with PBS. Wash by submerging the plate into a buffer tank containing PBS and removing the air bubbles in the wells by agitation.

3. Decant PBS, block with 120 μl of 2% MPBS per well ≥30 min at room temperature. Wash wells three times with 0.1% (w/v) Tween 20 in PBS, and twice with PBS. Decant final solution. Add 50 μl of 4% MPBS to all wells.

4. Add 50 μl culture supernatant containing the soluble antibody fragment to the appropriate wells. If possible, include positive controls of supernatant from a clone whose antibody fragment binds the coating antigen (i.e., the antigen coated on the plate). Also include negative controls of supernatant from a nonbinding scFv/Fab to determine nonspecific binding of a nonrelevant ScFv/Fab to the coated Ag. Mix by pipetting up and down. Incubate ~1.5 hr at room temperature with mild shaking.

5. Discard the solution, and wash out the wells three times with Tween in PBS and three times with PBS. Add 100 μl of 1.6 μg/ml 9E10 mAb in 2% MPBS to each well. Incubate 1 hr at room temperature.

6. Discard the antibody, and wash out the wells three times with Tween 20 in PBS, and three times with PBS by immersing in a tank of buffer or by using a squeeze bottle containing buffer.

7. Add 100 μl of diluted HRP- or AP-conjugated anti-mouse secondary antibody to each well. Incubate 1 hr at room temperature. Discard the secondary antibody, and wash the wells three times with Tween 20 in PBS and three times with PBS.

8. (*Optional*) For alkaline phosphatase detection, rinse plates twice with 0.9% (w/v) NaCl to remove the PBS.

For peroxidase staining

9a. *To develop with ABTS:* Dissolve one 10-mg ABTS substrate tablet in 20 ml of 50 mM citrate buffer, pH 4.5. Add 10 μl of 30% hydrogen peroxide to the substrate solution immediately before dispensing 100 μl to each well.

9b. *To develop with TMB:* Dispense 100 μl freshly prepared TMB solution per well. Proceed with development in the dark. Add hydrogen peroxide just before dispensing.

For alkaline phosphatase staining

9c. *To develop with pNPP:* Dissolve one 20-mg pNPP tablet in 20 ml pNPP buffer and add 100 μl per well.

10. Incubate 10 to 30 min (or longer if necessary) at room temperature. Time the development period if comparing data with data from other plates.

11. Quench the reaction by adding 50 to 100 μl of either 0.02% sodium azide solution or 3.2 mg/ml sodium fluoride in water for ABTS, or 2 N H_2SO_4 for TMB. Read absorbance at 405 nm for ABTS or pNPP, or 450 nm for TMB.

In general, negative control antibodies will produce signals <0.1, while positive control antibodies will yield signals >0.25, and sometimes as high as 3.0.

BASIC PROTOCOL 6

AMPLIFICATION AND FINGERPRINTING OF SELECTED CLONES

Materials *(see* APPENDIX 1 *for items with ✓)*

- 10× PCR buffer (provided by *Taq* manufacturer)
- ✓ 5 mM dNTPs
- 10 μM 5′ and 3′ primers
- 5 U/μl *Taq* polymerase
- 25 mM $MgCl_2$ (optional)
- Single positive colony or bacterial growth identified by either phage ELISA (see Basic Protocol 4) or scFv ELISA (see Basic Protocol 5)
- Purified recombinant yielding a similar sized fragment as the target in the same vector as investigated (positive control)
- Light white mineral oil (Sigma; optional)
- 2% agarose gel
- ✓ 10× TBE
- 10 mg/ml acetylated BSA
- 10× *Bst*NI buffer
- 10 U/μl *Bst*NI
- Purified DNA known to contain *Bst*NI sites
- 4% Nuseive agarose (BMA) gel
- ✓ Ethidium bromide staining solution
- 10× Ficoll Orange: 0.4% (w/v) Orange G in 25% (w/v) Ficoll 400 in water
- 100- to 500-bp ladder

- 96-well PCR microplates (optional)
- Toothpicks, sterile
- UV transilluminator

1. Prepare a PCR mastermix at 20 μl per clone containing:

   15.0 μl H_2O
   2.0 μl 10× PCR buffer
   1.0 μl 5 mM dNTPs
   1.0 μl 10 μM 5′ primer
   1.0 μl 10 μM 3′ primer
   0.05 μl 5 U/μl *Taq* polymerase.

   If the PCR buffer does not contain Mg^{2+}, add to a final concentration of 1.5 mM (e.g., 1.6 μl of 25 mM $MgCl_2$).

2. Pipet 20 μl of the mastermix into 0.5-ml microcentrifuge tubes or into 96-well PCR microplates. Using a sterile toothpick, gently touch a single positive colony or bacterial growth identified by either phage ELISA or scFv ELISA and swirl in the PCR reaction mix, taking care not to take too much material as excess bacteria in the PCR reaction can cause inhibition of *Taq* polymerase. Discard the toothpick or use it to rescue the clone into selective medium or onto a selective plate.

3. Include a purified recombinant sample yielding a similar sized fragment as the target in the same vector as investigated as a positive control, and reaction mix without DNA as a negative control.

4. (*Optional*) If the PCR apparatus does not have a heated lid, overlay the reactions with one droplet of light white mineral oil.

5. Heat to 94°C for 10 min using the PCR block to break open the bacteria and release the template DNA. Perform the following PCR program:

30 cycles:	1 min	94°C	(denaturation)
	1 min	55° to 60°C	(annealing)
	1 min/kb	72°C	(extension) *or*
	2 min/kb	72°C	(extension for Fabs).

6. (*Optional*) Electrophorese 5 µl of the PCR reaction on a 2% agarose gel in 0.5× TBE to determine how many clones lack the insert.

7. After the PCR, prepare a master mix, scaling the following recipe for one reaction according to the total number of reactions:

   0.2 µl 10 mg/ml acetylated BSA
   4 µl 10× *Bst*NI buffer
   15.5 µl H_2O
   0.5 µl 10 U/µl *Bst*NI.

   Add 20 µl to each PCR tube underneath the mineral oil if present. In addition, include controls of purified DNA known to contain *Bst*NI sites (to evaluate the restriction digest). Digest samples 2 to 3 hr at 60°C.

8. Combine 2 µl of 10× Ficoll Orange (sample dye) with 8 µl reaction mix (from under the oil if present) and analyze the product by agarose gel electrophoresis on a 4% Nuseive agarose gel cast in 0.5× TBE buffer containing 0.05 µg/ml ethidium bromide staining solution. Also run a 100- to 500-bp ladder. Electrophorese at 100 V (10 V/cm) and compare the banding patterns of individual clones on a UV transilluminator.

SUPPORT PROTOCOL 5

PREPARATION OF PERIPLASMIC PROTEINS

5

Materials *(see* APPENDIX 1 *for items with* ✓*)*

Antibody library
2× TYAG medium (see Basic Protocol 2)
✓ 2× TY medium containing 100 µg/ml ampicillin and 0.1% (w/v) glucose
2× TY medium containing 100 µg/ml ampicillin and 1 mM IPTG
1 M IPTG: store in 100-µl aliquots for up to a year or more at −20°C
1 mM EDTA in PBS (see *APPENDIX 1* for PBS), ice cold
1 M $MgCl_2$

250-ml culture flask
50-ml conical tube
End-over-end rotator (optional)
0.2-µm filter or 10 kDa-cut-off dialysis tubing

1. Grow a 50-ml overnight culture of antibody library in 2× TYAG medium at 30°C in a 250-ml culture flask.

2. Add 500 µl of the overnight culture to 50 ml of 2× TY medium containing 100 µg/ml ampicillin and 0.1% (w/v) glucose in a culture flask, and incubate 2 hr at 37°C with good aeration (i.e., shaking at 270 rpm).

3a. *For normal induction:* At an OD_{600} of 0.8 to 1.0, add up to 1 mM IPTG from a 1 M stock, and continue the incubation for 4 hr at 30°C for scFv.

3b. *For improved induction:* Grow cells in 50 ml 2× TYAG medium to an OD_{600} of 0.6 to 0.8. Centrifuge cells 15 min at 4000 × *g*, room temperature. Decant the supernatant and resuspend the pellet in 50 ml prewarmed 2× TY medium containing 100 μg/ml ampicillin and 1 mM IPTG. Incubate 4 hr at 30°C.

4. Centrifuge cells 15 min at 4000 × *g*, room temperature. Resuspend the pellet in 1/50 the original volume ice-cold 1 mM EDTA in PBS in a 50-ml conical tube to permeabilize the outer membrane. Incubate 30 min on ice with occasional but careful shaking. If available, use an end-over-end rotator at 4°C for 10 to 15 min.

5. Centrifuge cells 15 min at 4000 × *g*, room temperature. Carefully remove the supernatant from the slightly sloppy cell pellet with a pipet (the supernatant is the periplasmic fraction). Remove residual cell contamination by centrifuging 20 min at 8000 × *g*, 4°C.

6. To the periplasmic fraction, add $MgCl_2$ to 2 mM final concentration from a 1 M stock. Filter sterilize through a 0.2-μm filter and store at 4°C. Alternatively, dialyze the periplasmic extract against PBS (with 10-kDa cut-off membrane) overnight at 4°C with three buffer changes to remove the EDTA. Use immediately.

Reference: Kay et al., 1996.

Contributors: Andrew Bradbury, Daniele Sblattero, Roberto Marzari, Louise Rem, and Hennie Hoogenboom

Table 5.9.1 Troubleshooting Phage Display Protocols

Problem	Possible cause	Solution
No phagemid after growth of individual clones or library	Helper phage not functional	Change helper phage preparation
	OD of bacteria at time of helper phage infection too high	Check OD carefully, and keep the OD_{600} <0.5
	Bacteria no longer have pilus	Check for presence of pilus by determining if bacteria can be infected with helper phage. With TG1 the bacteria need to be grown on minimal plates to maintain the pilus.
	Wrong antibiotic used	Check antibiotic
No ELISA signals	Problems with substrate or secondary antibodies	Always include appropriate positive controls. Change substrates or antibodies.
	Growth temperature too high, resulting in proteolytic degradation	Growth should be conducted at 30°C, but in some cases, even this is too high, so reduce to 25°C and perform pilot experiments
	Phage not produced	See "no phagemid after growth of individual clones or library." Try scFv ELISA instead.
	scFv not produced	Check for induction on a western blot to see if IPTG is still active. Try phage ELISA instead. Some scFv work as phage but not as soluble scFv.
No output after selection	Bacterial stock contaminated	Check for contamination by streaking out a colony and looking for plaques in top agar. Alternatively, pick a fresh colony.
	Wrong selection used	Check the antibiotic resistance of the vector used
	Glucose omitted in plate	Add glucose (this will help the growth of some colonies)

continued

Table 5.9.1 Troubleshooting Phage Display Protocols, continued

Problem	Possible cause	Solution
	Poor infection of bacteria	See "no phagemid after growth of individual clones or library" above
	Elution conditions too severe	Check concentrations of reagents. Try adding bacteria at OD_{600} 0.5 for 60 min to elute. Test to see if elution conditions are killing phage
No positive clones	Selection has failed	Use a different selection strategy (e.g., biotinylated antigen)
	Antigen did not coat well	If other antibodies or ligands are available, use these to test coating conditions
	Library is not functional (i.e., there is no displayed protein/antibody)	Check display levels by performing an immunoblot using an anti-tag or anti-g3 antibody. The displayed protein should be 5% to 10% with the anti-tag antibody and 1% to 10% with the anti-gene 3 antibody. If display level is poor, remake or reobtain library.
Selected antibody not recognized by tag antibody	Tags can be proteolyzed	Perform an immunoblot to test presence of tag. If it is not present, prepare scFv again. Some scFv are recognized by protein A or protein L and these can be used as secondary reagents.
	scFv functional on phage, but not as soluble scFv	This occasionally happens for unknown reasons. Usually it is best to test another clone.
Antibody not purified well	Buffer conditions are wrong	Check buffer conditions. Ensure EDTA is not in the binding buffer when using His-tag purification.

UNIT 5.10

Detection of Protein Phosphorylation in Tissues and Cells

When planning phosphorylation experiments, it is important to choose a cell preparation that will provide the most relevant results. It is often useful to combine different cell preparations to confirm results.

CAUTION: Safety equipment is important when using [^{32}P]orthophosphate. Utmost care must be exercised to avoid radioactive contamination and exposure.

BASIC PROTOCOL 1

LABELING OF PHOSPHOPROTEINS IN SITU

Materials *(see* APPENDIX 1 *for items with ✓)*

Cultured cells or tissues in 10-cm dishes
Phosphate-free MEM (Life Technologies), 37°C
Neurotransmitters, protein kinases, or phosphatase activators/inhibitors of interest
[^{32}P]orthophosphate (9000 mCi/mol; NEN Life Sciences)
✓ PBS, 4°C
✓ Lysis buffer with or without detergents

Antibody against protein of interest, attached to protein A–Sepharose beads according to manufacturer's recommendations (Amersham Pharmacia Biotech)

✓ Lysis buffer containing 2% (v/v) Triton X-100 (4°C)

✓ Lysis buffer containing 500 to 750 mM NaCl (4°C)

✓ SDS sample buffer

✓ Super Stain

✓ Destaining solution

Cellophane sheets (BioRad)

10-cm culture dishes

Disposable cell scrapers

Tabletop ultracentrifuge (e.g., Optima TL with TLA 100.2 rotor; Beckman)

Probe sonicator

Platform rocker

Phosphor imager (optional)

1. Wash cells (or tissue) once in 37°C phosphate-free MEM.

2. Add [^{32}P]orthophosphate to phosphate-free MEM to a final concentration of 1 mCi/ml. Add 3 ml of this medium per 10-cm dish of adherent cells, and incubate in a 37°C, 5% CO_2 incubator until the [^{32}P]orthophosphate reaches labeling equilibrium with intracellular ATP and the phosphoproteins (as determined by pilot experiments; see annotation below).

To accurately measure changes in the state of phosphorylation, it is important to reach labeling equilibrium of cell phosphate pools before analyzing the phosphorylation of proteins. Conduct pilot experiments to determine the appropriate length of prelabeling with [^{32}P]orthophosphate by labeling the cells for 1, 2, 4, and 8 hr. Collect the cells at the different time points and analyze the incorporation of ^{32}P into the protein of interest (following the remainder of the steps in this protocol). The incorporation of ^{32}P into proteins should increase with incubation up to a point where the labeling levels off and becomes stable. This is the labeling equilibrium at which subsequent experiments should be conducted. The preincubation period may take 3 to 4 hr or longer in some preparations.

3. Subject cells to appropriate treatments with neurotransmitters, protein kinases, or phosphatase activators or inhibitors.

4. Upon completion of the incubation, carefully use a pipet to remove the radioactive medium. Wash cells carefully but rapidly twice with PBS (3 to 5 ml per 10-cm dish) at 4°C.

5a. *For combined analysis of membrane and soluble fractions:* Solubilize cells in 1 ml of 4°C lysis buffer containing nondenaturing detergents (e.g., Triton X-100) by adding buffer to the plate, rocking gently, and then using a cell scraper to collect the cells. Keep on ice throughout.

To lower background, pretreat extracts with SDS; however, SDS can also destroy native epitopes for some antibodies, so perform pilot studies to determine the best detergent conditions for immunoprecipitation.

5b. *For separate analysis of membrane and soluble fractions:* Harvest cells in 1 ml lysis buffer in the absence of detergent using a disposable cell scraper. Sonicate on ice for 20 to 30 sec on a medium setting using a probe sonicator reserved for radioactive use. Separate the membrane and soluble fraction by centrifugation in a tabletop ultracentrifuge for 20 min at 50,000 × *g*, 4°C. Keep lysed cells on ice as much as possible.

6. Remove any residual particulate material from the cell extract by microcentrifuging 15 min at maximum speed, 4°C, or at 50,000 × *g*, 4°C. Retain the supernatant.

7. Immunoprecipitate the protein of interest by incubating the cell extract for 2 hr with rocking at 4°C with 80 µl of 1:1 slurry of specific antibody that has been preattached to protein A–Sepharose beads in the buffer provided by the manufacturer.
8. Wash the Sepharose beads with 800 µl of 4°C lysis buffer containing 2% Triton X-100, centrifuge 1 min at 10,000 × *g*, room temperature. Discard supernatant and repeat wash. Discard supernatant and wash twice with 800 µl of 4°C lysis buffer plus high salt (500 to 750 mM NaCl), and three times with 800 µl of 4°C PBS.
9. Elute the immunoprecipitated protein from the beads by boiling the beads for 2 to 5 min in 50 to 150 µl of 1× SDS sample buffer. Resolve the samples using SDS-PAGE. Include lanes of radioactive standards or fluorescent markers to facilitate alignment of the film with the gel.
10. Immerse gel in Super Stain for 20 to 60 min, remove stain, then wash by immersing successively in a series of changes of destaining solution over several hours. Finally, dry on a gel dryer between two sheets of cellophane.
11. Expose dried gel to film with an intensifying screen in a −80°C freezer (i.e., perform autoradiography) or alternatively use a phosphor imager to reveal labeled phosphoproteins. Proceed with phosphopeptide mapping if required (see Basic Protocol 2).

BASIC PROTOCOL 2

PHOSPHOPEPTIDE MAP ANALYSIS

Materials *(see* APPENDIX 1 *for items with* ✓*)*

Dried acrylamide gel containing the radiolabeled protein of interest (see Basic Protocol 1)
50% (v/v) methanol
0.3 mg/ml L-tosylamide-2-phenylethyl chloromethyl ketone (TPCK)-treated trypsin or thermolysin (both compounds available from Sigma) in 0.4% (w/v) NH_4HCO_3
1 mg/ml basic fuschin (Sigma)
1 mg/ml phenol red
✓ Peptide map electrophoresis buffer, pH 3.5
✓ Ascending chromatography buffer

Glass vials
65°C water bath
20-cm cellulose TLC plates (Analtech)
Hair dryer
Whatman 3MM filter paper
Paint thinner (Varsol)
Electrophoresis tank designed to accommodate TLC plates (Fisher) or Hunter thin-layer peptide mapping electrophoresis system (CBS Scientific)
500-V power supply
Ascending chromatography tank
Phosphor imager (optional)

1. Cut the phosphorylated, radioactive band out of the dried acrylamide gel using a clean razor blade and place slice in a small glass vial. Wash dried gel slice three times at room temperature in the vial, each time with 20 ml of destaining solution, rocking the vial slowly on a platform rocker for 30 min per wash.

2. Wash twice with 20 ml of 50% methanol at room temperature, rocking the vial slowly on a platform rocker for 30 min per wash. Using tweezers, remove the gel piece from the 50% methanol and place in a microcentrifuge tube. Dry in a Speedvac evaporator for 2 hr.

3. Incubate the dried gel slice overnight in a 37°C water bath with 1 ml of 0.3 mg/ml TPCK-treated trypsin or TPCK-treated thermolysin dissolved in 0.4% NH_4HCO_3. Remove the supernatant and save it in a separate microcentrifuge tube.

4. Add 0.5 ml of water to the gel piece and incubate another 1 to 2 hr. During this incubation, begin drying down the saved supernatant from step 3 by centrifugation in a Speedvac evaporator.

5. Remove the supernatant from the vial containing the gel slice and add to the dried pellet from step 4. Repeat drying. Once completely dry, add 0.5 ml of water to the pellet and redry the sample. Repeat as necessary to remove residual volatile salts.

6. Measure the counts per minute of ^{32}P in a scintillation counter in the absence of scintillation fluid using Cerenkov radiation.

   Cerenkov radiation is photons given off by high-energy particles moving through a clear solution faster than the speed of light.

7. Resuspend the dried pellet in 10 µl water, heat to 65°C for 1 min, vortex, microcentrifuge 5 min at maximum speed, room temperature, and transfer the supernatant to a second microcentrifuge tube.

8. Spot the sample onto a cellulose TLC plate 1 µl at a time. Apply the sample to the center of the plate 4 cm from the bottom and let the sample dry completely before applying the next microliter (using a hair dryer). Apply one 1-µl sample (>1000 cpm as monitored by Cerenkov counting, or entire sample if not many counts) per plate.

9. After the radioactive sample has been completely loaded and dried, apply 1 µl of 1 mg/ml basic fuschin dye on top of dried sample and let dry. Apply 1 µl of 1 mg/ml phenol red dye on top of the sample. Mark the plate 4 cm from the side edges with a pencil.

10. Cut a piece of Whatman 3MM filter paper to fit the TLC plate and cut a 3-cm hole in it to surround the sample origin. Prewet the filter paper with peptide map electrophoresis buffer, pH 3.5, and cover the TLC plate with it. Pipet more buffer onto the filter paper, allowing the solution to wick onto the TLC plate evenly. Finally, pipet buffer from the outer edge of the plate toward the 3-cm hole and let the buffer diffuse slowly into the uncovered area containing the sample, which prevents the sample from diffusing from the tight spot of application and can actually concentrate the sample if done correctly.

11. Place the wet TLC plate into the electrophoresis tank containing peptide map electophoresis buffer, pH 3.5, with the sample positioned in the middle, equidistant between the two electrodes. Make sure that both edges of the TLC plate are submerged in buffer. Layer Varsol (paint thinner) on top of the pH 3.5 buffer to cover all portions of the TLC plate and prevent drying out of any exposed portions during the procedure. Electrophorese at 500 V until the basic fuschin (red) dye reaches the premarked indicator 4 cm from the edge (usually ~1.5 hr).

12. Remove the TLC plate in a fume hood and allow to dry completely (30 min to 1 hr). Avoid scratching the plates. Place the dried TLC plate into the ascending chromatography tank containing ascending chromatography buffer with the sample at the bottom. Allow ascending chromatography to proceed until the buffer approaches the top of the plate to resolve the peptides in a second dimension. Remove the plate and leave in a fume hood to dry.

13. Autoradiograph the dried TLC plate or expose to a phosphor imager screen for 1 to 7 days to reveal the resolved phosphopeptides.

BASIC PROTOCOL 3

PHOSPHOAMINO ACID ANALYSIS

Materials *(see* APPENDIX 1 *for items with ✓)*

- 6 N HCl
- N_2 gas
- ✓ 10 mg/nm phosphorylated amino acid standards
- ✓ Peptide map electrophoresis buffer, pH 1.9
- 1% (w/v) ninhydrin (Sigma) in acetone

- 16 × 75–mm Kimax screw-top tube with Teflon top
- 105°C oven
- 20-cm cellulose TLC plates (Analtech)
- Electrophoresis tanks (2) designed to accommodate TLC plates (Fisher) or Hunter thin-layer peptide mapping electrophoresis system (CBS Scientific)

1. Extract the phosphopeptides from the gel (see Basic Protocol 2, steps 1 to 6). Resuspend the dried pellet in 0.5 ml of 6 N HCl to hydrolyze the phosphopeptides to individual phosphoamino acids. Transfer the solution to a 16 × 75–mm Kimax Teflon screw-cap tube. Blow N_2 gently over the liquid and screw the lid on tightly.

2. Incubate for 1 to 2 hr in a 105°C oven. Allow the tube to cool and transfer the liquid to a microcentrifuge tube. Remove the liquid by centrifugation in a Speedvac evaporator. Resuspend the pellet in 0.5 ml water and redry in a Speedvac evaporator.

3. Resuspend the pellet in 10 μl water, vortex, and microcentrifuge 10 min at maximum speed, room temperature. Transfer the supernatant to a new microcentrifuge tube and discard the pellet.

4. Spot the sample onto a cellulose TLC plate, 4 cm from the bottom of the plate, 1 μl at a time. Let the sample dry completely before applying the next microliter (using a hair dryer). If processing multiple samples, spot each sample 4 cm from the bottom of the plate allowing 2 cm between samples and 2 cm from the side edges of the plate. Gently make two marks on the plate 5 and 14 cm from the origin with a pencil.

5. Spot 1 μl of each phosphorylated amino acid standard on top of the sample spot. Dry the spot completely between the application of each standard. Apply 1 μl of 1 mg/ml phenol red directly on top of the sample and allow the spot to dry completely.

6. Cut a piece of Whatman 3MM filter paper to fit the TLC plate and cut a 3-cm hole in it to surround the sample origin. Prewet the filter paper with peptide map electrophoresis buffer, pH 1.9 and cover the TLC plate with it. Pipet more peptide map electrophoresis buffer, pH 1.9, onto the filter paper and allow the solution to wick onto the TLC plate evenly. Finally, pipet buffer from the outer edge of the plate toward the 3-cm plate and let the buffer diffuse slowly into the uncovered area containing the sample, which prevents the sample from diffusing from the tight spot of application and can actually concentrate the sample if done correctly.

7. While wet, remove filter paper and place the TLC plate in the electrophoresis tank containing peptide map electrophoresis buffer, pH 1.9. Make certain that the sample is on the side of the TLC plate nearest the anode and electrophorese at 500 V until the dye reaches the first mark, 5 cm above the origin.

8. Without allowing the plate to dry, transfer it in the same orientation (sample side nearest the anode) into the other electrophoresis tank, containing peptide map electrophoresis

buffer, pH 3.5. Electrophorese at 500 V until the dye reaches the second mark 14 cm from the origin.

9. Remove the TLC plate, place in a fume hood and allow to dry completely. Submerge the plate in 1% ninhydrin in acetone until the standards are visibly stained purple (~15 min), then remove the plate and allow to dry.

10. Autoradiograph the dried TLC plate or expose to a phosphor imager screen for 1 to 7 days to reveal the resolved phosphoamino acids. Align the autoradiogram with the stained phosphoamino acid standards on the TLC plates to determine if the phosphoamino acids are phosphoserine, phosphothreonine, and/or phosphotyrosine.

SUPPORT PROTOCOL

PHOSPHORYLATION OF FUSION PROTEINS IN VITRO

Materials *(see APPENDIX 1 for items with ✓)*

Purified fusion protein (*UNIT 5.6*)
Protein kinase, e.g.:
  cAMP-dependent protein kinase (PKA)
  Protein kinase C (PKC)
  Calcium/calmodulin-dependent protein kinase II (CAM-KII)
Appropriate kinase buffer, e.g.:
  ✓ PKA buffer
  ✓ PKC buffer
  ✓ CAM-KII buffer
10 Ci/mmol [γ-^{32}P]ATP
✓ 3× SDS sample buffer
✓ Super Stain
✓ Destaining solution

Cellophane sheets (BioRad)
Phosphor imager (optional)

1. Prepare the following reaction mixture:

   ~1 μg purified fusion protein
   5 μCi [γ-^{32}P]ATP
   Appropriate quantity of protein kinase (45 ng PKA, 6 ng PKC, or 10 ng CAM-KII)
   Appropriate kinase buffer to 100 μl total reaction volume.

   Incubate 30 min at 30°C. Stop the phosphorylation reaction by adding 50 μl of 3× SDS sample buffer (1× final concentration).

2. Resolve the very radioactive fusion proteins using SDS-PAGE. Immerse gel in Super Stain for 20 to 60 min, remove stain, then wash by immersing successively in a series of changes of destaining solution over several hours. Finally, dry on a gel dryer between two sheets of cellophane.

3. Expose dried gel to film with an intensifying screen in a −80°C freezer for several hours, or longer if needed (i.e., perform autoradiography), or alternatively use a phosphor imager to reveal labeled phosphoproteins. Proceed with phosphopeptide mapping if required (see Basic Protocol 2).

References: Czernik et al., 1991; Pearson and Kemp, 1991

Contributors: Katherine W. Roche and Richard L. Huganir

Overview of Membrane Protein Solubilization

Selection of the proper solubilizing agent is largely a matter of trial and error that depends on the protein of interest and the particular goals of the investigator. Excellent reviews describing approaches to the solubilization of membrane proteins have been published (Hjelmeland and Chrambach, 1984; Hjelmeland, 1990). The purpose of this overview is to describe the methodology for finding the optimal conditions for detergent solubilization and to relate the selection of a specific detergent to explicit experimental goals.

SELECTION OF DETERGENT

Choice of detergent may involve several considerations. First, the goals for using the solubilized preparation should be delineated. For example, if the goal is to estimate a molecular weight for the solubilized protein using hydrodynamic techniques such as gel filtration chromatography or sucrose gradient centrifugation, a detergent such as digitonin is unsuitable since it has the same partial specific volume as amino acids and one cannot compensate for bound detergent by using H_2O/D_2O systems (Clarke, 1975; O'Brien et al., 1978). Because this type of experiment does not require a homogeneous sample and can be completed in 24 to 48 hr, protein purity and long-term stability are not important. Alternatively, if the goal is to purify the protein in a manner in which it retains its enzymatic or ligand binding activity, long-term stability of the protein, effective solubilization, and compatibility with anticipated purification technologies and assay methods are crucial. If the ultimate goal is to reconstitute the purified protein into lipids, then ease of detergent removal and avidity of detergent binding must be considered. Similarly, if a preparation is to be used to grow crystals suitable for X-ray diffraction, one must be able to either solubilize and purify the protein in an appropriate detergent or use a detergent that can be readily exchanged for one that is compatible with crystallization strategies and criteria (Garavito et al., 1996). Finally, if there are several equally effective detergents, expense and availability may also be worth considering.

The several types of detergents that are commercially available are generally classified as either zwitterionic, nonionic, or ionic molecules. Table 5.11.1 lists some representative detergents from each of these classes. An excellent summary of the structures and properties of commercially available detergents is available from Calbiochem (Neugebauer, 1992). Most commercially available detergents are relatively pure and can be used as obtained from the manufacturer; some detergents, however, such as cholate, can be purified further without much additional effort by recrystalization or chromatography (Ross and Schatz, 1978). Ionic impurities can be removed from nonionic detergents such as Lubrol-PX by passing a concentrated detergent solution over a mixed-bed ion-exchange resin.

Because there are few rules for selection of a detergent for purification of a membrane protein, it is often advantageous to search the literature to determine whether similar proteins have been purified. For example, many G-protein-coupled receptors have been purified in digitonin solutions, G-proteins in cholate, and Na^+/K^+-ATPases in $C_{12}E_9$ (nonaethylene glycol monododecyl ether). If the protein to be purified is unique, it is advisable to purchase a small amount of several different types of detergents for small-scale experiments to evaluate which class of detergent might be effective, and then screen several members of that class to further narrow possibilities.

Table 5.11.1 Representative Detergents[a]

Type	Name	Chemical name (if different)
Nonionic	Big CHAP	*N,N*-bis(3-D-gluconamidopropyl)cholamide
	Digitonin	
	$C_{12}E_8$	Octaethylene glycol monododecyl ether
	n-Octyl-β-D-glucopyranoside	
	Triton X-100	Nonaethylene glycol octylphenol ether
	MEGA-8	Octanoyl-*N*-methylglucamide
Ionic	CTAB	Cetyltrimethylammonium bromide
	Sodium cholate	
	Sodium deoxycholate	
Zwitterionic	CHAPS	3-[(3-cholamidopropyl)dimethylammonio]-1-propanesulfonate
	CHAPSO	3-[(3-cholamidopropyl)dimethylammonio]-2-hydroxypropane-1-sulfonate
	LDAO	Dodecyldimethylamine oxide
	Zwittergent 3-08	3-(Octyldimethylammonio)propane-1-sulfonate
	EMPIGEN BB	*N*-Dodecyl-*N,N*-dimethylglycine

[a]Adapted from Neugebauer (1992).

EVALUATION OF SOLUBILIZATION CONDITIONS

Prior to initiating trial solubilizations, the investigator should have a highly purified preparation of membranes containing the protein of interest. This can often be obtained by differential centrifugation to prepare a microsomal fraction followed by further purification using continuous or discontinuous sucrose gradient centrifugation (e.g., see Castle, 1995).

Secondly, a simple functional assay should be available. This might be the measurement of enzymatic activity or the binding of a radiolabeled ligand measured by a spin column gel filtration assay, precipitation assay, or binding of the protein-ligand complex to a filter disc. A last resort is to use equilibrium dialysis, but this method is cumbersome and often requires 24 to 48 hr to reach completion. An alternative approach for trial experiments assayed by ligand binding is to prebind the ligand to membrane-bound protein and attempt to solubilize the complex. The advantage is that, presumably, the conditions required to obtain high specific binding have already been evaluated. This approach has the disadvantage, however, that the protein-ligand complex may be more stable than the free protein in detergent, resulting in deceptively high measures of stability and recovery for the solubilized protein.

Third, an assay method for determining the protein concentration should be available. There are several options, including absorbance at 280 nm, modified Lowry methods (Peterson, 1983), and several dye-binding assays that are commercially available in kit form from vendors such as Pierce Bio-Rad. It is important to note that many detergents are not compatible with particular protein assays. For example, detergents containing aromatic groups, such as Triton X-100, give high background readings at 280 nm. Triton X-100 also gives a high background reading in Coomassie brilliant blue G-250 dye-binding assays for protein (Bradford, 1976). Protein determinations using bicinchonic acid (Smith et al., 1985) appear to be less sensitive to many detergents than the Lowry assays. As a last resort, it may be possible to quantitatively precipitate protein from solution prior to assay (Wessel and Flugge, 1984; Minamide and Bamburg, 1990).

Finally, the approximate critical micelle concentration (CMC) of the detergent should be known. The CMC is defined as the concentration of free detergent at which the transition from disperse detergent molecules to a micellar structure occurs. Since solubilization corresponds to removal of the protein from the membrane into the detergent micelle, the CMC is the minimal concentration of detergent necessary to form the required micellar structure for protein extraction. Although these data are generally available from the manufacturer, the CMC varies with temperature and buffering conditions. If the CMC for a detergent is not known, it can easily be estimated by simple spectrophotometric procedures (Rosenthal and Kousale, 1983; Vulliez-Le Normand and Eisele, 1993).

Prepare concentrated buffered stock solutions containing the membrane preparation and the detergent. The membrane solution should contain protease inhibitors. These usually include EDTA and EGTA (calcium-activated neutral proteases) and phenylmethyl sulfonyl fluoride (serine proteases). Several additional inhibitors specific for other types of proteases may also be included (for a summary see Hulme and Birdsall, 1992). The membranes and detergent are then combined in a volume of 1 to 5 ml such that the final concentration of membrane protein is in the range of 1 to 5 mg/ml and the detergent concentration varies from below to well above its CMC (often in the range of 0.05% to 5% w/v). Gently stir the solution for 30 to 60 min at room temperature or at 4°C, then centrifuge 1 hr at 100,000 × *g*, 4°C. Remove the supernatant and resuspend the pellet in nondetergent buffer (the pellet may be washed and recentrifuged to remove residual detergent). Quantify the recovery of protein and its activity for both membrane pellet and detergent extract. Graph the results as total or specific activity solubilized and remaining in the membrane pellet versus detergent-to-protein ratio.

Several outcomes are possible. First, no activity may be measurable in either extract or pellet, in which case the protein was inactivated by treatment with the detergent. Second, the plot of specific activity versus detergent-to-protein ratio may go through a maximum and then decrease. If the recovery of total activity is acceptable at this point, those solubilization conditions can be further optimized. Finally, all or most of the activity may remain in the pellet, indicating that the detergent is not an effective solubilizing agent for the protein of interest.

Further optimization of solubilization conditions can be undertaken by independently varying temperature, time, ionic strength, and pH. Higher recoveries might be obtained by a second detergent extraction of the pellet after centrifugation and higher specific activities by using a double extraction technique. In the latter method, membranes are first treated with detergent at the highest detergent-to-protein ratio where no activity is solubilized. The supernatant, which contains protein but no activity, is discarded and the pellet resuspended and treated with detergent at a sufficient detergent-to-protein ratio to solubilize the desired activity. If this method is to be used, the detergent-to-protein ratio for the second extraction should be redetermined, because it is likely to have changed after the first detergent treatment.

SOLUBILIZATION OF m2 MUSCARINIC RECEPTORS

Early work from the author's laboratory on solubilization of m2 muscarinic receptors may serve as an example of some of the approaches and problems described above (Cremo et al., 1981; Peterson and Schimerlik, 1984). The goal of those studies was to solubilize the muscarinic receptor from porcine atrial membranes so that the protein could be purified. An initial literature search indicated that other laboratories had used saturated digitonin solutions to solubilize muscarinic binding sites from brain in relatively low yield. Digitonin preparations were heterogeneous, containing digitonin, gitonin, and minor saponins in a variable ratio and giving highly variable results depending on supplies and lot number. The detergent preparations were also poorly soluble in aqueous solution, with relative solubility highly dependent on lot number.

To obtain more reproducible detergent solutions, a small amount of a second detergent was added. Although sodium cholate caused irreversible loss of muscarinic binding sites, addition of a low concentration dissolved digitonin, and the detergent mixture was able to solubilize >80% of the muscarinic receptor binding sites from atrial microsomes at final protein concentrations of 3 mg/ml and detergent concentrations of 0.4% (w/v) digitonin and 0.08% (w/v) cholate (Cremo et al., 1981).

These initial results were improved upon in later work (Peterson and Schimerlik, 1984) by using sucrose-gradient-purified atrial membranes and a double extraction technique. The first detergent treatment was done at 5.5 mg/ml protein and a final concentration of 0.4% (w/v) digitonin and 0.08% (w/v) cholate. After centrifugation, the supernatant was discarded, and the membrane pellet was resuspended and extracted a second time at 10 mg/ml protein, 0.8% (w/v) digitonin, and 0.16% (w/v) cholate. The result was a 70% recovery of binding sites and a two- to three-fold purification. Receptor prepared in this manner was stable when stored for several months at 4°C.

To conduct hydrodynamic studies, different detergent systems had to be used. For this purpose, purified protein was stabilized as a complex with a radiolabeled antagonist (which also facilitated analysis for receptor sites) and diluted into Triton X-405. Although the protein was unstable in this system, sufficient recoveries were obtained for analysis of gel filtration and sucrose gradient profiles in H_2O and D_2O (Peterson et al., 1986).

Because of the expense of digitonin and the variability in commercially available preparations, it seemed worthwhile to continue to examine other detergents. Although octyl-β-D-glucopyranoside was ineffective, a related detergent, dodecyl-β-D-maltoside, did solubilize ~50% of the receptor sites, yielding a three-fold purification, using a double extraction protocol (Peterson et al., 1988). The preparation was sufficiently stable ($t_{1/2}$ at 4°C of 20 days) and suitable for hydrodynamic studies, but not nearly as stable as receptor solubilized in digitonin/cholate mixtures. In purifying recombinant porcine m2 muscarinic receptors expressed in Chinese hamster ovary (CHO) cells (Peterson et al., 1995), it became clear that the optimal procedure for digitonin/cholate extraction of CHO cell membranes differed significantly from that developed for porcine atrial membranes. Although suppliers now offer digitonin as either completely or partially water-soluble preparations, the results remain lot-dependent and it is necessary to test all available lots of both types of detergent to find one that works best. When a suitable lot is found, it is a good idea to purchase as much as one can afford.

5

CONCLUSIONS

As the above discussion makes clear, the methodology for solubilizing membrane proteins is poorly defined. This is due in part to the heterogeneous nature of membrane systems, which are composed of protein(s), lipid, and detergent, and in part to the difficulty of predicting what interactions or forces are required to stabilize the native, membrane-bound state. For example, in some cases detergents may substitute successfully for lipids that are normally required to maintain proper protein conformation; in others they may solubilize the protein with a tightly bound lipid annulus intact. To purify a protein, the detergent must be able to disrupt interactions between that protein and other membrane-bound proteins; however, these interactions may be required to maintain protein stability. Finally, other purely physical parameters, such as lateral pressure, micelle size, and curvature, may also be important. These parameters most likely differ for different proteins, lipids, and detergents. Although defining successful solubilization conditions can be an arduous task, the possibility of developing new and creative approaches is always open.

Contributor: Michael I. Schimerlik

UNIT 5.12

Immunoprecipitation

See Table 5.12.2 at the end of this unit for a troubleshooting guide to immunoprecipitation.

NOTE: All solutions should be ice cold and procedures should be carried out at 4°C or on ice.

BASIC PROTOCOL 1

IMMUNOPRECIPITATION USING CELLS IN SUSPENSION LYSED WITH A NONDENATURING DETERGENT SOLUTION

This procedure results in the release of both soluble and membrane proteins; however, many cytoskeletal and nuclear proteins, as well as a fraction of membrane proteins, are not efficiently extracted under these conditions. The procedure allows immunoprecipitation with antibodies to epitopes that are exposed in native proteins. See Figure 5.12.1.

Materials *(see* APPENDIX 1 *for items with ✓)*

Unlabeled or labeled cells in suspension
✓ PBS, ice cold
✓ Nondenaturing lysis buffer, ice cold
50% (v/v) protein A–Sepharose (Sigma, Amersham Pharmacia Biotech) slurry in PBS containing 0.1% (w/v) BSA and 0.01% (w/v) NaN_3
Specific polyclonal antibody (antiserum or affinity-purified immunoglobulin) or monoclonal antibody (ascites, culture supernatant, or purified immunoglobulin)
Control antibody of same type as specific antibody (e.g., preimmune serum or purified irrelevant immunoglobulin for specific polyclonal antibody; irrelevant ascites, hybridoma culture supernatant, or purified immunoglobulin for specific monoclonal antibody)
10% (w/v) BSA
✓ Wash buffer, ice cold
End-over-end tube rotator

1. Collect 0.5–2 × 10^7 cells in suspension by centrifuging 5 min at 400 × g, 4°C, in a 15- or 50-ml capped conical tube. Place tube on ice. Aspirate supernatant and resuspend cells by gently tapping the bottom of the tube.

2. Rinse cells twice with ice-cold PBS, using the same volume of PBS as in the initial culture and centrifuging as in step 1.

3. Add 1 ml ice-cold nondenaturing lysis buffer and resuspend pellet by gentle agitation for 3 sec with a vortex mixer set at medium speed. Do not shake vigorously.

4. Keep suspension on ice 15 to 30 min. Transfer to a 1.5-ml conical microcentrifuge tube and microcentrifuge 15 min at maximum speed, 4°C.

5. Transfer supernatant to a fresh microcentrifuge tube. Do not disturb the pellet, and leave the last 20 to 40 μl of supernatant behind. Keep cleared lysate on ice.

   Cell extracts can be frozen at −70°C, but it is best to lyse cells immediately before immunoprecipitation. Frozen lysates should be thawed and microcentrifuged 15 min at maximum speed, 4°C, before preclearing (step 12).

5

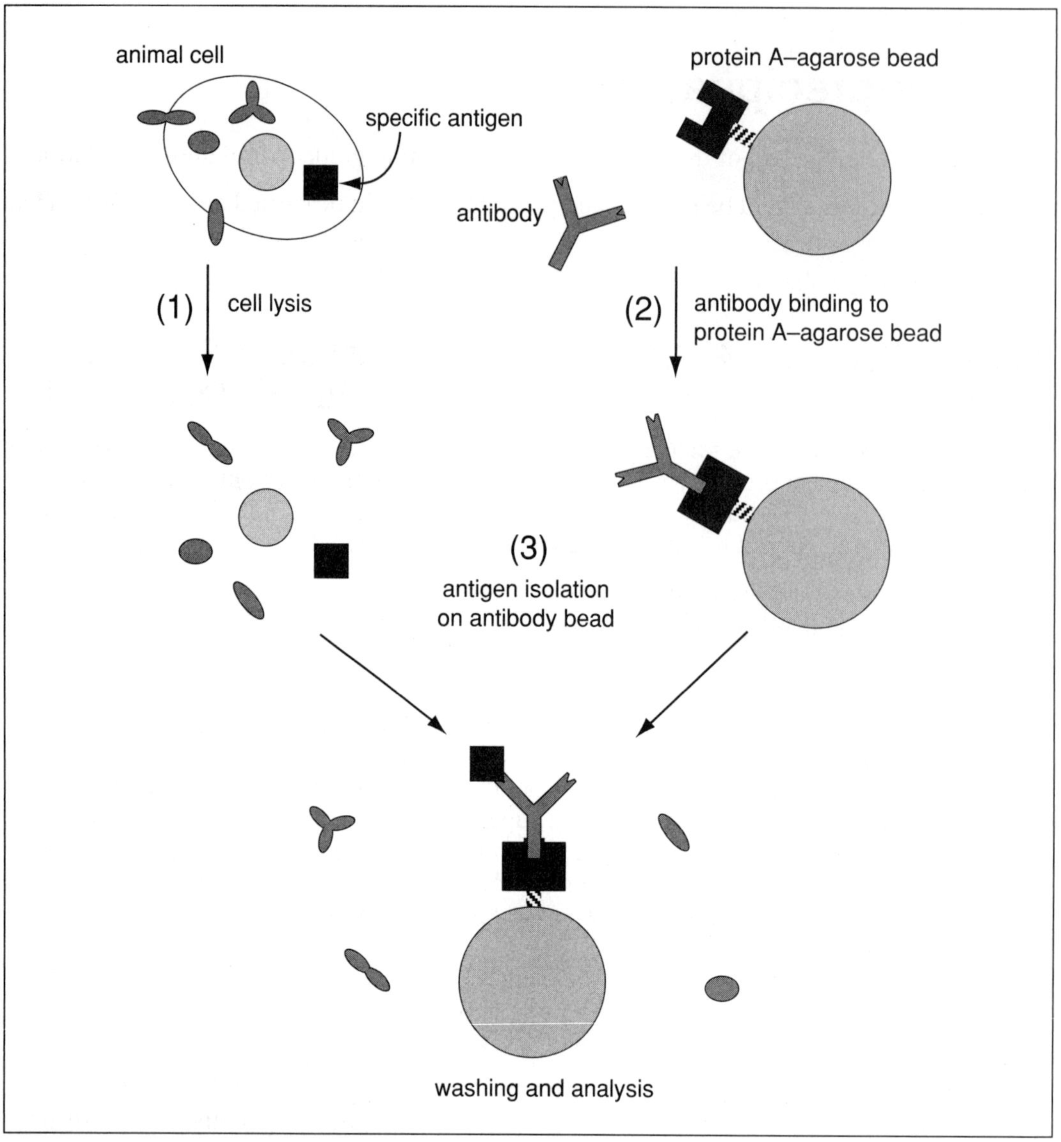

Figure 5.12.1 Schematic representation of the stages of the immunoprecipitation protocol presented in Basic Protocol 1. (**1**) Cell lysis: antigens are solublilized by extraction of the cells in the presence or absence of detergents. To increase specificity, the cell lysate can be precleared with protein A-agarose beads. (**2**) Antibody immobilization: a specific antibody is bound to protein A-agarose beads. (**3**) Antigen capture: the solubilized antigen is isolated on antibody-conjugated beads. Basic Protocol 1 uses beads made of Sepharose (a more stable cross-linked form of agarose).

6. In a 1.5-ml conical microcentrifuge tube, combine 30 μl of 50% protein A–Sepharose slurry, 0.5 ml ice-cold PBS, and the specific antibody (select one):

   1 to 5 μl polyclonal antiserum
   1 μg affinity-purified polyclonal antibody
   0.2 to 1 μl ascitic fluid containing monoclonal antibody
   1 μg purified monoclonal antibody
   20 to 100 μl culture supernatant containing monoclonal antibody.

   Substitute protein G for protein A if antibodies are of a species or subclass that does not bind to protein A (see Table 5.12.1).

Table 5.12.1 Binding of Antibodies to Protein A and Protein G[a,b,c]

Antibody	Protein A binding	Protein G binding[d]
Monoclonal antibodies[e]		
Human IgG1	++	++
Human IgG2	++	++
Human IgG3	–	++
Human IgG4	++	++
Mouse IgG1	+	++
Mouse IgG2a	++	++
Mouse IgG2b	++	++
Mouse IgG3	++	++
Rat IgG1	+	+
Rat IgG2a	–	++
Rat IgG2b	–	++
Rat IgG2c	++	++
Polyclonal antibodies		
Chicken	–	–
Donkey	–	++
Goat	+	++
Guinea pig	++	+
Hamster	+	++
Human	++	++
Monkey	++	++
Mouse	++	++
Rabbit	++	++
Rat	+	+
Sheep	+	++

[a]++, moderate to strong binding; +, weak binding; –, no binding.

[b]A hybrid protein A/G molecule that combines the features of protein A and protein G, coupled to a solid-phase matrix, is available from Pierce.

[c]Information from Harlow and Lane (1999), and from Amersham Pharmacia Biotech, Pierce, and Jackson Immunoresearch.

[d]Native protein G binds albumin from several animal species. Recombinant variants of protein G have been engineered for better binding to rat, mouse, and guinea pig IgG, as well as for avoiding binding to serum albumin.

[e]Protein A binds some IgM, IgA, and IgE antibodies in addition to IgG, whereas protein G binds only IgG.

Antibody-conjugated beads can be prepared prior to preparation of the cell lysate (steps 1 to 5) in order to minimize the time that the lysate is kept on ice.

7. Set up a nonspecific immunoprecipitation control using the appropriate control antibody:

 1 to 5 µl preimmune serum
 1 µg purified irrelevant polyclonal antibody
 0.2 to 1 µl ascitic fluid containing irrelevant monoclonal antibody
 1 µg purified irrelevant monoclonal antibody
 20 to 100 µl hybridoma culture supernatant containing irrelevant monoclonal antibody.

 The irrelevant antibody should be to an epitope not present in the cell lysate. The amount should match that of the specific antibody, and the antibody should be from the same species as the specific antibody.

8. Mix thoroughly and tumble end over end ≥1 hr (up to 24 hr) at 4°C in a tube rotator.

9. Microcentrifuge 2 sec at maximum speed, 4°C. Aspirate supernatant using a fine-tipped Pasteur pipet connected to a vacuum aspirator.

10. Add 1 ml nondenaturing lysis buffer to beads and resuspend by inverting three or four times.

11. Wash by repeating steps 9 and 10, and then step 9 once more. Store antibody-bound beads up to 6 hr at 4°C.

12. *Optional:* To preclear lysate, combine 1 ml cell lysate (step 5) and 30 μl of 50% protein A–Sepharose slurry in a microcentrifuge tube. Tumble end over end 30 min at 4°C in a tube rotator. Microcentrifuge 5 min at maximum speed, 4°C.

    If lysate was prepared from cells expressing immunoglobulins—such as spleen cells or cultured B cells—the preclearing step should be repeated at least three times to ensure complete removal of endogenous immunoglobulins.

13. Add 10 μl of 10% BSA to tube containing antibody-bound beads (step 11). Add the entire volume of lysate (step 5 or 12). In order to avoid carryover of beads with precleared material, leave 20 to 40 μl supernatant on top of the pellet in the preclearing tube. If a nonspecific immunoprecipitation control is performed, divide lysate in two ~0.4-ml aliquots, one for the specific antibody and the other for the nonspecific control. Incubate 1 to 2 hr at 4°C while mixing end over end in a tube rotator.

14. Microcentrifuge 5 sec at maximum speed, 4°C. Aspirate supernatant using a fine-tipped Pasteur pipet connected to a vacuum aspirator.

    The supernatant can be kept up to 8 hr at 4°C or up to 1 month at −70°C for sequential immunoprecipitation of other antigens or for analysis of total proteins.

15. Add 1 ml ice-cold wash buffer to beads and resuspend by inverting 3 or 4 times. Microcentrifuge 2 sec at maximum speed, 4°C. Aspirate supernatant, leaving ~20 μl supernatant on top of the beads. Repeat wash three more times. The total wash time should be ~30 min; keep samples on ice for 3 to 5 min between washes if necessary.

16. Wash beads once more using 1 ml ice-cold PBS and aspirate supernatant completely with a drawn-out Pasteur pipet or an adjustable pipet fitted with a disposable tip. The final product should be 15 μl of settled beads containing bound antigen. Analyze immediately or store at −20°C.

17. Analyze by electrophoresis or immunoblotting (*UNIT 5.7*).

    For troubleshooting, see Table 5.12.2 at the end of this unit.

5

ALTERNATE PROTOCOL 1

IMMUNOPRECIPITATION USING ADHERENT CELLS LYSED WITH A NONDENATURING DETERGENT SOLUTION

Additional Materials *(also see Basic Protocol 1)*

Unlabeled or labeled cells grown as a monolayer on a tissue culture plate
Rubber policeman

1. Rinse cells attached to a tissue culture plate twice with ice-cold PBS. Remove PBS by aspiration with a Pasteur pipet attached to a vacuum trap. Place plate on ice.

2. Add ice-cold nondenaturing lysis buffer to the tissue culture plate (e.g., 1 ml lysis buffer for an 80% to 90% confluent 100-mm-diameter tissue culture plate). Depending on the cell type, a confluent 100-mm dish will contain 0.5–2 × 10^7 cells.

3. Scrape cells off the plate with a rubber policeman. Transfer suspension to a 1.5-ml conical microcentrifuge tube, vortex gently for 3 sec, and keep on ice for 15 to 30 min.
4. Clear the lysate and perform immunoprecipitation (Basic Protocol 1, steps 4 to 17).

ALTERNATE PROTOCOL 2

IMMUNOPRECIPITATION USING CELLS LYSED WITH DETERGENT UNDER DENATURING CONDITIONS

If epitopes of native proteins are not accessible to antibodies, or if the antigen cannot be extracted from the cell with nonionic detergents, cells should be solubilized under denaturing conditions. Only antibodies that react with denatured proteins can be used to immunoprecipitate proteins solubilized by this protocol. This protocol can be adapted for adherent cells (see Alternate Protocol 1).

Additional Materials *(also see Basic Protocol 1; see* APPENDIX 1 *for items with ✓)*

✓ Denaturing lysis buffer
Heating block set at 95°C (Eppendorf Thermomixer 5436 or equivalent)
25-G needle attached to 1-ml syringe

1. Collect cells in suspension culture (Basic Protocol 1, steps 1 and 2). Place tubes on ice. Add 100 μl denaturing lysis buffer and resuspend cells by vortexing vigorously 2 to 3 sec at maximum speed.
2. Transfer suspension to a 1.5-ml conical microcentrifuge tube and heat 5 min in a 95°C heating block. Dilute the suspension with 0.9 ml nondenaturing lysis buffer. Mix gently.

   This sequesters SDS into Triton X-100 micelles.

3. Shear DNA by passing the suspension five to ten times through a 25-G needle attached to a 1-ml syringe. Incubate 5 min on ice.
4. Clear the lysate and perform immunoprecipitation (Basic Protocol 1, steps 4 to 17).

5

ALTERNATE PROTOCOL 3

IMMUNOPRECIPITATION USING CELLS LYSED WITHOUT DETERGENT

Immunoprecipitation of proteins that are already soluble within cells (e.g., cytosolic or lumenal organellar proteins) may not require the use of detergents.

Additional Materials *(also see Basic Protocol 1; see* APPENDIX 1 *for items with ✓)*

✓ Detergent-free lysis buffer
25-G needle attached to 3-ml syringe

1. Collect and wash cells in suspension (Basic Protocol 1, steps 1 and 2). Add 1 ml of ice-cold detergent-free lysis buffer and resuspend cells by gentle agitation for 3 sec with a vortex mixer set at medium speed.
2. Break cells by passing the suspension 15 to 20 times through a 25-G needle attached to a 3-ml syringe. Monitor cell breakage under a bright-field or phase-contrast microscope and repeat procedure until >90% cells are broken.

If the cells are particularly resistant to mechanical breakage, they can be swollen for 10 min at 4°C with a hypotonic solution containing 10 mM Tris·Cl, pH 7.4, before mechanical disruption.

3. Clear lysate and perform immunoprecipitation (Basic Protocol 1, steps 4 to 17), *except* use detergent-free lysis buffer in step 10.

ALTERNATE PROTOCOL 4

IMMUNOPRECIPITATION WITH ANTIBODY-SEPHAROSE

This protocol, which follows the steps presented in Figure 5.12.2, relies on the formation of an insoluble immune complex between a protein antigen and an antigen-specific monoclonal (or polyclonal) antibody covalently bound to Sepharose.

Materials *(also see Basic Protocol 1; see* APPENDIX 1 *for items with ✓)*

Unlabeled cells, surface-labeled cells (e.g., with ^{125}I or biotin) *or* biosynthetically ^{35}S-, ^{3}H-, or ^{14}C-labeled cells
✓ Triton X-100 lysis buffer
✓ Dilution buffer
Antibody (Ab)–Sepharose (see Support Protocol)
Activated, quenched (control) Sepharose, prepared as for Ab-Sepharose but eliminating Ab or substituting irrelevant Ab during coupling
✓ Tris/saline/azide (TSA) solution
✓ 0.05 M Tris·Cl, pH 6.8
✓ 2× SDS sample buffer

1. Incubate cells in Triton X-100 lysis buffer (5 × 10^7 cells/ml) for 1 hr at 4°C. Centrifuge 10 min at 3000 × *g* to remove nuclei. Centrifuge the supernatant 1 hr at 100,000 × *g* and save the supernatant. Use within several days or store at −70°C.

2. Preclear 200 μl supernatant by adding 10 μl activated, quenched (control) Sepharose. Shake on an orbital shaker 2 hr at room temperature or overnight at 4°C. Centrifuge 1 min at 200 × *g* and save supernatant.

3. Precoat 1.5-ml microcentrifuge tubes by filling with Triton X-100 lysis buffer 10 min at room temperature. Remove the solution by aspiration.

4. Add 10^5 to 10^6 cpm of radiolabeled (^{125}I or ^{35}S) supernatant containing antigen (step 2) to a precoated microcentrifuge tube and bring to 200 μl with dilution buffer. For nonradiolabled samples, use 0.2 to 1 ml precleared lysate.

   The recommended amount of radioactivity is appropriate for eukaryotic cells with >1000 molecules of antigen/cell.

5. Add ~10 μl of a 1:1 slurry of Ab-Sepharose/dilution buffer and shake 1.5 hr (up to 3 hr) at 4°C on an orbital shaker, keeping Sepharose suspended.

6. Wash the Ab-Sepharose with 1 ml of the buffers listed below. After each wash, centrifuge 1 min at 200 × *g* or microcentrifuge 5 sec. Carefully aspirate supernatant with a fine-tipped Pasteur pipet, leaving 10 μl of fluid above the pellet. After the fourth wash, centrifuge again to bring down any residual drops on the side of the tube, aspirate, and leave 10 μl over the pellet.

   First wash: dilution buffer
   Second wash: dilution buffer
   Third wash: TSA solution
   Fourth wash: 0.05 M Tris·Cl, pH 6.8.

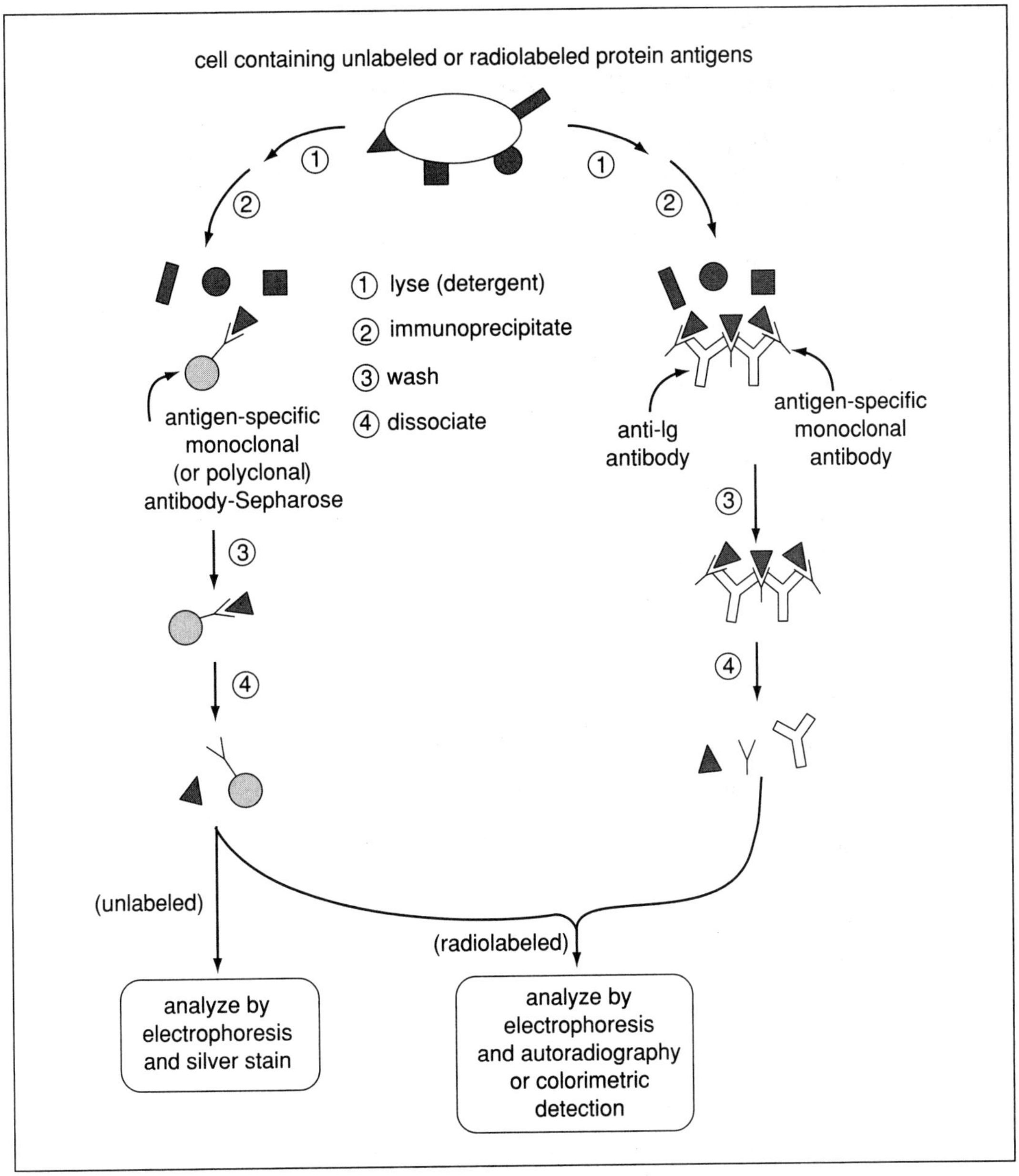

Figure 5.12.2 Schematic representation of the stages of the immunoprecipitation protocols using either antibody-Sepharose (left, see Alternate Protocol 4) or anti-Ig serum (right, see Alternate Protocol 5). (**1**) Cell lysis. (**2**) Immunoprecipitation using specific antibodies coupled covalently to Sepharose beads (left) or specific antibodies combined with anti-Ig serum (right). (**3**) Washing. (**4**) Dissociation of the antigen-antibody complex in sample buffer for electrophoresis.

7. Add 20 to 50 µl of 2× SDS sample buffer. Because the sample buffer has a higher density than the wash solution, it will sink into the Sepharose; do not vortex, because Sepharose may stick to side of tube above buffer level. Cap the tube securely and incubate 5 min at 100°C.

8. Vortex and centrifuge 1 min at 200 × *g* or microcentrifuge 5 sec. Analyze supernatant by SDS-PAGE. Detect labeled proteins by autoradiography with an enhancing screen (^{125}I), by fluorography (^{35}S, ^{14}C, or ^{3}H), or by colorimetric or chemiluminescent detection (biotinylated proteins).

SUPPORT PROTOCOL

PREPARATION OF ANTIBODY-SEPHAROSE

Materials *(see* APPENDIX 1 *for items with ✓)*

1 to 30 mg/ml antigen-specific monoclonal or polyclonal antibody
0.1 M $NaHCO_3$/0.5 M NaCl
Sepharose CL-4B (or Sepharose CL-2B for high-molecular-weight antigens; Amersham Pharmacia Biotech)
0.2 M Na_2CO_3
✓ Cyanogen bromide (CNBr)/acetonitrile
1 mM and 0.1 mM HCl, ice-cold
0.05 M glycine (or ethanolamine), pH 8.0
✓ Tris/saline/azide (TSA) solution

Dialysis tubing (MWCO >10,000)
Whatman no. 1 filter paper
Buchner funnel
Erlenmeyer filtration flask
Water aspirator

1. Dialyze 1 to 30 mg/ml antibody against 0.1 M $NaHCO_3$/0.5 M NaCl at 4°C with three buffer changes over 24 hr. Use a volume of dialysis solution that is 500 times the volume of antibody solution.
2. Centrifuge dialysate 1 hr at 100,000 × *g*, 4°C, to remove aggregates. Save the supernatant.
3. Measure the A_{280} of an aliquot of the solution and determine the concentration of the antibody (mg/ml IgG = A_{280}/1.44). Dilute with 0.1 M $NaHCO_3$/0.5 M NaCl to 5 mg/ml (or to the desired concentration for Ab-Sepharose) and keep at 4°C. Measure the A_{280} of this solution for later use in step 11.
4. Allow Sepharose slurry to settle in a beaker. Decant and discard the supernatant. Weigh out the desired quantity of Sepharose (assume density = 1.0).
5. Set up a filter apparatus using Whatman no. 1 filter paper in a Buchner funnel and an Erlenmeyer filtration flask attached to a water aspirator. Wash the Sepharose on the filter apparatus with 10 vol water.
6. Transfer Sepharose to 50-ml beaker and add an equal volume of 0.2 M Na_2CO_3.
7. Activate Sepharose in a fume hood at room temperature using 3.2 ml CNBr/acetonitrile per 100 ml Sepharose. Add CNBr/acetonitrile dropwise with a Pasteur pipet over 1 min, while slowly stirring the slurry with a magnetic stirrer. Continue stirring slowly for 5 min (excessive and vigorous stirring may fracture the Sepharose beads).
8. Rapidly filter the CNBr-activated Sepharose as in step 5. Aspirate to semidryness (i.e., until the Sepharose cake cracks and loses its sheen).
9. Wash with 10 vol ice-cold 1 mM HCl, then with 2 vol of ice-cold 0.1 mM HCl. Hydrate the cake with enough ice-cold 0.1 mM HCl so the cake regains its sheen, but so there is no excess liquid above the cake.

   CNBr-activated Sepharose can be purchased premade from Amersham Pharmacia Biotech, but the coupling capacity will be lower.

10. Immediately transfer a weighed amount of Sepharose (assume density = 1.0) to a beaker. Add an equal volume of a solution of antibody dissolved in 0.1 M $NaHCO_3$/0.5 M NaCl

(step 2). Stir gently with a magnetic stirrer or rotate end over end 2 hr at room temperature or overnight at 4°C.

11. Add 0.05 M glycine (or ethanolamine), pH 8.0, to saturate the remaining reactive groups on the Sepharose and allow the slurry to settle. Remove an aliquot of the supernatant, centrifuge to remove any residual Sepharose, and measure A_{280}. Compare absorbance to that of the A_{280} of the antibody solution (measured in step 3) to determine the percentage coupling. Store the Ab-Sepharose in TSA solution.

ALTERNATE PROTOCOL 5

IMMUNOPRECIPITATION OF RADIOLABELED ANTIGEN WITH ANTI-IG SERUM

Additional Materials *(also see Alternate Protocol 4)*

Normal serum
Anti-Ig serum (Zymed Laboratories)
Antigen-specific antiserum *or* antigen-specific purified monoclonal antibody *or* antigen-specific hybridoma culture supernatant

Follow the procedures in Alternate Protocol 4, with the following modifications at the indicated steps:

2a. Preclear by adding normal serum at a concentration of 2 μl/ml radiolabeled antigen. Add the proper amount of anti-Ig serum and let stand 12 to 18 hr at 4°C. Centrifuge 10 min at 1000 × *g* and reserve supernatant.

Normal serum is the source of carrier Ig. The proper amount of anti-Ig serum must be determined by titration with radiolabeled antigen or Ig. For high-titered anti-Ig serum, this amount would be 20× to 40× the volume of antigen-specific antiserum, 2 to 4 μl/μg purified MAb, or one-third the volume of hybridoma culture supernatant.

5a. Add 1 μl antigen-specific antiserum, 3 μg antigen-specific purified MAb, or antigen-specific hybridoma culture supernatant (30 μl cloned line or 100 μl uncloned line). Vortex and allow to stand 2 hr at 4°C. Then add the proper amount of anti-Ig serum, vortex, and allow to stand 12 to 18 hr at 4°C.

6a. Wash the immunoprecipitate as described, but centrifuge 7 min at 1000 × *g*.

7a. Add SDS sample buffer as described, but incubate 1 hr at 56°C and then 5 min at 100°C.

The initial 56°C incubation enhances the dissolution of the immunoprecipitates by reducing irreversible aggregation that occurs when precipitated protein is rapidly heated to 100°C.

BASIC PROTOCOL 2

IMMUNOPRECIPITATION-RECAPTURE

Once an antigen has been isolated by immunoprecipitation, it can be dissociated from the beads and reimmunoprecipitated ("recaptured") either with the same antibody or with a different antibody. Immunoprecipitation-recapture with the same antibody allows identification of a specific antigen in cases where the first immunoprecipitation contains too many bands to allow unambiguous identification. By using a different antibody in the second immunoprecipitation, immunoprecipitation-recapture can be used to analyze the subunit composition of multi-protein complexes. The feasibility of this approach depends on the ability of the second antibody to recognize denatured antigens.

Materials *(see* APPENDIX 1 *for items with ✓)*

- ✓ Elution buffer
- Beads containing bound antigen (see Basic Protocol 1)
- 10% (w/v) BSA
- ✓ Nondenaturing lysis buffer
- Heating block set at 95°C (Eppendorf Thermomixer 5436 or equivalent)

1. Add 50 μl elution buffer to 15 μl beads containing bound antigen and vortex. Incubate 5 min at room temperature and 5 min in a 95°C heating block. Cool to room temperature.
2. Add 10 μl of 10% BSA and mix by gentle vortexing.
3. Add 1 ml nondenaturing lysis buffer and incubate 10 min at room temperature.
4. Clear the lysate and perform second immunoprecipitation (Basic Protocol 1, steps 4 to 17).

References: Harlow and Lane, 1999; Hjelmeland and Chrambach, 1984.

Contributors: Jsuan S. Bonifacino, Esteban C. Dell'Angelica, and Timothy A. Springer

Table 5.12.2 Troubleshooting Guide for Immunoprecipitation

Problem	Cause	Solution
No specific radiolabeled antigen band		
Gel is completely blank after prolonged autoradiographic exposure	Poorly labeled cells: too little radiolabeled precursor, too few cells labeled, lysis/loss of cells during labeling, too much cold amino acid in labeling mix, wrong labeling temperature	Check incorporation of label by TCA precipitation; troubleshoot the labeling procedure
Only nonspecific bands present	Antigen does not contain the amino acid used for labeling	Label cells with another radiolabeled amino acid or, for glycoproteins, with tritiated sugar
	Antigen expressed at very low levels	Substitute cells known to express higher levels of antigens as detected by other methods; transfect cells for higher expression
	Protein has high turnover rate and is not well labeled by long-term labeling	Use pulse labeling
	Protein has a low turnover rate and is not well labeled by short-term labeling	Use long-term labeling
	Protein is not extracted by lysis buffer used to solubilize cells	Solublize with a different nondenaturing detergent or under denaturing conditions
	Antibody is nonprecipitating	Identify and use antibody that precipitates antigen
	Epitope is not exposed in native antigen	Extract cells under denaturing conditions
	Antibody does not recognize denatured antigen	Extract cells under nondenaturing conditions
	Antibody does not bind to immunoadsorbent	Use a different immunoadsorbent; use intermediate antibody
	Antigen is degraded during immunoprecipitation	Ensure that fresh protease inhibitors are present

continued

Table 5.12.2 Troubleshooting Guide for Immunoprecipitation, continued

Problem	Cause	Solution
High background of nonspecific bands		
Isolated lanes on gel with high background	Random carryover of detergent-insoluble proteins	Remove supernatant immediately after centrifugation, leaving a small amount with pellet; if resuspension occurs, recentrifuge
High background in all lanes	Incomplete washing	Cap tubes and invert several times during washes
	Poorly radiolabeled protein	Optimize duration of labeling to maximize signal-to-noise ratio
	Incomplete removal of detergent-insoluble proteins	Centrifuge lysate 1 hr at 100,000 × g
	Insufficient unlabeled protein to quench nonspecific binding	Increase concentration of BSA
	Antibody contains aggregates	Microcentrifuge antibody 15 min at maximum speed before binding to beads
	Antibody solution contains nonspecific antibodies	Use affinity-purified antibodies; absorb antibody with acetone extract of cultured cells that do not express antigen
	Too much antibody	Use less antibody
	Incomplete preclearing	Preclear with irrelevant antibody of same species of origin and immunoglobulin subclass bound to immunoadsorbent
	Nonspecifically immunoprecipitated proteins	Fractionate cell lysate (e.g, ammonium sulfate precipitation, lectin absorption, or gel filtration) prior to immunoprecipitation; after washes in wash buffer, wash beads once with 0.1% SDS in wash buffer or 0.1% SDS/0.1% sodium deoxycholate
Immunoprecipitating antibody detected in immunoblots		
Complete immunoglobulin or heavy and/or light chains visible in immunoblot	Protein A conjugate or secondary antibody recognizes immunoprecipitating antibody	Use antibody coupled covalently to solid-phase matrix for immunoprecipitation; probe blots with primary antibody from a different species and the appropriate secondary antibody specific for immunoblotting primary antibody

5

UNIT 5.13

Detection of Protein-Protein Interactions by Coprecipitation

Coprecipitation (Fig. 5.13.1) may be the single method of choice, or may be used in combination with other methods that detect protein-protein interactions, such as two-hybrid analysis and copurification schemes, and tests of physical association using purified proteins.

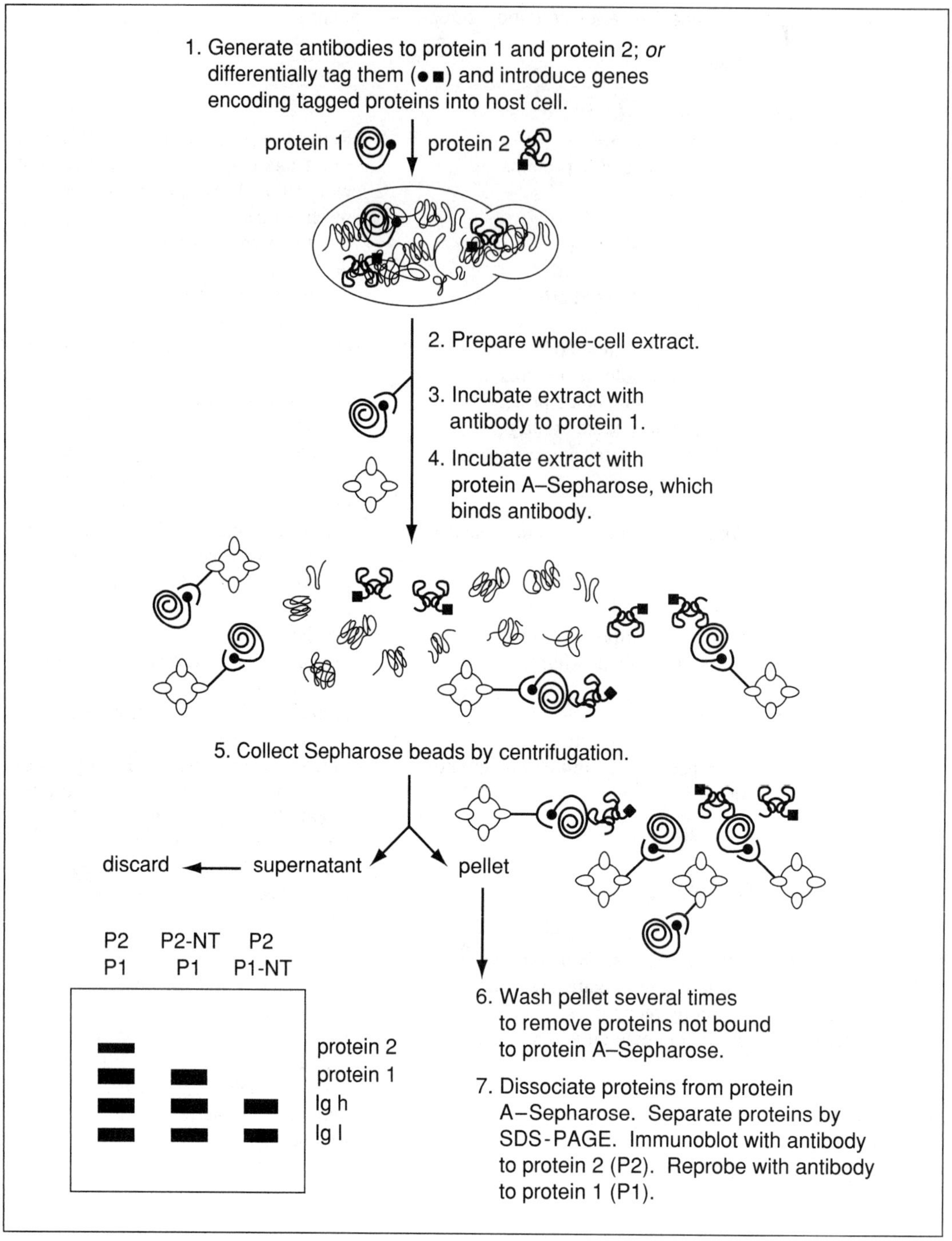

Figure 5.13.1 Flow chart for the coprecipitation of two proteins that have been differentially tagged and introduced into the host organism. Ig h and Ig l, immunoglobulin heavy and light chains, respectively; NT, no tag.

BASIC PROTOCOL

COPRECIPITATING PROTEINS WITH PROTEIN A– OR PROTEIN G–SEPHAROSE

It is essential to keep all buffers and tubes cold by using an ice bath and a refrigerated centrifuge. The conditions of coprecipitation match the conditions of the lysis buffer used to prepare whole-cell extracts.

Materials *(see APPENDIX 1 for items with ✓)*

Whole-cell extract
Antibody
5 M NaCl
✓ Co-immunoprecipitation buffer, adjusted as needed for activity of protein of interest
✓ Protein A– or protein G–Sepharose slurry
✓ 2× SDS sample buffer, made with 1 mM sodium azide and sterile stock solutions if sample is to be stored

Microcentrifuge, 4°C
Test tube rotator
20-ml syringe and 18-G needle
Hamilton syringe

1. Prepare duplicate samples in microcentrifuge tubes on ice:

    0.5 to 1 mg whole-cell extract
    1 μg antibody
    5 M NaCl to equalize at 100 mM NaCl
    Co-immunoprecipitation buffer to 0.5 ml final volume.

    Invert tube gently several times and incubate 90 min on ice, inverting occasionally.

2. Microcentrifuge 10 min at maximum speed, 4°C. Transfer supernatant to a new microcentrifuge tube and add 50 μl protein A– or protein G–Sepharose slurry (25 to 30 μl bead volume). Evenly suspend slurry before distributing it to samples. Rotate tube gently (do not rock) 30 to 60 min at 4°C.

3. Gently pellet Sepharose by centrifuging 30 sec at 1000 rpm in a tabletop centrifuge, 4°C. Wash pellet three times with 1 ml co-immunoprecipitation buffer. Use a 20-ml syringe with an 18-G needle to aspirate and remove supernatant after pelleting. Aspirate as much liquid as possible without touching the beads and add 25 μl of 2× SDS sample buffer.

    Sample can be frozen up to several months at −80°C prior to SDS-PAGE (step 4).

4. Boil sample 5 min, vortex, and pellet beads. Use a Hamilton syringe to load eluates onto an SDS-polyacrylamide gel, arranging duplicate samples to allow preparation of duplicate blots. Separate by electrophoresis. Include aliquots of the whole-cell extract for comparison and as a positive control for the immunoblot.

5. Immunoblot duplicate samples separately with antibodies for each of the two proteins. Reprobe blots with antibody to the other protein if needed.

ALTERNATE PROTOCOL

COPRECIPITATING A GST FUSION PROTEIN

This procedure might be used when the protein in question comigrates with immunoglobulin heavy or light chain in an SDS-PAGE gel or if the antibodies being used precipitate too many cross-reacting proteins.

Additional Materials *(also see Basic Protocol; see* APPENDIX 1 *for items with ✓)*

✓ Glutathione-agarose or glutathione-Sepharose slurry

1. Prepare duplicate samples as described (see Basic Protocol, step 1), omitting antibody and incubation. Microcentrifuge 10 min at maximum speed, 4°C and transfer supernatant to new microcentrifuge tube.
2. Add 30 μl glutathione-agarose or glutathione-Sepharose slurry (25- to 30-μl bead volume). Evenly suspend slurry before distributing it to samples. Rotate tube gently 30 to 60 min at 4°C.
3. Complete precipitation and immunoblot analysis as described (see Basic Protocol, steps 3 to 5).

References: BioSupplyNet Source Book, 1999; Phizicky and Fields, 1995

Contributor: Elaine A. Elion

UNIT 5.14

Imaging Protein-Protein Interactions by Fluorescence Resonance Energy Transfer (FRET) Microscopy

5

NOTE: All incubations are performed in a humidified 37°C, 5% CO_2 incubator unless otherwise specified. Some media (e.g., DMEM) may require altered levels of CO_2 to maintain pH 7.4. Use aseptic technique for tissue culture.

BASIC PROTOCOL

FRET MICROSCOPY OF FIXED CELLS

Materials *(see* APPENDIX 1 *for items with ✓)*

Cells of interest
Plasmid for GFP-tagged protein
Transfection reagent (e.g., Fugene 5 from Boehringer Mannheim, Lipofectin from Life Technologies, Effectene from Qiagen, or Superfect from Qiagen)
Serum-free medium
✓ Low-background fluorescence CO_2-independent medium (Life Technologies)
✓ PBS, pH 7.4
✓ 4% (w/v) formaldehyde fixative solution
Quench solution: 50 mM Tris·Cl (pH 8.0)/100 mM NaCl
0.1% (v/v) Triton X-100 in PBS

Antibody (e.g., PY72 monoclonal anti-phosphotyrosine antibody) labeled with Cy3 (see Support Protocol 2)
1% (w/v) BSA (fraction V) in PBS
✓ Mowiol mounting medium

6- and 12-well tissue culture plates
Coverslips
Microscope slides
Confocal laser scanning microscope (e.g., Zeiss LSM 510), equipped with argon (488 nm) and He/Ne (543 nm) lasers selected by the HFT 488/543 double dichroic filter, GFP fluorescence selected by the NFT 545 dichroic and BP 505-530 emission filter, and Cy3 fluorescence selected by the LP560 emission filter
Imaging software package (e.g., NIH-image or IPLab Spectrum from Scanalytics)

1. Seed the cells onto culture dishes containing coverslips. For live-cell FRET experiments, seed the cells onto glass-bottom MatTek culture dishes. Transfect the adherent monolayer cells with the plasmid for GFP-tagged protein of interest using calcium phosphate precipitation, lipofection, or the activated dendrimer reagent, Superfect.

2. Grow the cells as as monolayer, typically 1 to 2 days post transfection, until expression levels of the GFP-tagged protein are high enough to detect with fluorescence microscopy. When needed, starve the cells overnight with serum-free medium to make them semi-quiescent.

3. On the day of the experiment, transfer the coverslips to wells of a 12-well tissue culture plate. Subject the cells to the desired experimental conditions (e.g., growth factor/hormone stimulation, or incubation with drugs or inhibitors).

4. Wash the cells with ice-cold PBS, aspirate, and add 1 ml 4% formaldehyde fixative solution. Allow cells to fix at room temperature for 10 min. Quench excess fixative with quench solution by briefly washing the cells, then changing to fresh quench solution and incubating 5 min at room temperature.

5. Permeabilize cellular membranes by incubating 5 min with 0.1% Triton X-100 to allow penetration of acceptor-labeled molecules into fixed cells. Wash the cells with PBS to remove the permeabilization solution.

6. Dilute Cy3-labeled antibodies appropriately in 1% BSA/PBS. Incubate the coverslips with appropriately diluted (e.g., 0.1 to 10 μg/ml determined by titration) Cy3-labeled antibodies for 1 hr, at room temperature.

7. Transfer the coverslips to wells of a 6-well tissue culture plates and wash four times, each time with 3 ml PBS, to remove excess antibody. Blot off excess PBS with tissue and mount the coverslips on slides with ~10 μl Mowiol mounting solution. Allow Mowiol to harden overnight at 4°C before imaging.

8. View the specimen on a confocal microscope using a 63× or 100× oil-immersion objective. Acquire an image in the GFP channel (excitation, 488 nm; emission, NFT 545, BP505-530). Either take an image of the entire field of vision or select a region of interest containing the cell that is to be imaged.

   Do not use the full dynamic range of the detector because unquenching will result in added fluorescence. Depending on the make of the confocal microscope, make sure that a second image can be made at exactly the same location. Do not adjust the settings (i.e., pinhole size, contrast, brightness, laser power, or averaging) for the GFP channel.

   This acquisition provides the FRET-quenched donor image (F_{DA}), since the acceptor is present throughout the cell, causing FRET with the GFP donor.

5

9. Change to the Cy3 channel (excitation, 543 nm; emission, LP560) and take an image of the cell, minimizing Cy3 photobleaching. Select a portion of the cell in which the FRET efficiency is to be determined.

   This region is where the acceptor will be photobleached, thereby revealing the unquenched donor intensities enabling FRET calculation.

   Consequently, any portion of the cell where the acceptor is not photobleached serves as control [i.e., $1 - F_{DA}/F_{DA} = 0$]. This area in the same cell will provide an essential control to judge the effects of photobleaching GFP, lateral movement, or focal mismatch between the two consecutive donor images.

10. Photobleach the selected acceptor region by repeated scanning with the 543-nm He/Ne laser line at full power. Follow the progress of photobleaching by monitoring the intensities of the respective images (emission: NFT 545, BP505-530). Continue until there is no more discernible Cy3 intensity (~1 to 20 min).

11. Return to the GFP channel and make the second acquisition, using identical settings and location of the prebleached image to provide the F_D reference in the region where the acceptor was photobleached.

12. To calculate the FRET efficiency in the bleached area, use an appropriate imaging software package. Subtract the prebleach donor image from the post-bleach donor image (i.e., $F_D - F_{DA}$). Divide this image by the postbleach donor image: $(F_D - F_{DA})/F_D$; this is identical to $1 - (F_{DA}/F_D) = E$, the image of FRET efficiencies.

    A software package capable of performing the abovementioned image processing is NIH Image. This package is freely available from http://rsb.info.nih.gov/nih-image/ and versions are available for Apple Macintosh, Windows (Scion image) and even a Java version is available that runs on any platform (Image J). A more versatile commercial package is IPLab Spectrum.

    Image arithmetic for calculating the FRET efficiency should be performed on a region of interest obtained by thresholding the images in order to prevent the amplification of background noise. Therefore, thresholds should be chosen in such a way that the background is omitted in the calculation. Accuracy can be improved by subtracting the average background intensity from the images before calculating the FRET efficiency.

SUPPORT PROTOCOL 1

NUCLEAR AND CYTOSOLIC MICROINJECTION

Materials

Cells of interest, cultured in MatTek glass-bottom 35-mm dishes (MatTek)
DNA (e.g., human EGFR cDNA in the Clontech pEGFP-N1 expression vector)
HPLC-grade water

Millex-GV4 0.22-μm filtration unit
GELloader tips (Eppendorf)
Needles for microinjection (e.g., Femtotip from Eppendorf)
Microinjector (e.g., Eppendorf model 5244)
Micromanipulator (e.g., Eppendorf model 5170)
Inverted microscope with 10× and 40× air objectives

1. Prepare DNA, to be microinjected, of the highest possible quality (e.g., double cesium chloride banded of Qiagen ion exchange purified DNA)

2a. *For optimal expression a few hours after microinjection:* Dilute DNA in HPLC-grade water to 100 μg/ml.

2b. *For expression overnight:* Dilute DNA in HPLC-grade water to 1 μg/ml.

3. Clear the DNA solution to prevent blocking of the glass needle during microinjection as follows. Place a 0.22 μm Millex filtration unit in a 0.5-ml microcentrifuge tube and place the entire unit in a 1.5-ml microcentrifuge tube to enable centrifugation. Filter 10 μl of the DNA solution by microcentrifuging 1 min at maximum speed, room temperature.

4. Load 2 μl of DNA solution (or Cy3-protein solution) using GELloader tips into the capillary glass needle of the microinjector.

5. On an inverted microscope, microinject the DNA solution (or Cy3-protein solution) into the nucleus (or perinuclear cytosol) of cells grown in MatTek culture dishes.

   Typical settings are: 0.3 sec, 150 to 400 hPa injection pressure with 20-hPa back-pressure to prevent medium from entering the needle. The injection pressure may be varied according to the needle opening and cell type.

   No major movement of the nucleus (or cell organelles) should be observed. A visual indication for excessive pressure is the separation of the nucleus from the surrounding cellular material (i.e., light ring around the nucleus) and leakage into the cytosol, visible by movement of the cellular organelles.

   Restrict the microinjection procedure to a maximum of 10 min. In the authors' laboratory, microinjection is performed at room temperature in normal CO_2-dependent medium. After 10 min, the medium starts to acidify significantly (i.e., purple medium). CO_2-independent (or HEPES-buffered) media can be used for longer periods.

SUPPORT PROTOCOL 2

PROTEIN LABELING WITH Cy3

Materials *(see* APPENDIX 1 *for items with* ✓*)*

Antibody (PY72 monoclonal anti-phosphotyrosine antibody)
✓ 1 M Tris·Cl, pH 8.0
10 mM and 100 mM Bicine/NaOH, pH 8.0
100 ml citric acid/NaOH, pH 2.8
1 M Bicine/NaOH, pH 9.0
1 M NaCl
Labeling buffer: 100 mM Bicine/NaOH (pH 8.0)/100 mM NaCl
Cy3.29-OSu monofunctional sulfoindocyanine succinimide ester (Amersham Pharmacia Biotech)
Dimethylformamide (DMF) dried by addition of 10 to 20 mesh 3-pore diameter molecular sieve dehydrate (Fluka)

1-ml Protein G HiTrap columns (Amersham Pharmacia Biotech)
Centricon YM30 concentrators (Amicon)
Biogel P6DG Econopac prepacked size-exclusion columns (5.5 × 1.5–cm, ~10 ml; Bio-Rad)
1- and 10-ml syringes with HPLC Luer-Lok fitted tubing

1. Resuspend antibody to 1 mg/ml in PBS. Prepare a syringe-operated 1-ml protein G HiTrap column. Equilibrate column with 10 ml PBS at a maximum flow of 4 ml/min.

2. Add 0.1 vol 1 M Tris·Cl, pH 8.0, to the antibody solution. Load antibody solution onto the column and wash column with 10 ml of 100 mM Bicine/NaOH, pH 8.0. Collect run-through.

3. Wash column with 10 ml of 10 mM Bicine/NaOH, pH 8.0. Collect run-through. Elute column with 5 ml of 100 mM citric acid/NaOH, pH 2.8, and collect 0.5-ml fractions (i.e., 8 drops) in 1.5-ml microcentrifuge tubes containing 100 µl of 1 M Bicine/NaOH, pH 9.0, to neutralize the pH. Mix immediately and store on ice.

4. Determine the A_{280} of the eluted fractions using a spectrophotometer and pool the fractions that contain protein (typically the first four fractions).

5. Add 0.1 vol of 1 M NaCl. Concentrate the solution in a YM30 Centricon to ~200 µl by centrifuging at 5000 × *g*, 4°C. Redilute to 2 ml with labeling buffer. Repeat this step one time.

6. Concentrate to ~50 to 100 µl as in step 5 and collect the concentrated protein solution.

7. Reconstitute Cy3.29-OSu in 20 µl dry DMF to give a ~10 mM Cy3 solution. Determine the exact concentration by measuring the absorption of a 10^4-fold diluted solution in PBS. From the ε_{550} of 150 $mM^{-1}cm^{-1}$, calculate the concentration.

   Cy3.29-OSu is supplied as a desiccated pellet in microcentrifuge tubes.

   DMF is dried by addition of hygroscopic beads to the container.

   The ε_{650} of Cy5 is 250 $mM^{-1}cm^{-1}$.

8. Determine the protein concentration of the antibody (or protein) solution to be labeled based on A_{280} reading.

   Antibody concentration at $A_{280} = 1.0$ is typically 1 mg/ml. At low protein concentrations (i.e., <0.1 mg/ml), or for smaller proteins (where the $A_{280} = 1.0 = 1$ mg/ml protein relation does not hold true), the protein concentration can be determined using the Bio-Rad Coomassie-based protein determination assay.

9. Slowly add a 10- to 20-fold molar excess of dye to the protein solution (the added volume of Cy3/DMF should not exceed 10% of the total volume) while simultaneously stirring the solution with the pipet tip. Incubate 30 min at room temperature, shaking the tube gently every 10 min.

10. Remove the unconjugated dye by size-exclusion chromatography on a Bio-Rad P6DG Econopac (6-kDa exclusion size) column. Equilibrate the column with 30 ml PBS (or a buffer that is specifically formulated for the protein to be labeled).

11. Load the labeling reaction mixture onto the column and wash with a small amount (i.e., ~200 µl) of PBS (or equilibration buffer). Discard the first 2.5 ml (void volume ~3.3 ml).

12. Collect the colored protein fraction (or 2 ml of eluate if the staining is too weak to be visible) in a YM30 Centricon concentrator and concentrate to approximately the volume of the labeling reaction.

13. Estimate the labeling ratio by either of the following formulas:

$$A_{554} \times M/[A_{280} - (0.05 \times A_{554}) \times 150] \text{ for Cy3}$$

$$A_{650} \times M/[A_{280} - (0.05 \times A_{650}) \times 250] \text{ for Cy5}$$

   where A_x is the absorption at wavelength x and M is the molecular weight of the protein in kDa.

   The extinction coefficients (in units of $mM^{-1}cm^{-1}$) 150 and 250 are for Cy3 and Cy5, respectively. These formulas assume the $A_{280} = 1.0 = 1$ mg/ml relationship as found for larger proteins (i.e., >50 kDa) and correct for the 5% absorption of Cy3 at 280 nm. For smaller proteins or lower protein concentrations (i.e., <0.1 mg/ml) that cannot be reliably estimated from the A_{280}, *determine the protein concentration by the Bio-Rad protein assay and estimate the labeling ratio from:*

$A_{554} \times M/(P \times 150)$ *for Cy3*
$A_{650} \times M/(P \times 250)$ *for Cy5*

where P is the protein concentration in mg/ml as determined by Bio-Rad assay.

As previously mentioned, IgG concentration obtained with BSA as standard has to be multiplied by 2.

14. Verify covalent labeling of the antibody by SDS-PAGE of the labeled product. Use transillumination with a UV source (302 nm) to visualize labeled protein.

References: Bastiaens and Squire, 1999; Clegg, 1996

Contributors: Fred S. Wouters and Philippe I.H. Bastiaens

UNIT 5.15

An Overview on the Generation of BAC Transgenic Mice for Neuroscience Research

Functional studies of the mammalian nervous system require tools to study the system at all levels: genes, cells, circuitry, and behavior. Because of the extraordinary complexity of the mammalian nervous system, many aspects of neuroscience research still heavily rely on studies of the intact nervous system in vivo and in live brain slices. The mouse nervous system has become a favorite mammalian model system to study because of the powerful genetic tools that are available to manipulate the mouse genome. One such tool is the generation of transgenic mice; a stable integration of foreign DNA into the mouse genome. Recent advances using large genomic DNA clones, called bacterial artificial chromosomes (BACs), to generate transgenic mice and apply the BAC transgenic technology in neuroscience research are discussed in this unit.

5

TRANSGENIC MICE: SOME GENERAL CONSIDERATIONS

Transgenic mice are usually generated by direct microinjection of DNA fragments into fertilized one-cell mouse embryos, followed by transfer of these embryos into recipient mothers that can carry the pregnancy to term. Transgenic founders, which have the foreign DNA stably and randomly integrated into the genome, can then germline transmit the integrated transgene to their offspring to establish a transgenic mouse line. Since the transgene integration is a random event, each transgenic line is unique in that the transgene is intergrated at a distinct chromosomal location and the line has certain copies of the transgene. In 10% to 30% of the cases, foreign DNA fragments may integrate in more than one genomic loci, and subsequent breeding can separate these independent integration sites into separate transgenic "sub-lines." Transgenes are usually integrated as multiple copies (the range is from one to several hundred), and these integrated DNA copies usually form head-to-tail repeats (called concatmers).

The transgenic mouse approach is widely used to study gene expression and gene regulation in vivo. For these applications, it has several distinct advantages over the gene targeting approach. The gene targeting approach requires embryonic stem cells (ES cells), which are currently only available for a very few strains of mice (i.e., 129Sv and C57/bL6J). In addition, ES cell work is usually time-consuming and costly. The transgenic approach involves direct microinjection of DNA into fertilized embryos and does not require ES cells. Therefore, the transgenic approach is faster, less expensive, and available for many different strains of mice (i.e., FvB, C57/BL6, CD1, and others), and for many different species for which ES cells are not available, such as zebrafish, rat, sheep, and cow. Moreover, since the transgene is integrated as multiple copies, in some cases, the transgenic approach can achieve higher transgene expression levels than those achieved by the gene targeting approach.

The transgenic approach also has its limitations. The most severe limitation using a conventional transgenic approach with <20 kb genomic DNA in the transgenic construct often results in positional effects, that is, expression of the transgene is influenced by its integration site. Position effects may manifest in different forms, including lack of transgene expression, ectopic transgene expression (unintended sites of expression), mosaic expression (only a subset of cells express the transgene), and extinction (diminishing transgene expression in successive generations). The most important cause for position effects is lack of important regulatory elements in the transgenic construct. Other causes include high integration copy number causing repeat-mediated gene silencing, and transgene integration into a heterochromatic region of the chromosome. Another limitation in the transgenic approach is that integration of the transgene may cause rearrangement or deletion on the endogenous chromosome near the integration site. Since most transgenic founders are heterozygous at the transgene integration site and do not exhibit obvious phenotypes, a small number of heterozygous transgenic mice and some of the homozygous transgenic mice may have phenotypes that arise from disruption of the murine endogenous genes near the transgene integration site. Because of these limitations, it is often necessary to perform phenotypic studies on at least two transgenic mouse lines for a given transgene, and only those phenotypes that are reproducible between different lines can be unambiguously attributed to the transgene expression, and to rule out that these phenotypes are due to positional effects or disruption of endogenous genes near the integration site.

Bacterial Artificial Chromosome (BAC)-Mediated Transgenesis in Mice

Bacterial artificial chromosomes (BACs) are large-insert DNA clones based on the *E.coli* fertility factor (Shizuya et al., 1992). BACs contain on average ~180 to 200 kb genomic DNA (BACs can accomodate DNA inserts up to 700 kb; Shizuya et al., 1992). BACs have many advantages as large genomic DNA clones for transgenic studies. BACs have very high stability and low chimerism (Marra et al., 1997) and are very easy to work with—one can easily isolate a large quantity of high quality, relatively intact BAC DNA. BAC DNA is suitable for direct sequencing and for microinjections to generate transgenic mice. BACs have served as the primary genomic DNA clones to most of the genome sequencing projects. As a result, almost all of the fully sequenced genomes, including the human, mouse, rat, and zebrafish genomes, also contain dense contigs of BACs that are mapped across the genome. Therefore, for any gene of interest, known or predicted, one may readily locate BACs that contain a specific gene from the database, and obtain these BACs from public repositories to immediately begin functional investigations. Finally, several methods have been developed to utilize homologous recombination in *E. coli* to rapidly modify BACs and to introduce desirable mutations in BACs, including insertion, deletion, and point mutations (Yang et al., 1997; Zhang et al., 1998; Lee et al., 2001; Gong et al., 2002). These methods are highly efficient and can be scaled up for large-scale BAC-mediated functional genomic studies in mice (Gong et al., 2003; Valenzuela et al., 2003).

The major advantage of using BACs for transgenic studies is that BAC transgenic constructs can overcome positional effects to produce integration-site independent, copy-number dependent, and accurate transgene expression in vivo. This is because the majority of mammalian genes span <100 kb of genomic DNA, and a correctly chosen BAC (with ~200 kb of genomic DNA insert) is likely to contain all the regulatory elements necessary to confer accurate transgene expression in vivo. The advantage of the BAC system was first demonstrated with the murine *Zipro1* gene, a zinc finger transcription factor specifically expressed in the postnatally proliferating granule cell populations in the brain (Yang et al., 1997). A conventional transgenic construct with a 10-kb promoter and a lacZ marker gene produced ten transgenic lines, which all exhibited positional effects—lack of transgene expression or ectopic expression. But with a 130-kb Zipro1 BAC driving the lacZ marker gene, the marker gene is expressed in all of the expected neuronal populations in the brain. This initial observation is further substantiated by many subsequent BAC transgenic studies, particularly the GENSAT study, which demonstrated, with ~100 different BACs, that 80% to 85% of the BAC transgenes produced accurate and reproducible transgene expression in vivo (Gong et al., 2003). In summary, BAC transgenesis is a highly reproducible method to achieve accurate transgene expression in mice, and it has become a method of choice in mammalian transgenic studies.

BAC TRANSGENIC CONSTRUCT DESIGN

A transgenic construct requires three basic components: a promoter to drive expression of the transgene, a transgenic open reading frame (Tg-ORF) encoding genes to be expressed (e.g., EGFP, lacZ, Cre), and a polyadenylation signal (PolyA) to terminate transcription. In a BAC transgenic construct, rather than using a small genomic DNA fragment as a promoter, the entire BAC containing the genomic locus of a chosen gene (called driver gene) is used to drive the transgene expression. In this section, four commonly used BAC transgenic constructs (Fig. 5.15.1) are described.

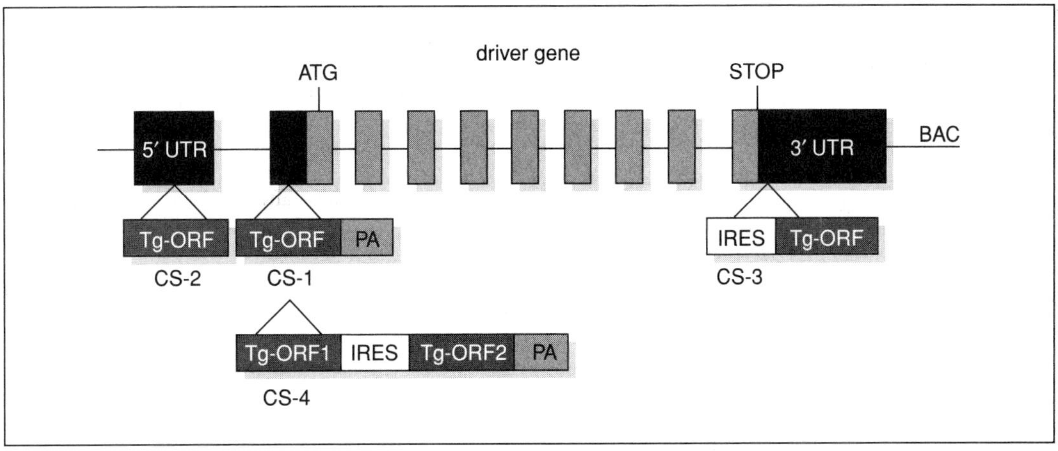

Figure 5.15.1 Four basic designs of BAC transgenic constructs. Construct 1 (CS-1) and construct 2 (CS-2) are designed to overexpress a transgenic open reading frame (Tg-ORF), such as the green fluorescent protein, lacZ, Cre, and others, but the driver gene itself is not overexpressed. Both CS-1 and CS-2 have Tg-ORF inserted in front of the translation initiation sequence (ATG) of the driver gene. The difference between these two constructs is that CS-1 has Tg-ORF inserted into exon 1 and uses an endogenous polyA signal from the driver gene, and Tg-ORF2 is not inserted into exon 1 and uses an exogenous polyA signal (PA). Construct 3 (CS-3) uses an internal ribosomal entry sequence (IRES) to drive overexpression of both the driver gene and a Tg-ORF. The IRES.Tg-ORF is often inserted into the 3′-untranslated region (3′-UTR) of the driver gene. Finally, construct-4 (CS-4) is designed to use IRES to overexpress two transgenes of interest, but the driver gene is not overexpressed.

Construct 1 and 2 (CS-1 and CS-2): Overexpression of an Exogenous Transgene but not the Driver Gene

Two different BAC transgenic constructs can achieve overexpression of an exogenous transgene but not the driver gene. In the first design (Fig. 5.15.1, CS-1), the Tg-ORF followed by a polyA signal (PA) is placed in the 5′-untranslated region (5′-UTR) immediately 5′ to the translation initiation site (ATG) of the driver gene. The polyadenylation signal is usually a mammalian polyA signal, such as the pGK-polyA. Viral polyA signals should be avoided if possible since they may induce transgene methylation and silencing. The mRNA transcript initiated at the driver gene promoter is terminated at the Poly A sequence, therefore, the coding region of the driver gene is not transcribed at all. To achieve a high level of transgene expression, the Tg-ORF should use an optimal Kozak translation initiation sequence (GCC ACC ATG G) (Kozak, 1999). This CS-1 construct design has been successfully used in the large-scale GENSAT BAC transgenic project (Gong et al., 2003). In the second design (Fig. 5.15.1, CS-2), the Tg-ORF, without an exogenous polyA signal, is inserted in the 5′-UTR portion in exon 1 of the driver gene. In this design, the mRNA transcript initiated at the driver gene promoter will transcribe through the Tg-ORF and also splice through the entire driver gene and use the endogenous polyA signal to terminate the fusion transcript. Since the mammalian ribosome only translates one ORF in the most 5′ end of the transcript, the Tg-ORF, but not the driver gene, will be translated from the fusion transcript. The Tg-ORF needs to be placed in the first exon of the driver gene because the internal exons of the mammalian genes are usually too small (180 to 200 bp) to accommodate the Tg-ORF (Berget et al., 1995). The first or the last exon, however, is usually relatively large (2 to 10 kb) and can accommodate a transgene insertion. This transgenic construct design (CS-2) has a distinct advantage over the CS-1. Since it incorporates the entire driver gene mRNA as part of the expressed transgene, the CS-2 construct preserves the mRNA processing of the driver gene, thus allowing crucial mRNA processing steps to be recapitulated and studied in these transgenic mice.

Construct 3 (CS-3): Overexpression of Both the Driver Gene and a Second Exogenous Gene

The CS-3 construct is designed to overexpress the driver gene and a second transgene (e.g., EGFP, lacZ, Cre) from the same mRNA transcript. In this design, the driver gene is overexpressed, and the second marker gene labels the cell types with driver gene overexpression. To translate two proteins from a single mRNA transcript, an internal ribosomal entry site (or IRES) is needed in front of the second Tg-ORF. Several IRES sequences (e.g., EMCV, FMDV, HCV, VEGF) have been successfully used in transgenic mice (Kim et al., 1992; Yang et al., 1997, 1999). In the CS-3 construct, the IRES-Tg-ORF is inserted into the 3′-UTR region of the driver gene and the endogenous polyA signal is used to terminate the transcription. If the Tg-ORF is not placed in the last exon of the driver gene, an exogenous polyA signal is needed to follow the Tg-ORF to properly terminate the transcription of the transgene.

Construct 4 (CS-4): Overexpression of Two Exogenous Genes from a Single Transcript

In the CS-4 design, two exogenous genes are expressed from a single transcript in the BAC. In this construct design (Fig. 5.15.1, CS-4), the first exogenous gene, or Tg-ORF1, is inserted into the 5′-UTR of the driver gene. Tg-ORF1 is immediately followed by an IRES sequence, the second transgene (Tg-ORF2), and a pGK-polyA signal (PA). This design allows co-expression of two exogenous genes, and the endogenous driver gene is not expressed from the BAC.

Selecting Appropriate BACs for Transgenic Studies

The next step in generating BAC transgenic mice is to select appropriate BACs containing the driver gene of interest. Several public genome databases, such as the Genome Browsers at the University of California at Santa Cruz (*http://www.genome.ucsc.edu*) or the mouse Clonefinders at the National Center for Biotechnology Information (NCBI) (*http://www.ncbi.nlm.nih.gov/genome/clone/clonefinder/CloneFinder.html*), can be used to readily locate human or mouse BACs that contain the gene of interest. For BAC transgenic studies, BACs that contain at least 50 kb of flanking genomic DNA sequence both 5′ and 3′ to the driver gene of interest are required, and the driver gene should be located essentially in the center of the BAC (Gong et al., 2003). For a mammalian gene that is <100 kb, which the vast majority of the genes are, one can routinely identify BACs that fulfill this criteria. BACs can be readily obtained from the public BAC clone repository (BACPAC Resource Center at Oakland Children's Hospital, *http://bacpac.chori.org/*), or from commercial sources (e.g., Invitrogen).

Other Considerations in BAC Transgenic Construct Design

Driver Gene Size

For driver genes that span <100 kb, 85% of BACs selected using the criteria listed above produce reproducible transgene expression (Gong et al., 2003). For the small number of mammalian genes that are >200 kb, experience thus far indicates that the BAC transgenes are relatively inconsistent in producing accurate transgene expression in vivo. BAC transgenic mice with reproducible and accurate expression with driver genes that are 100 to 170 kb in size have been generated. Therefore, BAC transgenic studies are most suitable for driver genes that are <200 kb in size.

Other Genes on the BAC

BACs often contain several other genes in addition to the driver gene. One concern for BAC transgenesis is that overexpression of these non-driver genes on the BAC may produce phenotypes that could confound intended studies. Since a large number of BAC transgenic studies use marker gene (e.g., EGFP and lacZ) expression as a phenotypic readout, the accuracy of marker gene expression patterns in these mice can be readily assessed and validated. To assess other phenotypes, such as pathological or behavioral phenotypes, it may be necessary to take additional steps to assess whether the phenotype is due to overexpression of the intended transgene or due to overexpression of the other genes on the BAC. First, the pattern and timing of gene expression may be used as criteria. For example, if other genes on the BAC are not expressed in the nervous system at all, they can be effectively ruled out as being responsible for the observed CNS phenotypes. For a more definitive answer, control transgenic mice containing BACs with deletion of the driver gene, or deletion of one or more of the non-driver genes in the BAC may be generated. Comparison of phenotypes in control BAC transgenic mice and original BAC transgenic mice (with similar transgene copy numbers) will help to determine whether the phenotypes are due to overexpression of the driver gene or overexpression of the non-driver genes. This approach has been used to determine that overexpression of the murine *Zipro1* gene, and not other genes on the BAC, is responsible for regulation of cerebellar granule cell proliferation in vivo (Yang et al., 1999).

BAC MODIFICATION BY HOMOLOGOUS RECOMBINATION IN *E. coli*

To facilitate the use of BACs for transgenic studies, several methods have been developed to introduce modifications, such as marker gene insertion, deletion, and point mutations, into

BACs. The first method to introduce a precise mutation into a BAC took advantage of the powerful homologous recombination system of *E. coli* (Yang et al., 1997). Since the BACs are maintained in a recombination-deficient *E. coli* host bacteria, which lacks the critical recombination enzyme RecA to modify a BAC, the recombination competence of these bacteria must be temporarily restored by re-introducing either RecA or other recombination enzymes that can rescue RecA deficiency (e.g., RecET, λ Red; Zhang et al., 1998; Lee et al., 2001). In this overview, focus is placed on a highly efficient RecA-based BAC modification method, which has been successfully applied to a large-scale BAC transgenic study in mice (Gong et al., 2002, 2003).

Shuttle Vector (PLD53-SC-AB)

To modify a BAC using the RecA-based method, a specialized shuttle vector called pLD53.SC-AB (Fig. 5.15.2) is used. This shuttle vector is derived from the PLD53 vector originally

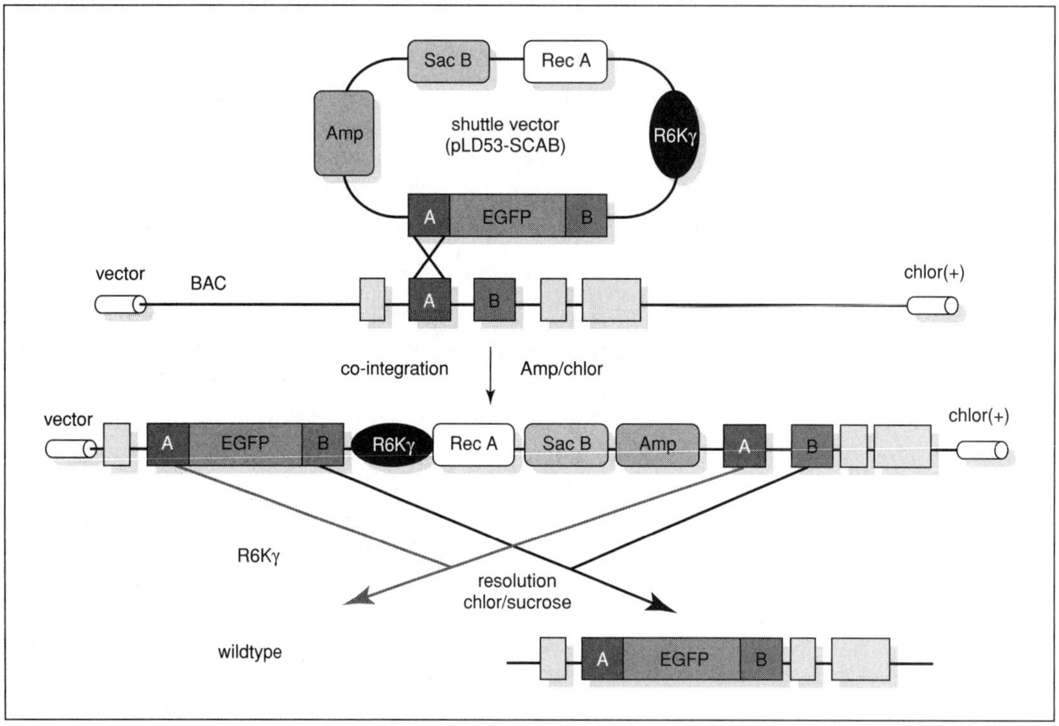

Figure 5.15.2 Modification of BAC by homologous recombination in *E. coli.* A shuttle vector (pLD53.SCAB) is used for modification. This vector contains an R6Kγ origin of replication, a RecA gene to support recombination, a SacB gene for negative selection, and a recombination cassette containing two small homology boxes A and B (~500-bp each) flanking a modification to be introduced (e.g., EGFP gene). The shuttle vector plasmid is electroporated into the BAC host bacteria. Some of the shuttle vectors can undergo homologous recombination with the BAC through either the A or B boxes, resulting in a complete integration of the shuttle vector into the BAC to form a co-integrate BAC. Bacteria containing co-integrate BACs can be selected by growth on chloramphenicol (chlor) and ampicillin (Amp). A small percentage of bacteria with co-integrates can undergo a second homologous recombination step through the A or B boxes (called resolution), resulting in the excision of the shuttle vector. The excised shuttle vector is automatically lost, and the bacteria with resolved BAC can be selected by growth on chloramphenicol and sucrose (sucrose selects for loss of the SacB gene). If co-integration and resolution occur through different homology boxes, the result is a correctly modified BAC with precise placement of the EGFP marker gene on a chosen locus on the BAC.

described by Metcalf et al. (1996). This plasmid has an R6Kγ origin of replication, which can replicate in the presence of the π protein. The π; protein is present in specialized bacterial cell lines such as the Pir2 cells (Invitrogen), but is absent in the BAC host bacteria strain, DH10B, or other conventional bacteria such as DH5α. In addition, the shuttle vector contains the *E. coli* RecA gene, which can restore the recombination competence of the BAC host bacteria. The shuttle vector also contains the SacB gene, which is used for negative selection against the shuttle vector. Since the SacB gene product, levasucrase, converts sucrose to levan, which is highly toxic to the host bacteria, SacB-expressing bacteria cannot grow on 5% sucrose plates (Gay et al., 1985).

BAC Modification Step 1: Construction of Shuttle Vectors with a Recombination Cassette

To modify a BAC by homologous recombination, the first step is to subclone a recombination cassette into the PLD53.SC-AB shuttle vector. This recombination cassette should contain two small genomic DNA fragments of ~200 to 500 bp, called homology boxes A and B. Boxes A and B are derived from the driver gene locus on the BAC, and their sequences and relative position should preserve those in the BAC. These boxes are designed so that after homologous recombination, the intended modification (i.e., gene insertion, deletion, or point mutation) is placed precisely between the chosen A and B boxes on the BAC. Unless the A and B boxes are immediately adjacent to each other on the BAC, after BAC modification, the DNA fragment between them is deleted and replaced by the intended modification. The DNA sequences used to design PCR primers for the homology boxes can be readily obtained from Genbank (NCBI) or the UCSC Genome Browsers. The authors generally use PCR to amplify the homology boxes from the BAC, and try to retain A and B boxes with similar lengths to ensure that the two subsequent homologous recombination steps occur with comparable efficiency.

BAC Modification Step 2: Co-integration

When the shuttle vector is electroporated into the BAC-host bacteria, the RecA gene is expressed from the shuttle vector and a small percentage of the shuttle vector undergoes homologous recombination with the BAC through either the A or B box and produces complete integration of the shuttle vector into the BAC, a step called co-integration. The resulting BAC with insertion of the entire shuttle vector is called a co-integrate BAC. Since non-integrated shuttle vectors cannot replicate and are automatically lost, the host bacteria containing co-integrate BACs can be readily selected by growing the electroporated BAC in Luria broth (LB) containing chloramphenicol (a selection marker for the BAC) and ampicillin (a selection marker for the shuttle vector). Bacteria with co-integrate BACs can be readily selected by colony PCR or Southern blot analyses. The efficiency for co-integration, that is the percentage of PCR-screened clones that contain correct co-integrate BACs, is on an average ~80% (Gong et al., 2002). Typically, only ten colonies need to be screened to obtain several bacteria clones containing correct co-integrate BACs.

BAC Modification Step 3: Resolution

Since the bacteria containing co-integrate BACs still express the RecA gene and have two pairs of homology boxes when they are plated on the chloramphenicol plates alone, a second homologous recombination may occur in some bacteria either between the two A boxes or between the two B boxes to result in the excision of the shuttle vector, a step called resolution. The excised shuttle vector plasmids cannot replicate and are automatically lost, therefore, the bacteria containing the resolved BACs no longer express RecA and return to the recombination-deficient state. Such bacteria can be selected by growing on chloramphenicol (to select for the

BACs) and 5% sucrose (to select for the loss of the SacB gene on the shuttle vector). These bacteria can be secondarily screened for UV sensitivity (due to loss of RecA) or for loss of the Amp gene. There are two possible end-products at the resolution step. If the co-integration and resolution both occur through the same homology box (i.e., both A boxes or both B boxes), the resolved BAC is the wildtype BAC. If the co-integration and resolution occur through two different boxes, the resolved BAC has the intended modification precisely placed between the A and B boxes in the BAC. Bacteria containing the correctly resolved BAC can be readily identified by colony PCR or Southern blot analyses. The resolution efficiency is also very high, ranging from 10% to 80%. Typically, only 10 to 20 resolved BAC clones need to be screened to obtain several bacteria clones containing correctly resolved BACs.

CHARACTERIZATION OF MODIFIED BACs AND PREPARATION OF BAC DNA FOR MICROINJECTIONS

Characterization of Modified BACs Prior to Microinjections

Several characterizations are typically performed to ensure that the modified BACs contain only the intended modification, and not any unwanted rearrangements or deletions. First, restriction digestion patterns of a modified BAC are compared to those of the parental BAC, a method called fingerprinting, which results in only a few differing bands. A large-scale fingerprinting of >100 modified BACs using the RecA-based method demonstrated that >95% of the BACs have only one to three visible band changes, suggesting that this method rarely introduces gross rearrangements or deletions into the BACs (Gong et al., 2003). Second, pulsed-field gel electrophoresis (PFGE) is routinely used to separate very large DNA fragments (a few kilobases to a few hundred kilobases) and to ensure that the modified BACs still contain the correct size insert after they are digested with *Not* I (which releases the insert) or with *Sce*I (which linearizes the BAC). If the modified BAC also contains rare-cutting sites (e.g., *Sal*I, *Mlu*, *Pme*I) within the modification region, PFGE can be used to map the modified BAC after digestion with these rare-cutting restriction enzymes. Fingerprinting can also be done using PFGE when the BACs are digested with *Pac*I, *Sal*I, or *Xho*I. If necessary, PCR may be used to amplify the modified genomic regions from the BAC and to sequence PCR products to ensure that the crucial sequences within the modified region of the BAC do not harbor any unwanted mutations.

Purification of BAC DNA for Transgenic Studies

A critical last step in the generation of BAC transgenic mice is the purification of intact BAC DNA. BAC DNA inserts are very large (ranging from a few kilobases to >100 kb) and prone to degradation, and degraded BAC DNAs may confound experiments by producing positional effects when they are integrated as transgenes. Two methods are used to prepare BAC DNA with minimal degradation. The first method involves cesium chloride gradient centrifugation followed by dialysis against the microinjection buffer. The second method employs conventional alkaline lysis to obtain DNA, followed by a phenol/chloroform extraction, and a final purification step using a Sepharose CL-4B column (Yang et al., 1997). The cesium chloride gradient method produces the highest and most consistent yield of transgenic founders, and the alkaline-CL-4B method also generates a relatively consistent yield of transgenic founders. Although both linear and circular DNA can produce transgenic founders with accurate transgene expression (Yang et al., 1999), the authors have found that linearization of BACs using either *Not*I digestion or *Sce*I digestion improves the rate of transgenic founder production. Therefore, if possible, linearized, purified BAC DNA at a concentration of ~2 ng/μl for microinjection into the mouse one-cell embryo should be used to generate BAC transgenic mice.

MOUSE STRAIN CONSIDERATIONS

Various inbred and hybrid mouse strains can be used to develop transgenic mice. Since these strains have different characteristics including behavioral profiles, neuropathology, and disease or toxin susceptibility, it is important to take these factors into account when choosing appropriate mouse strains for BAC transgenic studies (Bothe et al., 2004). The most commonly used strains are FvB, C57BL/6J, and F1 hybrids between C57 and DBA or CBA. The FvB mice are frequently used for transgenesis since their embryos have large pronuclei, which facilitate microinjection. Usually, FvB mice produce more transgenic founders than other inbred strains, and these mice have a relatively large litter size. There are several disadvantages of the FvB mice, e.g., these mice are blind, have partial dysgenesis of corpus callosum, and are relatively hyperactive. C57BL/6J mice are advantageous when used in a variety of cognitive behavioral studies, however, they are relatively poor breeders and it is difficult to use their embryos for pronuclear injections. It is necessary to carefully choose appropriate mouse strains for a given BAC transgenic study.

APPLICATIONS OF BAC TRANSGENIC MICE IN NEUROSCIENCE RESEARCH

The BAC transgenic approach has been increasingly applied to the study of the mammalian nervous system. Some of the applications that are being used to investigate gene expression, gene function, and to study the neuronal circuitry in vivo are described below.

Mapping Gene Expression Patterns and Studying Gene Regulation

For the majority of mammalian genes that are <100 kb, a BAC containing a gene of this size can be used to express a marker gene, such as EGFP, in transgenic mice to quickly map the in vivo expression of this gene. Since BACs have a success rate of ~85% in reproducing an accurate and endogenous-like transgene expression pattern, the BAC transgenic approach is a highly reliable way to assess gene expression in vivo (Gong et al., 2003).

Once a gene expression pattern is recapitulated by a BAC transgene, the regulatory elements that are crucial to gene regulation, RNA processing, and other processes may be systematically dissected (Yu et al., 1999). Such elements may include transcription enhancers, repressors, locus control regions, imprinting signals, or splicing signals. Using comparative genomics, a method comparing genomic DNA sequences of a given gene from different species (Boffelli et al., 2004), one may readily locate highly conserved, putative regulatory elements in the noncoding region near the gene of interest, and then experimentally test these putative elements in vivo using BAC modification and BAC transgenesis (Nobrega et al., 2003).

Labeling Live Neurons of Interest for Imaging, Electrophysiology, and Purification of Neurons

Large numbers of neuronal cell types in the brain cannot be identified unless they are fixed, sectioned, and stained with appropriate antibodies or by in situ hybridization. Using a fluorescent marker gene such as EGFP, distinct types of neurons and their processes can be labeled in BAC transgenic mice while these neurons are still alive in the mouse brain. These BAC-EGFP mice allow for the use of other approaches, such as in vivo imaging and electrophysiology, to study genetically identified neuronal types in the brain (Roseberry et al., 2004). These labeled neurons may also be purified, either by laser-capture microdissection, or by dissociation followed by fluorescent-activated cell sorting (or FACS). Purified neurons from the brain can be used for gene expression profiling using microarrays (X.W. Yang, unpub.

observ.). Furthermore, FACS-sorted neurons can be used to develop primary neuronal cultures for distinct neuronal types of interest. The public GENSAT project has developed hundreds of lines of BAC-EGFP transgenic mice, which contain labeled distinct types of neurons in the brain (*http://www.gensat.org*), and this tremendous resource will tremendously enhance the ability to study these different neuron types in the mammalian brain.

Developing Cre Mouse Lines for Conditional Gene Targeting

Conditional gene targeting offers a powerful tool to generate cell type- or regional-specific loss-of-function alleles using the Cre-LoxP system (*UNIT 4.19*). To achieve this, Cre recombinase expression is necessary in specific neuronal types in the brain. BAC transgenesis has been increasingly applied to develop Cre mouse lines with desirable patterns of Cre expression in vivo (Casanova et al., 2001; Aller et al., 2003; Ohyama and Groves, 2004). The BAC vectors most commonly used in the development of BAC-Cre mice usually have 1 or 2 LoxP sites and these sites need to be removed prior to microinjection, either by *Not*I digestion, or by BAC modification to delete the LoxP sites. Retaining these LoxP sites in the BAC-Cre construct may cause transgene instability and subsequent loss of Cre expression.

Neuronal Circuitry Mapping

Traditional methods to map neuronal circuitry rely on stereotactic injection of anterograde- or retrograde-chemical tracers into distinct regions of the brain. The injection sites usually involve many different types of neurons, thus the tracers often label different types of neurons. Genetically-encoded anterograde tracers, such as wheat germ agglutinin (WGA; Yoshihara et al., 1999) and barley lectin (BL), can be expressed in distinct neuronal subpopulations by BAC transgenesis to map the connectivity of these neurons. Since BAC transgenic mice can be engineered so that only specific neuronal types in the brain are labeled, the BAC transgenic system is highly suitable for expressing these genetic tracers to map different neuronal circuits in the brain.

In addition to genetic tracers, another elegant approach has been developed to perform retrograde and trans-synaptic mapping of the neuronal circuitry (DeFalco et al., 2001). In this study, BAC transgenic mice were developed to express Cre in the NPY(+) or POMC(+) neurons in the hypothalamus. Genetically engineered pseudorabies viruses were stereotactically injected into the hypothalamic region, and these viruses expressed EGFP only upon Cre-mediated excision of a transcription termination sequence. Upon retrograde trans-synaptic transfer of the pseudorabies virus, only neurons presynaptic to the Cre-expressing NPY or POMC neurons were green due to EGFP expression. Subsequent trans-synaptic transfers allowed further visualization of higher order neurons further upstream to the original NPY or POMC neurons. These genetically based neuronal tracing methods, combined with the BAC transgenic technology, allow systematic mapping of the complex neuronal circuitry in the mammalian brain.

Epitope-Tagging of Proteins In Vivo for Subcellular Localization and Proteomic Studies

Epitope tags, such as Flag tag, Myc tag, or HA tag, and protein tags, such as EGFP, Protein A, or alkaline phosphatase, can be engineered in the BAC to form fusion proteins with a gene-of-interest on the BAC. These tags can be fused at either the N terminal or the C terminal of the protein. Transgenic mice derived from these constructs can be used to study protein expression using immunoblot analysis with antibodies against these tags, or to visualize the subcellular localization by immunohistochemical staining. Furthermore, the epitope tags (such as Flag, EGFP, and protein A tags) may also allow immunopurification or immunoprecipitation of protein complexes containing the tagged protein from extracts of transgenic mouse brain for further proteomic studies.

Gain-of-Function Studies: Gene Dosage Study, Dominant-Negative Studies

Genetic studies in multiple model organisms, from yeast, *Drosophila*, *C. elegans* to mammals, have demonstrated that only about one-third of the genes have obvious loss-of-function phenotypes in laboratory experimental conditions. Therefore, in most organisms, the gain-of-function genetic approach is a powerful tool to assess gene function. BACs have been used for gain-of-function studies, e.g., gene dosage studies of the *Zipro1* gene (Yang et al., 1999) and *Acha9* gene (Zuo et al., 1999). Although the loss of function of these genes does not appear to have obvious ramifications in mice, a BAC-mediated increase of gene dosages (with appropriate controls) has provided crucial functional insight on these genes.

Genetic Complementation

A relatively straightforward application of BAC transgenesis is to complement a recessive mutation in mice. BAC-mediated complementation has been used in positional cloning to identify mutant genes (Antoch et al., 1997). If a BAC transgene can complement a recessive mutation, then it implies that the transgene is expressing a functional protein in correct expression patterns in vivo.

Development of Mouse Models of Human Disease

BACs are powerful tools for developing mouse models for human diseases that result from dominant genetic mechanisms. A number of human neuropsychiatric diseases are caused by dominant genetic mutations acting as a dominant gain of function, therefore, BAC transgenesis can be used to create models for these diseases. Neurodegenerative diseases, including Huntington's disease (HD) and other polyglutamine diseases, and some familial forms of Parkinson's disease, familial Alzheimer's disease (FAD), and amyotrophic lateral sclerosis (ALS), are all due to dominant mechanisms. In HD (Hodgson et al., 1999), FAD (Lamb et al., 1993), and spinobulbar muscular atrophy (Sopher et al., 2004), YAC transgenic models have all been shown to recapitulate human neuropathology better than conventional transgenic models. Because BAC transgenesis has multiple advantages over the YAC transgenic system (see above), the BAC transgenic system is believed to be an ideal system for the development of robust mouse models for human diseases that are caused by dominant gain-of-function mutations. One important consideration in modeling human disease is the choice between using human BACs versus mouse BACs containing the disease gene. Since human disease occurs in the context of the human disease protein, which may have sequence differences from the mouse homolog of the disease protein, it is very common to use human disease genes rather than the mouse genes to develop disease models in mice.

BAC Transgenesis in Other Organisms

BAC transgenesis can also be applied to other organisms including zebrafish, rat, rabbit, sheep, and cow. In these organisms, ES cells are not available but transgenesis by pronuclear microinjection is possible. BAC transgenic technology has been successfully applied to the zebrafish system (Jessen et al., 1998). Future applications of BAC transgenic technology in other mammalian models, such as rat, may lead to the development of better models to study the mammalian brain.

Contributors: X. William Yang and Shiaoching Gong

Modification of Bacterial Artificial Chromosomes (BACs) and Preparation of Intact BAC DNA for Generation of Transgenic Mice

BASIC PROTOCOL

MODIFICATION OF BACs USING THE pLD53.SC-AB SHUTTLE VECTOR

Materials *(see APPENDIX 1 for items with ✓)*

pLD53.SC-AB shuttle vector (Fig. 5.16.1; available from N. Heintz, Rockefeller University; Email: walshj@mail.rockefeller.edu)
Pir2 competent cells
Plasmid DNA isolation kit (miniprep and midiprep kits, Qiagen) containing:
 Buffer P1
 Buffer P2
 Buffer P3
*Asc*I restriction enzyme (New England Biolabs)
*Not*I restriction enzyme (New England Biolabs)
Shrimp alkaline phosphatase (SAP; Roche Diagnostics)
SeaPlaque low-melting agarose (Cambrex)
DNA gel purification kit (e.g., Geneclean, Bio101)
High DNA mass ladder (Invitrogen)
BAC clone (e.g., BACPAC Resource Center at Oakland Children's Hospital)
DNA ligation kit (e.g., Version 2, Takara)
✓ LB/Amp plates
✓ Alkaline lysis solution
✓ Luria broth (LB) medium
10% (v/v) glycerol, cold
Liquid nitrogen or 100% ethanol
SOC medium (BD Diagnostics)
Ampicillin (Amp) and chloramphenicol (Chl)
LB/Chl medium: LB containing 20 μg/ml Chl
LB/Chl/Amp medium: LB containing 20 μg/ml Chl and 30 μg/ml Amp
LB/Chl/sucrose plates: LB-agar plates supplemented with 20 μg/ml Chl and 5% sucrose
LB/Chl plate: LB-agar plates supplemented with 20 μg/ml Chl
LB/Chl/Amp plates: LB-agar plates supplemented with 20 μg/ml Chl and 30 μg/ml Amp
10 mM EDTA, pH 8.0
✓ 2 M potassium acetate, cold
100% isopropanol
✓ TE buffer
70% and 100% ethanol
Cesium chloride (CsCl)
10 mg/ml ethidium bromide
✓ NaCl-saturated butanol
Embryo transfer water (Sigma)
3 M sodium acetate

PI-*Sce*I enzyme and 10× buffer (New England Biolabs)
✓ Injection buffer

50-ml centrifuge tubes
Electroporator (Micropulser, BioRad)
0.1-cm electroporator cuvette
17× 100–mm polypropylene round-bottom tube
Incubator shaker
Pulsed-field gel system
250- and 500-ml bottles
5/8 × 3–in. Beckman Quickseal tube
10-ml syringe
18- and 23-G needles
Ultracentrifuge and NVT65 rotor (Beckman)
2.0-ml microcentrifuge tubes
25-mm, 0.025-μm filters (Millipore)
100-mm petri dishes
Kodak Gel Logic 100 imaging system

1. Transform 1 μl of pLD53.SC-AB shuttle vector into Pir2 cells using standard chemical transformation, and isolate DNA using a plasmid DNA maxi isolation kit.

2. Digest 5 to 10 μg of the shuttle vector with *Asc*I/*Not*I by incubating overnight with appropriate amounts of enzyme at 37°C. Treat the digested DNA with 1 μl (1 U) shrimp alkaline phosphatase (SAP) for 30 min at 37°C and inactivate the SAP for 30 min at 65°C.

3. Gel-purify the digested vector fragment on a low-melting-temperature agarose gel using a DNA gel purification kit, and determine the concentration of a small aliquot of the purified DNA by running it against a DNA standard. For this protocol, use 100 to 200 ng of purified *Asc*I/*Not*I–digested shuttle vector fragment for subcloning into the recombination cassette.

4. Construct a recombination cassette containing two homology boxes (A and B) and a modification (gene insertion, deletion, or point mutation) located between boxes A and B.

   The recombination boxes are ~200- to 500-bp each and are usually amplified from the BAC by PCR. The recombination cassette is usually constructed in a high-copy cloning plasmid, such as pBluescript (Stratagene). The AscI and NotI sites are usually located in the A and B boxes so that double digestion with these enzymes will allow for the excision of the recombination cassette.

5. Digest the plasmid containing the recombination cassette with *Asc*I/*Not*I, and gel-purify the fragment (as in step 3). Ligate 100 ng of *Asc*I/*Not*I–digested shuttle vector with 100 to 300 ng *Asc*I/*Not*I–digested recombination cassette fragment using a DNA ligation kit for 1 hr at room temperature or overnight at 15°C.

6. Transform the ligated DNA into Pir2 cells by chemical transformation and plate the transformed cells on LB/Amp plates. Individually pick a few colonies and test by PCR to amplify the A or B boxes, or the modification regions, and select colonies with the correct insert.

7. Prepare shuttle vector DNA from the positive clones using a conventional alkaline lysis method (*APPENDIX 2F*) or a miniprep kit, digest the plasmids with *Asc*I/*Not*I, and electrophorese to ensure the presence of the proper insert in the shuttle vector.

8. (*Optional*) Sequence the DNA prep to confirm the sequences with the modification cassette in the shuttle vector.

9. Prepare a midiprep of the shuttle vector DNA using a plasmid purification midiprep kit. Store DNA at −20°C.

10. Inoculate 200 ml of LB medium with 1/1000 volume of a fresh overnight BAC culture. Grow cells at 37°C with vigorous shaking to an OD_{600} of ∼0.7 (5 to 6 hr).

11. Harvest cells by centrifuging in a cold rotor 10 min at 4000 × *g*, −5°C. Decant supernatant and resuspend pellets in an equal volume of 10% cold glycerol (autoclaved and stored at 4°C). Transfer cells to a new 50-ml centrifuge tube.

12. Repeat step 11 two times. Decant the supernatant and gently resuspend cells to a final volume of 400 μl with 10% cold glycerol.

13. Dispense 40-μl aliquots of the electrocompetent cells into sterile 1.5-ml microcentrifuge tubes and snap-freeze in liquid nitrogen or 100% ethanol and dry ice and store up to 6 months at −80°C.

14. Thaw one 40-μl aliquot of BAC electrocompetent cells on ice, mix with 2 μl of 0.5 μg/μl DNA, and place the mixture on ice for 1 min. Transfer each sample to an ice-cold 0.1-cm electroporation cuvette. Set the electroporator capacitance at 25 μF, the voltage at 1.8 kV, and resistance to 200 W. Electroporate sample.

15. After electroporation, add 1 ml of SOC medium to each cuvette. Gently resuspend the cells, and transfer the cell suspension to a 17 × 100–mm polypropylene tube, and incubate 1 hr at 37°C with shaking at 225 rpm to allow cells to recover from electroporation. Add 5 ml of LB/Chl/Amp medium and shake overnight at 37°C at 300 rpm.

16. Dilute the overnight culture 1:1000 by adding 5 μl of the overnight culture into 5 ml LB/Chl/Amp medium and grow for an additional 8 to 14 hr at 37°C. Make a serial dilution (e.g., 1:1000; 1:2500; 1:5000), and plate 10 μl and 100 μl of each dilution on LB/Chl/Amp plates, and incubate overnight at 37°C.

17. Pick ten single colonies per plate and inoculate each colony into 5 ml of LB LB/Chl/Amp medium. Streak the same colonies onto LB/Chl/Amp master plates and grow overnight at 37°C. Identify clones with proper co-integrate BACs by colony PCR or by Southern blot analysis.

18. Pick two to three independent bacteria clones containing co-integrate BACs from the LB/Chl/Amp master plates, inoculate each colony into 1 ml LB/Chl medium and incubate for 1 hr at 37°C with shaking at 225 rpm.

The following steps describe a final step of BAC modification, which is to select bacteria that have undergone a second homologous recombination to excise the shuttle vectors and yield a correctly modified BAC (termed resolution). Bacteria containing correctly resolved BACs can be readily selected by growth on plates containing chloramphenicol and sucrose, and can be further confirmed by colony PCR reactions.

19. Spread 100 μl and 10 μl of each mixture onto LB/Chl/sucrose plates and incubate overnight at 37°C. Pick both the large and small colonies, plate them onto LB/Chl plates, and also streak same colonies onto LB/Chl/Amp plates. Incubate the plates overnight at 37°C.

20. To identify the correctly resolved clones, perform PCR. To further confirm that the resolved BACs are correct, perform Southern blot analysis using probes specific to the modification (e.g., EGFP) on the DNA prepared from resolved BAC clones.

21. Use one of the following methods to analyze the resolved BACs prior to preparation of BAC DNA for microinjection:

a. *Pulsed-field gel analysis*. Prepare BAC DNA, digest with *Not*I and other rare-cutting enzymes that are introduced to the modification region of the BAC (e.g., *Mlu*I, *Pme*), separate using pulsed-field gel electrophoresis (PFGE; BioRad, CHEF II system). Use the following pulsed-field gel running conditions:

0.5× TBE gel in 0.5× TBE buffer;
initial time: 5 sec;
final time: 15 sec;
field strength: 6 V/cm;
running time: 18 hr (with current at 140 to 200 mA).

b. *Fingerprinting of resolved BACs*. To determine whether any gross rearrangements or deletions were generated during the modification, perform unique restriction enzyme digestion, i.e., with *Eco*RI, *Hind*III, *Xba*I, and *Xho*I, followed by high-resolution fingerprinting of the original and modified BACs.

22. Pick a single colony from a freshly streaked LB/Chl/Amp plate and inoculate 3 ml LB/Chl/Amp medium. Incubate 8 hr at 37°C. Transfer 0.4 to 1.0 ml (depending on cell density) into 500 ml of LB/Chl/Amp medium and incubate overnight at 30°C.

23. Centrifuge bacteria 30 min at 6000 × *g*, 4°C. Decant the supernatant by inverting the centrifuge tube, resuspend cells in 40 ml of 10 mM EDTA, pH 8.0 and transfer to a 250-ml bottle. Add 80 ml of alkaline lysis solution. Mix by gently swirling and incubate 5 min at room temperature.

24. Add 60 ml of cold 2 M potassium acetate. Mix by gently swirling and incubate on ice 5 min. Centrifuge 30 min at 20,000 × *g*, 4°C. Transfer supernatant into a 500-ml bottle, add 180 ml isopropanol. Mix by gently swirling and centrifuge 30 min at 3850 × *g*, 4°C. Decant the supernatant and dissolve DNA pellet in 18 ml of 10:50 TE buffer.

25. Add 9 ml of 7.5 M potassium acetate, mix, and incubate 30 min at −70°C. Thaw solution at room temperature and centrifuge 10 min at 3000 × *g*, 4°C. Transfer supernatant to a new tube and add 2.5 vol 100% ethanol. Centrifuge 30 min at 3000 × *g*, 4°C, to precipitate the DNA.

26. Carefully decant supernatant and gently resuspend pellet (while still wet) in 4.4 ml TE buffer. Dissolve 10.2 g of CsCl in another 4.4 ml of TE buffer and gently add this solution to the 4.4 ml of DNA until the CsCl is completely dissolved.

27. Add 0.2 ml of 10 mg/ml ethidium bromide and immediately mix. Using a 10-ml syringe equipped with an 18-G needle, remove solution and load into a 5/8 × 3–in. Beckman Quickseal tube. Seal tubes and place in an ultracentrifuge (carefully equilibrate tubes to be centrifuged in opposing positions by weighing on a precision balance and assuring a weight differential of <0.05 g). Centrifuge overnight (>8 hr) at 340,000 × *g*, 18°C.

28. Remove centrifuge tubes from the rotor very gently so as not to perturb the gradients. Use a hand-held UV light to visualize the ethidium bromide-stained band. Using a 23-G needle, poke a hole in the top of the tube. With an 18-G needle, carefully remove the band (bottom band if there are two) with the needle bevel up. Remove only the band (usually ~200 μl). Place in a clean tube and bring up to 2 ml with TE buffer.

29. Extract four to five times with NaCl-saturated butanol until the pink color is gone by adding 10 ml NaCl-saturated butanol, mixing gently using a wide-bore pipet tip, allowing the solution to sit 30 sec to allow for separation, and removing and discarding the top layer.

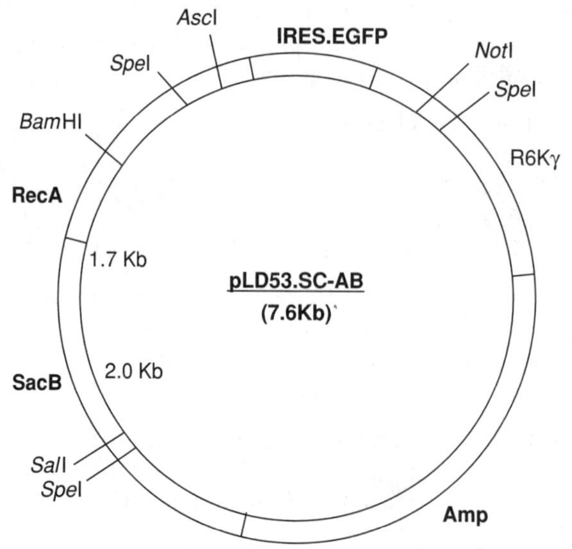

Figure 5.16.1 Map of the pLD53.SC-AB vector. This shuttle vector contains a conditional R6Kγ origin of replication, a RecA gene to supports homologous recombination, SacB for negative selection on sucrose, and an Amp gene for positive selection. The recombination cassette can be cloned between *Asc*I and *Not*I sites.

30. Add 1 ml embryo transfer water and 2.5 to 3.0 vol of 100% ethanol and mix. Place 30 min at −20°C. Centrifuge DNA 30 min at 3850 × *g*, 4°C, and resuspend in 0.5 ml of 0.3 M sodium acetate.

31. Transfer to a 2.0-ml microcentrifuge tube (while still wet) and add 1 ml 100% ethanol. Centrifuge DNA 30 min at 3850 × *g*, 4°C. Discard supernatant, fill tube with 70% ethanol, and let sit 5 min at room temperature. Centrifuge again, decant supernatant, and dry the DNA pellet 1 min at room temperature. Use paper towels to remove trace amounts of ethanol. Resuspend gently in 20 to 40 μl of TE buffer. Place DNA for 20 to 30 min in 37°C incubator. Store BAC DNA up to 6 months at 4°C.

32. Linearize 5 to 10 μl (~10 μg) BAC DNA with 2 μl PI-*Sce*I enzyme, 5 μl of 10× buffer, and water to 50 μl. Incubate 3 to 4 hr at 37°C.

33. Dialyze by loading the DNA on top of a 25-mm, 0.025-μm filter and floating the filter on 20 ml of injection buffer in a 100-mm petri dish with the shiny side up for at least 4 hr at room temperature. Transfer the dialyzed DNA to a microcentrifuge tube and add injection buffer to attain the original volume.

34. Dilute a standard BAC preparation to 1 ng/μl, 2 ng/μl, 4 ng/μl, 8 ng/μl, and 16 ng/μl. Electrophorese the sample BAC preparation along with these standards on a pulsed-field gel and estimate the concentration of sample BAC DNA by comparing the intensity of its band to those of the BAC standards using a UV gel-imaging system (e.g., Kodak Gel Logic 100 imaging system). Dilute the sample BAC DNA to 2 μg/ml using injection buffer and use for pronuclear microinjection within 1 week of preparation.

ALTERNATE PROTOCOL

PREPARATION OF BAC DNA BY ALKALINE LYSIS AND SEPHAROSE CL-4B CHROMATOGRAPHY

Additional Materials *(also see Basic Protocol)*

Freshly streaked LB/Chl plates (see Basic Protocol)
100% isopropanol (Sigma)
1:1 (v/v) phenol (pH 8.0)/chloroform
Sepharose CL-4B resin
10× DNA dye
Ethidium bromide: 30 μl of 10 mg/ml stock in 500 ml water

70-μm cell strainer
Vacuum
30-ml centrifuge tubes
5-ml sterile plastic pipet (Falcon)
24-well plate
Parafilm

1. Pick a single colony from a freshly streaked LB/Chl plate (20 μg Chl/ml). Grow 4-ml starter culture in LB/Chl medium (20 μg Chl/ml) and incubate with shaking overnight in a 30°C incubator.
2. Pour the 4-ml starter culture into 1 liter (1:250 dilution) of LB/Chl medium (12.5 μg Chl/ml) and incubate 3 to 4 hr at 37°C until OD_{600} reaches ~1.2 to 1.4.
3. Centrifuge bacteria 20 min at 3500 × *g*, 4°C (use 500-ml centrifuge tubes and centrifuge two times per sample to remove residual LB). Carefully remove all traces of LB medium by inverting the centrifuge tubes onto a paper towel. Resuspend bacterial pellet in 30 ml P1 buffer and gently pipet up and down to ensure all clumps are resuspended.
4. Add 30 ml P2 buffer, mix gently by inverting tube ten times, and incubate 5 min at room temperature. Add 30 ml ice-cold P3 buffer, mix gently by inverting tube ten times, incubate on ice for 30 min, and centrifuge 30 min at 11,000 × *g*, 4°C, to remove debris.
5. Filter the supernantant through a sterile 70-μm cell strainer into a new 250-ml centrifuge tube. Precipitate the DNA by adding 63 ml (0.7 × vol) 100% isopropanol and centrifuge 45 min at 11,000 × *g*, 4°C. Carefully decant the supernantant and invert the tube on top of a paper towel for 5 min.
6. Dry the pellet in a vacuum for 2 min. Dissolve the pellet in 1 ml TE buffer and carefully pipet up and down a few times using a wide-bore pipet tip.
7. Transfer the DNA into a new 30-ml centrifuge tube. Add 9 ml TE buffer to make a total volume of 10 ml. Add 10 ml of freshly prepared 1:1 phenol (pH 8.0)/chloroform and mix gently by inverting tube ten times. Centrifuge tubes for 12 min at 11,000 × *g*, 4°C.
8. Carefully remove the supernatant using a 10-ml pipet (remove the top 80% of the supernatant and do not touch the interface) and transfer the DNA into a new 30-ml centrifuge tube. Add 0.1 vol of 3 M sodium acetate, pH 7.0, 2× vol of 100% ethanol, mix and centrifuge 40 min at 11,000 × *g*, 4°C.

9. Decant supernatant and wash the pellet in cold 70% ethanol. Dry the pellet by first inverting the tube on a paper towel for 5 min and then air drying or drying in a vacuum for 2 min. Dissolve the DNA in 0.8 ml injection buffer. Incubate 30 min at 65°C to ensure that the DNA is completely dissolved.

The DNA can be stored up to 1 month at 4°C before further analysis and purification.

10. Digest 1/3 to 1/2 of the BAC DNA sample with PI-*Sce*I as described in Basic Protocol, step 32.

11. Place ~20 ml of Sepharose CL-4B resin into a 50-ml centrifuge tube and centrifuge 5 min at 3500 × *g*, 4°C. Carefully decant supernatant.

12. Add 20 ml injection buffer and resuspend the Sepharose CL-4B by inverting tube several times. Centrifuge 5 min at 3500 × *g*, 4°C. Carefully decant the supernatant. Repeat two additional times. Resuspend the Sepharose CL-4B by adding 10 ml injection buffer.

13. Load the Sepharose CL-4B column by taking a 5-ml sterile plastic pipet, blowing air into it so that the cotton moves to the tip, and clamping the pipet to a stand. Add the Sepharose CL-4B until the packed resin reaches the −2-ml mark (a total of 7 ml). Never let the resin dry. Place a 10-ml syringe on top of the column to act as the buffer resevoir. Run 1 to 2 ml of injection buffer to ensure the column flows properly.

14. Add 5 µl of 10× DNA dye into the 0.5-ml circular or digested BAC DNA. Carefully remove the reservoir and gently add the DNA onto the top of the Sepharose CL-4B column with a wide-bore tip Pasteur pipet. Allow the DNA to enter the column, then gently add 0.5 ml of injection buffer on top of the column.

15. Reset the reservoir containing 10 ml injection buffer on top of the column. Start collecting 0.36 ml per well (~6 drops, which takes ~3 min per well) into wells of a sterile 24-well plate. Collect ~18 fractions (until the DNA dye is at the 2-ml mark). Seal the plate with Parafilm and store the BAC DNA up to 1 month at 4°C.

16. Electrophorese 40 µl of fractions 3 to 16 on a pulsed-field gel to identify the appropriate fractions with most of the intact BAC DNA and the least amount of degradation. Stain the gel with ethidium bromide for 30 min with gentle shaking.

17. Dialyze the DNA on a 25-mm, 0.025-µm filter floating on 20 ml of injection buffer with the shiny side up for 3 to 4 hr. Store purified BAC DNA fraction up to 1 month at 4°C.

18. Determine DNA concentration as described in Basic Protocol, step 34. Dilute the purified BAC DNA with injection buffer to a final concentration of 1 to 2 µg/ml prior to pronuclear microinjection into murine zygotes.

Contributors: Shiaoching Gong and X. William Yang

APPENDIX 1

Reagents and Solutions

This appendix includes recipes for all reagents, media, and solutions used in *Short Protocols in Neuroscience: Cellular and Molecular Methods*. Recipes are listed alphabetically, with the unit(s) in which the solution is used listed parenthetically. No unit numbers are indicated for commonly used solutions. In some cases (e.g., PBS), there is a recipe for the commonly used version followed by other formulations that are unique to specific units. Recipes for some solutions include reagents for which separate recipes are included elsewhere in this appendix; these are indicated by a ✓ or (see recipe). It is critical that the correct recipe be used for each protocol and that the correct base or stock solutions be used in preparing the final solution.

When preparing solutions, use deionized, distilled water and reagents of the highest available grade. Sterilization—by filtration through a 0.22-μm filter or by autoclaving—is recommended for most applications, especially for cell culture. Discard any reagent that shows evidence of contamination, precipitation, or discoloration or has exceeded the recommended shelf life. General information about the molarities and specific gravities of acids and bases can be found in Table A.1.1.

CAUTION: Follow standard laboratory safety guidelines and heed manufacturers' precautions when working with hazardous chemicals.

Acid, concentrated stock solutions
See Table A.1.1.

Acridine orange staining solution (*UNIT 5.3*)
Dissolve 0.1 g acridine orange in 100 ml H_2O. Add 3 ml of this solution to 97 ml CPBS (see recipe). Make fresh and protect from light.

Adenovirus storage buffer, 2× (*UNIT 4.11*)
✓ 10 mM Tris·Cl, pH 8.0
100 mM NaCl
1 mM $MgCl_2$
50% (v/v) glycerol
0.1% (w/v) BSA
Filter sterilize
Store up to 1 year at 4°C

Alkaline lysis solution (*UNIT 5.16*)
0.2 N NaOH
✓ 1% (w/v) SDS
Prepare fresh

A medium, 5× (*UNIT 4.2*)
5 g $(NH_4)_2SO_4$
22.5 g KH_2PO_4
52.5 g K_2HPO_4
2.5 g sodium citrate.$2H_2O$
H_2O to 1 liter

Table A.1.1 Molarities and Specific Gravities of Concentrated Acids and Bases[a]

Acid/base	Molecular weight	% by weight	Molarity (approx.)	1 M solution (ml/liter)	Specific gravity
Acetic acid (glacial)	60.05	99.6	17.4	57.5	1.05
Ammonium hydroxide	35.0	28	14.8	67.6	0.90
Formic acid	46.03	90	23.6	42.4	1.205
		98	25.9	38.5	1.22
Hydrochloric acid	36.46	36	11.6	85.9	1.18
Nitric acid	63.01	70	15.7	63.7	1.42
Perchloric acid	100.46	60	9.2	108.8	1.54
		72	12.2	82.1	1.70
Phosphoric acid	98.00	85	14.7	67.8	1.70
Sulfuric acid	98.07	98	18.3	54.5	1.835

[a] *CAUTION:* Handle strong acids and bases carefully.

Sterilize by autoclaving
Store indefinitely at room temperature
Before use, dilute to 1× with sterile water and add the following sterile solutions, per liter:
1 ml 1 M $MgSO_4.7H_2O$
10 ml 20% carbon source (sugar or glycerol)

Ammonium hydroxide, concentrated stock solution
See Table A.1.1.

Ammonium sulfate ($[NH_4]_2SO_4$), saturated, pH 7.0 *(UNIT 4.11)*

Add 450 g $(NH_4)_2SO_4$ to 500 ml water. Heat on a stir plate until $(NH_4)_2SO_4$ dissolves completely. Filter through Whatman paper while still warm and allow to cool (upon cooling crystals will form which should not be removed). Adjust pH to pH 7.0 with ammonium hydroxide. Store up to 1 year at 4°C.

Antibody buffer *(UNIT 3.3)*

Prepare 0.5% (v/v) Tween 20 in PBS (see recipe for PBS) containing 3% (v/v) normal serum from species in which secondary antibody was produced. Prepare immediately before use.

Antibody dilution buffer *(UNIT 2.2)*
✓ *PBS, containing:*
10% (v/v) normal goat serum
0.05% (v/v) Triton X-100 *or* saponin
Prepare fresh

Use same detergent as that used in blocking/permeabilization buffer (see below).

AP (alkaline phosphatase) buffer *(UNIT 4.14)*
✓ 100 mM Tris·Cl, pH 8.5
150 mM NaCl
5 mM $MgCl_2$
Store up to 1 year at room temperature

APH (aqueous prehybridization/hybridization) solution *(UNIT 4.3)*
✓ 0.5 M sodium phosphate buffer, pH 7.4
1% (w/v) BSA
✓ 4% (w/v) SDS
Store up to 1 month at 4°C

If SDS precipitates during storage, warm to 37°C until dissolved.

AP substrate solution *(UNIT 4.14)*

Prepare fresh: Dissolve one tablet BCIP in 0.5 ml 100% dimethylformamide (final BCIP 50 mg/ml). Dissolve one tablet NBT in 1 ml water (final NBT 10 mg/ml). Add 330 µl NBT solution to 10 ml AP buffer (see recipe). Mix and then add 33 µl BCIP solution. Prepare fresh.

Artificial cerebrospinal fluid (aCSF)
Monkey
Dissolve in 900 ml H_2O:
268 mg $Na_2HPO_4{\cdot}7H_2O$ (1.0 mM final)
8.6 g NaCl (147 mM final)
0.22 g KCl (3 mM final)
0.19 g $CaCl_2{\cdot}2H_2O$ (1.3 mM final)
0.20 g $MgCl_2{\cdot}6H_2O$ (1.0 mM final)
35 mg ascorbic acid (weigh accurately; 0.20 mM final)
Add H_2O to 1 liter
Filter sterilize
Make fresh the day of the experiment

If this solution has been prepared correctly, the pH should be 7.4 ± 0.1. To achieve the desired pH, the ascorbic acid must be weighed accurately, as even small variations may affect pH drastically. If necessary, adjust pH to 7.4 with small volumes of 100 mM mono- or dibasic sodium phosphate. If pH is significantly different from 7.4, prepare new solution.

Rodent
Dissolve in 900 ml H_2O:
7.25 g NaCl (124 mM final)
0.37 g KCl (5 mM final)
0.015 g $CaCl_2{\cdot}2H_2O$ (0.1 mM final)
0.65 g $MgCl_2{\cdot}6H_2O$ (3.2 mM final)
2.18 g $NaHCO_3$ (26 mM final)
1.8 g glucose (10 mM final)
Adjust pH to 7.4 with 0.1 M NaOH or HCl
H_2O to 1 liter
Filter sterilize
Prepare fresh on day of experiment

aCSF is a balanced salt solution which can vary according to species and cell type. A balanced salt solution is an isotonic solution containing appropriate ions and buffers suitable for maintaining neurons in vitro. Consult a physiology reference text to formulate an appropriate solution for the animal being studied. A general Ringer's solution (e.g., Life Technologies) is often substituted for aCSF, but the requirement for serum should be considered.

Ascending chromatography buffer *(UNIT 5.10)*
15 parts (v/v) pyridine
10 parts (v/v) 1-butanol
3 parts (v/v) glacial acetic acid
12 parts (v/v) H_2O
Store up to 6 months at room temperature

Bio948, 500 pM *(UNIT 4.17)*
500 pM biotinylated control oligonucleotide Bio948 (5'-GTCAA-GATGCTACCGT-TCAG-3') (custom synthesis)
6× SSPE (Bio-Whittaker)
0.1 mg/ml herring sperm DNA (Promega)
0.005% (v/v) Triton X-100 (Sigma; molecular biology grade)
✓ Prepare in RNase-free H_2O (DEPC-treated)
Store up to 1 year at −20°C

Blocking buffer appropriate for membrane and detection protocol *(UNIT 5.7)*
Colorimetric detection

For nitrocellulose and PVDF: 0.1% (v/v) Tween 20 in TBS (TTBS). Store <1 week at 4°C.

For neutral and positively charged nylon: Tris-buffered saline (TBS; see recipe) containing 10% (w/v) nonfat dry milk. Prepare just before use.

Luminescence detection

For nitrocellulose, PVDF, and neutral nylon (e.g., Pall Biodyne A): 0.2% casein (e.g., Hammarsten grade or I-Block; Applied Biosystems) in TTBS. Prepare just before use.

For positively charged nylon: 6% (w/v) casein/ 1% (v/v) polyvinyl pyrrolidone (PVP) in TTBS. Prepare just before use. With constant mixing, add casein and PVP to warm (65°C) TTBS. Stir for 5 min. Cool before use.

Blocking/permeabilization buffer *(UNIT 2.2)*

PBS, containing:
10% (v/v) normal goat serum
0.2% (v/v) Triton X-100 *or* saponin
Prepare fresh

Triton X-100 is commonly used as the detergent in this step; however, for some proteins, especially plasma membrane proteins, saponin is preferred.

Blocking solution *(UNIT 3.3)*

Prepare 0.5% (v/v) Triton X-100 in PBS (see recipe for PBS) containing 3% (v/v) normal serum from species in which secondary antibody was produced. Prepare immediately before use.

Borate buffers, pH 8.5, 9.0, and 10 *(UNIT 5.5)*

pH 8.5 (0.1 M):
18.55 g boric acid
2850 ml H_2O
10 M NaOH to pH 8.5
H_2O to 3000 ml

pH 9.0 (0.16 M):
6.1 g $Na_2B_4O_7$ (sodium borate)
0.76 g NaCl
95 ml H_2O
10 M NaOH to pH 9.0
H_2O to 100 ml

pH 10 (0.1 M):
0.618 g boric acid
95 ml H_2O
10 M NaOH to pH 10
H_2O to 100 ml

BrdU labeling reagent *(UNIT 3.3)*

Add 20 mg/ml BrdU to sterile 0.007 N NaOH/ 0.9% NaCl. Warm gently in the hand and vortex to dissolve BrdU. Make fresh on the day of injection.

Buffer A *(UNIT 5.6)*

✓ 50 mM Tris·Cl, pH 8.0
25% (w/v) sucrose
✓ 10 mM EDTA
Store up to 1 year at 4°C

Buffer C *(UNIT 5.6)*

20 mM HEPES, pH 7.6
100 mM KCl
✓ 0.2 mM EDTA
20% (v/v) glycerol
1 mM DTT
Store up to 1 year at 4°C
Just prior to use add the following protease inhibitors:
1 μg/ml aprotinin
1 μg/ml leupeptin
1 μg/ml pepstatin A

Buffers A and B *(UNIT 4.8)*

Buffer A:
✓ 10 mM Tris·Cl, pH 7.5
1 mM $MgCl_2$
135 mM NaCl
Make fresh on day of use
Buffer B:
Add 10% (v/v) glycerol to Buffer A
Make fresh on day of use

$CaCl_2$, 0.5 M *(UNIT 4.10)*

36.7 g $CaCl_2 \cdot 2H_2O$
H_2O to 500 ml
Filter sterilize through a 0.45-μm nitrocellulose filter
Store up to 2 months at 4°C in 50-ml aliquots

$CaCl_2$, 2.5 M *(UNIT 4.11)*

183.7 g $CaCl_2 \cdot 2H_2O$ (Sigma; tissue culture grade)
H_2O to 500 ml
Filter sterilize through a 0.45-μm nitrocellulose filter (Nalgene)
Store at −20°C in 10-ml aliquots

This solution can be frozen and thawed repeatedly.

$CaCl_2$ solution *(APPENDIX 2H)*

60 mM $CaCl_2$
15% (v/v) glycerol
10 mM PIPES [piperazine-*N,N′*-bis(2-hydroxypropanesulfonic acid)], pH 7
Filter sterilize using a disposable filter unit, or autoclave
Store at room temperature (stable for years)

Calcium-sensitive membrane-impermeant indicator dye solution *(UNIT 2.4)*

Dissolve 1 mg of the calcium-sensitive indicator (calcium green 1 or fura-2 coupled to dextran or membrane-impermeant salts (see Table 2.4.1; Molecular Probes) in 2 to 4 μl of distilled water or Tyrode's or Ringers solution.

The dyes are very water soluble and will dissolve easily. The solution will be between 25 and 50 mM.

Addition of 0.2% to 2.5% Triton X-100 or DMSO to the Tyrode's or Ringers solution can facilitate retrograde loading.

Calcium-sensitive membrane-permeant indicator dye solution *(UNIT 2.4)*

Dissolve 50 μg calcium indicator (fura-2 AM (acetoxymethyl), calcium green 1, indo-1 AM, or fluo-3 AM (Molecular Probes.) in 50 μl DMSO.

Pipet vigorously to mix, taking care to wash the indicator from the side walls of the tube.

The solution will be ~1 mM.

Stock solutions can be stored at −20°C in anhydrous DMSO for a year or more, but in moist environments, spontaneous hydrolysis can occur.

Typically, the stock solution is diluted in physiological saline to a final concentration of 1 to 10 μM immediately before use. In certain applications, higher concentrations (100 μM) may be used (Dani et al, 1992).

CAM-KII buffer *(UNIT 5.10)*
10 mM HEPES-free acid
10 mM $MgCl_2$
800 μM $CaCl_2$
10 μg/ml calmodulin
50 μM ATP
Ajust pH to 7.0 with 10 N NaOH
Store up to 6 months at 4°C

Carbonate buffer, pH 9.6 *(UNITS 5.5 & 5.9)*
160 mg Na_2CO_3 (15 mM final)
294 mg $NaHCO_3$ (35 mM final)
20 mg NaN_3, pH 9.6 (0.2 g/liter)
H_2O to 100 ml

cDNA synthesis buffer *(UNIT 5.3)*
✓ 1× IST reaction buffer
7 mM dithiothreitol (DTT)
250 μM 4dNTP mix
Filter sterilize
0.12 U/μl RNasin (Promega)
Store up to 1 year at −20°C

Chase medium *(UNIT 4.12)*
Eagle's MEM (Life Technologies) containing:
2 mM L-glutamine
✓ 20 mM HEPES
150 μg/ml unlabeled (cold) methionine
Store up to 1 month at 4°C

Chopping buffer *(UNIT 2.2)*

Prepare DNase Type II-S (2000 U/mg protein; Sigma) according to the manufacturer's instructions and store at −20°C in 1-ml aliquots up to 6 months. On day of dissection, thaw a 1-ml aliquot and dilute 1:9 in dissection buffer (see recipe). Filter and sterilize through a 0.2-μm filter and store at 4°C until needed for preparation of cells.

The DNase is used to help prevent clumping of tissue.

Chopping solution *(UNIT 3.4)*
0.01 g DNase I (0.01% w/v final)
✓ 100 ml dissection solution
Adjust pH to 7.2 with 0.1 M NaOH
Filter sterilize through 0.2-μm filter
Divide into 10-ml aliquots
Store up to 6 months at −20°C

Citrate buffer, 50 mM *(UNIT 5.9)*
1 ml 1 M (29.41 g/100 ml H_2O) trisodium citrate
1 ml 1 M (21 g/100 ml H_2O) citric acid
18 ml water
Adjust pH to 4.5 with 1 M citric acid
Store up to 1 month at 4°C

CM (complete minimal) dropout medium *(UNIT 4.2)*
1.3 g dropout powder (Table A.1.2)
1.7 g YNB −AA/AS
5 g $(NH_4)_2SO_4$
20 g dextrose

(Alternatively, replace last three ingredients with 27 g minimal medium premix)

CM dropout powder, also known as minus or omission powder, lacks a single nutrient but contains the other nutrients listed in Table A.1.2. Complete minimal (CM) dropout medium is used to test for genes involved in biosynthetic pathways and to select for gene function in transformation experiments. To test for a gene involved in histidine biosynthesis one would determine if the yeast strain in question can grow on CM minus histidine (−His) or "histidine dropout" plates. It is convenient to make several dropout powders, each lacking a single nutrient, to avoid weighing each component separately for all the different dropout plates required in the laboratory.

It may be preferable to use a 10× solution of dropout powder (i.e., 13 g of dropout powder in 100 ml water) that has been "sterilized" separately (and added to the other ingredients after autoclaving) to improve the growth rate in this medium.

CMF-HBSS (calcium- and magnesium-free Hank's buffered salt solution)

HBSS (see recipe) prepared omitting all calcium and magnesium salts.

4CN visualization solution *(UNIT 5.7)*

Mix 20 ml ice-cold methanol with 60 mg 4CN. Separately mix 60 ul of 30% H_2O_2 with 100 ml TBS at room temperature. Rapidly mix the two solutions and use immediately.

CNBr/acetonitrile *(UNIT 5.12)*

To 25 g of cyanogen bromide *(CNBr should be white, not yellow, crystals)*, add 50 ml acetonitrile to make a 62.5% (w/v) solution. This may be stored indefinitely at −20°C in a desiccator over silica. Allow to warm before opening.

CAUTION: CNBr is a highly toxic lachrymator; handle in fume hood.

Co-immunoprecipitation buffer *(UNIT 5.13)*
✓ 50 mM Tris·Cl, pH 7.5
15 mM EGTA
100 mM NaCl
0.1% (w/v) Triton X-100
Store at 4°C
Immediately before use add:
✓ 1× protease inhibitor mix
1 mM dithiothreitol (DTT)
1 mM phenylmethylsulfonyl fluoride (PMSF; from fresh 250 mM solution in 95% ethanol)

Table A.1.2 Nutrient Concentrations for Dropout Powders[a]

Nutrient[b]	Amount in dropout powder(g)[c]	Final conc. in prepared media (μg/ml)	Liquid stock conc. (mg/100 ml)[d]
Adenine (hemisulfate salt)	2.5	40	500
L-arginine (HCl)	1.2	20	240
L-aspartic acid[e]	6.0	100	1200
L-glutamic acid (monosodium salt)	6.0	100	1200
L-histidine	1.2	20	240
L-leucine[f]	3.6	60	720
L-lysine (mono-HCl)	1.8	30	360
L-methionine	1.2	20	240
L-phenylalanine	3.0	50	600
L-serine	22.5	375	4500
L-threonine[e]	12.0	200	2400
L-tryptophan	2.4	40	480
L-tyrosine	1.8	30	180[f]
L-valine	9.0	150	1800
Uracil	1.2	20	240

[a]CM dropout powder lacks a single nutrient but contains the other nutrients listed in this table. Nomenclature in this manual refers to, e.g., a preparation that omits histidine as histidine dropout powder. Conditions from Sherman et al., 1979.

[b]Amino acids not listed here can be added to a final concentration of 40 μg/ml (40 mg/liter).

[c]Grind powders into a homogeneous mixture with a clean, dry mortar and pestle. Store in a clean, dry bottle or a covered flask.

[d]Use 8.3 ml/liter of each stock for special nutritional requirements. Store adenine, aspartic acid, glutamic acid, leucine, phenylalanine, tyrosine, and uracil solutions at room temperature. All others should be stored at 4°C.

[e]While these amino acids can be used reliably when included in autoclaved media, they supplement growth better when added after autoclaving.

[f]Use 16.6 ml/liter for L-tyrosine nutritional requirement.

The protease inhibitor mix, PMSF, and DTT should be added fresh at the time of experimentation. The mixture without those components can be stored for months at 4°C with the addition of 1 mM sodium azide. PMSF is labile in aqueous buffer and should be added at the last minute.

Color buffer 1 *(UNIT 4.8)*
- ✓ 100 mM Tris-HCl
- 150 mM NaCl
- Adjust pH to 7.5 with 1 M NaOH
- Store at room temperature up to 6 months

Color buffer 2 *(UNIT 4.8)*

At least 1 hr prior to use, dissolve blocking reagent from the Dig DNA labeling and detection kit (Roche) in color buffer 1(see recipe). Because the reagent does not dissolve rapidly, use mild heat to assist dissolution (50° to 70°C). Store up to 6 months at −20°C.

Color buffer 3 *(UNIT 4.8)*
- ✓ 100 mM Tris-HCl
- 100 mM NaCl
- 50 mM $MgCl_2$
- Adjust pH to 9.5 with 1 M NaOH
- Store up to 6 months at room temperature

Color buffer 4 *(UNIT 4.8)*
- ✓ 10 mM Tris-HCl
- ✓ 1 mM EDTA
- Adjust pH to 8.0 with 1 M NaOH
- Store at room temperature up to 6 months

Complete BHK-21 cell medium *(UNIT 4.12)*
- 45% (v/v) DMEM (Life Technologies)
- 45% (v/v) Iscove's modified Dulbecco's medium (IMDM; Life Technologies)
- 10% (v/v) fetal bovine serum (FBS), heat-inactivated (15 min at 55°C)
- 4 mM L-glutamine
- Store up to 1 month at 4°C

Complete DMEM/10% (v/v) FBS and DMEM/2% (v/v) FBS media *(UNIT 4.11)*
- *Dulbecco's Modified Eagle Medium (DMEM; Life Technologies) containing:*
- 4.5 g/liter glucose
- 10% or 2% (v/v) fetal bovine serum (FBS)
- 100 U/ml penicillin
- 100 μg/ml streptomycin sulfate
- Store up to 2 months at 4°C

Complete IMDM/10% (v/v) FBS medium (*UNIT 4.11*)
Iscove's Modified Dulbecco's Medium (Life Technologies) containing:
10% (v/v) fetal bovine serum (FBS)
100 U/ml penicillin
100 μg/ml streptomycin sulfate
Store up to 2 months at 4°C

Complete RPMI (*APPENDIX 2I*)
RPMI 1640 medium (e.g., Life Technologies) containing:
2%, 5%, 10%, 15%, or 20% FBS, heat-inactivated (optional)
✓ 2 mM L-glutamine
100 U/ml penicillin
100 μg/ml streptomycin sulfate
Filter sterilize and store ≤1 month at 4°C

Coverslips, poly-L-lysine or poly-L-ornithine treated (*UNIT 4.8*)

Clean coverslips: Prepare a solution of sulfo-chromic acid by dissolving 100 g of potassium dichromate in 850 ml distilled water. Add 50 ml sulfuric acid, allow to cool for several minutes then add an additional 50 ml of sulfuric acid. Adjust volume to 1000 ml with distilled water. Place 22 × 22–mm or 16-mm diameter glass coverslips in a polytetrafluoroethylene (PTFE) rack and wash twice in distilled water, each time for 10 min. Place racks into sulfo-chromic acid for 18 to 36 hr. Wash off all traces of sulfo-chromic acid under running tap water and leave in distilled water for 2 hr. Wash again with distilled water for an additional 2 hr. Wrap the rack in aluminum foil and bake for 1 hr at 180°C, and then allow to cool to room temperature. In a laminar flow cabinet, place coverslips in 35-mm or 22-mm diameter tissue culture plates using sterile, fine, curved forceps.

Coat cleaned coverslips in 0.1 mg/ml poly-L-lysine solution for neocortical neuronal or glial cultures or 0.1 mg/ml poly-L-ornithine solution for ventral-mesencephalic cultures. If the coverslips are not to be used immediately, they should be stored dry after washing in distilled water. They can be stored up to 2 months in a sterile container (Lowenstein et al., 1996).

To prepare poly-L-lysine coated coverslips: Prepare a filter-sterilized stock solution of poly-L-lysine by dissolving 10 mg poly-L-lysine in 0.15 M sodium tetraborate (pH 8.5). Dilute the poly-L-lysine stock solution 10-fold with sterile distilled water to obtain a working solution of 0.1 mg/ml. Using a sterile Pasteur pipet, add 5 to 7 drops of the working solution of poly-L-lysine to each coverslip. If cleaned correctly, the poly-L-lysine will spread out to cover the entire coverslip. Allow to dry in laminar flow cabinet overnight. Wash with sterile distilled water twice for 2 hr. Add 1 ml of plating medium (see recipe) and incubate at 37°C in an atmosphere of 5% CO_2 until used.

To prepare poly-L-ornithine/laminin coated coverslips: Prepare a filter-sterilized stock solution of poly-L-ornithine by dissolving 10 mg poly-L-ornithine hydrobromide in 0.15 M sodium tetraborate (pH 8.5). Dilute the poly-L-ornithine stock solution 10-fold with sterile distilled water to obtain a working solution of 0.1 mg/ml. Add 1 ml/well of poly-L-ornithine working solution to 24-well plates or glass coverslips in 12-well plates, and allow to dry at room temperature for 1 to 4 hours in a laminar flow hood. After poly-L-ornithine coating, wash the plates twice with distilled water for 5 min. Add 0.5 ml of laminin (dissolved in sterile water to a final concentration of 10 mg/ml) to each well, and allow to stand for 2 to 24 hr at 37°C. Before seeding the cells, wash the plates 3 times, each time for 5 min, in Dulbecco's PBS (Life Technologies) and once in VM culture medium (see recipe).

CPBS (citric acid/sodium phosphate buffer) (*UNIT 5.3*)
80 ml of 0.2 M sodium phosphate (dibasic)
120 ml of 0.1 ml citric acid
Adjust pH to 4.1 with HCl
Filter sterilize
Store up to 1 year at room temperature

Crystal violet stain (*UNIT 4.14*)

Prepare a solution containing 0.5% (w/v) crystal violet and 0.2% (w/v) sodium acetate. Adjust the pH to 3.6 with acetic acid before bringing it to final volume. Filter through Whatman 1MM paper and store up to several months at room temperature.

The stain can be reused several times.

CsCl, 5.7 M, DEPC-treated (*APPENDIX 2E*)

Dissolve CsCl in 0.1 M EDTA, pH 8.0 (see recipe). Add 0.002 vol DEPC, shake 20 to 30 min, and autoclave. Weigh the bottle of solution before and after autoclaving and make up the weight lost to evaporation during autoclaving with DEPC-treated H_2O (see recipe). This ensures that the solution is actually 5.7 M when used.

CsCl buffer (*UNIT 4.8*)
✓ 5 mM Tris-HCl
✓ 1 mM EDTA
Adjust pH to 7.8 with 1 M NaOH
Filter sterilize
Store up to 6 months at room temperature

CsCl gradient solutions (*UNIT 4.11*)
1.2 g/ml density:
27.5 g CsCl
✓ PBS to 100 ml
1.3 g/ml density:
40.5 g CsCl
✓ PBS to 100 ml
1.37 g/ml density:
50 g CsCl
✓ PBS to 100 ml
1.4 g/ml density:
54.5 g CsCl
✓ PBS to 100 ml

1.5 g/ml density:
67.5 g CsCl
✓ PBS to 100 ml

Check density of each solution by weighing 1 ml. Filter sterilize. Store the gradient solutions up to 1 year at room temperature.

CsCl solution, 1.33 and 1.45 g/ml *(UNIT 4.8)*
1.33 g/ml solution:
8.31 g CsCl
✓ 16 ml CsCl buffer
1.45 g/ml CsCl solution:
8.31 g CsCl
✓ 11 ml of CsCl buffer
Make these solutions fresh just prior to use

Cupric sulfate solution *(UNIT 1.2)*
10 mM $CuSO_4$
50 mM ammonium acetate, pH, 5.0
Store indefinitely at 4°C

Cutting medium *(UNIT 4.12)*
Roller tube culture medium (see recipe) containing:
10 mM $MgCl_2$
0.5 μM tetrodotoxin (e.g., Latoxan)
Filter sterilize and store up to 2 months at 4°C

DAB staining solution *(UNIT 4.9)*
DAB solution 1:
7.25 g ammonium nickel sulfate
600 mg glucose
120 mg ammonium chloride
150 ml 0.2 M sodium acetate, pH 6.0
Store up to several months in the dark at 4°C
DAB solution 2:
15 mg diaminobenzidine (DAB; Sigma)
1 mg glucose oxidase (Sigma)
15 ml distilled H_2O
Prepare fresh, directly before use

Mix equal volumes of DAB solutions 1 and 2 immediately before use, and discard unused solution (DAB is photosensitive).

DAB substrate solution *(UNIT 1.2)*

Dissolve 5 mg of 3,3′-diaminobenzidine tetrahydrochloride (DAB) in 10 ml of 50 mM Tris·Cl, pH 7.6 (see recipe). Prepare a fresh 0.01% stock solution of hydrogen peroxide by adding 33 μl of 30% hydrogen peroxide to 1 ml distilled water. Add 100 μl of the diluted hydrogen peroxide per 10 ml of DAB solution, shake well, and use immediately.

CAUTION: *Diaminobenzidine (DAB) is carcinogenic; wear appropriate protective gear such as gloves and work in a chemical fume hood. Place all supplies that come into contact with DAB into a disposable beaker containing bleach and leave overnight to detoxify. The DAB itself should be placed in a container of 2 M sulfuric acid and 0.2 M aqueous potassium permanganate and disposed of according to institutional procedures.*

DAB/NiCl₂ visualization solution *(UNIT 5.7)*
✓ 5 ml 100 mM Tris·Cl, pH 7.5
100 μl DAB stock (40 mg/ml in H_2O, stored in 100-μl aliquots at −20°C)
25 μl $NiCl_2$ stock (80 mg/ml in H_2O, stored in 100-μl aliquots at −20°C)
15 μl 3% H_2O_2
Mix just before use

CAUTION: Handle DAB carefully, wearing gloves and mask; it is a carcinogen.

Suppliers of peroxidase substrates are Sigma, Kirkegaard & Perry, Moss, and Vector Labs.

Defined minimal A medium, 5× *(UNIT 4.2)*
5 g $(NH_4)_2SO_4$
22.5 g KH_2PO_4
52.5 g K_2HPO_4
2.5 g sodium citrate·$2H_2O$

Before they are used, concentrated media should be diluted to 1× with sterile water and the following sterile solutions, per liter:
1 ml 1 M $MgSO_4 \cdot 7H_2O$
10 ml 20% carbon source (sugar or glycerol)
and, if required:
0.1 ml 0.5% vitamin B1 (thiamine)
5 ml 20% Casamino acids *or* L amino acids to 40 ug/ml *or* DL amino acids to 80 ug/ml
Antibiotic as indicated

Degenerate PCR amplification buffer, 10× *(UNIT 4.1)*
✓ 200 mM Tris·Cl, pH 8.8
100 mM KCl
100 mM $(NH_4)_2SO_4$
20 mM $MgSO_4$
1% (v/v) Triton X-100
1 mg/ml nuclease-free BSA
Store indefinitely at −20°C

Denaturing lysis buffer *(UNIT 5.12)*
1% (w/v) SDS
✓ 50 mM Tris·Cl, pH 7.4
✓ 5 mM EDTA
Store up to 1 week at room temperature (SDS precipitates at 4°C)
Add the following fresh before use:
10 mM dithiothreitol (DTT; from powder)
1 mM phenylmethylsulfonyl fluoride (PMSF; store 100 mM stock in 100% ethanol up to 6 months at −20°C)
2 μg/ml leupeptin (store 10 mg/ml stock in H_2O up to 6 months at −20°C)
15 U/ml DNase I (store 15,000 U/ml stock solution up to 2 years at −20°C)

1 mM 4-(2-aminoethyl)benzenesulfonyl fluoride (AEBSF), added fresh from a 0.1 M stock solution in H_2O, can be used in place of PMSF. AEBSF stock can be stored up to 1 year at −20°C.

Denaturing solution *(UNIT 5.1)*
500 μl formamide
162 μl 12.3 M (37%) formaldehyde
✓ 100 μl MOPS buffer
Make fresh from stock solutions immediately before use

If formamide has a yellow color, deionize as follows: add 5 g of mixed-bed ion-exchange resin [e.g., Bio-Rad AG 501-X8 or X8(D) resins] per 100 ml formamide, stir 1 hr at room temperature, and filter through Whatman #1 filter paper.

CAUTION: Formamide is a teratogen. Handle with care.

Denaturing solution *(APPENDIX 2E)*

4 M guanidinium thiocyanate
25 mM sodium citrate, pH 7
0.1 M 2-ME (added as noted below)
0.5% (v/v) *N*-lauroylsarcosine (Sarkosyl)

Prepare a stock solution by dissolving 250 g guanidinium thiocyanate in a solution of 293 ml H_2O, 17.6 ml of 0.75 M sodium citrate, pH 7, and 26.4 ml of 10% Sarkosyl at 60° to 65°C with stirring. The stock solution can be stored up to 3 months at room temperature.

Prepare working solution by adding 0.35 ml of 2-ME per 50 ml of stock solution. The working denaturing solution can be stored 1 month at room temperature.

Denhardt's solution, 100×

10 g Ficoll 400
10 g polyvinylpyrrolidone
10 g bovine serum albumin (Pentax Fraction V; Miles Laboratories)
H_2O to 500 ml
Filter sterilize
Store at −20°C in 25-ml aliquots

DEPC (diethylpyrocarbonate) treatment of solutions

Add 0.2 ml DEPC to 100 ml of the solution to be treated (excluding Tris solutions). Shake vigorously to get the DEPC into solution. Autoclave the solution to inactivate the remaining DEPC.

Many investigators keep the solutions they use for RNA work separate to ensure that "dirty" pipets do not go into them.

CAUTION: *Wear gloves and use a fume hood when using DEPC, as it is a suspected carcinogen.*

Destaining solution *(UNIT 5.10)*

50% (v/v) methanol
10% (v/v) acetic acid
Balance H_2O
Store up to 6 months at room temperature

Detection buffer *(UNIT 1.3)*

✓ 100 mM Tris·Cl, pH 7.5
150 mM NaCl
5% (v/v) normal goat serum (Vector Labs)
0.6% (v/v) Triton X-100
Store indefinitely at −20°C

Detergent-free lysis buffer *(UNIT 5.12)*

✓ *PBS containing:*
✓ 5 mM EDTA
0.02% (w/v) sodium azide
Store up to 6 months at 4°C
Immediately before use add:
10 mM iodoacetamide (from powder)
1 mM PMSF (store 100 mM stock in 100% ethanol up to 6 months at −20°C)
2 μg/ml leupeptin (store 10 mg/ml stock in H_2O up to 6 months at −20°C)

1 mM AEBSF, added fresh from a 0.1 M stock solution in H_2O, can be used in place of PMSF. AEBSF stock can be stored up to 1 year at −20°C.

Development buffer *(UNIT 1.3)*

✓ 100 mM Tris·Cl, pH 9.5
100 mM NaCl
50 mM $MgCl_2$
Store up to 1 year at room temperature

Diethanolamine buffer, pH 9.8, 0.1 M *(UNIT 5.5)*

50 ml 0.2 M diethanolamine
5.74 ml 0.2 M HCl
44.26 ml H_2O

Digestion buffer *(UNIT 4.9)*

✓ 10 mM Tris-Cl, pH 8.0
10 mM NaCl
✓ 25 mM EDTA
✓ 1% (w/v) SDS
4 mg/ml proteinase K
Prepare fresh before each use

Digestion buffer *(APPENDIX 2D)*

100 mM NaCl
✓ 10 mM Tris·Cl, pH 8
✓ 25 mM EDTA, pH 8
✓ 0.5% (w/v) SDS
0.1 mg/ml proteinase K
Store at room temperature

The proteinase K is labile and must be added fresh with each use.

Dilution buffer *(UNIT 5.12)*

✓ *TSA solution containing:*
0.1% (w/v) Triton X-100 (store at room temperature in dark)
0.1% (w/v) bovine hemoglobin (store frozen)

Dioxetane phosphate substrate buffer *(UNIT 5.7)*

1 mM $MgCl_2$
0.1 M diethanolamine
0.02% sodium azide (optional)
Adjust to pH 10 with HCl and use fresh

Traditionally, the AMPPD substrate buffer has been a solution containing 1 mM $MgCl_2$ and 50 mM sodium carbonate/bicarbonate, pH 9.6. The use of diethanolamine results in better light output (Applied Biosystems Western Light instructions).

Alternatively, 100 mM Tris·Cl (pH 9.5)/100 mM NaCl/5 mM $MgCl_2$ can be used.

Dioxetane phosphate visualization system *(UNIT 5.7)*

Prepare 0.1 mg/ml AMPPD or CSPD (Applied Biosystems) *or* Lumigen-PPD (Lumigen) substrate in dioxetane phosphate substrate buffer (see recipe). Prepare just before use. Lumi-Phos 530 (Boehringer Mannheim or Lumigen) is a ready-to-use solution and can be applied directly to the membrane.

This concentration (240 uM) of AMPPD substrate is the minimum recommended by Applied Biosystems Western Light. Ten-fold lower concentrations can be used but require longer exposures.

Dissection buffer *(UNIT 2.2)*

490 ml 1× Hanks' Balanced Salt Solution (Invitrogen)
5 ml 1 M HEPES (Invitrogen)
5 ml penicillin/streptomycin (10,000 U/ml penicillin/10,000 μg/ml streptomycin; Invitrogen)
Store up to one month at 4°C.

Dissection medium *(UNIT 3.5)*

100 ml 10× calcium- and magnesium-free (CMF)-HBSS (e.g., Life Technologies)
800 ml H_2O
3.7 g HEPES
Adjust to pH 7.2 to 7.4 with 1 N HCl
Add H_2O to 1 liter
Filter sterilize using a 0.2-μm filter
Store up to 1 month at 4°C

Dissection solution *(UNIT 3.4)*

100 ml 10× CMF-HBSS (e.g., Life Technologies)
800 ml H_2O
3.9 g HEPES (15 mM final)
0.84 g $NaHCO_3$ (10 mM final; cell culture tested, Sigma)
10 ml 100× penicillin/streptomycin (e.g., Life Technologies; 100 U/ml penicillin/100 μg/ml streptomycin final)
Adjust pH to 7.2 with 1 N HCl
Add H_2O to 1 liter
Filter sterilize through 0.2-μm sterile filter
Store up to 1 month at 4°C

Dissociation solutions (Dis A and Dis B) *(UNIT 3.2)*

4% (w/v) EDTA stock solution:
2 g EDTA (Sigma)
45 ml tissue culture-grade water
Adjust pH to 7.2 with 1 N NaOH
Adjust volume to 50 ml
Filter through 0.22-μm Millipore filter
Store up to 12 months at 4°C
Dis A:
5 ml 4% EDTA solution (see above; 0.04% final)
✓ 495 ml 1.25× PBS
Store up to 12 months at 4°C
Dis B:
Dilute trypsin stock solution (see recipe) five-fold with Dis A solution (0.05% trypsin final). Store up to 12 months at −20°C.

DMEM/F-12/N2, supplemented *(UNIT 3.4)*

✓ 900 ml DMEM/F-12 with L-glutamine and antibiotics
3.9 g HEPES
3.7 g $NaHCO_3$ (cell culture tested, Sigma)
0.11 g sodium pyruvate
N2 supplements (see recipe or purchase from Life Technologies)
Adjust pH to 7.2 or 7.3 with 1 N HCl
Adjust volume to 1 liter with DMEM/F-12
Filter sterilize through 0.2-μm filter
Store protected from light up to 1 month at 4°C
This recipe is derived from Bottenstein (1985).

DMEM/F-12/N2/10% (w/v) FBS, supplemented *(UNIT 3.4)*

Add 50 ml heat-inactivated (30 min at 56°C) FBS (Sigma) to 450 ml supplemented DMEM/F-12/N2 (see recipe). Store protected from light up to 1 month at 4°C.

Check several lots of serum and then choose and reserve the one that best supports cell growth for your cultures.

DMEM/HBSS/10% FBS, supplemented *(UNIT 3.5)*

550 ml DMEM
✓ 325 ml HBSS
15 ml 20% (w/v) glucose
100 ml FBS (10% v/v final)
2.38 g HEPES
10 ml penicillin/streptomycin
Adjust to pH 7.2 to 7.4 with 1 N HCl
Filter sterilize using a 0.2-μm filter
Store protected from light up to 1 month at 4°C

DMEM, HBSS, FBS, and antibiotics may be obtained from Life Technologies.

DNase digestion buffer *(UNIT 4.11)*

✓ 10 mM Tris·Cl, pH 7.5
10 mM $MgCl_2$
2 mM $CaCl_2$
50 U/ml DNase I
Prepare fresh

DNase I, RNase-free, 1 mg/ml

Prepare a solution of 0.1 M iodoacetic acid plus 0.15 M sodium acetate and adjust pH to 5.3. Filter sterilize. Add sterile solution to lyophilized RNase-free DNase I (e.g., Worthington) to give a final concentration of 1 mg/ml. Heat 40 min at 55°C and then cool. Add 1 M $CaCl_2$ to a final concentration of 5 mM. Store at −80°C in small aliquots.

DNase solution *(UNIT 3.6)*

7.5 mg DNase (e.g., DP DNase, Worthington)
15 ml DMEM
Do not adjust pH
Store up to 12 months at −20°C

DNase stop mix *(APPENDIX 2E)*

✓ 50 mM EDTA
✓ 1.5 M sodium acetate
✓ 1% (v/v) SDS

The SDS may precipitate at room temperature. Heat briefly to redissolve.

Dowex AG50W-X8 cation exchange resin (100- to 200-mesh; Bio-Rad) *(APPENDIX 2F)*

The resin is prepared in large batches (200 to 400 ml packed resin) by the following series of washing steps. These washes can be conveniently performed using a large Buchner funnel and filter paper to collect the resin between changes of wash solution.

1. Wash resin in ≥10 vol of 0.5 N NaOH until no color is observed in wash solution (resin will retain its buff color).
2. Wash with 5 to 10 vol of 0.5 N HCl.
3. Wash with 5 to 10 vol of 0.5 M NaCl.
4. Wash with 5 to 10 vol of distilled H_2O.
5. Wash with 5 to 10 vol of 0.5 N NaOH.
6. Wash with distilled H_2O until pH = 9.
7. Store prepared resin indefinitely in 0.5 M NaCl/0.1 M Tris, pH 7.5, at 4°C.

Downstream PCR amplification buffer, 10×
(UNIT 4.1)

500 mM KCl
✓ 100 mM Tris·Cl, pH 8.8
15 mM $MgCl_2$
30 mM DTT
1 mg/ml BSA
Store at −20°C for ≤3 months

DPBS (Dulbecco's phosphate-buffered saline)
(UNIT 4.14)

8.00 g NaCl (137 mM)
0.20 g KCl (2.7 mM)
0.20 g KH_2PO_4 (1.1 mM)
0.10 g $MgCl_2 \cdot 6H_2O$ (0.5 mM)
2.16 g $Na_2HPO_4 \cdot 7H2O$ (8.1 mM)
0.10 g anhydrous $CaCl_2$ (0.9 mM)
H_2O to 1 liter
Store up to 6 months at room temperature

Omit $MgCl_2 \cdot 6H_2O$ and $CaCl_2$ for calcium- and magnesium-free DPBS (CMF-DPBS)

DPBS and CMF-DPBS are also commercially available (e.g., Fisher). For packaging procedures, commercial DPBS is strongly recommended to ensure the utmost purity.

DTNB solution *(UNIT 5.5)*

Dissolve 39.6 mg 5, 5′-dithio-bis(2-nitrobenzoic acid) (DTNB; Fisher) in 10 ml of 0.05 M potassium phosphate buffer, pH 8.0 (see recipe). Adjust to pH 8.0 with 0.1 M NaOH if necessary. Prepare fresh for each assay.

Dulbecco's modified Eagle medium (DMEM), supplemented

Dulbecco's modified Eagle medium, high-glucose formulation (e.g., Life Technologies), containing:
5%, 10%, or 20% (v/v) FBS, heat-inactivated (optional; see recipe)
1% (v/v) nonessential amino acids
2 mM L-glutamine
100 U/ml penicillin
100 μg/ml streptomycin sulfate
Filter sterilize and store up to 1 month at 4°C

DMEM containing this set of additives is sometimes called "complete DMEM." The percentage of serum used is indicated after the medium name—e.g., "DMEM/5% FBS." Absence of a number indicates no serum is used. DMEM is also known as Dulbecco's minimum essential medium.

Two common nutrient mixtures, Ham's F-12 nutrient mixture and N2 supplements (both available commercially, e.g., from Life Technologies), are often added to DMEM either individually or in combination; the resulting media are known as DMEM/F-12, DMEM/N2, and DMEM/F-12/N2.

Dulbecco's modified essential medium (DMEM), supplemented *(UNIT 4.14)*

Dulbecco's modified essential medium (e.g., Fisher) containing:
1% (w/v) penicillin/streptomycin
4 mM L-glutamine
500 μg/ml G418 (neomycin analog)
Store up to 2 months at 4°C.

A 50× (200 mM) aqueous glutamine stock solution can be stored for a year or more at −20°C. A 100 mg/ml G418 stock solution in 100 mM HEPES, pH 7.3, can also be stored for a year or more at −20°C.

It may not be necessary to maintain the cells in the presence of G418 at all times. The authors have grown the cells without selective pressure for at least short periods of time with good results. Because G418 may affect cellular metabolism in ways other than conferring G418 resistance, and because it is expensive, the authors often include it in the medium only for the first two passages of the cells after they are thawed, after which the cells can be maintained without G418 for the remainder of the packaging procedure (UNIT 4.14, Basic Protocol 2).

EDTA (ethylenediaminetetraacetic acid), 0.5 M (pH 8.0)

Dissolve 186.1 g disodium EDTA dihydrate in 700 ml water. Adjust pH to 8.0 with 10 M NaOH (~50 ml; add slowly). Add water to 1 liter and filter sterilize. Store up to 6 months at room temperature.

Begin titrating before the sample is completely dissolved. EDTA, even in the disodium salt form, is difficult to dissolve at this concentration unless the pH is increased to between 7 and 8.

Electrode buffer, 10× *(UNIT 5.3)*

100 mM HEPES, pH 7.4
1.2 M KCl
10 mM $MgCl_2$
Filter sterilize and store 1-ml aliquots up to 1 year at −20°C

Electrode solution with T7-oligo(dT)$_{24}$ primer
(UNIT 5.3)

31.25 μl H_2O
✓ 5 μl 10× electrode buffer
5 μl 4dNTP mix (2.5 mM each)
1 μl 100 ng/μl T7-oligo(dT)$_{24}$

1.25 μl 20 U/μl avian myeloblastosis virus reverse transcriptase (AMV-RT)
Prepare fresh the day of the experiment

The promoter sequence is 5′-AAA CGA CGG CCA GTG AAT TGT AAT ACG ACT CAC TAT AGG GCG T_{24}-3′.

The final concentration of T7-oligo(dT)$_{24}$ in the electrode solution described here is 2 ng/μl, but 0.5 to 1 ng/μl also works well. The AMV-RT should be stored at −70°C and repeated freeze-thaw cycles avoided.

Elution buffer *(UNIT 5.12)*
✓ 1% (w/v) SDS
✓ 100 mM Tris·Cl, pH 7.4
Store up to 1 week at room temperature
10 mM DTT (add fresh from powder before use)

ES DIF (ES differentiation) medium *(UNIT 3.2)*
450 ml DMEM, high-glucose formulation of DMEM containing glutamine and sodium pyruvate (e.g., Life Technologies)
50 ml FBS
5 ml 100× nonessential amino acids (Life Technologies)
✓ 5 ml 100× nucleoside stock
✓ 5 ml 100× 2-mercaptoethanol stock
13 ml 1 M HEPES, pH 7.4
Store up to 1 month at 4°C
Supplement with 5 ml of 100 mM L-glutamine every 2 weeks

ES PRO (ES proliferation) medium *(UNIT 3.2)*
400 ml DMEM, high-glucose formulation of DMEM containing glutamine and sodium pyruvate (e.g., Life Technologies)
100 ml FBS
5 ml 100× nonessential amino acids (Life Technologies)
✓ 5 ml 100× nucleoside stock
✓ 5 ml 100× 2-mercaptoethanol stock
13 ml 1 M HEPES, pH 7.4
Store up to 1 month at 4°C
Supplement with 5 ml of 100 mM L-glutamine every 2 weeks

Immediately before use add murine leukemia inhibitory factor (LIF; ESGRO from Life Technologies) at a final concentration of 10^3 U/ml.

Ethidium bromide assay solution *(APPENDIX 2G)*

Add 10 ml of 10× TNE buffer (see recipe) to 89.5 ml H_2O. Filter through a 0.45-μm filter, then add 0.5 ml of 1 mg/ml ethidium bromide.

Add the dye after filtering, as ethidium bromide will bind to most filtration membranes.

Ethidium bromide staining solution

Concentrated stock (10 mg/ml): Dissolve 0.2 g ethidium bromide in 20 ml H_2O. Mix well and store at 4°C in dark or in a foil-wrapped bottle. Do not sterilize.

Working solution: Dilute stock to 0.5 μg/ml or other desired concentration in electrophoresis buffer (e.g., 1× TBE or TAE) or in H_2O.

Ethidium bromide working solution is used to stain agarose gels to permit visualization of nucleic acids under UV light. Gels should be placed in a glass dish containing sufficient working solution to cover them and shaken gently or allowed to stand for 10 to 30 min. If necessary, gels can be destained by shaking in electrophoresis buffer or H_2O for an equal length of time to reduce background fluorescence and facilitate visualization of small quantities of DNA.

Alternatively, a gel can be run directly in ethidium bromide by using working solution (made with electrophoresis buffer) as the solvent and running buffer for the gel.

CAUTION: *Ethidium bromide is a mutagen and must be handled carefully.*

FACS permeabilizing buffer *(UNIT 4.8)*
D-PBS (Life Technologies) containing:
1% (v/v) FBS (Life Technologies) decomplemented by heating at 56°C
0.1% (v/v) Triton X-100
Store up to 6 months at 4°C

FACS staining buffer *(UNIT 4.8)*

D-PBS containing 1% (v/v) FBS (Life Technologies) decomplemented by heating at 56°C. Store up to 6 months at 4°C.

Fe solution *(UNIT 4.14)*
✓ 200 ml PBS
332 mg potassium ferricyanide
424 mg potassium ferrocyanide
0.2 ml 1 M $MgCl_2$
0.2 ml 20% (w/v) Nonidet P40
0.2 ml 10% (w/v) sodium deoxycholate
Store up to 6 months at 4°C in a darkened container
This reagent is light sensitive.

Fetal bovine serum (FBS)

Thaw purchased fetal bovine serum (shipped on dry ice and kept frozen until needed). Store 3 to 4 weeks at 4°C. If FBS is not to be used within this time, aseptically divide into smaller aliquots and refreeze until used. Store ≤1 year at −20°C. To inactivate FBS, heat serum 30 min to 1 hr in a 56°C water bath.

Repeated thawing and refreezing should be avoided as it may cause denaturation of the serum.

Inactivated FBS (FBS that has been treated with heat to inactivate complement protein and thus prevent an immunological reaction against cultured cells) is useful for a variety of purposes. It can be purchased commercially or made in the lab as described above.

Fibronectin, 1 µg/ml *(UNIT 3.4)*

Resuspend 1 mg sterile, lyophilized bovine fibronectin (Life Technologies) in 1 ml sterile deionized, distilled water according to the manufacturer's instructions. Store up to 3 to 4 weeks at 4°C. On day of use, dilute 1:1000 to 1 µg/ml with sterile water.

Fixative solution for perfusion *(UNIT 1.1)*

✓ 500 ml 8% (w/v) formaldehyde stock solution
500 ml H_2O
3.2 g NaH_2PO_4
10.9 g Na_2HPO_4
9 g NaCl
pH should be ~7.4
Store indefinitely at 4°C

Fixing solution *(UNIT 4.12)*

10% (v/v) acetic acid
30% (v/v) methanol
Store up to 6 months at room temperature

Fluorescent mounting medium *(UNIT 4.11)*

Dissolve 0.24 to 0.36 g *n*-propyl gallate in 12 g glycerol (this will take some time). Slowly add 4.8 g Mowiol to form a slurry. Slowly add 12 ml distilled deionized water while heating over low heat. Add 24 ml of 0.2 M Tris·Cl, pH 8.2 to 8.5. Continue stirring for several hours. Cover tightly with paraffin. Allow undissolved Mowiol to settle overnight at 4°C. Store in aliquots up to 6 months at −20°C.

The solution should have a slight tan color. If the color darkens significantly or turns yellow, discard solution.

Formaldehyde fixative, 4% (w/v) *(UNIT 5.14)*

Dissolve 4 g of paraformaldehyde (Sigma) in 50 ml water. Add 1 ml 1 M NaOH solution and stir gently on a heating block (>60°C) until the paraformaldehyde is dissolved. Add 10 ml of 10× PBS (see recipe) and allow to cool to room temperature. Adjust the pH to 7.4 using 1 M HCl (~1 ml). Adjust to 100 ml with water and filter through a Millipore 0.45-µM filter using a syringe to remove traces of undissolved paraformaldehyde. Store up to several months at −20°C.

Formaldehyde in saline, 4% *(UNIT 1.1)*

✓ 500 ml 8% (w/v) formaldehyde stock solution
500 ml H_2O
9 g NaCl
Store indefinitely at 4°C

Formaldehyde loading buffer *(UNIT 5.1)*

✓ 1 mM EDTA, pH 8.0
0.25% (w/v) bromphenol blue
0.25% (w/v) xylene cyanol
50% (v/v) glycerol
Store up to 3 months at room temperature

Formaldehyde stock solution, 8% *(UNIT 1.1)*

Heat 800 ml H_2O to 65°C, then add 80 g paraformaldehyde granules (EM grade, Electron Microscopy Sciences) in a fume hood. Stir for 1 min, then slowly add 2 to 4 ml 5 N NaOH (until solution is clear). Let solution cool, then filter through Whatman no. 1 filter paper. Finally, add H_2O to 1 liter. Store up to 2 weeks at 4°C.

CAUTION: *Formaldehyde is toxic, and the preparation process involving heating results in considerable vaporization, which increases the hazard. It is essential to use appropriate safety procedures, such as working in a fume hood.*

Formamide loading dye *(UNIT 5.1)*

98% (v/v) deionized formamide
✓ 10 mM EDTA pH 8.0
0.025% (w/v) xylene cyanol
0.025% (w/v) bromphenol blue
Store indefinitely at −20°C

FPH (formamide prehybridization/hybridization) solution *(UNIT 5.1)*

✓ 5× SSPE
✓ 5× Denhardt solution
50% (v/v) formamide
✓ 0.5% (w/v) SDS
72 ug/ml denatured herring sperm DNA (Promega)
Make fresh from stock solutions immediately before use

Fragmentation buffer, 5x *(UNIT 4.17)*

Dissolve 6.06 g Tris base (Sigma; molecular biology grade) in 175 ml Rnase-free H_2O (see recipe for DEPC-treated H_2O). Adjust pH to 8.1 with glacial acetic acid. Add 12.3 g potassium acetate (from an unopened or dedicated bottle) and 8.04 g magnesium acetate (Sigma; molecular biology grade), and adjust volume to 250 ml (final pH ~8.4). Filter sterilize with a 0.2-µm filter. Store up to 6 months at −20°C.

Frozen stock (FS) solution, 2× *(UNIT 3.2)*

57 ml DMEM
25 ml FBS
18 ml dimethyl sulfoxide (DMSO)
Total 100 ml
Aliquot and store up to 12 months at −20°C

Gelatin-subbing solution *(UNIT 1.1)*

Dissolve 3.0 g gelatin in 150 ml distilled water and heat to 50°C with stirring until gelatin is completely dissolved. Add 0.3 g chromium potassium sulfate dodecahydrate and 450 ml distilled water, then mix 15 sec and filter through Whatman no. 1 filter paper. Prepare fresh.

Gel loading buffer, 6×

0.25% (w/v) bromphenol blue
0.25% (w/v) xylene cyanol FF
40% (w/v) sucrose *or* 15% (w/v) Ficoll 400 *or* 30% (v/v) glycerol
Store at 4°C (room temperature if Ficoll is used)

This buffer does not need to be sterilized. Sucrose, Ficoll 400, and glycerol are essentially interchangeable in this recipe.

Other concentrations (e.g., 10×) can be prepared if more convenient.

Glass bead suspension *(APPENDIX 2C)*

Transfer 200 to 300 μl of 200-μm acid-washed glass beads (National Scientific Supply) into a 1.5-ml microcentrifuge tube and add an equal volume of water. Vortex briefly to suspend just before using.

If glass beads do not come acid-washed, prepare as follows: Wash by soaking 1 hr in concentrated nitric acid. Rinse thoroughly with water. Dry in a baking oven, cool to room temperature, and store at 4°C until needed.

Glass beads, chilled and acid-washed *(UNIT 4.2)*

Wash 150- to 212-μm glass beads by soaking 1 hr in concentrated nitric acid. Rinse thoroughly with water. Dry the beads in a baking oven, cool to room temperature, and store at 4°C until needed.

Glass coverslips, cleaned and sterilized *(UNIT 3.5)*

Soak precleaned 12 × 24-mm glass coverslips (Becton Dickinson Primary Care Diagnostics) in 0.1 N HCl for 24 hr. Wash for an additional 24 hr in water, wash an additional three times, 5 min each wash, in water, and soak 30 min in isopropanol. Dry in an oven at ∼80°C, then sterilize with dry heat for 1 hr at 200°C. Store up to 1 month at room temperature.

Glial feed *(UNIT 2.2)*

Prepare the following in a 50-ml conical tube:
44.9 ml Neurobasal Medium (Invitrogen)
5 ml FBS, heat-inactivated 30 min at 55°C
100 μl 100× G-5 Supplement (Invitrogen)
Prepare fresh each time; warm in 37°C tissue culture incubator before use.

Glucose/Tris/EDTA (GTE) solution *(APPENDIX 2F)*

50 mM glucose
✓ 25 mM Tris·Cl, pH 8.0
✓ 10 mM EDTA

L-Glutamine, 0.2 M (100×) *(APPENDIX 2I)*

Thaw commercially prepared frozen L-glutamine or prepare an 0.2 M solution in water, aliquot aseptically into usable portions, then refreeze. For convenience, L-glutamine can be stored in 1-ml aliquots if 100-ml bottles of medium are used and in 5-ml aliquots if 500-ml bottles are used. Store ≤1 year at −20°C.

Many laboratories supplement medium with 2 mM L-glutamine—1% (v/v) of 100× stock—just prior to use.

Glutaraldehyde, 0.3% (v/v) *(UNIT 5.5)*

1.2 ml 25% glutaraldehyde (Sigma)
✓ 98.8 ml borate buffer, pH 10
Prepare immediately before use

Glutathione-agarose bead slurry *(UNIT 5.6)*

Pour glutathione-agarose beads (Sigma) into a 50-ml conical screw-cap polypropylene centrifuge tube. Fill tube with ice-cold buffer C (see recipe) and allow beads to swell overnight at 4°C. Wash beads three times with cold buffer C, each time by centrifuging 5 min at 800 × *g*, 4°C, and carefully removing supernatant. After the last wash, dilute 1 volume of washed beads with 1 volume of buffer C. Store at 4°C; stable at least 1 year.

After elution, the glutathione-agarose beads can be recycled by washing them in 3 M NaCl overnight followed by several rinses/washes in buffer C.

Glutathione-agarose or glutathione-Sepharose slurry *(UNIT 5.13)*

Swell 1.5 g glutathione-agarose or glutathione-Sepharose beads (e.g., Pierce, Sigma) in 30 ml of 50 mM Tris·Cl, pH 7.5 (see recipe), for 1 to 2 hr on ice. Pellet beads by gravity or very gentle centrifugation (1 min at 1000 rpm in a tabletop centrifuge) and then wash four times with co-immunoprecipitation buffer (see recipe) that lacks protease inhibitor mix and contains 1 mM sodium azide. Resuspend beads in 15 ml of this buffer to yield a final slurry concentration of ∼100 mg/ml. Store at 4°C (stable for months).

Glycerol solution *(UNIT 4.2)*

65% (v/v) glycerol, sterile
0.1 M $MgSO_4$
✓ 25 mM Tris-Cl, pH 8.0
Store up to 1 year at room temperature

Glyoxal, 6 M, deionized *(UNIT 5.1)*

Immediately before use, deionize glyoxal by passing through a small column of mixed-bed ion-exchange resin [e.g., Bio-Rad AG 501-X8 or X8(D) resins] until the pH is >5.0.

Glyoxal loading buffer *(UNIT 5.1)*

10 mM sodium phosphate, pH 7.0
0.25% (w/v) bromphenol blue
0.25% (w/v) xylene cyanol
50% (v/v) glycerol
Store up to 3 months at room temperature

GST solution *(UNIT 4.15)*

✓ PBS
2% (v/v) goat serum
0.2% (v/v) Triton X-100
Store up to 1 month at 4°C

Guanidinium solution *(APPENDIX 2G)*

4 M guanidinium isothiocyanate
✓ 20 mM sodium acetate, pH 5.2
0.1 mM DTT
0.5% *N*-lauroylsarcosine (Sarkosyl)

Dissolve the guanidinium isothiocyanate in water and the appropriate amount of sodium acetate. Heating the solution slightly (65°C) may be necessary to dissolve the guanidinium. Add the DTT and Sarkosyl. Check the pH—it should be ∼5.5. If not, adjust with acetic acid. Bring to volume and filter the solution through a Nalgene filter. Store at room temperature.

Hanker-Yates solution B (*UNIT 4.9*)

25 ml 0.1 M sodium cacodylate, pH 5.1, containing
100 mg of ammonium nickel sulfate
12.5 mg of *p*-phenylenediamine
25 mg of catechol
50 μl 30% hydrogen peroxide
Prepare immediately before use
CAUTION: *Solution is toxic!*

Hank's balanced salt solution, HEPES-buffered (HBSS/HEPES) (*UNIT 3.1*)

800 ml H_2O
100 ml 10× Ca^{2+}- and Mg^{2+}-free (CMF-) HBSS (e.g., Life Technologies)
3.7 g $NaHCO_3$
3.9 g HEPES
Adjust pH to 7.1 or 7.2 with 1 N HCl
Add H_2O to 1 liter
Filter sterilize

HBSS (Hanks' balanced salt solution)

0.40 g KCl (5.4 mM final)
0.09 g $Na_2HPO_4 \cdot 7H_2O$ (0.3 mM final)
0.06 g KH_2PO_4 (0.4 mM final)
0.35 g $NaHCO_3$ (4.2 mM final)
0.14 g $CaCl_2$ (1.3 mM final)
0.10 g $MgCl_2 \cdot 6H_2O$ (0.5 mM final)
0.10 g $MgSO_4 \cdot 7H_2O$ (0.6 mM final)
8.0 g NaCl (137 mM final)
1.0 g D-glucose (5.6 mM final)
0.2 g phenol red (0.02%; optional)
Add H_2O to l liter and adjust pH to 7.4 with 1 M HCl or 1 M NaOH
Filter sterilize and store up to 1 month at 4°C

HBSS may be made or purchased without Ca^{2+} and Mg^{2+} (CMF-HBSS). These components are optional and usually have no effect on an experiment; in a few cases, however, their presence may be detrimental. Consult individual protocols to see if the presence or absence of these components is recommended.

HEPES-buffered saline (HeBS), 2× (*UNIT 4.8*)

50 mM HEPES, sodium salt
280 mM NaCl
1.5 mM disodium orthophosphate
Adjust pH to 7.12 with 1 M NaOH
Filter sterilize
Store up to 6 months at −20°C

HEPES-buffered saline (HeBS) solution, 2× (*UNITS 4.10 & 411*)

16.4 g NaCl (0.28 M final)
11.9 g *N*-2-hydroxyethylpiperazine-*N'*-2-ethanesulfonic acid (HEPES; 0.05 M final)
0.21 g Na_2HPO_4 (1.5 mM final)
800 ml H_2O
Titrate to pH 7.05 with 5 N NaOH
Add H_2O to 1 liter
Filter sterilize through a 0.45-μm nitrocellulose filter
Test for transfection efficiency
Store at −20°C in 50-ml aliquots

An exact pH is extremely important for efficient transfection. The optimal pH range is 7.05 to 7.12.

There can be wide variability in the efficiency of transfection obtained between batches of 2× HeBS. Efficiency should be checked with each new batch. The 2× HeBS solution can be rapidly tested by mixing 0.5 ml of 2× HeBS with 0.5 ml of 250 mM $CaCl_2$, and vortexing. A fine precipitate should develop that is readily visible in the microscope. Transfection efficiency must still be confirmed, but if the solution does not form a precipitate in this test, there is something wrong.

Hoechst 33258 assay solutions (*APPENDIX 2G*)

Stock solution: Dissolve in H_2O at 1 mg/ml. Stable for ~6 months at 4°C.

Working solution: Add 10 ml of 10× TNE buffer (see recipe) to 90 ml H_2O. Filter through a 0.45-μm filter, then add 10 μl of 1 mg/ml Hoechst 33258.

Hoechst 33258 is a fluorochrome dye with a molecular weight of 624 and a molar extinction coefficient of 4.2×10^4 $M^{-1}cm^{-1}$ at 338 nm.

The dye is added after filtering because it will bind to most filtration membranes.

CAUTION: *Hoechst 33258 is hazardous; use appropriate care in handling, storage, and disposal.*

Hybridization buffer (*UNIT 4.18*)

25% (v/v) formamide
✓ 5× SSC
✓ 0.1% (w/v) SDS
1 μg/μl Cot1-DNA (Invitrogen Life Technologies)
1 μg/μl poly(A)$^+$ DNA (Amersham Pharmacia Biotech)
Store up to 6 months at −20°C
This reagent is also known as DL-Klenow labeling buffer.

Hybridization buffer (*UNIT 5.2*)

Dissolve the following in 150 mlDEPC-treated H_2O:
23.4 g NaCl (400 mM final)
13.85 g piperazine-*N,N'*-bis(2-hydroxypropanesulfonic acid) (PIPES) disodium salt (40 mM final)
0.37 g EDTA disodium salt (1 mM final)
Adjust pH to 6.4 with 1 M NaOH
Add DEPC-treated H_2O to 200 ml
Add 800 ml formamide
Stable at least 6 months at −20°C

Hybridization solution (*UNIT 1.3*)

Prepare the following using DEPC-treated H_2O (see recipe)
Solution A:
100 μg/ml salmon sperm DNA (Sigma)
250 μg/ml yeast total RNA (Sigma)
250 μg/ml yeast tRNA (Sigma)
Store indefinitely at −20°C

Solution B:
23.8 ml formamide (Ultrapure; Invitrogen)
✓ 0.95 ml 1 M Tris·Cl, pH 7.4
✓ 0.19 ml 250 mM EDTA, pH 8.0
3.75 ml 4 M NaCl
9.52 ml 50% (w/v) dextran sulfate (Sigma)
✓ 0.95 ml 50× Denhardt solution
H_2O to 40 ml
Store indefinitely at −20°C

Solution C: Combine appropriate quantity of labeled probe with 4 μl solution A and water to 12 μl. Mix, then heat 5 min at 65°C and cool rapidly on ice to room temperature.
Working solution:
Add to solution C as prepared above:
84 μl solution B
2 μl 5 M DTT
1 μl 10% (w/v) sodium thiosulfate
✓ 1 μl 10% (w/v) sodium dodecyl sulfate
Mix well and use immediately

6-Hydroxydopamine solution (6-OHDA) *(UNIT 3.7)*

Dissolve ascorbic acid in saline (0.9% NaCl) to a final concentration of 0.2 mg ascorbic acid/ml. Dissolve 6-hydroxydopamine bromide (Sigma) in the ascorbic acid/saline solution to a final concentration of 3.6 mg/ml.

The ascorbic acid inhibits auto-oxidation of the 6-hydroxydopamine and the final solution is stable for 2 to 3 hr at room temperature, after which it undergoes a brownish discoloration. It should be discarded at that point, as it may have lost some of its potency.

Wear protective mask and gloves when handling 6-hydroxydopamine.

Ibotenic acid solution *(UNIT 3.7)*

Dissolve ibotenic acid (Sigma) in 1 × PBS, pH 7.4 (see recipe) to a final concentration of 10 μg/μl (63 mM).

Freeze in small aliquots. Aliquots can be stored for several months at −20°C. Repeated freezing and thawing does not appear to affect the compound's toxicity.

Wear protective mask and gloves when handling ibotenic acid.

Injection buffer *(UNIT 5.16)*
✓ 10 mM Tris·Cl, pH 7.5
✓ 0.1 mM EDTA
100 mM NaCl
Prepare using embryo transfer water (Sigma)
Vacuum filter with a 0.45-μm filter system
Prepare fresh for each microinjection use and keep on ice

Insulin stock, 5 mg/ml *(UNIT 3.2)*
100 mg insulin (Sigma; tissue culture grade, from bovine pancreas)
20 ml 0.1 N NaOH
Filter through 0.22-μm Millipore filter
Store 100- or 500-μl aliquots up to 12 months at −80°C

Iodixanol gradient solutions *(UNIT 4.11)*
15% iodixanol (with 1 M NaCl):
✓ 50 ml 10× PBS
0.5 ml 1 M $MgCl_2$
0.5 ml 2.5 M KCl
100 ml 5 M NaCl (1 M final concentration)
125 ml Optiprep (iodixanol; Nycomed)
0.75 ml 0.5% (w/v) phenol red stock
H_2O to 500 ml
Filter sterilize
25% iodixanol:
✓ 50 ml 10× PBS
0.5 ml 1 M $MgCl_2$
0.5 ml 2.5 M KCl
200 ml Optiprep (iodixanol; Nycomed)
1.0 ml 0.5% (w/v) phenol red stock
H_2O to 500 ml
Filter sterilize
40% iodixanol:
✓ 50 ml 10× PBS
0.5 ml 1 M $MgCl_2$
0.5 ml 2.5 M KCl
333 ml Optiprep (iodixanol; Nycomed)
H_2O to 500 ml
Filter sterilize

The 40% solution contains does not contain phenol red because this is where the virus will be after centrifugation. Phenol red is omitted to clearly distinguish this solution from the others.
60% iodixanol:

500 ml Optiprep (iodixanol; Nycomed)
0.25 ml 0.5% (w/v) phenol red stock
0.5 ml 1 M $MgCl_2$
0.5 ml 2.5 M KCl

Because the Optiprep is 60% iodixanol, this recipe actually yields a 54% iodixanol solution when prepared.

Store all of the iodixanol solutions up to 6 months at 4°C.

Isoperc *(UNIT 3.6)*
90 ml Percoll (Pharmacia Biotech)
✓ 10 ml 10× PBS
Store up to 6 months at 4°C

IST buffer, 10× *(UNIT 5.3)*
✓ 500 mM Tris·Cl, pH 8.3
60 mM $MgCl_2$
1.2 M KCl
Filter sterilize and store 1-ml aliquots up to 1 year at −20°C

ITSFn medium *(UNIT 3.2)*
100 ml DMEM/F-12 (e.g., Life Technologies)
✓ 2.5 ml 2 mg/ml transferrin stock
✓ 100 μl 5 mg/ml insulin stock
✓ 10 μl 300 μM sodium selenite stock
1 ml 100× penicillin/streptomycin (Life Technologies)
Store up to 1 week at 4°C

Klenow filling in (KFI) buffer, 10× *(UNIT 5.3)*
✓ 200 mM Tris·Cl, pH 7.5
100 mM $MgCl_2$
50 mM NaCl
50 mM dithiothreitol (DTT)
Filter sterilize and store 1-ml aliquots up to 1 year at −20°C

KPBS *(UNIT 1.1)*
0.45 g KH_2PO_4
3.81 g $K_2HPO_4 \cdot 3H_2O$
9 g NaCl
H_2O to 1 liter
Stir to dissolve
pH should be ~7.4 at room temperature
Store up to several weeks at 4°C (prepare fresh every month)

Laemmli sample buffer, 2× *(UNIT 4.2)*
10% (v/v) 2-ME
✓ 6% (w/v) SDS
20% (v/v) glycerol
0.2 mg/ml bromphenol blue
✓ 0.025× Laemmli stacking buffer (optional)
Store up to 2 months at room temperature
This reagent can conveniently be prepared 10 ml at a time.

Laemmli stacking buffer, 2.5× *(UNIT 4.2)*
✓ 0.3 M Tris-Cl, pH 6.8
✓ 0.25% (w/v) SDS
Store up to 1 month at 4°C

LB medium (Luria broth) and LB plates
Per liter add:
10 g tryptone
5 g yeast extract
5 g NaCl
1 ml 1 M NaOH
15 g agar or agarose (for plates only)

Autoclave solution. Add filter-sterilized additives (see below) after the solution has cooled to 55°C. For LB plates, pour agar-containing solution into sterile petri dishes in a tissue culture hood. Store autoclaved medium up to 1 month at room temperature. Store antibiotic-containing plates in the dark up to 2 weeks at 4°C.

Contamination is easily identified by the clouding of solutions or the presence of growth on plates.

Additives:
Antibiotics (if required):
Ampicillin to 50 μg/ml
Tetracycline to 12 μg/ml
Galactosides (if required):
5-bromo-4-chloro-3-indolyl-β-D-galactoside (Xgal) to 20 μg/ml
Isopropyl-1-thio-β-D-galactoside (IPTG) to 0.1 mM

Low-background-fluorescence CO_2-independent medium *(UNIT 5.14)*

Adjust the formulation of the standard medium by omitting the pH indicator phenol red, the antibiotics penicillin and streptomycin, folic acid, and riboflavin. Before use, supplement the medium with 50 mM HEPES/NaOH, pH 7.4. Store 1 to 2 months at 4°C.

Low-Tris/EDTA buffer *(UNIT 4.8)*
Prepare the following two solutions:
✓ 100 ml 1 M Tris·Cl, pH 7.5
✓ 10 ml 0.5 M EDTA, pH 8.0
Combine the two solutions
Filter sterilize
Store up to 6 months at −20°C

Luxol fast blue (LFB) solution *(UNIT 4.9)*
Prepare in 95% methylated spirits:
0.1% (w/v) Luxol fast blue (Sigma)
10% (v/v) acetic acid
Store at room temperature and filter before each use

Lysis buffer *(UNIT 4.12)*
1% (v/v) Nonidet P-40 (NP-40)
✓ 50 mM Tris·Cl, pH 7.6
150 mM NaCl
✓ 2 mM EDTA
Store up to 2 months at 4°C

Lysis buffer *(UNIT 5.10)*
✓ *PBS containing:*
✓ 1 mM EDTA
1 mM EGTA
10 U/ml aprotinin
50 mM NaF
10 mM sodium pyrophosphate
1 mM sodium vanadate
1% (v/v) nondenaturing detergent (Triton X-100 or NP-40; omit if membranes and solubilized fractions are to be analyzed separately)
Make fresh day of use and keep on ice.

Lysis buffer *(APPENDIX 2E)*
✓ 50 mM Tris·Cl, pH 8.0
100 mM NaCl
5 mM $MgCl_2$
0.5% (v/v) Nonidet P-40
✓ Prepare with DEPC-treated H_2O
Filter sterilize

If the RNA is to be used for northern blot analysis or the cells are particularly rich in ribonuclease, add ribonuclease inhibitors to the lysis buffer: 1000 U/ml placental ribonuclease inhibitor (e.g., RNasin; Promega) plus 1 mM DTT or 10 mM vanadyl-ribonucleoside complex.

Lysis solution *(UNIT 4.2)*
✓ 50 mM Tris-Cl, pH 7.5
✓ 10 mM EDTA
0.3% (v/v) 2-ME, added just before use
2% (v/v) ß-glucuronidase from *Helix pomatia* (Type HP-2; Sigma), added just before use

Maintenance medium *(UNIT 4.8)*
D-MEM with Earle's salts
10% (v/v) FBS
1× MEM, nonessential amino acids
2 mM L-glutamine
100 U/ml penicillin

1 mg/ml streptomycin
Store up to 1 month at 4°C
All of the above ingredients are available from Life Technologies.

293 Medium *(UNIT 4.8)*
D-MEM with Earle's salts
10% (v/v) FBS
1× MEM, nonessential amino acids
2 mM L-glutamine
100 U/ml penicillin
1 mg/ml streptomycin
Store up to 1 month at 4°C
All of the above ingredients are available from Life Technologies.

Medium L, pH 8.0 *(UNIT 2.2)*
0.174 g K_2HPO_4 (1 mM final)
✓ 2 ml 0.5 M EDTA (0.1 mM final)
1 liter H_2O
Store up to several months at 4°C.

2-Mercaptoethanol stock, 100× *(UNIT 3.2)*
10 ml DMEM (e.g., Life Technologies)
7 μl 2-mercaptoethanol
Filter through 0.22-μm Millipore filter
Use immediately

Microelectrode solution *(UNIT 5.3)*
✓ 1× electrode buffer
dATP, dCTP, dGTP, and dTTP (250 μM each)
T7-oligo(dT) primer
100 ng/μl
✓ DEPC-treated water
0.5 U/μl AMV-RT (Seikagaku America)
Prepare fresh the day of the experiment

Middle wash buffer (APPENDIX 2E)
0.15 M LiCl
✓ 10 mM Tris·Cl, pH 7.5
✓ 1 mM EDTA
✓ 0.1% (w/v) SDS

Moloney murine leukemia virus reverse transcriptase (MMLV RT) buffer, 5× *(UNIT 4.1)*
✓ 250 mM Tris·Cl, pH 8.3
375 mM KCl
50 mM DTT
15 mM $MgCl_2$
Store at −20°C for ≤3 months

MOPS, 10× *(UNIT 5.3)*
200 mM MOPS [3-(*N*-morpholino) propanesulfonic acid], pH 7.0
50 mM sodium acetate
✓ 10 mM EDTA
Filter sterilize and store 1-ml aliquots up to 1 year at −20°C

MOPS buffer *(UNIT 5.1)*
0.2 M MOPS [3-(*N*-morpholino)-propanesulfonic acid], pH 7.0
0.5 M sodium acetate
✓ 0.01 M EDTA
Store up to 3 months at 4°C in the dark
Discard if it turns yellow.

MOPS running buffer, 10× *(UNIT 5.1)*
0.4 M MOPS, pH 7.0
0.1 M sodium acetate
✓ 0.01 M EDTA
Store up to 3 months at 4°C

Mowiol 4-88 mounting solution *(UNIT 4.8)*

Dispense 6.9 g of 99% to 100% glycerol in a 15-ml polypropylene conical tube. Add 2.4 g of Mowiol 4-88 (Calbiochem) and mix. Add 6 ml of distilled water and leave for 2 hr at room temperature. Transfer to a 50-ml polypropylene conical tube and add 12 ml of 0.2 M Tris·Cl, pH 8.5 (see recipe), and incubate for 10 min at 50°C with occasional stirring (a shaking water bath with occasional stirring may be useful because Mowiol can take a very long time to dissolve). Transfer the tube to a shaking platform and leave overnight. Centrifuge 15 min at 1000 × *g*. Remove supernatant and aliquot into plastic pop-cap microcentrifuge tubes; store at −20°C. Thaw just before use and do not refreeze.

Mowiol mounting medium *(UNIT 5.14)*

Mix 6 ml glycerol, 2.4 g Mowiol 4-88 (Calbiochem), and 6 ml water. Shake for 2 hr. Add 12 ml of 200 mM Tris·Cl, pH 8.5 (see recipe), and incubate at 50°C with occasional mixing until the Mowiol dissolves (i.e., ~3 hr). Filter through 0.45-μM Millipore filtration unit and store in aliquots up to several weeks at 4°C or up to several months at −20°C.

N2 medium *(UNIT 3.1)*
✓ 500 ml DMEM/F-12 (or Life Technologies)
100 mg apotransferrin
1.55 g glucose
0.073 g glutamine
1.69 g $NaHCO_3$
Adjust pH to 7.1 or 7.2 with 1 N HCl
Add H_2O to 1 liter
Filter sterilize

In a separate tube, dissolve 25 mg insulin in 5 ml of 10 mM NaOH. Add 5 to 10 ml sterile medium (from above) to neutralize the solution, then add the following:
200 μl 100 μM progesterone (dissolved in 100% ethanol; 20 nM final)
100 μl 1 M putrescine (dissolved in H_2O; 100 μM final)
60 μl 500 μM selenium (dissolved in H_2O; 30 nM final)
Filter sterilize, and add to sterile medium
Add 50 U/ml penicillin and 50 μg/ml streptomycin sulfate
Store up to 4 weeks at 4°C in the dark

N2 supplements *(UNIT 3.4)*
0.005 to 0.025 g bovine insulin (Sigma or Intergen) in 1 ml 0.1 N NaOH
0.1 g apotransferrin (human, iron-poor; Sigma or Intergen)
100 μl 1 M putrescine (Sigma)
20 μl 1 mM progesterone (Sigma)

60 μl 0.5 mM sodium selenite (Sigma)
This recipe is sufficient for 1 liter of medium.

Stock solutions: Dissolve 1.64 g putrescine in 10 ml H_2O and store at −20°C in aliquots. Dissolve 0.032 g progesterone in 100 ml 100% ethanol and store at −20°C in aliquots. For sodium selenite, prepare 0.5 M solution by dissolving 0.88 g in 10 ml H_2O, and store at room temperature protected from light; on the day of use, dilute 1:1000 with H_2O to achieve 0.5 mM final concentration.

N3FL medium *(UNIT 3.2)*
100 ml DMEM/F-12 (e.g., Life Technologies)
✓ 2.5 ml 2 mg/ml transferrin stock
✓ 0.5 ml 5 mg/ml insulin stock
✓ 10 μl 300 μM sodium selenite stock
✓ 100 μl 20 μM progesterone stock
✓ 100 μl 100 μM putrescine stock
1 ml 100× penicillin/streptomycin (Life Technologies)
Store at 4°C; use within 1 week

Immediately before use, add basic fibroblast growth factor (bFGF) and laminin to final concentrations of 50 ng/ml and 1 μg/ml, respectively.

NaCl-saturated butanol *(UNIT 5.16)*
20 ml 3 M NaCl
100 ml butanol
Store up to 6 months at room temperature with the bottle wrapped in aluminum foil

NAS/avidin blocking solution *(UNIT 1.2)*
✓ *Tris-buffered saline (TBS), pH 7.2, containing:*
4% (v/v) normal animal serum (NAS) from species in which secondary antibody was made
0.1% (v/v) Triton X-100 (omit when staining for electron microscopy)
10 (g/ml avidin
Prepare fresh

NBT/BCIP substrate solution *(UNIT 1.3)*
Stock solutions:
75 mg/ml nitroblue tetrazolium chloride (NBT) in dimethylformamide (Invitrogen)
50 mg/ml 5-bromo-4-chloro-3-indolyl phosphate (BCIP) in dimethylformamide (Invitrogen)
Working solution:
Prepare in development buffer (see recipe):
0.34 mg/ml NBT (from 75 mg/ml stock)
0.18 mg/ml BCIP (from 50 mg/ml stock)
Prepare fresh just prior to use

Levamisole (Vector Labs) may be added to a concentration of 1 mM to block peripheral-type endogenous alkaline phosphatase.

Neurobasal medium/2% B-27 (Neurobasal/B-27) *(UNIT 3.5)*
960 ml Neurobasal medium
40 ml B27 supplement
73.55 mg (0.5 mM) L-glutamine
10 ml penicillin/streptomycin
Adjust to pH 7.2 to 7.4 with 1 N HCl
Filter sterilize with a 0.2-μm filter
Store protected from light 1 month at 4°C
All ingredients may be obtained from Life Technologies.

Neurobasal medium/B-27/5% FBS *(UNIT 3.2)*
50 ml Neurobasal medium (Life Technologies)
1 ml 50× B-27 supplement (Life Technologies)
2.5 ml FBS
0.5 ml 100 mM L-glutamine
0.5 ml 100× penicillin/streptomycin (Life Technologies)

Neuronal feed *(UNIT 2.2)*
Prepare the following in a 50-ml conical tube:
48.5 ml Neurobasal Medium (Invitrogen)
1 ml 50× B-27 Serum-Free Supplement (Invitrogen)
500 μl GlutaMAX Supplement (Invitrogen; final 25 μM glutamic acid)
Warm in 37°C tissue culture incubator until use
Prepare fresh each time

Use above recipe on plating day and until day 3 in culture (i.e., the first medium change). For that change, and subsequent ones, omit GlutaMAX and increase Neurobasal Medium to maintain a constant volume.

Nondenaturing lysis buffer *(UNIT 5.12)*
1% (w/v) Triton X-100 (store at room temperature in dark)
✓ 50 mM Tris·Cl, pH 7.4
300 mM NaCl
✓ 5 mM EDTA
0.02% (w/v) sodium azide
Store up to 6 months at 4°C
Immediately before use add:
10 mM iodoacetamide (from powder)
1 mM PMSF (store 100 mM stock in 100% ethanol up to 6 months at −20°C)
2 μg/ml leupeptin (store 10 mg/ml stock in H_2O up to 6 months at −20°C)

1 mM AEBSF, added fresh from a 0.1 M stock solution in H_2O, can be used in place of PMSF. AEBSF stock can be stored up to 1 year at −20°C.

pNPP buffer *(UNIT 5.9)*
0.1 M (0.75 g/liter) glycine
1 mM (0.2 g/liter) $MgCl_2$
1 mM (0.14 g/liter) $ZnCl_2$
Adjust pH to 10.4 with 1 M NaOH
Store up to 6 months at 4°C

3NTP mix *(UNIT 5.2)*

Prepare a mixture of equal volumes of 10 mM stock solutions either of ATP, GTP, and CTP (when using radiolabeled UTP), or ATP, GTP, and UTP (when using radiolabeled CTP). Store up to 6 months at −20°C.

3dNTP mix, 10× *(UNIT 5.2)*
✓ 10 mM Tris·Cl, pH 7.2
✓ 10uM EDTA
0.15 mM dATP (omit if radiolabeled dATP is used)

0.15 mM dCTP (omit if radiolabeled dCTP is used)
0.15 mM dGTP
0.15 mM dTTP
Store up to 6 months at −20°C

dNTPs, 5 mM *(UNIT 5.9)*

Mix equal volumes of the four different 100 mM dNTPs stock solutions, and then dilute five-fold in water. Store up to 1 year at −20°C.

rNTP mix, 10x *(UNIT 4.17)*

30mM rGTP
15 mM rATP
12 mM rCTP
12 mM UTP
Prepare using RNase-free H_2O (see recipe for DEPC-treated H_2O) and Ultrapure reagents from Pharmacia Biotech.
Store up to 3 months at (20°C in small aliquots (e.g., 50 to 100 μl) to prevent multiple freeze/thaw cycles.

Nucleoside stock, 100× *(UNIT 3.2)*

Dissolve in 100 ml H_2O:
80 mg adenosine
85 mg guanosine
73 mg cytidine
73 mg uridine
24 mg thymidine
Warm to 37°C with gentle shaking to dissolve the contents
Filter through 0.22-μm Millipore filter
Divide into aliquots and store up to 12 months at −20°C

Osmium tetroxide, 1% (w/v) *(UNIT 1.2)*

Prepare 2% (w/v) osmium tetroxide stock solution in water (stable several months at 4°C). Dilute with an equal volume of 0.2 M sodium phosphate, pH 7.4 (see recipe), immediately before use.

Paraformaldehyde, 4% (w/v) *(UNIT 3.3)*

Heat 900 ml water to 55° to 60°C on a stirring hot plate in a fume hood. Add 40 g paraformaldehyde powder and stir for 30 min. If powder has not dissolved, add a few NaOH pellets one a time (waiting a few minutes between pellets) until paraformaldehyde dissolves. Add 100 ml of 10× PBS (see recipe), filter, cool to room temperature, and adjust pH to 7.4 with HCl. Store up to 1 week at 4°C for routine perfusions for immunohistochemistry.

The fume hood is used because paraformaldehyde fumes are toxic. The "Prill" form of paraformaldehyde from EM Sciences is safest to use, as it pours without creating a cloud.

Paraformaldehyde-picric acid solution *(UNIT 1.3)*

Add 40 g paraformaldehyde to 500 ml water and heat to 60°C. Add 10 N NaOH until paraformaldehyde dissolves (e.g., until the mixture clears). Cool to below 30°C and add 100 ml of 10× PBS (see recipe), 150 ml saturated picric acid, and water to 1 liter. Dilute 1 part of this solution with 3 parts PBS, to obtain 1% paraformaldehyde working solution. Prepare fresh or freeze at −20°C until use.

Paraformaldehyde solution, 4% (w/v) *(UNIT 4.8)*

Solution A: Mix 8 g paraformaldehyde with 80 ml distilled water. Heat to 50° to 60°C and add drops of 1 M NaOH until the paraformaldehyde dissolves. Cool and filter into 100 ml of solution B.

Solution B: Prepare 100 ml 0.2 M PBS (see recipe) containing 0.12 M sucrose.

Adjust pH of the solution A/solution B mixture to 7.5 and bring volume to 200 ml. Store at 4°C and use within 24 hr.

Paraformaldehyde solution, 4% (w/v) *(UNIT 4.14)*

Add 20 g paraformaldehyde to 300 ml H_2O and heat to 55° to 60°C. Slowly add 1 M NaOH dropwise over ~10 min until the solution becomes clear, then cool the solution to room temperature. Use pH paper to check that the pH is 7.0 to 7.5 (add more NaOH if necessary). Add 100 ml of 0.5 M sodium phosphate buffer, pH 7.0 (see recipe), and then add water to a final volume of 500 ml (final 0.1 M phosphate; final pH 7.0 to 7.5). Store up to 1 week at 4°C.

If the paraformaldehyde is old (e.g., >7 days), or if cells are not fixed for a sufficient period of time, the cells will collapse and shrink, causing background staining problems.

CAUTION: *Paraformaldehyde is toxic, and the preparation process involving heating results in considerable vaporization, which increases the hazard. It is essential to use appropriate safety procedures, such as working in a fume hood.*

PBS (phosphate-buffered saline)

10× stock solution (per liter):
80 g NaCl
2 g KCl
11.5 g $Na_2HPO_4 \cdot 7H_2O$
2 g KH_2PO_4
1× working solution:
137 mM NaCl final
2.7 mM KCl final
4.3 mM $Na_2HPO_4 \cdot 7H_2O$ final
1.4 mM KH_2PO_4 final
If needed, adjust to desired pH (usually 7.2 to 7.4) with 1 M NaOH or 1 M HCl
Filter sterilize and store up to 1 month at 4°C
Without adjustment, the pH will generally be ~7.3.

PBS, 0.2 M *(UNIT 4.8)*

Solution A: Prepare 800 ml of 0.2 M disodium orthophosphate in 0.9% sodium chloride.

Solution B: Prepare 200 ml of 0.2 M sodium dihydrogen orthophosphate in 0.9% sodium chloride.

Add solution B to solution A with stirring until the pH is 7.5. Store up to 6 months at room temperature.

PBS with antibiotics, 10× *(UNIT 3.4)*

Dilute 10× PBS (see recipe) to 990 ml with water. Add 10 ml of 100× penicillin/streptomycin (Life Technologies; 100 U/ml penicillin/100 μg/ml streptomycin final) and adjust pH to 7.4 with 1 N NaOH. Filter sterilize through a 0.2-μm filter and store up to 1 month at 4°C.

Store 10× PBS at room temperature, as phosphates will precipitate at 4°C.

PBS/glucose *(UNIT 3.6)*

✓ 900 ml 1× PBS, calcium- and magnesium-free
2 g glucose (2% w/v final)
2 ml 0.5% (w/v) phenol red
Adjust pH to 7.4 with 1 M HCl or 1 N NaOH, as needed
Adjust volume to 1 liter with H_2O
Store up to 2 years at 4°C

PBS-MK *(UNIT 4.11)*

✓ 50 ml 10× PBS
0.5 ml 1 M $MgCl_2$
0.5 ml 2.5 M KCl
H_2O to 500 ml
Filter sterilize
Store up to 2 years at 4°C

PCR buffer, 10× *(UNIT 5.2)*

Dissolve the following in 80 ml H_2O:
1.21 g Tris base (100 mM Tris·Cl final)
3.73 g KCl (500 mM final)
0.30 g $MgCl_2 \cdot 6\ H_2O$ (15 mM final)
10 mg gelatin (0.01% w/v final)
Adjust pH to 8.3 with 1 M HCl
Add H_2O to 100 ml
Stable at least 6 months at −20°C

Peptide map electrophoresis buffer, pH 1.9 *(UNIT 5.10)*

10 parts glacial acetic acid
1 part formic acid
89 parts H_2O
Store up to 6 months at 4°C

Peptide map electrophoresis buffer, pH 3.5 *(UNIT 5.10)*

19 parts glacial acetic acid
1 part pyridine
89 parts H_2O
Store up to 6 months at room temperature

Percoll, 30% (v/v) *(UNIT 3.6)*

✓ 30 ml Isoperc
✓ 60 ml PBS/glucose
Store up to 6 months at 4°C

Percoll, 60% (v/v) *(UNIT 3.6)*

✓ 60 ml Isoperc
✓ 30 ml PBS/glucose
0.1 ml 0.4% (w/v) trypan blue (e.g., Life Technologies), filter sterilized
Store up to 6 months at 4°C

Percoll gradients *(UNIT 2.2)*

23% Percoll:
12 ml 100% Percoll
14 ml H_2O
✓ 26 ml 2× sucrose buffer
15% Percoll:
30 ml 23% Percoll
✓ 16 ml 1× sucrose buffer
10% Percoll:
20 ml 15% Percoll
✓ 10 ml 1× sucrose buffer
3% Percoll:
6 ml 10% Percoll
✓ 14 ml 1× sucrose buffer
Adjust solutions to pH 7.4
Store up to 5 days at 4°C

To create the discontinuous gradient, carefully layer 2 ml of each of these Percoll solutions on top of each other, starting with 23%, in 10-ml polycarbonate centrifuge tubes. Prepare the gradients on the day of the experiment or one day before. Keep at 4°C until use.

Phenol, buffered

Add 0.5 g of 8-hydroxyquinoline to a 2-liter glass beaker containing a stir bar. Gently pour in 500 ml of liquefied phenol or crystals of redistilled phenol that have been melted in a water bath at 65°C (the phenol should turn yellow). Add 500 ml of 50 mM Tris base (unadjusted pH ~10.5). Cover the beaker with aluminum foil. Stir 10 min at low speed with magnetic stirrer at room temperature. Stop stirring and let phases separate at room temperature. Gently decant the top (aqueous) phase into a suitable waste receptacle. Using a pipet and suction bulb, remove residual aqueous phase that cannot be decanted. Add 500 ml of 50 mM Tris·Cl, pH 8.0 (see recipe). Repeat two successive equilibrations with 500 ml of 50 mM Tris·Cl, pH 8.0 (i.e., the same procedure as with the Tris base). Check phenol phase with pH paper to verify that pH is 8.0; if not, perform additional equilibrations. Add 250 ml of 50 mM Tris·Cl, pH 8.0, or TE buffer, pH 8.0 (see recipe), and store up to 2 months at 4°C in brown glass bottles or clear glass bottles wrapped in aluminum foil.

CAUTION: *Phenol can cause severe burns to skin and damage clothing. Gloves, safety glasses, and a lab coat should be worn whenever working with phenol, and all manipulations should be carried out in a fume hood. A glass receptacle should be available exclusively for disposing of used phenol and chloroform.*

In this recipe, the 8-hydroxyquinoline acts as an antioxidant.

Phenol/chloroform/isoamyl alcohol, 25:24:1 (v/v/v)

✓ 25 vol buffered phenol (bottom yellow phase of stored solution)
24 vol chloroform
1 vol isoamyl alcohol
Store up to 2 months at 4°C

Phosphorylated amino acid standards *(UNIT 5.10)*

Make individually:
10 mg/ml phosphoserine
10 mg/ml phosphothreonine
10 mg/ml phosphotyrosine
Store up to 6 months at −20°C

Phosphoserine, phosphothreonine, and phosphotyrosine are all available from Sigma.

PKA buffer *(UNIT 5.10)*

40 mM HEPES free acid
20 mM $MgCl_2$
50 μM ATP
Adjust pH to 7.0 with 10 N NaOH
Store up to 6 months at 4°C

PKC buffer *(UNIT 5.10)*

10 mM HEPES free acid
800 μM $CaCl_2$
500 μg/ml phosphatidylserine (Sigma)
5 μg/ml diolein (Sigma)
50 μM ATP
Adjust pH to 7.0 with 10 N NaOH
Store up to 6 months at 4°C

Plaque agarose *(UNIT 4.14)*

Prepare supplemented DMEM (see recipe) that is double-strength (2×), and place it and a bottle of FBS into a 42°C water bath. Prepare a solution of molten 2% (w/v) tissue culture-grade agarose (e.g., SeaPlaque, FMC Bioproducts) in water and place it into the 42°C water bath. Allow temperatures to equilibrate. Mix reagents together at a ratio of 25:25:1 (v/v/v) agarose/2× DMEM/FBS. Prepare fresh, remove just before required, and use when the temperature is <40°C.

Plaque overlay solution *(UNIT 4.11)*

Prepare 2% (w/v) SeaPlaque low-melting/gelling temperature agarose (FMC Bioproducts) in distilled water. Autoclave and cool to 39°C. Also prepare 2× DMEM (without phenol red) from powder (Life Technologies) and supplement with 4% (v/v) FBS, 25 mM $MgCl_2$, 0.3% (w/v) sodium bicarbonate, 40 mM HEPES pH 7.5, 100 U penicillin and 100 μg/ml streptomycin sulfate. Warm the medium to 37°C. Mix equal amounts of the agarose and cell culture medium solutions and promptly add to cells. Make fresh with each use.

Plating medium *(UNIT 4.8)*

Eagle MEM without Earle's salts (Sigma)
10% (v/v) FBS (Life Technologies)
1× MEM, nonessential amino acids (Life Technologies)
2 mM L-glutamine (Life Technologies)
100 U/ml penicillin (Life Technologies)
1 mg/ml streptomycin (Life Technologies)
Store up to 1 month at 4°C

Poly(A) loading buffer *(APPENDIX 2E)*

0.5 M LiCl
✓ 10 mM Tris·Cl, pH 7.5
✓ 1 mM EDTA
✓ 0.1% (w/v) SDS

Polylysine-coated tissue culture surfaces

To coat coverslips: Prepare a stock solution by dissolving 25 mg/ml polylysine in 4.73 ml water (both poly-L-lysine and poly-D-lysine are used to coat tissue culture surfaces; check specific protocol for choice of isomer) and filter sterilize through a 0.22-μm filter. Store in 100-μl aliquots at −20°C. When ready to use, dilute one aliquot in 40 ml water to prepare 13 μg/ml working solution. Sterilize coverslips by autoclaving prior to coating. Dip coverslips in the working solution, then incubate 15 min to several hours in a humidified 37°C, 5% CO_2 incubator. Allow surface to dry.

To coat culture dishes or 8-well chamber slides: Prepare a stock solution by dissolving 100 mg polylysine in 100 ml water (both poly-L-lysine and poly-D-lysine are used to coat tissue culture surfaces; check specific protocol for choice of isomer) and filter sterilize through a 0.22-μm filter. Store in 5-ml aliquots at −20°C. When ready to use, dilute 1 part stock solution with 9 parts water to prepare 100 μg/ml working solution. Fill tissue culture dishes or slide wells with the working solution and incubate 1 hr in a humidified 37°C, 5% CO_2 incubator, then remove solution by vacuum aspiration and allow surface to dry.

Store coated tissue culture ware up to 3 months at 4°C. Use diluted solutions only once, but unused diluted aliquots can be stored up to 3 months at 4°C.

Poly-D-lysine solution, 20 μg/ml *(UNIT 4.14)*

Prepare a 1 mg/ml stock solution of poly-D-lysine (mol. wt. 70,000 to 150,000; Sigma) in water. Filter sterilize and store up to a year or more at −20°C. Dilute to 20 μg/ml in water immediately prior to use.

Polyornithine, 15 μg/ml *(UNITS 3.1 & 3.4)*

100× stock (1.5 mg/ml):
0.150 g polyornithine hydrobromide (MW 55,000; Sigma)
100 ml H_2O
Divide into aliquots and store (stable at least 6 months) at −20°C

Working solution (1 μg/ml):
When needed, dilute 1:100 with H_2O
Filter sterilize through 0.2-μm filter
Store up to 1 month at 4°C

Ponceau S solution *(UNIT 5.7)*

Dissolve 0.5 g Ponceau S in 1 ml glacial acetic acid. Bring to 100 ml with water. Prepare just before use.

Potassium acetate, 2 M *(UNIT 5.16)*

50 ml 7.5 M potassium acetate
23 ml glacial acetic acid
127 ml ddH_2O
Store at 4°C up to 6 months

Table A.1.3 Preparation of 0.1 M Sodium and Potassium Phosphate Buffers[a]

Desired pH	Solution A (ml)	Solution B (ml)	Desired pH	Solution A (ml)	Solution B (ml)
5.7	93.5	6.5	6.9	45.0	55.0
5.8	92.0	8.0	7.0	39.0	61.0
5.9	90.0	10.0	7.1	33.0	67.0
6.0	87.7	12.3	7.2	28.0	72.0
6.1	85.0	15.0	7.3	23.0	77.0
6.2	81.5	18.5	7.4	19.0	81.0
6.3	77.5	22.5	7.5	16.0	84.0
6.4	73.5	26.5	7.6	13.0	87.0
6.5	68.5	31.5	7.7	10.5	90.5
6.6	62.5	37.5	7.8	8.5	91.5
6.7	56.5	43.5	7.9	7.0	93.0
6.8	51.0	49.0	8.0	5.3	94.7

[a]Adapted by permission from CRC (1975).

Potassium acetate solution (APPENDIX 2F)
For miniprep (pH ~4.8):
29.5 ml glacial acetic acid
KOH pellets (several) to pH 4.8
H_2O to 100 ml
For large-scale prep (pH ~5.5):
294 g potassium acetate (3 M final)
50 ml 90% formic acid (1.18 M final)
H_2O to 1 liter
Store solutions at room temperature (do not autoclave).

Potassium phosphate buffer, 0.1 M

Solution A: 27.2 g KH_2PO_4 per liter (0.2 M final)

Solution B: 34.8 g K_2HPO_4 per liter (0.2 M final)

Referring to Table A.1.3 for desired pH, mix the indicated volumes of solutions A and B, then dilute with water to 200 ml. Filter sterilize if necessary. Store up to 3 months at room temperature.

This buffer may be made as a 5- or 10-fold concentrate simply by scaling up the amount of potassium phosphate in the same final volume. Phosphate buffers show concentration-dependent changes in pH, so check the pH of the concentrate by diluting an aliquot to the final concentration.

Pre-hybridization solution (UNIT 4.8)
✓ 5× SSC
1% (w/v) blocking reagent (from Dig DNA labeling and detection kit)
0.1% (w/v) N-lauroylsarcosine
✓ 0.02% (w/v) SDS

Blocking reagent does not dissolve readily. Heat to 60°C for 30 min. Store up to 6 months at −20°C.

Primary antibody diluent (UNIT 1.2)
✓ *Tris-buffered saline (TBS), pH 7.2, containing:*
1% normal animal serum (NAS) from species in which secondary antibody was made
50 μg/ml biotin
Prepare fresh

Primer-extension buffer, 10× (UNIT 5.2)
Dissolve in the following in 80 ml H_2O:
23.4 g HEPES sodium salt (900 mM final)
2.03 g $MgCl_2 \cdot 6\,H_2O$ (100 mM final)
Adjust pH to 6.6 with 4 M NaOH
Add DEPC-treated H_2O to 100 ml
Stable at least 6 months at −20°C

Probe elution buffer (UNIT 5.2)
38.54 g ammonium acetate (0.5 M final)
0.37 g EDTA disodium salt (1 mM final)
1.0 g SDS (0.1% w/v final)
DEPC-treated H_2O to 1 liter
Stable at least 6 months at −20°C

Progesterone, 20 μM (UNIT 3.2)
2 mM stock solution:
6.3 mg progesterone (Sigma; tissue culture grade)
10 ml ethanol
Store up to 12 months at −20°C
20 μM working solution:
100 μl 2 mM stock solution
10 ml H_2O
Filter through 0.22-μm Millipore filter
Store 100-μl aliquots up to 12 months at −80°C

Protease inhibitor mix (UNIT 4.7)
1mM phenylmethylsulfonyl fluoride (PMSF)
0.5 mM iodoacetamide
0.2 mM Na-*p-tosyl*-L-lysine chloromethyl ketone (TLCK), prepared fresh
2 mM benzamidine-HCl

CAUTION: *Prepare stock solutions in a chemical fume hood.*

Protease inhibitor mix, 1000× *(UNIT 5.13)*

Dissolve in DMSO:
5 mg/ml chymostatin
5 mg/ml pepstatin A
5 mg/ml leupeptin
5 mg/ml antipain
Store in aliquots up to 1 year at −20°C

Protease inhibitors *(UNIT 2.2)*

Aprotinin: Prepare 10 ml of aqueous solution containing 20 mg (2 mg/ml final) aprotinin. Store up to 1 year at 4°C. Where indicated, add immediately before use at a 1:1000 dilution (2 µg/ml final).

Leupeptin: Prepare a stock solution of 1 mg leupeptin/ml aqueous solution. Store up to 1 year at −20°C. Where indicated, add immediately before use at a 1:1000 dilution (1 µg/ml final).

Phenylmethylsulfonyl fluoride (PMSF): Prepare 10 ml of 100% ethanol containing 0.435 g PMSF (250 mM final). Store indefinitely at −20°C. Where indicated, add immediately before use at a 1:250 dilution (1 mM final).

Protein A/G–Sepharose slurry *(UNIT 5.13)*

Swell 1.5 g protein A– or protein G–Sepharose beads (e.g., Pierce, Sigma) in 30 ml of 50 mM Tris·Cl, pH 7.5 (see recipe), for 1 to 2 hr on ice. Pellet beads by gravity or very gentle centrifugation (1 min at 1000 rpm in a tabletop centrifuge) and then wash four times with co-immunoprecipitation buffer (see recipe) that lacks protease inhibitor mix and contains 1 mM sodium azide. Resuspend beads in 15 ml of this buffer to yield a final slurry concentration of ~100 mg/ml. Store at 4°C (stable for months).

Proteinase K/SDS solution *(UNIT 5.2)*

Dissolve the following in 40 ml DEPC-treated H_2O:
0.605 g Tris base (50 mM Tris·Cl final)
50 ml glycerol (50% v/v final)
33 mg $CaCl_2$ (3 mM final)
Adjust pH to 8.0 with 1 M HCl
Add DEPC-treated H_2O to 100 ml
Dissolve 5 mg proteinase K per ml of the above buffer
Add an equal volume of 20% SDS (see recipe)
Stable at least 6 months at −20°C

Proteinase solution *(UNIT 4.11)*

1 M NaCl
1% (w/v) *N*-lauroylsarcosine (Sarkosyl)
100 g/ml proteinase K
Make fresh each time

Putrescine, 100 µM *(UNIT 3.2)*

1.61 g putrescine (Sigma; cell culture tested)
100 ml H_2O
Filter through 0.22-µm Millipore filter
Store 100-µl aliquots 12 months at −80°C

Release-agent-coated microscope slides *(UNIT 1.2)*

Dip microscope slides briefly in liquid release agent (Electron Microscopy Sciences) and allow to dry overnight at room temperature in a dust-free area.

The liquid release agent used here is soluble in water. Take care that no moisture comes in contact with the release-agent-coated slides, otherwise any resin-embedded tissue that has been cured on such a slide may become permanently bound to the slide.

Resin stubs *(UNIT 1.2)*

Place epoxy resin (e.g., Epon from Ted Pella or Durcupan from Electron Microscopy Sciences) into size 00 BEEM capsules (Ted Pella). Place the capsules, with the pointed sides up, in a 60°C oven for 2 to 3 days to cure the resin. Store indefinitely at room temperature.

Restriction endonuclease buffers *(UNIT 5.6)*

10× sodium chloride-based buffers
✓ 100 mM Tris·Cl, pH 7.5
100 mM $MgCl_2$
10 mM dithiothreitol (DTT)
1 mg/ml bovine serum albumin (BSA)
0, 0.5, 1.0, or 1.5 M NaCl

The concentration of NaCl depends upon the restriction endonuclease. The four different NaCl concentrations listed above are sufficient to cover the range for essentially all commercially available enzymes except those requiring a buffer containing KCl instead of NaCl (see recipe below). Autoclaved gelatin (at 1 mg/ml) can be used instead of BSA. Note that most restriction enzymes are provided by the suppliers with the appropriate buffers.

10× potassium acetate-based buffer
660 mM potassium acetate
330 mM Tris·acetate, pH 7.9
100 mM magnesium acetate
5 mM DTT
1 mg/ml BSA (optional)

10× potassium chloride-based buffer
✓ 60 mM Tris·Cl, pH 8.0
60 mM $MgCl_2$
200 mM KCl
60 mM 2-mercaptoethanol (2-ME)
1 mg/ml BSA

2× potassium glutamate-based buffer
200 mM potassium glutamate
50 mM Tris·acetate, pH 7.5
100 µg/ml BSA
1 mM 2-ME

Ringer's solution *(UNIT 2.4)*

119 mM NaCl (6.95 g/liter)
2.5 mM KCl (0.19 g/liter)
1.3 mM $MgSO_4$ (0.16 g/liter)
2.5 mM $CaCl_2$ (0.37 g/liter)
26 mM $NaHCO_3$ (2.18 g/liter)
1.0 mM NaH_2PO_4 (1.14 g/liter)

11mM glucose (1.98 g/liter)
Prepare fresh

RNA amplification buffer, 10× *(UNIT 5.3)*
✓ 400 mM Tris·Cl, pH 7.5
70 mM $MgCl_2$
100 mM NaCl
20 mM spermidine
Filter sterilize and store 1-ml aliquots up to 1 year at −20°C

RNase A/RNase T1 stock solution *(UNIT 5.2)*
Dissolve the following in 1 ml RNase storage buffer (see recipe):
0.5 mg RNase A (500 μg/ml final)
10,000 U RNase T1 (10,000 U/ml final)
Stable at least 6 months at −20°C

RNase A solution *(UNIT 1.3)*
Buffer stock solution:
62.5 ml 4 M NaCl
✓ 2.5 ml 2 M Tris·Cl, pH 8.0
✓ 0.5 ml 0.25 M EDTA, pH 8.0
Store up to 1 year at room temperature

Working solution: Just prior to use, prewarm 500 ml of buffer stock solution to 37°C and add 1 ml of 10 mg/ml RNase A. Use immediately.

RNase A stock solution, DNase-free, 2 mg/ml

Dissolve RNase A (e.g., Sigma) in DEPC-treated H_2O (see recipe) to 2 mg/ml. Boil 10 min in a 100°C water bath. Store up to 1 year at 4°C.

The activity of the enzyme varies from lot to lot; therefore, prepare several 10-ml aliquots of each dilution to facilitate standardization.

RNase digestion buffer *(UNIT 5.2)*
Dissolve the following in 80 ml DEPC-treated H_2O:
0.12 g Tris base (10 mM Tris·Cl final)
1.76 g NaCl (300 mM final)
0.19 g EDTA disodium salt (5 mM final)
Adjust pH to 7.5 with 1 M HCl
Add DEPC-treated H_2O to 100 ml
Stable at least 6 months at −20°C

RNase storage buffer *(UNIT 5.2)*
Dissolve the following in 40 ml DEPC-treated H_2O:
0.12 g Tris base (10 mM Tris·Cl final)
0.12 g NaCl (20 mM final)
50 ml glycerol (50% v/v final)
Adjust pH to 7.5 with 1 M HCl
Add DEPC-treated H_2O to 100 ml
Stable at least 6 months at −20°C

Roller tube culture medium *(UNIT 4.12)*
50 ml Eagle's basal medium (based on Earle's or Hanks' balanced salt solution; without glutamine, which tends to precipitate; store up to 6 months at 4°C), containing 10 μg/ml phenol red
25 ml Earle's or Hanks' balanced salt solution (BSS; store up to 6 months at room temperature), containing 10 μg/ml phenol red
25 ml horse serum, mycoplasma screened (heat-inactivate complement 30 min at 56°C; store up to 18 months at −20°C; thaw slowly at room temperature)
27.7 mM glucose
1 mM L-glutamine
Store up to ∼1 month at 4°C

BSS with Earle's salts contains more bicarbonate than Hanks' and will offer more buffering capacity against the greater amounts of CO_2 and lactic acid produced by larger cultures or cocultures.

RT buffer, 10× *(UNIT 5.3)*
✓ 500 mM Tris·Cl, pH 8.3
1.2 M KCl
100 mM $MgCl_2$
0.5 mM sodium pyrophosphate
Filter sterilize and store 1-ml aliquots up to 1 year at −20°C

S1 buffer, 10× *(UNIT 5.3)*
2 M NaCl
500 mM sodium acetate, pH 4.5
10 mM $ZnSO_4$
Filter sterilize and store 1-ml aliquots up to 1 year at −20°C

S1 nuclease digestion buffer, 2× *(UNIT 5.2)*
Dissolve the following in 80 ml DEPC-treated H_2O:
3.28 g NaCl (0.56 M final)
1.36 g sodium acetate trihydrate (0.1 M final)
0.26 g $ZnSO_4$ heptahydrate (9.0 mM final)
Adjust pH of final solution to 4.6 to 5.0 with 4 M NaOH
Add DEPC-treated H_2O to 100 ml
Stable at least 6 months at −20°C

S1 nuclease stop buffer *(UNIT 5.2)*
30.8 g ammonium acetate (4 M final)
✓ 10 ml 0.5 M EDTA, pH 8.0 (50 mM final)
5 mg (carrier) RNA (50 μg/ml final)
DEPC-treated H_2O to 100 ml
Stable at least 6 months at −20°C

SDS, 20% (w/v) *(UNIT 5.2)*
20 g sodium dodecyl sulfate (SDS)
DEPC-treated H_2O to 100 ml
Stable at least 3 months at room temperature

SDS-PAGE sample buffer, 2× *(UNIT 4.12)*
✓ 62.5 mM Tris·Cl, pH 8.8
20% (v/v) glycerol
✓ 4% (v/v) SDS
0.26 mg/ml bromphenol blue
Store up to 1 year at room temperature

SDS sample buffer (for discontinuous systems)

Prepare the 2× or 4× mixture described in Table A.1.4. Mix well, divide into ten 1-ml aliquots, and store at −80°C.

To avoid reducing proteins to subunits (if desired), omit 2-ME (reducing agent) and add 10 mM iodoacetamide to prevent disulfide interchange.

Table A.1.4 Preparation of SDS Sample Buffer

Ingredient	2×	4×	Final conc. in 1× buffer
0.5 M Tris·Cl, pH 6.8[a]	2.5 ml	5.0 ml	62.5 mM
SDS	0.4 g	0.8 g	2% (w/v)
Glycerol	2.0 ml	4.0 ml	10% (v/v)
Bromphenol blue	20 mg	40 mg	0.1% (w/v)
2-Mercaptoethanol	400 μl	800 μl	~300 mM
H_2O	to 10 ml	to 10 ml	—

[a]See recipe.

Secondary antibody diluent *(UNIT 1.2)*

Prepare Tris-buffered saline (TBS; see recipe), pH 7.2, containing 1% (v/v) normal animal serum (NAS) from species in which secondary antibody was made. Make fresh solution for each use.

Second-strand buffer, 10× *(UNIT 5.3)*

✓ 1 M Tris·Cl, pH 7.4
200 mM KCl
100 mM $MgCl_2$
50 mM dithiothreitol (DTT)
Filter sterilize and store 1-ml aliquots up to 1 year at −20°C

SOB/amp medium, per liter *(UNIT 4.15)*

20 g tryptone
5 g yeast extract
0.5 g NaCl
H_2O to 950 ml
10 ml 250 mM KCl
Adjust pH to 7.0 with 5 N NaOH
Autoclave and store up to 6 months at room temperature
Just before use, add 5 ml sterile 2 M $MgCl_2$ and 1 ml of 50 mg/ml ampicillin

Sodium acetate, 3 M

Dissolve 408 g sodium acetate trihydrate ($NaC_2H_3O_2 \cdot 3H_2O$) in 800 ml H_2O
Adjust pH to 4.8, 5.0, or 5.2 (as desired) with 3 M acetic acid (see Table A.1.1)
Add H_2O to 1 liter
Filter sterilize
Store up to 6 months at room temperature

Sodium acetate buffer *(APPENDIX 2E)*

✓ 100 mM sodium acetate, pH 5.2
✓ 1 mM EDTA

Sodium acetate–saturated phenol *(APPENDIX 2E)*

Melt 100 grams ultrapure phenol (redistilled nucleic acid grade) and mix with 100 ml sodium acetate buffer (see recipe). Shake vigorously, and let phases separate; phenol is the lower phase. Preheat to 60°C before using.

Sodium iodide (NaI) solution, 6 M *(APPENDIX 2C)*

Dissolve 0.75 g Na_2SO_3 in 40 ml H_2O. Add 45 g NaI (Sigma) and stir until dissolved (~30 min). Filter through Whatman paper or nitrocellulose and store 3 to 4 months in the dark (in aluminum foil) at 4°C. Discard if precipitate is observed.

Sodium phosphate, pH 7.0, 100 mM and 10 mM *(UNIT 5.1)*

100 mM stock solution:
5.77 ml 1 M Na_2HPO_4
4.23 ml 1 M NaH_2PO_4
H_2O to 100 ml
Store up to 3 months at room temperature

Sodium phosphate buffer, 0.1 M

Solution A: 27.6 g $NaH_2PO_4 \cdot H_2O$ per liter (0.2 M final)

Solution B: 53.65 g $Na_2HPO_4 \cdot 7H_2O$ per liter (0.2 M final)

Referring to Table A.1.3 for desired pH, mix the indicated volumes of solutions A and B, then dilute with water to 200 ml. Filter sterilize if necessary. Store up to 3 months at room temperature.

This buffer may be made as a 5- or 10-fold concentrate simply by scaling up the amount of sodium phosphate in the same final volume. Phosphate buffers show concentration-dependent changes in pH, so check the pH of the concentrate by diluting an aliquot to the final concentration.

Sodium selenite, 300 μM *(UNIT 3.2)*

0.5 M stock solution:
86.5 g sodium selenite (Sigma)
1 liter H_2O
Store up to 12 months at 4°C

300 μM working solution:
6 μl 0.5 M stock solution
10 ml H_2O
Filter through 0.22-μm Millipore filter
Store up to 12 months at 4°C

SSC (sodium chloride/sodium citrate), 20×

Dissolve in 900 ml H_2O:
175 g NaCl (3 M final)
88 g trisodium citrate dihydrate (0.3 M final)
Adjust pH to 7.0 with 1 M HCl
Adjust volume to 1 liter
Filter sterilize
Store up to 6 months at room temperature

SSPE (sodium chloride/sodium phosphate/EDTA), 20×

175.2 g NaCl
27.6 g $NaH_2PO_4 \cdot H_2O$
7.4 g disodium EDTA
800 ml H_2O
Adjust pH to 7.4 with 6 M NaOH, then bring volume to 1 liter with H_2O
Filter sterilize
Store up to 6 months at room temperature
The final sodium concentration of 20× SSPE is 3.2 M.

Starvation medium *(UNIT 4.12)*

Eagle's MEM (Life Technologies) containing:
2 mM L-glutamine
✓ 20 mM HEPES
Store up to 1 month at 4°C

Stock primer mix *(UNIT 4.3)*

10 µl each of 100 mM dATP, dCTP, dGTP, dTTP solutions (12.5 mM final)
10 µl 50 OD/ml or 820 µg/ml poly $(dT)_{19-24}$
30 µl H_2O
Store 80-µl aliquots for up to 1 year at −20°C

Stripping solution *(UNIT 5.1)*

✓ 1% (w/v) SDS
✓ 0.1× SSC
✓ 40 mM Tris·Cl, pH 7.5 to 7.8
Store up to 1 year at room temperature

Where formamide stripping is desired, prepare the above solution and add an equal volume of formamide just before use.

Substrate buffer *(UNIT 5.7)*

✓ 50 mM Tris·Cl, pH 7.5 (HRP) or
✓ Dioxetane phosphate substrate buffer (AP)

Sucrose buffer *(UNIT 2.2)*

2× sucrose buffer:
21.9 g sucrose (0.64 M)
✓ 2 ml 0.1 M EDTA (2 mM)
240 mg HEPES (10 mM)
100 ml H_2O

1× sucrose buffer:
Dilute 2× sucrose buffer 1:1 with H_2O.
Adjust pH to 7.4 with NaOH.
Store up to 5 days at 4°C.

When performing multiple brain dissections, use freshly prepared 1× sucrose buffer containing protease inhibitors (see recipe) and/or 0.25 mM DTT.

Sucrose gradient *(UNIT 2.2)*

7 ml 1.2 M sucrose
6 ml 1.0 M sucrose
6 ml 0.8 M sucrose
6 ml 0.6 M sucrose
6 ml 0.4 M sucrose

Prepare above solutions in Medium L (see recipe) and store up to 5 days at 4°C. On day of experiment, layer them on top of each other in a thin-walled, 38.5-ml, 25 × 89–mm ultracentrifuge tube. Keep at 4°C until use.

Sucrose-infiltration solution *(UNIT 1.1)*

✓ 0.1 M sodium phosphate buffer, pH 7.4
0.9% (w/v) saline
20% to 30% (w/v) sucrose
Store indefinitely at 4°C

Sucrose solution, 10%, 30%, and 60% (w/v) *(UNIT 4.14)*

Prepare 100 ml of a 10× stock solution of DPBS (see recipe; with calcium and magnesium). Add 600 g sucrose, mix, and add water to 1 liter. Adjust pH to 7.4 with 1 M NaOH. Autoclave for no more than 20 min. Use this 60% sucrose stock solution to make 30% and 10% sucrose solutions by diluting appropriately with 1× DPBS. Store up to 1 year at 4°C. Warm the solutions to room temperature before use.

When 60% sucrose is autoclaved, it will turn yellow and cloudy. However, it will clarify as it cools, and any remaining precipitate can be dissolved by swirling the bottle.

Super stain *(UNIT 5.10)*

50% (v/v) methanol
10% (v/v) acetic acid
0.1% (w/v) Coomassie brilliant blue
Store up to 6 months at room temperature

T4 polynucleotide kinase buffer, 10× *(UNIT 5.2)*

Dissolve the following in 80 ml DEPC-treated H_2O:
6.05 g Tris base (0.5 M Tris·Cl final)
2.03 g $MgCl_2$ hexahydrate (0.1 M final)
25 mg spermidine trihydrochloride (1.0 mM final)
✓ 200 µl 0.5 M EDTA, pH 8.0 (1.0 mM final)
Adjust pH to 7.6 with concentrated HCl
Add 0.77 g dithiothreitol (DTT; 50 mM final)
Add DEPC-treated H_2O to 100 ml
Stable at least 6 months at −20°C

TAE (Tris/acetate/EDTA) electrophoresis buffer, 10× *(UNIT 4.15)*

24.2 Tris base
5.71 ml glacial acetic acid
3.72 g $Na_2EDTA \cdot 2H_2O$
H_2O to 1 liter
Store indefinitely at room temperature

Tailing buffer, 5× *(UNIT 1.3)*

500 mM potassium cacodylate, pH 7.2
10 mM $CoCl_2$
1.0 mM DTT
Store indefinitely at −20°C

TBE (Tris/borate/EDTA) electrophoresis buffer, 10×

108 g Tris base (890 mM)
55 g boric acid (890 mM)
960 ml H_2O
✓ 40 ml 0.5 M EDTA, pH 8.0 (20 mM final)
Store up to 6 months at room temperature

TBS (Tris-buffered saline)

✓ 100 mM Tris·Cl, pH 7.5
0.9% (w/v) NaCl
Store up to several months at 4°C

TBS (Tris-buffered saline) *(UNIT 4.8)*
50 mM Tris base
0.9% (w/v) NaCl
Adjust pH to 7.4 with 1 M HCl
Store up to 6 months at room temperature

TBS (Tris-buffered saline) *(UNIT 4.11)*
✓ 25 mM Tris·Cl pH 7.4
140 mM NaCl
30 mM KCl
7 mM Na_2HPO_4
6 mM dextrose
Filter sterilize
Store up to 1 year at room temperature

TE (Tris/EDTA) buffer
✓ 10 mM Tris·Cl, pH 7.4, 7.5, or 8.0 (or other pH)
✓ 1 mM EDTA, pH 8.0
Store up to 6 months at room temperature

Terminal deoxynucleotidyl transferase (TdT) buffer, 5× *(UNIT 4.1)*
1 M potassium cacodylate
✓ 125 mM Tris·Cl, pH 7.4
1.25 μg/μl BSA
Store at −20°C for ≤6 weeks

TES solution *(APPENDIX 2E)*
✓ 10 mM Tris·Cl, pH 7.4
✓ 5 mM EDTA
✓ 1% (w/v) SDS

Thionin solution *(UNIT 1.1)*
9 ml glacial acetic acid
1.88 g thionin (Sigma)
1.08 g NaOH
H_2O to 750 ml
Store in a dark bottle indefinitely at room temperature and filter before each use

Thrombin, 200 U/ml *(UNIT 3.5)*

Dissolve 1000 U thrombin (Sigma; lyophilized powder, 50 to 100 NIH U/mg) in 5 ml Gey's balanced buffer solution (GBSS; e.g., Life Technologies) or HBSS (see recipe or, e.g., Life Technologies).

TMB buffer, 10× *(UNIT 5.9)*

Dissolve 37.4 g sodium citrate·3 H_2O in 250 ml water. Adjust pH to 5.5 with saturated citric acid solution (92.5 g sodium citrate·3 H_2O in 50 ml water). Store up to 6 months at 4°C.

This buffer is also known as 1.1 M citrate buffer.

TMB stock *(UNIT 5.9)*

Dissolve 1 mg 3,3′,5,5′-tetramethylbenzidine dihydrochloride (TMB; e.g., Sigma, Pierce) in 0.1 ml DMSO. Store up to 6 months at −20°C.

TMB working solution *(UNIT 5.9)*
✓ 0.1 ml TMB (3,3′,5,5′-tetramethylbenzidine dihydrochloride) stock
✓ 1 ml 10× TMB buffer
9 ml H_2O
Use immediately

TNB-G buffer *(UNIT 4.18)*
✓ 0.1 M Tris·Cl, pH 7.5
0.15 M NaCl
0.5% (v/v) blocking reagent (MICROMAX TSA kit, NEN Life Science Products)
10% (v/v) goat serum (Invitrogen Life Technologies)

Prepare Tris/NaCl and then add blocking reagent in small portions with continuous stirring. Heat gradually to 65°C with continuous stirring until the blocking reagent is completely dissolved (may take several hours). The solution should be slightly opaque, without visible particles. Do not filter the solution. Cool to room temperature. Aliquot and store up to 6 months at −20°C. Add 10% goat serum before use.

TNE buffer, 10× *(APPENDIX 2G)*
100 mM Tris base
✓ 10 mM EDTA
2.0 M NaCl
Adjust pH to 7.4 with concentrated HCl
As needed, dilute with H_2O to desired concentration

TNT buffer *(UNIT 4.18)*
✓ 0.1 M Tris·Cl, pH 7.5
0.15 M NaCl
0.05% (v/v) Tween 20
Store up to several weeks at room temperature

Toluidine blue, 1% (w/v) *(UNIT 4.9)*
1 g toluidine blue
1 g disodium tetraborate
100 ml distilled water

Dissolve the disodium tetraborate in the water. Add the toluidine blue and dissolve by sonication. Filter before use. Store at room temperature.

Transcription buffer, 5× *(UNIT 1.3)*
Prepare using DEPC-treated water (see recipe):
✓ 200 mM Tris·Cl, pH 7.9
30 mM $MgCl_2$
50 mM NaCl
10 mM spermidine
Store indefinitely at −20°C

Transcription buffer, 10× *(UNIT 4.12)*
✓ 400 mM HEPES, pH 7.4 (adjusted with KOH)
60 mM magnesium acetate (Sigma)
20 mM spermidine (Sigma)
Store up to 1 month at −20°C

Transcription buffer, 10× *(UNIT 5.2)*
Dissolve the following in 80 ml DEPC-treated H_2O:
4.84 g Tris base (400 mM Tris·Cl final)
1.42 g $MgCl_2$·6 H_2O (70 mM final)
0.51 g spermidine trihydrochloride (20 mM final)
1.46 g NaCl (250 mM final)
Adjust pH to 7.5 with 1 M HCl
Add 1.54 g dithiothreitol (DTT; 100 mM final)
Add DEPC-treated H_2O to 100 ml
Stable at least 6 months at −20°C

Transfer buffer *(UNIT 5.7)*

Add 18.2 g Tris base and 86.5 g glycine to 4 liters of water. Add 1200 ml methanol and bring to 6 liters with water. The pH of the solution is ~8.3 to 8.4. For use with PVDF filters, decrease methanol concentration to 15%; for nylon filters, omit methanol altogether.

CAPS transfer buffer can also be used. Add 2.21 g cyclohexylaminopropane sulfonic acid (CAPS; free acid), 0.5 g DTT, 150 ml methanol, and water to 1 liter. Adjust to pH 10.5 with NaOH and chill to 4°C. For proteins >60 kDa, reduce methanol content to 1%.

Transferrin stock, 2 mg/ml *(UNIT 3.2)*

100 mg human transferrin (Sigma)
50 ml H_2O
Filter through 0.22-μm Millipore filter
Store in 2.5-ml aliquots up to 12 months at −80°C

Triethanolamine/saline solution *(UNIT 1.1)*

Dissolve 74 g triethanolamine and 36 g NaCl in 3500 ml distilled water, then adjust pH to 8.0 with 5 N NaOH. Add water to 4 liters. Store indefinitely at room temperature. If solution is to be in phosphate-buffered saline add 0.122 g KH_2PO_4 and 0.815 g Na_2PO_4.

Tris·Cl, 1 M

Dissolve 121 g Tris base in 800 ml H_2O
Adjust to desired pH with concentrated HCl
Adjust volume to 1 liter with H_2O
Filter sterilize if necessary
Store up to 6 months at 4°C or room temperature

Approximately 70 ml HCl is needed to achieve a pH 7.4 solution, and ~42 ml for a solution that is pH 8.0.

IMPORTANT NOTE: *The pH of Tris buffers changes significantly with temperature, decreasing approximately 0.028 pH units per 1°C. Tris-buffered solutions should be adjusted to the desired pH at the temperature at which they will be used. Because the pK_a of Tris is 8.08, Tris should not be used as a buffer below pH ~7.2 or above pH ~9.0.*

Triton X-100 lysis buffer *(UNIT 5.12)*

✓ *TSA solution containing:*
1% (w/v) Triton X-100 (store at room temperature in dark)
1% (w/v) bovine hemoglobin (store frozen)
1 mM iodoacetamide (from powder)
Aprotinin (0.2 trypsin inhibitor U/ml)
1 mM PMSF (store 100 mM stock in 100% ethanol up to 6 months at −20°C)
Prepare fresh

1 mM AEBSF, added fresh from a 0.1 M stock solution in H_2O, can be used in place of PMSF. AEBSF stock can be stored up to 1 year at −20°C.

Trituration solution *(UNIT 3.4)*

0.1 g DNase I (from bovine pancreas; 1700 to 3300 U/mg dry weight; Worthington; 1% w/v final)
✓ 10 ml 1× CMF-HBSS
Adjust pH to 7.2 with 0.1 N NaOH
Filter sterilize through 0.2-μm filter
Divide into 1.5-ml aliquots
Store up to 6 months at −20°C

Trypsin buffer *(UNIT 3.3)*

✓ 0.1 M Tris·Cl, pH 7.5
0.1% (w/v) $CaCl_2$
0.1% (w/v) trypsin (e.g., type II-S, Sigma)

Trypsin/DNase solution *(UNIT 3.6)*

150 mg trypsin (TRL trypsin, Worthington)
15 mg DNase (DP DNase, Worthington)
✓ 15 ml PBS/glucose containing 5 mg/ml $MgSO_4$
90 μl 1 N NaOH
Store up to 1 year at −20°C

Trypsin/EDTA solution *(UNIT 3.2)*

Stock solution:
0.25 g trypsin (trypsin 1:250, Difco; 0.25% w/v final)
✓ 100 ml Dis A
Filter through 0.22-μm Millipore filter
Aliquot and store up to 12 months at −20°C
Working solution:
Dilute trypsin stock solution (see recipe) 1:1 with Dis A solution before use.

Trypsin/EDTA solution *(APPENDIX 2I)*

✓ *Prepare in sterile HBSS or 0.9% (w/v) NaCl:*
0.25% (w/v) trypsin
✓ 0.2% (w/v) EDTA
Store ≤1 year (until needed) at −20°C

Specific applications may require different concentrations of trypsin.

Trypsin/EDTA solution is commercially available in various concentrations including 10×, 1×, and 0.25% (w/v). It is received frozen from the manufacturer and can be thawed and aseptically divided into smaller volumes. Preparing trypsin/EDTA from powdered stocks may reduce its cost; however, most laboratories prefer commercially prepared solutions for convenience.

EDTA (disodium ethylenediamine tetraacetic acid) is added as a chelating agent to bind Ca^{2+} and Mg^{2+} ions that can interfere with the action of trypsin.

Trypsinization solution *(UNIT 3.4)*

0.025 g trypsin (from bovine pancreas; 200 to 300 U/mg protein; Worthington; 0.1% w/v final)
0.1 g DNase I (from bovine pancreas; 1700 to 3300 U/mg dry weight; Worthington; 0.4% w/v final)
✓ 25 ml dissection solution
Adjust pH to 7.2 with 0.1 N NaOH
Filter sterilize through 0.2-μm filter
Divide into 2.5-ml aliquots
Store up to 6 months at −20°C

TSA (Tris/saline/azide) solution (*UNIT 5.12*)
✓ 10 mM Tris·Cl, pH 8.0
140 mM NaCl
0.025% NaN_3
CAUTION: Sodium azide (NaN_3) is poisonous; wear gloves.

TY medium, 2× (*UNIT 5.9*)
Per liter:
16 g tryptone
10 g yeast extract
5 g NaCl
Store at room temperature

TY and TYAG plates, 2× (*UNIT 5.9*)

Dissolve 15 g bacto agar in 1 liter 2× TY medium (see recipe). Sterilize by autoclaving and allow to cool to 55°C. At this temperature add antibiotics and glucose (e.g., 100 µg/ml ampicillin and 2% w/v glucose for TYAG plates) just prior to pouring into plates. Store plates up to 1 month at 4°C.

Tyrode's solution (*UNIT 2.4*)
For 1 liter:
8 g NaCl (132 mM final)
1 ml 26.5% (w/v) $CaCl_2·2H_2O$ (1.8 mM final)
1 ml 5% (w/v) $NaH_2PO_4·2H_2O$ (0.32 mM final)
1 g glucose (5.56 mM final)
1 g $NaHCO_3$ (11.6 mM final)
0.2 g KCl (2.68 mM final)
Prepare fresh

Prior to use, add 100 µl of 100 U/ml heparin solution to ech liter of Tyrode's solution and gas with 95% O_2/5% CO_2 for 30 min.

Tyrode's solution (*UNIT 4.9*)
For 1 liter:
8 g NaCl (132 mM)
1 ml 26.5% (w/v) $CaCl_2.2H_2O$ (1.8 mM)
1 ml 5% (w/v) $NaH_2PO_4.2H_2O$ (0.32 mM)
1 g glucose (5.56 mM)
1 g $NaHCO_3$ (11.6 mM)
0.2 g KCl (2.68 mM)
Prepare fresh immediately before use

Prior to using, add 100 µl of 1000 U/ml heparin solution to each liter of Tyrode's and gas with 95% O_2/5% CO_2 for 30 min.

Virion lysis buffer (*UNIT 4.8*)
✓ 20 mM Tris·Cl, pH 7.8
✓ 10 mM EDTA, pH 7.8
✓ 1% (w/v) SDS
1 mg/ml proteinase K (Sigma)
Store up to 6 months at room temperature

VM culture medium (*UNIT 4.8*)
1:1 mixture of D-MEM, supplemented and Ham's F12 medium (both available from Sigma)
10% (v/v) fetal bovine serum (FBS)
4.0 mM L-glutamine (Life Technologies)
100 U/ml penicillin (Life Technologies)
1 mg/ml streptomycin (Life Technologies)
Store up to 1 month at 4°C

Wash buffer (*UNIT 2.2*)
6.142 ml 2 M NaCl (122 mM final)
500 µl 1 M KCl (5 mM final)
460 µl 0.25 M NaH_2PO_4 (1.15 mM final)
4 ml 0.5 M PIPES (20 mM final)
0.1 g D(+)-glucose (1 mg/ml final)
100 ml H_2O
Adjust to pH 6.8.
Prepare buffer fresh on day of experiment and keep at 4°C until use.

Wash buffer (*UNIT 5.12*)
0.1% (w/v) Triton X-100 (store at room temperature in dark)
✓ 50 mM Tris·Cl, pH 7.4
300 mM NaCl
✓ 5 mM EDTA
0.02% (w/v) sodium azide
Store up to 6 months at 4°C

Wash solution (*APPENDIX 2C*)
✓ 20 mM Tris·Cl, pH 7.4
✓ 1 mM EDTA
100 mM NaCl

Prepare above reagents as indicated and then dilute with an equal volume of 100% ethanol. Store 3 to 4 months at 0°C.

Water-saturated phenol (*APPENDIX 2E*)

Dissolve 100 g phenol crystals in H_2O at 60° to 65°C. Aspirate the upper water phase and store up to 1 month at 4°C.

Do not use buffering phenol in place of water-saturated phenol.

Xgal staining solution (*UNIT 4.5*)
Inorganic salt mix
✓ *PBS, pH 7.4 containing:*
4 mM $K_3Fe(CN)_6$ (from 400 mM stock)
4 mM $K_4Fe(CN)_6·3H_2O$ (from 400 mM stock)
2 mM $MgCl_2·6H_2O$ (from 200 mM stock)

Working solution: Immediately before use, add 20 mg/ml Xgal stock solution in dimethylformamide (store at −20°C) to a final concentration of 1 mg/ml.

Prepare the inorganic salt mix fresh for each experiment by adding appropriate amounts of the three stock solutions (all stored frozen at −20°C) to 10× PBS at ten times the final concentrations listed above, diluting this 10× mix to 1× with water, and then adding Xgal stock as described above.

Xgal staining solution (*UNIT 4.8*)
5 mM potassium ferrocyanide
5 mM potassium ferricyanide
2 mM $MgCl_2$
20 mg Xgal dissolved in 500 µl DMSO
Add ddH_2O to 20 ml
Keep solution in the dark at room temperature and use immediately

X-gal staining solution *(UNIT 4.14)*

✓ PBS, pH 7.5
20 mM $K_3Fe(CN)_6$
20 mM $K_4Fe(CN)_6 \cdot 3H_2O$
2 mM $MgCl_2$
Filter sterilize and store up to 1 year at 4°C

Before use, equilibrate solution to 37°C and add 20 μl/ml of 50 mg/ml of 5-bromo-4-chloro-3-indolyl-β-D-galactopyranoside (X-gal) in DMSO. Store X-gal solution in 1-ml aliquots up to several years at −20°C in the dark.

YPD medium and plate *(UNIT 4.2)*

Per liter:
10 g yeast extract (1% final)
20 g peptone (2% w/v final)
20 g dextrose (2% w/v final)

For plates:
20 g agar
1 NaOH pellet

This rich, complex medium—also known as YEPD medium—is widely used for the growth of yeast when special conditions are not required.

It is preferable to use a 20% (10×) solution of dextrose that has been filter sterilized or autoclaved separately (and added to the other ingredients after autoclaving) to prevent darkening of the medium and to promote optimal growth.

2YT medium *(UNIT 5.6)*

To 900 ml H_2O add:
16 g Bacto tryptone (Difco)
10 g Bacto yeast extract (Difco)
5 g NaCl

Shake until solutes have dissolved. Adjust pH to 7.0 with NaOH. Adjust volume of solution to 1 liter with H_2O. Sterilize by autoclaving 20 min at 15 psi on liquid cycle. Add antibiotics after solutions have cooled below 50°C or just prior to use.

2YT medium (Difco) is also available in powdered form from Fisher.

2YT plates *(UNIT 5.6)*

Add 15 g/liter Bacto agar (Difco) to 2YT medium (see recipe). Dissolve and autoclave. To pour plates, melt 2YT agar in microwave oven, add appropriate antibiotics when solution cools below 50°C, and pour ~25 ml into each 10-cm culture plate to be prepared. Allow to cool and solidify prior to use.

Z buffer *(UNIT 4.2)*

16.1 g $NaH_2PO_4.7H_2O$ (60 mM final)
5.5 g $NaH_2PO_4.H_2O$ (40 mM final)
0.75 g KCl (10 mM final)
0.246 g $MgSO_4.7H_2O$ (1 mM final)
2.7 ml 2-ME (50 mM final)
Adjust to pH 7.0 and bring to 1 liter with H_2O; do not autoclave

APPENDIX 2

Molecular Biology Techniques

APPENDIX 2A

Molecular Biology References

Many of the units in this manual (e.g., those found in Chapters 1, 4, and 5) describe methods and techniques for the cloning, expression, and structural analysis of neural genes and proteins. We assume that users of these protocols have at least some introductory background in recombinant DNA technology (or are working with a collaborator who does); therefore, we have not provided comprehensive coverage of all of these topics, but rather have concentrated on presenting selected techniques that will be of the most interest and use to the general neuroscience laboratory. More comprehensive coverage of these topics can be found in *Current Protocols in Molecular Biology* (CPMB), which is extensively cross-referenced throughout this manual. These cross-references are summarized in Table A.2A.1; alternatively, the reader may wish to

Table A.2A.1 Molecular Biology Techniques

Technique	CPMB reference
Antibody purification	*UNIT 11.11*
Autoradiography	*APPENDIX 3A*
Baculovirus	*UNITS 16.9–16.11*
Blotting	
dot and slot	*UNIT 2.9B*
northern	*UNIT 4.9*
Southern	*UNIT 2.9A*
western (immuno-)	*UNIT 10.8*
Chemiluminescence detection using streptavidin/HRP	*UNIT 10.8*
Cloning	
of PCR products	*UNIT 15.7*
blunt ends, using Klenow fragment	*UNIT 3.5*
subcloning, DNA fragments	*UNIT 3.16*
Database searching, BLAST	*UNIT 19.3*
Dialysis	*APPENDIX 3C*
DNA	
ethanol precipitation	*UNIT 2.1*
extraction from mammalian tissue	*UNIT 2.2*
phenol extraction	*UNIT 2.1*
preparation, miniprep	*UNIT 1.6*
quantitation, spectrophotometric	*APPENDIX 3D*
recovery from agarose gels	*UNIT 2.6*
sequencing	Chapter 7

continued

Table A.2A.1 Molecular Biology Techniques, continued

Technique	CPMB reference
E. coli	
growth in liquid medium	*UNIT 1.2*
growth on solid medium	*UNIT 1.3*
lysate preparation	*UNIT 1.7*
transformation	*UNIT 1.8*
Embryonic stem (ES) cells	
culture	*UNIT 23.5*
feeder cell preparation	*UNIT 23.2*
Expression systems	
baculovirus	*UNITS 16.9–16.11*
inducible, tet-regulated	*UNIT 16.21*
mammalian	*UNITS 16.12–16.14*
tagged proteins	*UNIT 16.12*
vaccinia	*UNITS 16.15–16.19*
Gel electrophoresis	
agarose	*UNIT 2.5A*
denaturing polyacrylamide	*UNIT 2.12*
gel drying	*UNIT 7.6*
low-gelling/melting temperature agarose	*UNIT 2.6*
nondenaturing polyacrylamide	*UNIT 2.7*
SDS–PAGE	*UNIT 10.2*
staining gels with Coomassie blue	*UNIT 10.6*
HPLC	*UNIT 10.12*
Hybridization conditions	
analysis	*UNIT 2.10*
bacterial filter	*UNITS 6.3, 6.4*
Immunoprecipitation	*UNIT 11.15*
In situ hybridization, in northern blot analysis	*UNIT 4.9*, Chapter 14
Libraries	
amplification of	
bacteriophage	*UNIT 5.10*
cosmid	*UNIT 5.11*
construction of	
cDNA	*UNITS 5.5, 5.8A*
cosmid	*UNIT 5.11*
genomic	*UNITS 5.7–5.9*
screening	
cDNA	*UNIT 6.3*
genomic	*UNIT 6.4*
Mass spectrometry	*UNIT 10.21*
Media	
E. coli	*UNITS 1.1–1.3*
selective	*UNIT 9.5*
tissue culture	*APPENDIX 3F*
yeast	*UNITS 13.1, 13.2*
mRNA, conversion into ds cDNA	*UNIT 5.5*
Mutagenesis	
site-directed	Chapter 8
deletion	*UNIT 3.17*

continued

Table A.2A.1 Molecular Biology Techniques, continued

Technique	CPMB reference
Oligonucleotides	
primer synthesis	*UNIT 2.11*
purification	*UNIT 2.12*
phosphorylation	*UNIT 3.10*
PCR	
general	*UNIT 15.1*
genotyping, using	*UNIT 15.1*
primer design	*UNIT 15.1*
primer synthesis	*UNIT 2.11*
Phage titering	*UNIT 1.11*
Phenol/chloroform extraction	*UNIT 4.1*
Plasmid preparation	
alkaline lysis miniprep	*UNIT 1.6*
cesium chloride density gradient	*UNIT 1.7*
phenol extraction	*UNIT 2.1A*
Protein determination	*UNIT 10.1A*
Random primer labeling	*UNIT 3.5*
Restriction endonuclease digestion	*UNIT 3.1*
Restriction mapping	*UNITS 3.1–3.3*
Retroviruses	
concentration of viral supernatants	*UNIT 9.12*
detection of helper virus in viral stocks	*UNIT 9.13*
life cycle	*UNIT 9.10*
preparation and titering	
in stable producer cell lines	*UNIT 9.10*
in transient producer cell lines	*UNIT 9.11*
RNA preparation and purification	
extraction with guanidinium isothiocyanate	*UNIT 4.2*
poly(A)$^+$	*UNIT 4.5*
total	*UNITS 4.1, 4.2*
Silanization of glassware	*APPENDIX 3B*
Spectrophotometry	*APPENDIX 3D*
T4 DNA ligase	*UNIT 3.14*
TCA precipitation	*UNIT 4.10*
Tissue culture, mammalian cell	*APPENDIX 3F*
Transfection, liposome-mediated	*UNIT 9.4*
Transformants, stable	*UNIT 9.5*
Transformation, by electroporation	*UNIT 1.8*
Trypan blue exclusion	*APPENDIX 3F*
Yeast	
allele replacement	*UNIT 13.10*
culture	*UNIT 13.6*
medium	*UNIT 13.1*
replica plating	*UNIT 13.3*
transformation with DNA	*UNIT 13.7*

consult any of a wide variety of textbooks that are specifically devoted to recombinant DNA techniques.

References: Ausubel et al., 2006; Sambrook et al., 1989

APPENDIX 2B

Optimization of Transfection

A2

When embarking upon any transfection procedure, a critical first step is to optimize conditions. Every mammalian cell type has a characteristic set of requirements for optimal introduction of foreign DNA; there is a tremendous degree of variability in the transfection conditions that work, even among cell types that are very similar to one another. Often, an experimenter must screen a wide variety of cell types for a desired regulatory trait, such as an appropriate response to a particular effector molecule. It is thus helpful to have a straightforward, systematic approach to optimizing transfection efficiency. Transient assay systems are particularly useful for this purpose. A fusion gene that is known to function in mammalian cells can be transfected into cells under a variety of conditions, and transfection efficiency can be easily monitored by assaying for the fusion gene product. The human growth hormone (hGH; see *CPMB UNIT 9.7A*) assay system is particularly useful for this purpose because both harvest and assay take very little time. However, any reporter system can be used to optimize transfection efficiency.

The single most important factor in optimizing transfection efficiency is selecting the proper transfection protocol. This usually comes down to a choice among calcium phosphate–mediated gene transfer (*CPMB UNIT 9.1*), DEAE-dextran-mediated gene transfer (*CPMB UNIT 9.2*), electroporation (*CPMB UNIT 9.3*), and liposome-mediated transfection (*CPMB UNIT 9.4*). Fusion techniques such as protoplast fusion and microinjection may also be considered. Cells are variable with respect to which transfection protocol is most efficient. It is recommended that any adherent cell line under investigation be tested for transfection ability with DEAE-dextran, calcium phosphate, and liposome-mediated transfection. Nonadherent cell lines can be transfected by electroporation and liposome-mediated transfection. Generally, if a cell can be grown in culture, it can be transfected.

CALCIUM PHOSPHATE TRANSFECTION

The primary factors that influence efficiency of calcium phosphate transfection are the amount of DNA in the precipitate, the length of time the precipitate is left on the cell, and the use and duration of glycerol or DMSO shock. A calcium phosphate optimization is shown in Table A.2B.1. Generally, higher concentrations of DNA (10 to 50 μg) are used in calcium phosphate transfection. Total DNA concentration in the precipitate can have a dramatic effect on efficiency of uptake of DNA with calcium phosphate-mediated transfection. With some cell lines, more than 10 to 15 μg of DNA added to a 10-cm dish results in excessive cell death and very little uptake of DNA. With other cell types, such as primary cells, a high concentration of DNA in the precipitate is necessary to get any DNA at all into the cell on a routine basis. For example, with human foreskin fibroblasts, transfection of 5 μg of a reporter plasmid with 5 μg of carrier DNA (e.g., pUC13) gives significantly less expression than does transfection of 5 μg of reporter plasmid with 35 μg of carrier DNA. Presumably, this is because the amount of DNA affects the nature of the precipitate and thus alters the fraction of the applied DNA that is taken up into cells.

The optimal length of time that the precipitate is left on cells varies with cell type. Some cell types, such as HeLa or BALB/c 3T3, are efficiently transfected by leaving the precipitate on for 16 hr. Other cell types cannot survive this length of exposure. Transfection efficiency of some

Table A.2B.1 Optimization of Calcium Phosphate Transfection

Dish (10-cm)	pXGH5 (μg)	pUC13 (μg)	Exposure to precipitate (hr)	Glycerol shock (min)
1	5	5	6	—
2	5	15	6	—
3	5	35	6	—
4	5	5	16	—
5	5	15	16	—
6	5	35	16	—
7	5	5	6	3
8	5	15	6	3
9	5	35	6	3
10	5	5	16	3
11	5	15	16	3
12	5	35	16	3

cell types, such as CHO DUKX BII, is dramatically increased by glycerol or DMSO shock. The pilot experiment listed will indicate whether the cell type is tolerant to long exposure to a calcium phosphate precipitate and whether glycerol shock should be used. Once the results of this experiment are in hand, finer experiments can be done to further optimize conditions. For example, if shocking with 10% glycerol for 3 min enhances transfection efficiency, an experiment varying the time of glycerol shock or also trying 10% and 20% DMSO shock might be done.

Once optimal conditions for transfection are found, extensive DNA curves varying the amount of reporter plasmid should be prepared. The total amount of DNA should be kept constant at the optimal level determined in the first experiment. The amount of reporter plasmid DNA (e.g., pXGH5) should be varied, and carrier DNA (e.g., pUC13) should be used to make up the difference. This is to ensure that transfections are performed under conditions where the amount of reporter plasmid in the cell is not saturating the cellular transcription and translation machinery.

DEAE-DEXTRAN TRANSFECTION

There are several factors that can be varied in DEAE-dextran transfection. The number of cells, concentration of DNA, and concentration of DEAE-dextran added to the dish are the most important to optimize. To a first approximation, most cell types that can be transfected using DEAE-dextran will have a preference for 1 to 10 μg DNA/10-cm dish and for 100 to 400 μg DEAE-dextran/ml of medium. Table A.2B.2 shows how the dishes in an optimization might be chosen. The 20-dish experiment consists of two sets of 10 dishes; one set is plated at 5×10^5 cells/dish, the other is plated at 2×10^6 cells/dish. Each set contains dishes that will be transfected with 1 to 10 μg of a reporter plasmid and 100 to 400 μg/ml DEAE-dextran. If an hGH expression vector such as pXGH5 is used, a time course of expression under each condition can be determined by removing 100-μl aliquots of the medium 2, 4, and 7 days posttransfection (with a medium change after the day 4 aliquot is removed).

With the results of this pilot experiment in hand, a second experiment using a narrower range of DEAE-dextran concentrations and a wider range of DNA doses should be undertaken. For example, if the cells appear to express more hGH at 100 μg/ml DEAE-dextran than at

Table A.2B.2 Optimization of DEAE-Dextran Transfection

5×10^5 cells/10-cm dish:			2×10^6 cells/10-cm dish:		
Dish	pXGH5 (μg)	DEAE-dextran (μg/ml)	Dish	pXGH5 (μg)	DEAE-dextran (μg/ml)
1	1	400	11	1	400
2	1	200	12	1	200
3	1	100	13	1	100
4	4	400	14	4	400
5	4	200	15	4	200
6	4	100	16	4	100
7	10	400	17	10	400
8	10	200	18	10	200
9	10	100	19	10	100
10	0	200	20	0	200

higher concentrations in the pilot experiment, the second experiment should cover from 25 to 150 μg/ml DEAE-dextran. Because DEAE-dextran is toxic to some cells, a brief exposure to small concentrations may be optimal. The wide range of added DNA in this experiment is crucial in two respects. First, it is valuable to know the smallest amount of the transfected reporter gene that can give a readily detectable signal. Second, the linearity of the dose of DNA with the amount of reporter gene expression generally decays for large amounts of input DNA. When excessive (i.e., nonlinear) amounts of DNA are used in transfection experiments, it is possible that the effects observed are dose-response effects rather than the phenomenon intended for study. This serious and common problem can be eliminated by doing a careful DNA dose-response curve as above.

ELECTROPORATION

Perhaps because it is not a chemically based protocol, electroporation tends to be less affected by DNA concentration than either DEAE-dextran- or calcium phosphate-mediated gene transfer. Generally, DNA amounts in the range of 10 to 40 μg/10^7 cells work well, and there is a good linear correlation between the amount of DNA present and the amount taken up. The parameter that can be varied to optimize electroporation is the amplitude and length of the electric pulse, the latter being determined by the capacitance of the power source. The extent to which this can be varied is determined by the electronics of the power supply used to supply the pulse. The objective is to find a pulse that kills between 20% and 60% of the cells. This generally is in the range of 1.5 kV at 25 μF. If excessive cell death occurs, the length of the pulse can be lowered by lowering the capacitance. Settings between 3 and 25 μF can be tried.

LIPOSOME-MEDIATED TRANSFECTION

Three primary parameters—the concentrations of lipid and DNA and incubation time of the liposome-DNA complex—affect the success of DNA transfection by cationic liposomes. These should be systematically examined to obtain optimal transfection frequencies.

Table A.2B.3 Optimization of Liposome-Mediated Transfection

Dish (35-mm)	pSV2CAT (μg)	Liposomes (μl)	Dish (35-mm)	pSV2CAT (μg)	Liposomes (μl)
1	0.1	1	11	5	5
2	0.1	2	12	5	10
3	0.1	4	13	5	15
4	0.1	8	14	5	20
5	0.1	12	15	5	30
6	0.5	1	16	10	5
7	0.5	2	17	10	10
8	0.5	4	18	10	15
9	0.5	8	19	10	20
10	0.5	12	20	10	30

Concentration of lipid. In general, increasing the concentrations of lipid improves transfection of four cell lines examined (CV-1 and COS-7 cells with Lipofectin, and HeLa and BHK-21 cells with TransfectACE, both available from Life Technologies). However, at high levels (>100 μg), the lipid can be toxic. For each particular liposome mixture tested, it is important to vary the amount as indicated in Table A.2B.3.

Concentration of DNA. In many of the cell types tested, relatively small amounts of DNA are effectively taken up and expressed. In fact, higher levels of DNA can be inhibitory in some cell types with certain liposome preparations. In the optimization protocol outlined in Table A.2B.3, the standard reporter vector pSV2CAT is used; however, any plasmid DNA whose expression can be easily monitored would be suitable.

Time of incubation. When the optimal amounts of lipid and DNA have been established, it is desirable to determine the length of time required for exposure of the liposome-DNA complex to the cells. In general, transfection efficiency increases with time of exposure to the liposome-DNA complex, although after 8 hr, toxic conditions can develop. HeLa or BHK-21 cells typically require ~3 hr incubation with the liposome-DNA complex for optimal tranfection, while CV-1 and COS-7 cells require 5 hr of exposure.

Contributor: John K. Rose

APPENDIX 2C

Purification and Concentration of DNA from Aqueous Solutions

IMPORTANT NOTE: The smallest amount of contamination of DNA preparations by recombinant phages or plasmids can be disastrous. All materials used for preparation of plasmid or phage DNA should be kept separate from those used for preparation of genomic DNA, and disposable items should be used wherever possible. Particular care should be taken to avoid contamination of commonly used rotors.

BASIC PROTOCOL

PHENOL EXTRACTION AND ETHANOL PRECIPITATION OF DNA

Materials *(see* APPENDIX 1 *for items with ✓)*

- $\leq$1 mg/ml DNA to be purified
- ✓ 25:24:1 (v/v/v) phenol/chloroform/isoamyl alcohol
- ✓ 3 M sodium acetate, pH 5.2
- 100% ethanol, ice cold
- 70% ethanol, room temperature
- ✓ TE buffer, pH 8.0
- Speedvac evaporator (Savant)

1. Add an equal volume of 25:24:1 phenol/chloroform/isoamyl alcohol to the DNA solution to be purified in a 1.5-ml microcentrifuge tube.

   DNA solutions containing $\leq$0.5 M monovalent cations can be used. Extracting volumes $\leq$100 μl is difficult; small volumes should be diluted to obtain a volume that is easy to work with.

   High salt concentrations can cause the inversion of the aqueous and organic phases. If this happens, the organic phase can be identified by its yellow color.

2. Vortex vigorously for 10 sec and microcentrifuge 15 sec at maximum speed, room temperature, to separate the phases. If DNA is viscous (containing a large amount of protein), microcentrifuge longer (1 to 2 min).
3. Carefully remove the top (aqueous) phase containing the DNA using a 200-μl pipettor and transfer to a new tube. If a white precipitate is present at the aqueous/organic interface, repeat steps 1 to 3. If starting with a small amount of DNA (<1 μg), improve recovery by reextracting the organic phase with 100 μl TE buffer, pH 8.0, then pool this aqueous phase with that from the first extraction.
4. Add 1/10 vol of 3 M sodium acetate, pH 5.2, to the solution of DNA. Mix by vortexing briefly or by flicking the tube several times with a finger. Make appropriate dilutions to keep NaCl and sodium acetate concentrations below 0.5 M.

   For high concentrations of DNA (>50 to 100 μg/ml), precipitation is essentially instantaneous at room temperature. If ethanol precipitation is not desirable, residual organic solvents can be removed by ether extraction (see Support Protocol 2). In this case, no salt should be added.

5. Add 2 to 2.5 vol (calculated *after* salt addition) of ice-cold 100% ethanol. Mix by vortexing and place tube in crushed dry ice for $\geq$5 min; or for $\geq$15 min at −70°C. Microcentrifuge 5 min at maximum speed and remove the supernatant.
6. Add 1 ml room-temperature 70% ethanol. Invert the tube several times and microcentrifuge as in step 5. If the DNA molecules being precipitated are very small (<200 bases), use 95% ethanol at this step. Remove the supernatant and dry the pellet in a desiccator under vacuum or in a Speedvac evaporator.
7. Dissolve the dry pellet in an appropriate volume of water if it is going to be used for further enzymatic manipulations requiring specific buffers. Dissolve in TE buffer, pH 8.0, if it is going to be stored indefinitely.

   DNA pellets will not dissolve well in high-salt buffers. To facilitate resuspension, the DNA concentration of the final solution should be kept at <1 mg/ml.

If DNA is resuspended in a volume of TE buffer or water to yield a DNA concentration of <1 mg/ml, small quantities (<25 μg) of precipitated plasmids or restriction fragments should dissolve quickly upon gentle vortexing or flicking of the tube. However, larger quantities of DNA may require vortexing and brief heating (5 min at 65°C) to resuspend. High-molecular-weight genomic DNA may require one to several days to dissolve and should be shaken gently (not vortexed) to avoid shearing, particularly if it is to be used for cosmid cloning or other applications requiring high-molecular-weight DNA. Gentle shaking on a rotating platform or a rocking apparatus is recommended.

A2

ALTERNATE PROTOCOL 1

PRECIPITATION OF DNA USING ISOPROPANOL

Equal volumes of isopropanol and DNA solution are used in precipitation. Note that the isopropanol volume is half that of the given volume of ethanol in precipitations. This allows precipitation from a large starting volume (e.g., 0.7 ml) in a single microcentrifuge tube. Isopropanol is less volatile than ethanol and takes longer to remove by evaporation. Some salts are less soluble in isopropanol (compared to ethanol) and will be precipitated along with nucleic acids. Extra washings may be necessary to eliminate these contaminating salts.

SUPPORT PROTOCOL 1

CONCENTRATION OF DNA USING BUTANOL

Additional Materials *(also see Basic Protocol)*

sec-butanol
Polypropylene tube

1. Add an equal volume of *sec*-butanol to the sample and mix well by vortexing or by gentle inversion (if the DNA is of high molecular weight). Perform extraction in a polypropylene tube, as butanol will damage polystyrene. Centrifuge 5 min at 1200 × *g*, room temperature, or in a microcentrifuge for 10 sec at maximum speed.
2. Remove and discard the upper (*sec*-butanol) phase. Repeat steps 1 and 2 until the desired volume of aqueous solution is obtained.
3. Extract the lower, aqueous phase with 25:24:1 phenol/chloroform/isoamyl alcohol and ethanol precipitate (see Basic Protocol) or remove *sec*-butanol by two ether extractions (see Support Protocol 2).

    Addition of too much sec-butanol can result in complete loss of the water phase into the sec-butanol layer. If this happens, add 1/2 vol water back to the sec-butanol, mix well, and spin. The DNA can be recovered in this new aqueous phase and can be further concentrated with smaller amounts of sec-butanol.

SUPPORT PROTOCOL 2

REMOVAL OF RESIDUAL PHENOL, CHLOROFORM, OR BUTANOL BY ETHER EXTRACTION

The final DNA solution will be free of organic solvents and will have salt concentrations that are roughly three-fourths of those that were in the aqueous solution before phenol extraction (solute concentrations are lowered in the two phenol/chloroform/isoamyl alcohol extractions steps).

Materials *(see* APPENDIX 1 *for items with ✓)*

- Diethyl ether
- DNA sample
- ✓ TE buffer, pH 8.0
- Polypropylene tube

1. Mix diethyl ether with an equal volume of water or TE buffer, pH 8.0, in a polypropylene tube. Vortex vigorously for 10 sec and let the phases separate (ether being the top phase).
2. Add an equal volume of hydrated ether to the DNA sample. Mix well by vortexing or by gentle inversion (if the DNA is of high molecular weight). Microcentrifuge 5 sec at maximum speed or let the phases separate by setting the tube upright in a test tube rack. Remove and discard the top (ether) layer. Repeat.
3. Remove ether by leaving the sample open under a fume hood for 15 min (small volumes, <100 μl) or under vacuum for 15 min (larger volumes).

ALTERNATE PROTOCOL 2

PURIFICATION OF DNA USING GLASS BEADS

Additional Materials *(also see Basic Protocol; see* APPENDIX 1 *for items with ✓)*

- ✓ 6 M sodium iodide (NaI) solution
- DNA in a 50- to 200-μl volume
- ✓ Glass bead suspension
- ✓ Wash solution
- ✓ TE buffer, pH 8.0
- Incubator or water bath at 45°C

NOTE: The above materials are also available as commercial kits (e.g., Glas-Pac, National Scientific Supply; GeneClean, Bio101; and Qiaex Gel Extraction Kit, Qiagen).

1. Add 3 vol of 6 M NaI solution to DNA in a 1.5-ml microcentrifuge tube. Add glass bead suspension as follows: for amounts of DNA <5 μg, use 5 μl glass bead suspension; for amounts of DNA >5 μg, use 5 μl plus an additional 1 μl for each 0.5-μg increment above 5 μg. Incubate 5 min at room temperature.
2. Microcentrifuge DNA/glass beads 5 sec at maximum speed. Remove and discard supernatant. To enhance yield, save the supernatant and reincubate with another sample of glass bead suspension as in step 1.
3. Wash the DNA/glass bead pellet three times with 500 μl wash solution. Lightly vortex the mixture to resuspend, then microcentrifuge briefly at maximum speed to pellet the beads.
4. Resuspend pellet in TE buffer, pH 8.0, at 0.5 μg/μl. Incubate 2 to 3 min at 45°C to elute DNA from the glass beads. Microcentrifuge 1 min and transfer the DNA-containing supernatant to a fresh tube. Store at 4°C until use.

ALTERNATE PROTOCOL 3

PURIFICATION AND CONCENTRATION OF RNA AND DILUTE SOLUTIONS OF DNA

The following adaptations to the purification procedure (see Basic Protocol) are used if RNA or dilute solutions of DNA are to be purified.

Purification and Concentration of RNA

The procedure outlined in the Basic Protocol is identical for purification of RNA, except that 2.5 vol ethanol should be used routinely for the precipitation (step 5). It is essential that all water used directly or in buffers be treated with diethylpyrocarbonate (DEPC) to inactivate RNases.

Dilute Solutions of DNA

When DNA solutions are dilute (<10 μg/ml) or when <1 μg of DNA is present, the ratio of ethanol to aqueous volume should be increased to 3:1 and the time on dry ice (step 5) extended to 30 min. Microcentrifugation should be carried out for 15 min in a cold room to ensure the recovery of DNA from these solutions.

Nanogram quantities of labeled or unlabeled DNA can be efficiently precipitated by the use of carrier nucleic acid. A convenient method is to add 10 μg of commercially available tRNA from *E. coli*, yeast, or bovine liver to the desired DNA sample. The DNA will be coprecipitated with the tRNA. The carrier tRNA will not interfere with most enzymatic reactions, but will be phosphorylated efficiently by polynucleotide kinase and should not be added if this enzyme will be used in subsequent radiolabeling reactions.

Recovery of small quantities of short DNA fragments and oligonucleotides can be enhanced by adding magnesium chloride to a concentration of <10 mM before adding ethanol (step 4). However, DNA precipitated from solutions containing >10 mM magnesium or phosphate ions is often difficult to redissolve and such solutions should be diluted prior to ethanol precipitation.

DNA in Large Aqueous Volumes (>0.4 to 10 ml)

Larger volumes can be accommodated by simply scaling up the amounts used in the Basic Protocol or by using butanol concentration (see Support Protocol 1). For the phenol extraction (see Basic Protocol, steps 1 through 3), tightly capped 15- or 50-ml polypropylene tubes should be used, as polystyrene tubes cannot withstand the phenol/chloroform mixture. Centrifugation steps should be performed for 5 min at speeds not exceeding 1200 × *g*, room temperature. The ethanol precipitate (step 6) should be centrifuged in thick-walled Corning glass test tubes (15- or 30-ml capacity) for 15 min in fixed-angle rotors at 8000 × *g*, 4°C. Glass tubes should be silanized to facilitate recovery of small amounts of DNA (<10 μg).

ALTERNATE PROTOCOL 4

REMOVAL OF LOW-MOLECULAR-WEIGHT OLIGONUCLEOTIDES AND TRIPHOSPHATES BY ETHANOL PRECIPITATION

The use of ammonium acetate in place of sodium acetate allows the preferential precipitation of longer DNA molecules. If the nucleic acid is to be phosphorylated, this protocol should not be used because T4 polynucleotide kinase is inhibited by ammonium ions.

Additional Materials *(also see Basic Protocol)*

4 M ammonium acetate, pH 4.8

1. Add an equal volume of 4 M ammonium acetate, pH 4.8, to the DNA solution. Mix well. Add 2 vol (calculated after salt addition) of ice-cold 100% ethanol. Vortex and set tube in crushed dry ice for 5 min or ≥15 min at −70°C.

2. Microcentrifuge 5 min at high speed, room temperature. Carefully remove supernatant and redissolve pellet in 100 μl TE buffer, pH 8.0. Repeat steps 1 and 2.

3. Wash and resuspend DNA (see Basic Protocol, steps 6 and 7).

Contributor: David Moore

A2

APPENDIX 2D

Preparation of Genomic DNA from Mammalian Tissue

BASIC PROTOCOL

PREPARATION OF GENOMIC DNA FROM MAMMALIAN TISSUE

To minimize the activity of endogenous nucleases, it is essential to rapidly isolate, mince, and freeze tissue. Tissue culture cells should be cooled and washed quickly.

Materials *(see* APPENDIX 1 *for items with ✓)*

Tissues, whole, or cultured cells
Liquid nitrogen
✓ Digestion buffer
✓ PBS, ice cold
✓ 25:24:1 (v/v/v) phenol/chloroform/isoamyl alcohol
7.5 M ammonium acetate
70% and 100% ethanol
✓ TE buffer, pH 8
Incubator or water bath at 50°C, with shaker

If beginning with whole tissue

1a. As soon as possible after excision, quickly mince tissue and freeze in liquid nitrogen. If working with liver, remove the gallbladder, which contains high levels of degradative enzymes.

2a. Starting with between 200 mg and 1 g, grind tissue with a prechilled mortar and pestle, or crush with a hammer to a fine powder (keep the tissue fragments, if crushing is incomplete).

3a. Suspend the powdered tissue (without clumps) in 1.2 ml digestion buffer per 100 mg tissue. Proceed to step 4.

If beginning with tissue culture cells

1b. Pellet suspension culture out of its serum-containing medium. Trypsinize adherent cells and collect cells from the flask. Centrifuge 5 min at 500 × *g*, 4°C, and discard supernatant.

2b. Resuspend cells with 1 to 10 ml ice-cold PBS. Centrifuge 5 min at 500 × *g* and discard supernatant. Repeat this resuspension and centrifugation step.

3b. Resuspend cells in 1 vol digestion buffer. For $<3 \times 10^7$ cells, use 0.3 ml digestion buffer. For larger numbers of cells use 1 ml digestion buffer/10^8 cells.

4. Incubate the samples with shaking at 50°C for 12 to 18 hr in tightly capped tubes.

   The samples will be viscous. After a 12-hr incubation, the tissue should be almost indiscernible, a sludge should be apparent from the organ samples, and tissue culture cells should be relatively clear.

5. Thoroughly extract the samples with an equal volume of phenol/chloroform/isoamyl alcohol (*APPENDIX 2C*). Centrifuge 10 min at 1700 × *g* in a swinging bucket rotor.

   If the phases do not resolve well, add another volume of digestion buffer, omitting proteinase K, and repeat the centrifugation.

   If there is a thick layer of white material at the interface between the phases, repeat the organic extraction.

6. Transfer the aqueous (top) layer to a new tube and add 1/2 vol of 7.5 M ammonium acetate and 2 vol (of original amount of top layer) of 100% ethanol (DNA should immediately form a stringy precipitate). Recover DNA by centrifuging 2 min at 1700 × *g*.

7. Rinse the pellet with 70% ethanol to remove residual salt and phenol. Decant ethanol and air dry the pellet. Resuspend DNA at ~1 mg/ml in TE buffer until dissolved (~2 mg DNA from 1 g mammalian cells). Shake gently at room temperature or at 65°C for several hours to facilitate solubilization. Store indefinitely at 4°C.

Reference: Gross-Bellard et al., 1973

Contributor: William M. Strauss

APPENDIX 2E

Preparation of RNA from Tissues and Cells

STRATEGIC PLANNING

The major source of failure in any attempt to produce RNA is contamination by ribonuclease. To avoid contamination problems, take the following precautions:

1. Solutions. Any water or salt solutions used in RNA preparation should be treated with diethylpyrocarbonate (DEPC). Solutions containing Tris cannot be effectively treated with DEPC because Tris reacts with DEPC to inactivate it.

2. Glassware and plastic. Autoclaving will not fully inactivate many RNases. Glassware can be baked at 300°C for 4 hr. Plasticware straight out of the package is generally free from contamination and can be used as is.

3. Hands are a major source of contaminating RNase. Wear gloves.

BASIC PROTOCOL 1

HOT PHENOL EXTRACTION OF RNA

Materials *(see APPENDIX 1 for items with ✓)*

Tissue or cells
✓ Sodium acetate buffer, ice cold
✓ PBS, ice cold *or* serum-free medium, ice cold
10% (w/v) SDS

- ✓ Sodium acetate-saturated phenol, at 60°C
- Chloroform
- 95% and 70% ethanol, −20°C
- ✓ TE buffer, pH 8.0

- 15- and 50-ml polypropylene tubes
- Tissue homogenizer (Brinkmann or Polytron type)
- 60°C water bath

A2

1a. *If starting with tissue:* Place tissue in a 15-ml polypropylene tube. Add ~5 ml ice-cold sodium acetate buffer. Keeping tube on ice, homogenize 10 sec with tissue homogenizer.

1b. *If starting with tissue culture cells:* Remove 10^6 to 10^8 cells in tissue culture medium (if monolayer cells, by trypsinization or scraping) to a 15-ml polypropylene tube. Pellet the cells by centrifuging 5 min at 800 × *g* and discard the supernatant. Rinse the cells with ice-cold PBS or ice-cold serum-free medium, centrifuge and discard the supernatant. Resuspend the cells in ~5 ml of ice cold sodium acetate buffer.

2. Add 0.05 vol of 10% SDS and immediately add 1 vol of 60°C sodium acetate–saturated phenol (lower phase). Immediately vortex 5 to 10 sec.

3. Place tube in 60°C water bath. For each milliliter of starting sample, incubate 1 min and vortex every 30 to 60 sec for 5 to 10 sec—replace in 60°C bath. Cool quickly on ice by swirling 5 min in ice-water.

4. Centrifuge 5 min at 800 × *g*. Remove aqueous (upper) phase to a new 15-ml polypropylene tube. Add equal volume chloroform, mix by inversion, and centrifuge again.

5. Repeat chloroform extraction as in step 4 and remove aqueous phase to a 50-ml polypropylene tube. Precipitate by adding 2.5 vol of 95% ethanol and letting stand overnight at −20°C.

6. Centrifuge 15 min at 1500 × *g* to obtain RNA pellet. Wash pellet briefly with 70% ethanol at −20°C. Allow pellet to air dry and resuspend in TE buffer, pH 8.0. Determine the RNA concentration, where A_{260} of 1.0 ≅ 40 µg/ml single-stranded RNA.

   The A_{260}/A_{280} ratio should be 1.9 to 2.0 for highly purified RNA (APPENDIX 2G).

BASIC PROTOCOL 2

PREPARATION OF CYTOPLASMIC RNA FROM TISSUE CULTURE CELLS

Materials *(see APPENDIX 1 for items with ✓)*

- Tissue culture cells
- DEPC
- ✓ PBS, ice cold
- ✓ Lysis buffer, ice-cold
- 20% (w/v) SDS
- 20 mg/ml proteinase K
- ✓ 25:24:1 (v/v/v) phenol/chloroform/isoamyl alcohol
- 24:1 (v/v) chloroform/isoamyl alcohol
- ✓ 3 M sodium acetate, pH 5.2

100% ethanol
75% ethanol/25% 0.1 M sodium acetate, pH 5.2
✓ DEPC-treated water

1. Wash cells free of medium with ice-cold PBS. For monolayer cultures, rinse three times. For suspension cultures, pellet by centrifuging 5 min at 300 × *g*, resuspend in half the original culture volume PBS, and pellet again.

   This procedure, as written, is used for up to 2 × 10^7 cells (two 10-cm dishes, two 75-cm^2 tissue culture flasks, or ~20 ml of suspension culture).

2. For monolayer cultures, scrape into a small volume of cold PBS with a rubber policeman. Transfer to a centrifuge tube on ice. Collect cells by centrifuging 5 min at 300 × *g*, 4°C, or 15 sec in a microcentrifuge, 4°C. Keep cells cold.

3. Resuspend cells by careful but vigorous vortexing without foaming in 375 µl ice-cold lysis buffer. Incubate 5 min on ice (suspension should clear rapidly, indicating cell lysis).

4. Microcentrifuge 2 min at 4°C. Remove supernatant (slightly cloudy, yellow-white cytoplasmic extract) to a clean tube containing 4 µl of 20% SDS. Mix immediately by vortexing. Add 2.5 µl of 20 mg/ml proteinase K. Incubate 15 min at 37°C.

5. Add 400 µl phenol/chloroform/isoamyl alcohol. Vortex thoroughly >1 min and microcentrifuge ≥5 min, room temperature. Remove the aqueous (upper) phase to a clean tube, avoiding precipitated material from the interface. Add 400 µl phenol/chloroform/isoamyl alcohol and repeat the extraction.

   If a very large precipitate forms after the first organic extraction and little or no aqueous phase can be recovered, first try spinning for a few minutes more. If the precipitate fails to collapse to the interface, remove and discard the organic phase from the bottom of the tube. Add 400 µl chloroform/isoamyl alcohol. Vortex well and spin ~2 min. The precipitate should have largely disappeared. Recover the upper aqueous phase and proceed.

6. Remove the aqueous phase to a clean tube. Add 400 µl chloroform/isoamyl alcohol. Vortex 15 to 30 sec and microcentrifuge 1 min. Again, remove the aqueous (upper) phase to a clean tube.

7. Add 40 µl of 3 M sodium acetate, pH 5.2, and 1 ml of 100% ethanol. Mix by inversion. Incubate 15 to 30 min on ice or store overnight at −20°C.

8. Microcentrifuge 15 min at 4°C and discard supernatant. Rinse the pellet with 1 ml of 75% ethanol/25% 0.1 M sodium acetate, pH 5.2. Dry the pellet and resuspend in 100 µl DEPC-treated water. Dilute 10 µl into 1 ml water to determine the A_{260} and A_{280}. Store the remaining RNA at −70°C.

SUPPORT PROTOCOL

REMOVAL OF CONTAMINATING DNA FROM AN RNA PREPARATION

Additional Materials *(see APPENDIX 1 for items with ✓)*

✓ TE buffer, pH 7.4
100 mM $MgCl_2$/10 mM DTT
✓ 2.5 mg/ml RNase-free DNase I
Placental ribonuclease inhibitor (e.g., RNasin from Promega)
✓ DNase stop mix

A2

1. Redissolve the RNA in 50 μl TE buffer, pH 7.4.
2. Prepare on ice a cocktail containing (per sample);

   10 μl of 100 mM $MgCl_2$/10 mM DTT
   0.2 μl of 2.5 mg/ml RNase-free DNase
   0.1 μl placental ribonuclease inhibitor (25 to 50 U/μl)
   39.7 μl TE buffer, pH 7.4.

   Add 50 μl of this cocktail to each RNA sample. Mix and incubate 15 min at 37°C.
3. Stop the DNase reaction by adding 25 μl DNase stop mix.
4. Extract once with phenol/chloroform/isoamyl alcohol and once with chloroform/ isoamyl alcohol.
5. Add 325 μl of 100% ethanol and precipitate 15 to 30 min on ice or overnight at −20°C. Resume Basic Protocol 2, step 8.

BASIC PROTOCOL 3

GUANIDINIUM METHOD FOR TOTAL RNA PREPARATION

Materials *(see* APPENDIX 1 *for items with ✓)*

✓ PBS
✓ Guanidinium solution
✓ 5.7 M cesium chloride (CsCl), DEPC-treated
✓ TES solution
✓ 3 M sodium acetate, pH 5.2, DEPC-treated
100% ethanol
✓ DEPC-treated water

Beckman JS-4.2 and SW-55 rotors (or equivalents)
Rubber policeman
6-ml syringe with 20-G needle
13 × 51–mm silanized and autoclaved polyallomer ultracentrifuge tube

1a. *If starting with a monolayer culture:* Wash cells at room temperature by adding 5 ml PBS per dish, swirling dishes, and pouring off. Repeat wash. Add 3.5 ml guanidinium solution for $\leq 10^8$ cells, dividing the solution equally between the dishes (cells should immediately lyse in place). Recover the viscous lysate by scraping the dishes with a rubber policeman. Remove lysate from dishes using a 20-G needle fitted on a 6-ml syringe. Combine lysates.

1b. *If starting with a suspension culture:* Pellet $\leq 10^8$ cells by centrifuging 5 min at 300 × *g* and discard the supernatant. Wash cells once at room temperature by resuspending the pellet in an amount of PBS equal to half the original volume and centrifuging. Discard supernatant. Add 3.5 ml guanidinium solution to the centrifuge tube.

2. Draw the resultant extremely viscous solution up and down four times through a 6-ml syringe with 20-G needle to shear the chromosomal DNA. Transfer the solution to a clean tube.

3. Place 1.5 ml of 5.7 M CsCl in a 13 × 51–mm silanized and autoclaved polyallomer ultracentrifuge tube. Layer 3.5 ml of cell lysate on top of CsCl cushion to create a step

gradient. The interface should be visible. Centrifuge 12 to 20 hr at 150,000 × g, 18°C. Set centrifuge for slow acceleration and deceleration to avoid disturbing the gradient.

4. Remove the supernatant very carefully. Place the end of the Pasteur pipet at the top of the solution and lower it as the level of the solution lowers. Leave ~100 μl in the bottom, invert the tube carefully, and pour the remaining liquid off. Carefully and completely remove the white band of DNA at the interface—it contains cellular DNA.

5. Drain the pellet 5 to 10 min, then resuspend it in 360 μl TES solution by repeatedly drawing the solution up and down in a pipet. Allow the pellet to resuspend 5 to 10 min at room temperature (allow ample time to completely resuspend pellet). Transfer to a clean microcentrifuge tube.

6. Add 40 μl of 3 M sodium acetate, pH 5.2, and 1 ml of 100% ethanol. Precipitate RNA 30 min on dry ice/ethanol. Microcentrifuge 10 to 15 min and discard supernatant. Resuspend the pellet in 360 μl water and repeat this step.

7. Drain the pellet 10 min and dissolve in ~200 μl water. Quantitate by diluting 10 μl to 1 ml and reading the A_{260} and A_{280}. Store RNA at −70°C either as an aqueous solution or as an ethanol precipitate.

A2

ALTERNATE PROTOCOL

SINGLE-STEP RNA ISOLATION FROM CULTURED CELLS OR TISSUES

NOTE: Carry out all steps at room temperature unless otherwise stated.

Additional Materials *(see* APPENDIX 1 *for items with ✓)*

- ✓ Denaturing solution
- ✓ 2 M sodium acetate, pH 4
- ✓ Water-saturated phenol
- 49:1 (v/v) chloroform/isoamyl alcohol
- 100% isopropanol
- 75% ethanol (prepared with DEPC-treated water)
- 0.5% (w/v) SDS, DEPC-treated

1a. *For tissue:* Add 1 ml denaturing solution/100 mg tissue and homogenize with a few strokes in a glass Teflon homogenizer.

1b. *For cultured cells:* Either centrifuge suspension cells and discard supernatant, *or* remove the culture medium from cells grown in monolayer cultures. Add 1 ml denaturing solution/10^7 cells and pass the lysate through a pipet seven to ten times. Do not wash cells with saline.

2. Transfer the homogenate into a 5-ml polypropylene tube. Add 0.1 ml of 2 M sodium acetate, pH 4, and mix thoroughly by inversion. Add 1 ml water-saturated phenol, mix thoroughly, and add 0.2 ml of 49:1 chloroform/isoamyl alcohol and cap tube tightly. Mix thoroughly and incubate the suspension 15 min at 0° to 4°C. Centrifuge 20 min at 10,000 × g, 4°C. Transfer the upper aqueous phase containing the RNA to a fresh tube.

3. Precipitate the RNA by adding 1 ml (1 vol) of 100% isopropanol. Place the samples 30 min at −20°C. Centrifuge 10 min at 10,000 × *g*, 4°C, and discard supernatant. For isolation of RNA from tissues with a high glycogen content (e.g., liver), follow this isopropanol precipitation and wash out glycogen from the RNA pellet by vortexing in 4 M LiCl. Sediment the insoluble RNA by centrifuging 10 min at 5000 × *g*.

4. Dissolve the RNA pellet in 0.3 ml denaturing solution and transfer into a 1.5-ml microcentrifuge tube. Precipitate the RNA with 0.3 ml of 100% isopropanol (1 vol) for 30 min at −20°C. Centrifuge 10 min at 10,000 × *g*, 4°C, and discard supernatant.

5. Resuspend the RNA pellet in 75% ethanol, vortex, and incubate 10 to 15 min at room temperature to dissolve residual amounts of guanidinium contaminating the pellet. Centrifuge 5 min at 10,000 × *g* and discard supernatant. Dry the RNA pellet in a vacuum for 5 to 15 min.

6. Dissolve the RNA pellet in 100 to 200 μl DEPC-treated water or in DEPC-treated 0.5% SDS, depending on the subsequent use of RNA. Quantitate as described in Basic Protocol 3, step 7. Store samples frozen at −70°C or in ethanol at −20°C.

BASIC PROTOCOL 4

PREPARATION OF POLY(A)$^+$ RNA

Materials *(see* APPENDIX 1 *for items with ✓)*

✓ DEPC-treated water
5 M NaOH
Oligo(dT) cellulose
0.1 M NaOH
✓ Poly(A) loading buffer
10 M LiCl, DEPC-treated
✓ Middle wash buffer
2 mM EDTA/0.1% SDS
✓ 3 M sodium acetate, DEPC-treated
✓ RNase-free TE buffer

Silanized column
Silanized SW-55 centrifuge tubes
Beckman SW-55 rotor (or equivalent)

1. Wash a silanized column (e.g., Pasteur pipet plugged with silanized glass wool or small disposable column with a 2-ml capacity) with 10 ml of 5 M NaOH, then rinse it with water.

2. Add 0.5 g dry oligo(dT) cellulose powder to 1 ml of 0.1 M NaOH. Pour the slurry into the column and rinse the column with ~10 ml water. Equilibrate the column with 10 to 20 ml of loading buffer to a pH ~7.5 at the end of the wash.

3. Heat ~2 mg total RNA in DEPC-treated water to 70°C for 10 min. Add 10 M DEPC-treated LiCl to 0.5 M final. Pass the RNA solution through the oligo(dT) column. Wash the column with 1 ml poly(A) loading buffer. Make certain to save the eluant from this loading step. Pass the eluant through the column two times more.

4. Rinse the column with 2 ml middle wash buffer. Elute the RNA into a fresh tube with 2 ml of 2 mM EDTA/0.1% SDS.

5. Reequilibrate the oligo(dT) column, as in step 2. Take the eluted RNA and repeat the poly(A)$^+$ selection, as described in steps 3 and 4.

6. Add 3 M DEPC-treated sodium acetate to 0.3 M final. Precipitate the RNA by adding 2.5 vol ethanol. Transfer the solution to two silanized SW-55 tubes. Incubate RNA overnight at −20°C or 30 min on dry ice/ethanol. Collect the precipitate by centrifuging 30 min at 304,000 × *g*, 4°C.

7. Pour off ethanol and allow pellets to air dry. Resuspend RNA in 150 µl of RNase-free TE buffer and pool the samples. Check the quality of the RNA by heating 5 µl at 70°C for 5 min and analyzing on a 1% agarose gel.

References: Chirgwin et al., 1979; Chomczynski and Sacchi, 1987

Contributors: Randall Ribaudo, Michael Gilman, Robert E. Kingston, Piotr Chomczynski, and Nicoletta Sacchi

APPENDIX 2F

Preparation of Bacterial Plasmid DNA

BASIC PROTOCOL 1

MINIPREP BY ALKALINE LYSIS

Materials *(see* APPENDIX 1 *for items with ✓)*

Plasmid-bearing *E. coli* strain
✓ LB medium containing ampicillin or other appropriate antibiotic
✓ Glucose/Tris/EDTA (GTE) solution
✓ TE buffer
NaOH/SDS solution: 0.2 N NaOH/1% (w/v) SDS (prepare just before use)
✓ Potassium acetate solution, pH 4.8
95% and 70% ethanol
✓ TE buffer containing 0.1 mg/ml RNase *or* 10 mg/ml DNase-free RNase
1.5-ml disposable microcentrifuge tubes

1. Inoculate 2 to 5 ml sterile LB medium (containing appropriate antibiotic) with a single bacterial colony. Grow to saturation—e.g., overnight (see Support Protocol 1).

2. Microcentrifuge 1.5 ml of cells 20 sec at maximum speed, 4°C or room temperature, to pellet. Completely remove the supernatant with a Pasteur pipet or a plastic pipettor tip and resuspend pellet completely in 100 µl GTE solution and let sit 5 min at room temperature.

3. Add 200 µl NaOH/SDS solution, mix by inversion, and place on ice for 5 min. Add 150 µl potassium acetate solution and vortex at maximum speed for 2 sec to completely mix. Place on ice for 5 min.

4. Microcentrifuge 3 min at maximum speed, 4°C or room temperature, to pellet cell debris and chromosomal DNA. Transfer 0.4 ml supernatant to a fresh tube, mix with 0.8 ml of 95% ethanol, and let sit 2 min at room temperature to precipitate nucleic acids.

5. Microcentrifuge 3 min at maximum speed, room temperature, to pellet plasmid DNA and RNA. Remove supernatant, wash pellet with 1 ml of 70% ethanol, and dry under vacuum.
6. Resuspend the pellet in 30 μl TE buffer/0.1 mg/ml RNase and store at −20°C. Use 2 to 5 μl of the resuspended DNA for a restriction digest.

BASIC PROTOCOL 2

A2

LARGE-SCALE CRUDE PREP BY ALKALINE LYSIS

Materials *(see* APPENDIX 1 *for items with ✓)*

Plasmid-bearing *E. coli* strain
✓ LB medium containing ampicillin or other appropriate antibiotic
✓ GTE solution
Hen egg white lysozyme
NaOH/SDS solution: 0.2 N NaOH/1% (w/v) SDS (prepare just before use)
✓ Potassium acetate solution, pH ~5.5
Isopropanol
70% ethanol

Sorvall GSA, GS-3, or Beckman JA-10 rotor (or equivalent)
Sorvall SS-34 or Beckman JA-17 rotor (or equivalent)

1. Inoculate 5 ml LB medium containing appropriate antibiotic with a single colony of *E. coli* containing the desired plasmid. Grow at 37°C with vigorous shaking overnight (see Support Protocol 1).
2. Inoculate 500 ml LB containing appropriate antibiotic in a 2-liter flask with ~5 ml of overnight culture. Grow at 37°C until culture is saturated ($OD_{600} \cong 4$; see Support Protocols 2 and 3).

   To increase yields, maximize aeration using a flask with high surface area and baffles; shake at >400 rpm.

3. Collect cells by centrifugation 10 min at 6000 × *g*, 4°C. Resuspend pellet from 500-ml culture in 4 ml GTE solution and transfer to high-speed centrifuge tube with ≥20-ml capacity.
4. Add 1 ml GTE solution containing hen egg white lysozyme added fresh to 25 mg/ml. Resuspend the pellet completely in this solution and allow it to stand 10 min at room temperature.
5. Add 10 ml freshly prepared NaOH/SDS solution and mix by stirring gently with pipet until solution becomes homogeneous and clears and very viscous. Let stand 10 min on ice.
6. Add 7.5 ml potassium acetate solution and again stir gently with pipet until viscosity is reduced and a large precipitate forms. Let stand 10 min on ice.
7. Centrifuge 10 min at 20,000 × *g*, 4°C. Decant the supernatant (plasmid DNA) into a clean centrifuge tube. Pour through several layers of cheesecloth if any floating material is visible. Add 0.6 vol isopropanol, mix by inversion, and let stand 5 to 10 min at room temperature.
8. Recover nucleic acids by centrifuging 10 min at 15,000 × *g*, room temperature. Wash the pellet with 2 ml of 70% ethanol; centrifuge briefly to collect pellet. Aspirate ethanol and dry pellet under vacuum. Store pellet indefinitely at 4°C.

BASIC PROTOCOL 3

PURIFICATION BY CsCl/ETHIDIUM BROMIDE EQUILIBRIUM CENTRIFUGATION

Materials (*see* APPENDIX 1 *for items with* ✓)

DNA pellet from crude prep (see Basic Protocol 2)
✓ TE buffer
Cesium chloride
10 mg/ml ethidium bromide
CsCl/TE solution: 100 ml TE buffer (*APPENDIX 1*) containing 100 g CsCl
✓ Dowex AG50W-X8 cation exchange resin
0.2 M NaCl prepared in TE buffer
100% and 70% ethanol

Beckman VTi65 or VTi80 rotor (or equivalent)
5-ml quick-seal ultracentrifuge tubes
3-ml syringes with 20-G needles

1. Resuspend the pellet obtained in the final step of the crude lysate preparation in 4 ml TE buffer. Add 4.4 g CsCl, dissolve, and add 0.4 ml of 10 mg/ml ethidium bromide.

   Ethidium bromide will form a complex with protein remaining in the solution to form a deep red flocculent precipitate. This can be removed by centrifuging the lysate/CsCl/ethidium bromide solution 5 min at ~2000 × g, room temperature. After this procedure, the protein/ethidium bromide complex will form a disc at the top of the solution. The solution can be pipetted out from beneath the disc or poured carefully, allowing the floating disc to adhere to the side of the tube.

2. Transfer the solution to a 5-ml ultracentrifuge tube. Top up the tube, if necessary, with CsCl/TE solution and seal tube. Band plasmid by centrifuging 3.5 hr at 500,000 × *g* or ≥14 hr at 350,000 × *g*, 20°C.

   It is very important that this centrifugation be done at a temperature no lower than 15°C. Because of the high concentration of cesium chloride and the high centrifugal force necessary to establish the gradient, lower temperatures will cause the cesium chloride at the bottom of the tube (where the density is highest) to precipitate during the run. The cesium chloride precipitate moves the center of mass towards the bottom of the tube. This can unbalance the rotor and cause breakage of the rotor and destruction of the centrifuge at least, and serious personal injury at worst. Also, warmer gradients achieve equilibrium more quickly than cold gradients.

3. Insert a 20-G needle gently into the top of the tube, and recover the plasmid band by suction with a 3-ml syringe with another 20-G needle attached (see Fig. A.2F.1). Do not allow the gradient to be mixed by rough handling or turbulence. Be certain not to cover the top of the first needle with a finger. Insert the second needle (collection needle) into the side of the tube ~1 cm below the plasmid band, which is the lower of the two bands (the beveled edge of the needle should face up, toward the DNA; see Fig. A.2F.1).

   Protein-ethidium bromide complexes will pellet on the outside edge of the tube if not removed earlier. To avoid contaminating the collection needle with this material do not put the collection needle through it. Occasionally the needle will become clogged if a piece of pellet enters it. Do not try to draw harder on the syringe as this will create turbulence in the tube and cause mixing of the gradient. Insert another needle and use it to draw off the band. Leave the clogged needle in place in the tube. (If the clogged needle is removed, the tube will empty through the hole that remains.) The air inlet needle can also become clogged; if it does, remove it and allow air to

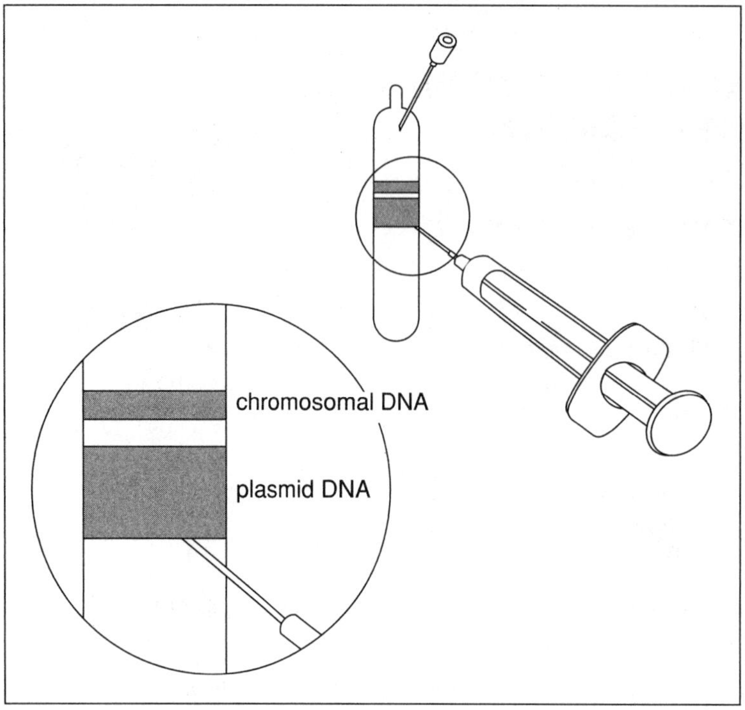

Figure A.2F.1 Collecting plasmid DNA from gradient tube.

A2

enter through the remaining hole. There should be no resistance felt in the syringe when drawing off the plasmid DNA. If resistance is felt, check that the needles are clear. Drawing the plasmid DNA through a very small opening remaining in a clogged needle may result in shearing of the DNA.

4. If higher-purity plasmid DNA is required, perform a second ultracentrifugation to eliminate any contaminating RNA or chromosomal DNA: add plasmid DNA band to another ultracentrifuge tube, top up with CsCl/TE solution containing 0.2 mg/ml ethidium bromide, and repeat steps 2 and 3.

5. Pour a Dowex AG50W-X8 column, 1.5 to 2 times the volume of the plasmid DNA/ethidium bromide solution, in a glass or plastic column. Pass several volumes of 0.2 M NaCl (prepared in TE buffer) through the column to wash and equilibrate.

6. Load the plasmid DNA/ethidium bromide solution directly from syringe to top of resin bed without disturbing the resin. Begin collecting the solution flowing through the column immediately after loading the plasmid solution. Wash the column with a volume of 0.2 M NaCl (prepared in TE buffer) twice that of the volume loaded (the final volume collected from the column should be three times that in which the plasmid DNA was removed from the gradient).

7. Precipitate the plasmid DNA with 2 vol of 100% ethanol at room temperature or at −20°C. Centrifuge 10 min at 10,000 × g, 4°C. Do not cool this solution below −20°C as this may cause the cesium chloride to precipitate.

8. Wash the pellet with 70% ethanol and dry under vacuum. Resuspend the pellet in TE buffer and store at 4°C.

SUPPORT PROTOCOL 1

GROWING AN OVERNIGHT CULTURE

Small freshly saturated cultures of *E. coli* are called overnights. To make an overnight, remove the cap from a sterile 16- or 18-mm culture tube. Working quickly to minimize contact of the tube with the possibly contaminated air, use a sterile pipet to transfer 5 ml of liquid medium into the tube. Inoculate the liquid with a single bacterial colony by touching a sterile inoculating loop to the colony, making certain that some of the cells have been transferred to the loop, and then dipping the loop into the liquid and shaking it a bit. Replace the tube's cap, and place the tube in a shaker or on a roller drum at 60 rpm, 37°C. Grow until the culture is freshly saturated (at a density of $1–2 \times 10^9$ cells/ml, which typically takes at least 12 hr). See Support Protocol 3 describing spectrophotometric measurement of cell density for corresponding OD readings.

A2

SUPPORT PROTOCOL 2

GROWING LARGER CULTURES

Larger cultures are generally inoculated with overnight cultures diluted 1:100. Use an Erlenmeyer or baffle flask whose volume is $\geq$5 times the volume of the culture. Grow the culture at 37°C with vigorous agitation ($\sim$300 rpm) to ensure proper aeration. If it is necessary to grow a culture without shaking (a rare occurrence; e.g., if the strain is temperature-sensitive for growth and no low-temperature shaker is available), grow the culture in an Erlenmeyer flask whose volume is $\geq$20 times that of the culture. See Support Protocol 3 describing spectrophotometric measurement of cell density for corresponding OD readings.

SUPPORT PROTOCOL 3

MONITORING GROWTH WITH A SPECTROPHOTOMETER

The concentration of cells in a culture can also be determined with a spectrophotometer by measuring the amount of 600-nm light scattered by the culture. The level of optical density at 600 nm will depend on the distance between the cuvette and the detector and will vary among spectrophotometers, often by a factor of 2. It is thus wise to calibrate each instrument by recording the OD_{600} (sometimes expressed as A_{600}) of a culture that contains a known number of cells determined by some other method, such as observation with a Petroff-Hausser chamber, counting on a Coulter counter, or titering for viable colonies.

If the culture is visibly turbid, also measure a ten-fold dilution of it. Calculate the number of cells/ml from whichever suspension (the undiluted or the diluted) has an $OD_{600} < 1$.

For a culture grown in rich medium, a good rule of thumb is that each 0.1 OD_{600} unit is roughly equivalent to 10^8 cells/ml. In general, saturated cultures have an OD_{600} reading of $\sim$4, while logarithmically growing cultures have an $OD_{600} < 0.7$.

Contributors: JoAnne Engebrecht, J.S. Heilig, and Roger Brent

APPENDIX 2G

Quantitation of DNA and RNA with Absorption and Fluorescence Spectroscopy

Reliable quantitation of nanogram and microgram amounts of DNA and RNA in solution is essential to researchers in molecular biology. Properties of the three assays described in this unit are listed in Table A.2G.1.

BASIC PROTOCOL

DETECTION OF NUCLEIC ACIDS USING ABSORPTION SPECTROSCOPY

A_{260} measurements are quantitative for relatively pure nucleic acid preparations in microgram quantities. Absorbance readings cannot discriminate between DNA and RNA; however, the ratio of A at 260 and 280 nm can be used as an indicator of nucleic acid purity. Proteins, for example, have a peak absorption at 280 nm that will reduce the A_{260}/A_{280} ratio. Absorbance at 325 nm indicates particulates in the solution or dirty cuvettes; contaminants containing peptide bonds or aromatic moieties such as protein and phenol absorb at 230 nm.

Materials *(see* APPENDIX 1 *for items with ✓)*

✓ 1× TNE buffer
DNA sample to be quantitated
✓ Calf thymus DNA standard solutions

Matched quartz semi-micro spectrophotometer cuvettes (1-cm pathlength)
Single- or dual-beam spectrophotometer (ultraviolet to visible)

1. Pipet 1.0 ml of 1× TNE buffer into a quartz cuvette. Place the cuvette in a single- or dual-beam spectrophotometer, read at 325 nm (note contribution of the blank relative to distilled water if necessary), and zero the instrument. Use this blank solution as the reference in double-beam instruments. For single-beam spectrophotometers, remove blank cuvette and insert cuvette containing DNA sample or standard suspended in the same solution as the blank. Take reading. Repeat this process at 280, 260, and 230 nm.

Table A.2G.1 Properties of Absorbance and Fluorescence Spectrophotometric Assays for DNA and RNA

Property	Absorbance (A_{260})	Fluorescence	
		H33258	EtBr
Sensitivity (μg/ml)			
DNA	1–50	0.01–15	0.1–10
RNA	1–40	n.a.	0.2–10
Ratio of signal (DNA/RNA)	0.8	400	2.2

2. To determine the concentration (C) of DNA present, use the A_{260} reading in conjunction with one of the following equations:

Single-stranded DNA: $$C\ (\text{pmol}/\mu\text{l}) = \frac{A_{260}}{10 \times S}$$

$$C\ (1\text{g/ml}) = \frac{A_{260}}{0.027}$$

Double-stranded DNA: $$C\ (\text{pmol}/\mu\text{l}) = \frac{A_{260}}{13.2 \times S}$$

$$C\ (\mu\text{g/ml}) = \frac{A_{260}}{0.020}$$

Single-stranded RNA: $$C\ (\mu\text{g/ml}) = \frac{A_{260}}{0.025}$$

Oligonucleotide: $$C\ (\text{pmol}/\mu\text{l}) = A_{260} \times \frac{100}{1.5\ N_A + 0.71\ N_C + 1.20\ N_G + 0.84\ N_T}$$

where S represents the size of the DNA in kilobases and N is the number or residues of base A, G, C, or T.

For double- or single-stranded DNA and single-stranded RNA: These equations assume a 1-cm-pathlength spectrophotometer cuvette and neutral pH. The calculations are based on the Lambert-Beer law, A = ECl, where A is the absorbance at a particular wavelength, C is the concentration of DNA, l is the pathlength of the spectrophotometer cuvette (typically 1 cm), and E is the extinction coefficient. For solution concentrations given in mol/liter and a cuvette of 1-cm pathlength, E is the molar extinction coefficient and has units of $M^{-1}cm^{-1}$. *If concentration units of μg/ml are used, then E is the specific absorption coefficient and has units of* $(\mu g/ml)^{-1}cm^{-1}$. *The values of E used here are as follows: ssDNA, 0.027* $(\mu g/ml)^{-1}cm^{-1}$; *dsDNA, 0.020* $(\mu g/ml)^{-1}cm^{-1}$; *ssRNA, 0.025* $(\mu g/ml)^{-1}cm^{-1}$. *Using these calculations, an* A_{260} *of 1.0 indicates 50 μg/ml double-stranded DNA, ~37 μg/ml single-stranded DNA, or ~40 μg/ml single-stranded RNA (adapted from Applied Biosystems, 1989).*

For oligonucleotides: Concentrations are calculated in the more convenient units of pmol/μl. The base composition of the oligonucleotide has significant effects on absorbance, because the total absorbance is the sum of the individual contributions of each base (Table A.2G.2).

3. Use the A_{260}/A_{280} ratio and readings at A_{230} and A_{325} to estimate the purity of the nucleic acid sample. For typical values at the four wavelengths for a highly purified preparation, see Table A.2G.3.

Table A.2G.2 Molar Extinction Coefficients of DNA Bases[a]

Base	$\varepsilon^{1M}_{260\ nm}$
Adenine	15,200
Cytosine	7,050
Guanosine	12,010
Thymine	8,400

[a]Measured at 260 nm; see Wallace and Miyada, 1987. Detailed spectrophotometric properties of nucleoside triphosphates are listed in *CPMB UNIT 3.4*.

Table A.2G.3 Spectrophotometric Measurements of Purified DNA[a]

Wavelength (nm)	Absorbance	A_{260}/A_{280}	Conc. (μg/ml)
325	0.01	—	—
280	0.28	—	—
260	0.56	2.0	28
230	0.30	—	—

[a]Typical absorbancy readings of highly purified calf thymus DNA suspended in 1× TNE buffer. The concentration of DNA was nominally 25 μg/ml.

A2

ALTERNATE PROTOCOL 1

DNA DETECTION USING THE DNA-BINDING FLUOROCHROME HOECHST 33258

Additional Materials *(also see Basic Protocol; see* APPENDIX 1 *for items with ✓)*

✓ Hoechst 33258 assay solution (working solution)
Dedicated filter fluorometer (Amersham Pharmacia Biotech DQ 200) or scanning fluorescence spectrophotometer (Shimadzu model RF-5000 or Perkin-Elmer model LS-5B or LS-3B)
Fluorometric square glass cuvettes *or* disposable acrylic cuvettes (Sarstedt)
Teflon stir rod

1. Prepare the scanning fluorescence spectrophotometer by setting the excitation wavelength to 365 nm and the emission wavelength to 460 nm.
2. Pipet 2.0 ml Hoechst 33258 assay solution into cuvette and place in sample chamber. Take a reading without DNA and use as background.
3. With the cuvette still in the sample chamber, add 2 μl DNA standard to the blank Hoechst 33258 assay solution. Mix in the cuvette with a Teflon stir rod or by capping and inverting the cuvette. Read emission in relative fluorescence units or set the concentration readout equal to the final DNA concentration. Repeat measurements with remaining DNA standards using fresh assay solution (take background zero reading and zero instrument if needed). Read samples in duplicate or triplicate, with a blank reading taken each time.
4. Repeat step 3 with test samples.

A dye concentration of 0.1 μg/ml is adequate for final DNA concentrations up to ~500 ng/ml. Increasing the working dye concentration to 1 μg/ml Hoechst 33258 will extend the assay's range to 15 μg/ml DNA, but will limit sensitivity at low concentrations (5 to 10 ng/ml). Sample volumes of ≤10 μl can be added to the 2.0-ml aliquot of Hoechst 33258 assay solution.

ALTERNATE PROTOCOL 2

DNA AND RNA DETECTION WITH ETHIDIUM BROMIDE FLUORESCENCE

Additional Materials *(also see Basic Protocol; see* APPENDIX 1 *for items with ✓)*

✓ Ethidium bromide assay solution

1. Pipet 2.0 ml ethidium bromide assay solution into cuvette and place in sample chamber. Set excitation wavelength to 302 nm (UV range) or 546 nm (visible range) and emission wavelength to 590 nm. Take an emission reading without DNA and use as background.
2. Read and calibrate these samples as described in Alternate Protocol 1, step 3.
3. Read emissions of the test samples as in Alternate Protocol 1, step 4.

A dye concentration of 5 μg/ml in the ethidium bromide assay solution is appropriate for final DNA concentrations up to 1000 ng/ml. 10 μg/ml ethidium bromide in the ethidium bromide assay solution will extend the assay's range to 10 μg/ml DNA, but is only used for DNA concentrations >1 μg/ml. Sample volumes of up to 10 μl can be added to the 2.0-ml aliquot of ethidium bromide assay solution.

Reference: Labarca and Paigen, 1980

Contributor: Sean R. Gallagher

A2

APPENDIX 2H

Introduction of Plasmid DNA into Cells

BASIC PROTOCOL

TRANSFORMATION USING CALCIUM CHLORIDE

Materials *(see APPENDIX 1 for items with ✓)*

Single colony of *E. coli* cells
✓ LB medium
✓ $CaCl_2$ solution, ice-cold
✓ LB plates containing ampicillin
Plasmid DNA (*APPENDIX 2F*)

Centrifuge with Beckman JS-5.2 rotor or equivalent and tubes
42°C water bath

1. Inoculate a single colony of *E. coli* cells into 50 ml LB medium. Grow overnight at 37°C with shaking (250 rpm).
2. Inoculate 4 ml of the culture into 400 ml LB medium in a 2-liter flask. Grow at 37°C, with shaking (250 rpm), to an OD_{590} of 0.375 (early- or mid-log phase). Dispense culture into eight 50-ml prechilled, sterile polypropylene tubes and leave the tubes on ice 5 to 10 min. Centrifuge cells 7 min at 1600 × *g*, 4°C.
3. Gently decant supernatant and resuspend each pellet in 10 ml ice-cold $CaCl_2$ solution. Centrifuge 5 min at 1100 × *g*, 4°C. Again, decant supernatant and resuspend each pellet in 10 ml cold $CaCl_2$ solution. Keep resuspended cells on ice for 30 min. Centrifuge 5 min at 1100 × *g*, 4°C.
4. Resuspend each pellet completely in 2 ml of ice-cold $CaCl_2$ solution. Aliquot 250 μl into prechilled, sterile polypropylene microcentrifuge tubes. Freeze immediately at −70°C (e.g., in liquid nitrogen).
5. Add 10 ng of purified plasmid DNA (10 to 25 μl) into a 15-ml sterile, round-bottom test tube and place on ice. Rapidly thaw competent cells from step 4 by warming between

hands. Dispense 100 μl immediately into test tubes containing DNA, swirl, and place on ice 10 min.

6. Heat shock cells by placing tubes 2 min in 42°C water bath. Add 1 ml LB medium to each tube. Incubate 1 hr at 37°C on roller drum (250 rpm).
7. Plate several dilutions (1, 10, and 25 μl) on appropriate antibiotic-containing LB plates, and incubate 12 to 16 hr at 37°C to determine transformation efficiency (10^6 to 10^8 transformants/μg DNA). Store remainder at 4°C for subsequent platings.

   The number of transformant colonies per aliquot volume (μl) × 10^5 is equal to the number of transformants per microgram of DNA.

ALTERNATE PROTOCOL

ONE-STEP PREPARATION AND TRANSFORMATION OF COMPETENT CELLS

***Additional Materials** (also see Basic Protocol)*

2× transformation and storage solution (TSS): ice-cold: dilute sterile 40% (w/v) polyethylene glycol (PEG) 3350 to 20% in sterile LB medium containing 100 mM $MgCl_2$; add DMSO to 10%

✓ LB medium containing 20 mM glucose

1. Dilute fresh overnight culture of bacteria 1:100 into LB medium and grow at 37°C to an OD_{600} of 0.3 to 0.4 (typically 2 to 4 hr).
2. Add a volume of ice-cold 2× TSS equal to the cell suspension, and gently mix on ice. For long-term storage, freeze small aliquots of the suspension in a dry ice/ethanol bath and store at −70°C. To use frozen cells for transformation, thaw slowly, then use immediately.
3. Add 100 μl competent cells and 1 to 5 μl DNA (0.1 to 100 ng) to an ice-cold polypropylene or glass tube. Incubate 5 to 60 min at 4°C.
4. Add 0.9 ml LB medium/20 mM glucose and incubate 30 to 60 min at 37°C with mild shaking. Select transformants on appropriate plates.

References: Dower et al., 1988; Hanahan, 1983

Contributors: Christine E. Seidman and Kevin Struhl

APPENDIX 2I

Techniques for Mammalian Cell Tissue Culture

In most cases, cells or tissues must be grown in culture for days or weeks to obtain sufficient numbers of cells for analysis. Maintenance of cells in long-term culture requires strict adherence to aseptic technique to avoid contamination and potential loss of valuable cell lines. The first section of this appendix discusses basic principles of sterile technique.

An important factor influencing the growth of cells in culture is the choice of tissue culture medium. Many different recipes for tissue culture media are available and each laboratory must determine which medium best suits their needs. Individual laboratories may elect to use commercially prepared medium or prepare their own. Commercially available medium can be obtained as a sterile and ready-to-use liquid, in a concentrated liquid form, or in a powdered form. Besides providing nutrients for growing cells, medium is generally supplemented with antibiotics, fungicides, or both to inhibit contamination. The second section of this appendix discusses medium preparation.

As cells grown in monolayer reach confluency, they must be subcultured or passaged. Failure to subculture confluent cells results in reduced mitotic index and eventually in cell death. The first step in subculturing is to detach cells from the surface of the primary culture vessel by trypsinization or mechanical means. The resultant cell suspension is then subdivided, or reseeded, into fresh cultures. When subculturing cells, add a sufficient number of cells to give a final concentration of $\sim 5 \times 10^4$ cells/ml in each new culture. Secondary cultures are checked for growth, fed periodically, and may be subsequently subcultured to produce tertiary cultures, etc. The time between passaging cells varies with the cell line and depends on the growth rate.

ASEPTIC TECHNIQUE

It is essential that aseptic technique be maintained when working with cell cultures. Aseptic technique involves a number of precautions to protect both the cultured cells and the laboratory worker from infection. The laboratory worker must realize that cells handled in the lab are potentially infectious and should be handled with caution. Protective apparel such as gloves, lab coats or aprons, and eyewear should be worn when appropriate (Knutsen, 1991). Care should be taken when handling sharp objects such as needles, scissors, scalpel blades, and glass that could puncture the skin. Sterile, disposable plastic supplies may be used to avoid the risk of broken or splintered glass (Rooney and Czepulkowski, 1992).

Frequently, specimens received in the laboratory are not sterile, and cultures prepared from these specimens may become contaminated with bacteria, fungus, or yeast. The presence of microorganisms can inhibit growth, kill cell cultures, or lead to inconsistencies in test results. The contaminants deplete nutrients in the medium and may produce substances that are toxic to cells. Antibiotics (penicillin, streptomycin, kanamycin, or gentamycin) and fungicides (amphotericin B or mycostatin) may be added to tissue culture medium to combat potential contaminants (see Table A.2I.1). An antibiotic/antimycotic solution or lyophilized powder that contains penicillin, streptomycin, and amphotericin B is available from Sigma. The solution can be used to wash specimens prior to culture and can be added to medium used for tissue culture. Similar preparations are available from other suppliers.

All materials that come into direct contact with cultures must be sterile. Sterile disposable dishes, flasks, pipets, etc., can be obtained directly from manufacturers. Reusable glassware must be washed, rinsed thoroughly, then sterilized by autoclaving or by dry heat before reusing. With dry heat, glassware should be heated 90 min to 2 hr at 160°C to ensure sterility. Materials that may be damaged by very high temperatures can be autoclaved 20 min at 120°C and 15 psi. All media, reagents, and other solutions that come into contact with the cultures must also be sterile; medium may be obtained as a sterile liquid from the manufacturer, autoclaved if not heat-sensitive, or filter sterilized. Supplements can be added to media prior to filtration, or they can be added aseptically after filtration. Filters with 0.20- to 0.22-μm pore size should be used to remove small gram-negative bacteria from culture media and solutions.

Contamination can occur at any step in handling cultured cells. Care should be taken when pipetting media or other solutions for tissue culture. The necks of bottles and flasks, as well as the tips of the pipets, should be flamed before the pipet is introduced into the bottle. If the pipet tip comes into contact with the benchtop or any other nonsterile surface, it should be discarded and a fresh pipet obtained. Forceps and scissors used in tissue culture can be rapidly sterilized by dipping in 70% alcohol and flaming.

Although tissue culture work can be done on an open bench if aseptic methods are strictly enforced, many labs prefer to perform tissue culture work in a room or low-traffic area reserved specifically for that purpose. At the very least, biological safety cabinets are recommended to protect the cultures as well as the laboratory worker. In a laminar flow hood, the flow of air protects the work area from dust and contamination and acts as a barrier between the work surface and the worker. Many different styles of safety hoods are available and the laboratory should consider the types of samples being processed and the types of potential pathogenic exposure in making their selection. Manufacturer recommendations should be followed regarding routine maintenance checks on air flow and filters. For day-to-day use, the cabinet should be turned on for at least 5 min prior to beginning work. All work surfaces both inside and outside of the hood should be kept clean and disinfected daily and after each use.

Some safety cabinets are equipped with ultraviolet (UV) lights for decontamination of work surfaces. However, use of UV lamps is no longer recommended, as it is generally ineffective (Knutsen, 1991). UV lamps may produce a false sense of security as they maintain a visible blue glow long after their germicidal effectiveness is lost. Effectiveness diminishes over time as the glass tube gradually loses its ability to transmit short UV wavelengths, and it may also be reduced by dust on the glass tube, distance from the work surface, temperature, and air movement. Even when the UV output is adequate, the rays must directly strike a microorganism in order to kill it; bacteria or mold spores hidden below the surface of a material or outside the direct path of the rays will not be destroyed. Another rule of thumb is that anything that can be seen cannot be killed by UV. UV lamps will only destroy microorganisms such as bacteria, virus, and mold spores; they will not destroy insects or other large organisms (Westinghouse Electric Company, 1976). The current recommendation is that work surfaces be wiped down with ethanol instead of relying on UV lamps, although some labs use the lamps in addition to ethanol wipes to decontaminate work areas. A special metering device is available to measure the output of UV lamps, and the lamps should be replaced when they fall below the minimum requirements for protection (Westinghouse Electric Company, 1976).

Cultures should be checked routinely for contamination. Indicators in the tissue culture medium change color when contamination is present: for example, medium that contains phenol red changes to yellow because of increased acidity. Cloudiness and turbidity are also observed in contaminated cultures. Once contamination is confirmed with a microscope, infected cultures are generally discarded. Keeping contaminated cultures increases the risk of contaminating other cultures. Sometimes a contaminated cell line can be salvaged by treating it with various combinations of antibiotics and antimycotics in an attempt to eradicate the infection (e.g., see Fitch et al., 1997). However, such treatment may adversely affect cell growth and is often unsuccessful in ridding cultures of contamination.

PREPARING CULTURE MEDIUM

Choice of tissue culture medium comes from experience. An individual laboratory must select the medium that best suits the type of cells being cultured. Chemically defined media are available in liquid or powdered form from a number of suppliers. Sterile, ready-to-use medium has the advantage of being convenient, although it is more costly than other forms. Powdered medium must be reconstituted with tissue culture grade water according to manufacturer's directions. Distilled or deionized water is not of sufficiently high quality for medium preparation; double- or triple-distilled water or commercially available tissue culture water should be used. The medium should be filter-sterilized and transferred to sterile bottles. Prepared medium can generally be stored ≤1 month at 4°C. Laboratories using large volumes of medium may choose to prepare their own medium from standard recipes. This is the most economical approach, but it is time-consuming.

Basic media such as Eagle's minimal essential medium (MEM), Dulbecco's modified Eagle medium (DMEM), Glasgow modified Eagle medium (GMEM), and RPMI 1640 and Ham

F10 nutrient mixtures (e.g., Life Technologies) are composed of amino acids, glucose, salts, vitamins, and other nutrients. A basic medium is supplemented by addition of L-glutamine, antibiotics (typically penicillin and streptomycin sulfate), and usually serum to formulate a "complete medium." Where serum is added, the amount is indicated as a percentage of fetal bovine serum (FBS) or other serum. Some media are also supplemented with antimycotics, nonessential amino acids, various growth factors, and/or drugs that provide selective growth conditions. Supplements should be added to medium prior to sterilization or filtration, or added aseptically just before use.

The optimum pH for most mammalian cell cultures is 7.2 to 7.4. Adjust pH of the medium as necessary after all supplements are added. Buffers such as bicarbonate and HEPES are routinely used in tissue culture medium to prevent fluctuations in pH that might adversely affect cell growth. HEPES is especially useful in solutions used for procedures that do not take place in a controlled CO_2 environment.

Most cultured cells will tolerate a wide range of osmotic pressure and an osmolarity between 260 and 320 mOsm/kg is acceptable for most cells. The osmolarity of human plasma is ~290 mOsm/kg, and this is probably the optimum for human cells in culture as well (Freshney, 1993).

Fetal bovine serum (FBS; sometimes known as fetal calf serum, FCS) is the most frequently used serum supplement. Calf serum, horse serum, and human serum are also used; some cell lines are maintained in serum-free medium (Freshney, 1993). Complete medium is supplemented with 5% to 30% (v/v) serum, depending on the requirements of the particular cell type being cultured. Serum that has been heat-inactivated (30 min to 1 hr at 56°C) is generally preferred, because heat treatment inactivates complement and is thought to reduce the number of contaminants. Serum is obtained frozen, then is thawed, divided into smaller portions, and refrozen until needed.

There is considerable lot-to-lot variation in FBS. Most suppliers will provide a sample of a specific lot and reserve a supply of that lot while the serum is tested for its suitability. The suitability of a serum lot depends upon the use. Frequently the ability of serum to promote cell growth equivalent to a laboratory standard is used to evaluate a serum lot. Once an acceptable lot is identified, enough of that lot should be purchased to meet the culture needs of the laboratory for an extended period of time.

Commercially prepared media containing L-glutamine are available, but many laboratories choose to obtain medium without L-glutamine, and then add it to a final concentration of 2 mM just before use. L-glutamine is an unstable amino acid that, upon storage, converts to a form cells cannot use. Breakdown of L-glutamine is temperature- and pH-dependent. At 4°C, 80% of the L-glutamine remains after 3 weeks, but at near incubator temperature (35°C) only half remains after 9 days (Barch et al., 1991). To prevent degradation, 100× L-glutamine (see *APPENDIX 1*) should be stored frozen in aliquots until needed.

As well as practicing good aseptic technique, most laboratories add antimicrobial agents to medium to further reduce the risk of contamination. A combination of penicillin and streptomycin is the most commonly used antibiotic additive; kanamycin and gentamycin are used alone. Mycostatin and amphotericin B are the most commonly used fungicides (Rooney and Czepulkowski, 1992). Table A.2I.1 lists the final concentrations for the most commonly used antibiotics and antimycotics. Combining antibiotics in tissue culture media can be tricky, as some antibiotics are not compatible, and one may inhibit the action of another. Furthermore, combined antibiotics may be cytotoxic at lower concentrations than is true for the individual antibiotics. In addition, prolonged use of antibiotics may cause cell lines to develop antibiotic resistance. For this reason, some laboratories add antibiotics and/or fungicides to medium when initially establishing a culture but eliminate them from medium used in later subcultures.

Table A.2I.1 Working Concentrations of Antibiotics and Fungicides for Mammalian Cell Culture

Additive	Final concentration
Penicillin	50–100 U/ml
Streptomycin sulfate	50–100 μg/ml
Kanamycin	100 μg/ml
Gentamycin	50 μg/ml
Mycostatin	20 μg/ml
Amphotericin B	0.25 μg/ml

All tissue culture medium, whether commercially prepared or prepared within the laboratory, should be tested for sterility prior to use. A small aliquot from each lot of medium is incubated 48 hr at 37°C and monitored for evidence of contamination such as turbidity (infected medium will be cloudy) and color change (if phenol red is the indicator, infected medium will turn yellow). Any contaminated medium should be discarded.

NOTE: All incubations are performed in a humidified 37°C, 5% CO_2 incubator unless otherwise specified.

BASIC PROTOCOL

TRYPSINIZING AND SUBCULTURING CELLS FROM A MONOLAYER

Materials *(see* APPENDIX 1 *for items with ✓)*

Primary cultures of cells
✓ HBSS *without* Ca^{2+} and Mg^{2+}, 37°C
✓ 0.25% (w/v) trypsin/0.2% EDTA solution, 37°C
✓ Complete medium with serum: e.g., DMEM supplemented with 10% to 15% (v/v) fetal bovine serum (complete DMEM-10), 37°C

Sterile Pasteur pipets
37°C warming tray *or* incubator
Tissue culture plasticware or glassware including pipets and 25-cm^2 flasks or 60-mm petri dishes, sterile

1. Remove all medium from primary culture with a sterile Pasteur pipet. Wash adhering cell monolayer once or twice with a small volume of 37°C HBSS without Ca^{2+} and Mg^{2+} to remove any residual FBS, which may inhibit the action of trypsin.

   If this is the first medium change, rather than discarding medium that is removed from primary culture, put it into a fresh dish or flask. The medium contains unattached cells that may attach and grow, thereby providing a backup culture.

2. Add enough 37°C trypsin/EDTA solution to culture to cover adhering cell layer. Place plate on a 37°C warming tray 1 to 2 min. Tap bottom of plate on the countertop to dislodge cells. Check culture with an inverted microscope to be sure that cells are rounded up and detached from the surface. If cells are not sufficiently detached, return plate to warming tray for an additional 1 or 2 min.

3. Add 2 ml of 37°C complete medium. Draw cell suspension into a Pasteur pipet and rinse cell layer two or three times to dissociate cells and to dislodge any remaining adherent cells. As soon as cells are detached, add serum or medium containing serum to inhibit further trypsin activity that might damage cells. If cultures are to be split 1/3 or 1/4 rather than 1/2, add sufficient medium such that 1 ml of cell suspension can be transferred into each fresh culture vessel.

4. Add an equal volume of cell suspension to fresh dishes or flasks that have been appropriately labeled. For primary cultures and early subcultures, use 60-mm petri dishes or 25-cm^2 flasks; larger petri dishes or flasks (e.g., 150-mm dishes or 75-cm^2 flasks) with proportionately more medium may be used for later subcultures.

5. Add 4 ml fresh medium to each new culture. Incubate in a humidified 37°C, 5% CO_2 incubator.

   Some laboratories now use incubators with 5% CO_2 and 4% O_2. The low oxygen concentration is thought to simulate the in vivo environment of cells and to enhance cell growth.

6. If necessary, feed subconfluent cultures after 3 or 4 days by removing old medium and adding fresh 37°C medium.

7. Passage secondary culture when it becomes confluent by repeating steps 1 to 5, and continue to passage as necessary.

SUPPORT PROTOCOL 1

FREEZING HUMAN CELLS GROWN IN MONOLAYER CULTURES

To preserve cells, avoid senescence, reduce the risk of contamination, and minimize effects of genetic drift, cell lines may be frozen for long-term storage. Without the use of a cryoprotective agent, freezing would be lethal to the cells in most cases. Generally, a cryoprotective agent such as DMSO is used in conjunction with complete medium for preserving cells at −70°C or lower. DMSO acts to reduce the freezing point and allows a slower cooling rate. Gradual freezing reduces the risk of ice crystal formation and cell damage.

Materials *(see APPENDIX 1 for items with ✓)*

Log-phase monolayer culture of cells in petri dish
Complete medium
Freezing medium: complete medium with 10% to 20% (v/v) FBS (e.g., complete DMEM/20% FBS; or complete RPMI/20% FBS) supplemented with 5% to 10% (v/v) DMSO, 4°C
Benchtop clinical centrifuge (e.g., Fisher Centrific or Clay Adams Dynac) with 45° fixed-angle or swinging-bucket rotor

1. Trypsinize log-phase monolayer culture of cells from plate (see Basic Protocol, steps 1 to 3).

2. Transfer cell suspension to a sterile centrifuge tube and add 2 ml complete medium with serum. Centrifuge 5 min at 300 to 350 × *g*, room temperature. Remove supernatant and add 1 ml of 4°C freezing medium. Resuspend pellet.

3. Add 4 ml of 4°C freezing medium, mix cells thoroughly, and place on wet ice. Count cells using a hemacytometer (see Support Protocol 4). Dilute with more freezing medium as necessary to get a final cell concentration of 10^6 or 10^7 cells/ml.

4. Pipet 1-ml aliquots of cell suspension into labeled 2-ml cryovials and tighten caps on vials. Place vials 1 hr to overnight in a −70°C freezer, then transfer to liquid nitrogen storage freezer. Keep accurate records of the identity and location of cells stored in liquid nitrogen freezers.

Cells may be stored for many years and proper information is imperative for locating a particular line for future use.

SUPPORT PROTOCOL 2

FREEZING CELLS GROWN IN SUSPENSION CULTURE

1. Transfer cell suspension to a centrifuge tube and centrifuge 10 min at 300 to 350 × *g*, room temperature.
2. Remove supernatant and resuspend pellet in 4°C freezing medium at a density of 10^6 to 10^7 cells/ml.

   Some laboratories freeze lymphoblastoid lines at the higher cell density because they plan to recover them in a larger volume of medium and because there may be a greater loss of cell viability upon recovery as compared to other types of cells (e.g., fibroblasts).

3. Transfer 1-ml aliquots of cell suspension into labeled cryovials and freeze as for monolayer cultures.

SUPPORT PROTOCOL 3

THAWING AND RECOVERING HUMAN CELLS

Materials *(see* APPENDIX 1 *for items with ✓)*

Cryopreserved cells stored in liquid nitrogen freezer
70% (v/v) ethanol
✓ Complete medium containing 20% FBS, 37°C: e.g., complete DMEM/20% FBS or complete RPMI/20% FBS, 37°C
Tissue culture dish or flask

1. Remove vial from liquid nitrogen freezer and immediately place it into a 37°C water bath. Agitate vial continuously until medium is thawed (usually <60 sec).
2. Wipe top of vial with 70% ethanol before opening. Transfer thawed cell suspension into a sterile centrifuge tube containing 2 ml warm complete medium containing 20% FBS. Centrifuge 10 min at 150 to 200 × *g*, room temperature. Discard supernatant.
3. Gently resuspend cell pellet in small amount (∼1 ml) of complete medium/20% FBS and transfer to properly labeled culture dish or flask containing the appropriate amount of medium.

   Cultures are reestablished at a higher cell density than that used for original cultures because there is some cell death associated with freezing. Generally, 1 ml cell suspension is reseeded in 5 to 20 ml medium.

4. Check cultures after ∼24 hr to ensure that cells have attached to the plate. Change medium after 5 to 7 days or when pH indicator (e.g., phenol red) in medium changes color. Keep cultures in medium with 20% FBS until cell line is reestablished.

SUPPORT PROTOCOL 4

DETERMINING CELL NUMBER AND VIABILITY WITH A HEMACYTOMETER AND TRYPAN BLUE STAINING

Determining the number of cells in culture is important in standardization of culture conditions and in performing accurate quantitation experiments. A hemacytometer is a thick glass slide with a central area designed as a counting chamber. Accompanying the hemacytometer slide is a thick, even-surfaced coverslip. Ordinary coverslips may have uneven surfaces, which can introduce errors in cell counting; therefore, it is imperative that the coverslip provided with the hemacytometer be used in determining cell number. Cell suspension is applied to a defined area and counted so cell density can be calculated.

A2

Materials

70% (v/v) ethanol
Cell suspension
0.4% (w/v) trypan blue *or* 0.4% (w/v) nigrosin, prepared in HBSS (*APPENDIX 1*)

Hemacytometer with coverslip (Improved Neubauer, Baxter Scientific; see Fig. A.2I.1)
Hand-held counter

1. Clean surface of hemacytometer slide and coverslip with 70% alcohol. Wet edge of coverslip slightly with tap water and press over grooves on hemacytometer (the coverslip should rest evenly over the silver counting area).

2. For cells grown in monolayer cultures, detach cells from surface of dish using trypsin (see Basic Protocol). Dilute cells as needed to obtain a uniform suspension. Disperse any clumps.

   When using the hemacytometer, a maximum cell count of 20 to 50 cells per 1-mm square is recommended.

3. Use a sterile Pasteur pipet to transfer cell suspension to edge of hemacytometer counting chamber. Hold tip of pipet under the coverslip and dispense one drop of suspension by capillary action. Fill second counting chamber.

   Hemacytometer should be considered nonsterile. If cell suspension is to be used for cultures, do not reuse the pipet and do not return any excess cell suspension in the pipet to the original suspension.

4. Allow cells to settle for a few minutes before beginning to count. Blot off excess liquid. View slide on microscope with 100× magnification.

   A 10× ocular with a 10× objective = 100× magnification.

   Position slide to view the large central area of the grid (section 3 in Fig. A.2I.1); this area is bordered by a set of three parallel lines. The central area of the grid should almost fill the microscope field. Subdivisions within the large central area are also bordered by three parallel lines and each subdivision is divided into sixteen smaller squares by single lines. Cells within this area should be evenly distributed without clumping. If cells are not evenly distributed, wash and reload hemacytometer.

5. Use a hand-held counter to count cells in each of the four corner and central squares (Fig. A.2I.1, squares numbered 1 to 5). Repeat counts for other counting chamber.

   Five squares (four corner and one center) are counted from each of the two counting chambers for a total of ten squares counted.

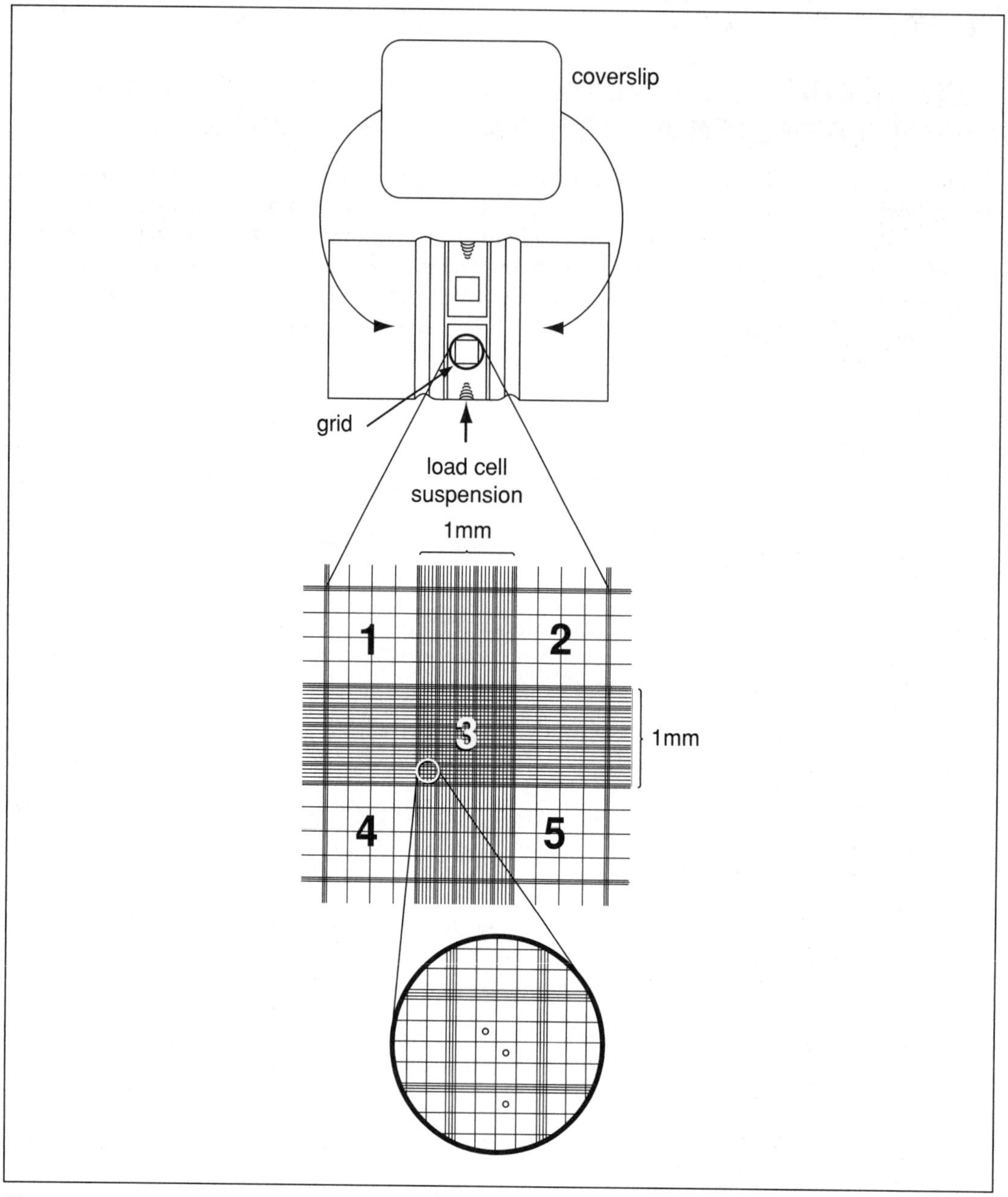

Figure A.2I.1 Hemacytometer slide (Improved Neubauer) and coverslip. Coverslip is applied to slide and cell suspension is added to counting chamber using a Pasteur pipet. Each counting chamber has a 3 × 3–mm grid (enlarged). The four corner squares (1, 2, 4, and 5) and the central square (3) are counted on each side of the hemacytometer (numbers added).

Count cells touching the middle line of the triple line on the top and left of the squares. Do not count cells touching the middle line of the triple lines on the bottom or right side of the square.

6. Determine cells per milliliter by the following calculations:

cells/ml = average count per square × dilution factor × 10^4

total cells = cells/ml × total original volume of cell suspension from which sample was taken

10^4 is the volume correction factor for the hemacytometer: each square is 1 × 1 mm and the depth is 0.1 mm.

7. Determine number of viable cells by adding 0.5 ml of 0.4% trypan blue, 0.3 ml HBSS, and 0.1 ml cell suspension to a small tube. Mix thoroughly and let stand 5 min before loading hemacytometer.

   Either 0.4% trypan blue or 0.4% nigrosin can be used to determine the viable cell number. Nonviable cells will take up the dye, whereas live cells will be impermeable to it.

8. Count total number of cells and total number of viable (unstained) cells. Calculate percent viable cells as follows:

$$\% \text{ viable cells} = \frac{\text{number of unstained cells}}{\text{total number of cells}} \times 100$$

9. Decontaminate coverslip and hemacytometer by rinsing with 70% ethanol and then deionized water. Air dry and store for future use.

A2

SUPPORT PROTOCOL 5

PREPARING CELLS FOR TRANSPORT

Both monolayer and suspension cultures can easily be shipped in 25-cm^2 tissue culture flasks. Cells are grown to near confluency in a monolayer or to desired density in suspension. Medium is removed from monolayer cultures and the flask is filled with fresh medium. Fresh medium is added to suspension cultures to fill the flask. *It is essential that the flasks be completely filled with medium to protect cells from drying if flasks are inverted during transport.* The cap is tightened and taped securely in place. The flask is sealed in a leak-proof plastic bag or other leak-proof container designed to prevent spillage in the event that the flask should become damaged. The primary container is then placed in a secondary insulated container to protect it from extreme temperatures during transport. A biohazard label is affixed to the outside of the package. Generally, cultures are transported by same-day or overnight courier.

Cells can also be shipped frozen. The vial containing frozen cells is removed from the liquid nitrogen freezer and placed immediately on dry ice in an insulated container to prevent thawing during transport.

Reference: Lee, 1991

Contributor: Mary C. Phelan

APPENDIX 3

Selected Suppliers of Reagents and Equipment

Listed below are addresses and phone numbers of commercial suppliers who have been recommended for particular items used in our manuals because: (1) the particular brand has actually been found to be of superior quality, or (2) the item is difficult to find in the marketplace. Consequently, this compilation may not include some important vendors of biological supplies. For comprehensive listings, see *Linscott's Directory of Immunological and Biological Reagents* (Santa Rosa, CA), *The Biotechnology Directory* (Stockton Press, New York), the annual Buyers' Guide supplement to the journal *Bio/Technology*, as well as various sites on the Internet.

A3

A.C. Daniels
72-80 Akeman Street
Tring, Hertfordshire, HP23 6AJ, UK
(44) 1442 826881
FAX: (44) 1442 826880

A.D. Instruments
5111 Nations Crossing Road #8
Suite 2
Charlotte, NC 28217
(704) 522-8415 FAX: (704) 527-5005
http://www.us.endress.com

A.J. Buck
11407 Cronhill Drive
Owings Mill, MD 21117
(800) 638-8673 FAX: (410) 581-1809
(410) 581-1800
http://www.ajbuck.com

A.M. Systems
131 Business Park Loop
P.O. Box 850
Carlsborg, WA 98324
(800) 426-1306 FAX: (360) 683-3525
(360) 683-8300
http://www.a-msystems.com

Aaron Medical Industries
7100 30th Avenue North
St. Petersburg, FL 33710
(727) 384-2323 FAX: (727) 347-9144
www.aaronmed.com

Abbott Laboratories
100 Abbott Park Road
Abbott Park, IL 60064
(800) 323-9100 FAX: (847) 938-7424
http://www.abbott.com

ABCO Dealers
55 Church Street Central Plaza
Lowell, MA 01852
(800) 462-3326 (978) 459-6101
http://www.lomedco.com/abco.htm

Aber Instruments
5 Science Park
Aberystwyth, Wales SY23 3AH, UK
(44) 1970 636300
FAX: (44) 1970 615455
http://www.aber-instruments.co.uk

ABI Biotechnologies
See Perkin-Elmer

ABI Biotechnology
See Apotex

Access Technologies
Subsidiary of Norfolk Medical
7350 N. Ridgeway
Skokie, IL 60076
(877) 674-7131 FAX: (847) 674-7066
(847) 674-7131
http://www.norfolkaccess.com

Accurate Chemical and Scientific
300 Shames Drive
Westbury, NY 11590
(800) 645-6264 FAX: (516) 997-4948
(516) 333-2221
http://www.accuratechemical.com

AccuScan Instruments
5090 Trabue Road
Columbus, OH 43228
(800) 822-1344 FAX: (614) 878-3560
(614) 878-6644
http://www.accuscan-usa.com

AccuStandard
125 Market Street
New Haven, CT 06513
(800) 442-5290 FAX: (877) 786-5287
http://www.accustandard.com

Ace Glass
1430 NW Boulevard
Vineland, NJ 08360
(800) 223-4524 FAX: (800) 543-6752
(609) 692-3333

ACO Pacific
2604 Read Avenue
Belmont, CA 94002
(650) 595-8588 FAX: (650) 591-2891
http://www.acopacific.com

Acros Organic
See Fisher Scientific

Action Scientific
P.O. Box 1369
Carolina Beach, NC 28428
(910) 458-0401 FAX: (910) 458-0407

AD Instruments
1949 Landings Drive
Mountain View, CA 94043
(888) 965-6040 FAX: (650) 965-9293
(650) 965-9292
http://www.adinstruments.com

Adaptive Biosystems
15 Ribocon Way
Progress Park
Luton, Bedsfordshire LU4 9UR, UK
(44)1 582-597676
FAX: (44)1 582-581495
http://www.adaptive.co.uk

Adobe Systems
1585 Charleston Road
P.O. Box 7900
Mountain View, CA 94039
(800) 833-6687 FAX: (415) 961-3769
(415) 961-4400
http://www.adobe.com

Advanced Bioscience Resources
1516 Oak Street, Suite 303
Alameda, CA 94501
(510) 865-5872 FAX: (510) 865-4090

Advanced Biotechnologies
9108 Guilford Road
Columbia, MD 21046
(800) 426-0764 FAX: (301) 497-9773
(301) 470-3220
http://www.abionline.com

Advanced ChemTech
5609 Fern Valley Road
Louisville, KY 40228
(502) 969-0000
http://www.peptide.com

Advance Machining and tooling
9850 Businesspark Avanue
San Diego, CA 92131
(858)530-0751 FAX:(858)530-611
http://www.amtmgf.com

Advanced Magnetics
See PerSeptive Biosystems

Advanced Process Supply
See Naz-Dar-KC Chicago

Advanced Separation Technologies
37 Leslie Court
P.O. Box 297
Whippany, NJ 07981
(973) 428-9080 FAX: (973) 428-0152
http://www.astecusa.com

Advanced Targeting Systems
11175-A Flintkote Avenue
San Diego, CA 92121
(877) 889-2288 FAX: (858) 642-1989
(858) 642-1988
http://www.ATSbio.com

Advent Research Materials
Eynsham, Oxford OX29 4JA, UK
(44) 1865-884440
FAX: (44) 1865-84460
www.advent-rm.com

Advet
Industrivagen 24
S-972 54 Lulea, Sweden
(46) 0920-211887
FAX: (46) 0920-13773

Aesculap
1000 Gateway Boulevard
South San Francisco, CA 94080
(800) 282-9000
http://www.aesculap.com

Affinity Chromatography
307 Huntingdon Road
Girton, Cambridge CB3 OJX, UK
(44) 1223 277192
FAX: (44) 1223 277502
http://www.affinity-chrom.com

Affinity Sensors
See Labsystems Affinity Sensors

Affymetrix
3380 Central Expressway
Santa Clara, CA 95051
(408) 731-5000 FAX: (408) 481-0422
(800) 362-2447
http://www.affymetrix.com

Agar Scientific
66a Cambridge Road
Stansted CM24 8DA, UK
(44) 1279-813-519
FAX: (44) 1279-815-106
http://www.agarscientific.com

A/G Technology
101 Hampton Avenue
Needham, MA 02494
(800) AGT-2535 FAX: (781) 449-5786
(781) 449-5774
http://www.agtech.com

Agen Biomedical Limited
11 Durbell Street
P.O. Box 391
Acacia Ridge 4110
Brisbane, Australia
61-7-3370-6300 FAX: 61-7-3370-6370
http://www.agen.com

Agilent Technologies
395 Page Mill Road
P.O. Box 10395
Palo Alto, CA 94306
(650) 752-5000
http://www.agilent.com/chem

Agouron Pharmaceuticals
10350 N. Torrey Pines Road
La Jolla, CA 92037
(858) 622-3000 FAX: (858) 622-3298
http://www.agouron.com

Agracetus
8520 University Green
Middleton, WI 53562
(608) 836-7300 FAX: (608) 836-9710
http://www.monsanto.com

AIDS Research and Reference
Reagent Program
U.S. Department of Health and Human Services
625 Lofstrand Lane
Rockville, MD 20850
(301) 340-0245 FAX: (301) 340-9245
http://www.aidsreagent.org

AIN Plastics
249 East Sanford Boulevard
P.O. Box 151
Mt. Vernon, NY 10550
(914) 668-6800 FAX: (914) 668-8820
http://www.tincna.com

Air Products and Chemicals
7201 Hamilton Boulevard
Allentown, PA 18195
(800) 345-3148 FAX: (610) 481-4381
(610) 481-6799
http://www.airproducts.com

ALA Scientific Instruments
1100 Shames Drive
Westbury, NY 11590
(516) 997-5780 FAX: (516) 997-0528
http://www.alascience.com

Aladin Enterprises
1255 23rd Avenue
San Francisco, CA 94122
(415) 468-0433 FAX: (415) 468-5607

Aladdin Systems
165 Westridge Drive
Watsonville, CA 95076
(831) 761-6200 FAX: (831) 761-6206
http://www.aladdinsys.com

Alcide
8561 154th Avenue NE
Redmond, WA 98052
(800) 543-2133 FAX: (425) 861-0173
(425) 882-2555
http://www.alcide.com

Aldrich Chemical
P.O. Box 2060
Milwaukee, WI 53201
(800) 558-9160 FAX: (800) 962-9591
(414) 273-3850 FAX: (414) 273-4979
http://www.aldrich.sial.com

Alexis Biochemicals
6181 Cornerstone Court East, Suite 103
San Diego, CA 92121
(800) 900-0065 FAX: (858) 658-9224
(858) 658-0065
http://www.alexis-corp.com

Alfa Laval
Avenue de Ble 5 - Bazellaan 5
BE-1140 Brussels, Belgium
32(2) 728 3811
FAX: 32(2) 728 3917 or
32(2) 728 3985
http://www.alfalaval.com

Alice King Chatham Medical Arts
11915-17 Inglewood Avenue
Hawthorne, CA 90250
(310) 970-1834 FAX: (310) 970-0121
(310) 970-1063

Allegiance Healthcare
800-964-5227
http://www.allegiance.net

Allelix Biopharmaceuticals
6850 Gorway Drive
Mississauga, Ontario
L4V 1V7 Canada
(905) 677-0831 FAX: (905) 677-9595
http://www.allelix.com

Allentown Caging Equipment
Route 526, P.O. Box 698
Allentown, NJ 08501
(800) 762-CAGE FAX: (609) 259-0449
(609) 259-7951
http://www.acecaging.com

Alltech Associates
Applied Science Labs
2051 Waukegan Road
P.O. Box 23
Deerfield, IL 60015
(800) 255-8324 FAX: (847) 948-1078
(847) 948-8600
http://www.alltechweb.com

Alomone Labs
HaMarpeh 5
P.O. Box 4287
Jerusalem 91042, Israel
972-2-587-2202 FAX: 972-2-587-1101
US: (800) 791-3904
FAX: (800) 791-3912
http://www.alomone.com

Alpha Innotech
14743 Catalina Street
San Leandro, CA 94577
(800) 795-5556 FAX: (510) 483-3227
(510) 483-9620
http://www.alphainnotech.com

Altec Plastics
116 B Street
Boston, MA 02127
(800) 477-8196 FAX: (617) 269-8484
(617) 269-1400

Alza
1900 Charleston Road P.O. Box 7210
Mountain View, CA 94043
(800) 692-2990 FAX: (650) 564-7070
(650) 564-5000
http://www.alza.com

Alzet
c/o Durect Corporation
P.O. Box 530
10240 Bubb Road
Cupertino CA 95015
(800)692-2990 (408)367-4036
FAX: (408)865-1406
http://www.alzet.com

Amac
160B Larrabee Road
Westbrook, ME 04092
(800) 458-5060 FAX: (207) 854-0116
(207) 854-0426

Amaresco
30175 Solon Industrial Parkway
Solon, Ohio 44139
(800) 366-1313 FAX: (440) 349-1182
(440) 349-1313

Ambion
2130 Woodward Street, Suite 200
Austin, TX 78744
(800) 888-8804 FAX: (512) 651-0190
(512) 651-0200
http://www.ambion.com

American Association of Blood Banks College of American Pathologists
325 Waukegan Road
Northfield, IL 60093
(800) 323-4040 FAX: (847) 8166
(847) 832-7000
http://www.cap.org

American Bio-Technologies
See Intracel Corporation

American Bioanalytical
15 Erie Drive
Natick, MA 01760
(800) 443-0600 FAX: (508) 655-2754
(508) 655-4336
http://www.americanbio.com

American Cyanamid
P.O. Box 400
Princeton, NJ 08543
(609) 799-0400 FAX: (609) 275-3502
http://www.cyanamid.com

American HistoLabs
7605-F Airpark Road
Gaithersburg, MD 20879
(301) 330-1200 FAX: (301) 330-6059

American International Chemical
17 Strathmore Road
Natick, MA 01760
(800) 238-0001 (508) 655-5805
http://www.aicma.com

American Laboratory Supply
See American Bioanalytical

American Medical Systems
10700 Bren Road West
Minnetonka, MN 55343
(800) 328-3881 FAX: (612) 930-6654
(612) 933-4666
http://www.visitams.com

American Qualex
920-A Calle Negocio
San Clemente, CA 92673
(949) 492-8298 FAX: (949) 492-6790
http://www.americanqualex.com

American Radiolabeled Chemicals
11624 Bowling Green
St. Louis, MO 63146
(800) 331-6661 FAX: (800) 999-9925
(314) 991-4545 FAX: (314) 991-4692
http://www.arc-inc.com

American Scientific Products
See VWR Scientific Products

American Society for Histocompatibility and Immunogenetics
P.O. Box 15804
Lenexa, KS 66285 (913) 541-0009
FAX: (913) 541-0156
http://www.swmed.edu/home_pages/ASHI/ashi.htm

American Type Culture Collection (ATCC)
10801 University Boulevard
Manassas, VA 20110
(800) 638-6597 FAX: (703) 365-2750
(703) 365-2700
http://www.atcc.org

Amersham
See Amersham Pharmacia Biotech

Amersham International
Amersham Place
Little Chalfont, Buckinghamshire
HP7 9NA, UK
(44) 1494-544100
FAX: (44) 1494-544350
http://www.apbiotech.com

Amersham Medi-Physics
Also see Nycomed Amersham
3350 North Ridge Avenue
Arlington Heights, IL 60004
(800) 292-8514 FAX: (800) 807-2382
http://www.nycomed-amersham.com

Amersham Pharmacia Biotech
800 Centennial Avenue
P.O. Box 1327
Piscataway, NJ 08855
(800) 526-3593 FAX: (877) 295-8102
(732) 457-8000
http://www.apbiotech.com

Amgen
1 Amgen Center Drive
Thousand Oaks, CA 91320
(800) 926-4369 FAX: (805) 498-9377
(805) 447-5725
http://www.amgen.com

Amicon
Scientific Systems Division
72 Cherry Hill Drive
Beverly, MA 01915
(800) 426-4266 FAX: (978) 777-6204
(978) 777-3622
http://www.amicon.com

Amika
8980F Route 108
Oakland Center
Columbia, MD 21045
(800) 547-6766 FAX: (410) 997-7104
(410) 997-0100
http://www.amika.com

Amoco Performance Products
See BPAmoco

AMPI
See Pacer Scientific

Amrad
576 Swan Street
Richmond, Victoria 3121, Australia
613-9208-4000
FAX: 613-9208-4350
http://www.amrad.com.au

Amresco
30175 Solon Industrial Parkway
Solon, OH 44139
(800) 829-2805 FAX: (440) 349-1182
(440) 349-1199

Anachemia Chemicals
3 Lincoln Boulevard
Rouses Point, NY 12979
(800) 323-1414 FAX: (518) 462-1952
(518) 462-1066
http://www.anachemia.com

Ana-Gen Technologies
4015 Fabian Way
Palo Alto, CA 94303
(800) 654-4671 FAX: (650) 494-3893
(650) 494-3894
http://www.ana-gen.com

Analox Instruments USA
P.O. Box 208
Lunenburg MA 01462
(978) 582-9368 FAX: (978) 582-9588
http://www.analox.com

Analytical Biological Services
Cornell Business Park 701-4
Wilmington, DE 19801
(800) 391-2391 FAX: (302) 654-8046
(302) 654-4492
http://www.ABSbioreagents.com

Analytical Genetics Testing Center
7808 Cherry Creek S. Drive, Suite 201
Denver, CO 80231
(800) 204-4721 FAX: (303) 750-2171
(303) 750-2023
http://www.geneticid.com

AnaSpec
2149 O'Toole Avenue, Suite F
San Jose, CA 95131
(800) 452-5530 FAX: (408) 452-5059
(408) 452-5055
http://www.anaspec.com

Ancare
2647 Grand Avenue
P.O. Box 814
Bellmore, NY 11710
(800) 645-6379 FAX: (516) 781-4937
(516) 781-0755
http://www.ancare.com

Ancell
243 Third Street North
P.O. Box 87
Bayport, MN 55033
(800) 374-9523 FAX: (651) 439-1940
(651) 439-0835
http://www.ancell.com

Anderson Instruments
500 Technology Court
Smyrna, GA 30082
(800) 241-6898 FAX: (770) 319-5306
(770) 319-9999
http://www.graseby.com

Andreas Hettich
Gartenstrasse 100
Postfach 260
D-78732 Tuttlingen, Germany
(49) 7461 705 0
FAX: (49) 7461 705-122
http://www.hettich-centrifugen.de

Anesthetic Vaporizer Services
10185 Main Street
Clarence, NY 14031
(719) 759-8490
www.avapor.com

Animal Identification and Marking Systems (AIMS)
13 Winchester Avenue
Budd Lake, NJ 07828
(908) 684-9105 FAX: (908) 684-9106
http://www.animalid.com

Annovis
34 Mount Pleasant Drive
Aston, PA 19014
(800) EASY-DNA FAX: (610) 361-8255
(610) 361-9224
http://www.annovis.com

Apotex
150 Signet Drive
Weston, Ontario M9L 1T9, Canada
(416) 749-9300 FAX: (416) 749-2646
http://www.apotex.com

Apple Scientific
11711 Chillicothe Road, Unit 2
P.O. Box 778
Chesterland, OH 44026
(440) 729-3056 FAX: (440) 729-0928
http://www.applesci.com

Applied Biosystems
See PE Biosystems

Applied Imaging
2380 Walsh Avenue, Bldg. B
Santa Clara, CA 95051
(800) 634-3622 FAX: (408) 562-0264
(408) 562-0250
http://www.aicorp.com

Applied Photophysics
203-205 Kingston Road
Leatherhead, Surrey, KT22 7PB
UK
(44) 1372-386537

Applied Precision
1040 12th Avenue Northwest
Issaquah, Washington 98027
(425) 557-1000
FAX: (425) 557-1055
http://www.api.com/index.html

Appligene Oncor
Parc d'Innovation
Rue Geiler de Kaysersberg, BP 72
67402 Illkirch Cedex, France
(33) 88 67 22 67
FAX: (33) 88 67 19 45
http://www.oncor.com/prod-app.htm

Applikon
1165 Chess Drive, Suite G
Foster City, CA 94404
(650) 578-1396 FAX: (650) 578-8836
http://www.applikon.com

Appropriate Technical Resources
9157 Whiskey Bottom Road
Laurel, MD 20723
(800) 827-5931 FAX: (410) 792-2837
http://www.atrbiotech.com

APV Gaulin
100 S. CP Avenue
Lake Mills, WI 53551
(888) 278-4321 FAX: (888) 278-5329
http://www.apv.com

Aqualon
See Hercules Aqualon

Aquebogue Machine and Repair Shop
Box 2055
Main Road
Aquebogue, NY 11931
(631) 722-3635 FAX: (631) 722-3106

Archer Daniels Midland
4666 Faries Parkway
Decatur, IL 62525
(217) 424-5200
http://www.admworld.com

Archimica Florida
P.O. Box 1466
Gainesville, FL 32602
(800) 331-6313 FAX: (352) 371-6246
(352) 376-8246
http://www.archimica.com

Arcor Electronics
1845 Oak Street #15
Northfield, IL 60093
(847) 501-4848

Arcturus Engineering
400 Logue Avenue
Mountain View, CA 94043
(888) 446 7911 FAX: (650) 962 3039
(650) 962 3020
http://www.arctur.com

Argonaut Technologies
887 Industrial Road, Suite G
San Carlos, CA 94070
(650) 998-1350 FAX: (650) 598-1359
http://www.argotech.com

Ariad Pharmaceuticals
26 Landsdowne Street
Cambridge, MA 02139
(617) 494-0400 FAX: (617) 494-8144
http://www.ariad.com

Armour Pharmaceuticals
See Rhone-Poulenc Rorer

Aronex Pharmaceuticals
8707 Technology Forest Place
The Woodlands, TX 77381
(281) 367-1666 FAX: (281) 367-1676
http://www.aronex.com

Artisan Industries
73 Pond Street
Waltham, MA 02254
(617) 893-6800
http://www.artisanind.com

ASI Instruments
12900 Ten Mile Road
Warren, MI 48089
(800) 531-1105 FAX: (810) 756-9737
(810) 756-1222
http://www.asi-instruments.com

Aspen Research Laboratories
1700 Buerkle Road
White Bear Lake, MN 55140
(651) 264-6000 FAX: (651) 264-6270
http://www.aspenresearch.com

Associates of Cape Cod
704 Main Street
Falmouth, MA 02540 (800) LAL-TEST
FAX: (508) 540-8680
(508) 540-3444
http://www.acciusa.com

Astra Pharmaceuticals
See AstraZeneca

AstraZeneca
1800 Concord Pike
Wilmington, DE 19850
(302) 886-3000 FAX: (302) 886-2972
http://www.astrazeneca.com

AT Biochem
30 Spring Mill Drive
Malvern, PA 19355
(610) 889-9300 FAX: (610) 889-9304

ATC Diagnostics
See Vysis

ATCC
See American Type Culture Collection

Athens Research and Technology
P.O. Box 5494
Athens, GA 30604
(706) 546-0207 FAX: (706) 546-7395

Atlanta Biologicals
1425-400 Oakbrook Drive
Norcross, GA 30093
(800) 780-7788 or (770) 446-1404
FAX: (800) 780-7374 or (770) 446-1404
http://www.atlantabio.com

Atomergic Chemical
71 Carolyn Boulevard
Farmingdale, NY 11735
(631) 694-9000 FAX: (631) 694-9177
http://www.atomergic.com

Atomic Energy of Canada
2251 Speakman Drive
Mississauga, Ontario
L5K 1B2 Canada
(905) 823-9040 FAX: (905) 823-1290
http://www.aecl.ca

ATR
P.O. Box 460
Laurel, MD 20725
(800) 827-5931 FAX: (410) 792-2837
(301) 470-2799
http://www.atrbiotech.com

Aurora Biosciences
11010 Torreyana Road
San Diego, CA 92121
(858) 404-6600 FAX: (858) 404-6714
http://www.aurorabio.com

Automatic Switch Company
A Division of Emerson Electric
50 Hanover Road
Florham Park, NJ 07932
(800) 937-2726 FAX: (973) 966-2628
(973) 966-2000
http://www.asco.com

Avanti Polar Lipids
700 Industrial Park Drive
Alabaster, AL 35007
(800) 227-0651 FAX: (800) 229-1004
(205) 663-2494 FAX: (205) 663-0756
http://www.avantilipids.com

Aventis
BP 67917
67917 Strasbourg Cedex 9, France
33 (0) 388 99 11 00
FAX: 33 (0) 388 99 11 01
http://www.aventis.com

Aventis Pasteur
1 Discovery Drive
Swiftwater, PA 18370
(800) 822-2463 FAX: (570) 839-0955
(570) 839-7187
http://www.aventispasteur.com/usa

Avery Dennison
150 North Orange Grove Boulevard
Pasadena, CA 91103
(800) 462-8379 FAX: (626) 792-7312
(626) 304-2000
http://www.averydennison.com

Avestin
2450 Don Reid Drive
Ottawa, Ontario K1H 1E1, Canada
(888) AVESTIN FAX: (613) 736-8086
(613) 736-0019
http://www.avestin.com

AVIV Instruments
750 Vassar Avenue
Lakewood, NJ 08701
(732) 367-1663 FAX: (732) 370-0032
http://www.avivinst.com

Axon Instruments
1101 Chess Drive
Foster City, CA 94404
(650) 571-9400 FAX: (650) 571-9500
http://www.axon.com

Azon
720 Azon Road
Johnson City, NY 13790
(800) 847-9374 FAX: (800) 635-6042
(607) 797-2368
http://www.azon.com

BAbCO
1223 South 47th Street
Richmond, CA 94804
(800) 92-BABCO FAX: (510) 412-8940
(510) 412-8930
http://www.babco.com

Bacharach
625 Alpha Drive
Pittsburgh, PA 15238
(800) 736-4666 FAX: (412) 963-2091
(412) 963-2000
http://www.bacharach-inc.com

Bachem Bioscience
3700 Horizon Drive
King of Prussia, PA 19406
(800) 634-3183 FAX: (610) 239-0800
(610) 239-0300
http://www.bachem.com

Bachem California
3132 Kashiwa Street
P.O. Box 3426
Torrance, CA 90510
(800) 422-2436 FAX: (310) 530-1571
(310) 539-4171
http://www.bachem.com

Baekon
18866 Allendale Avenue
Saratoga, CA 95070
(408) 972-8779 FAX: (408) 741-0944

Baker Chemical
See J.T. Baker

Bangs Laboratories
9025 Technology Drive
Fishers, IN 46038
(317) 570-7020 FAX: (317) 570-7034
www.bangslabs.com

Bard Parker
See Becton Dickinson

Barnstead/Thermolyne
P.O. Box 797
2555 Kerper Boulevard
Dubuque, IA 52004
(800) 446-6060 FAX: (319) 589-0516
http://www.barnstead.com

Barrskogen
4612 Laverock Place N
Washington, DC 20007
(800) 237-9192 FAX: (301) 464-7347

BAS
See Bioanalytical Systems

BASF
Specialty Products
3000 Continental Drive North
Mt. Olive, NJ 07828
(800) 669-2273 FAX: (973) 426-2610
http://www.basf.com

Baum, W.A.
620 Oak Street
Copiague, NY 11726
(631) 226-3940 FAX: (631) 226-3969
http://www.wabaum.com

Bausch & Lomb
One Bausch & Lomb Place
Rochester, NY 14604
(800) 344-8815 FAX: (716) 338-6007
(716) 338-6000
http://www.bausch.com

Baxter
Fenwal Division
1627 Lake Cook Road
Deerfield, IL 60015
(800) 766-1077 FAX: (800) 395-3291
(847) 940-6599 FAX: (847) 940-5766
http://www.powerfulmedicine.com

Baxter Healthcare
One Baxter Parkway
Deerfield, IL 60015
(800) 777-2298 FAX: (847) 948-3948
(847) 948-2000
http://www.baxter.com

Baxter Scientific Products
See VWR Scientific

Bayer
Agricultural Division
Animal Health Products
12707 Shawnee Mission Pkwy.
Shawnee Mission, KS 66201
(800) 255-6517 FAX: (913) 268-2803
(913) 268-2000
http://www.bayerus.com

Bayer
Diagnostics Division (Order Services)
P.O. Box 2009
Mishiwaka, IN 46546
(800) 248-2637 FAX: (800) 863-6882
(219) 256-3390
http://www.bayer.com

Bayer Diagnostics
511 Benedict Avenue
Tarrytown, NY 10591
(800) 255-3232 FAX: (914) 524-2132
(914) 631-8000
http://www.bayerdiag.com

Bayer Plc
Diagnostics Division
Bayer House, Strawberry Hill
Newbury, Berkshire RG14 1JA, UK
(44) 1635-563000
FAX: (44) 1635-563393
http://www.bayer.co.uk

BD Immunocytometry Systems
2350 Qume Drive
San Jose, CA 95131
(800) 223-8226 FAX: (408) 954-BDIS
http://www.bdfacs.com

BD Labware
Two Oak Park
Bedford, MA 01730
(800) 343-2035 FAX: (800) 743-6200
http://www.bd.com/labware

BD PharMingen
10975 Torreyana Road
San Diego, CA 92121
(800) 848-6227 FAX: (858) 812-8888
(858) 812-8800
http://www.pharmingen.com

BD Transduction Laboratories
133 Venture Court
Lexington, KY 40511
(800) 227-4063 FAX: (606) 259-1413
(606) 259-1550
http://www.translab.com

BDH Chemicals
Broom Road
Poole, Dorset BH12 4NN, UK
(44) 1202-745520
FAX: (44) 1202- 2413720

BDH Chemicals
See Hoefer Scientific Instruments

BDIS
See BD Immunocytometry Systems

Beckman Coulter
4300 North Harbor Boulevard
Fullerton, CA 92834
(800) 233-4685 FAX: (800) 643-4366
(714) 871-4848
http://www.beckman-coulter.com

Beckman Instruments
Spinco Division/Bioproducts Operation
1050 Page Mill Road
Palo Alto, CA 94304
(800) 742-2345 FAX: (415) 859-1550
(415) 857-1150
http://www.beckman-coulter.com

Becton Dickinson Immunocytometry & Cellular Imaging
2350 Qume Drive
San Jose, CA 95131
(800) 223-8226 FAX: (408) 954-2007
(408) 432-9475
http://www.bdfacs.com

Becton Dickinson Labware
1 Becton Drive
Franklin Lakes, NJ 07417
(888) 237-2762 FAX: (800) 847-2220
(201) 847-4222
http://www.bdfacs.com

Becton Dickinson Labware
2 Bridgewater Lane
Lincoln Park, NJ 07035
(800) 235-5953 FAX: (800) 847-2220
(201) 847-4222
http://www.bdfacs.com

Becton Dickinson Primary
Care Diagnostics
7 Loveton Circle
Sparks, MD 21152
(800) 675-0908 FAX: (410) 316-4723
(410) 316-4000
http://www.bdfacs.com

Behringwerke Diagnostika
Hoechster Strasse 70
P-65835 Liederback, Germany
(49) 69-30511 FAX: (49) 69-303-834

Bellco Glass
340 Edrudo Road
Vineland, NJ 08360
(800) 257-7043 FAX: (856) 691-3247
(856) 691-1075
http://www.bellcoglass.com

Bender Biosystems
See Serva

Beral Enterprises
See Garren Scientific

Berkeley Antibody
See BAbCO

Bernsco Surgical Supply
25 Plant Avenue
Hauppague, NY 11788
(800) TIEMANN FAX: (516) 273-6199
(516) 273-0005
http://www.bernsco.com

Beta Medical and Scientific (Datesand Ltd.)
2 Ferndale Road
Sale, Manchester M33 3GP, UK
(44) 1612 317676
FAX: (44) 1612 313656

Bethesda Research Laboratories (BRL)
See Life Technologies

Biacore
200 Centennial Avenue, Suite 100
Piscataway, NJ 08854
(800) 242-2599 FAX: (732) 885-5669
(732) 885-5618
http://www.biacore.com

Bilaney Consultants
St. Julian's
Sevenoaks, Kent TN15 0RX, UK
(44) 1732 450002
FAX: (44) 1732 450003
http://www.bilaney.com

Binding Site
5889 Oberlin Drive, Suite 101
San Diego, CA 92121
(800) 633-4484 FAX: (619) 453-9189
(619) 453-9177
http://www.bindingsite.co.uk

BIO 101
See Qbiogene

Bio Image
See Genomic Solutions

Bioanalytical Systems
2701 Kent Avenue
West Lafayette, IN 47906
(800) 845-4246 FAX: (765) 497-1102
(765) 463-4527
http://www.bioanalytical.com

Biocell
2001 University Drive
Rancho Dominguez, CA 90220
(800) 222-8382 FAX: (310) 637-3927
(310) 537-3300
http://www.biocell.com

Biocoat
See BD Labware

BioComp Instruments
650 Churchill Road
Fredericton, New Brunswick
E3B 1P6 Canada
(800) 561-4221 FAX: (506) 453-3583
(506) 453-4812
http://131.202.97.21

BioDesign
P.O. Box 1050
Carmel, NY 10512
(914) 454-6610 FAX: (914) 454-6077
http://www.biodesignofny.com

BioDiscovery
4640 Admiralty Way, Suite 710
Marina Del Rey, CA 90292
(310) 306-9310 FAX: (310) 306-9109
http://www.biodiscovery.com

Bioengineering AG
Sagenrainstrasse 7
CH8636 Wald, Switzerland
(41) 55-256-8-111
FAX: (41) 55-256-8-256

Biofluids
Division of Biosource International
1114 Taft Street
Rockville, MD 20850
(800) 972-5200 FAX: (301) 424-3619
(301) 424-4140
http://www.biosource.com

BioFX Laboratories
9633 Liberty Road, Suite S
Randallstown, MD 21133
(800) 445-6447 FAX: (410) 498-6008
(410) 496-6006
http://www.biofx.com

BioGenex Laboratories
4600 Norris Canyon Road
San Ramon, CA 94583
(800) 421-4149 FAX: (925) 275-0580
(925) 275-0550
http://www.biogenex.com

Bioline
2470 Wrondel Way
Reno, NV 89502
(888) 257-5155 FAX: (775) 828-7676
(775) 828-0202
http://www.bioline.com

Bio-Logic Research & Development
1, rue de l-Europe
A.Z. de Font-Ratel
38640 CLAIX, France
(33) 76-98-68-31
FAX: (33) 76-98-69-09

Biological Detection Systems
See Cellomics or Amersham

Biomeda
1166 Triton Drive, Suite E
P.O. Box 8045
Foster City, CA 94404
(800) 341-8787 FAX: (650) 341-2299
(650) 341-8787
http://www.biomeda.com

BioMedic Data Systems
1 Silas Road
Seaford, DE 19973
(800) 526-2637 FAX: (302) 628-4110
(302) 628-4100
http://www.bmds.com

Biomedical Engineering
P.O. Box 980694
Virginia Commonwealth University
Richmond, VA 23298
(804) 828-9829 FAX: (804) 828-1008

Biomedical Research Instruments
12264 Wilkins Avenue
Rockville, MD 20852
(800) 327-9498
(301) 881-7911
http://www.biomedinstr.com

Bio/medical Specialties
P.O. Box 1687
Santa Monica, CA 90406
(800) 269-1158 FAX: (800) 269-1158
(323) 938-7515

BioMerieux
100 Rodolphe Street
Durham, North Carolina 27712
(919) 620-2000
http://www.biomerieux.com

BioMetallics
P.O. Box 2251
Princeton, NJ 08543
(800) 999-1961 FAX: (609) 275-9485
(609) 275-0133
http://www.microplate.com

Biomol Research Laboratories
5100 Campus Drive
Plymouth Meeting, PA 19462
(800) 942-0430 FAX: (610) 941-9252
(610) 941-0430
http://www.biomol.com

Bionique Testing Labs
Fay Brook Drive
RR 1, Box 196
Saranac Lake, NY 12983
(518) 891-2356 FAX: (518) 891-5753
http://www.bionique.com

Biopac Systems
42 Aero Camino
Santa Barbara, CA 93117
(805) 685-0066 FAX: (805) 685-0067
http://www.biopac.com

Bioproducts for Science
See Harlan Bioproducts for Science

Bioptechs
3560 Beck Road Butler, PA 16002
(877) 548-3235 FAX: (724) 282-0745
(724) 282-7145
http://www.bioptechs.com

BIOQUANT-R&M Biometrics
5611 Ohio Avenue
Nashville, TN 37209
(800) 221-0549 (615) 350-7866
FAX: (615) 350-7282
http://www.bioquant.com

Bio-Rad Laboratories
2000 Alfred Nobel Drive
Hercules, CA 94547
(800) 424-6723 FAX: (800) 879-2289
(510) 741-1000 FAX: (510) 741-5800
http://www.bio-rad.com

Bio-Rad Laboratories
Maylands Avenue
Hemel Hempstead, Herts HP2 7TD, UK
http://www.bio-rad.com

BioRobotics
3-4 Bennell Court
Comberton, Cambridge CB3 7DS, UK
(44) 1223-264345
FAX: (44) 1223-263933
http://www.biorobotics.co.uk

BIOS Laboratories
See Genaissance Pharmaceuticals

Biosearch Technologies
81 Digital Drive
Novato, CA 94949
(800) GENOME1 FAX: (415) 883-8488
(415) 883-8400
http://www.biosearchtech.com

BioSepra
111 Locke Drive
Marlborough, MA 01752
(800) 752-5277 FAX: (508) 357-7595
(508) 357-7500
http://www.biosepra.com

Bio-Serv
1 8th Street, Suite 1
Frenchtown, NJ 08825
(908) 996-2155 FAX: (908) 996-4123
http://www.bio-serv.com

BioSignal
1744 William Street, Suite 600
Montreal, Quebec H3J 1R4, Canada
(800) 293-4501 FAX: (514) 937-0777
(514) 937-1010
http://www.biosignal.com

Biosoft
P.O. Box 10938
Ferguson, MO 63135
(314) 524-8029 FAX: (314) 524-8129
http://www.biosoft.com

Biosource International
820 Flynn Road
Camarillo, CA 93012
(800) 242-0607 FAX: (805) 987-3385
(805) 987-0086
http://www.biosource.com

BioSpec Products
P.O. Box 788
Bartlesville, OK 74005
(800) 617-3363 FAX: (918) 336-3363
(918) 336-3363
http://www.biospec.com

Biosure
See Riese Enterprises

Biosym Technologies
See Molecular Simulations

Biosys
21 quai du Clos des Roses
602000 Compiegne, France
(33) 03 4486 2275
FAX: (33) 03 4484 2297

Bio-Tech Research Laboratories
NIAID Repository
Rockville, MD 20850
http://www.niaid.nih.gov/ncn/repos.htm

Biotech Instruments
Biotech House
75A High Street
Kimpton, Hertfordshire SG4 8PU, UK
(44) 1438 832555
FAX: (44) 1438 833040
http://www.biotinst.demon.co.uk

Biotech International
11 Durbell Street
Acacia Ridge, Queensland 4110
Australia
61-7-3370-6396
FAX: 61-7-3370-6370
http://www.avianbiotech.com

Biotech Source
Inland Farm Drive
South Windham, ME 04062
(207) 892-3266 FAX: (207) 892-6774

Bio-Tek Instruments
Highland Industrial Park
P.O. Box 998
Winooski, VT 05404
(800) 451-5172 FAX: (802) 655-7941
(802) 655-4040
http://www.biotek.com

Biotecx Laboratories
6023 South Loop East
Houston, TX 77033
(800) 535-6286 FAX: (713) 643-3143
(713) 643-0606
http://www.biotecx.com

BioTherm
3260 Wilson Boulevard
Arlington, VA 22201
(703) 522-1705 FAX: (703) 522-2606

Bioventures
P.O. Box 2561
848 Scott Street
Murfreesboro, TN 37133
(800) 235-8938 FAX: (615) 896-4837
http://www.bioventures.com

BioWhittaker
8830 Biggs Ford Road
P.O. Box 127
Walkersville, MD 21793
(800) 638-8174 FAX: (301) 845-8338
(301) 898-7025
http://www.biowhittaker.com

Biozyme Laboratories
9939 Hibert Street, Suite 101
San Diego, CA 92131
(800) 423-8199 FAX: (858) 549-0138
(858) 549-4484
http://www.biozyme.com

Bird Products
1100 Bird Center Drive
Palm Springs, CA 92262
(800) 328-4139 FAX: (760) 778-7274
(760) 778-7200
http://www.birdprod.com/bird

B & K Universal
2403 Yale Way
Fremont, CA 94538
(800) USA-MICE FAX: (510) 490-3036

BLS Ltd.
Zselyi Aladar u. 31
1165 Budapest, Hungary
(36) 1-407-2602 FAX: (36)
1-407-2896
http://www.bls-ltd.com

Blue Sky Research
3047 Orchard Parkway
San Jose, CA 95134
(408) 474-0988 FAX: (408) 474-0989
http://www.blueskyresearch.com

Blumenthal Industries
7 West 36th Street, 13th floor
New York, NY 10018
(212) 719-1251 FAX: (212) 594-8828

BOC Edwards
One Edwards Park
301 Ballardvale Street
Wilmington, MA 01887
(800) 848-9800 FAX: (978) 658-7969
(978) 658-5410
http://www.bocedwards.com

Boehringer Ingelheim
900 Ridgebury Road
P.O. Box 368
Ridgefield, CT 06877
(800) 243-0127 FAX: (203) 798-6234
(203) 798-9988
http://www.boehringer-ingelheim.com

Boehringer Mannheim
Biochemicals Division
See Roche Diagnostics

Boekel Scientific
855 Pennsylvania Boulevard
Feasterville, PA 19053
(800) 336-6929 FAX: (215) 396-8264
(215) 396-8200
http://www.boekelsci.com

Bohdan Automation
1500 McCormack Boulevard
Mundelein, IL 60060
(708) 680-3939 FAX: (708) 680-1199

BPAmoco
4500 McGinnis Ferry Road
Alpharetta, GA 30005
(800) 328-4537 FAX: (770) 772-8213
(770) 772-8200
http://www.bpamoco.com

Brain Research Laboratories
Waban P.O. Box 88
Newton, MA 02468
(888) BRL-5544 FAX: (617) 965-6220
(617) 965-5544
http://www.brainresearchlab.com

Braintree Scientific
P.O. Box 850929
Braintree, MA 02185
(781) 843-1644 FAX: (781) 982-3160
http://www.braintreesci.com

Brandel
8561 Atlas Drive
Gaithersburg, MD 20877
(800) 948-6506 FAX: (301) 869-5570
(301) 948-6506
http://www.brandel.com

Branson Ultrasonics
41 Eagle Road
Danbury, CT 06813
(203) 796-0400 FAX: (203) 796-9838
http://www.plasticsnet.com/branson

B. Braun Biotech
999 Postal Road
Allentown, PA 18103
(800) 258-9000 FAX: (610) 266-9319
(610) 266-6262
http://www.bbraunbiotech.com

B. Braun Biotech International
Schwarzenberg Weg 73-79
P.O. Box 1120
D-34209 Melsungen, Germany
(49) 5661-71-3400
FAX: (49) 5661-71-3702
http://www.bbraunbiotech.com

B. Braun-McGaw
2525 McGaw Avenue
Irvine, CA 92614
(800) BBRAUN-2 (800) 624-2963
http://www.bbraunusa.com

B. Braun Medical
Thorncliffe Park
Sheffield S35 2PW, UK
(44) 114-225-9000
FAX: (44) 114-225-9111
http://www.bbmuk.demon.co.uk

Brenntag
P.O. Box 13788
Reading, PA 19612-3788
(610) 926-4151 FAX: (610) 926-4160
http://www.brenntagnortheast.com

Bresatec
See GeneWorks

Bright/Hacker Instruments
17 Sherwood Lane
Fairfield, NJ 07004
(973) 226-8450 FAX: (973) 808-8281
http://www.hackerinstruments.com

Brinkmann Instruments
Subsidiary of Sybron
1 Cantiague Road
P.O. Box 1019
Westbury, NY 11590
(800) 645-3050 FAX: (516) 334-7521
(516) 334-7500
http://www.brinkmann.com

Bristol-Meyers Squibb
P.O. Box 4500
Princeton, NJ 08543
(800) 631-5244 FAX: (800) 523-2965
http://www.bms.com

Broadley James
19 Thomas
Irvine, CA 92618
(800) 288-2833 FAX: (949) 829-5560
(949) 829-5555
http://www.broadleyjames.com

Brookhaven Instruments
750 Blue Point Road
Holtsville, NY 11742
(631) 758-3200 FAX: (631) 758-3255
http://www.bic.com

Brownlee Labs
See Applied Biosystems
Distributed by Pacer Scientific

Bruel & Kjaer
Division of Spectris Technologies
2815 Colonnades Court
Norcross, GA 30071
(800) 332-2040 FAX: (770) 847-8440
(770) 209-6907
http://www.bkhome.com

Bruker Analytical X-Ray Systems
5465 East Cheryl Parkway
Madison, WI 53711
(800) 234-XRAY FAX: (608) 276-3006
(608) 276-3000
http://www.bruker-axs.com

Bruker Instruments
19 Fortune Drive
Billerica, MA 01821
(978) 667-9580 FAX: (978) 667-0985
http://www.bruker.com

BTX
Division of Genetronics
11199 Sorrento Valley Road
San Diego, CA 92121
(800) 289-2465 FAX: (858) 597-9594
(858) 597-6006
http://www.genetronics.com/btx

Buchler Instruments
See Baxter Scientific Products

Buckshire
2025 Ridge Road
Perkasie, PA 18944
(215) 257-0116

Burdick and Jackson
Division of Baxter Scientific Products
1953 S. Harvey Street
Muskegon, MI 49442
(800) 368-0050 FAX: (231) 728-8226
(231) 726-3171
http://www.bandj.com/mainframe.htm

Burleigh Instruments
P.O. Box E
Fishers, NY 14453
(716) 924-9355 FAX: (716) 924-9072
http://www.burleigh.com

Burns Veterinary Supply
1900 Diplomat Drive
Farmer's Branch, TX 75234
(800) 92-BURNS FAX: (972) 243-6841
http://www.burnsvet.com

Burroughs Wellcome
See Glaxo Wellcome

The Butler Company
5600 Blazer Parkway
Dublin, OH 43017
(800) 551-3861 FAX: (614) 761-9096
(614) 761-9095
http://www.wabutler.com

Butterworth Laboratories
54-56 Waldegrave Road
Teddington, Middlesex
TW11 8LG, UK
(44)(0)20-8977-0750
FAX: (44)(0)28-8943-2624
http://www.butterworth-labs.co.uk

Buxco Electronics
95 West Wood Road #2
Sharon, CT 06069
(860) 364-5558 FAX: (860) 364-5116
http://www.buxco.com

C/D/N Isotopes
88 Leacock Street Pointe-Claire, Quebec
Canada H9R 1H1
(800) 697-6254 FAX: (514) 697-6148

A3

C.M.A./Microdialysis AB
73 Princeton Street
North Chelmsford, MA 01863
(800) 440-4980 FAX: (978) 251-1950
(978) 251-1940
http://www.microdialysis.com

Calbiochem-Novabiochem
P.O. Box 12087-2087
La Jolla, CA 92039
(800) 854-3417 FAX: (800) 776-0999
(858) 450-9600
http://www.calbiochem.com

California Fine Wire
338 South Fourth Street
Grover Beach, CA 93433
(805) 489-5144 FAX: (805) 489-5352
http://www.california.com

Calorimetry Sciences
155 West 2050 North
Spanish Fork, UT 84660
(801) 794-2600 FAX: (801) 794-2700
http://www.calscorp.com

Caltag Laboratories
1849 Bayshore Highway, Suite 200
Burlingame, CA 94010
(800) 874-4007 FAX: (650) 652-9030
(650) 652-0468
http://www.caltag.com

Cambridge Electronic Design
Science Park, Milton Road
Cambridge CB4 0FE, UK
44 (0) 1223-420-186
FAX: 44 (0) 1223-420-488
http://www.ced.co.uk

Cambridge Isotope Laboratories
50 Frontage Road
Andover, MA 01810
(800) 322-1174 FAX: (978) 749-2768
(978) 749-8000
http://www.isotope.com

Cambridge Research Biochemicals
See Zeneca/CRB

Cambridge Technology
109 Smith Place
Cambridge, MA 02138
(6l7) 441-0600 FAX: (617) 497-8800
http://www.camtech.com

Camlab
Nuffield Road
Cambridge CB4 1TH, UK
(44) 122-3424222
FAX: (44) 122-3420856
http://www.camlab.co.uk/home.htm

Campden Instruments
Park Road
Sileby Loughborough
Leicestershire LE12 7TU, UK
(44) 1509-814790
FAX: (44) 1509-816097
http://www.campden-inst.com/home.htm

Cappel Laboratories
See Organon Teknika Cappel

Carl Roth GmgH & Company
Schoemperlenstrasse 1-5
76185 Karlsrube
Germany
(49) 72-156-06164
FAX: (49) 72-156-06264
http://www.carl-roth.de

Carl Zeiss
One Zeiss Drive
Thornwood, NY 10594
(800) 233-2343 FAX: (914) 681-7446
(914) 747-1800
http://www.zeiss.com

Carlo Erba Reagenti
Via Winckelmann 1
20148 Milano
Lombardia, Italy
(39) 0-29-5231
FAX: (39) 0-29-5235-904
http://www.carloerbareagenti.com

Carolina Biological Supply
2700 York Road
Burlington, NC 27215
(800) 334-5551 FAX: (336) 584-76869
(336) 584-0381
http://www.carolina.com

Carolina Fluid Components
9309 Stockport Place
Charlotte, NC 28273
(704) 588-6101 FAX: (704) 588-6115
http://www.cfcsite.com

Cartesian Technologies
17851 Skypark Circle, Suite C
Irvine, CA 92614
(800) 935-8007
http://cartesiantech.com

Cayman Chemical
1180 East Ellsworth Road
Ann Arbor, MI 48108
(800) 364-9897 FAX: (734) 971-3640
(734) 971-3335
http://www.caymanchem.com

CB Sciences
One Washington Street, Suite 404
Dover, NH 03820
(800) 234-1757 FAX: (603) 742-2455
http://www.cbsci.com

CBS Scientific
P.O. Box 856
Del Mar, CA 92014
(800) 243-4959 FAX: (858) 755-0733
(858) 755-4959
http://www.cbssci.com

CCR (Coriell Cell Repository)
See Coriell Institute for Medical Research

Cedarlane Laboratories
5516 8th Line, R.R. #2
Hornby, Ontario L0P 1E0, Canada
(905) 878-8891 FAX: (905) 878-7800
http://www.cedarlanelabs.com

CE Instruments
Grand Avenue Parkway
Austin, TX 78728
(800) 876-6711 FAX: (512) 251-1597
http://www.ceinstruments.com

CEL Associates
P.O. Box 721854
Houston, TX 77272
(800) 537-9339 FAX: (281) 933-0922
(281) 933-9339
http://www.cel-1.com

Cel-Line Associates
See Erie Scientific

Celite World Minerals
130 Castilian Drive
Santa Barbara, CA 93117
(805) 562-0200 FAX: (805) 562-0299
http://www.worldminerals.com/celite

Cell Genesys
342 Lakeside Drive
Foster City, CA 94404
(650) 425-4400 FAX: (650) 425-4457
http://www.cellgenesys.com

Cell Systems
12815 NE 124th Street, Suite A
Kirkland, WA 98034
(800) 697-1211 FAX: (425) 820-6762
(425) 823-1010

Cellmark Diagnostics
20271 Goldenrod Lane
Germantown, MD 20876
(800) 872-5227 FAX: (301) 428-4877
(301) 428-4980
http://www.cellmark-labs.com

Cellomics
635 William Pitt Way
Pittsburgh, PA 15238
(888) 826-3857 FAX: (412) 826-3850
(412) 826-3600
http://www.cellomics.com

Celltech
216 Bath Road
Slough, Berkshire SL1 4EN, UK
(44) 1753 534655
FAX: (44) 1753 536632
http://www.celltech.co.uk

Cellular Products
872 Main Street
Buffalo, NY 14202
(800) CPI-KITS FAX: (716) 882-0959
(716) 882-0920
http://www.zeptometrix.com

CEM
P.O. Box 200
Matthews, NC 28106
(800) 726-3331

Centers for Disease Control
1600 Clifton Road NE
Atlanta, GA 30333
(800) 311-3435 FAX: (888) 232-3228
(404) 639-3311
http://www.cdc.gov

CERJ
Centre d'Elevage Roger Janvier
53940 Le Genest Saint Isle
France

Cetus
See Chiron

Chance Propper
Warly, West Midlands B66 1NZ, UK
(44)(0)121-553-5551
FAX: (44)(0)121-525-0139

Charles River Laboratories
251 Ballardvale Street
Wilmington, MA 01887
(800) 522-7287 FAX: (978) 658-7132
(978) 658-6000
http://www.criver.com

Charm Sciences
36 Franklin Street
Malden, MA 02148
(800) 343-2170 FAX: (781) 322-3141
(781) 322-1523
http://www.charm.com

Chase-Walton Elastomers
29 Apsley Street
Hudson, MA 01749
(800) 448-6289 FAX: (978) 562-5178
(978) 568-0202
http://www.chase-walton.com

ChemGenes
Ashland Technology Center
200 Homer Avenue
Ashland, MA 01721
(800) 762-9323 FAX: (508) 881-3443
(508) 881-5200
http://www.chemgenes.com

Chemglass
3861 North Mill Road
Vineland, NJ 08360
(800) 843-1794 FAX: (856) 696-9102
(800) 696-0014
http://www.chemglass.com

Chemicon International
28835 Single Oak Drive
Temecula, CA 92590
(800) 437-7500 FAX: (909) 676-9209
(909) 676-8080
http://www.chemicon.com

Chem-Impex International
935 Dillon Drive
Wood Dale, IL 60191
(800) 869-9290 FAX: (630) 766-2218
(630) 766-2112
http://www.chemimpex.com

Chem Service
P.O. Box 599
West Chester, PA 19381-0599
(610) 692-3026 FAX: (610) 692-8729
http://www.chemservice.com

Chemsyn Laboratories
13605 West 96th Terrace
Lenexa, Kansas 66215
(913) 541-0525 FAX: (913) 888-3582
http://www.tech.epcorp.com/ChemSyn/chemsyn.htm

Chemunex USA
1 Deer Park Drive, Suite H-2
Monmouth Junction, NJ 08852
(800) 411-6734
http://www.chemunex.com

Cherwell Scientific Publishing
The Magdalen Centre
Oxford Science Park
Oxford OX44GA, UK
(44)(1) 865-784-800
FAX: (44)(1) 865-784-801
http://www.cherwell.com

ChiRex Cauldron
383 Phoenixville Pike
Malvern, PA 19355
(610) 727-2215 FAX: (610) 727-5762
http://www.chirex.com

Chiron Diagnostics
See Bayer Diagnostics

Chiron Mimotopes Peptide Systems
See Multiple Peptide Systems

Chiron
4560 Horton Street
Emeryville, CA 94608
(800) 244-7668 FAX: (510) 655-9910
(510) 655-8730
http://www.chiron.com

Chrom Tech
P.O. Box 24248
Apple Valley, MN 55124
(800) 822-5242 FAX: (952) 431-6345
http://www.chromtech.com

Chroma Technology
72 Cotton Mill Hill, Unit A-9
Brattleboro, VT 05301
(800) 824-7662 FAX: (802) 257-9400
(802) 257-1800
http://www.chroma.com

Chromatographie
ZAC de Moulin No. 2
91160 Saulx les Chartreux
France
(33) 01-64-54-8969
FAX: (33) 01-69-0988091
http://www.chromatographie.com

Chromogenix
Taljegardsgatan 3
431-53 Mlndal, Sweden
(46) 31-706-20-70
FAX: (46) 31-706-20-80
http://www.chromogenix.com

Chrompack USA
c/o Varian USA
2700 Mitchell Drive
Walnut Creek, CA 94598
(800) 526-3687 FAX: (925) 945-2102
(925) 939-2400
http://www.chrompack.com

Chugai Biopharmaceuticals
6275 Nancy Ridge Drive
San Diego, CA 92121
(858) 535-5900 FAX: (858) 546-5973
http://www.chugaibio.com

Ciba-Corning Diagnostics
See Bayer Diagnostics

Ciba-Geigy
See Ciba Specialty Chemicals or Novartis Biotechnology

Ciba Specialty Chemicals
540 White Plains Road
Tarrytown, NY 10591
(800) 431-1900 FAX: (914) 785-2183
(914) 785-2000
http://www.cibasc.com

Ciba Vision
Division of Novartis AG
11460 Johns Creek Parkway
Duluth, GA 30097
(770) 476-3937
http://www.cvworld.com

Cidex
Advanced Sterilization Products
33 Technology Drive
Irvine, CA 92618
(800) 595-0200 (949) 581-5799
http://www.cidex.com

Cinna Scientific
Subsidiary of Molecular Research Center
5645 Montgomery Road
Cincinnati, OH 45212
(800) 462-9868 FAX: (513) 841-0080
(513) 841-0900
http://www.mrcgene.com

Cistron Biotechnology
10 Bloomfield Avenue
Pine Brook, NJ 07058
(800) 642-0167 FAX: (973) 575-4854
(973) 575-1700
http://www.cistronbio.com

Clark Electromedical Instruments
See Harvard Apparatus

Clay Adam
See Becton Dickinson Primary Care Diagnostics

CLB (Central Laboratory of the Netherlands)
Blood Transfusion Service
P.O. Box 9190
1006 AD Amsterdam, The Netherlands
(31) 20-512-9222
FAX: (31) 20-512-3332

Cleveland Scientific
P.O. Box 300
Bath, OH 44210
(800) 952-7315 FAX: (330) 666-2240
http:://www.clevelandscientific.com

Clonetics
Division of BioWhittaker
http://www.clonetics.com
Also see BioWhittaker

Clontech Laboratories
1020 East Meadow Circle
Palo Alto, CA 94303
(800) 662-2566 FAX: (800) 424-1350
(650) 424-8222 FAX: (650) 424-1088
http://www.clontech.com

Closure Medical Corporation
5250 Greens Dairy Road
Raleigh, NC 27616
(919) 876-7800 FAX: (919) 790-1041
http://www.closuremed.com

CMA Microdialysis AB
73 Princeton Street
North Chelmsford, MA 01863
(800) 440-4980 FAX: (978) 251-1950
(978) 251 1940
http://www.microdialysis.com

Cocalico Biologicals
449 Stevens Road
P.O. Box 265
Reamstown, PA 17567
(717) 336-1990 FAX: (717) 336-1993

Coherent Laser
5100 Patrick Henry Drive
Santa Clara, CA 95056
(800) 227-1955 FAX: (408) 764-4800
(408) 764-4000
http://www.cohr.com

Cohu
P.O. Box 85623
San Diego, CA 92186
(858) 277-6700 FAX: (858) 277-0221
http://www.COHU.com/cctv

Cole-Parmer Instrument
625 East Bunker Court
Vernon Hills, IL 60061
(800) 323-4340 FAX: (847) 247-2929
(847) 549-7600
http://www.coleparmer.com

Collaborative Biomedical Products and Collaborative Research
See Becton Dickinson Labware

Collagen Aesthetics
1850 Embarcadero Road
Palo Alto, CA 94303
(650) 856-0200 FAX: (650) 856-0533
http://www.collagen.com

Collagen Corporation
See Collagen Aesthetics

College of American Pathologists
325 Waukegan Road
Northfield, IL 60093
(800) 323-4040 FAX: (847) 832-8000
(847) 446-8800
http://www.cap.org/index.cfm

Colonial Medical Supply
504 Wells Road
Franconia, NH 03580
(603) 823-9911 FAX: (603) 823-8799
http://www.colmedsupply.com

Colorado Serum
4950 York Street
Denver, CO 80216
(800) 525-2065 FAX: (303) 295-1923
http://www.colorado-serum.com

Columbia Diagnostics
8001 Research Way
Springfield, VA 22153
(800) 336-3081 FAX: (703) 569-2353
(703) 569-7511
http://www.columbiadiagnostics.com

Columbus Instruments
950 North Hague Avenue
Columbus, OH 43204
(800) 669-5011 FAX: (614) 276-0529
(614) 276-0861
http://www.columbusinstruments.com

Computer Associates International
One Computer Associates Plaza
Islandia, NY 11749
(631) 342-6000 FAX: (631) 342-6800
http://www.cai.com

Connaught Laboratories
See Aventis Pasteur

Connectix
2955 Campus Drive, Suite 100
San Mateo, CA 94403
(800) 950-5880 FAX: (650) 571-0850
(650) 571-5100
http://www.connectix.com

Contech
99 Hartford Avenue
Providence, RI 02909
(401) 351-4890 FAX: (401) 421-5072
http://www.iol.ie/~burke/contech.html

Continental Laboratory Products
5648 Copley Drive
San Diego, CA 92111
(800) 456-7741 FAX: (858) 279-5465
(858) 279-5000
http://www.conlab.com

ConvaTec
Professional Services
P.O. Box 5254
Princeton, NJ 08543
(800) 422-8811
http://www.convatec.com

Cooper Instruments & Systems
P.O. Box 3048
Warrenton, VA 20188
(800) 344-3921 FAX: (540) 347-4755
(540) 349-4746
http://www.cooperinstruments.com

Cora Styles Needles 'N Blocks
56 Milton Street
Arlington, MA 02474
(781) 648-6289 FAX: (781) 641-7917

Coriell Cell Repository (CCR)
See Coriell Institute for Medical Research

Coriell Institute for Medical Research
Human Genetic Mutant Repository
401 Haddon Avenue
Camden, NJ 08103
(856) 966-7377 FAX: (856) 964-0254
http://arginine.umdnj.edu

Corion
8 East Forge Parkway
Franklin, MA 02038
(508) 528-4411 FAX: (508) 520-7583
(800) 598-6783
http://www.corion.com

Corning and Corning Science Products
P.O. Box 5000
Corning, NY 14831
(800) 222-7740 FAX: (607) 974-0345
(607) 974-9000
http://www.corning.com

Costar
See Corning

Coulbourn Instruments
7462 Penn Drive
Allentown, PA 18106
(800) 424-3771 FAX: (610) 391-1333
(610) 395-3771
http://www.coulbourninst.com

Coulter Cytometry
See Beckman Coulter

Covance Research Products
465 Swampbridge Road
Denver, PA 17517
(800) 345-4114 FAX: (717) 336-5344
(717) 336-4921
http://www.covance.com

Coy Laboratory Products
14500 Coy Drive
Grass Lake, MI 49240
(734) 475-2200 FAX: (734) 475-1846
http://www.coylab.com

CPG
3 Borinski Road
Lincoln Park, NJ 07035
(800) 362-2740 FAX: (973) 305-0884
(973) 305-8181

CPL Scientific
43 Kingfisher Court
Hambridge Road
Newbury RG14 5SJ, UK
(44) 1635-574902
FAX: (44) 1635-529322
http://www.cplscientific.co.uk

CraMar Technologies
8670 Wolff Court, #160
Westminster, CO 80030
(800) 4-TOMTEC
http://www.cramar.com

Crescent Chemical
1324 Motor Parkway
Hauppauge, NY 11788
(800) 877-3225 FAX: (631) 348-0913
(631) 348-0333
http://www.creschem.com

Crist Instrument
P.O. Box 128
10200 Moxley Road
Damascus, MD 20872
(301) 253-2184 FAX: (301) 253-0069
http://www.cristinstrument.com

Cruachem
See Annovis
http://www.cruachem.com

CS Bio
1300 Industrial Road
San Carlos, CA 94070
(800) 627-2461 FAX: (415) 802-0944
(415) 802-0880
http://www.csbio.com

CS-Chromatographie Service
Am Parir 27
D-52379 Langerwehe, Germany
(49) 2423-40493-0
FAX: (49) 2423-40493-49
http://www.cs-chromatographie.de

Cuno
400 Research Parkway
Meriden, CT 06450
(800) 231-2259 FAX: (203) 238-8716
(203) 237-5541
http://www.cuno.com

Curtin Matheson Scientific
9999 Veterans Memorial Drive
Houston, TX 77038
(800) 392-3353 FAX: (713) 878-3598
(713) 878-3500

CWE
124 Sibley Avenue
Ardmore, PA 19003
(610) 642-7719 FAX: (610) 642-1532
http://www.cwe-inc.com

Cybex Computer Products
4991 Corporate Drive
Huntsville, AL 35805
(800) 932-9239 FAX: (800) 462-9239
http://www.cybex.com

Cygnus
400 Penobscot Drive
Redwood City, CA
(650) 369-4300 FAX: (650) 599-2503
http://www.cygn.com/homepage.html

Cygnus Technology
P.O. Box 219
Delaware Water Gap, PA 18327
(570) 424-5701 FAX: (570) 424-5630
http://www.cygnustech.com

Cymbus Biotechnology
Eagle Class, Chandler's Ford
Hampshire SO53 4NF, UK
(44) 1-703-267-676
FAX: (44) 1-703-267-677
http://www.biotech.cymbus.com

Cytogen
600 College Road East
Princeton, NJ 08540
(609) 987-8200 FAX: (609) 987-6450
http://www.cytogen.com

Cytogen Research and Development
89 Bellevue Hill Road
Boston, MA 02132
(617) 325-7774 FAX: (617) 327-2405

CytRx
154 Technology Parkway
Norcross, GA 30092
(800) 345-2987 FAX: (770) 368-0622
(770) 368-9500
http://www.cytrx.com

Dade Behring
Corporate Headquarters
1717 Deerfield Road
Deerfield, IL 60015
(847) 267-5300 FAX: (847) 267-1066
http://www.dadebehring.com

Dagan
2855 Park Avenue
Minneapolis, MN 55407
(612) 827-5959 FAX: (612) 827-6535
http://www.dagan.com

Dako
6392 Via Real
Carpinteria, CA 93013
(800) 235-5763 FAX: (805) 566-6688
(805) 566-6655
http://www.dakousa.com

Dako A/S
42 Produktionsvej
P.O. Box 1359
DK-2600 Glostrup, Denmark
(45) 4492-0044 FAX: (45) 4284-1822

Dakopatts
See Dako A/S

Damon, IEC
See Thermoquest

Dan Kar Scientific
150 West Street
Wilmington, MA 01887
(800) 942-5542 FAX: (978) 658-0380
(978) 988-9696
http://www.dan-kar.com

DataCell
Falcon Business Park
40 Ivanhoe Road
Finchampstead, Berkshire
RG40 4QQ, UK
(44) 1189 324324
FAX: (44) 1189 324325
http://www.datacell.co.uk
In the US:
(408) 446-3575 FAX: (408) 446-3589
http://www.datacell.com

DataWave Technologies
380 Main Street, Suite 209
Longmont, CO 80501
(800) 736-9283 FAX: (303) 776-8531
(303) 776-8214

Datex-Ohmeda
3030 Ohmeda Drive
Madison, WI 53718
(800) 345-2700 FAX: (608) 222-9147
(608) 221-1551
http://www.us.datex-ohmeda.com

DATU
82 State Street
Geneva, NY 14456
(315) 787-2240 FAX: (315) 787-2397
http://www.nysaes.cornell.edu/datu

David Kopf Instruments
7324 Elmo Street
P.O. Box 636
Tujunga, CA 91043
(818) 352-3274 FAX: (818) 352-3139

Decagon Devices
P.O. Box 835
950 NE Nelson Court
Pullman, WA 99163
(800) 755-2751 FAX: (509) 332-5158
(509) 332-2756
http://www.decagon.com

Decon Labs
890 Country Line Road
Bryn Mawr, PA 19010
(800) 332-6647 FAX: (610) 964-0650
(610) 520-0610
http://www.deconlabs.com

Decon Laboratories
Conway Street
Hove, Sussex BN3 3LY, UK
(44) 1273 739241
FAX: (44) 1273 722088

Degussa
Precious Metals Division
3900 South Clinton Avenue
South Plainfield, NJ 07080
(800) DEGUSSA FAX: (908) 756-7176
(908) 561-1100
http://www.degussa-huls.com

Deneba Software
1150 NW 72nd Avenue
Miami, FL 33126
(305) 596-5644 FAX: (305) 273-9069
http://www.deneba.com

Deseret Medical
524 West 3615 South
Salt Lake City, UT 84115
(801) 270-8440 FAX: (801) 293-9000

Devcon Plexus
30 Endicott Street
Danvers, MA 01923
(800) 626-7226 FAX: (978) 774-0516
(978) 777-1100
http://www.devcon.com

Developmental Studies Hybridoma Bank
University of Iowa
436 Biology Building
Iowa City, IA 52242
(319) 335-3826 FAX: (319) 335-2077
http://www.uiowa.edu/~dshbwww

DeVilbiss
Division of Sunrise Medical Respiratory
100 DeVilbiss Drive
P.O. Box 635 Somerset, PA 15501
(800) 338-1988 FAX: (814) 443-7572
(814) 443-4881
http://www.sunrisemedical.com

Dharmacon Research
3200 Valmont Road, #5
Boulder, CO 80301
(800) 235-9880 FAX: (303) 415-9879
(303) 415-9880
http://www.dharmacon.com

DiaCheM
Triangle Biomedical
Gardiners Place
West Gillibrands, Lancashire
WN8 9SP, UK
(44) 1695-555581
FAX: (44) 1695-555518
http://www.diachem.co.uk

Diagen
Max-Volmer Strasse 4
D-40724 Hilden, Germany
(49) 2103-892-230
FAX: (49) 2103-892-222

Diagnostic Concepts
6104 Madison Court
Morton Grove, IL 60053
(847) 604-0957

Diagnostic Developments
See DiaCheM

Diagnostic Instruments
6540 Burroughs
Sterling Heights, MI 48314
(810) 731-6000 FAX: (810) 731-6469
http://www.diaginc.com

Diamedix
2140 North Miami Avenue
Miami, FL 33127
(800) 327-4565 FAX: (305) 324-2395
(305) 324-2300

DiaSorin
1990 Industrial Boulevard
Stillwater, MN 55082
(800) 328-1482 FAX: (651) 779-7847
(651) 439-9719
http://www.diasorin.com

Diatome US
321 Morris Road
Fort Washington, PA 19034
(800) 523-5874 FAX: (215) 646-8931
(215) 646-1478
http://www.emsdiasum.com

Difco Laboratories
See Becton Dickinson

Digene
1201 Clopper Road
Gaithersburg, MD 20878
(301) 944-7000 (800) 344-3631
FAX: (301) 944-7121
www.digene.com

Digi-Key
701 Brooks Avenue South
Thief River Falls, MN 56701
(800) 344-4539 FAX: (218) 681-3380
(218) 681-6674
http://www.digi-key.com

Digitimer
37 Hydeway
Welwyn Garden City, Hertfordshire
AL7 3BE, UK
(44) 1707-328347
FAX: (44) 1707-373153
http://www.digitimer.com

Dimco-Gray
8200 South Suburban Road
Dayton, OH 45458
(800) 876-8353 FAX: (937) 433-0520
(937) 433-7600
http://www.dimco-gray.com

Dionex
1228 Titan Way
P.O. Box 3603
Sunnyvale, CA 94088
(408) 737-0700 FAX: (408) 730-9403
http://dionex2.promptu.com

Display Systems Biotech
1260 Liberty Way, Suite B
Vista, CA 92083
(800) 697-1111 FAX: (760) 599-9930
(760) 599-0598
http://www.displaysystems.com

Diversified Biotech
1208 VFW Parkway
Boston, MA 02132
(617) 965-8557 FAX: (617) 323-5641
(800) 796-9199
http://www.divbio.com

DNA ProScan
P.O. Box 121585
Nashville, TN 37212
(800) 841-4362 FAX: (615) 292-1436
(615) 298-3524
http://www.dnapro.com

DNAStar
1228 South Park Street
Madison, WI 53715
(608) 258-7420 FAX: (608) 258-7439
http://www.dnastar.com

DNAVIEW
Attn: Charles Brenner
http://www.wco.com
~cbrenner/dnaview.htm

Doall NYC
36-06 48th Avenue
Long Island City, NY 11101
(718) 392-4595 FAX: (718) 392-6115
http://www.doall.com

Dojindo Molecular Technologies
211 Perry Street Parkway, Suite 5
Gaitherbusburg, MD 20877
(877) 987-2667
http://www.dojindo.com

Dolla Eastern
See Doall NYC

Dolan Jenner Industries
678 Andover Street
Lawrence, MA 08143
(978) 681-8000 (978) 682-2500
http://www.dolan-jenner.com

Dow Chemical
Customer Service Center
2040 Willard H. Dow Center
Midland, MI 48674
(800) 232-2436 FAX: (517) 832-1190
(409) 238-9321
http://www.dow.com

Dow Corning
Northern Europe
Meriden Business Park
Copse Drive
Allesley, Coventry CV5 9RG, UK
(44) 1676 528 000
FAX: (44) 1676 528 001

Dow Corning
P.O. Box 994
Midland, MI 48686
(517) 496-4000
http://www.dowcorning.com

Dow Corning (Lubricants)
2200 West Salzburg Road
Auburn, MI 48611
(800) 248-2481 FAX: (517) 496-6974
(517) 496-6000

Dremel
4915 21st Street
Racine, WI 53406
(414) 554-1390
http://www.dremel.com

Drummond Scientific
500 Parkway
P.O. Box 700
Broomall, PA 19008
(800) 523-7480 FAX: (610) 353-6204
(610) 353-0200
http://www.drummondsci.com

Duchefa Biochemie BV
P.O. Box 2281
2002 CG Haarlem, The Netherlands
31-0-23-5319093
FAX: 31-0-23-5318027
http://www.duchefa.com

Duke Scientific
2463 Faber Place
Palo Alto, CA 94303
(800) 334-3883 FAX: (650) 424-1158
(650) 424-1177
http://www.dukescientific.com

DuPont Biotechnology Systems
See NEN Life Science Products

DuPont Medical Products
See NEN Life Science Products

DuPont Merck Pharmaceuticals
331 Treble Cove Road
Billerica, MA 01862
(800) 225-1572 FAX: (508) 436-7501
http://www.dupontmerck.com

DuPont NEN Products
See NEN Life Science Products

Dyets
2508 Easton Avenue
P.O. Box 3485
Bethlehem, PA 18017
(800) 275-3938 FAX: (800) 329-3938

Dynal
5 Delaware Drive
Lake Success, NY 11042
(800) 638-9416 FAX: (516) 326-3298
(516) 326-3270
http://www.dynal.net

Dynal AS
Ullernchausen 52,
0379 Oslo, Norway
47-22-06-10-00 FAX: 47-22-50-70-15
http://www.dynal.no

Dynalab
P.O. Box 112
Rochester, NY 14692
(800) 828-6595 FAX: (716) 334-9496
(716) 334-2060
http://www.dynalab.com

Dynarex
1 International Boulevard
Brewster, NY 10509
(888) DYNAREX FAX: (914) 279-9601
(914) 279-9600
http://www.dynarex.com

Dynatech
See Dynex Technologies

Dynex Technologies
14340 Sullyfield Circle
Chantilly, VA 22021
(800) 336-4543 FAX: (703) 631-7816
(703) 631-7800
http://www.dynextechnologies.com

Dyno Mill
See Willy A. Bachofen

E.S.A.
22 Alpha Road
Chelmsford, MA 01824
(508) 250-7000 FAX: (508) 250-7090

E.W. Wright
760 Durham Road
Guilford, CT 06437
(203) 453-6410 FAX: (203) 458-6901
http://www.ewwright.com

E-Y Laboratories
107 N. Amphlett Boulevard
San Mateo, CA 94401
(800) 821-0044 FAX: (650) 342-2648
(650) 342-3296
http://www.eylabs.com

Eastman Kodak
1001 Lee Road
Rochester, NY 14650
(800) 225-5352 FAX: (800) 879-4979
(716) 722-5780 FAX: (716) 477-8040
http://www.kodak.com

ECACC
See European Collection of Animal Cell Cultures

EC Apparatus
See Savant/EC Apparatus

Ecogen, SRL
Gensura Laboratories
Ptge. Dos de Maig
9(08041) Barcelona (34) 3-450-2601
FAX: (34) 3-456-0607
http://www.ecogen.com

Ecolab
370 North Wabasha Street
St. Paul, MN 55102
(800) 35-CLEAN FAX: (651) 225-3098
(651) 352-5326
http://www.ecolab.com

ECO PHYSICS
3915 Research Park Drive, Suite A-3
Ann Arbor, MI 48108
(734) 998-1600 FAX: (734) 998-1180
http://www.ecophysics.com

Edge Biosystems
19208 Orbit Drive
Gaithersburg, MD 20879-4149
(800) 326-2685 FAX: (301) 990-0881
(301) 990-2685
http://www.edgebio.com

Edmund Scientific
101 E. Gloucester Pike
Barrington, NJ 08007
(800) 728-6999 FAX: (856) 573-6263
(856) 573-6250
http://www.edsci.com

EG&G
See Perkin-Elmer

Ekagen
969 C Industry Road
San Carlos, CA 94070
(650) 592-4500 FAX: (650) 592-4500

Elcatech
P.O. Box 10935
Winston-Salem, NC 27108
(336) 544-8613 FAX: (336) 777-3623
(910) 777-3624
http://www.elcatech.com

Electron Microscopy Sciences
321 Morris Road
Fort Washington, PA 19034
(800) 523-5874 FAX: (215) 646-8931
(215) 646-1566
http://www.emsdiasum.com

Electron Tubes
100 Forge Way, Unit F
Rockaway, NJ 07866
(800) 521-8382 FAX: (973) 586-9771
(973) 586-9594
http://www.electrontubes.com

Elicay Laboratory Products, (UK) Ltd.
4 Manborough Mews
Crockford Lane
Basingstoke, Hampshire
RG 248NA, England
(256) 811-118 FAX: (256) 811-116
http://www.elkay-uk.co.uk

Eli Lilly
Lilly Corporate Center
Indianapolis, IN 46285
(800) 545-5979 FAX: (317) 276-2095
(317) 276-2000
http://www.lilly.com

ELISA Technologies
See Neogen

Elkins-Sinn
See Wyeth-Ayerst

EMBI
See European Bioinformatics Institute

EM Science
480 Democrat Road
Gibbstown, NJ 08027
(800) 222-0342 FAX: (856) 423-4389
(856) 423-6300
http://www.emscience.com

EM Separations Technology
See R & S Technology

Endogen
30 Commerce Way
Woburn, MA 01801
(800) 487-4885 FAX: (617) 439-0355
(781) 937-0890
http://www.endogen.com

ENGEL-Loter
HSGM Heatcutting Equipment
& Machines
1865 E. Main Street, No. 5
Duncan, SC 29334
(888) 854-HSGM FAX: (864) 486-8383
(864) 486-8300
http://www.engelgmbh.com

Enzo Diagnostics
60 Executive Boulevard
Farmingdale, NY 11735
(800) 221-7705 FAX: (516) 694-7501
(516) 694-7070
http://www.enzo.com

Enzogenetics
4197 NW Douglas Avenue
Corvallis, OR 97330
(541) 757-0288

The Enzyme Center
See Charm Sciences

Enzyme Systems Products
486 Lindbergh Avenue
Livermore, CA 94550
(888) 449-2664 FAX: (925) 449-1866
(925) 449-2664
http://www.enzymesys.com

Epicentre Technologies
1402 Emil Street
Madison, WI 53713
(800) 284-8474 FAX: (608) 258-3088
(608) 258-3080
http://www.epicentre.com

Erie Scientific
20 Post Road
Portsmouth, NH 03801
(888) ERIE-SCI FAX: (603) 431-8996
(603) 431-8410
http://www.eriesci.com

ES Industries
701 South Route 73
West Berlin, NJ 08091
(800) 356-6140 FAX: (856) 753-8484
(856) 753-8400
http://www.esind.com

ESA
22 Alpha Road
Chelmsford, MA 01824
(800) 959-5095 FAX: (978) 250-7090
(978) 250-7000
http://www.esainc.com

Ethicon
Route 22, P.O. Box 151
Somerville, NJ 08876
(908) 218-0707
http://www.ethiconinc.com

Ethicon Endo-Surgery
4545 Creek Road
Cincinnati, OH 45242
(800) 766-9534 FAX: (513) 786-7080

Eurogentec
Parc Scientifique du Sart Tilman
4102 Seraing, Belgium
32-4-240-76-76 FAX: 32-4-264-07-88
http://www.eurogentec.com

European Bioinformatics Institute
Wellcome Trust Genomes Campus
Hinxton, Cambridge CB10 1SD, UK
(44) 1223-49444
FAX: (44) 1223-494468

European Collection of Animal Cell Cultures (ECACC)
Centre for Applied Microbiology & Research
Salisbury, Wiltshire SP4 0JG, UK
(44) 1980-612 512
FAX: (44) 1980-611 315
http://www.camr.org.uk

Evergreen Scientific
2254 E. 49th Street
P.O. Box 58248
Los Angeles, CA 90058
(800) 421-6261 FAX: (323) 581-2503
(323) 583-1331
http://www.evergreensci.com

Exalpha Biologicals
20 Hampden Street
Boston, MA 02205
(800) 395-1137 FAX: (617) 969-3872
(617) 558-3625
http://www.exalpha.com

Exciton
P.O. Box 31126
Dayton, OH 45437
(937) 252-2989 FAX: (937) 258-3937
http://www.exciton.com

Extrasynthese
ZI Lyon Nord
SA-BP62
69730 Genay, France
(33) 78-98-20-34
FAX: (33) 78-98-19-45

Factor II
1972 Forest Avenue
P.O. Box 1339
Lakeside, AZ 85929
(800) 332-8688 FAX: (520) 537-8066
(520) 537-8387
http://www.factor2.com

Falcon
See Becton Dickinson Labware

Fenwal
See Baxter Healthcare

Filemaker
5201 Patrick Henry Drive
Santa Clara, CA 95054
(408) 987-7000 (800) 325-2747

Fine Science Tools
202-277 Mountain Highway
North Vancouver, British Columbia
V7J 3P2 Canada
(800) 665-5355 FAX: (800) 665 4544
(604) 980-2481 FAX: (604) 987-3299

Fine Science Tools
373-G Vintage Park Drive
Foster City, CA 94404
(800) 521-2109 FAX: (800) 523-2109
(650) 349-1636 FAX: (630) 349-3729

Fine Science Tools
Fahrtgasse 7-13
D-69117 Heidelberg, Germany
(49) 6221 905050
FAX: (49) 6221 600001
http://www.finescience.com

Finn Aqua
AMSCO Finn Aqua Oy
Teollisuustiez, FIN-04300
Tuusula, Finland
358 025851 FAX: 358 0276019

Finnigan
355 River Oaks Parkway
San Jose, CA 95134
(408) 433-4800 FAX: (408) 433-4821
http://www.finnigan.com

Fisher Chemical Company
Fisher Scientific Limited
112 Colonnade Road Nepean
Ontario K2E 7L6, Canada
(800) 234-7437 FAX: (800) 463-2996
http://www.fisherscientific.com

W.F. Fisher & Son
220 Evans Way, Suite #1
Somerville, NJ 08876
(908) 707-4050 FAX: (908) 707-4099

Fisher Scientific
2000 Park Lane
Pittsburgh, PA 15275
(800) 766-7000 FAX: (800) 926-1166
(412) 562-8300
http://www3.fishersci.com

Fitzco
5600 Pioneer Creek Drive
Maple Plain, MN 55359
(800) 367-8760 FAX: (612) 479-2880
(612) 479-3489
http://www.fitzco.com

5 Prime → 3 Prime
See 2000 Eppendorf-5 Prime
http://www.5prime.com

Fleisch (Rusch)
2450 Meadowbrook Parkway
Duluth, GA 30096
(770) 623-0816 FAX: (770) 623-1829
http://ruschinc.com

Flow Cytometry Standards
P.O. Box 194344
San Juan, PR 00919
(800) 227-8143 FAX: (787) 758-3267
(787) 753-9341
http://www.fcstd.com

Flow Labs
See ICN Biomedicals

Flow-Tech Supply
P.O. Box 1388
Orange, TX 77631
(409) 882-0306 FAX: (409) 882-0254
http://www.flow-tech.com

Fluid Marketing
See Fluid Metering

Fluid Metering
5 Aerial Way, Suite 500
Sayosett, NY 11791
(516) 922-6050 FAX: (516) 624-8261
http://www.fmipump.com

Fluorochrome
1801 Williams, Suite 300
Denver, CO 80264
(303) 394-1000 FAX: (303) 321-1119

Fluka Chemical
See Sigma-Aldrich

FMC BioPolymer
1735 Market Street
Philadelphia, PA 19103
(215) 299-6000 FAX: (215) 299-5809
http://www.fmc.com

FMC BioProducts
191 Thomaston Street
Rockland, ME 04841
(800) 521-0390 FAX: (800) 362-1133
(207) 594-3400 FAX: (207) 594-3426
http://www.bioproducts.com

Forma Scientific
Milcreek Road
P.O. Box 649
Marietta, OH 45750
(800) 848-3080 FAX: (740) 372-6770
(740) 373-4765
http://www.forma.com

Fort Dodge Animal Health
800 5th Street NW
Fort Dodge, IA 50501
(800) 685-5656 FAX: (515) 955-9193
(515) 955-4600
http://www.ahp.com

Fotodyne
950 Walnut Ridge Drive
Hartland, WI 53029
(800) 362-3686 FAX: (800) 362-3642
(262) 369-7000 FAX: (262) 369-7013
http://www.fotodyne.com

Fresenius HemoCare
6675 185th Avenue NE, Suite 100
Redwood, WA 98052
(800) 909-3872
(425) 497-1197
http://www.freseniusht.com

Fresenius Hemotechnology
See Fresenius HemoCare

Fuji Medical Systems
419 West Avenue
P.O. Box 120035
Stamford, CT 06902
(800) 431-1850 FAX: (203) 353-0926
(203) 324-2000
http://www.fujimed.com

Fujisawa USA
Parkway Center North
Deerfield, IL 60015-2548
(847) 317-1088 FAX: (847) 317-7298

Ernest F. Fullam
900 Albany Shaker Road
Latham, NY 12110
(800) 833-4024 FAX: (518) 785-8647
(518) 785-5533
http://www.fullam.com

Gallard-Schlesinger Industries
777 Zechendorf Boulevard Garden City,
NY 11530
(516) 229-4000 FAX: (516) 229-4015
http://www.gallard-schlessinger.com

Gambro
Box 7373
SE 103 91 Stockholm, Sweden
(46) 8 613 65 00
FAX: (46) 8 611 37 31
In the US: COBE Laboratories
225 Union Boulevard
Lakewood, CO 80215
(303) 232-6800 FAX: (303) 231-4915
http://www.gambro.com

Garner Glass
177 Indian Hill Boulevard
Claremont, CA 91711
(909) 624-5071 FAX: (909) 625-0173
http://www.garnerglass.com

Garon Plastics
16 Byre Avenue
Somerton Park, South Australia 5044
(08) 8294-5126 FAX: (08) 8376-1487
http://www.apache.airnet.com.au/~garon

Garren Scientific
9400 Lurline Avenue, Unit E
Chatsworth, CA 91311
(800) 342-3725 FAX: (818) 882-3229
(818) 882-6544
http://www.garren-scientific.com

GATC Biotech AG
Jakob-Stadler-Platz 7
D-78467 Constance, Germany
(49) 07531-8160-0
FAX: (49) 07531-8160-81
http://www.gatc-biotech.com

Gaussian
Carnegie Office Park
Building 6, Suite 230
Carnegie, PA 15106
(412) 279-6700 FAX: (412) 279-2118
http://www.gaussian.com

G.C. Electronics/A.R.C. Electronics
431 Second Street
Henderson, KY 42420
(270) 827-8981 FAX: (270) 827-8256
http://www.arcelectronics.com

GDB (Genome Data Base, Curation)
2024 East Monument Street, Suite 1200
Baltimore, MD 21205
(410) 955-9705 FAX: (410) 614-0434
http://www.gdb.org

GDB (Genome Data Base, Home)
Hospital for Sick Children
555 University Avenue
Toronto, Ontario
M5G 1X8 Canada
(416) 813-8744 FAX: (416) 813-8755
http://www.gdb.org

Gelman Sciences
See Pall-Gelman

Gemini BioProducts
5115-M Douglas Fir Road
Calabasas, CA 90403
(818) 591-3530 FAX: (818) 591-7084

Gen Trak
5100 Campus Drive
Plymouth Meeting, PA 19462
(800) 221-7407 FAX: (215) 941-9498
(215) 825-5115
http://www.informagen.com

Genaissance Pharmaceuticals
5 Science Park
New Haven, CT 06511
(800) 678-9487 FAX: (203) 562-9377
(203) 773-1450
http://www.genaissance.com

GENAXIS Biotechnology
Parc Technologique
10 Avenue Ampère
Montigny le Bretoneux
78180 France
(33) 01-30-14-00-20
FAX: (33) 01-30-14-00-15
http://www.genaxis.com

GenBank
National Center for Biotechnology Information
National Library of Medicine/NIH
Building 38A, Room 8N805
8600 Rockville Pike
Bethesda, MD 20894
(301) 496-2475 FAX: (301) 480-9241
http://www.ncbi.nlm.nih.gov

Gene Codes
640 Avis Drive
Ann Arbor, MI 48108
(800) 497-4939 FAX: (734) 930-0145
(734) 769-7249
http://www.genecodes.com

Genemachines
935 Washington Street
San Carlos, CA 94070
(650) 508-1634 FAX: (650) 508-1644
(877) 855-4363
http://www.genemachines.com

Genentech
1 DNA Way
South San Francisco, CA 94080
(800) 551-2231 FAX: (650) 225-1600
(650) 225-1000
http://www.gene.com

General Scanning/GSI Luminomics
500 Arsenal Street
Watertown, MA 02172
(617) 924-1010 FAX: (617) 924-7327
http://www.genescan.com

General Valve
Division of Parker Hannifin Pneutronics
19 Gloria Lane
Fairfield, NJ 07004
(800) GVC-VALV
FAX: (800) GVC-1-FAX
http://www.pneutronics.com

Genespan
19310 North Creek Parkway, Suite 100
Bothell, WA 98011
(800) 231-2215 FAX: (425) 482-3005
(425) 482-3003
http://www.genespan.com

Généthon Human Genome Research Center
1 bis rue de l'Internationale
91000 Evry, France
(33) 169-472828
FAX: (33) 607-78698
http://www.genethon.fr

Genetic Microsystems
34 Commerce Way
Wobum, MA 01801
(781) 932-9333 FAX: (781) 932-9433
http://www.genticmicro.com

Genetic Mutant Repository
See Coriell Institute for Medical Research

Genetic Research Instrumentation
Gene House
Queenborough Lane
Rayne, Braintree, Essex CM7 8TF, UK
(44) 1376 332900
FAX: (44) 1376 344724
http://www.gri.co.uk

Genetics Computer Group
575 Science Drive
Madison, WI 53711
(608) 231-5200 FAX: (608) 231-5202
http://www.gcg.com

Genetics Institute/American Home Products
87 Cambridge Park Drive
Cambridge, MA 02140
(617) 876-1170 FAX: (617) 876-0388
http://www.genetics.com

Genetix
63-69 Somerford Road
Christchurch, Dorset BH23 3QA, UK
(44) (0) 1202 483900
FAX: (44)(0) 1202 480289
In the US: (877) 436 3849
US FAX: (888) 522 7499
http://www.genetix.co.uk

Gene Tools
One Summerton Way
Philomath, OR 97370
(541) 9292-7840 FAX: (541) 9292-7841
http://www.gene-tools.com

GeneWorks
P.O. Box 11, Rundle Mall
Adelaide, South Australia 5000, Australia
1800 882 555 FAX: (08) 8234 2699
(08) 8234 2644
http://www.geneworks.com

Genome Systems (INCYTE)
4633 World Parkway Circle
St. Louis, MO 63134
(800) 430-0030 FAX: (314) 427-3324
(314) 427-3222
http://www.genomesystems.com

Genomic Solutions
4355 Varsity Drive, Suite E
Ann Arbor, MI 48108
(877) GENOMIC FAX: (734) 975-4808
(734) 975-4800
http://www.genomicsolutions.com

Genomyx
See Beckman Coulter

Genosys Biotechnologies
1442 Lake Front Circle, Suite 185
The Woodlands, TX 77380
(281) 363-3693 FAX: (281) 363-2212
http://www.genosys.com

Genotech
92 Weldon Parkway
St. Louis, MO 63043
(800) 628-7730 FAX: (314) 991-1504
(314) 991-6034

GENSET
876 Prospect Street, Suite 206
La Jolla, CA 92037
(800) 551-5291 FAX: (619) 551-2041
(619) 515-3061
http://www.genset.fr

Gensia Laboratories Ltd.
19 Hughes
Irvine, CA 92718
(714) 455-4700 FAX: (714) 855-8210

Genta
99 Hayden Avenue, Suite 200
Lexington, MA 02421
(781) 860-5150 FAX: (781) 860-5137
http://www.genta.com

GENTEST
6 Henshaw Street
Woburn, MA 01801
(800) 334-5229 FAX: (888) 242-2226
(781) 935-5115 FAX: (781) 932-6855
http://www.gentest.com

Gentra Systems
15200 25th Avenue N., Suite 104
Minneapolis, MN 55447
(800) 866-3039 FAX: (612) 476-5850
(612) 476-5858
http://www.gentra.com

Genzyme
1 Kendall Square
Cambridge, MA 02139
(617) 252-7500 FAX: (617) 252-7600
http://www.genzyme.com
See also R&D Systems

Genzyme Genetics
One Mountain Road
Framingham, MA 01701
(800) 255-7357 FAX: (508) 872-9080
(508) 872-8400
http://www.genzyme.com

George Tiemann & Co.
25 Plant Avenue
Hauppauge, NY 11788
(516) 273-0005 FAX: (516) 273-6199

GIBCO/BRL
A Division of Life Technologies
1 Kendall Square
Grand Island, NY 14072
(800) 874-4226 FAX: (800) 352-1968
(716) 774-6700
http://www.lifetech.com

Gilmont Instruments
A Division of Barnant Company
28N092 Commercial Avenue
Barrington, IL 60010
(800) 637-3739 FAX: (708) 381-7053
http://barnant.com

Gilson
3000 West Beltline Highway
P.O. Box 620027
Middletown, WI 53562
(800) 445-7661
(608) 836-1551
http://www.gilson.com

Glas-Col Apparatus
P.O. Box 2128
Terre Haute, IN 47802
(800) Glas-Col FAX: (812) 234-6975
(812) 235-6167
http://www.glascol.com

Glaxo Wellcome
Five Moore Drive
Research Triangle Park, NC 27709
(800) SGL-AXO5 FAX: (919) 248-2386
(919) 248-2100
http://www.glaxowellcome.com

Glen Mills
395 Allwood Road
Clifton, NJ 07012
(973) 777-0777 FAX: (973) 777-0070
http://www.glenmills.com

Glen Research
22825 Davis Drive
Sterling, VA 20166
(800) 327-4536 FAX: (800) 934-2490
(703) 437-6191 FAX: (703) 435-9774
http://www.glenresearch.com

Glo Germ
P.O. Box 189
Moab, UT 84532
(800) 842-6622 FAX: (435) 259-5930
http://www.glogerm.com

Glyco
11 Pimentel Court
Novato, CA 94949
(800) 722-2597 FAX: (415) 382-3511
(415) 884-6799
http://www.glyco.com

Gould Instrument Systems
8333 Rockside Road
Valley View, OH 44125
(216) 328-7000 FAX: (216) 328-7400
http://www.gould13.com

Gralab Instruments
See Dimco-Gray

GraphPad Software
5755 Oberlin Drive #110
San Diego, CA 92121
(800) 388-4723 FAX: (558) 457-8141
(558) 457-3909
http://www.graphpad.com

Graseby Anderson
See Andersen Instruments
http://www.graseby.com

Grass Instrument
A Division of Astro-Med
600 East Greenwich Avenue
W. Warwick, RI 02893
(800) 225-5167 FAX: (877) 472-7749
http://www.grassinstruments.com

Greenacre and Misac Instruments
Misac Systems
27 Port Wood Road
Ware, Hertfordshire SF12 9NJ, UK
(44) 1920 463017
FAX: (44) 1920 465136

Greer Labs
639 Nuway Circle
Lenois, NC 28645
(704) 754-5237
http://greerlabs.com

Greiner
Maybachestrasse 2
Postfach 1162
D-7443 Frickenhausen, Germany
(49) 0 91 31/80 79 0
FAX: (49) 0 91 31/80 79 30
http://www.erlangen.com/greiner

GSI Lumonics
130 Lombard Street Oxnard, CA 93030
(805) 485-5559 FAX: (805) 485-3310
http://www.gsilumonics.com

GTE Internetworking
150 Cambridge Park Drive
Cambridge, MA 02140
(800) 472-4565 FAX: (508) 694-4861
http://www.bbn.com

GW Instruments
35 Medford Street
Somerville, MA 02143
(617) 625-4096 FAX: (617) 625-1322
http://www.gwinst.com

H & H Woodworking
1002 Garfield Street
Denver, CO 80206
(303) 394-3764

Hacker Instruments
17 Sherwood Lane
P.O. Box 10033
Fairfield, NJ 07004
800-442-2537 FAX: (973) 808-8281
(973) 226-8450
http://www.hackerinstruments.com

Haemenetics
400 Wood Road
Braintree, MA 02184
(800) 225-5297 FAX: (781) 848-7921
(781) 848-7100
http://www.haemenetics.com

Halocarbon Products
P.O. Box 661
River Edge, NJ 07661
(201) 242-8899 FAX: (201) 262-0019
http://halocarbon.com

Hamamatsu Photonic Systems
A Division of Hamamatsu
360 Foothill Road
P.O. Box 6910
Bridgewater, NJ 08807
(908) 231-1116 FAX: (908) 231-0852
http://www.photonicsonline.com

Hamilton Company
4970 Energy Way
P.O. Box 10030
Reno, NV 89520
(800) 648-5950 FAX: (775) 856-7259
(775) 858-3000
http://www.hamiltoncompany.com

Hamilton Thorne Biosciences
100 Cummings Center, Suite 102C
Beverly, MA 01915
http://www.hamiltonthorne.com

Hampton Research
27631 El Lazo Road
Laguna Niguel, CA 92677
(800) 452-3899 FAX: (949) 425-1611
(949) 425-6321
http://www.hamptonresearch.com

Harlan Bioproducts for Science
P.O. Box 29176
Indianapolis, IN 46229
(317) 894-7521 FAX: (317) 894-1840
http://www.hbps.com

Harlan Sera-Lab
Hillcrest, Dodgeford Lane
Belton, Loughborough
Leicester LE12 9TE, UK
(44) 1530 222123
FAX: (44) 1530 224970
http://www.harlan.com

Harlan Teklad
P.O. Box 44220
Madison, WI 53744
(608) 277-2070 FAX: (608) 277-2066
http://www.harlan.com

Harrick Scientific Corporation
88 Broadway
Ossining, NY 10562
(914) 762-0020 FAX: (914) 762-0914
http://www.harricksci.com

Harrison Research
840 Moana Court
Palo Alto, CA 94306
(650) 949-1565 FAX: (650) 948-0493

Harvard Apparatus
84 October Hill Road
Holliston, MA 01746
(800) 272-2775 FAX: (508) 429-5732
(508) 893-8999
http://harvardapparatus.com

Harvard Bioscience
See Harvard Apparatus

Haselton Biologics
See JRH Biosciences

Hazelton Research Products
See Covance Research Products

Health Products
See Pierce Chemical

Heat Systems-Ultrasonics
1938 New Highway
Farmingdale, NY 11735
(800) 645-9846 FAX: (516) 694-9412
(516) 694-9555

Heidenhain Corp
333 East State Parkway
Schaumberg, IL 60173
(847) 490-1191 FAX: (847) 490-3931
http://www.heidenhain.com

Hellma Cells
11831 Queens Boulevard
Forest Hills, NY 11375
(718) 544-9166 FAX: (718) 263-6910
http://www.helmaUSA.com

Hellma
Postfach 1163
D-79371 Müllheim/Baden, Germany
(49) 7631-1820
FAX: (49) 7631-13546
http://www.hellma-worldwide.de

Henry Schein
135 Duryea Road, Mail Room 150
Melville, NY 11747
(800) 472-4346 FAX: (516) 843-5652
http://www.henryschein.com

Heraeus Kulzer
4315 South Lafayette Boulevard
South Bend, IN 46614
(800) 343-5336
(219) 291-0661
http://www.kulzer.com

Heraeus Sepatech
See Kendro Laboratory Products

Hercules Aqualon
Aqualon Division
Hercules Research Center, Bldg. 8145
500 Hercules Road
Wilmington, DE 19899
(800) 345-0447 FAX: (302) 995-4787
http://www.herc.com/aqualon/pharma

Heto-Holten A/S
Gydevang 17-19
DK-3450 Allerod, Denmark
(45) 48-16-62-00
FAX: (45) 48-16-62-97
Distributed by ATR

Hettich-Zentrifugen
See Andreas Hettich

Hewlett-Packard
3000 Hanover Street
Mailstop 20B3
Palo Alto, CA 94304
(650) 857-1501 FAX: (650) 857-5518
http://www.hp.com

HGS Hinimoto Plastics
1-10-24 Meguro-Honcho
Megurouko
Tokyo 152, Japan
3-3714-7226 FAX: 3-3714-4657

Hitachi Scientific Instruments
Nissei Sangyo America
8100 N. First Street
San Elsa, CA 95314
(800) 548-9001 FAX: (408) 432-0704
(408) 432-0520
http://www.hii.hitachi.com

Hi-Tech Scientific
Brunel Road
Salisbury, Wiltshire, SP2 7PU
UK
(44) 1722-432320
(800) 344-0724 (US only)
http://www.hi-techsci.co.uk

Hoechst AG
See Aventis Pharmaceutical

Hoefer Scientific Instruments
Division of Amersham-Pharmacia
Biotech
800 Centennial Avenue
Piscataway, NJ 08855 (800) 227-4750
FAX: (877) 295-8102
http://www.apbiotech.com

Hoffman-LaRoche
340 Kingsland Street
Nutley, NJ 07110
(800) 526-0189 FAX: (973) 235-9605
(973) 235-5000
http://www.rocheUSA.com

Holborn Surgical and Medical Instruments
Westwood Industrial Estate
Ramsgate Road Margate, Kent CT9 4JZ UK
(44) 1843 296666
FAX: (44) 1843 295446

Honeywell
101 Columbia Road
Morristown, NJ 07962
(973) 455-2000 FAX: (973) 455-4807
http://www.honeywell.com

Honeywell Specialty Films
P.O. Box 1039
101 Columbia Road
Morristown, NJ 07962
(800) 934-5679 FAX: (973) 455-6045
http://www.honeywell-specialtyfilms.com

Hood Thermo-Pad Canada
Comp. 20, Site 61A, RR2
Summerland, British Columbia
V0H 1Z0 Canada
(800) 665-9555 FAX: (250) 494-5003
(250) 494-5002
http://www.thermopad.com

Horiba Instruments
17671 Armstrong Avenue
Irvine, CA 92714
(949) 250-4811 FAX: (949) 250-0924
http://www.horiba.com

Hoskins Manufacturing
10776 Hall Road
P.O. Box 218
Hamburg, MI 48139
(810) 231-1900 FAX: (810) 231-4311
http://www.hoskinsmfgco.com

Hosokawa Micron Powder Systems
10 Chatham Road
Summit, NJ 07901
(800) 526-4491 FAX: (908) 273-7432
(908) 273-6360
http://www.hosokawamicron.com

HT Biotechnology
Unit 4
61 Ditton Walk
Cambridge CB5 8QD, UK
(44) 1223-412583

Hugo Sachs Electronik
Postfach 138
7806 March-Hugstetten, Germany
D-79229(49) 7665-92000
FAX: (49) 7665-920090

Human Biologics International
7150 East Camelback Road, Suite 245
Scottsdale, AZ 85251
(480) 990-2005 FAX: (480)-990-2155
http://www.humanbiological.com

Human Genetic Mutant Cell Repository
See Coriell Institute for Medical Research

HVS Image
P.O. Box 100
Hampton, Middlesex TW12 2YD, UK
FAX: (44) 208 783 1223
In the US: (800) 225-9261
FAX: (888) 483-8033
http://www.hvsimage.com

Hybaid
111-113 Waldegrave Road
Teddington, Middlesex TW11 8LL, UK
(44) 0 1784 42500
FAX: (44) 0 1784 248085
http://www.hybaid.co.uk

Hybaid Instruments
8 East Forge Parkway
Franklin, MA 02028
(888)4-HYBAID FAX: (508) 541-3041
(508) 541-6918
http://www.hybaid.com

Hybridon
155 Fortune Boulevard
Milford, MA 01757
(508) 482-7500 FAX: (508) 482-7510
http://www.hybridon.com

HyClone Laboratories
1725 South HyClone Road
Logan, UT 84321
(800) HYCLONE FAX: (800) 533-9450
(801) 753-4584 FAX: (801) 750-0809
http://www.hyclone.com

Hyseq
670 Almanor Avenue
Sunnyvale, CA 94086
(408) 524-8100 FAX: (408) 524-8141
http://www.hyseq.com

IBF Biotechnics
See Sepracor

IBI (International Biotechnologies)
See Eastman Kodak
For technical service (800) 243-2555
(203) 786-5600

ICN Biochemicals
See ICN Biomedicals

ICN Biomedicals
3300 Hyland Avenue
Costa Mesa, CA 92626
(800) 854-0530 FAX: (800) 334-6999
(714) 545-0100 FAX: (714) 641-7275
http://www.icnbiomed.com

ICN Flow and Pharmaceuticals
See ICN Biomedicals

ICN Immunobiochemicals
See ICN Biomedicals

ICN Radiochemicals
See ICN Biomedicals

ICONIX
100 King Street West, Suite 3825
Toronto, Ontario M5X 1E3 Canada
(416) 410-2411 FAX: (416) 368-3089
http://www.iconix.com

ICRT (Imperial Cancer Research Technology)
Sardinia House
Sardinia Street
London WC2A 3NL, UK
(44) 1712-421136
FAX: (44) 1718-314991

Idea Scientific Company
P.O. Box 13210
Minneapolis, MN 55414
(800) 433-2535 FAX: (612) 331-4217
http://www.ideascientific.com

IEC
See International Equipment Co.

IITC
Life Sciences
23924 Victory Boulevard
Woodland Hills, CA 91367
(888) 414-4482 (818) 710-1556
FAX: (818) 992-5185
http://www.iitcinc.com

IKA Works
2635 N. Chase Parkway, SE
Wilmington, NC 28405
(910) 452-7059 FAX: (910) 452-7693
http://www.ika.net

Ikegami Electronics
37 Brook Avenue
Maywood, NJ 07607
(201) 368-9171 FAX: (201) 569-1626

Ikemoto Scientific Technology
25-11 Hongo
3-chome, Bunkyo-ku
Tokyo 101-0025, Japan
(81) 3-3811-4181
FAX: (81) 3-3811-1960

Imagenetics
See ATC Diagnostics

Imaging Research
c/o Brock University
500 Glenridge Avenue
St. Catharines, Ontario
L2S 3A1 Canada
(905) 688-2040 FAX: (905) 685-5861
http://www.imaging.brocku.ca

Imclone Systems
180 Varick Street
New York, NY 10014
(212) 645-1405 FAX: (212) 645-2054
http://www.imclone.com

A3

IMCO Corporation LTD., AB
P.O. Box 21195
SE-100 31
Stockholm, Sweden
46-8-33-53-09 FAX: 46-8-728-47-76
http://www.imcocorp.se

IMICO
Calle Vivero, No. 5-4a Planta
E-28040, Madrid, Spain
(34) 1-535-3960 FAX: (34) 1-535-2780

Immunex
51 University Street
Seattle, WA 98101
(206) 587-0430 FAX: (206) 587-0606
http://www.immunex.com

Immunotech
130, av. Delattre de Tassigny
B.P. 177
13276 Marseilles Cedex 9
France
(33) 491-17-27-00
FAX: (33) 491-41-43-58
http://www.immunotech.fr

Imperial Chemical Industries
Imperial Chemical House
Millbank, London SW1P 3JF, UK
(44) 171-834-4444
FAX: (44)171-834-2042
http://www.ici.com

Inceltech
See New Brunswick Scientific

Incstar
See DiaSorin

Incyte
6519 Dumbarton Circle
Fremont, CA 94555
(510) 739-2100 FAX: (510) 739-2200
http://www.incyte.com

Incyte Pharmaceuticals
3160 Porter Drive
Palo Alto, CA 94304
(877) 746-2983 FAX: (650) 855-0572
(650) 855-0555
http://www.incyte.com

Individual Monitoring Systems
6310 Harford Road
Baltimore, MD 21214

Indo Fine Chemical
P.O. Box 473
Somerville, NJ 08876
(888) 463-6346 FAX: (908) 359-1179
(908) 359-6778
http://www.indofinechemical.com

Industrial Acoustics
1160 Commerce Avenue
Bronx, NY 10462
(718) 931-8000 FAX: (718) 863-1138
http://www.industrialacoustics.com

Inex Pharmaceuticals
100-8900 Glenlyon Parkway
Glenlyon Business Park
Burnaby, British Columbia
V5J 5J8 Canada
(604) 419-3200 FAX: (604) 419-3201
http://www.inexpharm.com

Ingold, Mettler, Toledo
261 Ballardvale Street
Wilmington, MA 01887
(800) 352-8763 FAX: (978) 658-0020
(978) 658-7615
http://www.mt.com

Innogenetics N.V.
Technologie Park 6
B-9052 Zwijnaarde
Belgium
(32) 9-329-1329 FAX: (32) 9-245-7623
http://www.innogenetics.com

Innovative Medical Services
1725 Gillespie Way
El Cajon, CA 92020
(619) 596-8600 FAX: (619) 596-8700
http://www.imspure.com

Innovative Research
3025 Harbor Lane N, Suite 300
Plymouth, MN 55447
(612) 519-0105 FAX: (612) 519-0239
http://www.inres.com

Innovative Research of America
2 N. Tamiami Trail, Suite 404
Sarasota, FL 34236
(800) 421-8171 FAX: (800) 643-4345
(941) 365-1406 FAX: (941) 365-1703
http://www.innovrsrch.com

Inotech Biosystems
15713 Crabbs Branch Way, #110
Rockville, MD 20855 (800) 635-4070
FAX: (301) 670-2859
(301) 670-2850
http://www.inotechintl.com

INOVISION
22699 Old Canal Road
Yorba Linda, CA 92887
(714) 998-9600 FAX: (714) 998-9666
http://www.inovision.com

Instech Laboratories
5209 Militia Hill Road
Plymouth Meeting, PA 19462
(800) 443-4227 FAX: (610) 941-0134
(610) 941-0132
http://www.instechlabs.com

Instron
100 Royall Street
Canton, MA 02021
(800) 564-8378 FAX: (781) 575-5725
(781) 575-5000
http://www.instron.com

Instrumentarium
P.O. Box 300
00031 Instrumentarium
Helsinki, Finland
(10) 394-5566
http://www.instrumentarium.fi

Instruments SA
Division Jobin Yvon
16-18 Rue du Canal
91165 Longjumeau, Cedex, France
(33)1 6454-1300
FAX: (33)1 6909-9319
http://www.isainc.com

Instrutech
20 Vanderventer Avenue, Suite 101E
Port Washington, NY 11050
(516) 883-1300 FAX: (516) 883-1558
http://www.instrutech.com

Integrated DNA Technologies
1710 Commercial Park
Coralville, Iowa 52241
(800) 328-2661 FAX: (319) 626-8444
http://www.idtdna.com

Integrated Genetics
See Genzyme Genetics

Integrated Scientific Imaging Systems
3463 State Street, Suite 431
Santa Barbara, CA 93105
(805) 692-2390 FAX: (805) 692-2391
http://www.imagingsystems.com

Integrated Separation Systems (ISS)
See OWL Separation Systems

IntelliGenetics
See Oxford Molecular Group

Interactiva BioTechnologie
Sedanstrasse 10
D-89077 Ulm, Germany
(49) 731-93579-290
FAX: (49) 731-93579-291
http://www.interactiva.de

Interchim
213 J.F. Kennedy Avenue
B.P. 1140
Montlucon
03103 France
(33) 04-70-03-83-55
FAX: (33) 04-70-03-93-60

Interfocus
14/15 Spring Rise
Falcover Road
Haverhill, Suffolk CB9 7XU, UK
(44) 1440 703460
FAX: (44) 1440 704397
http://www.interfocus.ltd.uk

Intergen
2 Manhattanville Road
Purchase, NY 10577
(800) 431-4505 FAX: (800) 468-7436
(914) 694-1700 FAX: (914) 694-1429
http://www.intergenco.com

Intermountain Scientific
420 N. Keys Drive
Kaysville, UT 84037
(800) 999-2901 FAX: (800) 574-7892
(801) 547-5047 FAX: (801) 547-5051
http://www.bioexpress.com

International Biotechnologies (IBI)
See Eastman Kodak

International Equipment Co. (IEC)
See Thermoquest

International Institute for the Advancement of Medicine
1232 Mid-Valley Drive
Jessup, PA 18434
(800) 486-IIAM FAX: (570) 343-6993
(570) 496-3400
http://www.iiam.org

International Light
17 Graf Road
Newburyport, MA 01950
(978) 465-5923 FAX: (978) 462-0759

International Market Supply (I.M.S.)
Dane Mill
Broadhurst Lane
Congleton, Cheshire CW12 1LA, UK
(44) 1260 275469
FAX: (44) 1260 276007

International Marketing Services
See International Marketing Ventures

International Marketing Ventures
6301 Ivy Lane, Suite 408
Greenbelt, MD 20770
(800) 373-0096 FAX: (301) 345-0631
(301) 345-2866
http://www.imvlimited.com

International Products
201 Connecticut Drive
Burlington, NJ 08016
(609) 386-8770 FAX: (609) 386-8438
http://www.mkt.ipcol.com

Intracel Corporation
Bartels Division
2005 Sammamish Road, Suite 107
Issaquah, WA 98027
(800) 542-2281 FAX: (425) 557-1894
(425) 392-2992
http://www.intracel.com

Invitrogen
1600 Faraday Avenue
Carlsbad, CA 92008
(800) 955-6288 FAX: (760) 603-7201
(760) 603-7200
http://www.invitrogen.com

In Vivo Metric
P.O. Box 249
Healdsburg, CA 95448
(707) 433-4819 FAX: (707) 433-2407

IRORI
9640 Towne Center Drive
San Diego, CA 92121
(858) 546-1300 FAX: (858) 546-3083
http://www.irori.com

Irvine Scientific
2511 Daimler Street
Santa Ana, CA 92705
(800) 577-6097 FAX: (949) 261-6522
(949) 261-7800
http://www.irvinesci.com

ISC BioExpress
420 North Kays Drive
Kaysville, UT 84037
(800) 999-2901 FAX: (800) 574-7892
(801) 547-5047
http://www.bioexpress.com

ISCO
P.O. Box 5347
4700 Superior
Lincoln, NE 68505
(800) 228-4373 FAX: (402) 464-0318
(402) 464-0231
http://www.isco.com

Isis Pharmaceuticals
Carlsbad Research Center
2292 Faraday Avenue
Carlsbad, CA 92008
(760) 931-9200
http://www.isip.com

Isolabs
See Wallac

ISS
See Integrated Separation Systems

J & W Scientific
See Agilent Technologies

J.A. Webster
86 Leominster Road
Sterling, MA 01564
(800) 225-7911 FAX: (978) 422-8959
http://www.jawebster.com

J.T. Baker
See Mallinckrodt Baker
222 Red School Lane
Phillipsburg, NJ 08865
(800) JTBAKER FAX: (908) 859-6974
http://www.jtbaker.com

Jackson ImmunoResearch Laboratories
P.O. Box 9
872 W. Baltimore Pike
West Grove, PA 19390
(800) 367-5296 FAX: (610) 869-0171
(610) 869-4024
http://www.jacksonimmuno.com

The Jackson Laboratory
600 Maine Street
Bar Harbor, ME 04059
(800) 422-6423 FAX: (207) 288-5079
(207) 288-6000
http://www.jax.org

Jaece Industries
908 Niagara Falls Boulevard
North Tonawanda, NY 14120
(716) 694-2811 FAX: (716) 694-2811
http://www.jaece.com

Jandel Scientific
See SPSS

Janke & Kunkel
See Ika Works

Janssen Life Sciences Products
See Amersham

Janssen Pharmaceutica
1125 Trenton-Harbourton Road
Titusville, NJ 09560
(609) 730-2577 FAX: (609) 730-2116
http://us.janssen.com

Jasco
8649 Commerce Drive
Easton, MD 21601
(800) 333-5272 FAX: (410) 822-7526
(410) 822-1220
http://www.jascoinc.com

Jena Bioscience
Loebstedter Str. 78
07749 Jena, Germany
(49) 3641-464920
FAX: (49) 3641-464991
http://www.jenabioscience.com

Jencons Scientific
800 Bursca Drive, Suite 801
Bridgeville, PA 15017
(800) 846-9959 FAX: (412) 257-8809
(412) 257-8861
http://www.jencons.co.uk

JEOL Instruments
11 Dearborn Road
Peabody, MA 01960
(978) 535-5900 FAX: (978) 536-2205
http://www.jeol.com/index.html

Jewett
750 Grant Street
Buffalo, NY 14213
(800) 879-7767 FAX: (716) 881-6092
(716) 881-0030
http://www.JewettInc.com

John's Scientific
See VWR Scientific

John Weiss and Sons
95 Alston Drive
Bradwell Abbey
Milton Keynes, Buckinghamshire
MK1 4HF UK
(44) 1908-318017
FAX: (44) 1908-318708

Johnson & Johnson Medical
2500 Arbrook Boulevard East
Arlington, TX 76004
(800) 423-4018
http://www.jnjmedical.com

Johnston Matthey Chemicals
Orchard Road
Royston, Hertfordshire SG8 5HE, UK
(44) 1763-253000
FAX: (44) 1763-253466
http://www.chemicals.matthey.com

Jolley Consulting and Research
683 E. Center Street, Unit H
Grayslake, IL 60030
(847) 548-2330 FAX: (847) 548-2984
http://www.jolley.com

Jordan Scientific
See Shelton Scientific

Jorgensen Laboratories
1450 N. Van Buren Avenue
Loveland, CO 80538
(800) 525-5614 FAX: (970) 663-5042
(970) 669-2500
http://www.jorvet.com

JRH Biosciences and JR Scientific
13804 W. 107th Street
Lenexa, KS 66215
(800) 231-3735 FAX: (913) 469-5584
(913) 469-5580

Jule Bio Technologies
25 Science Park, #14, Suite 695
New Haven, CT 06511
(800) 648-1772 FAX: (203) 786-5489
(203) 786-5490
http://hometown.aol.com/precastgel/index.htm

K.R. Anderson
2800 Bowers Avenue
Santa Clara, CA 95051
(800) 538-8712 FAX: (408) 727-2959
(408) 727-2800
http://www.kranderson.com

Kabi Pharmacia Diagnostics
See Pharmacia Diagnostics

Kanthal H.P. Reid
1 Commerce Boulevard
P.O. Box 352440
Palm Coast, FL 32135
(904) 445-200 FAX: (904) 446-2244
http://www.kanthal.com

Kapak
5305 Parkdale Drive
St. Louis Park, MN 55416
(800) KAPAK-57 FAX: (612) 541-0735
(612) 541-0730
http://www.kapak.com

Karl Hecht
Stettener Str. 22-24
D-97647 Sondheim
Rhön, Germany
(49) 9779-8080 FAX: (49) 9779-80888

Karl Storz
Köningin-Elisabeth Str. 60
D-14059 Berlin, Germany
(49) 30-30 69 09-0
FAX: (49) 30-30 19 452
http://www.karlstorz.de

KaVo EWL
P.O. Box 1320
D-88293 Leutkirch im Allgäu, Germany
(49) 7561-86-0 FAX: (49) 7561-86-371
http://www.kavo.com/english/startseite.htm

Keithley Instruments
28775 Aurora Road
Cleveland, OH 44139
(800) 552-1115 FAX: (440) 248-6168
(440) 248-0400
http://www.keithley.com

Kemin
2100 Maury Street, Box 70
Des Moines, IA 50301
(515) 266-2111 FAX: (515) 266-8354
http://www.kemin.com

Kemo
3 Brook Court, Blakeney Road
Beckenham, Kent BR3 1HG, UK
(44) 0181 658 3838
FAX: (44) 0181 658 4084
http://www.kemo.com

Kendall
15 Hampshire Street
Mansfield, MA 02048
(800) 962-9888 FAX: (800) 724-1324
http://www.kendallhq.com

Kendro Laboratory Products
31 Pecks Lane
Newtown, CT 06470
(800) 522-SPIN FAX: (203) 270-2166
(203) 270-2080
http://www.kendro.com

Kendro Laboratory Products
P.O. Box 1220
Am Kalkberg
D-3360 Osterod, Germany
(55) 22-316-213
FAX: (55) 22-316-202
http://www.heraeus-instruments.de

Kent Laboratories
23404 NE 8th Street
Redmond, WA 98053
(425) 868-6200 FAX: (425) 868-6335
http://www.kentlabs.com

Kent Scientific
457 Bantam Road, #16
Litchfield, CT 06759
(888) 572-8887 FAX: (860) 567-4201
(860) 567-5496
http://www.kentscientific.com

Keuffel & Esser
See Azon

Keystone Scientific
Penn Eagle Industrial Park 320 Rolling Ridge Drive
Bellefonte, PA 16823
(800) 437-2999 FAX: (814) 353-2305
(814) 353-2300 Ext 1
http://www.keystonescientific.com

Kimble/Kontes Biotechnology
1022 Spruce Street
P.O. Box 729
Vineland, NJ 08360
(888) 546-2531 FAX: (856) 794-9762
(856) 692-3600
http://www.kimble-kontes.com

Kinematica AG
Luzernerstrasse 147a
CH-6014 Littau-Luzern, Switzerland
(41) 41 2501257 FAX: (41) 41 2501460
http://www.kinematica.ch

Kin-Tek
504 Laurel Street
LaMarque, TX 77568
(800) 326-3627
FAX: (409) 938-3710
http://www.kin-tek.com

Kipp & Zonen
125 Wilbur Place
Bohemia, NY 11716
(800) 645-2065 FAX: (516) 589-2068
(516) 589-2885
http://www.kippzonen.thomasregister.com/olc/kippzonen

Kirkegaard & Perry Laboratories
2 Cessna Court
Gaithersburg, MD 20879
(800) 638-3167 FAX: (301) 948-0169
(301) 948-7755
http://www.kpl.com

Kodak
See Eastman Kodak

Kontes Glass
See Kimble/Kontes Biotechnology

Kontron Instruments AG
Postfach CH-8010
Zurich, Switzerland
41-1-733-5733 FAX: 41-1-733-5734

David Kopf Instruments
P.O. Box 636
Tujunga, CA 91043
(818) 352-3274 FAX: (818) 352-3139

Kraft Apparatus
See Glas-Col Apparatus

Kramer Scientific Corporation
711 Executive Boulevard
Valley Cottage, NY 10989
(845) 267-5050 FAX: (845) 267-5550

Kulite Semiconductor Products
1 Willow Tree Road
Leonia, NJ 07605
(201) 461-0900 FAX: (201) 461-0990
http://www.kulite.com

Lab-Line Instruments
15th & Bloomingdale Avenues
Melrose Park, IL 60160
(800) LAB-LINE FAX: (708) 450-5830
FAX: (800) 450-4LAB
http://www.labline.com

Lab Products
742 Sussex Avenue
P.O. Box 639
Seaford, DE 19973
(800) 526-0469 FAX: (302) 628-4309
(302) 628-4300
http://www.labproductsinc.com

LabRepco
101 Witmer Road, Suite 700
Horsham, PA 19044
(800) 521-0754 FAX: (215) 442-9202
http://www.labrepco.com

Lab Safety Supply
P.O. Box 1368
Janesville, WI 53547
(800) 356-0783 FAX: (800) 543-9910
(608) 754-7160 FAX: (608) 754-1806
http://www.labsafety.com

Lab-Tek Products
See Nalge Nunc International

Labconco
8811 Prospect Avenue
Kansas City, MO 64132
(800) 821-5525 FAX: (816) 363-0130
(816) 333-8811
http://www.labconco.com

Labindustries
See Barnstead/Thermolyne

Labnet International
P.O. Box 841
Woodbridge, NJ 07095
(888) LAB-NET1 FAX: (732) 417-1750
(732) 417-0700
http://www.nationallabnet.com

LABO-MODERNE
37 rue Dombasle
Paris
75015 France
(33) 01-45-32-62-54
FAX: (33) 01-45-32-01-09
http://www.labomoderne.com/fr

Laboratory of Immunoregulation
National Institute of Allergy and
Infectious Diseases/NIH
9000 Rockville Pike
Building 10, Room 11B13
Bethesda, MD 20892
(301) 496-1124

Laboratory Supplies
29 Jefry Lane
Hicksville, NY 11801
(516) 681-7711

Labscan Limited
Stillorgan Industrial Park
Stillorgan
Dublin, Ireland
(353) 1-295-2684
FAX: (353) 1-295-2685
http://www.labscan.ie

Labsystems
See Thermo Labsystems

Labsystems Affinity Sensors
Saxon Way, Bar Hill
Cambridge CB3 8SL, UK
44 (0) 1954 789976
FAX: 44 (0) 1954 789417
http://www.affinity-sensors.com

Labtronics
546 Governors Road
Guelph, Ontario N1K 1E3, Canada
(519) 763-4930 FAX: (519) 836-4431
http://www.labtronics.com

Labtronix Manufacturing
3200 Investment Boulevard
Hayward, CA 94545
(510) 786-3200 FAX: (510) 786-3268
http://www.labtronix.com

Lafayette Instrument
3700 Sagamore Parkway North
P.O. Box 5729
Lafayette, IN 47903
(800) 428-7545 FAX: (765) 423-4111
(765) 423-1505
http://www.lafayetteinstrument.com

Lambert Instruments
Turfweg 4
9313 TH Leutingewolde
The Netherlands
(31) 50-5018461 FAX: (31)
50-5010034
http://www.lambert-instruments.com

Lancaster Synthesis
P.O. Box 1000
Windham, NH 03087
(800) 238-2324 FAX: (603) 889-3326
(603) 889-3306
http://www.lancastersynthesis-us.com

Lancer
140 State Road 419
Winter Springs, FL 32708
(800) 332-1855 FAX: (407) 327-1229
(407) 327-8488
http://www.lancer.com

LaVision GmbH
Gerhard-Gerdes-Str. 3
D-37079
Goettingen, Germany
(49) 551-50549-0
FAX: (49) 551-50549-11
http://www.lavision.de

Lawshe
See Advanced Process Supply

LC Laboratories
165 New Boston Street
Woburn, MA 01801
(781) 937-0777 FAX: (781) 938-5420
http://www.lclaboratories.com

LC Packings
80 Carolina Street
San Francisco, CA 94103
(415) 552-1855 FAX: (415) 552-1859
http://www.lcpackings.com

LC Services
See LC Laboratories

LECO
3000 Lakeview Avenue
St. Joseph, MI 49085
(800) 292-6141 FAX: (616) 982-8977
(616) 985-5496
http://www.leco.com

Lederle Laboratories
See Wyeth-Ayerst

Lee Biomolecular Research Laboratories
11211 Sorrento Valley Road, Suite M
San Diego, CA 92121
(858) 452-7700

The Lee Company
2 Pettipaug Road
P.O. Box 424
Westbrook, CT 06498
(800) LEE-PLUG FAX: (860) 399-7058
(860) 399-6281
http://www.theleeco.com

Lee Laboratories
1475 Athens Highway
Grayson, GA 30017
(800) 732-9150 FAX: (770) 979-9570
(770) 972-4450
http://www.leelabs.com

Leica
111 Deer Lake Road
Deerfield, IL 60015
(800) 248-0123 FAX: (847) 405-0147
(847) 405-0123
http://www.leica.com

Leica Microsystems
Imneuenheimer Feld 518
D-69120
Heidelberg, Germany
(49) 6221-41480
FAX: (49) 6221-414833
http://www.leica-microsystems.com

Leinco Technologies
359 Consort Drive
St. Louis, MO 63011
(314) 230-9477 FAX: (314) 527-5545
http://www.leinco.com

Leltz U.S.A.
See Leica

LenderKing Metal Products
8370 Jumpers Hole Road
Millersville, MD 21108
(410) 544-8795 FAX: (410) 544-5069
http://www.lenderking.com

Letica Scientific Instruments
Panlab s.i., c/Loreto 50
08029 Barcelona, Spain
(34) 93-419-0709
FAX: (34) 93-419-7145
www.panlab-sl.com

Leybold-Heraeus Trivac DZA
5700 Mellon Road
Export, PA 15632
(412) 327-5700

LI-COR
Biotechnology Division
4308 Progressive Avenue
Lincoln, NE 68504
(800) 645-4267 FAX: (402) 467-0819
(402) 467-0700
http://www.licor.com

Life Science Laboratories
See Adaptive Biosystems

Life Science Resources
Two Corporate Center Drive
Melville, NY 11747
(800) 747-9530 FAX: (516) 844-5114
(516) 844-5085
http://www.astrocam.com

Life Sciences
2900 72nd Street North
St. Petersburg, FL 33710
(800) 237-4323 FAX: (727) 347-2957
(727) 345-9371
http://www.lifesci.com

Life Technologies
9800 Medical Center Drive
P.O. Box 6482
Rockville, MD 20849
(800) 828-6686 FAX: (800) 331-2286
http://www.lifetech.com

Lifecodes
550 West Avenue
Stamford, CT 06902
(800) 543-3263 FAX: (203) 328-9599
(203) 328-9500
http://www.lifecodes.com

Lightnin
135 Mt. Read Boulevard
Rochester, NY 14611
(888) MIX-BEST FAX: (716) 527-1742
(716) 436-5550
http://www.lightnin-mixers.com

Linear Drives
Luckyn Lane, Pipps Hill
Basildon, Essex SS14 3BW, UK
(44) 1268-287070
FAX: (44) 1268-293344
http://www.leardrives.com

Linscott's USA
6 Grove Street
Mill Valley, CA 94941
(415) 389-9674 FAX: (415) 389-6025
http://www.linscottsdirectory.com

Linton Instrumentation
Unit 11, Forge Business Center
Upper Rose Lane
Palgrave, Diss, Norfolk IP22 1AP, UK
(44) 1-379-651-344
FAX: (44) 1-379-650-970
http://www.lintoninst.co.uk

List Biological Laboratories
501-B Vandell Way
Campbell, CA 95008
(800) 726-3213 FAX: (408) 866-6364
(408) 866-6363
http://www.listlabs.com

LKB Instruments
See Amersham Pharmacia Biotech

Lloyd Laboratories
604 West Thomas Avenue
Shenandoah, IA 51601
(800) 831-0004 FAX: (712) 246-5245
(712) 246-4000
http://www.lloydinc.com

Loctite
1001 Trout Brook Crossing
Rocky Hill, CT 06067
(860) 571-5100 FAX: (860)571-5465
http://www.loctite.com

Lofstrand Labs
7961 Cessna Avenue
Gaithersburg, MD 20879
(800) 541-0362 FAX: (301) 948-9214
(301) 330-0111
http://www.lofstrand.com

Lomir Biochemical
99 East Main Street
Malone, NY 12953
(877) 425-3604 FAX: (518) 483-8195
(518) 483-7697
http://www.lomir.com

LSL Biolafitte
10 rue de Temara
7810C St.-Germain-en-Laye, France
(33) 1-3061-5260 FAX: (33)
1-3061-5234

Ludl Electronic Products
171 Brady Avenue
Hawthorne, NY 10532
(888) 769-6111 FAX: (914) 769-4759
(914) 769-6111
http://www.ludl.com

Lumigen
24485 W. Ten Mile Road
Southfield, MI 48034
(248) 351-5600 FAX: (248) 351-0518
http://www.lumigen.com

Luminex
12212 Technology Boulevard
Austin, TX 78727
(888) 219-8020 FAX: (512) 258-4173
(512) 219-8020
http://www.luminexcorp.com

LYNX Therapeutics
25861 Industrial Boulevard
Hayward, CA 94545 (510) 670-9300
FAX: (510) 670-9302
http://www.lynxgen.com

Lyphomed
3 Parkway North
Deerfield, IL 60015
(847) 317-8100 FAX: (847) 317-8600

M.E.D. Associates
See Med Associates

Macherey-Nagel
6 South Third Street, #402
Easton, PA 18042
(610) 559-9848 FAX: (610) 559-9878
http://www.macherey-nagel.com

Macherey-Nagel
Valencienner Strasse 11
P.O. Box 101352
D-52313 Dueren, Germany
(49) 2421-969141
FAX: (49) 2421-969199
http://www.macherey-nagel.ch

Mac-Mod Analytical
127 Commons Court
Chadds Ford, PA 19317
800-441-7508 FAX: (610) 358-5993
(610) 358-9696
http://www.mac-mod.com

Mallinckrodt Baker
222 Red School Lane
Phillipsburg, NJ 08865
(800) 582-2537 FAX: (908) 859-6974
(908) 859-2151
http://www.mallbaker.com

Mallinckrodt Chemicals
16305 Swingley Ridge Drive
Chesterfield, MD 63017
(314) 530-2172 FAX: (314) 530-2563
http://www.mallchem.com

Malven Instruments
Enigma Business Park
Grovewood Road
Malven, Worchestershire
WR 141 XZ, United Kingdom

Marinus
1500 Pier C Street
Long Beach, CA 90813
(562) 435-6522 FAX: (562) 495-3120

Markson Science
c/o Whatman Labs Sales
P.O. Box 1359
Hillsboro, OR 97123
(800) 942-8626 FAX: (503) 640-9716
(503) 648-0762

Marsh Biomedical Products
565 Blossom Road
Rochester, NY 14610
(800) 445-2812 FAX: (716) 654-4810
(716) 654-4800
http://www.biomar.com

Marshall Farms USA
5800 Lake Bluff Road
North Rose, NY 14516
(315) 587-2295
e-mail: info@marfarms.com

Martek
6480 Dobbin Road
Columbia, MD 21045
(410) 740-0081 FAX: (410) 740-2985
http://www.martekbio.com

Martin Supply
Distributor of Gerber Scientific
2740 Loch Raven Road
Baltimore, MD 21218
(800) 282-5440 FAX: (410) 366-0134
(410) 366-1696

Mast Immunosystems
630 Clyde Court
Mountain View, CA 94043
(800) 233-MAST FAX: (650) 969-2745
(650) 961-5501
http://www.mastallergy.com

Matheson Gas Products
P.O. Box 624
959 Route 46 East
Parsippany, NJ 07054
(800) 416-2505 FAX: (973) 257-9393
(973) 257-1100
http://www.mathesongas.com

Mathsoft
1700 Westlake Avenue N., Suite 500
Seattle, WA 98109
(800) 569-0123 FAX: (206) 283-8691
(206) 283-8802
http://www.mathsoft.com

Matreya
500 Tressler Street
Pleasant Gap, PA 16823
(814) 359-5060 FAX: (814) 359-5062
http://www.matreya.com

Matrigel
See Becton Dickinson Labware

Matrix Technologies
22 Friars Drive
Hudson, NH 03051
(800) 345-0206 FAX: (603) 595-0106
(603) 595-0505
http://www.matrixtechcorp.com

MatTek Corp.
200 Homer Avenue
Ashland, Massachusetts 01721
(508) 881-6771 FAX: (508) 879-1532
http://www.mattek.com

Maxim Medical
89 Oxford Road
Oxford OX2 9PD
United Kingdom
44 (0)1865-865943
FAX: 44 (0)1865-865291
http://www.maximmed.com

Mayo Clinic
Section on Engineering
Project #ALA-1, 1982
200 1st Street SW
Rochester, MN 55905
(507) 284-2511 FAX: (507) 284-5988

McGaw
See B. Braun-McGaw

McMaster-Carr
600 County Line Road
Elmhurst, IL 60126
(630) 833-0300 FAX: (630) 834-9427
http://www.mcmaster.com

McNeil Pharmaceutical
See Ortho McNeil Pharmaceutical

MCNC
3021 Cornwallis Road
P.O. Box 12889
Research Triangle Park, NC 27709
(919) 248-1800 FAX: (919) 248-1455
http://www.mcnc.org

MD Industries
5 Revere Drive, Suite 415
Northbrook, IL 60062
(800) 421-8370 FAX: (847) 498-2627
(708) 339-6000
http://www.mdindustries.com

MDS Nordion
447 March Road
P.O. Box 13500
Kanata, Ontario K2K 1X8, Canada (800)
465-3666 FAX: (613) 592-6937
(613) 592-2790
http://www.mds.nordion.com

MDS Sciex
71 Four Valley Drive
Concord, Ontario
Canada L4K 4V8
(905) 660-9005 FAX: (905) 660-2600
http://www.sciex.com

Mead Johnson
See Bristol-Meyers Squibb

Med Associates
P.O. Box 319
St. Albans, VT 05478
(802) 527-2343 FAX: (802) 527-5095
http://www.med-associates.com

Medecell
239 Liverpool Road
London N1 1LX, UK
(44) 20-7607-2295
FAX: (44) 20-7700-4156
http://www.medicell.co.uk

Media Cybernetics
8484 Georgia Avenue, Suite 200
Silver Spring, MD 20910
(301) 495-3305 FAX: (301) 495-5964
http://www.mediacy.com

Mediatech
13884 Park Center Road
Herndon, VA 20171
(800) cellgro
(703) 471-5955
http://www.cellgro.com

Medical Systems
See Harvard Apparatus

Medifor
647 Washington Street
Port Townsend, WA 98368
(800) 366-3710 FAX: (360) 385-4402
(360) 385-0722
http://www.medifor.com

MedImmune
35 W. Watkins Mill Road
Gaithersburg, MD 20878
(301) 417-0770 FAX: (301) 527-4207
http://www.medimmune.com

Medoc Advanced Medical Systems
1502 West Highway 54, Suite 404
Durham, NC 27707
(919) 402-9600 FAX: (919) 402-9607
http://www.medoc-web.com

MedProbe AS
P.O. Box 2640
St. Hanshaugen
N-0131 Oslo, Norway
(47) 222 00137 FAX: (47) 222 00189
http://www.medprobe.com

Megazyme
Bray Business Park
Bray, County Wicklow
Ireland
(353) 1-286-1220
FAX: (353) 1-286-1264
http://www.megazyme.com

Melles Griot
4601 Nautilus Court South
Boulder, CO 80301
(800) 326-4363 FAX: (303) 581-0960
(303) 581-0337
http://www.mellesgriot.com

Menzel-Glaser
Postfach 3157
D-38021 Braunschweig, Germany
(49) 531 590080
FAX: (49) 531 509799

E. Merck
Frankfurterstrasse 250
D-64293 Darmstadt 1, Germany
(49) 6151-720

Merck
See EM Science

Merck & Company
Merck National Service Center
P.O. Box 4
West Point, PA 19486
(800) NSC-MERCK
(215) 652-5000
http://www.merck.com

Merck Research Laboratories
See Merck & Company

Merck Sharpe Human Health Division
300 Franklin Square Drive
Somerset, NJ 08873
(800) 637-2579 FAX: (732) 805-3960
(732) 805-0300

Merial Limited
115 Transtech Drive
Athens, GA 30601
(800) MERIAL-1 FAX: (706) 548-0608
(706) 548-9292
http://www.merial.com

Meridian Instruments
P.O. Box 1204
Kent, WA 98035
(253) 854-9914 FAX: (253) 854-9902
http://www.minstrument.com

Meta Systems Group
32 Hammond Road
Belmont, MA 02178
(617) 489-9950 FAX: (617) 489-9952

Metachem Technologies
3547 Voyager Street, Bldg. 102
Torrance, CA 90503
(310) 793-2300 FAX: (310) 793-2304
http://www.metachem.com

Metallhantering
Box 47172
100-74 Stockholm, Sweden
(46) 8-726-9696

MethylGene
7220 Frederick-Banting, Suite 200
Montreal, Quebec H4S 2A1, Canada
http://www.methylgene.com

Metro Scientific
475 Main Street, Suite 2A
Farmingdale, NY 11735
(800) 788-6247 FAX: (516) 293-8549
(516) 293-9656

Metrowerks
980 Metric Boulevard
Austin, TX 78758
(800) 377-5416
(512) 997-4700
http://www.metrowerks.com

Mettler Instruments
Mettler-Toledo
1900 Polaris Parkway
Columbus, OH 43240
(800) METTLER FAX: (614) 438-4900
http://www.mt.com

Miami Serpentarium Labs
34879 Washington Loop Road
Punta Gorda, FL 33982
(800) 248-5050 FAX: (813) 639-1811
(813) 639-8888
http://www.miamiserpentarium.com

Michrom BioResources
1945 Industrial Drive
Auburn, CA 95603
(530) 888-6498 FAX: (530) 888-8295
http://www.michrom.com

Mickle Laboratory Engineering
Gomshall, Surrey, UK
(44) 1483-202178

Micra Scientific
A division of Eichrom Industries
8205 S. Cass Ave, Suite 111
Darien, IL 60561
(800) 283-4752 FAX: (630) 963-1928
(630) 963-0320
http://www.micrasci.com

MicroBrightField
74 Hegman Avenue
Colchester, VT 05446
(802) 655-9360 FAX: (802) 655-5245
http://www.microbrightfield.com

Micro Essential Laboratory
4224 Avenue H
Brooklyn, NY 11210
(718) 338-3618 FAX: (718) 692-4491

Micro Filtration Systems
7-3-Chome, Honcho
Nihonbashi, Tokyo, Japan
(81) 3-270-3141

Micro-Metrics
P.O. Box 13804
Atlanta, GA 30324
(770) 986-6015 FAX: (770) 986-9510
http://www.micro-metrics.com

Micro-Tech Scientific
140 South Wolfe Road
Sunnyvale, CA 94086
(408) 730-8324 FAX: (408) 730-3566
http://www.microlc.com

MicroCal
22 Industrial Drive East
Northampton, MA 01060
(800) 633-3115 FAX: (413) 586-0149
(413) 586-7720
www.microcalorimetry.com

Microfluidics
30 Ossipee Road
P.O. Box 9101
Newton, MA 02164
(800) 370-5452 FAX: (617) 965-1213
(617) 969-5452
http://www.microfluidicscorp.com

Microgon
See Spectrum Laboratories

Microlase Optical Systems
West of Scotland Science Park
Kelvin Campus, Maryhill Road
Glasgow G20 0SP, UK
(44) 141-948-1000
FAX: (44) 141-946-6311
http://www.microlase.co.uk

Micron Instruments
4509 Runway Street
Simi Valley, CA 93063
(800) 638-3770 FAX: (805) 522-4982
(805) 552-4676
http://www.microninstruments.com

Micron Separations
See MSI

Micro Photonics
4949 Liberty Lane, Suite 170
P.O. Box 3129
Allentown, PA 18106
(610) 366-7103 FAX: (610) 366-7105
http://www.microphotonics.com

MicroTech
1420 Conchester Highway
Boothwyn, PA 19061
(610) 459-3514

Midland Certified Reagent Company
3112-A West Cuthbert Avenue
Midland, TX 79701
(800) 247-8766 FAX: (800) 359-5789
(915) 694-7950 FAX: (915) 694-2387
http://www.mcrc.com

Midwest Scientific
280 Vance Road
Valley Park, MO 63088
(800) 227-9997 FAX: (636) 225-9998
(636) 225-9997
http://www.midsci.com

Miles
See Bayer

Miles Laboratories
See Serological

Miles Scientific
See Nunc

Millar Instruments
P.O. Box 230227
6001-A Gulf Freeway
Houston, TX 77023
(713) 923-9171 FAX: (713) 923-7757
http://www.millarinstruments.com

MilliGen/Biosearch
See Millipore

Millipore
80 Ashbury Road
P.O. Box 9125
Bedford, MA 01730
(800) 645-5476 FAX: (781) 533-3110
(781) 533-6000
http://www.millipore.com

Miltenyi Biotec
251 Auburn Ravine Road, Suite 208
Auburn, CA 95603
(800) 367-6227 FAX: (530) 888-8925
(530) 888-8871
http://www.miltenyibiotec.com

Miltex
6 Ohio Drive
Lake Success, NY 11042
(800) 645-8000 FAX: (516) 775-7185
(516) 349-0001

Milton Roy
See Spectronic Instruments

Mini-Instruments
15 Burnham Business Park
Springfield Road
Burnham-on-Crouch, Essex CM0 8TE, UK
(44) 1621-783282
FAX: (44) 1621-783132
http://www.mini-instruments.co.uk

Mini Mitter
P.O. Box 3386
Sunriver, OR 97707
(800) 685-2999 FAX: (541) 593-5604
(541) 593-8639
http://www.minimitter.com

Misonix
1938 New Highway
Farmingdale, NY 11735
(800) 645-9846 FAX: (516) 694-9412
http://www.misonix.com

Mitutoyo (MTI)
See Dolla Eastern

MJ Research
Waltham, MA 02451
(800) PELTIER FAX: (617) 923-8080
(617) 923-8000
http://www.mjr.com

Modular Instruments
228 West Gay Street
Westchester, PA 19380
(610) 738-1420 FAX: (610) 738-1421
http://www.mi2.com

Molecular Biology Insights
8685 US Highway 24
Cascade, CO 80809-1333
(800) 747-4362 FAX: (719) 684-7989
(719) 684-7988
http://www.oligo.net

Molecular Biosystems
10030 Barnes Canyon Road
San Diego, CA 92121
(858) 452-0681 FAX: (858) 452-6187
http://www.mobi.com

Molecular Devices
1312 Crossman Avenue
Sunnyvale, CA 94089
(800) 635-5577 FAX: (408) 747-3602
(408) 747-1700
http://www.moldev.com

Molecular Designs
1400 Catalina Street
San Leandro, CA 94577
(510) 895-1313 FAX: (510) 614-3608

Molecular Dynamics
928 East Arques Avenue
Sunnyvale, CA 94086
(800) 333-5703 FAX: (408) 773-1493
(408) 773-1222
http://www.apbiotech.com

Molecular Probes
4849 Pitchford Avenue
Eugene, OR 97402
(800) 438-2209 FAX: (800) 438-0228
(541) 465-8300 FAX: (541) 344-6504
http://www.probes.com

Molecular Research Center
5645 Montgomery Road
Cincinnati, OH 45212
(800) 462-9868 FAX: (513) 841-0080
(513) 841-0900
http://www.mrcgene.com

Molecular Simulations
9685 Scranton Road
San Diego, CA 92121
(800) 756-4674 FAX: (858) 458-0136
(858) 458-9990
http://www.msi.com

Monoject Disposable Syringes & Needles/Syrvet
16200 Walnut Street
Waukee, IA 50263
(800) 727-5203 FAX: (515) 987-5553
(515) 987-5554
http://www.syrvet.com

Monsanto Chemical
800 North Lindbergh Boulevard
St. Louis, MO 63167
(314) 694-1000 FAX: (314) 694-7625
http://www.monsanto.com

Moravek Biochemicals
577 Mercury Lane
Brea, CA 92821
(800) 447-0100 FAX: (714) 990-1824
(714) 990-2018
http://www.moravek.com

Moss
P.O. Box 189
Pasadena, MD 21122
(800) 932-6677 FAX: (410) 768-3971
(410) 768-3442
http://www.mosssubstrates.com

Motion Analysis
3617 Westwind Boulevard
Santa Rosa, CA 95403
(707) 579-6500 FAX: (707) 526-0629
http://www.motionanalysis.com

Mott
Farmington Industrial Park
84 Spring Lane
Farmington, CT 06032
(860) 747-6333 FAX: (860) 747-6739
http://www.mottcorp.com

MSI (Micron Separations)
See Osmonics

Multi Channel Systems
Markwiesenstrasse 55
72770 Reutlingen, Germany
(49) 7121-503010
FAX: (49) 7121-503011
http://www.multichannelsystems.com

Multiple Peptide Systems
3550 General Atomics Court
San Diego, CA 92121
(800) 338-4965 FAX: (800) 654-5592
(858) 455-3710 FAX: (858) 455-3713
http://www.mps-sd.com

Murex Diagnostics
3075 Northwoods Circle
Norcross, GA 30071
(707) 662-0660 FAX: (770) 447-4989

MWG-Biotech
Anzinger Str. 7
D-85560 Ebersberg, Germany
(49) 8092-82890 FAX: (49)
8092-21084
http://www.mwg_biotech.com

Myriad Industries
3454 E Street
San Diego, CA 92102
(800) 999-6777 FAX: (619) 232-4819
(619) 232-6700
http://www.myriadindustries.com

Nacalai Tesque
Nijo Karasuma, Nakagyo-ku
Kyoto 604, Japan
81-75-251-1723
FAX: 81-75-251-1762
http://www.nacalai.co.jp

Nalge Nunc International
Subsidiary of Sybron International
75 Panorama Creek Drive
P.O. Box 20365
Rochester, NY 14602
(800) 625-4327 FAX: (716) 586-8987
(716) 264-9346
http://www.nalgenunc.com

Nanogen
10398 Pacific Center Court
San Diego, CA 92121
(858) 410-4600 FAX: (858) 410-4848
http://www.nanogen.com

Nanoprobes
95 Horse Block Road
Yaphank, NY 11980
(877) 447-6266 FAX: (631) 205-9493
(631) 205-9490
http://www.nanoprobes.com

Narishige USA
1710 Hempstead Turnpike
East Meadow, NY 11554
(800) 445-7914 FAX: (516) 794-0066
(516) 794-8000
http://www.narishige.co.jp

National Bag Company
2233 Old Mill Road
Hudson, OH 44236
(800) 247-6000 FAX: (330) 425-9800
(330) 425-2600
http://www.nationalbag.com

National Band and Tag
Department X 35, Box 72430
Newport, KY 41032
(606) 261-2035 FAX: (800) 261-8247
https://www.nationalband.com

National Biosciences
See Molecular Biology Insights

National Diagnostics
305 Patton Drive
Atlanta, GA 30336
(800) 526-3867 FAX: (404) 699-2077
(404) 699-2121
http://www.nationaldiagnostics.com

National Institute of Standards and Technology
100 Bureau Drive
Gaithersburg, MD 20899
(301) 975-NIST FAX: (301) 926-1630
http://www.nist.gov

National Instruments
11500 North Mopac Expressway
Austin, TX 78759
(512) 794-0100 FAX: (512) 683-8411
www.ni.com

National Labnet
See Labnet International

National Scientific Instruments
975 Progress Circle
Lawrenceville, GA 300243
(800) 332-3331 FAX: (404) 339-7173
http://www.nationalscientific.com

National Scientific Supply
1111 Francisco Bouldvard East
San Rafael, CA 94901
(800) 525-1779 FAX: (415) 459-2954
(415) 459-6070
http://www.nat-sci.com

Naz-Dar-KC Chicago
Nazdar
1087 N. North Branch Street
Chicago, IL 60622
(800) 736-7636 FAX: (312) 943-8215
(312) 943-8338
http://www.nazdar.com

NB Labs
1918 Avenue A
Denison, TX 75021
(903) 465-2694 FAX: (903) 463-5905
http://www.nblabslarry.com

NEB
See New England Biolabs

NEN Life Science Products
549 Albany Street
Boston, MA 02118
(800) 551-2121 FAX: (617) 451-8185
(617) 350-9075
http://www.nen.com

NEN Research Products, Dupont (UK)
Diagnostics and Biotechnology Systems
Wedgewood Way
Stevenage, Hertfordshire SG1 4QN, UK
44-1438-734831
44-1438-734000
FAX: 44-1438-734836
http://www.dupont.com

Neogen
628 Winchester Road
Lexington, KY 40505 (800) 477-8201
FAX: (606) 255-5532
(606) 254-1221
http://www.neogen.com

Neosystems
380, 11012 Macleod Trail South
Calgary, Alberta T2J 6A5 Canada
(403) 225-9022 FAX: (403) 225-9025
http://www.neosystems.com

Neuralynx
2434 North Pantano Road
Tucson, AZ 85715
(520) 722-8144 FAX: (520) 722-8163
http://www.neuralynx.com

Neuro Probe
16008 Industrial Drive
Gaithersburg, MD 20877
(301) 417-0014 FAX: (301) 977-5711
http://www.neuroprobe.com

Neurocrine Biosciences
10555 Science Center Drive
San Diego, CA 92121
(619) 658-7600 FAX: (619) 658-7602
http://www.neurocrine.com

Nevtek
HCR03, Box 99
Burnsville, VA 24487
(540) 925-2322 FAX: (540) 925-2323
http://www.nevtek.com

New Brunswick Scientific
44 Talmadge Road
Edison, NJ 08818
(800) 631-5417 FAX: (732) 287-4222
(732) 287-1200
http://www.nbsc.com

New England Biolabs (NEB)
32 Tozer Road
Beverly, MA 01915 (800) 632-5227
FAX: (800) 632-7440
http://www.neb.com

New England Nuclear (NEN)
See NEN Life Science Products

New MBR
Gubelstrasse 48
CH8050 Zurich, Switzerland
(41) 1-313-0703

Newark Electronics
4801 N. Ravenswood Avenue
Chicago, IL 60640
(800) 4-NEWARK FAX: (773) 907-5339
(773) 784-5100
http://www.newark.com

Newell Rubbermaid
29 E. Stephenson Street
Freeport, IL 61032
(815) 235-4171 FAX: (815) 233-8060
http://www.newellco.com

Newport Biosystems
1860 Trainor Street
Red Bluff, CA 96080
(530) 529-2448 FAX: (530) 529-2648

Newport
1791 Deere Avenue
Irvine, CA 92606
(800) 222-6440 FAX: (949) 253-1800
(949) 253-1462
http://www.newport.com

Nexin Research B.V.
P.O. Box 16
4740 AA Hoeven, The Netherlands
(31) 165-503172
FAX: (31) 165-502291

NIAID
See Bio-Tech Research Laboratories

Nichols Institute Diagnostics
33051 Calle Aviador
San Juan Capistrano, CA 92675
(800) 286-4NID FAX: (949) 240-5273
(949) 728-4610
http://www.nicholsdiag.com

Nichols Scientific Instruments
3334 Brown Station Road
Columbia, MO 65202
(573) 474-5522 FAX: (603) 215-7274
http://home.beseen.com technology/nsi_technology

Nicolet Biomedical Instruments
5225 Verona Road, Building 2
Madison, WI 53711
(800) 356-0007 FAX: (608) 441-2002
(608) 273-5000
http://nicoletbiomedical.com

N.I.G.M.S. (National Institute of General Medical Sciences)
See Coriell Institute for Medical Research

Nikon
Science and Technologies Group
1300 Walt Whitman Road
Melville, NY 11747
(516) 547-8500 FAX: (516) 547-4045
http://www.nikonusa.com

Nippon Gene
1-29, Ton-ya-machi
Toyama 930, Japan
(81) 764-51-6548
FAX: (81) 764-51-6547

Noldus Information Technology
751 Miller Drive
Suite E-5
Leesburg, VA 20175
(800) 355-9541 FAX: (703) 771-0441
(703) 771-0440
http://www.noldus.com

Nordion International
See MDS Nordion

North American Biologicals (NABI)
16500 NW 15th Avenue
Miami, FL 33169
(800) 327-7106 (305) 625-5305
http://www.nabi.com

North American Reiss
See Reiss

Northwestern Bottle
24 Walpole Park South
Walpole, MA 02081
(508) 668-8600 FAX: (508) 668-7790

NOVA Biomedical
Nova Biomedical 200
Prospect Street Waltham, MA 02454
(800) 822-0911 FAX: (781) 894-5915
http://www.novabiomedical.com

Novagen
601 Science Drive
Madison, WI 53711
(800) 526-7319 FAX: (608) 238-1388
(608) 238-6110
http://www.novagen.com

Novartis
59 Route 10
East Hanover, NJ 07936
(800)526-0175 FAX: (973) 781-6356
http://www.novartis.com

Novartis Biotechnology
3054 Cornwallis Road
Research Triangle Park, NC 27709
(888) 462-7288 FAX: (919) 541-8585
http://www.novartis.com

Nova Sina AG
Subsidiary of Airflow Lufttechnik GmbH
Kleine Heeg 21
52259 Rheinbach, Germany
(49) 02226 920-0
FAX: (49) 02226 9205-11

Novex/Invitrogen
1600 Faraday
Carlsbad, CA 92008
(800) 955-6288 FAX: (760) 603-7201
http://www.novex.com

Novo Nordisk Biochem
77 Perry Chapel Church Road
Franklington, NC 27525
(800) 879-6686 FAX: (919) 494-3450
(919) 494-3000
http://www.novo.dk

Novo Nordisk BioLabs
See Novo Nordisk Biochem

Novocastra Labs
Balliol Business Park West
Benton Lane
Newcastle-upon-Tyne
Tyne and Wear NE12 8EW, UK
(44) 191-215-0567
FAX: (44) 191-215-1152
http://www.novocastra.co.uk

Novus Biologicals
P.O. Box 802
Littleton, CO 80160
(888) 506-6887 FAX: (303) 730-1966
http://www.novus-biologicals.com/main.html

NPI Electronic
Hauptstrasse 96
D-71732 Tamm, Germany
(49) 7141-601534
FAX: (49) 7141-601266
http://www.npielectronic.com

NSG Precision Cells
195G Central Avenue
Farmingdale, NY 11735
(516) 249-7474 FAX: (516) 249-8575
http://www.nsgpci.com

Nu Chek Prep
109 West Main
P.O. Box 295
Elysian, MN 56028
(800) 521-7728 FAX: (507) 267-4790
(507) 267-4689

Nuclepore
See Costar

Numonics
101 Commerce Drive
Montgomeryville, PA 18936
(800) 523-6716 FAX: (215) 361-0167
(215) 362-2766
http://www.interactivewhiteboards.com

NYCOMED AS Pharma
c/o Accurate Chemical & Scientific
300 Shames Drive
Westbury, NY 11590
(800) 645-6524 FAX: (516) 997-4948
(516) 333-2221
http://www.accuratechemical.com

Nycomed Amersham
Health Care Division
101 Carnegie Center
Princeton, NJ 08540
(800) 832-4633 FAX: (800) 807-2382
(609) 514-6000
http://www.nycomed-amersham.com

Nyegaard
Herserudsvagen 5254
S-122 06 Lidingo, Sweden
(46) 8-765-2930

Ohmeda Catheter Products
See Datex-Ohmeda

Ohwa Tsusbo
Hiby Dai Building
1-2-2 Uchi Saiwai-cho
Chiyoda-ku
Tokyo 100, Japan
03-3591-7348 FAX: 03-3501-9001

Oligos Etc.
29970 SW Town Centre
Loop West, Suite B419
Wilsonville, OR 97070
(800) 888-2358 FAX: (800) 869-0813

Olis Instruments
130 Conway Drive
Bogart, GA 30622
(706) 353-6547 (800) 852-3504
http://www.olisweb.com

Olympus America
2 Corporate Center Drive
Melville, NY 11747
(800) 645-8160 FAX: (516) 844-5959
(516) 844-5000
http://www.olympusamerica.com

Omega Engineering
One Omega Drive
P.O. Box 4047
Stamford, CT 06907
(800) 848-4286 FAX: (203) 359-7700
(203) 359-1660
http://www.omega.com

Omega Optical
3 Grove Street
P.O. Box 573
Brattleboro, VT 05302
(802) 254-2690 FAX: (802) 254-3937
http://www.omegafilters.com

Omnetics Connector Corporation
7260 Commerce Circle
East Minneapolis, MN 55432
(800) 343-0025 (763) 572-0656
FAX: (763) 572-3925
http://www.omnetics.com

Omni International
6530 Commerce Court
Warrenton, VA 20187
(800) 776-4431 FAX: (540) 347-5352
(540) 347-5331
http://www.omni-inc.com

Omnion
2010 Energy Drive
P.O. Box 879
East Troy, WI 53120
(262) 642-7200 FAX: (262) 642-7760
http://www.omnion.com

Omnitech Electronics
See AccuScan Instruments

Oncogene Research Products
P.O. Box Box 12087
La Jolla, CA 92039-2087
(800) 662-2616 FAX: (800) 766-0999
http://www.apoptosis.com

Oncogene Science
See OSI Pharmaceuticals

Oncor
See Intergen

Operon Technologies
1000 Atlantic Avenue
Alameda, CA 94501
(800) 688-2248 FAX: (510) 865-5225
(510) 865-8644
http://www.operon.com

Optiscan
P.O. Box 1066
Mount Waverly MDC, Victoria
Australia 3149
61-3-9538 3333 FAX: 61-3-9562 7742
http://www.optiscan.com.au

Optomax
9 Ash Street
P.O. Box 840
Hollis, NH 03049
(603) 465-3385 FAX: (603) 465-2291

Opto-Line Associates
265 Ballardvale Street
Wilmington, MA 01887
(978) 658-7255 FAX: (978) 658-7299
http://www.optoline.com

Orbigen
6827 Nancy Ridge Drive
San Diego, CA 92121
(866) 672-4436 (858) 362-2030
(858) 362-2026
http://www.orbigen.com

Oread BioSaftey
1501 Wakarusa Drive
Lawrence, KS 66047
(800) 447-6501 FAX: (785) 749-1882
(785) 749-0034
http://www.oread.com

Organomation Associates
266 River Road West
Berlin, MA 01503
(888) 978-7300 FAX: (978)838-2786
(978) 838-7300
http://www.organomation.com

Organon
375 Mount Pleasant Avenue
West Orange, NJ 07052
(800) 241-8812 FAX: (973) 325-4589
(973) 325-4500
http://www.organon.com

Organon Teknika (Canada)
30 North Wind Place
Scarborough, Ontario
M1S 3R5 Canada
(416) 754-4344 FAX: (416) 754-4488
http://www.organonteknika.com

Organon Teknika Cappel
100 Akzo Avenue
Durham, NC 27712
(800) 682-2666 FAX: (800) 432-9682
(919) 620-2000 FAX: (919) 620-2107
http://www.organonteknika.com

Oriel Corporation of America
150 Long Beach Boulevard
Stratford, CT 06615
(203) 377-8282 FAX: (203) 378-2457
http://www.oriel.com

OriGene Technologies
6 Taft Court, Suite 300
Rockville, MD 20850
(888) 267-4436 FAX: (301) 340-9254
(301) 340-3188
http://www.origene.com

OriginLab
One Roundhouse Plaza
Northhampton, MA 01060
(800) 969-7720 FAX: (413) 585-0126
http://www.originlab.com

Orion Research
500 Cummings Center
Beverly, MA 01915
(800) 225-1480 FAX: (978) 232-6015
(978) 232-6000
http://www.orionres.com

Ortho Diagnostic Systems
Subsidiary of Johnson & Johnson
1001 U.S. Highway 202
P.O. Box 350
Raritan, NJ 08869
(800) 322-6374 FAX: (908) 218-8582
(908) 218-1300

Ortho McNeil Pharmaceutical
Welsh & McKean Road
Spring House, PA 19477
(800) 682-6532
(215) 628-5000
http://www.orthomcneil.com

Oryza
200 Turnpike Road, Unit 5
Chelmsford, MA 01824
(978) 256-8183 FAX: (978) 256-7434
http://www.oryzalabs.com

OSI Pharmaceuticals
106 Charles Lindbergh Boulevard
Uniondale, NY 11553
(800) 662-2616 FAX: (516) 222-0114
(516) 222-0023
http://www.osip.com

Osmonics
135 Flanders Road
P.O. Box 1046
Westborough, MA 01581
(800) 444-8212 FAX: (508) 366-5840
(508) 366-8212
http://www.osmolabstore.com

Oster Professional Products
150 Cadillac Lane
McMinnville, TN 37110
(931) 668-4121 FAX: (931) 668-4125
http://www.sunbeam.com

Out Patient Services
1260 Holm Road
Petaluma, CA 94954
(800) 648-1666 FAX: (707) 762-7198
(707) 763-1581

OWL Scientific Plastics
See OWL Separation Systems

OWL Separation Systems
55 Heritage Avenue
Portsmouth, NH 03801
(800) 242-5560 FAX: (603) 559-9258
(603) 559-9297
http://www.owlsci.com

Oxford Biochemical Research
P.O. Box 522
Oxford, MI 48371
(800) 692-4633 FAX: (248) 852-4466
http://www.oxfordbiomed.com

Oxford GlycoSystems
See Glyco

Oxford Instruments
Old Station Way
Eynsham
Witney, Oxfordshire OX8 1TL, UK
(44) 1865-881437
FAX: (44) 1865-881944
http://www.oxinst.com

Oxford Labware
See Kendall

Oxford Molecular Group
Oxford Science Park
The Medawar Centre
Oxford OX4 4GA, UK
(44) 1865-784600
FAX: (44) 1865-784601
http://www.oxmol.co.uk

Oxford Molecular Group
2105 South Bascom Avenue, Suite 200
Campbell, CA 95008
(800) 876-9994 FAX: (408) 879-6302
(408) 879-6300
http://www.oxmol.com

OXIS International
6040 North Cutter Circle
Suite 317
Portland, OR 97217
(800) 547-3686 FAX: (503) 283-4058
(503) 283-3911
http://www.oxis.com

Oxoid
800 Proctor Avenue
Ogdensburg, NY 13669
(800) 567-8378 FAX: (613) 226-3728
http://www.oxoid.ca

Oxoid
Wade Road
Basingstoke, Hampshire RG24 8PW, UK
(44) 1256-841144
FAX: (4) 1256-814626
http://www.oxoid.ca

Oxyrase
P.O. Box 1345
Mansfield, OH 44901
(419) 589-8800 FAX: (419) 589-9919
http://www.oxyrase.com

Ozyme
10 Avenue Ampère
Montigny de Bretoneux
78180 France
(33) 13-46-02-424
FAX: (33) 13-46-09-212
http://www.ozyme.fr

PAA Laboratories
2570 Route 724
P.O. Box 435
Parker Ford, PA 19457
(610) 495-9400 FAX: (610) 495-9410
http://www.paa-labs.com

Pacer Scientific
5649 Valley Oak Drive
Los Angeles, CA 90068
(323) 462-0636 FAX: (323) 462-1430
http://www.pacersci.com

Pacific Bio-Marine Labs
P.O. Box 1348
Venice, CA 90294
(310) 677-1056 FAX: (310) 677-1207

Packard Instrument
800 Research Parkway
Meriden, CT 06450
(800) 323-1891 FAX: (203) 639-2172
(203) 238-2351
http://www.packardinst.com

Padgett Instrument
1730 Walnut Street
Kansas City, MO 64108
(816) 842-1029

Pall Filtron
50 Bearfoot Road
Northborough, MA 01532
(800) FILTRON FAX: (508) 393-1874
(508) 393-1800

Pall-Gelman
25 Harbor Park Drive
Port Washington, NY 11050
(800) 289-6255 FAX: (516) 484-2651
(516) 484-3600
http://www.pall.com

PanVera
545 Science Drive
Madison, WI 53711
(800) 791-1400 FAX: (608) 233-3007
(608) 233-9450
http://www.panvera.com

Parke-Davis
See Warner-Lambert

A3

Parr Instrument
211 53rd Street
Moline, IL 61265
(800) 872-7720 FAX: (309) 762-9453
(309) 762-7716
http://www.parrinst.com

Partec
Otto Hahn Strasse 32
D-48161 Munster, Germany
(49) 2534-8008-0
FAX: (49) 2535-8008-90

PCR
See Archimica Florida

PE Biosystems
850 Lincoln Centre Drive
Foster City, CA 94404
(800) 345-5224 FAX: (650) 638-5884
(650) 638-5800
http://www.pebio.com

Pel-Freez Biologicals
219 N. Arkansas
P.O. Box 68
Rogers, AR 72757
(800) 643-3426 FAX: (501) 636-3562
(501) 636-4361
http://www.pelfreez-bio.com

Pel-Freez Clinical Systems
Subsidiary of Pel-Freez Biologicals
9099 N. Deerbrook Trail
Brown Deer, WI 53223
(800) 558-4511 FAX: (414) 357-4518
(414) 357-4500
http://www.pelfreez-bio.com

Peninsula Laboratories
601 Taylor Way
San Carlos, CA 94070
(800) 650-4442 FAX: (650) 595-4071
(650) 592-5392
http://www.penlabs.com

Pentex
24562 Mando Drive
Laguna Niguel, CA 92677
(800) 382-4667 FAX: (714) 643-2363
http://www.pentex.com

PeproTech
5 Crescent Avenue
P.O. Box 275
Rocky Hill, NJ 08553
(800) 436-9910 FAX: (609) 497-0321
(609) 497-0253
http://www.peprotech.com

Peptide Institute
4-1-2 Ina, Minoh-shi
Osaka 562-8686, Japan
81-727-29-4121 FAX: 81-727-29-4124
http://www.peptide.co.jp

Peptide Laboratory
4175 Lakeside Drive
Richmond, CA 94806
(800) 858-7322 FAX: (510) 262-9127
(510) 262-0800
http://www.peptidelab.com

Peptides International
11621 Electron Drive
Louisville, KY 40299
(800) 777-4779 FAX: (502) 267-1329
(502) 266-8787
http://www.pepnet.com

Perceptive Science Instruments
2525 South Shore Boulevard, Suite 100
League City, TX 77573
(281) 334-3027 FAX: (281) 538-2222
http://www.persci.com

Perimed
4873 Princeton Drive
North Royalton, OH 44133
(440) 877-0537 FAX: (440) 877-0534
http://www.perimed.se

Perkin-Elmer
761 Main Avenue
Norwalk, CT 06859
(800) 762-4002 FAX: (203) 762-6000
(203) 762-1000
http://www.perkin-elmer.com
See also PE Biosystems

PerSeptive Bioresearch Products
See PerSeptive BioSystems

PerSeptive BioSystems
500 Old Connecticut Path
Framingham, MA 01701
(800) 899-5858 FAX: (508) 383-7885
(508) 383-7700
http://www.pbio.com

PerSeptive Diagnostic
See PE Biosystems
(800) 343-1346

Pettersson Elektronik AB
Tallbacksvagen 51
S-756 45 Uppsala, Sweden
(46) 1830-3880 FAX: (46) 1830-3840
http://www.bahnhof.se/~pettersson

PGC Scientifics
7311 Governors Way
Frederick, MD 21704
(800) 424-3300 FAX: (800) 662-1112
(301) 620-7777 FAX: (301) 620-7497
http://www.pgcscientifics.com

Pharmacia Biotech
See Amersham Pharmacia Biotech

Pharmacia Diagnostics
See Wallac

Pharmacia LKB Biotech
See Amersham Pharmacia Biotech

Pharmacia LKB Biotechnology
See Amersham Pharmacia Biotech

Pharmacia LKB Nuclear
See Wallac

Pharmaderm Veterinary Products
60 Baylis Road
Melville, NY 11747
(800) 432-6673
http://www.pharmaderm.com

Pharmed (Norton)
Norton Performance Plastics
See Saint-Gobain Performance Plastics

PharMingen
See BD PharMingen

Phenomex
2320 W. 205th Street
Torrance, CA 90501
(310) 212-0555 FAX: (310) 328-7768
http://www.phenomex.com

PHLS Centre for Applied Microbiology and Research
See European Collection of Animal Cell Cultures (ECACC)

Phoenix Flow Systems
11575 Sorrento Valley Road, Suite 208
San Diego, CA 92121
(800) 886-3569 FAX: (619) 259-5268
(619) 453-5095
http://www.phnxflow.com

Phoenix Pharmaceutical
4261 Easton Road, P.O. Box 6457
St. Joseph, MO 64506
(800) 759-3644 FAX: (816) 364-4969
(816) 364-5777
http://www.phoenixpharmaceutical.com

Photometrics
See Roper Scientific

Photon Technology International
1 Deerpark Drive, Suite F
Monmouth Junction, NJ 08852
(732) 329-0910 FAX: (732) 329-9069
http://www.pti-nj.com

Physik Instrumente
Polytec PI
23 Midstate Drive, Suite 212
Auburn, MA 01501
(508) 832-3456 FAX: (508) 832-0506
http://www.polytecpi.com

Physitemp Instruments
154 Huron Avenue
Clifton, NJ 07013
(800) 452-8510 FAX: (973) 779-5954
(973) 779-5577
http://www.physitemp.com

Pico Technology
The Mill House, Cambridge Street
St. Neots, Cambridgeshire
PE19 1QB, UK
(44) 1480-396-395
FAX: (44) 1480-396-296
www.picotech.com

Pierce Chemical
P.O. Box 117
3747 Meridian Road
Rockford, IL 61105
(800) 874-3723 FAX: (800) 842-5007
FAX: (815) 968-7316
http://www.piercenet.com

Pierce & Warriner
44, Upper Northgate Street
Chester, Cheshire CH1 4EF, UK
(44) 1244 382 525
FAX: (44) 1244 373 212
http://www.piercenet.com

Pilling Weck Surgical
420 Delaware Drive
Fort Washington, PA 19034
(800) 523-2579 FAX: (800) 332-2308
www.pilling-weck.com

PixelVision
A division of Cybex Computer Products
14964 NW Greenbrier Parkway
Beaverton, OR 97006
(503) 629-3210 FAX: (503) 629-3211
http://www.pixelvision.com

P.J. Noyes
P.O. Box 381
89 Bridge Street
Lancaster, NH 03584
(800) 522-2469 FAX: (603) 788-3873
(603) 788-4952
http://www.pjnoyes.com

Plas-Labs
917 E. Chilson Street
Lansing, MI 48906
(800) 866-7527 FAX: (517) 372-2857
(517) 372-7177
http://www.plas-labs.com

Plastics One
6591 Merriman Road, Southwest
P.O. Box 12004
Roanoke, VA 24018
(540) 772-7950 FAX: (540) 989-7519
http://www.plastics1.com

Platt Electric Supply
2757 6th Avenue South
Seattle, WA 98134
(206) 624-4083 FAX: (206) 343-6342
http://www.platt.com

Plexon
6500 Greenville Avenue
Suite 730
Dallas, TX 75206
(214) 369-4957 FAX: (214) 369-1775
http://www.plexoninc.com

Polaroid
784 Memorial Drive
Cambridge, MA 01239
(800) 225-1618 FAX: (800) 832-9003
(781) 386-2000
http://www.polaroid.com

Polyfiltronics
136 Weymouth St.
Rockland, MA 02370
(800) 434-7659 FAX: (781) 878-0822
(781) 878-1133
http://www.polyfiltronics.com

Polylabo Paul Block
Parc Tertiare de la Meinau
10, rue de la Durance
B.P. 36
67023 Strasbourg Cedex 1
Strasbourg, France
33-3-8865-8020
FAX: 33-3-8865-8039

PolyLC
9151 Rumsey Road, Suite 180
Columbia, MD 21045
(410) 992-5400 FAX: (410) 730-8340

Polymer Laboratories
Amherst Research Park
160 Old Farm Road
Amherst, MA 01002
(800) 767-3963 FAX: (413) 253-2476
http://www.polymerlabs.com

Polymicro Technologies
18019 North 25th Avenue
Phoenix, AZ 85023
(602) 375-4100 FAX: (602) 375-4110
http://www.polymicro.com

Polyphenols AS
Hanabryggene Technology Centre
Hanaveien 4-6
4327 Sandnes, Norway
(47) 51-62-0990
FAX: (47) 51-62-51-82
http://www.polyphenols.com

Polysciences
400 Valley Road
Warrington, PA 18976
(800) 523-2575 FAX: (800) 343-3291
http://www.polysciences.com

Polyscientific
70 Cleveland Avenue
Bayshore, NY 11706
(516) 586-0400 FAX: (516) 254-0618

Polytech Products
285 Washington Street
Somerville, MA 02143
(617) 666-5064 FAX: (617) 625-0975

Polytron
8585 Grovemont Circle
Gaithersburg, MD 20877
(301) 208-6597 FAX: (301) 208-8691
http://www.polytron.com

Popper and Sons
300 Denton Avenue
P.O. Box 128
New Hyde Park, NY 11040
(888) 717-7677 FAX: (800) 557-6773
(516) 248-0300 FAX: (516) 747-1188
http://www.popperandsons.com

Porphyrin Products
P.O. Box 31
Logan, UT 84323
(435) 753-1901 FAX: (435) 753-6731
http://www.porphyrin.com

Portex
See SIMS Portex Limited

Powderject Vaccines
585 Science Drive
Madison, WI 53711
(608) 231-3150 FAX: (608) 231-6990
http://www.powderject.com

Praxair
810 Jorie Boulevard
Oak Brook, IL 60521
(800) 621-7100
http://www.praxair.com

Precision Dynamics
13880 Del Sur Street
San Fernando, CA 91340
(800) 847-0670 FAX: (818) 899-4-45
http://www.pdcorp.com

Precision Scientific Laboratory Equipment
Division of Jouan
170 Marcel Drive
Winchester, VA 22602
(800) 621-8820 FAX: (540) 869-0130
(540) 869-9892
http://www.precisionsci.com

Primary Care Diagnostics
See Becton Dickinson Primary Care Diagnostics

Primate Products
1755 East Bayshore Road, Suite 28A
Redwood City, CA 94063
(650) 368-0663 FAX: (650) 368-0665
http://www.primateproducts.com

5 Prime → 3 Prime
See 2000 Eppendorf-5 Prime
http://www.5prime.com

Princeton Applied Research
PerkinElmer Instr.: Electrochemistry
801 S. Illinois
Oak Ridge, TN 37830
(800) 366-2741 FAX: (423) 425-1334
(423) 481-2442
http://www.eggpar.com

Princeton Instruments
A division of Roper Scientific
3660 Quakerbridge Road
Trenton, NJ 08619
(609) 587-9797 FAX: (609) 587-1970
http://www.prinst.com

Princeton Separations
P.O. Box 300
Aldephia, NJ 07710
(800) 223-0902 FAX: (732) 431-3768
(732) 431-3338

Prior Scientific
80 Reservoir Park Drive
Rockland, MA 02370
(781) 878-8442 FAX: (781) 878-8736
http://www.prior.com

PRO Scientific
P.O. Box 448
Monroe, CT 06468
(203) 452-9431 FAX: (203) 452-9753
http://www.proscientific.com

Professional Compounding Centers of America
9901 South Wilcrest Drive
Houston, TX 77099
(800) 331-2498 FAX: (281) 933-6227
(281) 933-6948
http://www.pccarx.com

Progen Biotechnik
Maass-Str. 30
69123 Heidelberg, Germany
(49) 6221-8278-0
FAX: (49) 6221-8278-23
http://www.progen.de

Prolabo
A division of Merck Eurolab
54 rue Roger Salengro
94126 Fontenay Sous Bois Cedex
France
33-1-4514-8500
FAX: 33-1-4514-8616
http://www.prolabo.fr

Proligo
2995 Wilderness Place Boulder, CO 80301
(888) 80-OLIGO FAX: (303) 801-1134
http://www.proligo.com

Promega
2800 Woods Hollow Road
Madison, WI 53711
(800) 356-9526 FAX: (800) 356-1970
(608) 274-4330 FAX: (608) 277-2516
http://www.promega.com

Protein Databases (PDI)
405 Oakwood Road
Huntington Station, NY 11746
(800) 777-6834 FAX: (516) 673-4502
(516) 673-3939

Protein Polymer Technologies
10655 Sorrento Valley Road
San Diego, CA 92121
(619) 558-6064 FAX: (619) 558-6477
http://www.ppti.com

Protein Solutions
391 G Chipeta Way
Salt Lake City, UT 84108
(801) 583-9301 FAX: (801) 583-4463
http://www.proteinsolutions.com

Prozyme
1933 Davis Street, Suite 207
San Leandro, CA 94577
(800) 457-9444 FAX: (510) 638-6919
(510) 638-6900
http://www.prozyme.com

PSI
See Perceptive Science Instruments

Pulmetrics Group
82 Beacon Street
Chestnut Hill, MA 02167
(617) 353-3833 FAX: (617) 353-6766

Purdue Frederick
100 Connecticut Avenue
Norwalk, CT 06850
(800) 633-4741 FAX: (203) 838-1576
(203) 853-0123
http://www.pharma.com

Purina Mills
LabDiet
P. O. Box 66812
St. Louis, MO 63166
(800) 227-8941 FAX: (314) 768-4894
http://www.purina-mills.com

Qbiogene
2251 Rutherford Road
Carlsbad, CA 92008
(800) 424-6101 FAX: (760) 918-9313
http://www.qbiogene.com

Qiagen
28159 Avenue Stanford
Valencia, CA 91355
(800) 426-8157 FAX: (800) 718-2056
http://www.qiagen.com

Quality Biological
7581 Lindbergh Drive
Gaithersburg, MD 20879
(800) 443-9331 FAX: (301) 840-5450
(301) 840-9331
http://www.qualitybiological.com

Quantum Appligene
Parc d'Innovation
Rue Geller de Kayserberg
67402 Illkirch, Cedex, France
(33) 3-8867-5425
FAX: (33) 3-8867-1945
http://www.quantum-appligene.com

Quantum Biotechnologies
See Qbiogene

Quantum Soft
Postfach 6613
CH-8023
Zürich, Switzerland
FAX: 41-1-481-69-51
profit@quansoft.com

Questcor Pharmaceuticals
26118 Research Road
Hayward, CA 94545
(510) 732-5551 FAX: (510) 732-7741
http://www.questcor.com

Quidel
10165 McKellar Court
San Diego, CA 92121
(800) 874-1517 FAX: (858) 546-8955
(858) 552-1100
http://www.quidel.com

R-Biopharm
7950 Old US 27 South
Marshall, MI 49068
(616) 789-3033 FAX: (616) 789-3070
http://www.r-biopharm.com

R. C. Electronics
6464 Hollister Avenue
Santa Barbara, CA 93117
(805) 685-7770 FAX: (805) 685-5853
http://www.rcelectronics.com

R & D Systems
614 McKinley Place NE
Minneapolis, MN 55413
(800) 343-7475 FAX: (612) 379-6580
(612) 379-2956
http://www.rndsystems.com

R & S Technology
350 Columbia Street
Peacedale, RI 02880
(401) 789-5660 FAX: (401) 792-3890
http://www.septech.com

RACAL Health and Safety
See 3M
7305 Executive Way
Frederick, MD 21704
(800) 692-9500 FAX: (301) 695-8200

Radiometer America
811 Sharon Drive
Westlake, OH 44145
(800) 736-0600 FAX: (440) 871-2633
(440) 871-8900
http://www.rameusa.com

Radiometer A/S
The Chemical Reference Laboratory
kandevej 21
DK-2700 Brnshj, Denmark
45-3827-3827 FAX: 45-3827-2727

Radionics
22 Terry Avenue
Burlington, MA 01803
(781) 272-1233 FAX: (781) 272-2428
http://www.radionics.com

Radnoti Glass Technology
227 W. Maple Avenue
Monrovia, CA 91016
(800) 428-l4l6 FAX: (626) 303-2998
(626) 357-8827
http://www.radnoti.com

Rainin Instrument
Rainin Road
P.O. Box 4026
Woburn, MA 01888
(800)-4-RAININ FAX: (781) 938-1152
(781) 935-3050
http://www.rainin.com

Rank Brothers
56 High Street
Bottisham, Cambridge
CB5 9DA UK
(44) 1223 811369
FAX: (44) 1223 811441
http://www.rankbrothers.com

Rapp Polymere
Ernst-Simon Strasse 9
D 72072 Tübingen, Germany
(49) 7071-763157
FAX: (49) 7071-763158
http://www.rapp-polymere.com

Raven Biological Laboratories
8607 Park Drive
P.O. Box 27261
Omaha, NE 68127
(800) 728-5702 FAX: (402) 593-0995
(402) 593-0781
http://www.ravenlabs.com

Razel Scientific Instruments
100 Research Drive
Stamford, CT 06906
(203) 324-9914 FAX: (203) 324-5568

Reagents International
See Biotech Source

Receptor Biology
10000 Virginia Manor Road, Suite 360
Beltsville, MD 20705
(888) 707-4200 FAX: (301) 210-6266
(301) 210-4700
http://www.receptorbiology.com

Regis Technologies
8210 N. Austin Avenue
Morton Grove, IL 60053
(800) 323-8144 FAX: (847) 967-1214
(847) 967-6000
http://www.registech.com

Reichert Ophthalmic Instruments
P.O. Box 123
Buffalo, NY 14240
(716) 686-4500 FAX: (716) 686-4545
http://www.reichert.com

Reiss
1 Polymer Place
P.O. Box 60 Blackstone, VA 23824
(800) 356-2829 FAX: (804) 292-1757
(804) 292-1600
http://www.reissmfg.com

Remel
12076 Santa Fe Trail Drive
P.O. Box 14428
Shawnee Mission, KS 66215
(800) 255-6730 FAX: (800) 621-8251
(913) 888-0939 FAX: (913) 888-5884
http://www.remelinc.com

Reming Bioinstruments
6680 County Route 17
Redfield, NY 13437
(315) 387-3414 FAX: (315) 387-3415

RepliGen
117 Fourth Avenue
Needham, MA 02494
(800) 622-2259 FAX: (781) 453-0048
(781) 449-9560
http://www.repligen.com

Research Biochemicals
1 Strathmore Road
Natick, MA 01760
(800) 736-3690 FAX: (800) 736-2480
(508) 651-8151 FAX: (508) 655-1359
http://www.resbio.com

Research Corporation Technologies
101 N. Wilmot Road, Suite 600
Tucson, AZ 85711
(520) 748-4400 FAX: (520) 748-0025
http://www.rctech.com

Research Diagnostics
Pleasant Hill Road
Flanders, NJ 07836
(800) 631-9384 FAX: (973) 584-0210
(973) 584-7093
http://www.researchd.com

Research Diets
121 Jersey Avenue
New Brunswick, NJ 08901
(877) 486-2486 FAX: (732) 247-2340
(732) 247-2390
http://www.researchdiets.com

Research Genetics
2130 South Memorial Parkway
Huntsville, AL 35801
(800) 533-4363 FAX: (256) 536-9016
(256) 533-4363
http://www.resgen.com

Research Instruments
Kernick Road Pernryn
Cornwall TR10 9DQ, UK
(44) 1326-372-753
FAX: (44) 1326-378-783
http://www.research-instruments.com

Research Organics
4353 E. 49th Street
Cleveland, OH 44125
(800) 321-0570 FAX: (216) 883-1576
(216) 883-8025
http://www.resorg.com

Research Plus
P.O. Box 324
Bayonne, NJ 07002
(800) 341-2296 FAX: (201) 823-9590
(201) 823-3592
http://www.researchplus.com

Research Products International
410 N. Business Center Drive
Mount Prospect, IL 60056
(800) 323-9814 FAX: (847) 635-1177
(847) 635-7330
http://www.rpicorp.com

Research Triangle Institute
P.O. Box 12194
Research Triangle Park, NC 27709
(919) 541-6000 FAX: (919) 541-6515
http://www.rti.org

Restek
110 Benner Circle
Bellefonte, PA 16823
(800) 356-1688 FAX: (814) 353-1309
(814) 353-1300
http://www.restekcorp.com

Rheodyne
P.O. Box 1909
Rohnert Park, CA 94927
(707) 588-2000 FAX: (707) 588-2020
http://www.rheodyne.com

Rhone Merieux
See Merial Limited

Rhone-Poulenc
2 T W Alexander Drive
P.O. Box 12014
Research Triangle Park, NC 08512
(919) 549-2000 FAX: (919) 549-2839
http://www.Rhone-Poulenc.com
Also see Aventis

Rhone-Poulenc Rorer
500 Arcola Road
Collegeville, PA 19426
(800) 727-6737 FAX: (610) 454-8940
(610) 454-8975
http://www.rp-rorer.com

Rhone-Poulenc Rorer
Centre de Recherche de Vitry-Alfortville
13 Quai Jules Guesde, BP14 94403
Vitry Sur Seine, Cedex, France
(33) 145-73-85-11
FAX: (33) 145-73-81-29
http://www.rp-rorer.com

Ribi ImmunoChem Research
563 Old Corvallis Road
Hamilton, MT 59840
(800) 548-7424 FAX: (406) 363-6129
(406) 363-3131
http://www.ribi.com

RiboGene
See Questcor Pharmaceuticals

Ricca Chemical
448 West Fork Drive
Arlington, TX 76012
(888) GO-RICCA FAX: (800) RICCA-93
(817) 461-5601
http://www.riccachemical.com

Richard-Allan Scientific
225 Parsons Street
Kalamazoo, MI 49007
(800) 522-7270 FAX: (616) 345-3577
(616) 344-2400
http://www.rallansci.com

Richelieu Biotechnologies
11 177 Hamon
Montral, Quebec
H3M 3E4 Canada
(802) 863-2567 FAX: (802) 862-2909
http://www.richelieubio.com

Richter Enterprises
20 Lake Shore Drive
Wayland, MA 01778
(508) 655-7632 FAX: (508) 652-7264
http://www.richter-enterprises.com

Riese Enterprises
BioSure Division
12301 G Loma Rica Drive
Grass Valley, CA 95945
(800) 345-2267 FAX: (916) 273-5097
(916) 273-5095
http://www.biosure.com

Robbins Scientific
1250 Elko Drive
Sunnyvale, CA 94086
(800) 752-8585 FAX: (408) 734-0300
(408) 734-8500
http://www.robsci.com

Roboz Surgical Instruments
9210 Corporate Boulevard, Suite 220
Rockville, MD 20850
(800) 424-2984 FAX: (301) 590-1290
(301) 590-0055

Roche Diagnostics
9115 Hague Road
P.O. Box 50457
Indianapolis, IN 46256
(800) 262-1640 FAX: (317) 845-7120
(317) 845-2000
http://www.roche.com

Roche Molecular Systems
See Roche Diagnostics

Rocklabs
P.O. Box 18-142
Auckland 6, New Zealand
(64) 9-634-7696
FAX: (64) 9-634-7696
http://www.rocklabs.com

Rockland
P.O. Box 316
Gilbertsville, PA 19525
(800) 656-ROCK FAX: (610) 367-7825
(610) 369-1008
http://www.rockland-inc.com

Rohm
Chemische Fabrik
Kirschenallee
D-64293 Darmstadt, Germany
(49) 6151-1801 FAX: (49) 6151-1802
http://www.roehm.com

Roper Scientific
3440 East Brittania Drive, Suite 100
Tucson, AZ 85706
(520) 889-9933 FAX: (520) 573-1944
http://www.roperscientific.com

Rosetta Inpharmatics
12040 115th Avenue NE
Kirkland, WA 98034
(425) 820-8900 FAX: (425) 820-5757
http://www.rii.com

ROTH-SOCHIEL
3 rue de la Chapelle
Lauterbourg
67630 France
(33) 03-88-94-82-42
FAX: (33) 03-88-54-63-93

Rotronic Instrument
160 E. Main Street
Huntington, NY 11743
(631) 427-3898 FAX: (631) 427-3902
http://www.rotronic-usa.com

Roundy's
23000 Roundy Drive
Pewaukee, WI 53072
(262) 953-7999 FAX: (262) 953-7989
http://www.roundys.com

RS Components
Birchington Road
Weldon Industrial Estate
Corby, Northants NN17 9RS, UK
(44) 1536 201234
FAX: (44) 1536 405678
http://www.rs-components.com

Rubbermaid
See Newell Rubbermaid

SA Instrumentation
1437 Tzena Way
Encinitas, CA 92024
(858) 453-1776 FAX: (800)-266-1776
http://www.sainst.com

Safe Cells
See Bionique Testing Labs

Sage Instruments
240 Airport Boulevard
Freedom, CA 95076
831-761-1000 FAX: 831-761-1008
http://www.sageinst.com

Sage Laboratories
11 Huron Drive
Natick, MA 01760
(508) 653-0844 FAX: 508-653-5671
http://www.sagelabs.com

Saint-Gobain Performance Plastics
P.O. Box 3660
Akron, OH 44309
(330) 798-9240 FAX: (330) 798-6968
http://www.nortonplastics.com

San Diego Instruments
7758 Arjons Drive
San Diego, CA 92126
(858) 530-2600 FAX: (858) 530-2646
http://www.sd-inst.com

Sandown Scientific
Beards Lodge
25 Oldfield Road
Hampden, Middlesex TW12 2AJ, UK
(44) 2089 793300
FAX: (44) 2089 793311
http://www.sandownsci.com

Sandoz Pharmaceuticals
See Novartis

Sanofi Recherche
Centre de Montpellier
371 Rue du Professor Blayac
34184 Montpellier, Cedex 04
France
(33) 67-10-67-10
FAX: (33) 67-10-67-67

Sanofi Winthrop Pharmaceuticals
90 Park Avenue
New York, NY 10016
(800) 223-5511 FAX: (800) 933-3243
(212) 551-4000
http://www.sanofi-synthelabo.com/us

Santa Cruz Biotechnology
2161 Delaware Avenue
Santa Cruz, CA 95060
(800) 457-3801 FAX: (831) 457-3801
(831) 457-3800
http://www.scbt.com

Sarasep
(800) 605-0267 FAX: (408) 432-3231
(408) 432-3230
http://www.transgenomic.com

Sarstedt
P.O. Box 468
Newton, NC 28658
(800) 257-5101 FAX: (828) 465-4003
(828) 465-4000
http://www.sarstedt.com

Sartorius
131 Heartsland Boulevard
Edgewood, NY 11717
(800) 368-7178 FAX: (516) 254-4253
http://www.sartorius.com

SAS Institute
Pacific Telesis Center
One Montgomery Street
San Francisco, CA 94104
(415) 421-2227 FAX: (415) 421-1213
http://www.sas.com

Savant/EC Apparatus
A ThermoQuest company
100 Colin Drive
Holbrook, NY 11741
(800) 634-8886 FAX: (516) 244-0606
(516) 244-2929
http://www.savec.com

Savillex
6133 Baker Road
Minnetonka, MN 55345
(612) 935-5427

Scanalytics
Division of CSP
8550 Lee Highway, Suite 400
Fairfax, VA 22031
(800) 325-3110 FAX: (703) 208-1960
(703) 208-2230
http://www.scanalytics.com

Schering Laboratories
See Schering-Plough

Schering-Plough
1 Giralda Farms
Madison, NJ 07940
(800) 222-7579 FAX: (973) 822-7048
(973) 822-7000
http://www.schering-plough.com

Schleicher & Schuell
10 Optical Avenue
Keene, NH 03431
(800) 245-4024 FAX: (603) 357-3627
(603) 352-3810
http://www.s-und-s.de/english-index.html

Science Technology Centre
1250 Herzberg Laboratories
Carleton University
1125 Colonel Bay Drive
Ottawa, Ontario, Canada K1S 5B6
(613) 520-4442 FAX: (613) 520-4445
http://www.carleton.ca/universities/stc

Scientific Instruments
200 Saw Mill River Road
Hawthorne, NY 10532
(800) 431-1956 FAX: (914) 769-5473
(914) 769-5700
http://www.scientificinstruments.com

Scientific Solutions
9323 Hamilton
Mentor, OH 44060
(440) 357-1400 FAX: (440) 357-1416
www.labmaster.com

Scion
82 Worman's Mill Court, Suite H
Frederick, MD 21701
(301) 695-7870 FAX: (301) 695-0035
www.scioncorp.com

Scott Specialty Gases
6141 Easton Road
P.O. Box 310
Plumsteadville, PA 18949
(800) 21-SCOTT FAX: (215) 766-2476
(215) 766-8861
http://www.scottgas.com

Scripps Clinic and Research Foundation
Instrumentation and Design Lab
10666 N. Torrey Pines Road
La Jolla, CA 92037
(800) 992-9962 FAX: (858) 554-8986
(858) 455-9100
http://www.scrippsclinic.com

SDI Sensor Devices
407 Pilot Court, 400A
Waukesha, WI 53188
(414) 524-1000 FAX: (414) 524-1009

Sefar America
111 Calumet Street
Depew, NY 14043
(716) 683-4050 FAX: (716) 683-4053
http://www.sefaramerica.com

Seikagaku America
Division of Associates of Cape Cod
704 Main Street
Falmouth, MA 02540
(800) 237-4512 FAX: (508) 540-8680
(508) 540-3444
http://www.seikagaku.com

Sellas Medizinische Gerate
Hagener Str. 393
Gevelsberg-Vogelsang, 58285
Germany
(49) 23-326-1225

Sensor Medics
22705 Savi Ranch Parkway
Yorba Linda, CA 92887
(800) 231-2466 FAX: (714) 283-8439
(714) 283-2228
http://www.sensormedics.com

Sensor Systems LLC
2800 Anvil Street, North
Saint Petersburg, FL 33710
(800) 688-2181 FAX: (727) 347-3881
(727) 347-2181
http://www.vsensors.com

SenSym/Foxboro ICT
1804 McCarthy Boulevard
Milpitas, CA 95035
(800) 392-9934 FAX: (408) 954-9458
(408) 954-6700
http://www.sensym.com

Separations Group
See Vydac

Sepracor
111 Locke Drive
Marlboro, MA 01752
(877)-SEPRACOR (508) 357-7300
http://www.sepracor.com

Sera-Lab
See Harlan Sera-Lab

Sermeter
925 Seton Court, #7
Wheeling, IL 60090
(847) 537-4747

Serological
195 W. Birch Street
Kankakee, IL 60901
(800) 227-9412 FAX: (815) 937-8285
(815) 937-8270

Seromed Biochrom
Leonorenstrasse 2-6
D-12247 Berlin, Germany
(49) 030-779-9060

Serotec
22 Bankside
Station Approach
Kidlington, Oxford OX5 1JE, UK
(44) 1865-852722
FAX: (44) 1865-373899
In the US: (800) 265-7376
http://www.serotec.co.uk

Serva Biochemicals
Distributed by Crescent Chemical

S.F. Medical Pharmlast
See Chase-Walton Elastomers

SGE
2007 Kramer Lane
Austin, TX 78758
(800) 945-6154 FAX: (512) 836-9159
(512) 837-7190
http://www.sge.com

Shandon/Lipshaw
171 Industry Drive
Pittsburgh, PA 15275
(800) 245-6212 FAX: (412) 788-1138
(412) 788-1133
http://www.shandon.com

Sharpoint
P.O. Box 2212
Taichung, Taiwan
Republic of China
(886) 4-3206320
FAX: (886) 4-3289879
http://www.sharpoint.com.tw

Shelton Scientific
230 Longhill Crossroads
Shelton, CT 06484
(800) 222-2092 FAX: (203) 929-2175
(203) 929-8999
http://www.sheltonscientific.com

Sherwood-Davis & Geck
See Kendall

Sherwood Medical
See Kendall

Shimadzu Scientific Instruments
7102 Riverwood Drive
Columbia, MD 21046
(800) 477-1227 FAX: (410) 381-1222
(410) 381-1227
http://www.ssi.shimadzu.com

Sialomed
See Amika

Siemens Analytical X-Ray Systems
See Bruker Analytical X-Ray Systems

Sievers Instruments
Subsidiary of Ionics
6060 Spine Road
Boulder, CO 80301 (800) 255-6964
FAX: (303) 444-6272
(303) 444-2009
http://www.sieversinst.com

SIFCO
970 East 46th Street
Cleveland, OH 44103
(216) 881-8600 FAX: (216) 432-6281
http://www.sifco.com

Sigma-Aldrich
3050 Spruce Street
St. Louis, MO 63103
(800) 358-5287 FAX: (800) 962-9591
(800) 325-3101 FAX: (800) 325-5052
http://www.sigma-aldrich.com

Silenus/Amrad
34 Wadhurst Drive
Boronia, Victoria 3155 Australia
(613)9887-3909 FAX: (613)9887-3912
http://www.amrad.com.au

Silicon Genetics
2601 Spring Street
Redwood City, CA 94063
(866) SIG SOFT FAX: (650) 365 1735
(650) 367 9600
http://www.sigenetics.com

SIMS Deltec
1265 Grey Fox Road
St. Paul, Minnesota 55112
(800) 426-2448 FAX: (615) 628-7459
http://www.deltec.com

SIMS Portex
10 Bowman Drive
Keene, NH 03431
(800) 258-5361 FAX: (603) 352-3703
(603) 352-3812
http://www.simsmed.com

SIMS Portex Limited
Hythe, Kent CT21 6JL, UK
(44)1303-260551
FAX: (44)1303-266761
http://www.portex.com

Siris Laboratories
See Biosearch Technologies

Skatron Instruments
See Molecular Devices

SLM Instruments
See Spectronic Instruments

SLM-AMINCO Instruments
See Spectronic Instruments

Small Parts
13980 NW 58th Court
P.O. Box 4650
Miami Lakes, FL 33014
(800) 220-4242 FAX: (800) 423-9009
(305) 558-1038 FAX: (305) 558-0509
http://www.smallparts.com

Smith & Nephew
11775 Starkey Road
P.O. Box 1970
Largo, FL 33779
(800) 876-1261
http://www.smith-nephew.com

SmithKline Beecham
1 Franklin Plaza, #1800
Philadelphia, PA 19102
(215) 751-4000 FAX: (215) 751-4992
http://www.sb.com

Solid Phase Sciences
See Biosearch Technologies

SOMA Scientific Instruments
5319 University Drive, PMB #366
Irvine, CA 92612
(949) 854-0220 FAX: (949) 854-0223
http://somascientific.com

Somatix Therapy
See Cell Genesys

SOMEDIC Sales AB
Box 194
242 22 Hörby, Sweden
(46) 415-165-50
FAX: (46) 415-165-60
http://www.somedic.com

Sonics & Materials
53 Church Hill Road
Newtown, CT 06470
(800) 745-1105 FAX: (203) 270-4610
(203) 270-4600
http://www.sonicsandmaterials.com

Sonosep Biotech
See Triton Environmental Consultants

Sorvall
See Kendro Laboratory Products

Southern Biotechnology Associates
P.O. Box 26221
Birmingham, AL 35260
(800) 722-2255 FAX: (205) 945-8768
(205) 945-1774
http://SouthernBiotech.com

SPAFAS
190 Route 165
Preston, CT 06365
(800) SPAFAS-1 FAX: (860) 889-1991
(860) 889-1389
http://www.spafas.com

Specialty Media
Division of Cell & Molecular Technologies
580 Marshall Street
Phillipsburg, NJ 08865
(800) 543-6029 FAX: (908) 387-1670
(908) 454-7774
http://www.specialtymedia.com

Spectra Physics
See Thermo Separation Products

Spectramed
See BOC Edwards

SpectraSource Instruments
31324 Via Colinas, Suite 114
Westlake Village, CA 91362
(818) 707-2655 FAX: (818) 707-9035
http://www.spectrasource.com

Spectronic Instruments
820 Linden Avenue
Rochester, NY 14625
(800) 654-9955 FAX: (716) 248-4014
(716) 248-4000
http://www.spectronic.com

Spectrum Medical Industries
See Spectrum Laboratories

Spectrum Laboratories
18617 Broadwick Street
Rancho Dominguez, CA 90220
(800) 634-3300 FAX: (800) 445-7330
(310) 885-4601 FAX: (310) 885-4666
http://www.spectrumlabs.com

Spherotech
1840 Industrial Drive, Suite 270
Libertyville, IL 60048
(800) 368-0822 FAX: (847) 680-8927
(847) 680-8922
http://www.spherotech.com

SPSS
233 S. Wacker Drive, 11th floor
Chicago, IL 60606
(800) 521-1337 FAX: (800) 841-0064
http://www.spss.com

SS White Burs
1145 Towbin Avenue
Lakewood, NJ 08701
(732) 905-1100 FAX: (732) 905-0987
http://www.sswhiteburs.com

Stag Instruments
16 Monument Industrial Park
Chalgrove, Oxon OX44 7RW, UK
(44) 1865-891116
FAX: (44) 1865-890562

Standard Reference Materials Program
National Institute of Standards and Technology
Building 202, Room 204
Gaithersburg, MD 20899
(301) 975-6776 FAX: (301) 948-3730

Starplex Scientific
50 Steinway
Etobieoke, Ontario
M9W 6Y3 Canada
(800) 665-0954 FAX: (416) 674-6067
(416) 674-7474
http://www.starplexscientific.com

State Laboratory Institute of Massachusetts
305 South Street
Jamaica Plain, MA 02130
(617) 522-3700 FAX: (617) 522-8735
http://www.state.ma.us/dph

Stedim Labs
1910 Mark Court, Suite 110
Concord, CA 94520
(800) 914-6644 FAX: (925) 689-6988
(925) 689-6650
http://www.stedim.com

Steinel America
9051 Lyndale Avenue
Bloomington, MN 55420
(800) 852-4343 FAX: (952) 888-5132
http://www.steinelamerica.com

Stem Cell Technologies
777 West Broadway, Suite 808
Vancouver, British Columbia
V5Z 4J7 Canada
(800) 667-0322 FAX: (800) 567-2899
(604) 877-0713 FAX: (604) 877-0704
http://www.stemcell.com

Stephens Scientific
107 Riverdale Road
Riverdale, NJ 07457
(800) 831-8099 FAX: (201) 831-8009
(201) 831-9800

Steraloids
P.O. Box 689
Newport, RI 02840
(401) 848-5422 FAX: (401) 848-5638
http://www.steraloids.com

Sterling Medical
2091 Springdale Road, Ste. 2
Cherry Hill, NJ 08003
(800) 229-0900 FAX: (800) 229-7854
http://www.sterlingmedical.com

Sterling Winthrop
90 Park Avenue
New York, NY 10016
(212) 907-2000 FAX: (212) 907-3626

Sternberger Monoclonals
10 Burwood Court
Lutherville, MD 21093
(410) 821-8505 FAX: (410) 821-8506
http://www.sternbergermonoclonals.com

Stoelting
502 Highway 67
Kiel, WI 53042
(920) 894-2293 FAX: (920) 894-7029
http://www.stoelting.com

Stovall Lifescience
206-G South Westgate Drive
Greensboro, NC 27407
(800) 852-0102 FAX: (336) 852-3507
http://www.slscience.com

Stratagene
11011 N. Torrey Pines Road
La Jolla, CA 92037
(800) 424-5444 FAX: (888) 267-4010
(858) 535-5400
http://www.stratagene.com

Strategic Applications
530A N. Milwaukee Avenue
Libertyville, IL 60048
(847) 680-9385 FAX: (847) 680-9837

Strem Chemicals
7 Mulliken Way
Newburyport, MA 01950
(800) 647-8736 FAX: (800) 517-8736
(978) 462-3191 FAX: (978) 465-3104
http://www.strem.com

StressGen Biotechnologies
Biochemicals Division
120-4243 Glanford Avenue
Victoria, British Columbia
V8Z 4B9 Canada
(800) 661-4978 FAX: (250) 744-2877
(250) 744-2811
http://www.stressgen.com

Structure Probe/SPI Supplies
(Epon-Araldite)
P.O. Box 656
West Chester, PA 19381
(800) 242-4774 FAX: (610) 436-5755
http://www.2spi.com

Süd-Chemie Performance Packaging
101 Christine Drive
Belen, NM 87002
(800) 989-3374 FAX: (505) 864-9296
http://www.uniteddesiccants.com

Sumitomo Chemical
Sumitomo Building
5-33, Kitahama 4-chome
Chuo-ku, Osaka 541-8550, Japan
(81) 6-6220-3891
FAX: (81)-6-6220-3345
http://www.sumitomo-chem.co.jp

Sun Box
19217 Orbit Drive
Gaithersburg, MD 20879
(800) 548-3968 FAX: (301) 977-2281
(301) 869-5980
http://www.sunboxco.com

Sunbrokers
See Sun International

Sun International
3700 Highway 421 North
Wilmington, NC 28401
(800) LAB-VIAL FAX: (800) 231-7861
http://www.autosamplervial.com

Sunox
1111 Franklin Boulevard, Unit 6
Cambridge, ON N1R 8B5, Canada
(519) 624-4413 FAX: (519) 624-8378
http://www.sunox.ca

Supelco
See Sigma-Aldrich

SuperArray
P.O. Box 34494
Bethesda, MD 20827
(888) 503-3187 FAX: (301) 765-9859
(301) 765-9888
http://www.superarray.com

A3

Surface Measurement Systems
3 Warple Mews, Warple Way
London W3 ORF, UK
(44) 20-8749-4900
FAX: (44) 20-8749-6749
http://www.smsuk.co.uk/index.htm

SurgiVet
N7 W22025 Johnson Road, Suite A
Waukesha, WI 53186
(262) 513-8500 (888) 745-6562
FAX: (262) 513-9069
www.surgivet.com

Sutter Instruments
51 Digital Drive
Novato, CA 94949
(415) 883-0128 FAX: (415) 883-0572
http://www.sutter.com

Swiss Precision Instruments
1555 Mittel Boulevard, Suite F
Wooddale, IL 60191
(800) 221-0198 FAX: (800) 842-5164

Synaptosoft
3098 Anderson Place
Decatur, GA 30033
(770) 939-4366 FAX: 770-939-9478
http://www.synaptosoft.com

SynChrom
See Micra Scientific

Synergy Software
2457 Perkiomen Avenue
Reading, PA 19606
(800) 876-8376 FAX: (610) 370-0548
(610) 779-0522
http://www.synergy.com

Synteni
See Incyte

Synthetics Industry
Lumite Division
2100A Atlantic Highway
Gainesville, GA 30501
(404) 532-9756 FAX: (404) 531-1347

Systat
See SPSS

Systems Planning and Analysis (SPA)
2000 N. Beauregard Street
Suite 400
Alexandria, VA 22311
(703) 931-3500
http://www.spa-inc.net

3M Bioapplications
3M Center
Building 270-15-01
St. Paul, MN 55144
(800) 257-7459 FAX: (651) 737-5645
(651) 736-4946

T Cell Diagnostics and T Cell Sciences
38 Sidney Street
Cambridge, MA 02139
(617) 621-1400

TAAB Laboratory Equipment
3 Minerva House
Calleva Park
Aldermaston, Berkshire RG7 8NA, UK
(44) 118 9817775
FAX: (44) 118 9817881

Taconic
273 Hover Avenue
Germantown, NY 12526
(800) TAC-ONIC FAX: (518) 537-7287
(518) 537-6208
http://www.taconic.com

Tago
See Biosource International

TaKaRa Biochemical
719 Alliston Way
Berkeley, CA 94710
(800) 544-9899 FAX: (510) 649-8933
(510) 649-9895
http://www.takara.co.jp/english

Takara Shuzo
Biomedical Group Division
Seta 3-4-1
Otsu Shiga 520-21, Japan
(81) 75-241-5100
FAX: (81) 77-543-9254
http://www.Takara.co.jp/english

Takeda Chemical Products
101 Takeda Drive
Wilmington, NC 28401
(800) 825-3328 FAX: (800) 825-0333
(910) 762-8666 FAX: (910) 762-6846
http://takeda-usa.com

TAO Biomedical
73 Manassas Court
Laurel Springs, NJ 08021
(609) 782-8622 FAX: (609) 782-8622

Tecan US
P.O. Box 13953
Research Triangle Park, NC 27709
(800) 33-TECAN FAX: (919) 361-5201
(919) 361-5208
http://www.tecan-us.com

Techne
University Park Plaza
743 Alexander Road
Princeton, NJ 08540
(800) 225-9243 FAX: (609) 987-8177
(609) 452-9275
http://www.techneusa.com

Technical Manufacturing
15 Centennial Drive
Peabody, MA 01960
(978) 532-6330 FAX: (978) 531-8682
http://www.techmfg.com

Technical Products International
5918 Evergreen
St. Louis, MO 63134
(800) 729-4451 FAX: (314) 522-6360
(314) 522-8671
http://www.vibratome.com

Technicon
See Organon Teknika Cappel

Techno-Aide
P.O. Box 90763
Nashville, TN 37209
(800) 251-2629 FAX: (800) 554-6275
(615) 350-7030
http://www.techno-aid.com

Ted Pella
4595 Mountain Lakes Boulevard
P.O. Box 492477
Redding, CA 96049
(800) 237-3526 FAX: (530) 243-3761
(530) 243-2200
http://www.tedpella.com

Tekmar-Dohrmann
P.O. Box 429576 Cincinnati, OH 45242
(800) 543-4461 FAX: (800) 841-5262
(513) 247-7000 FAX: (513) 247-7050

Tektronix
142000 S.W. Karl Braun Drive
Beaverton, OR 97077
(800) 621-1966 FAX: (503) 627-7995
(503) 627-7999
http://www.tek.com

Tel-Test
P.O. Box 1421
Friendswood, TX 77546
(800) 631-0600 FAX: (281)482-1070
(281)482-2672
http://www.isotex-diag.com

TeleChem International
524 East Weddell Drive, Suite 3
Sunnyvale, CA 94089
(408) 744-1331 FAX: (408) 744-1711
http://www.gst.net/~telechem

Terrachem
Mallaustrasse 57
D-68219 Mannheim, Germany
0621-876797-0 FAX: 0621-876797-19
http://www.terrachem.de

Terumo Medical
2101 Cottontail Lane
Somerset, NJ 08873
(800) 283-7866 FAX: (732) 302-3083
(732) 302-4900
http://www.terumomedical.com

Tetko
333 South Highland Manor
Briarcliff, NY 10510
(800) 289-8385 FAX: (914) 941-1017
(914) 941-7767
http://www.tetko.com

TetraLink
4240 Ridge Lea Road
Suite 29
Amherst, NY 14226
(800) 747-5170 FAX: (800) 747-5171
http://www.tetra-link.com

TEVA Pharmaceuticals USA
1090 Horsham Road
P.O. Box 1090
North Wales, PA 19454
(215) 591-3000 FAX: (215) 721-9669
http://www.tevapharmusa.com

Texas Fluorescence Labs
9503 Capitol View Drive
Austin, TX 78747
(512) 280-5223 FAX: (512) 280-4997
http://www.teflabs.com

The Nest Group
45 Valley Road
Southborough, MA 01772
(800) 347-6378 FAX: (508) 485-5736
(508) 481-6223
http://world.std.com/~nestgrp

ThermoCare
P.O. Box 6069
Incline Village, NV 89450
(800) 262-4020
(775) 831-1201

Thermo Labsystems
8 East Forge Parkway
Franklin, MA 02038
(800) 522-7763 FAX: (508) 520-2229
(508) 520-0009
http://www.finnpipette.com

Thermometric
Spjutvagen 5A
S-175 61 Jarfalla, Sweden
(46) 8-564-72-200

Thermoquest
IEC Division
300 Second Avenue
Needham Heights, MA 02194
(800) 843-1113 FAX: (781) 444-6743
(781) 449-0800
http://www.thermoquest.com

Thermo Separation Products
Thermoquest
355 River Oaks Parkway
San Jose, CA 95134
(800) 538-7067 FAX: (408) 526-9810
(408) 526-1100
http://www.thermoquest.com

Thermo Shandon
171 Industry Drive
Pittsburgh, PA 15275
(800) 547-7429 FAX: (412) 899-4045
http://www.thermoshandon.com

Thomas Scientific
99 High Hill Road at I-295
Swedesboro, NJ 08085
(800) 345-2100 FAX: (800) 345-5232
(856) 467-2000 FAX: (856) 467-3087
http://www.wheatonsci.com/html/nt/Thomas.html

Thomson Instrument
354 Tyler Road
Clearbrook, VA 22624
(800) 842-4752 FAX: (540) 667-6878
(800) 541-4792 FAX: (760) 757-9367
http://www.hplc.com

Thorn EMI
See Electron Tubes

Thorlabs
435 Route 206
Newton, NJ 07860
(973) 579-7227 FAX: (973) 383-8406
http://www.thorlabs.com

Tiemann
See Bernsco Surgical Supply

Timberline Instruments
1880 South Flatiron Court, H-2
P.O. Box 20356
Boulder, CO 80308
(800) 777-5996 FAX: (303) 440-8786
(303) 440-8779
http://www.timberlineinstruments.com

Tissue-Tek
A Division of Sakura Finetek USA
1750 West 214th Street
Torrance, CA 90501
(800) 725-8723 FAX: (310) 972-7888
(310) 972-7800
http://www.sakuraus.com

Tocris Cookson
114 Holloway Road, Suite 200
Ballwin, MO 63011
(800) 421-3701 FAX: (800) 483-1993
(636) 207-7651 FAX: (636) 207-7683
http://www.tocris.com

Tocris Cookson
Northpoint, Fourth Way
Avonmouth, Bristol BS11 8TA, UK
(44) 117-982-6551
FAX: (44) 117-982-6552
http://www.tocris.com

Tomtec
See CraMar Technologies

TopoGen
P.O. Box 20607
Columbus, OH 43220
(800) TOPOGEN
FAX: (800) ADD-TOPO
(614) 451-5810 FAX: (614) 451-5811
http://www.topogen.com

Toray Industries, Japan
Toray Building 2-1
Nihonbash-Muromach
2-Chome, Chuo-Ku
Tokyo, Japan 103-8666
(03) 3245-5115 FAX: (03) 3245-5555
http://www.toray.co.jp

Toray Industries, U.S.A.
600 Third Avenue
New York, NY 10016
(212) 697-8150 FAX: (212) 972-4279
http://www.toray.com

Toronto Research Chemicals
2 Brisbane Road
North York, Ontario M3J 2J8, Canada
(416) 665-9696 FAX: (416) 665-4439
http://www.trc-canada.com

TosoHaas
156 Keystone Drive
Montgomeryville, PA 18036
(800) 366-4875 FAX: (215) 283-5035
(215) 283-5000
http://www.tosohaas.com

Towhill
647 Summer Street
Boston, MA 02210
(617) 542-6636 FAX: (617) 464-0804

Toxin Technology
7165 Curtiss Avenue
Sarasota, FL 34231
(941) 925-2032 FAX: (9413) 925-2130
http://www.toxintechnology.com

Toyo Soda
See TosoHaas

Trace Analytical
3517-A Edison Way
Menlo Park, CA 94025
(650) 364-6895 FAX: (650) 364-6897
http://www.traceanalytical.com

Transduction Laboratories
See BD Transduction Laboratories

Transgenomic
2032 Concourse Drive
San Jose, CA 95131
(408) 432-3230 FAX: (408) 432-3231
http://www.transgenomic.com

Transonic Systems
34 Dutch Mill Road
Ithaca, NY 14850
(800) 353-3569 FAX: (607) 257-7256
http://www.transonic.com

Travenol Lab
See Baxter Healthcare

Tree Star Software
20 Winding Way
San Carlos, CA 94070
800-366-6045
http://www.treestar.com

Trevigen
8405 Helgerman Court
Gaithersburg, MD 20877
(800) TREVIGEN FAX: (301) 216-2801
(301) 216-2800
http://www.trevigen.com

Trilink Biotechnologies
6310 Nancy Ridge Drive
San Diego, CA 92121
(800) 863-6801 FAX: (858) 546-0020
http://www.trilink.biotech.com

Tripos Associates
1699 South Hanley Road, Suite 303
St. Louis, MO 63144
(800) 323-2960 FAX: (314) 647-9241
(314) 647-1099
http://www.tripos.com

Triton Environmental Consultants
120-13511 Commerce Parkway
Richmond, British Columbia
V6V 2L1 Canada
(604) 279-2093 FAX: (604) 279-2047
http://www.triton-env.com

Tropix
47 Wiggins Avenue
Bedford, MA 01730
(800) 542-2369 FAX: (617) 275-8581
(617) 271-0045
http://www.tropix.com

TSI Center for Diagnostic Products
See Intergen

2000 Eppendorf-5 Prime
5603 Arapahoe Avenue
Boulder, CO 80303
(800) 533-5703 FAX: (303) 440-0835
(303) 440-3705

Tyler Research
10328 73rd Avenue
Edmonton, Alberta
T6E 6N5 Canada
(403) 448-1249 FAX: (403) 433-0479

UBI
See Upstate Biotechnology

Ugo-Basile
Via G. Borghi 43
21025 Comerio, Varese, Italy
(39) 332 744 574
FAX: (39) 332 745 488
http://www.ugobasile.com

UltraPIX
See Life Science Resources

Ultrasonic Power
239 East Stephenson Street
Freeport, IL 61032
(815) 235-6020 FAX: (815) 232-2150
http://www.upcorp.com

Ultrasound Advice
23 Aberdeen Road
London N52UG, UK
(44) 020-7359-1718
FAX: (44) 020-7359-3650
http://www.ultrasoundadvice.co.uk

UNELKO
14641 N. 74th Street
Scottsdale, AZ 85260
(480) 991-7272 FAX: (480)483-7674
http://www.unelko.com

Unifab Corp.
5260 Lovers Lane
Kalamazoo, MI 49002 (800) 648-9569
FAX: (616) 382-2825
(616) 382-2803

Union Carbide
10235 West Little York Road, Suite 300
Houston, TX 77040
(800) 568-4000 FAX: (713) 849-7021
(713) 849-7000
http://www.unioncarbide.com

United Desiccants
See Süd-Chemie Performance Packaging

United States Biochemical
See USB

United States Biological (US Biological)
P.O. Box 261
Swampscott, MA 01907
(800) 520-3011 FAX: (781) 639-1768
www.usbio.net

Universal Imaging
502 Brandywine Parkway
West Chester, PA 19380
(610) 344-9410 FAX: (610) 344-6515
http://www.image1.com

Upchurch Scientific
619 West Oak Street
P.O. Box 1529
Oak Harbor, WA 98277
(800) 426-0191 FAX: (800) 359-3460
(360) 679-2528 FAX: (360) 679-3830
http://www.upchurch.com

Upjohn
Pharmacia & Upjohn
http://www.pnu.com

Upstate Biotechnology (UBI)
1100 Winter Street, Suite 2300
Waltham, MA 02451
(800) 233-3991 FAX: (781) 890-7738
(781) 890-8845
http://www.upstatebiotech.com

USA/Scientific
346 SW 57th Avenue
P.O. Box 3565
Ocala, FL 34478
(800) LAB-TIPS FAX: (352) 351-2057
(3524) 237-6288
http://www.usascientific.com

USB
26111 Miles Road
P.O. Box 22400
Cleveland, OH 44122
(800) 321-9322 FAX: (800) 535-0898
FAX: (216) 464-5075
http://www.usbweb.com

USCI Bard
Bard Interventional Products
129 Concord Road
Billerica, MA 01821
(800) 225-1332 FAX: (978) 262-4805
http://www.bardinterventional.com

UVP (Ultraviolet Products)
2066 W. 11th Street
Upland, CA 91786
(800) 452-6788 FAX: (909) 946-3597
(909) 946-3197
http://www.uvp.com

V & P Scientific
9823 Pacific Heights Boulevard, Suite T
San Diego, CA 92121
(800) 455-0644 FAX: (858) 455-0703
(858) 455-0643
http://www.vp-scientific.com

Valco Instruments
P.O. Box 55603
Houston, TX 77255
(800) FOR-VICI FAX: (713) 688-8106
(713) 688-9345
http://www.vici.com

Valpey Fisher
75 South Street
Hopkin, MA 01748
(508) 435-6831 FAX: (508) 435-5289
http://www.valpeyfisher.com

Value Plastics
3325 Timberline Road
Fort Collins, CO 80525
(800) 404-LUER FAX: (970) 223-0953
(970) 223-8306
http://www.valueplastics.com

Vangard International
P.O. Box 308
3535 Rt. 66, Bldg. #4
Neptune, NJ 07754
(800) 922-0784 FAX: (732) 922-0557
(732) 922-4900
http://www.vangard1.com

Varian Analytical Instruments
2700 Mitchell Drive
Walnut Creek, CA 94598
(800) 926-3000 FAX: (925) 945-2102
(925) 939-2400
http://www.varianinc.com

Varian Associates
3050 Hansen Way
Palo Alto, CA 94304
(800) 544-4636 FAX: (650) 424-5358
(650) 493-4000
http://www.varian.com

Vector Core Laboratory/National Gene Vector Labs
University of Michigan
3560 E MSRB II
1150 West Medical Center Drive
Ann Arbor, MI 48109
(734) 936-5843 FAX: (734) 764-3596

Vector Laboratories
30 Ingold Road
Burlingame, CA 94010
(800) 227-6666 FAX: (650) 697-0339
(650) 697-3600
http://www.vectorlabs.com

Vedco
2121 S.E. Bush Road
St. Joseph, MO 64504
(888) 708-3326 FAX: (816) 238-1837
(816) 238-8840
http://database.vedco.com

Ventana Medical Systems
3865 North Business Center Drive
Tucson, AZ 85705
(800) 227-2155 FAX: (520) 887-2558
(520) 887-2155
http://www.ventanamed.com

Verity Software House
P.O. Box 247
45A Augusta Road
Topsham, ME 04086
(207) 729-6767 FAX: (207) 729-5443
http://www.vsh.com

Vernitron
See Sensor Systems LLC

Vertex Pharmaceuticals
130 Waverly Street
Cambridge, MA 02139
(617) 577-6000 FAX: (617) 577-6680
http://www.vpharm.com

Vetamac
Route 7, Box 208
Frankfort, IN 46041
(317) 379-3621

Vet Drug
Unit 8
Lakeside Industrial Estate
Colnbrook, Slough SL3 0ED, UK

Vetus Animal Health
See Burns Veterinary Supply

Viamed
15 Station Road
Cross Hills, Keighley
W. Yorkshire BD20 7DT, UK
(44) 1-535-634-542
FAX: (44) 1-535-635-582
http://www.viamed.co.uk

Vical
9373 Town Center Drive, Suite 100
San Diego, CA 92121
(858) 646-1100 FAX: (858) 646-1150
http://www.vical.com

Victor Medical
2349 North Watney Way, Suite D
Fairfield, CA 94533
(800) 888-8908 FAX: (707) 425-6459
(707) 425-0294

Virion Systems
9610 Medical Center Drive, Suite 100
Rockville, MD 20850
(301) 309-1844 FAX: (301) 309-0471
http://www.radix.net/~virion

VirTis Company
815 Route 208
Gardiner, NY 12525
(800) 765-6198 FAX: (914) 255-5338
(914) 255-5000
http://www.virtis.com

Visible Genetics
700 Bay Street, Suite 1000
Toronto, Ontario M5G 1Z6, Canada
(888) 463-6844 (416) 813-3272
http://www.visgen.com

Vitrocom
8 Morris Avenue
Mountain Lakes, NJ 07046
(973) 402-1443 FAX: (973) 402-1445

VTI
7650 W. 26th Avenue
Hialeah, FL 33106
(305) 828-4700 FAX: (305) 828-0299
http://www.vticorp.com

VWR Scientific Products
200 Center Square Road
Bridgeport, NJ 08014
(800) 932-5000 FAX: (609) 467-5499
(609) 467-2600
http://www.vwrsp.com

Vydac
17434 Mojave Street
P.O. Box 867 Hesperia, CA 92345
(800) 247-0924 FAX: (760) 244-1984
(760) 244-6107
http://www.vydac.com

Vysis
3100 Woodcreek Drive
Downers Grove, IL 60515
(800) 553-7042 FAX: (630) 271-7138
(630) 271-7000
http://www.vysis.com

W&H Dentalwerk Bürmoos
P.O. Box 1
A-5111 Bürmoos, Austria
(43) 6274-6236-0
FAX: (43) 6274-6236-55
http://www.wnhdent.com

Wako BioProducts
See Wako Chemicals USA

Wako Chemicals USA
1600 Bellwood Road
Richmond, VA 23237
(800) 992-9256 FAX: (804) 271-7791
(804) 271-7677
http://www.wakousa.com

Wako Pure Chemicals
1-2, Doshomachi 3-chome
Chuo-ku, Osaka 540-8605, Japan
81-6-6203-3741 FAX: 81-6-6222-1203
http://www.wako-chem.co.jp/egaiyo/index.htm

Wallac
See Perkin-Elmer

Wallac
A Division of Perkin-Elmer
3985 Eastern Road
Norton, OH 44203
(800) 321-9632 FAX: (330) 825-8520
(330) 825-4525
http://www.wallac.com

Waring Products
283 Main Street
New Hartford, CT 06057
(800) 348-7195 FAX: (860) 738-9203
(860) 379-0731
http://www.waringproducts.com

Warner Instrument
1141 Dixwell Avenue
Hamden, CT 06514
(800) 599-4203 FAX: (203) 776-1278
(203) 776-0664
http://www.warnerinstrument.com

Warner-Lambert
Parke-Davis
201 Tabor Road
Morris Plains, NJ 07950
(973) 540-2000 FAX: (973) 540-3761
http://www.warner-lambert.com

Washington University Machine Shop
615 South Taylor
St. Louis, MO 63310
(314) 362-6186 FAX: (314) 362-6184

Waters Chromatography
34 Maple Street
Milford, MA 01757
(800) 252-HPLC FAX: (508) 478-1990
(508) 478-2000
http://www.waters.com

Watlow
12001 Lackland Road
St. Louis, MO 63146 (314) 426-7431
FAX: (314) 447-8770
http://www.watlow.com

Watson-Marlow
220 Ballardvale Street
Wilmington, MA 01887
(978) 658-6168 FAX: (978) 988 0828
http://www.watson-marlow.co.uk

Waukesha Fluid Handling
611 Sugar Creek Road
Delavan, WI 53115
(800) 252-5200 FAX: (800) 252-5012
(414) 728-1900 FAX: (414) 728-4608
http://www.waukesha-cb.com

WaveMetrics
P.O. Box 2088
Lake Oswego, OR 97035
(503) 620-3001 FAX: (503) 620-6754
http://www.wavemetrics.com

Weather Measure
P.O. Box 41257
Sacramento, CA 95641
(916) 481-7565

Weber Scientific
2732 Kuser Road
Hamilton, NJ 08691
(800) FAT-TEST FAX: (609) 584-8388
(609) 584-7677
http://www.weberscientific.com

Weck, Edward & Company
1 Weck Drive
Research Triangle Park, NC 27709
(919) 544-8000

Wellcome Diagnostics
See Burroughs Wellcome

Wellington Laboratories
398 Laird Road, Guelph
Ontario, Canada, N1G 3X7
(800) 578-6985 FAX: (519) 822-2849
http://www.well-labs.com

Wesbart Engineering
Daux Road
Billingshurst, West Sussex
RH14 9EZ, UK
(44) 1-403-782738
FAX: (44) 1-403-784180
http://www.wesbart.co.uk

Whatman
9 Bridewell Place
Clifton, NJ 07014
(800) 631-7290 FAX: (973) 773-3991
(973) 773-5800
http://www.whatman.com

Wheaton Science Products
1501 North 10th Street
Millville, NJ 08332
(800) 225-1437 FAX: (800) 368-3108
(856) 825-1100 FAX: (856) 825-1368
http://www.algroupwheaton.com

Whittaker Bioproducts
See BioWhittaker

Wild Heerbrugg
Juerg Dedual Gaebrisstrasse 8 CH
9056 Gais, Switzerland
(41) 71-793-2723
FAX: (41) 71-726-5957
http://www.homepage.swissonline.net/dedual/wild_heerbrugg

Willy A. Bachofen AG Maschinenfabrik
Utengasse 15/17
CH4005 Basel, Switzerland
(41) 61-681-5151
FAX: (41) 61-681-5058
http://www.wab.ch

Winthrop
See Sterling Winthrop

Wolfram Research
100 Trade Center Drive
Champaign, IL 61820
(800) 965-3726 FAX: (217) 398-0747
(217) 398-0700
http://www.wolfram.com

World Health Organization
Microbiology and Immunology Support
20 Avenue Appia
1211 Geneva 27, Switzerland
(41-22) 791-2602
FAX: (41-22) 791-0746
http://www.who.org

World Precision Instruments
175 Sarasota Center Boulevard
International Trade Center
Sarasota, FL 34240
(941) 371-1003 FAX: (941) 377-5428
http://www.wpiinc.com

Worthington Biochemical
Halls Mill Road
Freehold, NJ 07728
(800) 445-9603 FAX: (800) 368-3108
(732) 462-3838 FAX: (732) 308-4453
http://www.worthington-biochem.com

WPI
See World Precision Instruments

Wyeth-Ayerst
2 Esterbrook Lane
Cherry Hill, NJ 08003
(800) 568-9938 FAX: (858) 424-8747
(858) 424-3700

Wyeth-Ayerst Laboratories
P.O. Box 1773
Paoli, PA 19301
(800) 666-7248 FAX: (610) 889-9669
(610) 644-8000
http://www.ahp.com

Xenotech
3800 Cambridge Street
Kansas City, KS 66103
(913) 588-7930 FAX: (913) 588-7572
http://www.xenotechllc.com

Xillix Technologies
300-13775 Commerce Parkway
Richmond, British Columbia
V6V 2V4 Canada
(800) 665-2236 FAX: (604) 278-3356
(604) 278-5000
http://www.xillix.com

Xomed Surgical Products
6743 Southpoint Drive N
Jacksonville, FL 32216
(800) 874-5797 FAX: (800) 678-3995
(904) 296-9600 FAX: (904) 296-9666
http://www.xomed.com

Yakult Honsha
1-19, Higashi-Shinbashi 1-chome
Minato-ku Tokyo 105-8660, Japan
81-3-3574-8960

Yamasa Shoyu
23-8 Nihonbashi Kakigaracho
1-chome, Chuoku
Tokyo, 103 Japan
(81) 3-479 22 0095
FAX: (81) 3-479 22 3435

Yeast Genetic Stock Center
See ATCC

Yellow Spring Instruments
See YSI

YMC
YMC Karasuma-Gojo Building
284 Daigo-Cho, Karasuma Nisihiirr
Gojo-dori Shimogyo-ku
Kyoto, 600-8106, Japan
(81) 75-342-4567
FAX: (81) 75-342-4568
http://www.ymc.co.jp

YSI
1725-1700 Brannum Lane
Yellow Springs, OH 45387
(800) 765-9744 FAX: (937) 767-9353
(937) 767-7241
http://www.ysi.com

Zeneca/CRB
See AstraZeneca
(800) 327-0125 FAX: (800) 321-4745

Zivic-Miller Laboratories
178 Toll Gate Road
Zelienople, PA 16063
(800) 422-LABS FAX: (724) 452-4506
(800) MBM-RATS FAX: (724) 452-5200
http://zivicmiller.com

Zymark
Zymark Center
Hopkinton, MA 01748
(508) 435-9500 FAX: (508) 435-3439
http://www.zymark.com

Zymed Laboratories
458 Carlton Court
South San Francisco, CA 94080
(800) 874-4494 FAX: (650) 871-4499
(650) 871-4494
http://www.zymed.com

Zymo Research
625 W. Katella Avenue, Suite 30
Orange, CA 92867
(888) 882-9682 FAX: (714) 288-9643
(714) 288-9682
http://www.zymor.com

Zynaxis Cell Science
See ChiRex Cauldron

References

Aboody-Guterman, K.S., Pechan, P.A., Rainov, N.G., Sena-Esteves, M., Jacobs, A., Snyder, E.Y., Wild, P., Schraner, E., Tobler, K., Breakefield, X.O., and Fraefel, C. 1997. Green fluorescent protein as a reporter for retrovirus and helper virus-free HSV-1 amplicon vector-mediated gene transfer into neural cells in culture and *in vivo*. *Neuroreport* 8:3801–3808.

Abuin, A. and Bradley, A. 1996. Recycling selectable markers in mouse embryonic stem cells. *Mol. Cell. Biol.* 16:1851–1856.

Agah, R., Frenkel, P.A., French, B.A., Michael, L.H., Overbeek, P.A., and Schneider, M.D. 1997. Gene recombination in postmitotic cells. Targeted expression of Cre recombinase provokes cardiac-restricted, site-specific rearrangement in adult ventricular muscle in vivo. *J. Clin. Invest.* 100:169–179.

Albertson, D.G., Fishpool, R.M., and Birchall, P.S. 1995. Fluorescence in situ hybridization for the detection of DNA and RNA. *Methods Cell Biol.* 48:339–364.

Alizadeh, A., Eisen, M., Botstein, D., Brown, P.O., and Staudt, L.M. 1998. Probing lymphocyte biology by genomic-scale gene expression analysis. *J. Clin. Immunol.* 18:373–379.

Aller, M.I., Jones, A., Merlo, D., Paterlini, M., Meyer, A.H., Amtmann, U., Brickley, S., Jolin, H.E., McKenzie, A.N., Monyer, H., Farrant, M., and Wisden, W. 2003. Cerebellar granule cell Cre recombinase expression. *Genesis* 36:97–103.

Altman, J. and Bayer, S. 1990. Prolonged sojourn of developing pyramidal cells in the intermediate zone of the hippocampus and their settling in the stratum pyramidale. *J. Comp. Neurol.* 301:343–364.

Amaral, D.G. and Witter, M.P. 1989. The three-dimensional organization of the hippocampal formation: A review of anatomical data. *Neuroscience* 31:571–591.

Amundson, S.A., Bittner, M., Chen, Y., Trent, J., Meltzer, P., and Fornace, A.J. Jr. 1999. Fluorescent cDNA microarray hybridization reveals complexity and heterogeneity of cellular genotoxic stress responses. *Oncogene* 18:3666–3672.

Angelichio, M.L., Beck, J.A., Johansen, H., and Ivey-Hoyle, M. 1991. Comparison of several promoters and polyadenylation signals for use in heterologous gene expression in cultured *Drosophila* cells. *Nucl. Acids Res.* 19:5037–5043.

Antoch, M.P., Song, E.J., Chang, A.M., Vitaterna, M.H., Zhao, Y., Wilsbacher, L.D., Sangoram, A.M., King, D.P., Pinto, L.H., and Takahashi, J.S. 1997. Functional identification of the mouse circadian clock gene by transgenic BAC rescue. *Cell* 89:655–667.

Applied Biosystems. 1989. User Bulletin Issue 11, Model No. 370. Applied Biosystems, Foster City, Calif.

Ashani, Y. and Catravas, G.N. 1980. Highly reactive impurities in Triton X-100 and Brij 35: Partial characterization and removal. *Anal. Biochem.* 109:55–62.

Ashburner, M. 1989. *Drosophila*: A laboratory manual. Cold Spring Harbor Laboratory Press, Cold Spring Harbor, N.Y.

Atwell, S., Ridgway, J.B., Wells, J.A., and Carter, P. 1997. Stable heterodimers from remodeling the domain interface of a homodimer using a phage display library. *J. Mol. Biol.* 270:26–35.

Ausubel, F.M., Brent, R., Kingston, R.E., Moore, D.D., Seidman, J.G., Smith, J.A., and Struhl, K. (eds.) 2006. Current Protocols in Molecular Biology. John Wiley & Sons, Hoboken.

Bacallao, R., Kiai, K., and Jesaitis, L. 1995. Guiding principles of specimen preservation for confocal fluorescence microscopy. *In* Handbook of Biological Confocal Microscopy, 2nd ed. (J. Pawley, ed.) pp. 311–326. Plenum Press, New York.

Baim, S.B., Labow, M.A., Levine, A.J., and Shenk, T. 1991. A chimeric mammalian transactivator based on the lac repressor that is regulated by temperature and isopropyl-β-D-thiogalactopyranoside. *Proc. Natl. Acad. Sci. U.S.A.* 88:5072–5076.

Balass, M., Katchalski-Katzir, E., and Fuchs, S. 1997. The alpha-bungarotoxin binding site on the nicotinic acetylcholine receptor: Analysis using a phage epitope library. *Proc. Natl. Acad. Sci. U.S.A.* 94:6054–6058.

Ballinger, M.D., Jones, J.T., Lofgren, J.A., Fairbrother, W.J., Akita, R.W., Sliwkowski, M.X., and Wells, J.A. 1998. Selection of heregulin variants having higher affinity for the *ErbB3* receptor by monovalent phage display. *J. Biol. Chem.* 273:11675–11684.

Banker, G. and Goslin, K. 1998. Culturing Nerve Cells, 2nd ed. MIT Press, Cambridge, Mass.

Bannigan, J. and Langman, J. 1979. The cellular effect of 5-bromodeoxyuridine on the mammalian embryo. *J. Embryol. Exp. Morphol.* 50:123–135.

Barbe, M.F. and Levitt, P. 1992. Attraction of specific thalamic input by cerebral grafts depends on the molecular identity of the implant. *Proc. Natl. Acad. Sci. U.S.A.* 89:3706–3710.

Barbe, M.F. and Levitt, P. 1995. Age-dependent specification of the corticocortical connections of cerebral grafts. *J. Neurosci.* 15:1819–1834.

Barch, M.J., Lawce, H.J., and Arsham, M.S. 1991. Peripheral Blood Culture. *In* The ACT Cytogenetics Laboratory Manual, 2nd ed. (M.J. Barch, ed.) pp. 17–30. Raven Press, New York.

Baserga, R. and Malamud, D. 1969. Autoradiography: Techniques and Application. Harper and Row, New York.

Bassett, D.E. Jr., Eisen, M.B., and Boguski, M.S. 1999. Gene expression informatics—it's all in your mine. *Nature Genet.* 21:51–55.

Bastiaens, P.I.H. and Squire, A. 1999. Fluorescence lifetime imaging microscopy: Spatial resolution of biochemical processes in the cell. *Trends Cell Biol.* 9:48–52.

Behr, M.A., Wilson, M.A., Gill, W.P., Salamon, H., Schoolnik, G.K., Rane, S., and Small, P.M. 1999. Comparative genomics of BCG vaccines by whole-genome DNA microarray. *Science* 284:1520–1523.

Bendixen, C., Gangloff, S., and Rothstein, R. 1994. A yeast mating-selection scheme for detection of protein-protein interactions. *Nucl. Acids Res.* 22:1778–1779.

Berget, S.M. 1995. Exon recognition in vertebrate splicing. *J. Biol. Chem.* 270:2411–2414.

Berkner, K.L. and Sharp, P.A., 1983. Generation of adenovirus by cotransfection of plasmids. *Nucl. Acids Res.* 11:6003–6020.

Berkner, K.L. and Sharp, P.A., 1984. Expression of dihydrofolate reductase, and of the adjacent E1b region, in an Ad5-dihydrofolate reductase recombinant virus. *Nucl. Acids Res.* 12:1925–1941.

Bernard, A., Payton, M., and Radford, K.R. 1999. Protein expression in the baculovirus system. *In* Current Protocols in Protein Science (J.E. Coligan, B.M. Dunn, H.L. Ploegh, D.W. Speicher, and P.T. Wingfield, eds.) pp. 5.5.1–5.5.18. John Wiley & Sons, New York.

Bernstein, E., Caudy, A.A., Hammond, S.M., and Hannon, G.J. 2001. Role for a bidentate ribonuclease in the initiation step of RNA interference. *Nature* 409:363–366.

BioSupplyNet Source Book. 1999. BioSupplyNet, Plainview, N.Y., and Cold Spring Harbor Laboratory Press, Cold Spring Harbor, N.Y.

Bishop, C.E. and Hatat, D. 1987. Molecular cloning and sequence analysis of a mouse Y chromosome RNA transcript expressed in the testis. *Nucl. Acids Res.* 15:2959–2969.

Bjerrum, O.J. and Schafer-Nielsen, C. 1986. Buffer systems and transfer parameters for semidry electroblotting with a horizontal apparatus. *In* Electrophoresis '86 (M.J. Dunn, ed.) pp. 315–327. VCH Publishers, Deerfield Beach, Fla.

Blomer, U., Naldini, L., Kafri, T., Trono, D., Verma, I.M., and Gage, FH. 1997. Highly efficient and sustained gene transfer in adult neurons with a lentivirus vector. *J. Virol.* 71:6641–6649.

Blond-Elguindi, S., Cwirla, S.E., Dower, W.J., Lipshutz, R.J., Sprang, S.R., Sambrook, J.F., and Gething, M.J.H. 1993. Affinity panning of a library of peptides displayed on bacteriophages reveals the binding specificity of BiP. *Cell* 75:717–728.

Blow, J.J. and Laskey, R.A. 1988. A role for the nuclear envelope in controlling DNA replication within the cell cycle. *Nature* 332:546–548.

Bluthner, M., Bautz, E.K., and Bautz, F.A. 1996. Mapping of epitopes recognized by PM/Scl autoantibodies with gene-fragment phage display libraries. *J. Immunol. Methods* 198:187–198.

Bobrow, M.N., Harris, T.D., Shaughnessy, K.J., and Litt, G.J. 1989. Catalyzed reporter deposition: A novel method of signal amplification. *J. Immunol. Methods* 125:279–285.

Boffelli, D., Nobrega, M.A., and Rubin, E.M. 2004. Comparative genomics at the vertebrate extremes. *Nat. Rev. Genet.* 5:456–465.

Bolam, J.P. (ed.) 1992. Experimental Neuroanatomy: A Practical Approach. Oxford University Press, Oxford.

Bonnycastle, L., Mehroke, J., Rashed, M., Gong, X., and Scott, J. 1996. Probing the basis of antibody reactivity with a panel of constrained peptide libraries displayed by filamentous phage. *J. Mol. Biol.* 258: 747–762.

Bothe, G.W., Molivar, V.J., Vedder, M.J., and Geistfeld, J.G. 2004. Genetics and behavioral differences among five inbred mouse strains commonly used in the production of transgenic and knockout mice. *Genes Brain Behav.* 3:149–157.

Bottenstein, J.E. 1985. Growth and differentiation of neural cells in defined media. *In* Cell Culture in the Neurosciences (J.E. Bottenstein and G. Sato, eds.) pp. 3–43. Plenum, New York.

Bradbury, A., Persic, L., Werge, T., and Cattaneo, A. 1993. From gene to antibody: The use of living columns to select specific phage antibodies. *BioTechniques* 11:1565–1569.

Bradford, M.M. 1976. A rapid and sensitive method for the quantitation of microgram quantities of protein utilizing the principle of protein-dye binding. *Anal. Biochem.* 72:248–254.

Bradley, D.J., Towle, H.C., and Young, W.S. III. 1992. Spatial and temporal expression of α and β thyroid hormone receptor mRNAs, including the β2 subtype, in the developing mammalian nervous system. *J. Neurosci.* 12:2288–2302.

Bredenbeek, P.J., Frolov, I., Rice, C.M., and Schlesinger, S. 1993. Sindbis virus expression vectors: Packaging of RNA replicons by using defective helper RNAs. *J. Virol.* 67:6439–6446.

Brelje, T.C., Wessendorf, M.W., and Sorenson, R.L. 1993. Multicolor laser scanning confocal immunofluorescence microsopy: Practical applications and limitations. *In* Cell Biological Applications of Confocal Microscopy (B. Matsumoto, ed.) pp. 98–182. Academic Press, San Diego.

Brent, R. and Ptashne, M. 1984. A bacterial repressor protein or a yeast transcriptional terminator can block upstream activation of a yeast gene. *Nature* 312:612–615.

Bridgman, P.C. and Reese, T.S. 1984. The structure of cytoplasm in dirctly frozen cultured cells. I:. Filamentous meshworks and the cytoplasmic ground substance. *J. Cell Biol.* 99:1655–1668.

Brocard, J., Warot, X., Wendling, O., Messaddeq, N., Vonesch, J.L., Chambon, P., and Metzger, D. 1997. Spatio-temporally controlled site-specific somatic mutagenesis in the mouse. *Proc. Natl. Acad. Sci. U.S.A.* 94:14559–14563.

Brocard, J., Feil, R., Chambon, P., and Metzger, D. 1998. A chimeric Cre recombinase inducible by synthetic, but not by natural ligands of the glucocorticoid receptor. *Nucl. Acids Res.* 26:4086–4090.

Brown, P.O. and Botstein, D. 1999. Exploring the new world of the genome with DNA microarrays. *Nat. Genet.* 21:33–37.

Buchli, P.J., Wu, Z., and Ciardelli, T.L. 1997. The functional display of interleukin-2 on filamentous phage. *Arch. Biochem. Biophys.* 339:79–84.

Bunch, T.A., Grinblat, Y., and Goldstein, S.B. 1988. Characterization and use of the *Drosophila* metallothionein promoter in cultured *Drosophila melanogaster* cells. *Nucl. Acids Res.* 16:1043–1061.

Burlingham, B.T., Brown, D.T., and Doerfler, W. 1974. Incomplete particles of adenovirus. I. Characteristics of the DNA associated with incomplete adenovirions of types 2 and 12. *Virology* 60:419–430.

Bursztajn, S., DeSouza, R., McPhie, D.L., Berman, S.A., Shioi, J., Robakis, N.K., and Neve, R.L. 1998. Overexpression in neurons of human presenilin-1 or a presenilin-1 familial Alzheimer disease mutant does not enhance apoptosis. *J. Neurosci.* 18:9790–9799.

Burton, R.A., Gibeaut, D.M., Bacic, A., Findlay, K., Roberts, K., Hamilton, A., Baulcombe, D.C., and Fincher, G.B. 2000. Virus-induced silencing of a plant cellulose synthase gene. *Plant Cell* 12:691–705.

Byrnes, A.P., MacLaren, R.D., and Charlton, H.M. 1996. Immunological instability of persistent adenovirus vectors in the brain: Peripheral exposure to vector leads to renewed inflammation, reduced gene expression, and demyelination. *J. Neurosci.* 16:3045–3055.

Cai, L., Hayes, N.L., and Nowakowski, R.S. 1997. Local homogeneity of cell cycle length in developing mouse cortex. *J. Neurosci.* 17:2079–2087.

Carlezon, W.A. Jr., Boundy, V.A., Haile, C.N., Kalb, R.G., Neve, R.L., and Nestler, E.J. 1997. Sensitization to morphine induced by viral-mediated gene transfer. *Science* 277:812–814.

Casanova, E., Fehsenfeld, S., Mantamadiotis, T., Lemberger, T., Greiner, E., Stewart, A.F., and Schutz, G. 2001. A CamKIIalpha iCre BAC allows brain-specific gene inactiviation. *Genesis* 31:37–42.

Castle, D.J. 1995. Purification of organelles from mammalian cells. *In* Current Protocols in Protein Science (J.E. Coligan, B.M. Dunn, H.L. Ploegh, D.W. Speicher, and P.T. Wingfield, eds.) pp. 4.2.1–4.2.56. John Wiley & Sons, New York.

Caudy, A.A., Myers, M., Hannon, G.J., and Hammond, S.M. 2002. Fragile X-related protein and VIG associate with the RNAi machinery. *Genes Dev.* 16:2491–2496.

Centonze, V. and Pawley, J. 1995. Tutorial on practical confocal microscopy and use of the confocal test specimen. In Handbook of Biological Confocal Microscopy, 2nd ed. (J. Pawley, ed.) pp. 549–570. Plenum Press, New York.

Chapot, M.P., Eshdat, Y., Marullo, S., Guillet, J.G., Charbit, A., Strosberg, A.D., and Delavier, K.C. 1990. Localization and characterization of three different beta-adrenergic receptors expressed in *Escherichia coli*. *Eur. J. Biochem.* 187:137–144.

Chartier, C., Degryse, E., Gantzer, M., Dieterle, A., Pavirani, A., and Mehtali, M., 1996. Efficient generation of recombinant adenovirus vectors by homologous recombination in *Escherichia coli*. *J. Virol.* 70:4805–4810.

Chemical Rubber Company. 1975. CRC Handbook of Biochemistry and Molecular Biology, Physical and Chemical Data, 3d ed., Vol. 1. CRC Press, Boca Raton, Fla.

Cheng, Y.C., Huang, E.S., Lin, J.C., Mar, E.C., Pagano, J.S., Dutschman, G.E., and Grill, S.P. 1983. Unique spectrum of activity of 9-[(1,3-dihydroxy-2-propoxy)methyl]guanine against herpesviruses in vitro and its mode of action against herpes simplex virus type 1. *Proc. Natl. Acad. Sci. U.S.A.* 80:2767–2770.

Cherbas, L., Moss, R., and Cherbas, P. 1994. Transformation techniques for *Drosophila* cell lines. *In* Methods in Cell Biology, (S. Lawrence, B. Goldstein, and E.A. Fyrberg, eds.) pp. 161–179. Academic Press, New York.

Cheung, V.G., Gregg, J.P., Gogolin-Ewens, K.J., Bandong, J., Stanley, C.A., Baker, L., Higgins, M.J., Nowak, N.J., Shows, T.B., Ewens, W.J., Nelson, S.F., and Spielman, R.S. 1998. Linkage-disequilibrium mapping without genotyping. *Nature Genet.* 18:225–230.

Chien, C.-T., Bartel, P.L., Sternglanz, R., and Fields, S. 1991. The two-hybrid system: A method to identify and clone genes for proteins that interact with a protein of interest. *Proc. Natl. Acad. Sci. U.S.A.* 88:9578–9582.

Chirgwin, J.J., Przbyla, A.E., MacDonald, R.J., and Rutter, W.J. 1979. Isolation of biologically active ribonucleic acid from sources enriched in ribonuclease. *Biochemistry* 18:5294.

Chiswell, D.J. and McCafferty, J. 1992. Phage antibodies: Will new "coliclonal" antibodies replace monoclonal antibodies? *Trends Biotech.* 10:80–84.

Cho, R.J., Campbell, M.J., Winzeler, E.A., Steinmetz, L., Conway, A., Wodicka, L., Wolfsberg, T.G., Gabrielian, A.E., Landsman, D., Lockhart, D.J., and Davis, R.W. 1998. A genome-wide transcriptional analysis of the mitotic cell cycle. *Mol. Cell* 2:65–73.

Choi-Lundberg, D.L., Lin, Q., Chang, Y.-N., Chiang, Y.L., Hay, C.M., Mohajeri, H., Davidson, B.L., and Bohn, M.C. 1997. Dopaminergic neurons protected from degeneration by GDNF gene therapy. *Science* 275:838–841.

Chomczynski, P. and Sacchi, N. 1987. Single-step method of RNA isolation by acid guanidinium thiocyanate-phenol-chloroform extraction. *Anal. Biochem.* 162:156–159.

Chou, P.Y. and Fasman, G.D. 1974. Prediction of protein conformation. *Biochemistry* 13:222–245.

Chu, S., DeRisi, J., Eisen, M., Mulholland, J., Botstein, D., Brown, P.O., and Herskowitz, I. 1998. The transcriptional program of sporulation in budding yeast. *Science* 282:699–705.

Chuang, C.-H. and Meyerowitz, E.M. 2000. Specific and heritable genetic interference by double-stranded RNA in *Arabidopsis thaliana*. *Proc. Natl. Acad. Sci. U.S.A.* 97:4985–4990.

Cisterni, C., Henderson, C.E., Aebischer, P., Pettmann, B., and Deglon, N. 2000. Efficient gene transfer and expression of biologically active glial cell line-derived neurotrophic factor in rat motoneurons transduced with lentiviral vectors. *J. Neurochem.* 74:1820–1828.

Clackson, T. and Wells, J.A. 1994. In vitro selection from protein and peptide libraries. *Trends Biotech.* 12:173–184.

Clarke, S. 1975. The size and detergent binding of membrane proteins. *J. Biol. Chem.* 250:5459–5469.

Clegg, R.M. 1996. Fluorescence resonance energy transfer spectroscopy and microscopy. *In* Fluorescence Imaging Spectroscopy and Microscopy (X.F. Wang and B. Herman eds.), pp. 179–251. John Wiley & Sons, New York.

Cogoni, C. and Macino, G. 1999. Gene silencing in *Neurospora crassa* requires a protein homologous to RNA-dependent RNA polymerase. *Nature* 399:166–169.

Cohen, L. 1993. Optical monitoring of physiological activity: A brief history. *Jpn. J. Physiol.* 43(S1–S6).

Colas, P., Cohen, B., Jessen, T., Grishina, I., McCoy, J., and Brent, R. 1996. Genetic selection of peptide aptamers that recognize and inhibit cyclin-dependent kinase 2. *Nature* 380:548–550.

Cole, R. and de Vellis, J. 1989. Preparation of astrocyte and oligodendrocyte cultures from primary rat glial cultures. *In* A Dissection and Tissue Culture Manual of the Nervous System (A. Shahar, J. de Vellis, A. Vernadakis, and B. Haber, eds.) pp. 121–133. Wiley-Liss, New York.

Coligan, J.E., Kruisbeek, A.M., Margulies, D.H., Shevach, E.M., and Strober, W. (eds.) 2006a. Current Protocols in Immunology. John Wiley & Sons, New York.

Coligan, J.E., Dunn, B.M., Ploegh, H.L., Speicher, D.W., and Wingfield, P.T. (eds.) 2006b. Current Protocols in Protein Science. John Wiley & Sons, New York.

Cooper, J.R., Bloom, F.E., and Roth, R.H. 1996. The Biochemical Basis of Neuropharmacology, 7th ed. Oxford University Press, New York.

Corrigan, A.J. and Huang, P.C. 1982. A basic microcomputer program for plotting the secondary structure of proteins. *Comput. Programs Biomed.* 3:163–168.

Cortese, I., Capone, S., Tafi, R., Grimaldi, L.M., Nicosia, A., and Cortese, R. 1998. Identification of peptides binding to IgG in the CSF of multiple sclerosis patients. *Mult. Scler.* 4:31–36.

Cortese, R., Felici, F., Galfre, G., Luzzago, A., Monaci, P., and Nicosia, A. 1994. Epitope discovery using peptide libraries displayed on phage. *Trends Biotech.* 12:262–267.

Cowsill, C., Southgate, T.D., Morrissey, G., Dewey, R.A., Morelli, A.E., Maleniak, T.C., Forrest, Z., Klatzmann, D., Wilkinson, G.W.G., Lowenstein, P.R., and Castro, M.G., 2000. Central nervous system toxicity of two adenoviral vectors encoding variants of the herpes simplex virus type 1 thymidine kinase: Reduced cytotoxicity of a truncated HSV1-TK. *Gene Ther.* 7:679–685.

Crameri, R. and Suter, M. 1993. Display of biologically active proteins on the surface of filamentous phages: A cDNA cloning system for selection of functional gene products linked to the genetic information responsible for their production. *Gene* 137:69–75.

Cremo, C.R., Herron, G.S., and Schimerlik, M.I. 1981. Solubilization of the atrial muscarinic receptor: A new detergent system and rapid assays. *Anal. Biochem.* 115:331–338.

Croen, K.D., Ostrove, J.M., Dragovic, L.J., Smialek, J.E., and Straus, S.E. 1987. Latent herpes simplex virus in human trigeminal ganglia. Detection of an immediate-early gene anti-sense transcript by *in situ* hybridization. *New Engl. J. Med.* 317: 1422–1432.

Cruz, A., Coburn, C.M., and Beverley, S.M. 1991. Double targeted gene replacement for creating null mutants. *Proc. Natl. Acad. Sci. U.S.A.* 88:7170–7174.

Cuello, A.C. 1993. Immunohistochemistry II (IBRO Handbook Series: Methods in the Neurosciences). John Wiley & Sons, Chichester, U.K.

Cunningham, C. and Davison, A.J. 1993. A cosmid-based system for constructing mutants of herpes simplex virus type 1. *Virology* 197:116–124.

Cunningham, M. and McKay, R.D.G. 1993. A hypothermic miniaturized stereotaxic instrument for surgery in newborn rats. *J. Neurosci. Methods* 47:105–114.

Cwirla, S.E., Balasubramanian, P., Duffin, D., Wagstrom, C., Gates, C., Singer, S., Davis, A., Tansik, R., Mattheakis, L., Boytos, C., Schatz, P., Baccanari, D., Wrighton, N., Barrett, R., and Dower, W. 1997. Peptide agonist of the thrombopoietin receptor as potent as the natural cytokine. *Science* 276:1696–1699.

Czernik, A.J., Girault, J.A., Nairn, A.C., Chen, J., Snyder, G., Kebabian, J. and Greengard, P. 1991. Production of phosphorylation state-specific antibodies. *Methods Enzymol.* 201:264–283.

Dalmay, T., Hamilton, A.J., Rudd, S., Angell, S., and Baulcombe, D.C. 2000. An RNA-dependent RNA polymerase gene in *Arabidopsis* is required for posttranscriptional gene silencing mediated by a transgene but not by a virus. *Cell* 101:543–553.

Davies, J. and Riechmann, L. 1996. Single antibody domains as small recognition units: Design and in vitro antigen selection of camelized, human VH domains with improved protein stability. *Protein Eng.* 9:531–537.

Dawes, C.J. 1971. Biological Techniques in Electron Microscopy (International Textbook Series). Barnes & Noble, New York.

DeAngelis, M.M., Wang, D.G., and Hawkins, T.L. 1995. Solid-phase reversible immobilization for the isolation of PCR products. *Nucl. Acids Res.* 23:4742–4743.

DeFalco, J., Tomishima, M., Liu, H., Zhao, C., Cai, X., Marth, J.D., Enquist, L., and Friedman, J.M. 2001. Virus-assisted mapping of neural inputs to a feeding center in the hypothalamus. *Science* 291: 2608–2613.

Deglon, N., Tseng, J.L., Bensadoun, J.C., Zurn, A.D., Arsenijevic, Y., Pereira de Almeida, L., Zufferey, R., Trono, D., and Aebischer, P. 2000. Self-inactivating lentiviral vectors with enhanced transgene expression as potential gene transfer system in Parkinson's disease. *Hum. Gene Ther.* 11:179–190.

de la Luna, S., Soria, I., Pulido, D., Ortin, J., and Jimenez, A. 1988. Efficient transformation of mammalian cells with constructs containing a puromycin-resistance marker. *Gene* 62:121–126.

DeLuca, N.A. and Schaffer, P.A. 1987. Activities of herpes simplex virus type 1 (HSV-1) ICP4 genes specifying nonsense peptides. *Nucl. Acids Res.* 15:4491–4511.

DeLuca, N.A., McCarthy, A.M. and Schaffer, P.A. 1985. Isolation and characterization of deletion mutants of herpes simplex virus type 1 in the gene encoding immediate-early regulatory protein ICP4. *J. Virol.* 56:558–570.

Deng, S.J., MacKenzie, C.R., Sadowska, J., Michniewicz, J., Young, N.M., Bundle, D.R., and Narang, S.A. 1994. Selection of antibody single-chain variable fragments with improved carbohydrate binding by phage display. *J. Biol. Chem.* 269: 9533–9538.

Dente, L., Vetriani, C., Zucconi, A., Pelicci, G., Lanfrancone, L., Pelicci, P.G., and Cesareni, G. 1997. Modified phage peptide libraries as a tool to study specificity of phosphorylation and recognition of tyrosine containing peptides. *J. Mol. Biol.* 269:694–703.

DeRisi, J.L., Iyer, V.R., and Brown, P.O. 1997. Exploring the metabolic and genetic control of gene expression on a genomic scale. *Science* 278:680–686.

Dewey, R.A., Morrissey, G., Cowsill, C.M., Stone, D., Bolognani, F., Dodd, N.J.F., Southgate, T.D., Katzmann, D., Lassmann, H., Castro, M.G., and Lowenstein, P.R. 1999. Chronic brain inflammation and persistent herpes simplex virus 1 thymidine kinase expression in survivors of syngenic glioma treated by adenovirus-mediated gene therapy: Implications for clinical trials. *Nature Med.* 5:1256–1263.

Diehn, M., Eisen, M., Brown, P., and Botstein, D. 2000. Large-scale identification of membrane-associated gene products using DNA microarrays. *Nature Genet.* 25:58–62.

Dion, L.D., Fang, J., and Garver, R.I. 1996. Supernatant rescue assay vs. polymerase chain reaction for detection of wild type adenovirus-containing recombinant adenovirus stocks. *J Virol. Method* 56:99–107.

Djikeng, A., Shi, H., Tschudi, C., and Ullu, E. 2001. RNA interference in *Trypanosoma brucei*: Cloning of small interfering RNAs provides evidence for retroposon-derived 24–26 nucleotide RNAs. *RNA* 7:1522–1530.

Donis, J.A., Michelman, M.V., and Neve, R.L. 1993. Comparison of expression of a series of mammalian vector promoters in the neuronal cell lines PC12 and HT4. *BioTechniques* 15:786–787.

Dower, W.J., Miller, J.F., and Ragdale, C.W. 1988. High efficiency transformation of *E. coli* by high voltage electroporation. *Nucl. Acids Res.* 16:6127–6145.

Dulac, C. and Axel, R. 1995. A novel family of genes encoding putative pheromone receptors in mammals. *Cell* 83:195–206.

Dulbecco, R. and Vogt, M. 1954. Plaque formation and isolation of pure lines with poliomyelitis viruses. *J. Exp. Med.* 99:167–182.

Dunkley, P.R., Heath, J.W., Harrison, S.M., Jarvie, P.E., Glenfield, P.J., and Rostas, J.A. 1988. A rapid Percoll gradient procedure for isolation of synaptosomes directly from an S1 fraction: Homogeneity and morphology of subcellular fractions. *Brain Res.* 441:59–71.

Dunnett, S.B. and Bjorklund, A. (eds.) 1992. Neural Transplantation. A Practical Approach. Oxford University Press, New York.

Durfee, T., Becherer, K., Chen, P.L., Yeh, S.H., Yang, Y., Kilburn, A.E., Lee, W.H., and Elledge, S.J. 1993. The retinoblastoma protein associates with the protein phosphatase type 1 catalytic subunit. *Genes Dev.* 7:555–569.

During, M.J., Naegele, J.R., O'Malley, K.L., and Geller, A.I. 1994. Long-term behavioral recovery in Parkinsonian rats by an HSV vector expressing tyrosine hydroxylase. *Science* 266:1399–1403.

Easton, R.M., Johnson, E.M., and Creedon, D.J. 1998. Analysis of events leading to neuronal death after infection with E1-deficient adenovirus vectors. *Mol. Cell. Neurosci.* 11:334–347

Eberwine, J. 1996. Amplification of mRNA populations using aRNA generated from immobilized oligo (dT)-T7 primed cDNA. *Biotechniques* 20:584–591.

Eberwine, J., Yeh, H., Miyashiro, K., Cao, Y., Nair, S., Finnell, R., Zettel, M., and Coleman, P. 1992. Analysis of gene expression in single live neurons. *Proc. Natl. Acad. Sci. U.S.A.* 89:3010–3014.

Eberwine, J., Crino, P., and Dichter, M. 1995. Single cell mRNA amplification: Basic science and clinical implications. *Neuroscientist* 1:200–211.

Ehrengruber, M.U., Lundstrom, K., Schweitzer, C., Heuss, C., Schlesinger, S., and Gahwiler, B.H. 1999. Recombinant Semliki Forest virus and Sindbis virus efficiently infect neurons in hippocampal slice cultures. *Proc. Natl. Acad. Sci. U.S.A.* 96:7041–7046.

Eisen, M.B., Spellman, P.T., Brown, P.O., and Botstein, D. 1998. Cluster analysis and display of genome-wide expression patterns. *Proc. Natl. Acad. Sci. U.S.A.* 95:14863–14868.

Elbashir, S.M., Martinez, J., Patkaniowska, A., Lendeckel, W., and Tuschl, T. 2001a. Functional anatomy of siRNAs for mediating efficient RNAi in Drosophila melanogaster embryo lysate. *EMBO J.* 20:6877–6888.

Elbashir, S.M., Harborth, J., Lendeckel, W., Yalcin, A., Weber, K., and Tuschl, T. 2001b. Duplexes of 21-nucleotide RNAs mediate RNA interference in cultured mammalian cells. *Nature* 411:494–498.

Fack, F., Hugle-Dorr, B., Song, D., Queitsch, I., Petersen, G., and Bautz, E.K. 1997. Epitope mapping by phage display: Random versus gene-fragment libraries. *J. Immunol. Methods* 206:43–52.

Fearon, E.R., Finkel, T., Gillison, M.L., Kennedy, S.P., Casella, J.F., Tomaselli, G.F., Morrow, J.S., and Dang, C.V. 1992. Karyoplasmic interaction selection strategy: A general strategy to detect protein-protein interactions in mammalian cells. *Proc. Nat. Acad. Sci. U.S.A.* 89:7958–7962.

Feil, R., Brocard, J., Mascrez, B., LeMeur, M., Metzger, D., and Chambon, P. 1996. Ligand-activated site-specific recombination in mice. *Proc. Natl. Acad. Sci. U.S.A.* 93:10887–10890.

Feil, R., Wagner, J., Metzger, D., and Chambon, P. 1997. Regulation of Cre recombinase activity by mutated estrogen receptor ligand-binding domains. *Biochem. Biophys. Res. Commun.* 237:752–757.

Feinendegen, L.E. 1967. Tritium-Labeled Molecules in Biology and Medicine. Academic Press, New York.

Ferea, T.L., Botstein, D., Brown, P.O., and Rosenzweig, R.F. 1999. Systematic changes in gene expression patterns following adaptive evolution in yeast. *Proc. Natl. Acad. Sci. U.S.A.* 96:9721–9726.

Fiering, S., Kim, C.G., Epner, E.M., and Groudine, M. 1993. An "in-out" strategy using gene targeting and FLP recombinase for the functional dissection of complex DNA regulatory elements: Analysis of the β-globin locus control region. *Proc. Natl. Acad. Sci. U.S.A.* 90:8469–8473.

Fiering, S., Epner, E., Robinson, K., Zhuang, Y., Telling, A., Hu, M., Martin, D.I., Enver, T., Ley, T.J., and Groudine, M. 1995. Targeted deletion of 5′HS2 of the murine β-globin LCR reveals that it is not essential for proper regulation of the β-globin locus. *Genes Dev.* 9:2203–2213.

Fiering S., Bender, M.A., and Groudine, M. 1999. Analysis of mammalian cis-regulatory DNA elements by homolgous recombination. *Methods Enzymol.* 306:42–66.

Fink, D.J., DeLuca, N.A., Goins, W.F., and Glorioso, J.C. 1996. Gene transfer to neurons using herpes simplex virus-based vectors. *Annu. Rev. Neurosci.* 19:265–287.

Finley, R.L., Jr., and Brent, R. 1994. Interaction mating reveals binary and ternary connections between *Drosophila* cell cycle regulators. *Proc. Natl. Acad. Sci. U.S.A.* 91:12980–12984.

Fire, A., Albertson, D., Harrison, S., and Moerman, D. 1991. Production of antisense RNA leads to effective and specific inhibition of gene expression in *C. elegans*. *Development* 113:503–514.

Fitch, F.W., Gajewski, T.F., and Yokoyama, W.M. 1997. Diagnosis and treatment of mycoplasma-contaminated cell cultures. *In* Current Protocols in Immunology (J.E. Coligan, A.M. Kruisbeek, D.H. Margulies, E.M. Shevach, and W. Strober, eds.) pp. A.3E.1–A.3E.4. John Wiley & Sons, New York.

Fodor, S.P., Read, J.L., Pirrung, M.C., Stryer, L., Lu, A.T., and Solas, D. 1991. Light-directed, spatially addressable parallel chemical synthesis. *Science* 251:767–773.

Follenzi, A., Ailles, L.E., Bakovic, S., Geuna, M., and Naldini, L. 2000. Gene transfer by lentiviral vectors is limited by nuclear translocation and rescued by HIV-1 *pol* sequences. *Nat. Genet.* 25:217–222.

Fraefel, C., Song, S., Lim, F., Lang, P., Yu, L., Wang, Y., Wild, P., and Geller, A.I. 1996. Helper virus-free transfer of herpes simplex virus type 1 plasmid vectors into neural cells. *J. Virol.* 70:7190–7197.

Fraefel, C., Jacoby, D.R., Lage, C., Hilderbrand, H., Chou, J.Y., Alt, F.W., Breakefield, X.O., and Majzoub, J.A. 1997. Gene transfer into hepatocytes mediated by helper virus free HSV/AAV hybrid vectors. *Mol. Med.* 3:813–825.

Fraefel, C., Breakefield, X.O., and Jacoby, D.R. 1998. HSV-1 amplicon. *In* Gene Therapy for Neurological Disorders and Brain Tumors (E.A. Chiocca and X.O. Breakefield, eds.) pp. 63–82. Humana Press, Totowa, N.J.

Freshney, R.I. 1993. Culture of Animal Cells. A Manual of Basic Techniques, 3rd ed. Wiley-Liss, New York.

Fuerst, T.R., Niles, E.G., Studier F.W., and Moss, B. 1986. Eukaryotic transient-expression system based on recombinant vaccinia virus that synthesizes bacteriophage T7 RNA polymerase. *Proc. Natl. Acad. Sci. U.S.A.* 83:8122–8126.

Gaffney, D.F., McLauchlin, J., Whitton, J.L., and Clements, J.B. 1985. A modular system for the assay of transcription regulatory signals: The sequence TAATGARAT is required for herpes simplex virus immediate early gene activation. *Nucl. Acids Res.* 13:7847–7863.

Gahwiler, B.H. 1981. Organotypic monolayer cultures of nervous tissue. *J. Neurosci. Methods* 4:329–342.

Garavito, M.R., Picot, D., and Loll, P.J. 1996. Strategies for crystallizing membrane proteins. *J. Bioenerg. Biomembr.* 28:13–27.

Gay, P., Le Coq, D., Steinmetz, M., Berkelman, T., and Kado, C.I. 1985. Positive selection procedure for entrapment of insertion sequence elements in gram-negative bacteria. *J. Bacteriol.* 164:918–921.

Geerligs, H.J., Weijer, W.J., Bloemhoff, W., Welling, G.W., and Welling-Wester, S. 1988. The influence of pH and ionic strength on the coating of peptides of herpes simplex virus type 1 in an enzyme-linked immunosorbent assay. *J. Immunol. Methods* 106:239–244.

Geller, A.I., and Breakefield, X.O. 1988. A defective HSV-1 vector expresses Escherichia coli beta-galactosidase in cultured peripheral neurons. *Science* 241:1667–1669.

Geller, A.I., Keyomarsi, K., Bryan, J., and Pardee, A.B. 1990. An efficient deletion mutant packaging system for defective herpes simplex virus vectors: Potential applications to human gene therapy and neuronal physiology. *Proc. Natl. Acad. Sci. U.S.A.* 87:8950–8954.

Geller, A.I., During, M.J., Haycock, J.W., Freese, A., and Neve, R. 1993. Long-term increases in neurotransmitter release from neuronal cells expressing a constitutively active adenylate cyclase from a herpes simplex virus type 1 vector. *Proc. Natl. Acad. Sci. U.S.A.* 90:7603–7607.

Gerdes, C.A., Castro, M.G., and Lowenstein, P.R. 2000. Strong promoters are the key to highly efficient, noninflammatory and noncytotoxic adenoviral-mediated transgene delivery into the brain in vivo. *Mol. Ther.* 2:330–338.

Geschwind, D.H. 2001. Sharing gene expression data: An array of options. *Nat. Rev. Neuro.* 2:436–438.

Geschwind, D.H., Ou, J., Easterday, M.C., Dougherty, J.D., Jackson, R.L., Chen, Z., Antoine, H., Terskikh, A., Weissman, I.L., Nelson, S.F., and Kornblum, H.I. 2001. A genetic analysis of neural progenitor differentiation. *Neuron* 29:325–339.

Gillespie, P.G. and Hudspeth, A.J. 1991. Chemiluminescence detection of proteins from single cells. *Proc. Natl. Acad. Sci. U.S.A.* 88:2563–2567.

Giloh, H. and Sedat, J.W. 1982. Fluorescence microscopy: Reduced photobleaching of rhodamine and fluorescein protein conjugates by n-propyl gallate. *Science* 217:1252–1255.

Giordano, E., Rendina, R., Peluso, I., and Furia, M. 2002. RNAi triggered by symmetrically transcribed transgenes in *Drosophila melanogaster*. *Genetics* 160:637–648.

Gluzman, Y. and Van Doren, K. 1983. Palindromic adenovirus type 5-SV40 hybrid. *J. Virol.* 45:91–103.

Gong, S., Yang, X.W., Li, J., and Heintz, N. 2002. Highly efficient modification of bacterial artificial chromosomes (BACs) using novel shuttle vectors containing the R6Kγ origin of replication. *Genome Res.* 12:1992–1998.

Gong, S., Zheng, C., Goughty, M.L., Losos, K., Didkovsky, N., Schambra, U.B., Nowak, N.J., Joyner, A., Leblanc, G., Hatten, M.E., and Heintz, N. 2003. A gene expression atlas of the central nervous system based on bacterial artificial chromosomes. *Nature* 425:917–925.

Gossen, M. and Bujard, H. 1992. Tight control of gene expression in mammalian cells by tetracycline-responsive promoters. *Proc. Natl. Acad. Sci. U.S.A.* 89:5547–5551.

Graham, F.L., Smiley, J., Russell, W.C., and Nairu, R., 1977. Characteristics of a human cell line transformed by DNA from human adenovirus type 5. *J. Gen. Virol.* 68:937–940.

Graus, Y.F., de Baets, M.H., Parren, P.W., Berrih-Aknin, S., Wokke, J., van Breda Vriesman, P.J. and Burton, D.R. 1997. Human anti-nicotinic acetylcholine receptor recombinant Fab fragments isolated from thymus-derived phage display libraries from myasthenia gravis patients reflect predominant specificities in serum and block the action of pathogenic serum antibodies. *J. Immunol.* 158:1919–1929.

Gravel, C., Gotz, R., Lorrain, A., and Sendtner, M. 1997. Adenoviral gene transfer of ciliary neurotrophic factor and brain-derived neurotrophic factor leads to long-term survival of axotomized motor neurons. *Nat. Med.* 3:765–770.

Greene, L.A. and Tischler, A.S. 1976. Establishment of a noradrenergic clonal line of rat adrenal pheochromocytoma cells which respond to nerve growth factor. *Proc. Natl. Acad. Sci. U.S.A.* 73:2424–2428.

Gritz, L. and Davies, J. 1983. Plasmid-encoded hygromycin-B resistance: The sequence of hygromycin-B-phosphotransferase gene and its expression in *E. coli* and *S. cerevisiae*. *Gene* 25:179–188.

Gross-Bellard, M., Oudet, P., and Chambon, P. 1973. Isolation of high-molecular-weight DNA from mammalian cells. *Eur. J. Biochem.* 36:32–38.

Gu, H., Marth, J.D., Orban, P.C., Mossmann, H., and Rajewsky, K. 1994. Deletion of a DNA polymerase β gene segment in T cells using cell type-specific gene targeting. *Science* 265:103–106.

Gu, H., Yi, Q., Bray, S.T., Riddle, D.S., Shiau, A.K., and Baker, D. 1995. A phage display system for studying the sequence determinants of protein folding. *Protein Sci.* 4:1108–1117.

Guo, S. and Kemphues, K. 1995. Par-1, a gene required for establishing polarity in C. elegans embryos, encodes a putative Ser/Thr kinase that is asymmetrically distributed. *Cell* 81:611–620.

Gyuris, J., Golemis, E.A., Chertkov, H., and Brent, R. 1993. Cdi1, a human G1- and S-phase protein phosphatase that associates with Cdk2. *Cell* 75:791–803.

Haase, G., Kennel, P., Petemann, B., Vigne, E., Akli, S., Revah, F., Schmalbruch, H., and Kahn, A. 1997. Gene therapy of murine motor neurons disease using adenoviral vectors for neurotrophic factors. *Nat. Med.* 3:429–436.

Hacia, J.G., Brody, L.C., Chee, M.S., Fodor, S.P., and Collins, F.S. 1996. Detection of heterozygous mutations in *BRCA1* using high density oligonucleotide arrays and two-colour fluorescence analysis. *Nature Genet.* 14:441–447.

Hamilton, A.J. and Baulcombe, D.C. 1999. A novel species of small antisense RNA in post-transcriptional gene silencing. *Science* 286:950–952.

Hamilton, A.J., Voinnet, O., Chappell, L., and Baulcombe, D.C. 2002. Two classes of short interfering RNA in RNA silencing. *EMBO J.* 21:4671–4679.

Hammond, S.M., Bernstein, E., Beach, D., and Hannon, G. 2000. An RNA-directed nuclease mediates post-transcriptional gene silencing in *Drosophila* cell extracts. *Nature* 404:293–296.

Hammond, S.M., Boettcher, S., Caudy, A.A., Kobayashi, R., and Hannon, G.J. 2001. Argonaute2, a link between genetic and biochemical analyses of RNAi. *Science* 293:1146–1150.

Hanahan, D. 1983. Studies on transformation of *Escherichi coli* with plasmids. *J. Mol. Biol.* 166:557–580.

Hanker, J.S., Yates, P.E., Metz, C.B., and Rustioni, A. 1977. A new specific, sensitive and non-carcinogenic reagent for the demonstration of horseradish peroxidase. *Histochem. J.* 9:789–792.

Hannas-Djebbara, Z., Bazes, M.D., Sacchettoni, S., Prod'hon, C., Jouvet, M., Belin, M.-F., and Jacquemont, B. 1997. Transgene expression of plasmid DNAs directed by viral or neural promoters in the rat brain. *Mol. Brain Res.* 46:91–99.

Harborth, J., Elbashir, S.M., Bechert, K., Tuschl, T., and Weber, K. 2001. Identification of essential genes in cultured mammalian cells using small interfering RNAs. *J. Cell Sci.* 114:4557–4565.

Harlow, E., and Lane, D. 1988. Antibodies: A Laboratory Manual. Cold Spring Harbor Laboratory Press, Cold Spring Harbor, N.Y.

Harlow, E. and Lane, D. 1988. Immunoblotting. *In* Antibodies: A Laboratory Manual, pp. 471–510. CSH Laboratory, Cold Spring Harbor, N.Y.

Harlow, E. and Lane, D. 1999. Antibodies: A Laboratory Manual. Cold Spring Harbor Laboratory Press, Cold Spring Harbor, N.Y.

Harris, S., Craig, L., Mehroke, J., Rashed, M., Zwick, M., Kenar, K., Toone, E., Greenspan, N., Auzanneau, F., Marino-Albernas, J., Pinto, B., and Scott, J. 1997. Exploring the basis of peptide-carbohydrate cross-reactivity: Evidence for discrimination by peptides between closely related anti-carbohydrate antibodies. *Proc. Natl. Acad. Sci. U.S.A.* 94:2454–2459.

Hartman, S.C. and Mulligan, R.C. 1988. Two dominant-acting selectable markers for gene transfer studies in mammalian cells. *Proc. Natl. Acad. Sci. U.S.A.* 85:8047–8051.

Hatten, M.E. 1985. Neuronal regulation of astroglial morphology and proliferation in vitro. *J. Cell Biol.* 100:384–396.

Haugland, R.P. 1996. Handbook of Fluorescent Probes and Research Chemicals Molecular Probes, Inc., Eugene, Ore.

Hayward, R., DeRisi, J., Alfadhli, S., Kaslow, D., Brown, P., and Rathod, P. 2000. Shotgun DNA microarrays and stage-specific gene expression in *Plasmodium falciparum* malaria. *Mol. Microbiol.* 35:6–14.

Hennecke, M., Kola, A., Baensch, M., Wrede, A., Klos, A., Bautsch, W., and Kohl, J. 1997. A selection system to study C5a-C5a-receptor interactions: Phage display of a novel C5a anaphylatoxin, Fos-C5aAla27. *Gene* 184:263–272.

Hjelmeland, L.M. 1990. Solubilization of native membrane proteins. *Methods Enzymol.* 182:253–264.

Hjelmeland, J.M. and Chrambach, A. 1984a. Solubilization of functional membrane proteins. *Methods Enzymol.* 104:305–318.

Hjelmeland, L.M. and Chrambach, A. 1984b. Solubilization of functional membrane-bound receptors. *In* Membranes, Detergents, and Receptor Solubilization (J.C. Venter and L.C. Harrison, eds.) pp. 35–46. Alan R. Liss, New York.

Hodgson, J.G., Agopyan, N., Gutekunst, C.A., Leavitt, B.R., LePiane, F., Singaraja, R., Smith, D.J., Bissada, N., McCutcheon, K., Nasir, J., Jamot, L., Li, X.J., Stevens, M.E., Rosemond, E., Roder, J.C., Phillips, A.G., Rubin, E.M., Hersch, S.M., and Hayden, M.R. 1999. A YAC mouse model for Huntington's disease with full-length mutant huntingtin, cytoplasmic toxicity, and selective striatal neurodegeneration. *Neuron* 23:181–192.

Hoffmann, S., and Willbold, D. 1997. A selection system to study protein-RNA interactions: Functional display of HIV-1 Tat protein on filamentous bacteriophage M13. *Biochem. Biophys. Res. Commun.* 235:806–811.

Honess, R.W. and Roizman, B. 1974. Regulation of herpes virus macromolecular synthesis. I. Cascade regulation of the synthesis of three groups of viral proteins. *J. Virol.* 14:8–19.

Hopp, T.P. and Woods, K.R. 1981. Prediction of protein antigenic determinants from amino acid sequences. *Proc. Natl. Acad. Sci. U.S.A.* 78:3824–3828.

Horikawa, K. and Armstrong W. 1988. A versatile means of intracellular labeling: Injection of biocytin and its detection with avidin conjugates. *J. Neurosci. Methods* 25:1–11.

Hulme, E.C. and Birdsall, N.J.M. 1992. Strategy and tactics in receptor-binding. *In* Studies in Receptor-Ligand Interactions. A Practical Approach. (E.C. Hulme, ed.) pp. 153–156. IRL Press, Oxford.

Hunter, A.J., Leslie, R.A., Gloger, I.S., and Lawrence, M. 1995. Probing the function of novel genes in the nervous system: Is antisense the answer? *Trends Neurosci.* 18:320–323.

Hunyady, B., Krempels, K., Harta, G., and Mezey, E. 1996. Immunohistochemical signal amplification by catalyzed reporter deposition and its application in double immunostainings. *J. Histochem. Cytochem.* 12:1353–1362.

Iannolo, G., Minenkova, O., Petruzzelli, R., and Cesareni, G. 1995. Modifying filamentous phage capsid: Limits in the size of the major capsid protein. *J. Mol. Biol.* 248:835–844.

Inoue, S. 1986. Video Microscopy. Plenum Press, New York.

Isacson, I. 1995. Behavioral effects and gene delivery in a rat model of Parkinson's disease. *Science* 269: 856–857.

Iyer, V.R., Eisen, M.B., Ross, D.T., Schuler, G.T., Moore, J.C., Lee, F., Trent, J.M., Staudt, L.M., Hudson, J. Jr., Boguski, M.S., Lashkari, D., Shalon, D., Botstein, D., and Brown, P.O. 1999. The transcriptional program in the response of human fibroblasts to serum. *Science* 283:83–87.

Jacobsson, K., and Frykberg, L. 1998. Gene VIII-based, phage-display vectors for selection against complex mixtures of ligands. *BioTechniques* 24:294–301.

Jacobsson, K., Jonsson, H., Lindmark, H., Guss, B., Lindberg, M., and Frykberg, L. 1997. Shot-gun phage display mapping of two streptococcal cell-surface proteins. *Microbiol. Res.* 152:121–128.

Jespers, L.S., Messens, J.H., De Keyser, A., Eeckhout, D., Van Den Brande, I., Gansemans, Y.G., Lauwereys, M.J., Vlasuk, G.P., and Stanssens, P.E. 1995. Surface expression and ligand-based selection of cDNAs fused to filamentous phage gene VI. *Biotechnology (N.Y.)* 13:378–382.

Jespers, L., Jenne, S., Lasters, I., and Collen, D. 1997. Epitope mapping by negative selection of randomized antigen libraries displayed on filamentous phage. *J. Mol. Biol.* 269:704–718.

Jessen, J.R., Meng, A., McFarlane, R.J., Paw, B.H., Zon, L.I., Smith, G.R., and Lin, S. 1998. Modification of bacterial artificial chromosomes through chi-stimulated homologous recombination and its application in zebra fish transgenesis. *Proc. Natl. Acad. Sci. U.S.A.* 95:5121–5126.

Jin, B.K., Belloni, M., Conti, B., Federoff, H.J., Starr, R., Son, J.H., Baker, H., and Joh, T.H. 1996. Prolonged in vivo gene expression driven by a tyrosine hydroxylase promoter in a defective herpes simplex virus amplicon vector. *Hum. Gene Ther.* 7:2015–2024.

Johe, K.K., Hazel, T.G., Muller, T., Dugich-Djordjevic, M.M., and McKay, R.D.G. 1996. Single factors direct the differentiation of stem cells from the fetal and adult central nervous system. *Genes Dev.* 10:3129–3140.

Johnson, G.D., Davidson, R.S., McNamee, K.C., Russell, G., Goodwin, D., and Holborow, E.J. 1982. Fading of immunofluorescence during microscopy: A study of the phenomenon and its remedy. *J. Immunol. Methods.* 55:231–242.

Johnson, P.A., Miyanohara, A., Levine, F., Cahill, T., and Friedmann, T. 1992. Cytotoxicity of a replication-defective mutant of herpes simplex virus type 1. *J. Virol.* 66:2952–2965.

Johnston, K.M., Jacoby, D., Pechan, P., Fraefel, C., Borghesani, P., Schuback, D., Dunn, R.J., Smith, F.I., and Breakefield, X.O. 1997. HSV/AAV hybrid amplicon vectors extend transgene expression in human glioma cells. *Hum. Gene Ther.* 8:359–370.

Kajiwara, K., Byrnes, A.P., Charlton, H.M., Wood, M.J.A., and Wood, K.J. 1997. Immune responses to adenoviral vectors during gene transfer in the brain. *Hum. Gene Ther.* 8:253–265.

Kallioniemi, A., Kallioniemi, O.P., Sudar, D., Rutovitz, D., Gray, J.W., Waldman, F., and Pinkel, D. 1992. Comparative genomic hybridization for molecular cytogenetic analysis of solid tumors. *Science* 258:818–821.

Kamath, R.S., Fraser, A.G., Dong, Y., Poulin, G., Durbin, R., Gotta, M., Kanapin, A., Le Bot, N., Moreno, S., Sohrmann, M., Welchman, D.P., Zipperlen, P., and Ahringer, J. 2003. Systemic functional analysis of the Caenorhabditis elegans genome using RNAi. *Nature* 421:231–237.

Kaplitt, M.G., Leone, P., Samulski, R.J., Xiao, X., Pfaff, D.W., O'Malley, K.L., and During, M.J. 1994. Long-term gene expression and phenotypic correction using adeno-associated virus vectors in the mammalian brain. *Nature Genet.* 8:148–154.

Katz, J., Bodin, T., and Coen, D.M. 1990. Quantitative polymerase chain reaction analysis of herpes simplex virus DNA in ganglia of mice infected with replication incompetent mutants. *J. Virol.* 64:4288–4295.

Kaufman, M.H. 1995. The Atlas of Mouse Development. Academic Press, San Diego.

Kaufman, R.J., Murtha, P., Ingolia, D.E., Yeung, C.-Y., and Kellems, R.E. 1986. Selection and amplification of heterologous genes encoding adenosine deaminase in mammalian cells. *Proc. Natl. Acad. Sci. U.S.A.* 83:3136–3140.

Kay, B., Winter, J., and McCafferty, J. 1996. Phage Display of Peptides and Proteins. Academic Press, New York.

Keller, E. 1995. Objective lenses for confocal microscopy. *In* Handbook of Biological Confocal Microscopy, 2nd ed. (J. Pawley, ed.) pp. 155–166. Plenum Press, New York.

Kermani, P., Peloquin, L., and Lagace, J. 1995. Production of ScFv antibody fragments following immunization with a phage-displayed fusion protein and analysis of reactivity to surface- exposed epitopes of the protein F of Pseudomonas aeruginosa by cytofluorometry. *Hybridoma* 14:323–328.

Ketner, G., Spencer, F., Tugendrreich, S., Connely, C., and Hieter, P., 1994. Efficient manipulation of the human adenovirus genome as an infectious yeast artificial chromosome clone. *Proc. Natl. Acad. Sci. U.S.A.* 91:6186–6190.

Ketting, R., Haverkamp, T., van Luenen, H., and Plasterk, R. 1999. *mut-7* of *C. elegans*, required for transposon silencing and RNA interference, is a homolog of Werner syndrome helicase and RNaseD. *Cell* 99:133–141.

Kiewitz, A. and Wolfes, H. 1997. Mapping of protein-protein interactions between c-myb and its coactivator CBP by a new phage display technique. *FEBS Lett.* 415:258–262.

Kim, D.G., Kang, H.M., Jang, S.K., and Shin, H.S. 1992. Construction of a bifunctional mRNA in the mouse by using the internal ribosomal entry site of the encephalomyocarditis virus. *Mol. Cell Biol.* 12:3636–3643.

Knutsen, T. 1991. *In* The ACT Cytogenetics Laboratory Manual, 2nd ed. (M.J. Barch, ed.) pp. 563–587. Raven Press, New York.

Kooter, J.M. and Mol, J.N.M. 1993. *Trans*-inactivation of gene expression in plants. *Plant Biotech.* 4:166–171.

Kozak, M. 1989. The scanning model for translation: An update. *J. Cell Biol.* 108:229–241.

Kozak, M. 1999. Initiation of translation in prokaryotes and eukaryotes. *Gene* 234:187–208.

Kretzschmar, T. and Geiser, M. 1995. Evaluation of antibodies fused to minor coat protein III and major coat protein VIII of bacteriophage M13. *Gene* 155:61–65.

Kuhn, R., Schwenk, F., Aguet, M., and Rajewsky, K. 1995. Inducible gene targeting in mice. *Science* 269:1427–1429.

Kyte, J. and Doolittle, R.F. 1982. A simple method for displaying the hydropathic character of a protein. *J. Mol. Biol.* 157:105–132.

Labarca, C. and Paigen, K. 1980. A simple, rapid, and sensitive DNA assay procedure. *Anal. Biochem.* 102:344–352.

Lagos-Quintana, M., Rauhut, R., Lendeckel, W., and Tuschl, T. 2001. Identification of novel genes coding for small expressed RNAs. *Science* 294:853–858.

Lamb, B.T., Sisodia, S.S., Lawler, A.M., Slunt, H.H., Kitt, C.A., Kearns, W.G., Pearson, P.L., Price, D.L., and Gearhart, J.D. 1993. Introduction and expression of the 400 kilobase amyloid precursor protein gene in transgenic mice. *Nat. Genet.* 5:22–30.

Latchman, D.S. 1990. Molecular biology of herpes simplex virus latency. *J. Exp. Pathol.* 71:133–141.

Lau, N.C., Lim, L.P., Weinstein, E.G., and Bartel, D.P. 2001. An abundant class of tiny RNAs with probable regulatory roles in *Caenorhabditis elegans*. *Science* 294:858–862.

Lauwereys, M., Arbabi Ghahroudi, M., Desmyter, A., Kinne, J., Holzer, W., De Genst, E., Wyns, L., and Muyldermans, S. 1998. Potent enzyme inhibitors derived from dromedary heavy-chain antibodies. *EMBO J.* 17:3512–3520.

Lee, E.C. 1991. Cytogenetic Analysis of Continuous Cell Lines. *In* The ACT Cytogenetics Laboratory Manual, 2nd ed. (M.J. Barch, ed.) pp. 107–148. Raven Press, New York.

Lee, E.C., Yu, D., Martinez de Velasco, J., Tessarollo, L., Swing, D.A., Court, D.L., Jenkins, N.A., and Copeland, N.G. 2001. A highly efficient *Escherichia coli*-based chromosome engineering system adapted for recombinogenic targeting and subcloning of BAC DNA. *Genomics* 73:56–65.

Lee, K.H. and Cotanche, D.A. 1995. Detection of β-actin mRNA by RT-PCR in normal and regenerating chicken cochlea. *Hearing Res.* 87:9–15.

Lee, R. and Ambros, V. 2001. An extensive class of small RNAs in *Caenorhabditis elegans*. *Science* 294:862–864.

Levey, A.I., Kitt, C.A., Simonds, W.F., Price, D.L., and Brann, M.R. 1991. Identification and localization of muscarinic acetylcholine receptor proteins in brain with subtype-specific antibodies. *J. Neurosci.* 11: 3218–3226.

Levey, A.I., Hersch, S.M., Rye, D.B., Sunahara, R.K., Niznik, H.B., Kitt, C.A., Price, D.L., Maggio, R., Brann, M.R., and Ciliax, B.J. 1993. Localization of D1 and D2 dopamine receptors in rat, monkey, and human brain with subtype-specific antibodies. *Proc. Natl. Acad. Sci. U.S.A.* 90:8861–8865.

Li, H., Li, W.X., and Ding, S.W. 2002. Induction and suppression of RNA silencing by an animal virus. *Science* 296:1319–1321.

Liebermann, H. and Mentel, R. 1994. Quantification of adenvirus particles. *J. Virol. Methods* 50:281–291.

Lim, F., Hartley, D., Starr, P., Lang, P., Song, S., Yu, L., Wang, Y., and Geller, A.I. 1996. Generation of high-titer defective HSV-1 vectors using an IE 2 deletion mutant and quantitative study of expression in cultured cortical cells. *BioTechniques* 20:460–470.

Linscott's Directory of Immunological and Biological Reagents, Mill Valley, Calif.

Littlefield, J.W. 1964. Selection of hybrids from matings of fibroblasts in vitro and their presumed recombinants. *Science* 145:709–710.

Littlewood, T.D., Hancock, D.C., Danielian, P.S., Parker, M.G., and Evan, G.I. 1995. A modified oestrogen receptor ligand-binding domain as an improved switch for the regulation of heterologous proteins. *Nucl. Acids Res.* 23:1686–1690.

Liu, X., Chen, S., Chen, N., Gao, L., and Zhao, Z. 1996b. Specific binding of human bone morphogenetic protein (2A) with mouse osteoblastic cells. *Chin. Med. Sci. J.* 11:97–99.

Llave, C., Xie, Z., Kasschau, K.D., and Carrington, J.C. 2002a. Cleavage of *Scarecrow-like* mRNA targets directed by a class of *Arabidopsis* miRNA. *Science* 297:2053–2056.

Llave, C., Kasschau, K.D., Rector, M.A., and Carrington, J.C. 2002b. Endogenous and silencing-associated small RNAs in plants. *Plant Cell* 14:1–15.

Lockhart, D.J., Dong, H., Byrne, M.C., Follettie, M.T., Gallo, M.V., Chee, M.S., Mittmann, M., Wang, C., Kobayashi, M., Horton, H., and Brown, E.L. 1996. Expression monitoring by hybridization to high-density oligonucleotide arrays. *Nature Biotechnol.* 14:1675–1680.

Louis, N., Evelegh, C., and Graham, F.L. 1997. Cloning and sequencing of the cellular-viral junctions from the human adenovirus type 5 transformed 293 cell line. *Virology* 233:423–429.

Lowenstein, P.R., Fournel, S., Bain, D., Tomasec, P., Clissold, P.M., Castro, M.G., and Epstein, A.L. 1995. Simultaneous detection of amplicon and HSV-1 helper encoded proteins reveals that neurons and astrocytoma cells do express amplicon-borne transgenes in the absence of synthesis of virus immediate early proteins. *Mol. Brain Res.* 30:169–175.

Lowenstein, P.R., Shering, A., Bain, D., Castro, M.G., and Wilkinson, G.W.G. 1996. The use of adenovirus vectors to transferr genes to identified target brain cells in vitro. *In* Protocols for Gene Transfer in Neuroscience: Towards Gene Therapy of Neurological Disorders. (P.R. Lowenstein and L.W. Enquist, eds.) pp. 93–114. John Wiley & Sons, Chichester, U.K.

Lowman, H.B. and Wells, J.A. 1993. Affinity maturation of human growth hormone by monovalent phage display. *J. Mol. Biol.* 234:564–578.

Lubkowski, J., Hennecke, F., Pluckthun, A., and Wlodawer, A. 1998. The structural basis of phage display elucidated by the crystal structure of the N-terminal domains of g3p. *Nat. Struct. Biol.* 5:140–147.

Lucic, M.R., Forbes, B.E., Grosvenor, S.E., Carr, J.M., Wallace, J.C., and Forsberg, G. 1998. Secretion in *Escherichia coli* and phage-display of recombinant insulin-like growth factor binding protein-2 [In Process Citation]. *J. Biotechnol.* 61:95–108.

Luckow, V.A. and Summers, M.D. 1988. Trends in the development of baculovirus expression vectors. *Bio/Technology* 6:47–55.

Luo, L., Salunga, R.C., Guo, H., Bittner, A., Joy, K.C., Galindo, J.E., Xiao, H., Rogers, K.E., Wan, J.S., Jackson, M.R., and Erlander, M.G. 1999. Gene expression profiles of laser-captured adjacent neuronal subtypes. *Nature Med.* 5:117–122.

Mackem, S. and Roizman, B. 1982a. Differentiation between alpha promoter and regulatory regions of herpes simplex virus type I: The functional domains and sequence of a movable alpha regulator. *Proc. Natl. Acad. Sci. U.S.A.* 79:4917–4921.

Mackem, S. and Roizman, B. 1982b. Structural features of the herpes simplex virus alpha gene 4, 0, and 27 promoter-regulatory sequences which confer alpha regulation on chimeric thymidine kinase. *J. Virol.* 44:939–949.

Mandel, R.J., Rendahl, K.G., Spratt, S.K., Snyder, R.O., Cohen, L.K., and Leff, S.E. 1998. Characterization of intrastriatal recombinant adeno-associated virus-mediated gene transfer of human tyrosine hydroxylase and human GFP-cyclohydrolase I in a rat model of Parkinson's disease. *J. Neurosci.* 18:4271–4284.

Mansour, S.L. 1990. Gene targeting in murine embryonic stem cells: Introduction of specific alterations into the mammalian genome. *Genet. Anal. Tech. Appl.* 7:219–227.

Mansour, S.L., Thomas, K.R., and Capecchi, M.R. 1988. Disruption of the proto-oncogene int-2 in mouse embryo-derived stem cells: A general strategy for targeting mutations to nonselectable genes. *Nature* 336:348–352.

Mansour, S.L., Thomas, K.R., Deng, C.X., and Capecchi, M.R. 1990.

Introduction of a lacZ reporter gene into the mouse int-2 locus by homologous recombination. *Proc. Natl. Acad. Sci. U.S.A.* 87:7688–7692.

Marconi, P., Krisky, D., Oligino, T., Poliani, P.L., Ramakrishnan, R., Goins, W.F., Fink, D.J., and Glorioso, J.C. 1996. Replication-defective herpes simplex virus vectors for gene transfer in vitro. *Proc. Natl. Acad. Sci. U.S.A.* 93:11319–11320.

Marra, M.A., Kucaba, T.A., Dietrich, N.L., Green, E.D., Brownstein, B., Wilson, R.K., McDonald, K.M., Hillier, L.W., McPherson, J.D., and Waterston, R.H. 1997. High throughput fingerprint analysis of large-insert clones. *Genome Res.* 7:1072–1084.

Martin, F., Toniatti, C., Salvati, A.L., Ciliberto, G., Cortese, R., and Sollazzo, M. 1996. Coupling protein design and in vitro selection strategies: Improving specificity and affinity of a designed beta-protein IL-6 antagonist. *J. Mol. Biol.* 255:86–97.

Marzari, R., Edomi, P., Bhatnagar, R.K., Ahmad, S., Selvapandiyan, A., and Bradbury, A. 1997. Phage display of *Bacillus thuringiensis* CryIA(a) insecticidal toxin. *Febs Lett.* 411:27–31.

Matsumoto, B. (ed.) 1993. Cell Biological Applications of Confocal Microscopy. Academic Press, San Diego.

Matthews, D.J. and Wells, J.A. 1993. Substrate phage: Selection of protease substrates by monovalent phage display. *Science* 260:1113–1117.

McAllister, L., Penland, L., and Brown, P.O. 1998. Enrichment for loci identical-by-descent between pairs of mouse or human genomes by genomic mismatch scanning. *Genomics* 47:7–11.

McCaffrey, A.P., Meuse, L., Pham, T.T.T., Conklin, D.S., Hannon, G.J., and Kay, M.A. 2002. Gene expression-RNAi in adult mice. *Nature* 418:38–39.

McCarthy, A.M., McMahan, L., and Schaffer, P.A. 1989. Herpes simplex virus type 1 ICP27 deletion mutants exhibit altered patterns of transcription and are DNA deficient. *J. Virol.* 63:18–27.

McCown, T.J., Xiao, X., Li, J., Breese, G.R., and Samulski, R.J. 1996. Differential and persistent expression patterns of CNS gene transfer by an adeno-associated virus (AAV) vector. *Brain Res.* 713(1–2):99–107.

McIntosh, R.S., Asghar, M.S., and Weetman, A.P. 1997. The antibody response in human autoimmune thyroid disease [editorial]. *Clin. Sci. (Colch.)* 92:529–541.

Merlin, S., Rowold, E., Abegg, A., Berglund, C., Klover, J., Staten, N., McKearn, J.P. and Lee, S.C. 1997. Phage presentation and affinity selection of a deletion mutant of human interleukin-3. *Appl. Biochem. Biotechnol.* 67:199–214.

Metcalf, W.W., Jiang, W., Daniels, L.L., Kim, S.K., Haldimann, A., and Wanner, B.L. 1996. Conditionally replicative and conjugative plasmids carrying lacZ alpha cloning, mutagenesis, and allele replacement in bacteria. *Plasmid* 35:1–13.

Mezey, E., Chandross, K.J., Harta G., Maki, R.A., and McKercher, S.R. 2000. Turning blood into brain: Cells bearing neuronal antigens generated in vivo from bone marrow. *Science* 290:1775–1779.

Mezey, E., Key, S., Vogelsang, G., Szalayova, I., Lange, G.D., and Crain, B. 2003. Transplanted bone marrow generates new neurons in human brains, *Proc. Nat. Acad. Sci. U.S.A.* 100:1364–1369.

Mirnics, K., Middleton, F.A., Marquez, A., Lewis, D.A., and Levitt, P. 2000. Molecular characterization of schizophrenia viewed by microarray analysis of gene expression in prefrontal cortex. *Neuron* 28:53–67.

Minamide, L.S. and Bamburg, J.R. 1990. A filter paper dye-binding assay for quantitative determination of protein without interference from reducing agents or detergents. *Anal. Biochem.* 190:66–70.

Miyoshi, H., Takahashi, M., Gage, F.H., and Verma, I.M. 1997. Stable and efficient gene transfer into the retina using an HIV-based lentiviral vector. *Proc. Natl. Acad. Sci. U.S.A.* 94:10319–10323.

Mombaerts, P., Clarke, A.R., Hooper, M.L., and Tonegawa, S. 1991. Creation of a large genomic deletion at the T-cell antigen receptor beta-subunit locus in mouse embryonic stem cells by gene targeting. *Proc. Natl. Acad. Sci. U.S.A.* 88:3084–3087.

Montgomery, M.K., Xu, S.Q., and Fire, A. 1998. RNA as a target of double-stranded RNA-mediated genetic interference in Caenorhabditis elegans. *Proc. Natl. Acad. Sci. U.S.A.* 95:15502–15507.

Montgomery, R.I., Warner, M.S., Lum, B.J., and Spear, P.G. 1996. Herpes simplex virus-1 entry into cells mediated by a novel member of the TNF/NGF receptor family. *Cell* 87:427–436.

Morelli, A.E., Larregina, A.T., Smith-Arica, J., Dewey, R.A., Southgate, T.D., Ambar, B., Fontana, A., Castro, M.G., and Lowenstein, P.R. 1999. Neuronal and glial cell type-specific promoters within adenovirus recombinants restrict the expression of the apoptosis-inducing molecule Fas ligand to predetermined brain cell types, and abolish peripheral liver toxicity. *J. Gen. Virol.* 80:571–583.

Mortensen, R.M., Zubiaur, M., Neer, E.J., and Seidman, J.G. 1991. Embryonic stem cells lacking a functional inhibitory G-protein subunit (alpha i2) produced by gene targeting of both alleles. *Proc. Natl. Acad. Sci. U.S.A.* 88:7036–7040.

Mortensen, R.M., Conner, D.A., Chao, S., Geisterfer, L.A., and Seidman, J.G. 1992. Production of homozygous mutant ES cells with a single targeting construct. *Mol. Cell. Biol.* 12:2391–2395.

Mourelatos, Z., Dostie, J., Paushkin, S., Sharma, A., Charroux, B., Abel, L., Rappsilber, J., Mann, M., and Dreyfuss, G. 2002. miRNPs: A novel class of ribonucleoproteins containing numerous microRNAs. *Genes Dev.* 16:720–728.

Mourrain, P., Beclin, C., Elmayan, T., Feuerbach, F., Godon, C., Morel, J.-B., Jouette, D., Lacombe, A.-M., Nikic, S., Picault, N., Remoue, K., Sanial, M., Vo, T.-A., and Vaucheret, H. 2000. *Arabidopsis SGS2* and *SGS3* genes are required for posttranscriptional gene silencing and natural virus resistance. *Cell* 101:533–542.

Mullen, C.A., Kilstrup, M., and Blaese, R.M. 1992. Transfer of the bacterial gene for cytosine deaminase to mammalian cells confers lethal sensitivity to 5-fluorocytosine: A negative selection system. *Proc. Natl. Acad. Sci. U.S.A.* 89:33–37.

Mulligan, R.C. and Berg, P. 1981. Selection for animal cells that express the *E. coli* gene coding for xanthine-guanine phosphoribosyltransferase. *Proc. Natl. Acad. Sci. U.S.A.* 78: 2072–2076.

Mulsant, P., Gatignol, A., Dalens, M., and Tiraby, G. 1988. Phleomycin resistance as a dominant selectable marker in CHO cells. *Somatic Cell Mol. Genet.* 14:243–252.

Murphy, C.I. and Piwnica-Worms, H. 1994a. Preparation of insect cell cultures and baculovirus stocks. *In*

Current Protocols in Molecular Biology (F.M. Ausubel, R. Brent, R.E. Kingston, D.D. Moore, J.G. Seidman, J.A. Smith, and K. Struhl, eds.) pp. 16.10.1–16.10.8. John Wiley & Sons, New York.

Murphy, C.I. and Piwnica-Worms, H. 1994b. Generation of recombinant baculoviruses and analysis of recombinant protein expression. *In* Current Protocols in Molecular Biology (F.M. Ausubel, R. Brent, R.E. Kingston, D.D. Moore, J.G. Seidman, J.A. Smith, and K. Struhl, eds.) pp. 16.11.1–16.11.19. John Wiley & Sons, New York.

Murphy, C.I. and Piwnica-Worms, H. 1999. Overview of the baculovirus expression system. *In* Current Protocols in Protein Science (J.E. Coligan, B.M. Dunn, H.L. Ploegh, D.W. Speicher, and P.T. Wingfield, eds.) pp. 5.4.1–5.4.4. John Wiley & Sons, New York.

Naldini, L., Blomer, U., Gallay, P., Ory, D., Mulligan, R., Gage, F.H., Verma, I.M., and Trono, D. 1996. In vivo gene delivery and stable transduction of nondividing cells by a lentiviral vector. *Science* 272:263–267.

Nelson, S.F., McCusker, J.H., Sander, M.A., Kee, Y., Modrich, P., and Brown, P.O. 1993. Genomic mismatch scanning: A new approach to genetic linkage mapping. *Nature Genet.* 4:11–18.

Neugebauer, J. 1992. A Guide to the Properties and Uses of Detergents in Biology and Biochemistry. Corp. Doc. No. CB0068-0892. Calbiochem-Novabiochem, La Jolla, Calif.

Neve, R.L. and Geller, A.I. 1996. A defective herpes simplex virus vector system for gene delivery into the brain: Comparison with alternative gene delivery systems and usefulness for gene therapy. *Clin. Neurosci.* 3:262–267.

Neve, R.L., Howe, J.R., Hong, S., and Kalb, R.G. 1997. Introduction of glutamate receptor subunit 1 into motor neurons in vivo using a recombinant herpes simplex virus alters the functional properties of AMPA receptors. *Neuroscience* 79:435–447.

No, D., Yao, T.P., and Evans, R.M. 1996. Ecdysone-inducible gene expression in mammalian cells and mice. *Proc. Natl. Acad. Sci. U.S.A.* 93:3346–3351.

Noble, M. and Mayer-Proschel, M. 1998. Culture of astrocytes, oligodendrocytes and O-2A progenitor cells. *In* Culturing Nerve Cells: Cellular and Molecular Neuroscience. (G. Banker and K. Goslin, eds.). MIT Press, Cambridge, Mass.

Nobrega, M.A., Ovcharenko, I., Afzal, V., and Rubin, E.M. 2003. Scanning human gene deserts for long-range enhancers. *Science* 302:413.

Novina, C.D., Murray, M.F., Dykxhoorn, D.M., Beresford, P.J., Riess, J., Lee, S.-K., Collman, R.G., Lieberman, J., Shankar, P., and Sharp, P.A. 2002. siRNA-directed inhibition of HIV-1 infection. *Nat. Med.* 8:681–686.

Nyberg-Hoffman, C. and Aguilar-Cordova, E. 1999. Instability of adenoviral vectors during transport and its implication for clinical studies. *Nature Medicine* 5:955–957.

Nykanen, A., Haley, B., and Zamore, P.D. 2001. ATP requirements and small interfering RNA structure in the RNAi pathway. *Cell* 107:309–321.

O'Brien, R.D., Timpone, C.A., and Gibson, R.E. 1978. The measurement of partial specific volumes and sedimentation coefficients by sucrose density centrifugation. *Anal. Biochem.* 86:602–615.

O'Donovan, M.J. and Ritter, A. 1995. Optical recording and lesioning of spinal neurones during rhythmic activity in the chick embryo spinal cord. *In* Alpha and Gamma Motor Systems (A. Taylor, M.H. Gladden, and R. Durbaba, eds.) pp. 557–563. Plenum, New York.

O'Donovan, M.J., Ho, S., Sholomenko, G., and Yee, W. 1993. Real-time imaging of neurons retrogradely and anterogradely labelled with calcium sensitive dyes. *J. Neurosci. Methods* 46:91–106.

O'Donovan, M., Ho, S., and Yee, W. 1994. Calcium imaging of rhythmic network activity in the developing spinal cord of the chick embryo. *J. Neurosci.* 14:6354–6369.

Ohyama, T. and Groves, A.K. 2004. Generation of Pax2-Cre mice by modification of a Pax2 bacteria artificial chromosome. *Genesis* 38: 195–199.

Okabe, S., Forsberg-Nilsson, K., Spiro, A.C., Segal, M., and McKay, R.D.G. 1996. Development of neuronal precursor cells and functional neurons from embryonic stem cells in vitro. *Mech. Dev.* 59:89–102.

O'Leary, D.D.M. and Stanfield, B.B. 1989. Selective elimination of axons extended by developing cortical neurons is dependent on regional locale: Experiments utilizing fetal cortical transplants. *J. Neurosci.* 9:2230–2246.

Olson, E.N., Arnold, H.H., Rigby, P.W., and Wold, B.J. 1996. Know your neighbors: Three phenotypes in null mutants of the myogenic bHLH gene MRF4. *Cell* 85:1–4.

Oroskar, A.A. and Read, G.S. 1989. Control of mRNA stability by the virion host shutoff function of herpes simplex virus. *J. Virol.* 63:1897–1906.

Paddison, P.J., Caudy, A.A., and Hannon, G.J. 2002. Stable suppression of gene expression by RNAi in mammalian cells. *Proc. Natl. Acad. Sci. U.S.A.* 99:1443–1448.

Palmer, T.D., Hock, R.A., Osborne, W.R.A., and Miller, A.D. 1987. Efficient retrovirus-mediated transfer and expression of a human adenosine deaminase gene in diploid skin fibroblasts from an adenosine-deficient human. *Proc. Natl. Acad. Sci. U.S.A.* 84:1055–1059.

Pasqualini, R. and Ruoslahti, E. 1996. Organ targeting in vivo using phage display peptide libraries. *Nature* 380:364–366.

Paterson, T. and Everett, R.D. 1990. A prominent serine-rich region in Vmw175, the major transcriptional regulator protein of herpes simplex virus type 1, is not essential for virus growth in tissue culture. J. Gen. Virol. 71:1775–1783.

Pawley, J.B. (ed.) 1995. Handbook of Biological Confocal Microscopy, 2nd ed. Plenum Press, New York.

Paxinos, G. and Watson, C. 1982. The Rat Brain in Stereotaxic Coordinates. Academic Press, New York.

Paxinos, G. and Watson, C. 1986. The Rat Brain in Stereotaxic Coordinates, 2nd ed. Academic Press, London.

Paxinos, G., Ashwell, K.W.S., and Tork, I. 1994. Atlas of the Developing Rat Nervous System. Academic Press, San Diego.

Pearson, R.B. and Kemp, B.E. 1991. Protein kinase phosphorylation site sequences and consensus specificity motifs: tabulations. *Methods Enzymol.* 200:62–81.

Peart, J.R., Cook, G., Feys, B.J., Parker, J.E., and Baulcombe, D.C. 2002a. An *EDS1* orthologue is required for *N*-mediated resistance against tobacco mosaic virus. *Plant J.* 29:569–579.

Peart, J.R., Lu, R., Sadanandom, A., Malcuit, I., Moffett, P., Brice, D.C., Schauser, L., Jaggard, D.A.W., Xiao,

S., Coleman, M., Dow, J.M., Jones, J.D.G., Shirasu, K., and Baulcombe, D.C. 2002b. Ubiquitin ligase-associated protein SGT1 is required for host and nonhost disease resistance in plants. *Proc. Natl. Acad. Sci. U.S.A.* 99:10865–10869.

Pechan, P.A., Fotaki, M., Thompson, R.L., Dunn, R.J., Chase, M., Chiocca, E.A., and Breakefield, X.O. 1996. A novel "piggyback" packaging system for herpes simplex virus amplicon vectors. *Hum. Gene Ther.* 7:2003–2013.

Perou, C.M., Jeffrey, S.S., van de Rijn, M., Rees, C.A., Eisen, M.B., Ross, D.T., Pergamenschikov, A., Williams, C.F., Zhu, S.X., Lee, J.C., Lashkari, D., Shalon, D., Brown, P.O., and Botstein, D. 1999. Distinctive gene expression patterns in human mammary epithelial cells and breast cancers. *Proc. Natl. Acad. Sci. U.S.A.* 96:9212–9217.

Perucho, M., Hanahan, D., and Wigler, M. 1980. Genetic and physical linkage of exogenous sequences in transformed cells. *Cell* 22:309–317.

Petersen, G., Song, D., Hugle-Dorr, B., Oldenburg, I., and Bautz, E.K. 1995. Mapping of linear epitopes recognized by monoclonal antibodies with gene-fragment phage display libraries. *Mol. Gen. Genet.* 249:425–431.

Peterson, G.L. 1983. Determination of total protein. *Methods Enzymol.* 91: 95–119.

Peterson, G.L. and Schimerlik, M.I. 1984. Large scale preparation and characterization of membrane-bound and detergent-solubilized muscarinic acetylcholine receptor from pig atria. *Prep. Biochem.* 14: 33–74.

Peterson, G.L., Rosenbaum, L.C., Broderick, D.J., and Schimerlik, M.I. 1986. Physical properties of the purified cardiac muscarinic acetylcholine receptor. *Biochemistry* 25: 3189–3202.

Peterson, G.L., Rosenbaum, L.C., and Schimerlik, M.I. 1988. Solubilization and hydrodynamic properties of pig atrial muscarinic acetylcholine receptor in dodecyl β-D-maltoside. *Biochem. J.* 255:553–560.

Peterson, G.L., Toumadje, A., Johnson, W.C., Jr., and Schimerlik, M.I. 1995. Purification of recombinant porcine m2 muscarinic acetylcholine receptor from Chinese hamster ovary cells. *J. Biol. Chem.* 270:17808–17814.

Pham, C.T., MacIvor, D.M., Hug, B.A., Heusel, J.W., and Ley, T.J. 1996. Long-range disruption of gene expression by a selectable marker cassette. *Proc. Natl. Acad. Sci. U.S.A.* 93:13090–13095.

Phizicky, E.M. and Fields, S. 1995. Protein-protein interactions: Methods for detection and analysis. *Microbiol. Rev.* 59:94–123.

Pinkel, D., Segraves, R., Sudar, D., Clark, S., Poole, I., Kowbel, D., Collins, C., Kuo, W.L., Chen, C., Zhai, Y., Dairkee, S.H., Ljung, B.M., Gray, J.W., and Albertson, D.G. 1998. High resolution analysis of DNA copy number variation using comparative genomic hybridization to microarrays. *Nature Genet.* 20:207–211.

Plenz, D. and Kitai, S.T. 1996. Organotypic cortex-striatum-mesencephalon cultures: The nigrostriatal pathway. *Neurosci. Lett.* 209:177–180.

Poeschla, E.M., Wong-Staal, F., and Looney, D.J. 1998. Efficient transduction of nondividing human cells by feline immunodeficiency virus lentiviral vectors. *Nat. Med.* 3:354–357.

Pollack, J.R., Perou, C.M., Alizadeh, A.A., Eisen, M.B., Pergamenschikov, A., Williams, C.F., Jeffrey, S.S., Botstein, D., and Brown, P.O. 1999. Genome-wide analysis of DNA copy-number changes using cDNA microarrays. *Nature Genet.* 23:41–46.

Pomeroy, S.L., Tamayo, P., Gaasenbeek, M., et al. 2002. Prediction of central nervous system embryonal tumour outcome based on gene expression. *Nature* 415:436–442.

Prickett, K.S., Amberg, D.C., and Hopp, T.P. 1989. A calcium-dependent antibody for identification and purification of recombinant proteins. *BioTechniques* 7:580–589.

Rajewsky, K., Gu, H., Kuhn, R., Betz, U.A., Muller, W., Roes, J., and Schwenk, F. 1996. Conditional gene targeting. *J. Clin. Invest.* 98:600–603.

Revah, F., Horellou, P., Vigne, E., Le Gal La Salle, G., Robert, J.J., Perricaudet, M., and Mallet, J. 1996. Gene trasfer into the central and peripheral nervous system using adenoviral vectors. *In* Protocols for Gene Transfer in Neuroscience: Towards Gene Therapy of Neurological Disorders. (P.R. Lowenstein and L.W. Enquist, eds.) pp. 81–92. John Wiley & Sons, Chichester, U.K.

Rheinhart, B.J., Weinstein, E.G., Rhoades, M., Bartel, B., and Bartel, D.P. 2002. MicroRNAs in plants. *Genes Dev.* 16:1616–1626.

Rhoades, M.W., Reinhart, B.J., Lim, L.P., Burge, C.B., Bartel, B., and Bartel, D.P. 2002. Prediction of plant microRNA targets. *Cell* 110:513–520.

Rivera, V.M., Clackson, T., Nateson, S., Pollock, R., Amara, J.F., Keenen, T., Magari, S.R., Phillips, T., Courage, N.L., Cerasoli, F., Holt, D.A. Jr., and Gilman, M. 1996. A humanized system for pharmacologic control of gene expression. *Nature Med.* 2:1028–1032.

Roberts, B.L., Markland, W., Ley, A.C., Kent, R.B., White, D.W., Guterman, S.K. and Ladner, R.C. 1992. Directed evolution of a protein: Selection of potent neutrophil elastase inhibitors displayed on M13 fusion phage. *Proc. Natl. Acad. Sci. U.S.A.* 89:2429–2433.

Robertson, E.J. 1987. Embryo-derived stem cell lines. *In* Teratocarcinoma and Embryonic Stem Cells: A Practical Approach (J.E. Robertson, ed.) pp. 71–112. IRL Press, Washington, D.C.

Robertson, E.J. 1991. Using embryonic stem cells to introduce mutations into the mouse germ line. *Biol. Reprod.* 44:238–245.

Robins, D.M., Ripley, S., Henderson, A.S., and Axel, R. 1981. Transforming DNA integrates into the host chromosome. *Cell* 23:29–39.

Rohlmann, A., Gotthardt, M., Willnow, T.E., Hammer, R.E., and Herz, J. 1996. Sustained somatic gene inactivation by viral transfer of Cre recombinase. *Nat. Biotechnol.* 14:1562–1565.

Roizman, B. and Jenkins, F.J. 1985. Genetic engineering of novel genomes of large DNA viruses. *Science* 229: 1208–1214.

Roizman, B. and Sears, A.E. 1990. Herpes simplex viruses and their replication. *In* Virology, 2nd ed. (B. Fields, D.M. Knipe, R.M. Chanock et al., eds.) pp. 1795–1841. Raven Press, New York.

Rooney, D.E. and Czepulkowski, B.H. (eds.) 1992. Human Cytogenetics: A Practical Approach, Vol. I. Constitutional Analysis, 2nd ed. IRL Press, Washington, D.C.

Roseberry, A.G., Liu, H., Jackson, A.C., Cai, X., and Griedman, J.M. 2004. Neuropeptide Y-mediated inhibition of proopiomelanocortin neurons in the arcuate nucleus shows enhanced desensitization in ob/ob mice. *Neuron* 41:711–722.

Rosenthal, K.S. and Kousale, F. 1983. Critical micelle determination of nonionic detergents with Coomassie Brilliant Blue G-250. *Anal. Chem.* 55:1115–1117.

Ross, E.M. and Schatz, G. 1978. Purification and subunit composition of cytochrome c_1 from baker's yeast *Saccharomyces cerevisiae*. *Methods Enzymol.* 53:222–229.

Rossenu, S., Dewitte, D., Vandekerckhove, J., and Ampe, C. 1997. A phage display technique for a fast, sensitive, and systematic investigation of protein-protein interactions. *J. Prot. Chem.* 16:499–503.

Rost, F.W.D. 1992. Fluorescence Microscopy. Cambridge University Press, Cambridge.

Ruden, D.M., Ma, J., Li, Y., Wood, K., and Ptashne, M. 1991. Generating yeast transcriptional activators containing no yeast protein sequences. *Nature* 350:426–430.

Russ, J. 1995. The Image Processing Handbook. CRC Press, Boca Raton, Fla.

Salcini, A.E., Confalonieri, S., Doria, M., Santolini, E., Tassi, E., Minenkova, O., Cesareni, G., Pelicci, P.G., and Di Fiore, P.P. 1997. Binding specificity and in vivo targets of the EH domain, a novel protein-protein interaction module. *Genes Dev.* 11:2239–2249.

Sambrook, J., Fritsch, E.F., and Maniatis, T. 1989. Molecular Cloning: A Laboratory Manual, 2nd edition. Cold Spring Harbor Laboratory, Cold Spring Harbor, N.Y.

Samulski, R.J., Sally, M., and Muzyczka, N. 1999. Adeno-associated viral vectors. *In* The Development of Human Gene Therapy. (T. Friedman, ed.) pp. 36:131–172. Cold Spring Harbor Laboratory Press. Cold Spring Harbor, N.Y.

Sandberg, R., Yasuda, R., Pankratz, D.G., Carter, T.A., Del Rio, J.A., Wodicka, L., Mayford, M., Lockhart, D.J., and Barlow, C. 2000. Regional and strain-specific gene expression mapping in the adult mouse brain. *Proc. Natl. Acad. Sci. U.S.A.* 97:11038–11043.

Santiago Vispo, N., Felici, F., Castagnoli, L., and Cesareni, G. 1993. Hybrid Rop-pIII proteins for the display of constrained peptides on filamentous phage capsids. *Ann. Biol. Clin.* 51:917–922.

Sauer, B. 1993. Manipulation of transgenes by site-specific recombination: Use of Cre recombinase. *Methods Enzymol.* 225:890–900.

Sauer, B. and Henderson, N. 1989. Cre-stimulated recombination at loxP-containing DNA sequences placed into the mammalian genome. *Nucl. Acids Res.* 17:147–161.

Schena, M., Shalon, D., Davis, R.W., and Brown, P.O. 1995. Quantitative monitoring of gene expression patterns with a complementary DNA microarray. *Science* 270:467–470.

Schetz, J.A., Kim, O.-J., and Sibley, D.R. 2003. Pharmacological characterization of mammalian D_1 and D_2 dopamine receptors expressed in *Drosophila* Schneider-2 cells. *J. Receptors Signal Transduct.* 23: 99–109.

Schier, R., Bye, J., Apell, G., McCall, A., Adams, G.P., Malmqvist, M., Weiner, L.M. and Marks, J.D. 1996a. Isolation of high-affinity monomeric human anti-c-*erbB-2* single chain Fv using affinity-driven selection. *J. Mol. Biol.* 255:28–43.

Schier, R., McCall, A., Adams, G.P., Marshall, K.W., Merritt, H., Yim, M., Crawford, R.S., Weiner, L.M., Marks, C., and Marks, J.D. 1996b. Isolation of picomolar affinity anti-c-*erB-2* single-chain Fv by molecular evolution of the complementarity determining regions in the center of the antibody binding site. *J. Mol. Biol.* 263:551–567.

Schmitz, R., Baumann, G., and Gram, H. 1996. Catalytic specificity of phosphotyrosine kinases *Blk, Lyn,* c-*Src* and *Syk* as assessed by phage display. *J. Mol. Biol.* 260:664–677.

Schneider, I. 1972. Cell lines derived from late embryonic stages of *Drosophila melanogaster. J. Embryol. Exp. Morphol.* 27:353–365.

Schneppenheim, R., Budde, U., Dahlmann, N. and Rautenberg, P. 1991. Luminography—a new, highly sensitive visualization method for electrophoresis. *Electrophoresis* 12: 367–372.

Schwenk, F., Baron, U., and Rajewsky, K. 1995. A cre-transgenic mouse strain for the ubiquitous deletion of *loxP*-flanked gene segments including deletion in germ cells. *Nucl. Acids Res.* 23:5080–5081.

Shering, A.F., Bain, D., Stewart, K., Epstein, A.L., Castro, M.G., Wilkinson, G.W.G., and Lowenstein, P.R. 1997. Cell type-specific expression in brain cell cultures from a short human cytomegalovirus major immediate early promoter depends on whether it is inserted into herpesvirus or adenovirus vectors. *J. Gen. Virol.* 78:445–459.

Sherman, F., Fink, G.R., and Lawrence, C.W. 1979. Methods in Yeast Genetics. Cold Spring Harbor Laboratory, Cold Spring Harbor, N.Y.

Shibata, H., Toyama, K., Shioya, H., Ito, M., Hirota, M., Hasegawa, S., Matsumoto, H., Takano, H., Akiyama, T., Toyoshima, K., Kanamaru, R., Kanegae, Y., Saito, I., Nakamura, Y., Shiba, K., and Noda, T. 1997. Rapid colorectal adenoma formation initiated by conditional targeting of the *Apc* gene. *Science* 278:120–123.

Shields, G. and Sang, J.H. 1977. Improved medium for culture of Drosophila embryonic cells. Drosophila Information Services 52: 161.

Shizuya, H., Birren, B., Kim, U.J., Mancino, V., Slepak, T., Tachiiri, Y., and Simon, M. 1992. Cloning and stable maintenance of 300-kilobase-pair fragments of human DNA in *Echerichia coli* using an F-factor based vector. *Proc. Natl. Acad. Sci. U.S.A* 89:8794–8797.

Shockett, P.E. and Schatz, D.G. 1996. Diverse strategies for tetracycline-regulated inducible gene expression. *Proc. Natl. Acad. Sci. U.S.A.* 93:5173–5176.

Shotton, D.M. (ed.) 1993. Electronic Light Microscopy. John Wiley and Sons, New York.

Sibille, P., Ternynck, T., Nato, F., Buttin, G., Strosberg, D., and Avrameas, A. 1997. Mimotopes of polyreactive anti-DNA antibodies identified using phage- display peptide libraries. *Eur. J. Immunol.* 27:1221–1228.

Sijen, T., Fleenor, J., Simmer, F., Thijssen, K.L., Parrish, S., Timmons, L., Plasterk, R.H.A., and Fire, A. 2001. On the role of RNA amplification in dsRNA-triggered gene silencing. *Cell* 107:465–476.

Sikorski, R.S. and Hieter, P. 1989. A system of shuttle vectors and yeast host strains designed for efficient manipulation of DNA in Saccharomyces cerevisiae. *Genetics* 122:19–27.

Simonsen, C.C. and Levinson, A.D. 1983. Isolation and expression of an altered mouse dihydrofolate reductase cDNA. *Proc. Natl. Acad. Sci. U.S.A.* 80:2495–2499.

Smith, G. 1993. Preface: Surface display and peptide libraries. *Gene* 128:1–2.

Smith, G. and Petrenko, V. 1997. Phage display. *Chem. Rev.* 97:391–410.

Smith, G.P. 1985. Filamentous fusion phage: Novel expression vectors that display cloned antigens on the virion surface. *Science* 228:1315–1317.

Smith, G.P., Patel, S.U., Windass, J.D., Thornton, J.M., Winter, G., and Griffiths, A.D. 1998. Small binding proteins selected from a combinatorial repertoire of knottins displayed on phage. *J. Mol. Biol.* 277:317–332.

Smith, I.L., Hardwicke, M.A., and Sandri-Goldin, R.M. 1992. Evidence that the herpes simplex virus immediate early protein ICP27 acts post-transcriptionally during infection to regulate gene expression. *Virology* 186:74–86.

Smith, P.K., Krohn, R.I., Hermanson, G.T., Mallia, A.K., Gartner, F.H., Provenzano, M.D., Fujimoto, E.K., Goeke, N.M., Olson, B.J., and Klenk, D.C. 1985. Measurement of protein using bicinchonic acid. *Anal. Biochem.* 150:76–85.

Solinas-Toldo, S., Lampel, S., Stilgenbauer, S., Nickolenko, J., Benner, A., Dohner, H., Cremer, T., and Lichter, P. 1997. Matrix-based comparative genomic hybridization: Biochips to screen for genomic imbalances. *Genes Chrom. Cancer* 20:399–407.

Songyang, Z. 1994. Specific motifs recognized by the SH2 domains of *CsK, 3BP2, fps/fes, GRB-2, HCP, SHC, Syk,* and *Vav*. *Mol. Cell. Biol.* 14:2777–2785.

Sopher, B.L., Thomas, P.S. Jr., LaFevre-Bernt, M.A., Holm, I.E., Wilke, S.A., Ware, C.B., Jin, L.W., Libby, R.T., Ellerby, L.M., and La Spada, A.R. 2004. Androgen receptor YAC transgenic mice recapitulate SBMA motor neuronopathy and implicate VEGF164 in the motor neuron degeneration. *Neuron* 41:687–699.

Souriau, C., Fort, P., P, R., Hartley, O., Lefranc, M., and Weill, M. 1997. A simple luciferase assay for signal transduction activity detection of epidermal growth factor displayed on phage. *Nucl. Acids Res.* 25:1585–1590.

Southern, P.J. and Berg, P. 1982. Transformation of mammalian cells to antibiotic resistance with a bacterial gene under control of the SV40 early region promoter. *J. Mol. Appl. Gen.* 1:327–341.

Spaete, R. and Frenkel, N. 1982. The herpes simplex virus amplicon: A new eucaryotic defective-virus cloning-amplifying vector. *Cell* 30:305–310.

Spaete, R. and Frenkel, N. 1985. The herpes simplex virus amplicon: Analyses of *cis*-acting replication functions. *Proc. Natl. Acad. Sci. U.S.A.* 82:694–698.

Sparks, A.B., Quilliam, L.A., Thorn, J.M., Der, C.J., and Kay, B.K. 1994. Identification and characterization of *Src* SH3 ligands from phage-displayed random peptide libraries. *J. Biol. Chem.* 269:23853–23856.

Spear, P.G. 1993. Entry of alphaherpesviruses into cells. *Semin. Virol.* 4:167–180.

Spellman, P.T., Sherlock, G., Zhang, M.Q., Iyer, V.R., Anders, K., Eisen, M.B., Brown, P.O., Botstein, D., and Futcher, B. 1998. Comprehensive identification of cell cycle-regulated genes of the yeast *Saccharomyces cerevisiae* by microarray hybridization. *Mol. Biol. Cell* 9:3273–3297.

Spenger, C., Studer, L., Evtouchenko, L., Egli, M., Burgunder, J.-M., Markwalder, R., and Seiler, R.W. 1994. Long-term survival of dopaminergic neurones in free-floating roller tube cultures of human fetal ventral mesencephalon. *J. Neurosci. Methods* 54:63–73.

Spessot, R., Inchley, K., Hupel, T.M., and Bacchetti, S., 1989. Cloning of the herpes-simplex virus ICP4 gene in an adenovirus vector-effects on adenovirus gene-expression and replication. *Virology* 168:378–387.

Stanfield, B.B. and O'Leary, D.D.M. 1985. Fetal occipital cortical neurons transplanted to rostral cortex can extend and maintain a pyramidal tract axon. *Nature* 313:135–137.

Staschke, K.A., Colacino, J.M., Mabry, T.E., and Jones, C.D. 1994. The in vitro anti-hepatitis B virus activity of FIAU [1-(2′-deoxy-2′-fluoro-1-β-D-arabinofuranosyl-5-iodo)uracil] is selective, reversible, and determined, at least in part, by the host cell. *Antivir. Res.* 23:45–61.

Stears, R.L., Getts, R.C., and Gullans, S.R. 2000. A novel, sensitive detection system for high-density microarrays using dendrimer technology. *Physiol. Genomics* 3:93–99.

Steller, H. and Pirrotta, V. 1985. A transposable P vector that confers selectable G418 resistance in Drosophila larvae. *EMBO J.* 4:167–171.

Stemmer, W.P. 1994. DNA shuffling by random fragmentation and reassembly: In vitro recombination for molecular evolution. *Proc. Natl. Acad. Sci. U.S.A.* 91:10747–10751.

Sternberger, L.A. 1979. Immunocytochemistry. 2nd ed. John Wiley & Sons, New York.

Sternberger, L.A., Hardy, P.H., Cuculis, J.J., and Meyer, H.G. 1970. The unlabeled antibody enzyme method of immunohistochemistry. *J. Histochem. Cytochem.* 18:315–333.

Stevens, J.G., Wagner, E.K., Devi-Rao, G.B., Cook, M.L., and Feldman, L.T. 1987. RNA complementary to a herpes virus alpha gene mRNA is prominent in latently infected neurons. *Science* 235:1056–1059.

St-Onge, L., Furth, P.A., and Gruss, P. 1996. Temporal control of the Cre recombinase in transgenic mice by a tetracycline responsive promoter. *Nucl. Acids Res.* 24:3875–3877.

Stoppini, L., Buchs, P.A., and Muller, D. 1991. A simple method for organotypic cultures of nervous tissue. *J. Neurosci. Methods* 37:173–182.

Stow, N. and McMonagle, E. 1982. Propagation of foreign DNA sequences linked to a herpes simplex virus origin of replication. *In* Eucaryotic Viral Vectors (Y. Gluzman, ed.) pp. 199–204. Cold Spring Harbor Laboratory Press, Cold Spring Harbor, N.Y.

Suck, R.W.L. and Krupinska, K. 1996. Repeated probing of Western blots obtained from Coomassie Brilliant Blue-stained or unstained polyacrylamide gels. *BioTechniques* 21:418–422.

Sugiyama, M., Thompson, C.J., Kumagai, T., Suzuki, K., Deblaere, R., Villarroel, R., and Davies, J. 1994. Characterisation by molecular cloning of two genes from *Streptomyces verticillus* encoding resistance to bleomycin. *Gene* 151:11–16.

Sui, G., Soohoo, C., Affar, E.B., Gay, F., Shi, Y., Forrester, W.C., and Shi, Y. 2002. A DNA vector-based RNAi technology to suppress gene expression in mammalian cells. *Proc. Natl. Acad. Sci. U.S.A.* 99:5515–5520.

Summers, M.D. and Smith, G.E. 1987. A manual of methods for baculovirus vectors and insect cell culture procedures. Texas Agricultural Experimental Station Bulletin No. 1555. College Station, Texas.

Szardenings, M., Tornroth, S., Mutulis, F., Muceniece, R., Keinanen, K., Kuusinen, A. and Wikberg, J.E. 1997. Phage display selection on whole cells yields a peptide specific for melanocortin receptor 1. *J. Biol. Chem.* 272:27943–27948.

Takeshima, T., Shimoda, K., Johnston, J.M., and Commissiong, J.W. 1996. Standardized methods to bioassay neurotrophic factors for dopaminergic neurons. *J. Neurosci. Methods* 67:27–41.

Tang, X.B., Dallaire, P., Hoyt, D.W., Sykes, B.D., O'Connor-McCourt, M. and Malcolm, B.A. 1997. Construction of transforming growth factor alpha (TGF-alpha) phage library and identification of high binders of epidermal growth factor receptor (EGFR) by phage display. *J. Biochem. (Tokyo)* 122:686–690.

Tanner, V.A., Ploug, T., and Tao-Cheng, J.-H. 1996. Subcellular localization of SV2 and other secretory vesicle components in PC12 cells by an efficient method of pre-embedding EM immunocytochemistry for cell cultures. *J. Histochem. Cytochem.* 44:1481–1488.

Terasaki, M. and Dailey, M.E. 1995. Confocal microscopy of living cells. *In* Handbook of Biological Confocal Microscopy, 2nd ed. (J. Pawley, ed.) pp. 327–346. Plenum Press, New York.

te Riele, H., Maandag, E.R., Clarke, A., Hooper, M., and Berns, A. 1990. Consecutive inactivation of both alleles of the pim-1 proto-oncogene by homologous recombination in embryonic stem cells. *Nature* 348:649–651.

Thomas, C.E., Birket, D., Anozie, I., Castro, M.G., and Lowenstein, P.R. 2001. Acute direct adenoviral vector cytotoxicity and chronic, but not acute inflammatory responses correlate with decreased vector-mediated transgene expression in the brain. *Mol. Ther.* 3:36–46.

Thomas, P.S. 1980. Hybridization of denatured RNA and small DNA fragments transferred to nitrocellulose. *Proc. Natl. Acad. Sci. U.S.A.* 77:5201–5205.

Timmons, L. and Fire, A. 1998. Specific interference by ingested dsRNA. *Nature* 395:854.

Tokashiki, M. and Takamatsu, H. 1993. Perfusion culture apparatus for suspended mammalian cells. *Cytotechnology* 13:149–159.

Tollefson, A.E., Hermiston, T.W., and Wold, W.S.M., 1999. Preparation and titration of CsCl-banded adenovirus stock. *In* Adenovirus Methods and Protocols. (W.S.M. Wold, ed.) pp. 1–9. Humana Press Inc., New Jersey.

Trent, J.M., Bittner, M., Zhang, J., Wiltshire, R., Ray, M., Su, Y., Gracia, E., Meltzer, P., De Risi, J., Penland, L., and Brown, P. 1997. Use of microgenomic technology for analysis of alterations in DNA copy number and gene expression in malignant melanoma. *Clin. Exp. Immunol.* 107:33–40.

Tripathy, S.K., Black, H.B., Goldwasser, E., and Leiden, J.M. 1996. Immune responses to transgene-encoded proteins limit the stability of gene expression after injection of replication-defective adenovirus vectors. *Nat. Med.* 2:545–550.

Tsau, Y., Wenner, P., O'Donovan, M.J., Cohen, L.B., Loew, L.M., and Wuskell, J.P. 1996. Dye screening and signal-to-noise ratio for retrogradely transported voltage-sensitive dyes. *J. Neurosci. Methods* 70:121–129.

Tsien, R.Y. and Waggoner, A. 1995. Fluorophores for confocal microscopy. *In* Handbook of Biological Confocal Microscopy, 2nd ed. (J. Pawley, ed.) pp. 267–280. Plenum Press, New York.

Valenzuela, D.M., Murphy, A.J., Frendewey, D., Gale, N.W., Economides, A.N., Auerbach, W., Poueymirous, W.T., Adams, N.C., Rojas, J., Yasenchak, J., Chernomorsky, R., Boucher, M., Elsasser, A.L., Esau, L., Zheng, J., Griffiths, J.A., Wang, X., Su, H., Xue, Y., Dominguez, M.G., Noguera, I., Torres, R., Macdonald, L.E., Stewart, A.F., DeChiara, T.M., and Yancopoulos, G.D. 2003. High-throughput engineering of the mouse genome coupled with high-resolution expression analysis. *Nat. Biotechnol.* 6:652–659.

Valyi-Nagi, T., Deshmane, S.L., Spivack, J.G., Steiner, I., Ace, C.I., Preston, C.M., and Fraser, N.W. 1991. Investigation of herpes simplex virus type 1 (HSV-1) gene expression and DNA synthesis during the establishment of latent infection by an HSV-1 mutant, in 1814, that does not replicate in mouse trigeminal ganglia. *J. Gen. Virol.* 72:641–649.

van der Neut, R. 1997. Targeted gene disruption: Applications in neurobiology. *J. Neurosci. Methods* 71:19–27.

van der Straten, A., Johansen, H., Sweet, R., and Rosenberg, M. 1987. Efficient expression of foreign genes in cultured Drosophila melanogaster cells using hygromycin B selection. In Invertebrate Cell Systems Applications, Vol. 1 (Jun Mitsuhashi, ed.) pp. 131–134. CRC Press, Boca Raton, Fla.

van Meijer, M., Roelofs, Y., Neels, J., Horrevoets, A., van Zonneveld, A., and Pannekoek, H. 1996. Selective screening of a large phage display library of plasminogen activator inhibitor 1 mutants to localize interaction sites with either thrombin or the variable region 1 of tissue-type plasminogen activator. *J. Biol. Chem.* 271:7423–7428.

Van Regenmortel, M.H.V., Briand, J.P., Muller, S., and Plaue, S. 1988. Synthetic polypeptides as antigens. *In* Laboratory Techniques in Biochemistry and Molecular Biology, Vol. 19 (R.H. Burdon and P.H. von Knippenberg, eds.). Elsevier, Amsterdam and New York.

Vasavada, H.A., Ganguly, S., Germino, F.J., Wang, Z.X., and Weissman, S.M. 1991. A contingent replication assay for the detection of protein-protein interactions in animal cells. *Proc. Nat. Acad. Sci. U.S.A.* 88: 10686–10690.

Vicario-Abejon, C., Collin, C., McKay, R.D.G., and Segal, M. 1998. Neurotrophins induce formation of functional excitatory and inhibitory synapses between cultured hippocampal neurons. *J. Neursci.* 18: 7256–7271.

Vispo, N.S., Callejo, M., Ojalvo, A.G., Santos, A., Chinea, G., Gavilondo, J.V. and Arana, M.J. 1997. Displaying human interleukin-2 on the surface of bacteriophage. *Immunotechnology* 3:185–193.

Vlazny, D., Kwong, A., and Frenkel, N. 1982. Site-specific cleavage/packaging of herpes simplex virus DNA and the selective maturation of nucleocapsids containing full-length viral DNA. *Proc. Natl. Acad. Sci. U.S.A.* 79:1423–1427.

Voinnet, O. 2001. RNA silencing as a plant immune system against viruses. *Trends Genet.* 17:449–459.

Vojtek, A.B., Hollenberg, S.M., and Cooper, J.A. 1993. Mammalian Ras interacts directly with the serine/threonine kinase Raf. *Cell* 74:205–214.

Vooijs, M., van der Valk, M., te Riele, H., and Berns, A. 1998. Flp-mediated tissue-specific inactivation of the retinoblastoma tumor suppressor gene in the mouse. *Oncogene* 17:1–12.

Vulliez-Le Normand, B. and Eisele, J.L. 1993. Determination of detergent critical micellar concentration by solubilization of a colored dye. *Anal. Biochem.* 208:241–243.

Wallace, R.B. and Miyada, C.G. 1987. Oligonucleotide probes for the screening of recombinant DNA libraries. *Meth. Enzymol.* 152:432–442.

Wang, L.F., Du Plessis, D.H., White, J.R., Hyatt, A.D., and Eaton, B.T. 1995. Use of a gene-targeted phage display random epitope library to map an antigenic determinant on the bluetongue virus outer capsid protein VP5. *J. Immunol. Methods* 178:1–12.

Wang, S. and Vos, J. 1996. A hybrid herpesvirus infectious vector based on Epstein-Barr virus and herpes simplex virus type 1 for gene transfer into human cells in vitro and in vivo. *J. Virol.* 70:8422–8430.

Wang, Y., Krushel, L.A., and Edelman, G.M. 1996. Targeted DNA recombination in vivo using an adenovirus carrying the cre recombinase gene. *Proc. Natl. Acad. Sci. U.S.A.* 93:3932–3936.

Waterhouse, P.M., Graham, H.W., and Wang, M.B. 1998. Virus resistance and gene silencing in plants can be induced by simultaneous expression of sense and antisense RNA. *Proc. Natl. Acad. Sci. U.S.A.* 95:13959–13964.

Waterman-Storer, C.M., Sanger, J.W., and Sanger, J.M. 1993. Dynamics of organelles in the mitotic spindles of living cells: Membrane and microtubule interactions. *Cell Motil. Cytoskeleton* 26:19–39.

Watson, M.A., Buckholz, R., and Weiner, M.P. 1996. Vectors encoding alternative antibiotic resistance for use in the yeast two-hybrid system. *BioTechniques* 21:255–259.

Webb, R.H. and Dorey, C.K. 1995. The pixilated image. *In* Handbook of Biological Confocal Microscopy, 2nd ed. (J. Pawley, ed.) pp. 55–68. Plenum Press, New York.

Wei, K. and Huber, B.E. 1996. Cytosine deaminase gene as a positive selection marker. *J. Biol. Chem.* 271:3812–3816.

Weinstein, D.E., Shelanski, M.L., and Liem, R.K. 1990. C17, a retrovirally immortalized neuronal cell line, inhibits the proliferation of astrocytes and astrocytoma cells by a contact-mediated mechanism. *Glia* 3:130–139.

Welford, S.M., Gregg, J., Chen, E., Garrison, D., Sorensen, P.H., Denny, C.T., and Nelson, S.F. 1998. Detection of differentially expressed genes in primary tumor tissues using representational differences analysis coupled to microarray hybridization. *Nucl. Acids Res.* 26:3059–3065.

Wenner, P., Tsau, Y., Cohen, L.B., O'Donovan, M.J., and Dan, Y. 1996. Voltage sensitive dye recording using retrogradely transported dye in the chicken spinal cord: Staining and signal characteristics. *J. Neurosci. Methods* 70:111–120.

Wesley, S.V., Helliwell, C.A., Smith, N.A., Wang, M., Rouse, D.T., Liu, Q., Gooding, P.S., Singh, S.P., Abbott, D., Stoutjesdijk, P.A., Robinson, S.P., Gleave, A.P., Green, A.G., and Waterhouse, P. 2001. Construct design for efficient, effective and high-throughput gene silencing in plants. *Plant J.* 27:581–590.

Wessel, D. and Flugge, U.I. 1984. A method for the quantitative recovery of protein in dilute solution in the presence of detergents and lipids. *Anal. Biochem.* 138:141–143.

Wessendorf, M.W. and Brelje, T.C. 1993. Multicolor fluorescence microscopy using the laser-scanning confocal microscope. *Neuroprotocols* 2:121–140.

West, R.W.J., Yocum, R.R., and Ptashne, M. 1984. *Saccharomyces cerevisiae GAL1-GAL10* divergent promoter region: Location and function of the upstream activator sequence UASG. *Mol. Cell Biol.* 4:2467–2478.

Westinghouse Electric Company. 1976. Westinghouse sterilamp germicidal ultraviolet tubes. Westinghouse Electric Corp., Bloomfield, N.J.

Wigler, M., Silverstein, S., Lee, L.-S., Pellicer, A., Cheng, Y.-C., and Axel, R. 1977. Transfer of purified herpes virus thymidine kinase gene to cultured mouse cells. *Cell* 11:223–232.

Wilkinson, D.G. (ed.) 1992. In Situ Hybridization. A Practical Approach. Oxford University Press, New York.

Wilkinson, M. 1991. Purification of RNA. *In* Essential Molecular Biology: A Practical Approach, Vol. 1 (T.A. Brown, ed.) pp. 69–87. IRL Press, Oxford.

Wilson, M., DeRisi, J., Kirstensen, H., Imboden, P., Rane, S., Brown, P., and Schoolnik, G. 1999. Exploring drug-induced alterations in gene expression in *Mycobacterium tuberculosis* by microarray hybridization. *Proc. Natl. Acad. Sci. U.S.A.* 96:12833–12838.

Winter, G., Griffiths, A.D., Hawkins, R.E., and Hoogenboom, H.R. 1994. Making antibodies by phage display technology. *Annu. Rev. Immunol.* 12:433–455.

Wodicka, L., Dong, H., Mittmann, M., Ho, M.H., and Lockhart, D.J. 1997. Genome-wide expression monitoring in *Saccharomyces cerevisiae*. *Nature Biotechnol.* 15:1359–1367.

Wong, R.O., Chernjavsky, A., Smith, S.J., and Shatz, C.J. 1995. Early functional neural networks in the developing retina. *Nature* 374:716–718.

Wood, M.J.A., Charlton, H.M., Wood, K.J., Kajiwara, K., and Byrnes, A.P. 1996a. Immune responses to adenovirus vectors in the nervous system. *Trends Neurosci.* 19:497–501.

Wood, M.J.A., Byrnes, A.P., McMenamin, M., Kajiwara, K., Vine, A., Gordon, I., Lang, J., Wood, K.J., and Charlton, H.M. 1996b. Immune responses to viruses: Practical implications for the use of viruses as vectors for experimental and clinical gene therapy. *In* Protocols for Gene Transfer in Neuroscience (P.R. Lowenstein and L.W. Enquist, eds.) pp. 365–376. John Wiley & Sons, New York.

Wray, S. 1992. Organotypic slice explant roller-tube cultures. *In* Practical Cell Culture Techniques (A.A. Boulton, G.B. Baker, and W. Walz, eds.) pp. 201–239. Humana Press, Totowa, N.J.

Xia, Z., Dudek, H., Miranti, C.K., and Greenberg, M.E. 1996. Calcium influx via the NMDA receptor induces immediate early gene transcription by a MAP kinase/ERK-dependent mechanism. *J. Neurosci.* 16:5425–5436.

Xiao, X., Li, J., and Samulski, R.J. 1998. Production of high-titer recombinant adeno-associated virus vectors in the absence of helper adenovirus. *J. Virol.* 72:2224–2232.

Yagi, T., Ikawa, Y., Yoshida, K., Shigetani, Y., Takeda, N., Mabuchi, I., Yamamoto, T., and Aizawa, S. 1990. Homologous recombination at c-fyn locus of mouse embryonic stem cells with use of diphtheria toxin A-fragment gene in negative selection. *Proc. Natl. Acad. Sci. U.S.A.* 87:9918–9922.

Yang, D., Buchholz, F., Huang, Z.D., Goga, A., Chen, C.Y., Brodsky, F.M., and Bishop, J.M. 2002. Short RNA duplexes produced by hydrolysis with *Escherichia coli* RNase III mediate effective RNAi in mammalian cells. *Proc. Natl. Acad. Sci. U.S.A.* 99:9942–9947.

Yang, X.W., Model, P., and Heintz, N. 1997. Homologous recombination based modification in *Escherichia coli* and germline transmission in transgenic mice of a bacterial artificial chromosome. *Nat. Biotechnol.* 15:859–865.

Yang, X.W., Wynder, C., Doughty, M.L., and Heintz, N. 1999. BAC-mediated gene-dosage analysis reveals a role for *Zipro1* (*Ru49/Zfp38*) in progenitor cell proliferation in cerebellum and skin. *Nat. Genet.* 22:327–335.

Yayon, A., Aviezer, D., Safran, M., Gross, J.L., Heldman, Y., Cabilly, S., Givol, D., and Katchalski-Katzir, E. 1993. Isolation of peptides that inhibit binding of basic fibroblast growth factor to its receptor from a random phage-epitope library. *Proc. Natl. Acad. Sci. U.S.A.* 90:10643–10647.

Yoshihara, Y., Mizuno, T., Nakahira, M., Kawasaki, M., Watanabe, Y., Kagamiyama, H., Jishage, K., Ueda, O., Suzuki, H., Tabuchi, K., Sawamoto, K., Okano., Noda, T., and Mori, K. 1999. A genetics approach to visualization of multisynaptic neural pathways using plant lectin transgene. *Neuron* 22:33–41.

Young, W.S. III 1992. In situ hybridization with oligodeoxyribonucleotide probes. *In* In Situ Hybridization: A Practical Approach (D.G. Wilkinson, ed.) pp. 33–44. Oxford University Press, New York.

Young, W.S. III, Mezey, E., and Siegel, R.E. 1986. Vasopressin and oxytocin mRNAs in adrenalectomized and Brattleboro rats: Analysis by quantitative in situ hybridization histochemistry. *Mol. Brain Res.* 1:231–241.

Yu, W., Misulovin, Z., Suh, H., Hardy, R.R., Jankovic, M., Yannoutsos, N., and Nussenzweig, M.C. 1999. Coordinate regulation of RAG1 and RAG2 by cell type-specific DNA elements 5' of RAG2. *Science* 285: 1080–1084.

Yue, H., Eastman, P.S., Wang, B.B., Minor, J., Doctolero, M.H., Nuttall, R.L., Stack, R., Becker, J.W., Montgomery, J.R., Vainer, M., and Johnston, R. 2001. An evaluation of the performance of cDNA microarrays for detecting changes in global mRNA expression. *Nucl. Acids Res.* 29:E41–1. (available online at *http://nar.oupjournals.org/cgi/content/full/29/8/e41*)

Yuste, R. and Katz, L.C. 1991. Control of postsynaptic Ca^{2+} influx in developing neocortex by excitatory and inhibitory neurotransmitters. *Neuron* 6:333–344.

Yuste, R., Peinado, A., and Katz, L.C. 1992. Neuronal domains in developing neocortex. *Science* 257:665–669.

Zamore, P.D. 2001. RNA interference: Listening to the sound of silence. *Nat. Struc. Biol.* 8:746–750.

Zhang, Y., Riesterer, C., Ayrall, A.M., Sablitzky, F., Littlewood, T.D., and Reth, M. 1996. Inducible site-directed recombination in mouse embryonic stem cells. *Nucl. Acids Res.* 24:543–548.

Zhang, Y., Buchholz, F., Muyrers, J.P., and Stewart, A.F. 1998. A new logic for DNA engineering using recombination in *Escherichia coli*. *Nat. Genet.* 20:123–128.

Zimmer, A. 1992. Manipulating the genome by homologous recombination in embryonic stem cells. *Annu. Rev. Neurosci.* 15:115–137.

Zuffery, R., Nagy, D., Mandel, R.J., Naldini, L., and Trono, D. 1997. Multiply attenuated lentiviral vector achieves efficient gene delivery in vivo. *Nat. Biotechnol.* 15:871–875.

Zuo, J., Treadaway, J., Buckner, T.W., and Fritzsch, B. 1999. Visualization of α9 acetylcholine receptor expression in hair cells of transgenic mice containing a modified bacterial artificial chromosome. *Proc. Natl. Acad Sci. U.S.A.* 96:14100–14105.

INDEX

Page numbers in this book are hyphenated: the number before the hyphen refers to the chapter and the number after the hyphen refers to the page within the chapter (e.g., 4-3 is page 3 of Chapter 4). A range of pages is indicated by an arrow connecting the page numbers (e.g., 4-3→4-5 refers to pages 3 through 5 of Chapter 4).

A

B

C

D

E

F

G

H

I

J

K

L

M

N

O

P

S